Green Oxidation in Organic Synthesis

Green Oxidation in Organic Synthesis

Edited by

Ning Jiao
Peking University
Beijing, China

Shannon S. Stahl
University of Wisconsin-Madison
Madison, WI, USA

WILEY

This edition first published 2019

Registered Offices
John Wiley & Sons, Inc., 111 River Street, Hoboken, NJ 07030, USA
John Wiley & Sons Ltd, The Atrium, Southern Gate, Chichester, West Sussex, PO19 8SQ, UK

Editorial Office
The Atrium, Southern Gate, Chichester, West Sussex, PO19 8SQ, UK

For details of our global editorial offices, customer services, and more information about Wiley products visit us at www.wiley.com.

Wiley also publishes its books in a variety of electronic formats and by print-on-demand. Some content that appears in standard print versions of this book may not be available in other formats.

Library of Congress Cataloging-in-Publication Data

Names: Jiao, Ning, 1976– editor. | Stahl, Shannon S., 1970– editor.
Title: Green oxidation in organic synthesis / edited by Ning Jiao (Peking University, Beijing, China), Shannon S. Stahl (University of Wisconsin-Madison, Madison, US).
Description: First edition. | Hoboken, NJ : Wiley, 2019. | Includes bibliographical references and index. |
Identifiers: LCCN 2019008186 (print) | LCCN 2019015930 (ebook) | ISBN 9781119304494 (Adobe PDF) | ISBN 9781119304043 (ePub) | ISBN 9781119304166 (hardback)
Subjects: LCSH: Organic compounds–Synthesis. | Green chemistry. | Oxidation.
Classification: LCC QD262 (ebook) | LCC QD262 .G66485 2019 (print) | DDC 547/.2–dc23
LC record available at https://lccn.loc.gov/2019008186

Cover design: Wiley
Cover image: © Sebastian Arning/EyeEm/Getty Images

Set in 10/12pt Warnock by SPi Global, Pondicherry, India

10 9 8 7 6 5 4 3 2 1

Contents

Preface

Oxidations represent a prominent class of reactions in organic synthesis. The development of oxidation methods that use O_2 or H_2O_2 as oxidants, in addition to oxidations that achieve good selectivity and/or generate minimal waste or by-products, have advanced considerably in recent years. These methods often address the priorities of green and sustainable chemistry, and they often match or surpass the utility and effectiveness of classical methods that employ stoichiometric oxidants, such as inorganic oxides and salts, peroxides, hypervalent iodines, and quinones. The present compilation of articles is intended to complement important previous compilations, such as *Modern Oxidation Methods* (2004, 2011) edited by Jan-Erling Bäckvall, with a particular attention to recent advances in green oxidation methods.

This multi-authored book provides an extensive survey of oxidative processes and serves as a valuable introduction to recent and historical literature on topic of green oxidation. Chapter 1 introduces the topics that will be covered in this book, which are organized according to the functional groups generated in the reactions. Chapter 1 surveys methods for the synthesis of alcohols and phenols, and Chapter 2 highlights recent advances in the synthesis of aldehydes and ketones. Oxidative methods for the synthesis of ethers, esters, and organic halides is the focus of Chapter 3, with the subtopic of epoxide formation reviewed in Chapter 4, with an emphasis on metal-catalyzed epoxidation reactions that use O_2 and H_2O_2. Oxidation methods for the synthesis of carboxylic acids are reviewed in Chapter 5, completing the series of chapters emphasizing carbon–oxygen bond formation. Carbon-nitrogen bonds are the focus of the next series of chapters. Chapter 6 addresses oxidation routes for the synthesis of amines, amides, and imines, Chapter 7 presents methods for the synthesis of nitriles, and Chapter 8 reviews the oxidative synthesis of azo, diazo, and azido compounds. Carbon–carbon bond formation is the focus of Chapter 9, with examples including oxidative Heck and Sonogashira reactions for the synthesis of substituted olefins and alkynes, and the subsequent two chapters focus on oxidative methods for the preparation of substituted arenes (Chapter 10) and the synthesis of heterocyclic compounds (Chapter 11). Methods include C—H functionalization, cross-dehydrogenative coupling, dehydrogenative aromatization, and aerobic oxygenation. Sulfur oxidation, emphasizing the preparation of sulfoxides and sulfones, is the topic of Chapter 12, and methods for the oxidative C—P and C—S bond formation are covered in Chapter 13. Various classes of photochemical methods are available for selective oxidation of organic molecules, including photoredox strategies and generation of singlet oxygen with photosensitizers. These and other photochemical reactions are the focus of

Chapter 14. Chapter 15 summarizes the developments of electrochemical oxidations, and the book concludes in Chapter 16 with a presentation of enzymatic oxidations.

The contents herein highlight the considerable growth of interest in the topic of green oxidation chemistry, and we hope that this book and the references cited herein will stimulate further expansion of this field and be of value to chemists interested in applying such methods.

April 2019 *Ning Jiao and Shannon S. Stahl*

List of Contributors

Marc Biermann
Faculty of Chemistry, Bielefeld University, Germany

Arun Jyoti Borah
State Key Laboratory of Coordination Chemistry, School of Chemistry and Chemical Engineering, Nanjing University, China

Qun Cao
Queen's University Belfast, United Kingdom

Tieqiao Chen
College of Materials and Chemical Engineering, Hainan University, Haikou, China

Shunsuke Chiba
Nanyang Technological University, Singapore

Ji Eun Choi
Faculty of Chemistry, Bielefeld University, Germany

Miquel Costas
Facultat de Ciències, Departament de Química I Institut de Química Computacional i Catàlisi, Universitat de Girona, Spain

Wei Ding
College of Chemistry, Central China Normal University, Wuhan, China

Laura M. Dornan
Queen's University Belfast, United Kingdom

Xiaoyu Geng
State Key Laboratory of Natural and Biomimetic Drugs, School of Pharmaceutical Sciences, Peking University, Beijing, China

Harald Gröger
Faculty of Chemistry, Bielefeld University, Germany

Li-Biao Han
National Institute of Advanced Industrial Science and Technology (AIST), Tsukuba, Japan

Ryutaro Hayashi
Department of Synthetic Chemistry and Biological Chemistry, Graduate School of Engineering, Kyoto University, Japan

Yoshiharu Iwabuchi
Department of Organic Chemistry, Graduate School of Pharmaceutical Sciences, Tohoku University, Sendai, Japan

Huanfeng Jiang
Key Laboratory of Functional Molecular Engineering of Guangdong Province, School of Chemistry and Chemical

Engineering, South China University of Technology, Guangzhou, China

Xuefeng Jiang
School of Chemistry and Molecular Engineering, East China Normal University, Shanghai, China

Ning Jiao
State Key Laboratory of Natural and Biomimetic Drugs, School of Pharmaceutical Sciences, Peking University, Beijing, China

Tianlong Lan
Tsinghua University, Beijing, China

Aiwen Lei
College of Chemistry and Molecular Sciences, the Institute for Advanced Studies (IAS), Wuhan University, China
State Key Laboratory for Oxo Synthesis and Selective Oxidation, Lanzhou Institute of Chemical Physics, Chinese Academy of Sciences, Lanzhou, China

Chao-Jun Li
Department of Chemistry and FRQNT Centre for Green Chemistry and Catalysis, McGill University, Montreal, Quebec, Canada

Tianlei Li
Tsinghua University, Beijing, China

Yiming Li
School of Chemistry and Molecular Engineering, East China Normal University, Shanghai, China

Yujie Liang
State Key Laboratory of Natural and Biomimetic Drugs, School of Pharmaceutical Sciences, Peking University, Beijing, China

Mingxin Liu
Department of Chemistry and FRQNT Centre for Green Chemistry and Catalysis, McGill University, Montreal, Quebec, Canada

Tianyi Liu
College of Chemistry and Molecular Sciences, the Institute for Advanced Studies (IAS), Wuhan University, China

Liang-Qiu Lu
College of Chemistry, Central China Normal University, Wuhan, China

Mark J. Muldoon
Queen's University Belfast, United Kingdom

Yasushi Obora
Department of Chemistry and Materials Engineering, Faculty of Chemistry, Materials and Bioengineering, Kansai University, Suita, Japan

Keiko Oike
Faculty of Chemistry, Bielefeld University, Germany

Lu Ouyang
Key Laboratory of Functional Molecular Engineering of Guangdong Province, School of Chemistry and Chemical Engineering, South China University of Technology, Guangzhou, China

Sreekumar Pankajakshan
Nanyang Technological University, Singapore

Yu Rao
Tsinghua University, Beijing, China

Zhuangzhi Shi
State Key Laboratory of Coordination Chemistry, School of Chemistry and

Chemical Engineering, Nanjing University, China

Akihiro Shimizu
Department of Synthetic Chemistry and Biological Chemistry, Graduate School of Engineering, Kyoto University, Japan

Atul K. Singh
College of Chemistry and Molecular Sciences, the Institute for Advanced Studies (IAS), Wuhan University, China

Zilu Tang
College of Chemistry and Molecular Sciences, The Institute for Advanced Studies (IAS), Wuhan University, China

Liguo Wang
Tsinghua University, Beijing, China

Severin Wedde
Faculty of Chemistry, Bielefeld University, Germany

Wanqing Wu
Key Laboratory of Functional Molecular Engineering of Guangdong Province, School of Chemistry and Chemical Engineering, South China University of Technology, Guangzhou, China

Wen-Jing Xiao
College of Chemistry, Central China Normal University, Wuhan, China

Hong Yi
College of Chemistry and Molecular Sciences, the Institute for Advanced Studies (IAS), Wuhan University, China

Jun-ichi Yoshida
Department of Synthetic Chemistry and Biological Chemistry, Graduate School of Engineering, Kyoto University, Japan

Nadine Zumbrägel
Faculty of Chemistry, Bielefeld University, Germany

1

Green Oxidative

Synthesis of Alcohols and Phenols

Tianlei Li, Tianlong Lan, Liguo Wang, and Yu Rao

Tsinghua University, Beijing, China

1.1 Introduction

Alcohol and phenol derivatives are among the most important and common organic chemicals used for industrial materials such as polymers and commodity chemicals, and are the core structures of many biologically active compounds, including pharmaceuticals, agrochemicals, and natural products. The development of synthetic methods for introducing hydroxyl groups into organic compounds is a fundamental task in organic chemistry. Due to the sustainable chemistry and environmentally friendly manufacturing techniques extensively demanded in society today, development of new synthetic methodologies and novel reaction conditions without compromising product selectivity, energy efficiency, and environmental safety has become the main theme of current chemical research. Over the past decades, numerous methods have been reported, the most important and straightforward strategy is the oxidation of various C—H, C—C, and C=C bonds to hydroxyl groups. In this review, we will focus on the latest developments adopting green oxidative strategies with aerobic (O_2 and air), hydrogen peroxide, enzymatic oxidation, and photocatalytic oxidation for synthesis of alcohol and phenol derivatives that have happened since 2008. Mechanistic studies of the novel transformations including the preparation and isolation of reaction intermediates will also be discussed. Our intention is to highlight here the interesting achievements in green oxidative synthesis of alcohols and phenols over the last 10 years. There may be important literature reports that have escaped our attention, for which we extend our sincere apologies.

1.2 C—H Hydroxylation via Aerobic Oxidant

C—H hydroxylation is considered as one of the most straightforward approaches to preparing alcohols and phenols. During this process, various oxidants, such as inorganic salt, TBHP, $PhI(OAc)_2$, $K_2S_4O_8$, and IBX, are used. For green and sustainable chemistry, molecular oxygen is considered as an ideal oxidant due to its natural, inexpensive, and environmentally friendly characters, and therefore offers attractive academic and industrial prospects. In addition, it exhibits a highly atom-efficient oxidant per weight.

Green Oxidation in Organic Synthesis, First Edition. Edited by Ning Jiao and Shannon S. Stahl.

In this context, various significant progress in catalytic C—H hydroxylation via aerobic oxidant has been achieved over the past 10 years, especially in transition metal- (such as palladium, copper, and cobalt) catalyzed oxidation reactions employing molecular oxygen as the sole oxidant. In the next section, we will illustrate the recent advances over the last 10 years in C—H hydroxylation using molecular oxygen or air as an oxidant.

1.2.1 Synthesis of Phenols by C(sp^2)-H Hydroxylation

An intriguing example of employing transition metals to catalyze *ortho*-hydroxylation involving strong coordination of the carbonyl group was reported by Yu and co-workers in 2009 [1]. Notably, hydroxylation proceeded using 1 atm of air as the sole oxidant under nonacidic conditions in high yields with variously substituted benzoic acids such as trifluoromethyl, acetyl, cyanide, and nitro. Preliminary mechanistic investigations strongly supported the proposal that the aryl-Pd species could be directly oxygenated by molecular O_2 (Scheme 1.1).

In 2012, Lei's group reported a copper-catalyzed "oxygenase-type" oxidation of arenes and heteroarenes at room temperature by the oxidation of O_2 from air (1 atm). The authors proposed a mechanism of a novel combination of single-electron transfer (SET) initiation and organometallic catalytic cycle for this copper-catalyzed aerobic oxidation reaction (Scheme 1.2) [2].

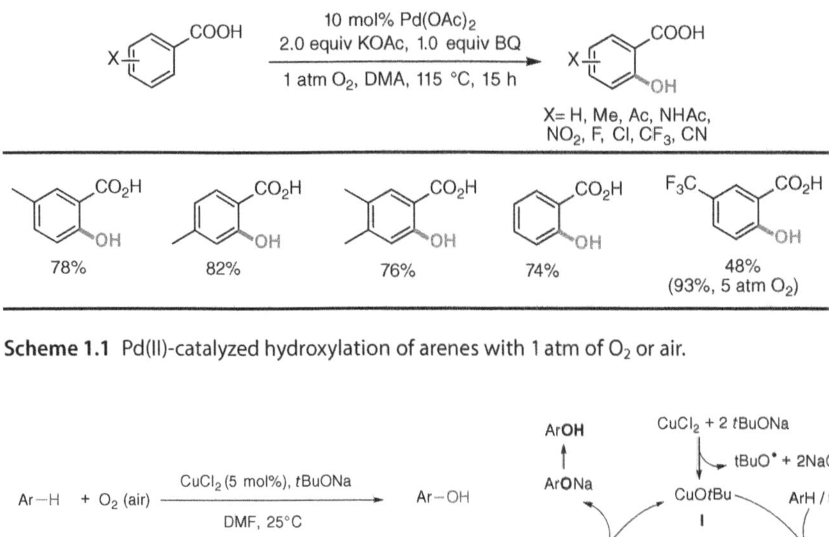

Scheme 1.1 Pd(II)-catalyzed hydroxylation of arenes with 1 atm of O_2 or air.

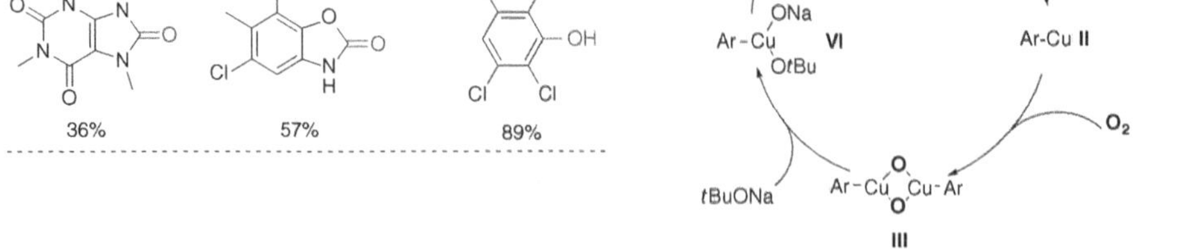

Scheme 1.2 Copper-catalyzed oxidation of arenes and heteroarenes.

In 2013, Jiao's group reported a novel $PdCl_2$ and *N*-hydroxyphthalimide co-catalyzed direct C(sp^2)-H hydroxylation of 2-phenyl-pyridines. Molecular oxygen was employed as a reagent and the sole oxidant under neutral conditions without the addition of any other stoichiometric oxidant and base. This transformation provided a green and practical method to synthesize a variety of substituted 2-(pyridin-2-yl)phenols. Its synthetic use has been exemplified in further applications in the preparation of a series of biologically active molecules from 2-(pyridin-2-yl)phenols. Based on extensive mechanistic studies, a novel combination of transition-metal-catalyzed C—H activation and a NHPI-initiated radical process was proposed as the mechanism (Scheme 1.3) [3].

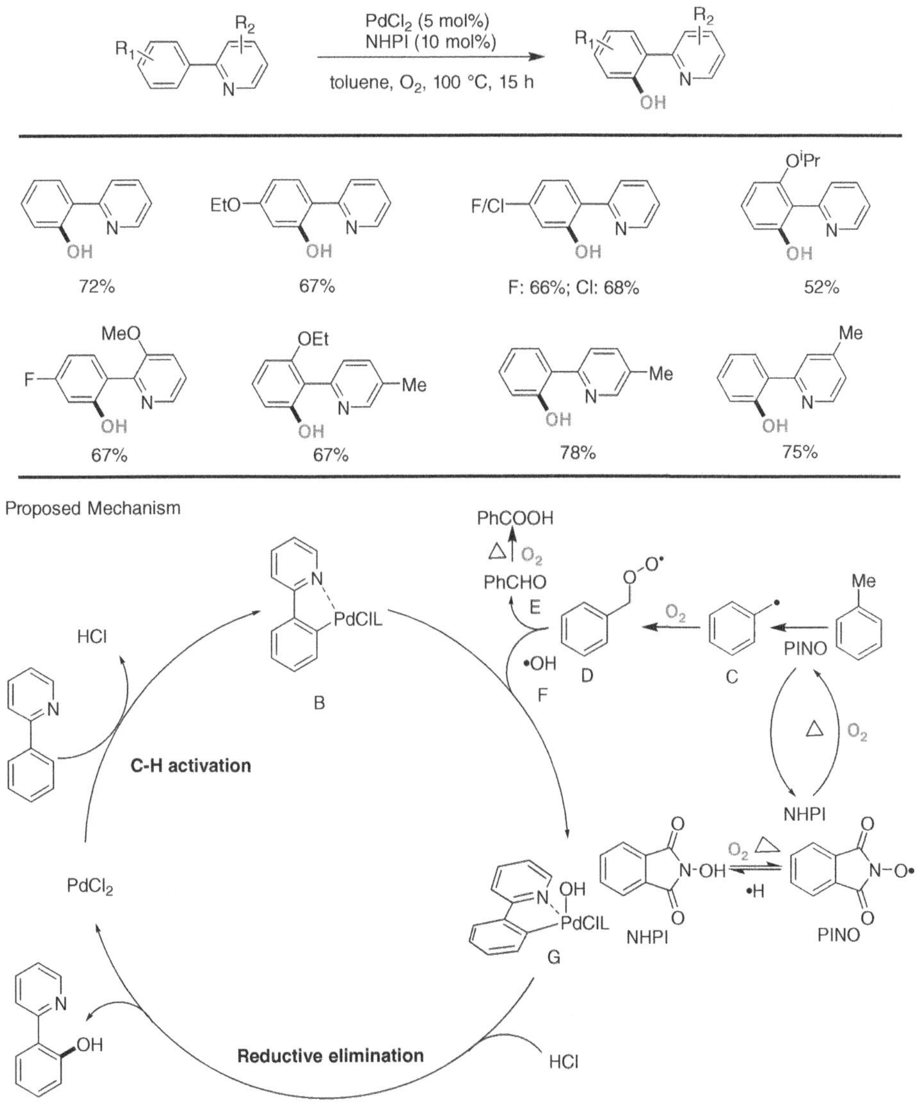

Scheme 1.3 $PdCl_2$ and *N*-hydroxyphthalimide co-catalyzed C(sp^2) by dioxygen activation.

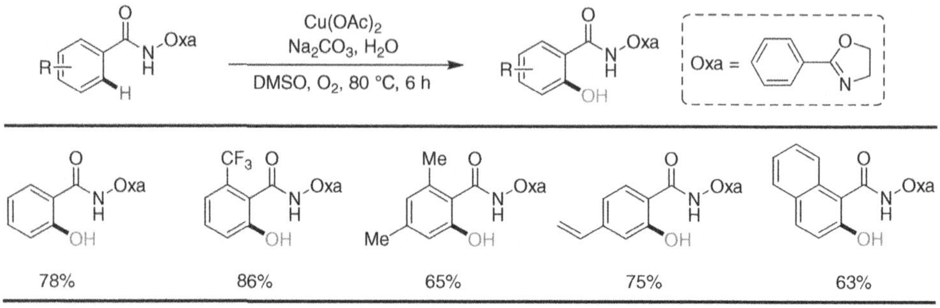

Scheme 1.4 Cu(II)-mediated ortho-C—H hydroxylation using a removable directing group.

In 2015, Yu and co-workers reported Cu(II)-mediated *ortho*-C–H hydroxylation using a removable directing group. The reaction tolerates a wide range of functional groups, and the use of O_2 as the oxidant presents a significant advantage. The practicality of this new methodology has been proven by gram-scale synthesis of 2-hydroxybenzamide [4]. Very recently, the author developed a similar reaction to generate the corresponding phenols via a weakly coordinating monodentate directing group with the assistance of an external oxazoline ligand (Scheme 1.4) [5].

In 2016, Guin and co-workers reported a Pd-catalyzed direct C—H hydroxylation of 2-arylpyridines. This method offered easy access to a broad range of substituted 2-(pyridin-2-yl)phenols in good isolated yields. The ^{18}O labeling experiment proved that the oxygen atom of the hydroxyl group originated from molecular oxygen, and the use of molecular oxygen as a sole oxidant makes the process environmentally attractive. A possible reaction pathway is depicted in Scheme 1.5. The catalytic cycle begins with the chelation-assisted C—H activation through *ortho* palladation of 1a with a palladium catalyst to form the complex E. Meanwhile, aerobic oxidation of *n*-butyraldehyde to *n*-butyric acid produces an active acyl peroxo-radical intermediate B. The oxidative addition of the radical B to Pd^{II}-complex E may lead to a transient peroxo-palladium species that decomposes to the most likely Pd^{IV}- intermediate, F. Finally, reductive elimination of F furnishes the hydroxylated product 2a and regenerates the catalyst [6].

Due to multicopper proteins that bind and activate O_2 exhibiting highly selective oxidative transformations and playing important roles in a number of biological processes, the multicopper(II) complex has received much interest from researchers in chemistry fields that range from coordination chemistry to synthetic methods. Arising from these studies, three basic Cu_2O_2 core structures have been widely described [7]. Recently, Ribas and co-workers systematically studied O_2 activation and selective phenolate *ortho*-hydroxylation by an unsymmetric di-copper μ-η^1: η^1-peroxido complex, which represented a novel di-copper complex based on a heptadentate ligand that gave rise to an asymmetric $N_3Cu^{II}N_4Cu^{II}(\mu\text{-}\eta^1\text{: }\eta^1\text{-}O_2)$ core and hitherto exhibited reactivity patterns not observed for symmetric analogs [8]. The combined experimental and computational evidence indicated that the *ortho*-hydroxylation of a phenolate by a Cu_2O_2 species can occur by adjacent binding of phenolate and O_2 at a common N_3Cu site without requiring the peroxide to be side-on bound, thus offering a conceptually new understanding of O_2 activation at di-copper sites (Scheme 1.6) [9].

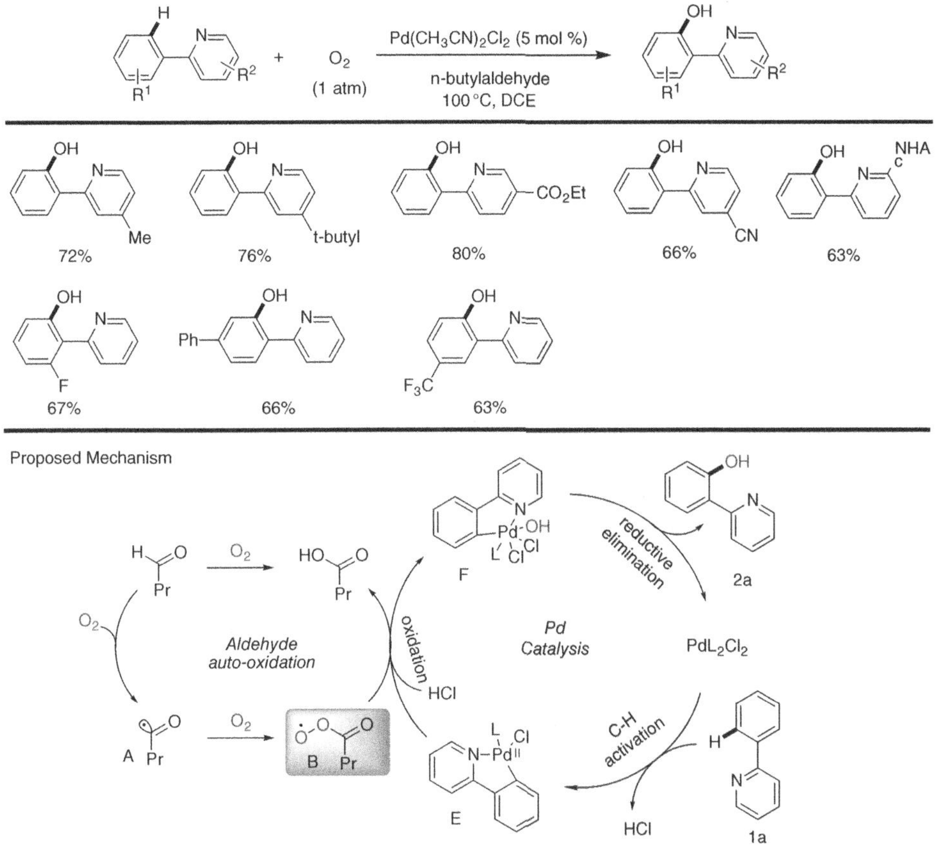

Scheme 1.5 Pd-catalyzed direct C—H hydroxylation of 2-arylpyridines.

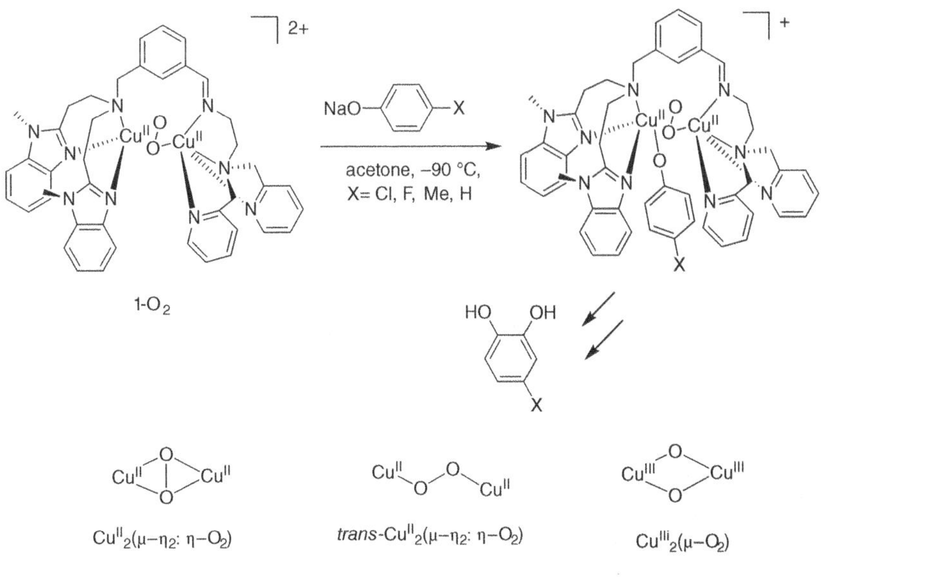

Scheme 1.6 O_2 activation and selective phenolate *ortho*-hydroxylation by an unsymmetric di-copper.

1.2.2 Synthesis of Alcohol by C(sp^3)-H Hydroxylation

Inspired by discovery and mechanism evaluation of previously unknown Pd-(III)-Pd(III) complexes [10], Ritter and co-workers developed a chemo- and regioselective α-hydroxylation reaction of carbonyl compounds by oxygen transfer from O_2, catalyzed by a dinuclear Pd(II) complex in 2011 (Scheme 1.7) [11]. The hydroxylation reaction was regioselective and compatible with other electrophilic and nucleophilic oxygen transfer reagents such as double bonds and sulfides. A broad range of carbonyl compounds were smoothly transformed to tertiary alcohols with good yields under very mild conditions. It is interesting to note that the dinuclear Pd(II) complex functions as an oxygen transfer catalyst, reminiscent of an oxygenase. On the basis of an oxygen uptake experiment, both oxygen atoms from O_2 were incorporated into the product. Hence, this oxidation proceeded without generation of stoichiometric amounts of waste except for solvent. This reaction represents the first example of the catalytic hydroxylation of C—H bonds with a bimetallic palladium(III) catalysis.

In 2012, Tan and co-workers described pentanidium-catalyzed α-hydroxylation of 3-substituted-2-oxindoles using molecular oxygen with good yields and enantioselectivities (Scheme 1.8) [12]. Several 3-hydroxyl-2-oxindoles with excellent enantioselectivities (85–98% *ee*) were obtained. The reaction was demonstrated to consist of two steps: an enantioselective formation of hydroperoxide oxindole and a kinetic resolution of the hydroperoxide oxindole via reduction with enolates generated from the oxindoles. This reaction does not require an additional reductant such as triethyl phosphite, which was typically added to reduce the peroxide intermediate. This chiral phase-transfer catalysis operated with a high degree of efficiency and selectivity that made it a valuable tool for constructing highly enantiopure 3-hydroxyl-2-oxindoles in organic synthesis.

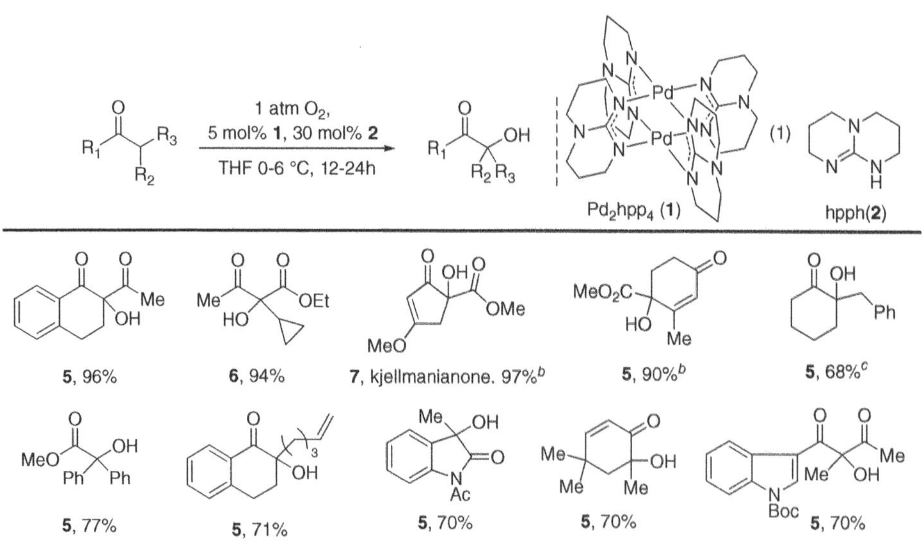

[a]A total of 2.5 mol% of **1** was used. [b]No **2** was added. [c]A total of 10 mol % of **1**, and 60 mol % of **2** were used.

Scheme 1.7 Dinuclear palladium catalyst for α-hydroxylation of carbonyls with O_2.

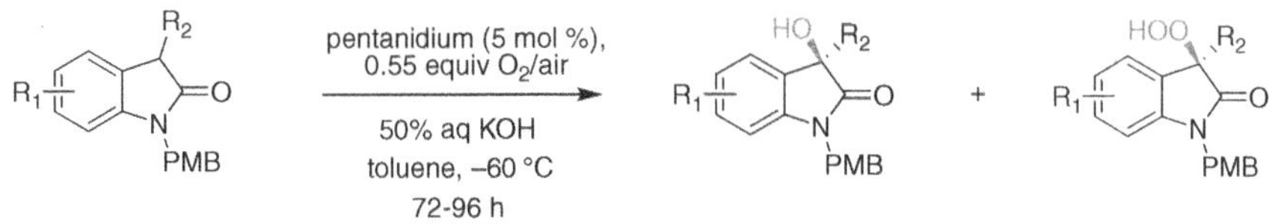

Entry	R_1	R_2	Yield (%)	ee (%)	ratio of 1/2
1	5-F	Me	84	86	90:10
2	5-OMe	Me	88	98	91:9
3	5-Me	Me	86	96	92:8
4	H	nBu	84	85	90:10
5	H	Bn	92	92	97:3
6	H	*p*-Bn-OMe	93	92	97:3
7	H	$CH_2C{\equiv}CH$	85	91	92:8
8	H	$CH_2CH{=}CH_2$	89	91	94:6
9	5-OMe	$CH_2CH{=}CH_2$	83	94	94:6

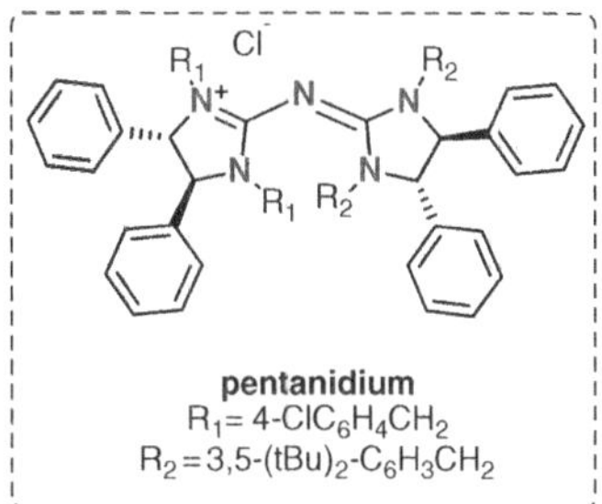

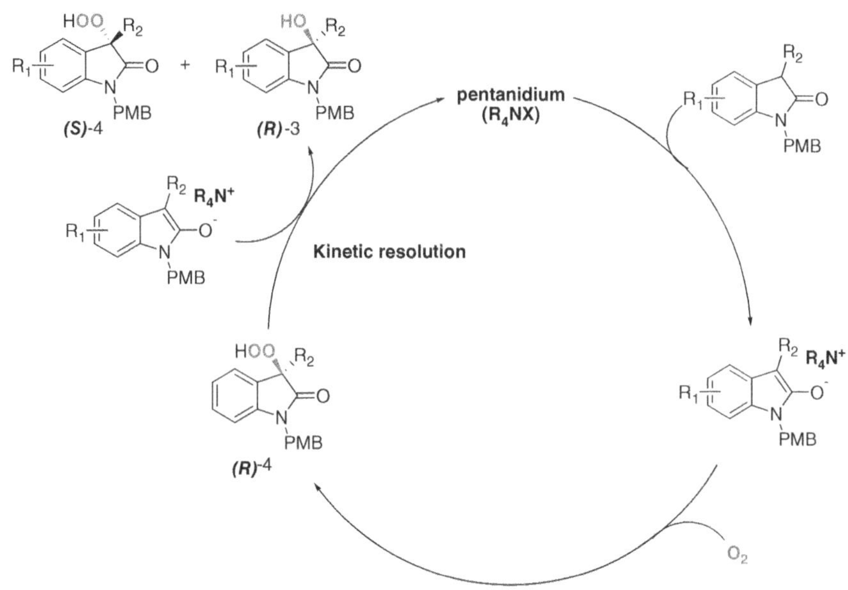

Scheme 1.8 Pentanidium-catalyzed enantioselective α-hydroxylation of oxindoles.

In 2015, Gu and co-workers reported efficient ionic liquid ([bmIm]OH)-catalyzed α-hydroxylation of phosphonates using O_2 as the oxygen source (Scheme 1.9) [13]. The procedure was environmentally friendly, avoiding hazardous organic solvents and toxic catalysts. This method provided a novel and convenient methodology for the construction of quaternary α-hydroxy phosphonates.

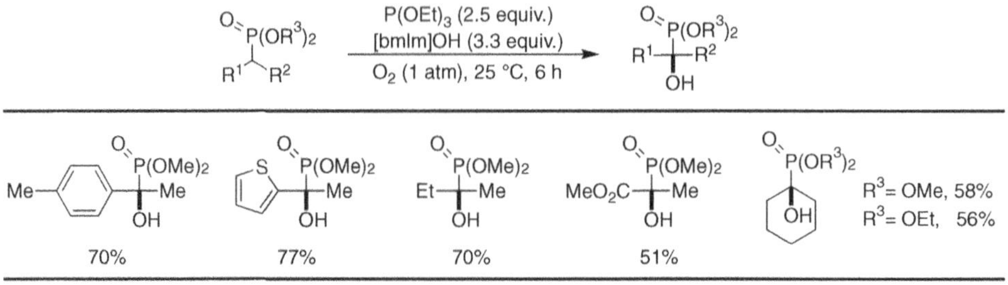

Scheme 1.9 Synthesis of quaternary α-hydroxy phosphonates via C—H hydroxylation.

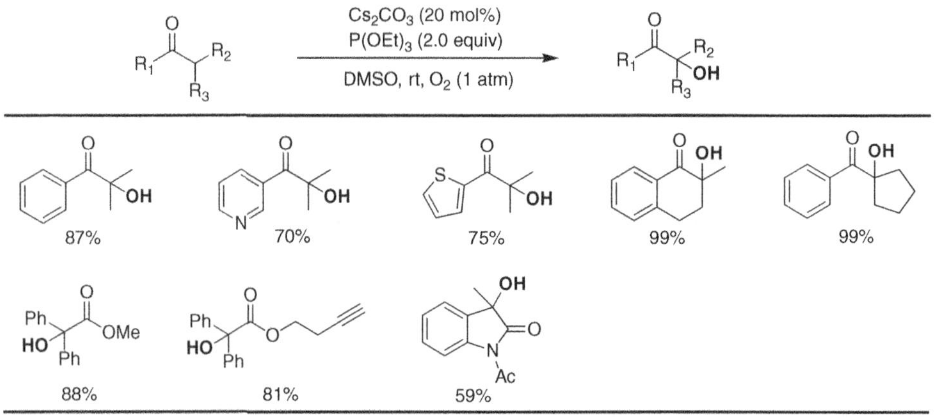

Scheme 1.10 Cs_2CO_3-initiated α-hydroxylation of ketones with O_2.

A significant number of transition-metal-catalyzed autoxidation and aerobic oxidative hydroxylation reactions have been proved to be environmentally friendly strategies for the preparation of various phenols and alcohols. However, main group metal-catalyzed C—H hydroxylation has been rarely reported. Recently, Jiao and co-workers described a transition-metal-free Cs_2CO_3-catalyzed α-hydroxylation of carbonyl compounds with O_2 as the oxygen source to give tertiary α-hydroxycarbonyl compounds, which are highly valued chemicals and widely used in the chemical and pharmaceutical industries (Scheme 1.10) [14]. The simple conditions and the use of molecular oxygen as both the oxidant and the oxygen source make this protocol very environmentally friendly and practical. This transformation is highly efficient and selective for tertiary $C(sp^3)$—H bond cleavage. Remarkably, its synthetic utilities have been exemplified in further late-stage modification of several drug substrates such as ketoprofen, ibuprofen, and naproxen. Shortly after this report, the authors applied a similar CsOH-catalyzed aerobic oxidation to efficiently access various *p*-quinols from multi-alkyl phenols with moderate to good yields [15].

In 2015, Zhao and co-workers demonstrated an efficient and enantioselective α-hydroxylation of acyclic as well as cyclic ketones using molecular oxygen (Scheme 1.11) [16]. Synthetically valuable acyclic ketones could be hydroxylated in modest to good yields with good enantioselectivities, whereas a wide range of cyclic α-hydroxy ketones could be obtained with excellent yields and enantiopurity.

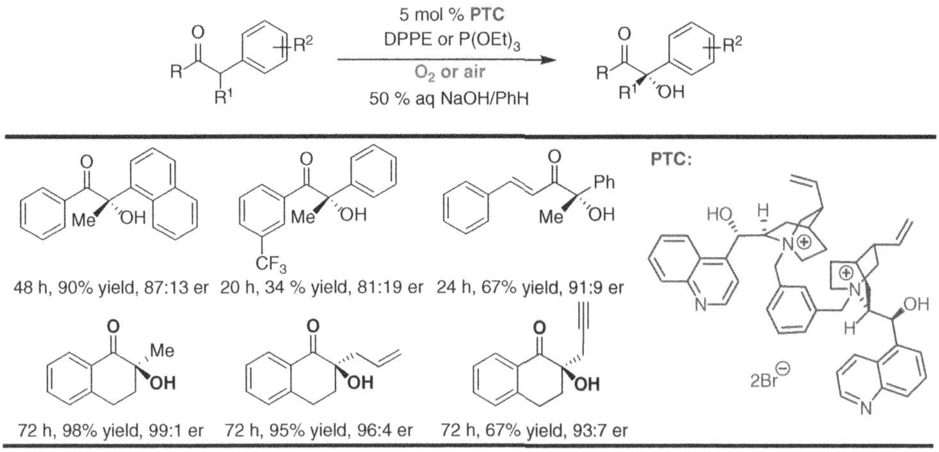

Scheme 1.11 Phase-transfer-catalyzed enantioselective α-hydroxylation of acyclic and cyclic ketones.

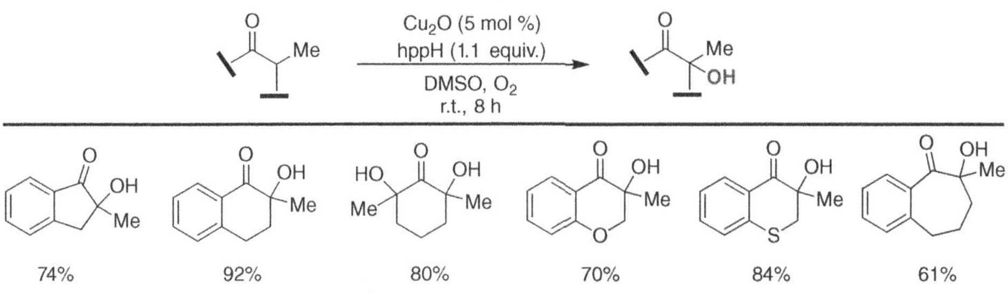

Scheme 1.12 Cu_2O-catalyzed hydroxylation of α- methyl ketones.

In 2016, Schoenebeck and co-workers developed a selective aerobic hydroxylation of tertiary α-C—H bonds in ketones without C—C cleavage (Scheme 1.12) [17]. They found that α-alkyl-substituted ketones selectively generated α-hydroxy ketones under Cu_2O/DMSO/hppH conditions, and the hydroxylated products were formed under these conditions, tolerating different ring sizes, heteroatoms, and substituents at the aromatic ring. Furthermore, the experimental and computational studies uncovered the role of hppH [1,3,4,6,7,8-hexahydro-2H-pyrimido[1,2-a]pyrimidine] to act not only as a base in the transformation but also as a reductant of the peroxide to the corresponding α-hydroxy ketone.

In 2016, Kanai and co-workers reported that a radical *N*-oxyl directing activator promoted the aerobic oxygenation of benzylic, propargylic, tertiary, and unactivated acyclic methylene C(sp^3)—H bonds in aliphatic alcohols with γ- (or δ-) selectivity under mild conditions (room temperature to 50 °C) (Scheme 1.13) [5]. Molecular oxygen was used as the stoichiometric oxidant, and the reactions proceeded regioselectively at the γ- (and δ-) position(s), whereas the α, β, and other positions beyond the d position remained intact. This regioselectivity can be explained in terms of the intramolecular accessibility of the reactive *N*-oxyl radical site, despite the low regioselectivity between γ- and δ-positions in electronically nonbiased substrates is a current limitation that

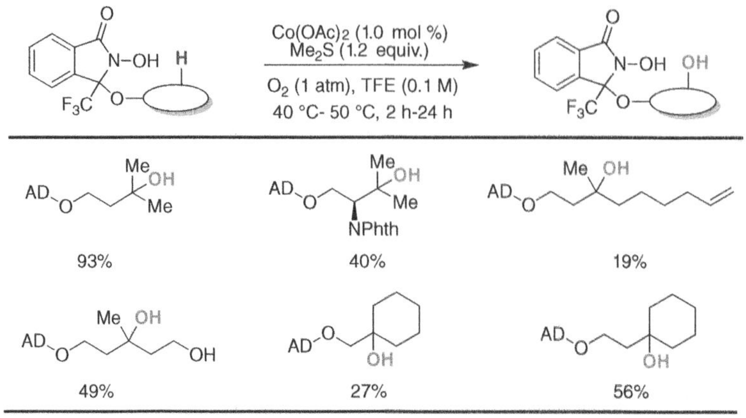

Scheme 1.13 Cobalt-catalyzed C—H hydroxylation for synthesis of alcohols.

must be solved in future studies. Preliminary structural tuning of the DA led to an alteration of the regioselectivity, providing a selective ultraremote aerobic C–H oxygenation.

Following the pioneering report by Schönecker and co-workers on the Cu-mediated C–H oxidation for the synthesis of 12β-hydroxy steroids [18], Baran and co-workers reinvestigated this fascinating Cu-mediated Schönecker oxidation; the only practical solution to the challenge of site-specific steroidal C12 functionalization was dramatically improved (Scheme 1.14) [19]. The new imine directing group and alternative reducing agent rendered this an operationally simple reaction that was no longer limited to a 50% maximum yield with long reaction times. Various types of ketone-derived substrates were prepared with acceptably isolated yields and short reaction time under the derived procedure.

A new mechanistic picture was proposed based on a series of NMR studies. Following initial Cu binding to give 16, additional uncoordinated Cu(I) and O_2 could complex to form the imine complex 17, a $[Cu_2O_2]$ species. The active Cu-species was likely the bis(μ-oxo)di-copper(III) complex 18, which could also be the mixed bis(μ-oxo)Cu(II)/Cu(III) complex. Oxidation of the proximal C—H bond then presumably occurred through an oxygen-rebound mechanism. The resulting Cu(II) that was not directly ligated to the substrate in the $[Cu_2O_2]$ complex 19 was then reduced by ascorbate to Cu(I) and released, allowing for further substrate engagement. In 2017, Holthausen and co-workers reported a detailed mechanistic DFT study of Schçnecker's reaction [20].

1.2.3 Oxidative Aromatization for Synthesis of Phenols

In 2011, Stahl and co-workers developed a palladium(II) catalyst system, incorporating an unconventional *ortho*-dimethylaminopyridine ligand, for the conversion of substituted cyclohexanones to the corresponding phenols (Scheme 1.15) [21]. This approach proved to be effective in the preparation of a number of substituted phenol derivatives. These reactions achieved high conversions and product yields, and the catalyst tolerated useful substrate functional groups, including aromatic and heteroatom substituents. This reactivity demonstrates a versatile and efficient strategy for the synthesis of substituted aromatic molecules with fundamentally different selectivity constraints from numerous known synthetic methods that rely on substitution of a pre-existing aromatic ring.

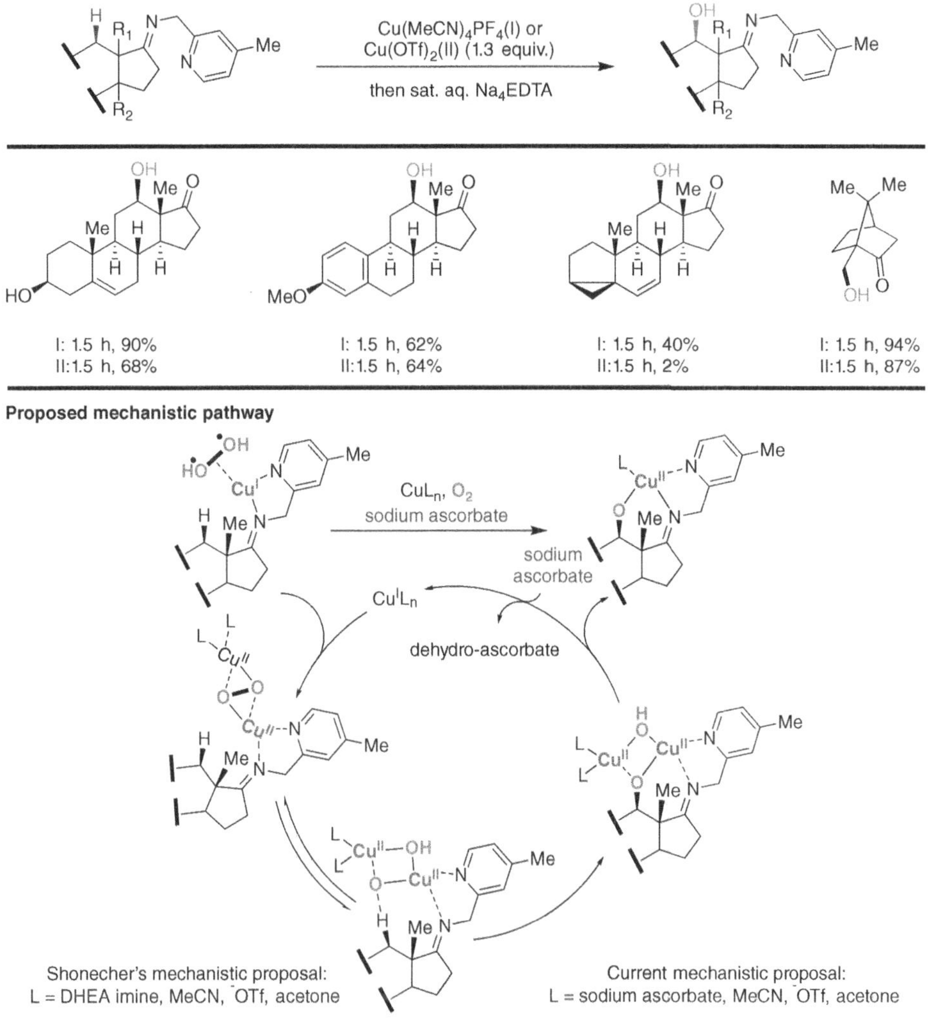

Scheme 1.14 Scalable C—H oxidation with copper for synthesis of polyoxypregnanes.

In palladium-catalyzed aerobic oxidation reactions, this transformation involves initial dehydrogenation of cyclohexanones via sequential Pd-mediated C—H activation and β-hydride elimination steps, followed by tautomerization of the resulting dienone product. This strategy is appealing because Pd^{II}–hydride intermediates formed in this mechanism could be oxidized by molecular oxygen, thereby enabling the overall process to be catalytic in Pd with water as the sole by-product (Scheme 1.16).

In 2013, Stahl's group disclosed that kinetic and mechanistic studies of these reactions revealed the key role of the dimethylsulfoxide (DMSO) ligand in controlling this chemoselectivity [22]. DMSO had minimal kinetic influence on the rate of $Pd(TFA)_2$-catalyzed dehydrogenation of cyclohexanone to cyclohexenone, while it strongly inhibited the second dehydrogenation step, the conversion of cyclohexenone to phenol. These contrasting kinetic effects of DMSO provided the basis for chemoselective formation of cyclohexenones.

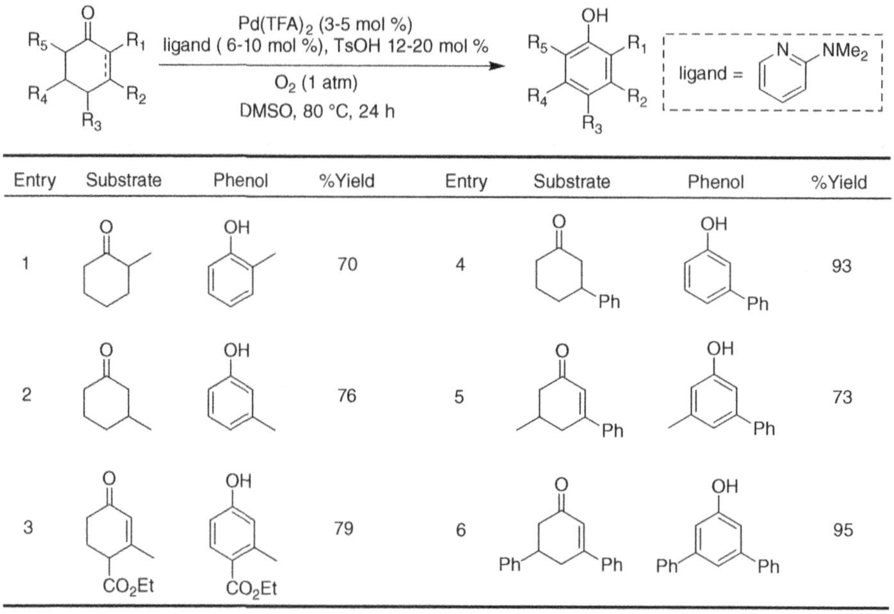

Scheme 1.15 Palladium-catalyzed aerobic dehydrogenation of substituted cyclohexanones to phenols.

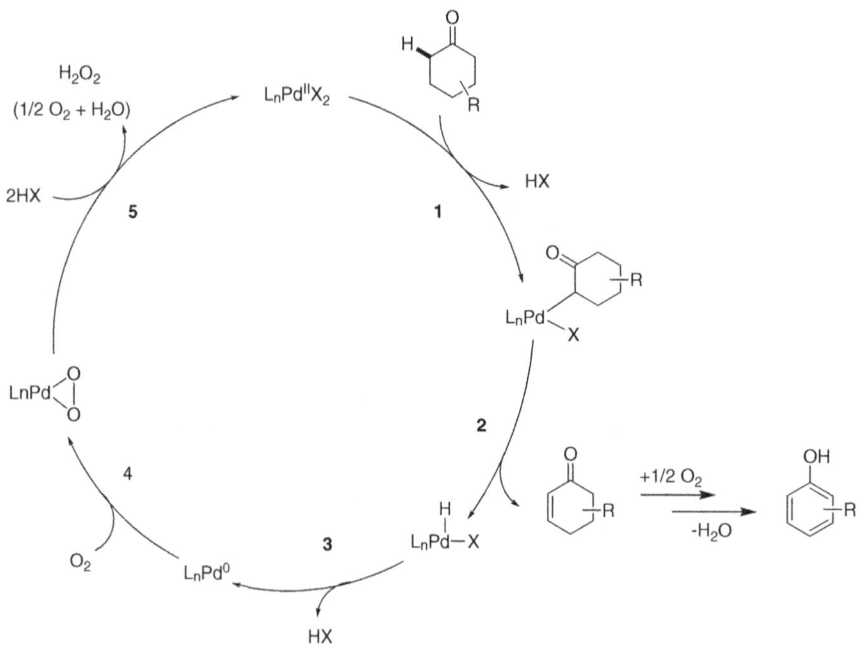

Scheme 1.16 Catalytic mechanism whereby cyclohexanone dehydrogenation can be achieved with O_2 as the terminal oxidant.

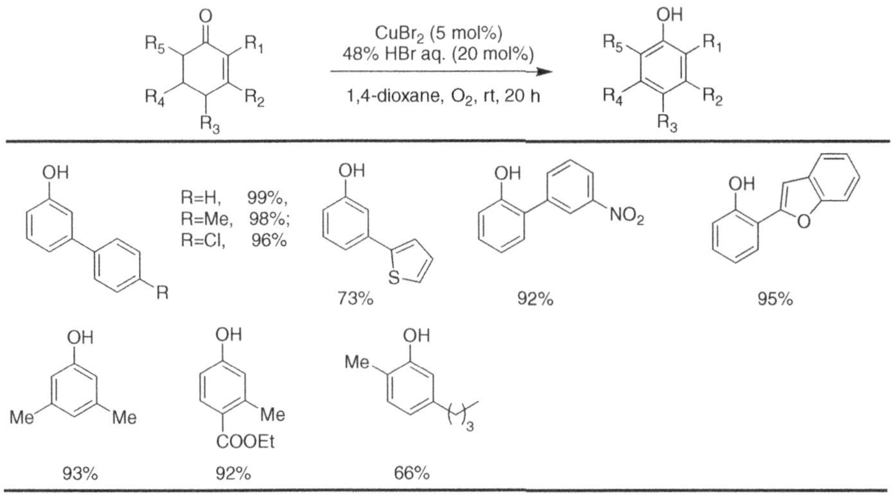

Scheme 1.17 Copper-catalyzed oxidative aromatization of 2-cyclohexen-1-ones for synthesis of phenol derivatives.

In the synthesis of phenol, various oxidative aromatizations of cyclohexenones are straightforward procedures for the preparation of multi-substituted phenol derivatives. Following this strategy, Nishina and co-workers developed a catalytic oxidative aromatization of 2-cyclohexen-1-ones for synthesis of phenol derivatives in the presence of a catalytic amount of copper salt and aqueous HBr under molecular oxygen in 2013 (Scheme 1.17) [23]. Various mono-, di-, and trisubstituted phenols with substituents at the desired positions could be synthesized under cheap and simple conditions. The authors proposed that copper(II) species involved generation of molecular bromine along with reduced copper(I) species, followed by bromination and dehydrobromination of 2-cyclohexen-1-one to give the desired phenol. The reduced copper(I) would be oxidized by molecular oxygen and HBr to regenerate the copper(II) species.

1.2.4 Oxygenation of Alkenes for Synthesis of Alcohols

Following the pioneering report by Feldman and Parvez on the thiol-olefin co-oxygenation reaction, Naito and co-workers developed a domino effect of reactions of vinylcyclopropyl oxime ethers involving thiyl radical addition, ring-opening, and hydroxylation reactions to form highly functionalized ε-thio-δ,γ-unsaturated-β-hydroxy oxime ethers in 2009, which are versatile intermediates in organic synthesis and play an important role in biological and chemical processes (Scheme 1.18) [24]. Importantly, this multicomponent reaction provided a single and efficient operation to form the allyl sulfide containing 1,3-amino alcohol compounds.

On the other hand, various saturated-β-hydroxy oxime ethers have been prepared by an intermolecular hydroxylation of α,β-unsaturated imines (Scheme 1.19) [25]. This reaction proceeds via two key transitional steps such as Et_3B-mediated regioselective alkyl radical addition and subsequent hydroxylation with molecular oxygen. This approach is characterized by mild conditions, is straightforward, and allows for the efficient and concomitant construction of a C—C bond and a C—O bond. In the proposed

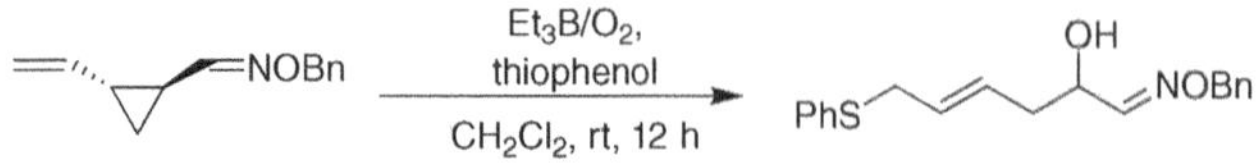

Entry	Substrate	Thiol (equiv.)	Yield
1	1E	PhSH	78%
2	1E	*p*-Cl-C_6H_4SH	88%
3	1E	*p*-MeO-C_6H_4SH	84%
4	1Z	*p*-Cl-C_6H_4SH	87%
5	1Z	*p*-MeO-C_6H_4SH	82%
6	1E/Z	*p*-Cl-C_6H_4SH	82%

Scheme 1.18 Hydroxysulfenylation of Vinylcyclopropyl Oxime.

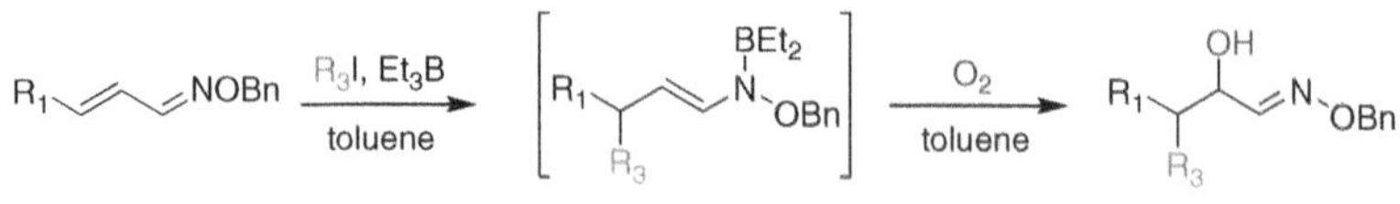

R_3 = i-Pr, c-pentyl, t-butyl, Et

Entry	Substrate	Conditions	Product	Yield (*anti/syn*)
1	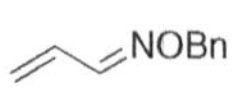	Et_3B (3.0 equiv), Me_3Al (2.2 equiv), O_2 bubbling in CH_2Cl_2	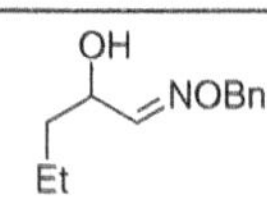	60%
2	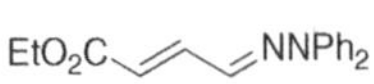	Et_3B (3.0 equiv), O_2 bubbling in toluene	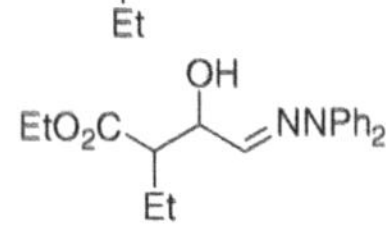	34% (3:2)
3	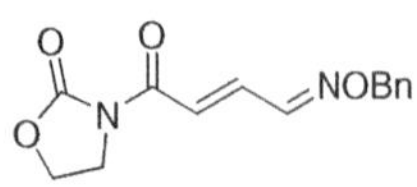	Et_3B (3.0 equiv), O_2 bubbling in toluene	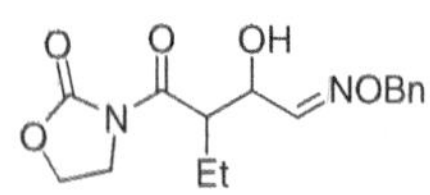	79% (17:1)
4	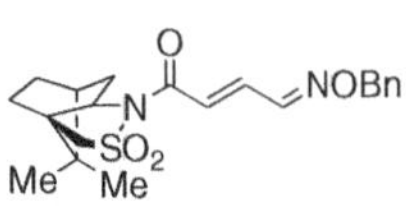	Et_3B (3.0 equiv), Me_3Al (2.2 equiv), O_2 bubbling in CH_2Cl_2	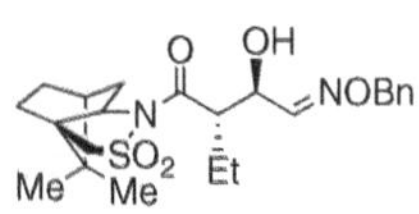	72% (3:1)

Scheme 1.19 Hydroxyalkylation of *N*-borylenamine.

reaction pathway, the first step involved regioselective alkyl radical addition assisting Et_3B for trapping of the enaminyl radical. The borylenamine was followed the oxidation by molecular oxygen to generate the desired alcohol.

In 2011, Taniguchi and co-workers reported a Fe-catalyzed oxidative arylation of alkenes by oxygen-mediated formation of aryl radicals from arylhydrazines (Scheme 1.20) [26]. Both peroxides and alcohols were smoothly obtained in satisfactory yields from a simple alkene in the presence of a catalytic amount of potassium ferrocyanide

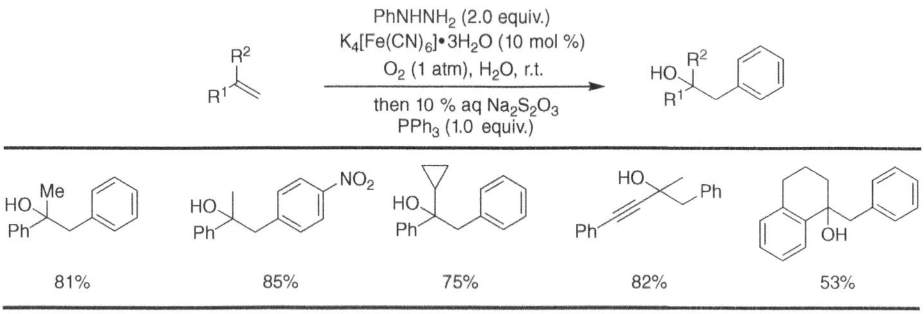

Scheme 1.20 Fe-catalyzed oxidative arylation of alkenes.

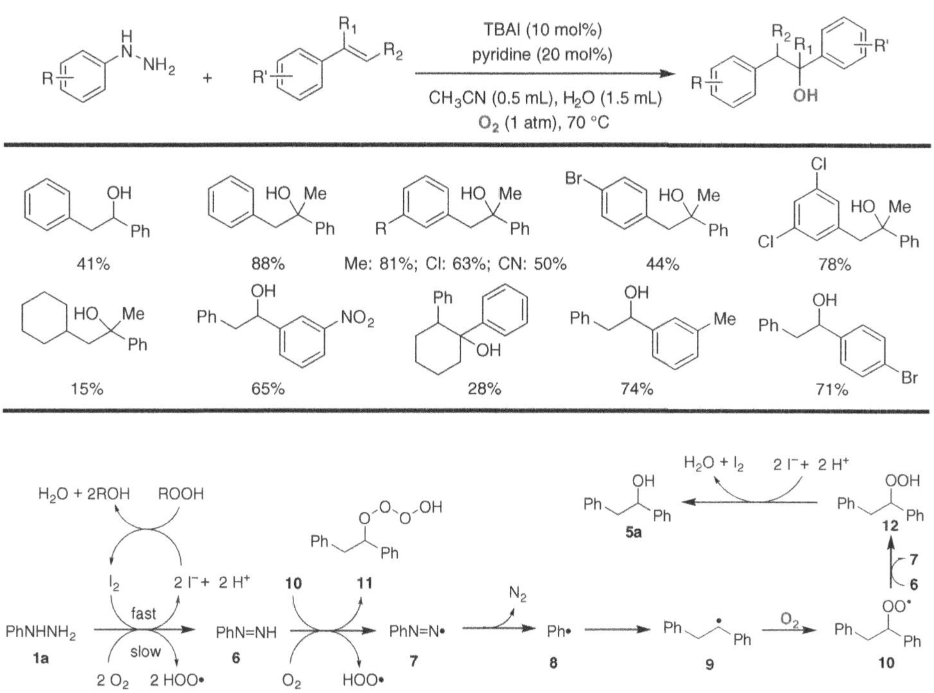

Scheme 1.21 Coupling and oxygenation of alkenes for the direct synthesis of alcohols.

trihydrate ($K_4[Fe(CN)_6]·3H_2O$) and water. This reaction showed excellently reactivity, regioselectivity, and good functional-group tolerance. Due to the involvement of environmentally friendly and inexpensive reagents ($K_4[Fe(CN)_6]·3H_2O$, oxygen gas, and water), this provided a mild and economical approach for the synthesis of peroxides and alcohols through C—C bond formation.

In 2013, Jiao and co-workers developed a novel highly chemoselective coupling and oxygenation of alkenes for the direct synthesis of alcohols (Scheme 1.21) [27]. Various 1,2-diarylethanols were obtained in moderate to good yields from the corresponding aryl hydrazines and substituted styrenes. Remarkably, molecular oxygen not only participated as an oxidant, but also underwent dioxygen activation through a radical process. The proposed mechanism for this transformation involved initial dehydrogenation of hydrazine via molecular oxygen and iodine to form phenyl radical **8** and a large amount of hydroperoxyl radical (HOO) with the release of N_2 in the presence of base, followed by the addition of alkene and molecular oxygen to generate the benzyl peroxy radical **10**, which captured a H atom from the strong H-donor **6** to provide hydroperoxide **12**, along with the formation of radical intermediate **7** to complete the chain propagation step. Subsequently, the hydroxylation product **5a** was generated by hydroperoxide **12** through the Landolt reaction with the regeneration of iodine.

In 2014, Taniguchi's group reported a unique 1,4-hydroxylation reaction of aliphatic alkenes by iron-catalyzed aerobic hydration and C(sp^3)-H hydroxylation (Scheme 1.22) [28]. Notably, this reaction enabled oxidation of all types of C(sp^3)—H bonds (methyl/primary,

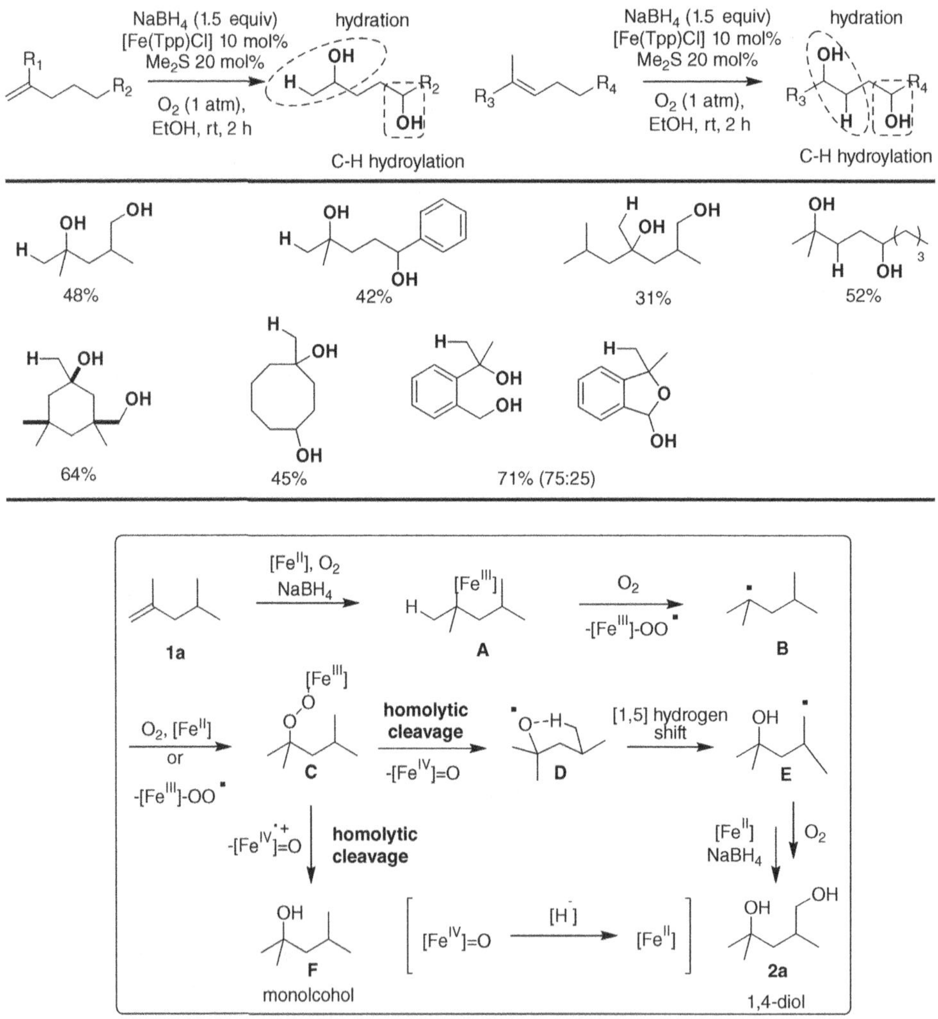

Scheme 1.22 Iron-catalyzed aerobic hydration and C—H hydroxylation.

methylene/secondary, and tertiary carbon centers) and allowed efficient generation of various 1,4-diols from simple alkenes using nontoxic and inexpensive reagents under mild conditions. The catalysis systems consisted of a user-friendly iron phthalocyanine complex, sodium borohydride. Molecular oxygen was the source for the two oxygen atoms of the 1,4-diols. This transformation of simple molecules into functionalized compounds was realized with a convenient and common reaction system. Based on their experiments, they proposed that a putative iron(III) hydride complex **A**, which was formed from iron phtholocyanine and sodium borohydride in the presence of oxygen, followed by the oxidation of molecular oxygen to initiate tertiary-carbon-centered radical **B** and iron peroxide complex **C**. Subsequently, highly reactive alkoxy radical **D** was able to undergo a [1,5]hydrogen shift to provide alkyl radical intermediate E, which was transformed into the resulting 1,4-diol through a similar pathway as the formation of complex C from alkyl radical **B**.

In 2015, Lei's group developed a selective radical dioxygenation of alkenes using hydroxamic acid and O_2, and cobalt was used as the catalyst without assistance from any additional ligands or bases (Scheme 1.23) [29]. Various α-oxo-tertiary alcohols were achieved from 1,1-disubstituted alkenes with good yields under mild conditions.

In 2015, Li and co-workers developed a general and practical method for hydroxysulfenylation of alkenes through an aerobic copper catalysis (Scheme 1.24) [30]. This method presented a selective and efficient synthesis of β-hydroxysulfides bearing

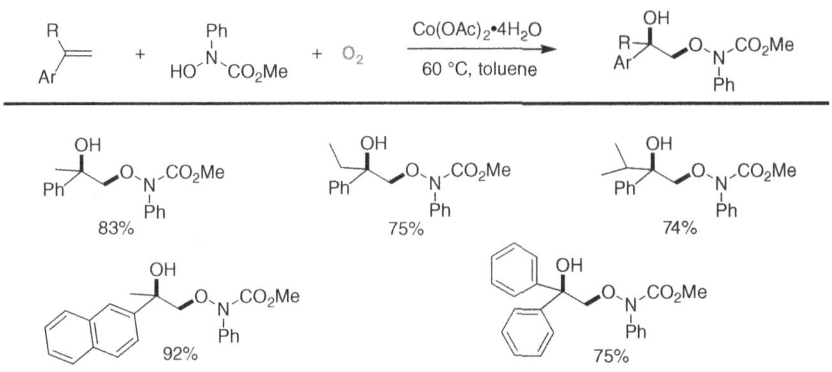

Scheme 1.23 Cobalt-catalyzed radical dioxygenation of 1,1-disubstituted alkenes.

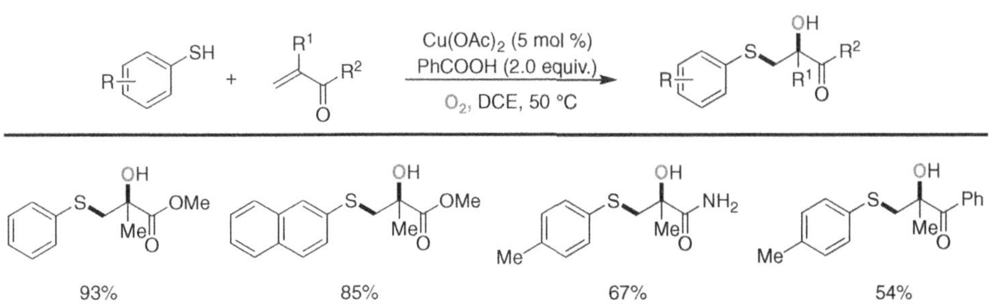

Scheme 1.24 Copper-catalyzed hydroxysulfenylation of alkenes.

electron-withdrawing groups, and a broad range of alkenes were applied smoothly for the hydroxysulfenylation under optimized conditions, including α,β-unsaturated esters, amides and ketone, styrenes, and the conjugated diene.

In 2010, Taniguchi and co-workers reported the Fe(III)-catalyzed oxidative addition of alkoxycarbonyl radicals to alkenes in air to give β-hydroxylesters (Scheme 1.25) [31]. This approach represents an environmentally friendly reaction, which employed the use of a cheap and nontoxic iron catalyst, readily available carbazate precursors, and an oxygen molecule green oxidant. Based on the knowledge that radical species could be generated from hydrazine compounds through the oxidative formation of diazenes, the alkoxycarbonyl radical is likely to be generated through a similar diazene pathway.

In 2015, Jiao and co-workers described the synthesis of β-azido alcohols through an efficient Mn-catalyzed aerobic oxidative hydroxyazidation of olefins at room temperature (Scheme 1.26). This approach showed broad substrate scope, inexpensive Mn

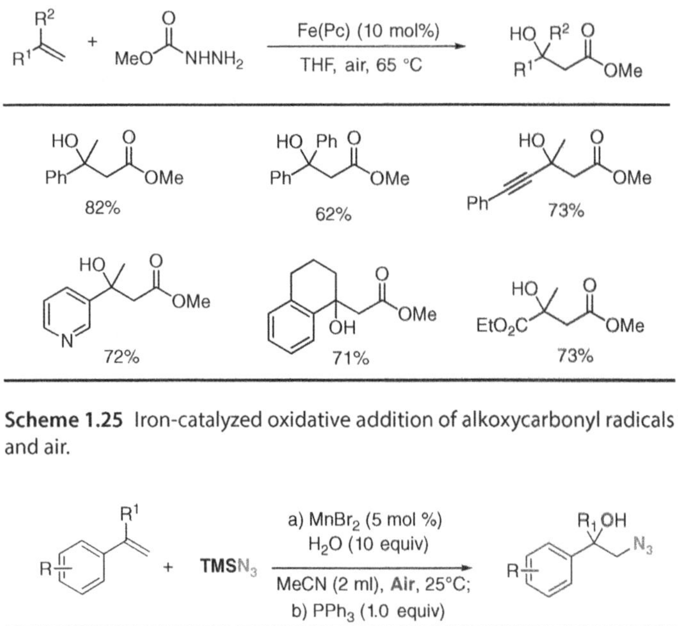

Scheme 1.25 Iron-catalyzed oxidative addition of alkoxycarbonyl radicals to alkenes with carbazates and air.

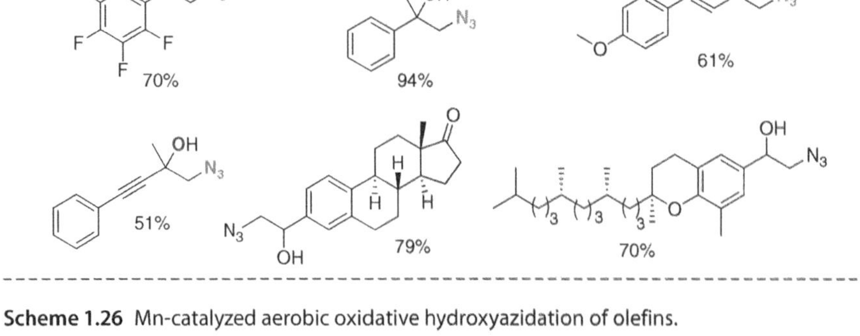

Scheme 1.26 Mn-catalyzed aerobic oxidative hydroxyazidation of olefins.

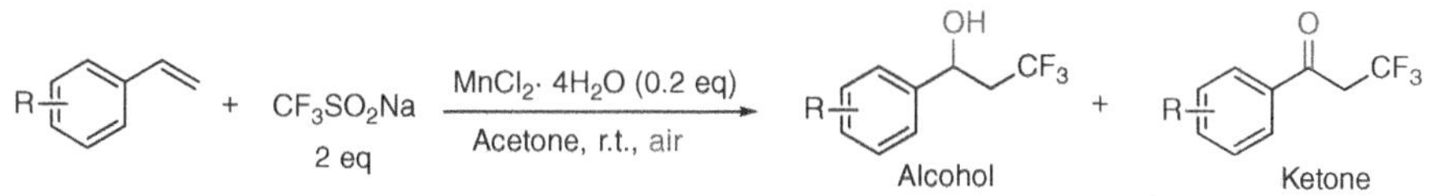

Entry	Substrate	ratio of A/K	Yield A+K
1	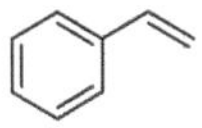	6:1	65+10%
2	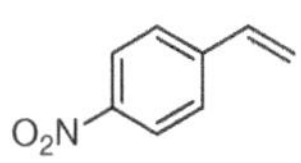	3:1	59+20%
3	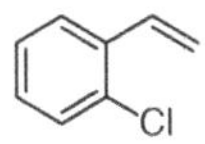	>10:1	60% (A)
4	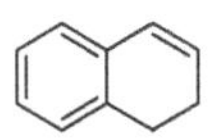	1.5:1	36+23%

Scheme 1.27 Mn-catalyzed aerobic oxytrifluoromethylation of styrene derivatives with CF_3SO_2Na.

catalyst, high efficiency, easy operation, and mild conditions. The authors applied this approach to the modification of complex bioactive molecules containing alkenyl group, such as (+)-δ-tocopherol.

In 2015, Vicic's group reported the activation of CF_3SO_2Na by Mn catalyst under aerobic conditions, leading to oxytrifluoromethylations of olefins [32]. For most alkene substrates examined, this approach gives the benzylic alcohol products in moderate yields accompanied by the formation of ketone as the minor products. It was proved that the ketone was not derived from the in situ oxidation of the benzylic alcohol. Based on experimental observations the authors proposed a free radical process as shown in Scheme 1.27.

In 2016, Kang's group reported the acetoxyhydroxylation of alkenes cooperatively catalyzed by *t*BuONO and palladium at room temperature using air as the sole oxidant (Scheme 1.28) [33]. Like other Pd-NO catalyzed aerobic processes, the NO generated in situ played a key role in the reaction. This protocol could be used to provide a divergent access to various vicinal diols, diacetoxyalkanes, and dihalogenoalkanes. However, this reaction could not proceed well with conjugate electron-withdrawing alkenes as well as internal alkenes.

1.3 C—H Hydroxylation via Hydrogen Peroxide

1.3.1 Synthesis of Phenols by C(sp^2)-H Hydroxylation

In 2009, Rybak-Akimova and co-workers described the regioselective hydroxylation of aromatic acids with hydrogen peroxide in the presence of iron(II) complexes with tetradentate aminopyridine ligands (Scheme 1.29) [34]. The hydroxylation of the aromatic ring occurred exclusively in the vicinity of the carboxylate functional group:

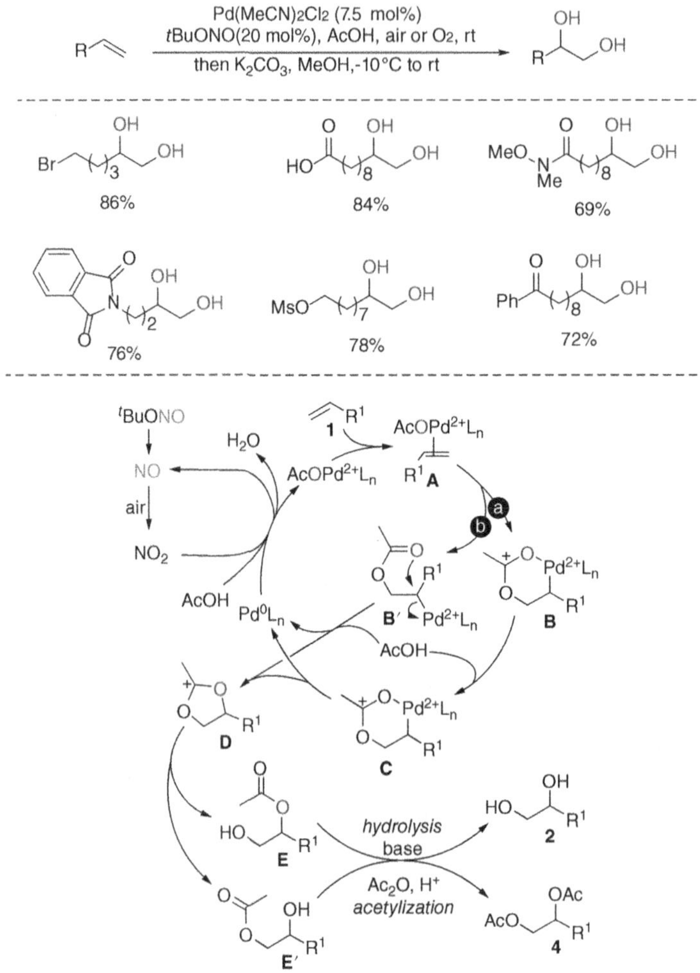

Scheme 1.28 Aerobic acetoxyhydroxylation of alkenes co-catalyzed by organic nitrite and palladium.

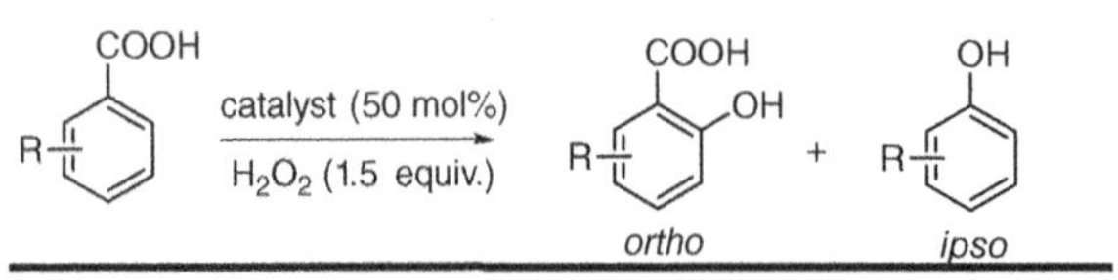

Substituent	Position	Yield of hydroxylated product [%] *ortho*	*ipso*
OMe	2-	-	70
	3-	52	-
	4-	-	22
Cl	2-	-	75
	3-	98	-
	4-	-	60
Cl	2-	52	14
	3-	56	-
	4-	68	-

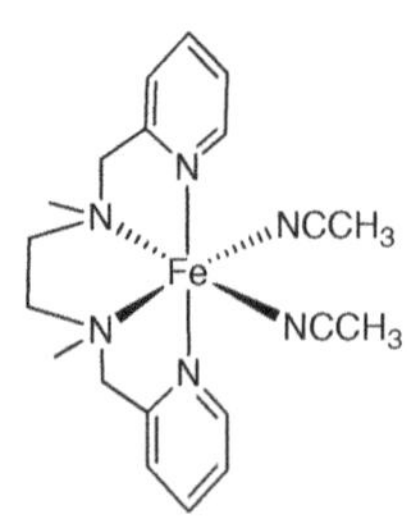

$[Fe^{II}(BPMEN)(CH_3CN)_2](ClO_4)_2$

Scheme 1.29 Iron-promoted *ortho*- and/or ipso-hydroxylation of benzoic acids with H_2O_2.

ortho-hydroxylation afforded salicylates, whereas 2- and, to a lesser extent, 4-substituted substrates tended to undergo ipso-hydroxylation through decarboxylation yielding phenolates.

As a continuous study of C—H hydroxylation, in 2012, Mizuno's group reported a divanadium-substituted phosphotungstate, $[\gamma\text{-}PW_{10}O_{38}V_2(m\text{-}OH)_2]^{3-}$ catalyzed oxidation system to direct hydroxylation of various structurally diverse arenes to phenols with H_2O_2 under mild conditions (Scheme 1.30) [35]. This study provided the first example of a synthetic catalyst that could chemoselectively hydroxylate the aromatic ring C(sp^2)—H bonds without causing the oxidation of the more reactive aromatic side-chain C(sp^2)—H bonds. With an excess of substrate relative to H_2O_2, the reaction exhibited high H_2O_2 efficiency.

In 2012, Zhao's group reported benzene hydroxylation catalyzed by vanadyl(IV) complex grafted PMOs using hydrogen peroxide as the terminal oxidant (Scheme 1.31). The PMO catalyst could be recovered and reused [36]. Under the best conditions, the benzene conversion could reach 27.4 with 100% selectivity to form phenol. A reversible V^{4+}/V^{5+} redox mechanism was proposed for the hydroxylation of benzene to phenol as depicted in Scheme 1.17.

In 2014, Kühn's group reported the hydroxylation of benzene and toluene by an Fe^{II} NHC complex using aqueous H_2O_2 as the oxidant (Scheme 1.32) [37]. The authors suspected that the reaction underwent an electrophilic mechanism by an Fe=O species. Under the reported conditions, benzene was hydroxylated with a high selectivity to phenol and toluene was hydroxylated to cresols with a good selectivity for the *ortho* and *para* positions. However, this reaction showed low H_2O_2 efficiency with less than 13% total conversion, even in the presence of 10 equiv. H_2O_2.

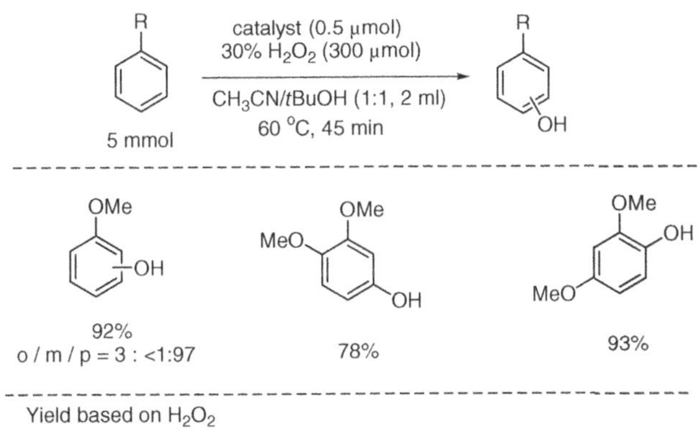

Scheme 1.30 Hydroxylation of arenes with H_2O_2 catalyzed by a divanadium-substituted phosphotungstate.

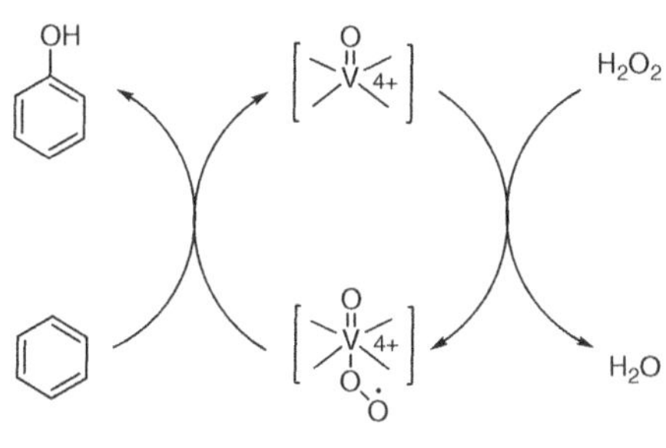

Scheme 1.31 Hydroxylation of benzene to phenol with H_2O_2 by vanadyl(IV) complex grafted PMOs.

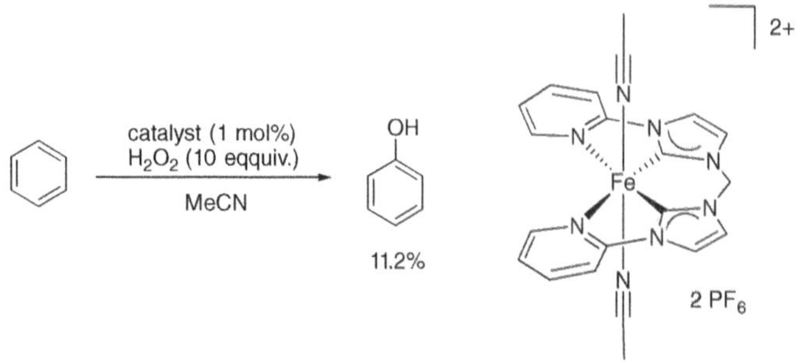

Scheme 1.32 Iron-complex-catalyzed hydroxylation of benzene with H_2O_2.

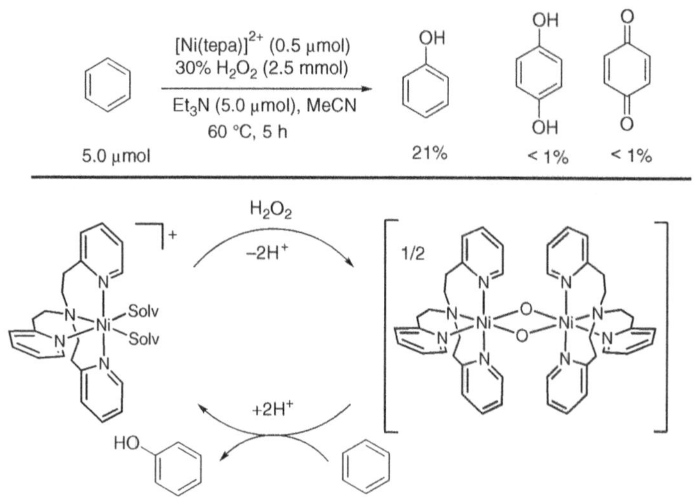

Scheme 1.33 Nickel-complex-catalyzed hydroxylation of benzene with H_2O_2.

In 2015, Itoh's group reported a benzene hydroxylation reaction employing nickel complexes supported by pyridylalkylamine ligands as the catalysts and H_2O_2 as the oxidant under mild conditions (Scheme 1.33) [38]. The maximum yield of phenol was 21% based on benzene without the formation of quinone or diphenol. However, high catalyst loading (10 mol%) and a large excess of H_2O_2 (500 equiv.) was employed. This catalyst could also be used for the hydroxylation of toluene, in which cresol was obtained as the major product with 90% *ortho* and *para* selectivity in the presence of 1 mol% catalyst and 5 equiv. H_2O_2.

In 2016, Fukuzumi reported the benzene hydroxylation with hydrogen peroxide in the presence of catalytic amounts of Cu(II) complexes ($[Cu(tmpa)]^{2+}$) in acetone to yield phenol at room temperature. The author also demonstrated the incorporation of the Cu(II) into mesoporous silica–alumina (Al-MCM-41) to make it a heterogeneous catalyst.

1.3.2 Synthesis of Alcohol by C(sp^3)-H Hydroxylation

As a great master of chemical synthesis, nature has devised many elegant metalloenzymes such as the heme-containing cytochrome P450 and the nonheme methane monooxygenase to enable biologically essential transformations under mild conditions.

Iron-containing cytochrome P450 selectively hydroxylates the long aliphatic side chain of cholesterol and thus plays an important role in the biosynthesis of the female hormone progesterone. Inspired by the extremely fine craftsmanship of enzyme active sites, continuous efforts have been devoted to the development of nonheme iron catalysts mimicking enzyme activities for selective C—H hydroxylations.

Based on the principles in oxidation catalysis with heme complexes and the seminal work reported by White, Costas' group in 2009 designed a chemically robust nonheme iron catalyst for the alkane hydroxylation using H_2O_2 as the oxidant (Scheme 1.34) [39]. The catalyst was proved to be highly efficient; by using 1 mol% of this iron complex, stereospecific, and site-predictable hydroxylation of alkanes was achieved in synthetically useful yields. The catalytic hydroxylation preferentially occurred at tertiary C—H bonds to give the 3° alcohol stereoselectively. The authors observed the formation of monomeric $[LFe^{III}(OR)(CF3SO3)]^+$ (OR = OH, OAc) species by mass spectroscopy, thus it was presumed that the formation of high-valent iron–oxo species was responsible for catalytic activity.

A small-molecule, nonheme iron hydroxylation catalyst possessing mixed hydroxylase/desaturase activity with carboxylic acid containing aliphatic substrates by using H_2O_2 as the terminal oxidant was reported by White's group in 2011 (Scheme 1.35) [40].

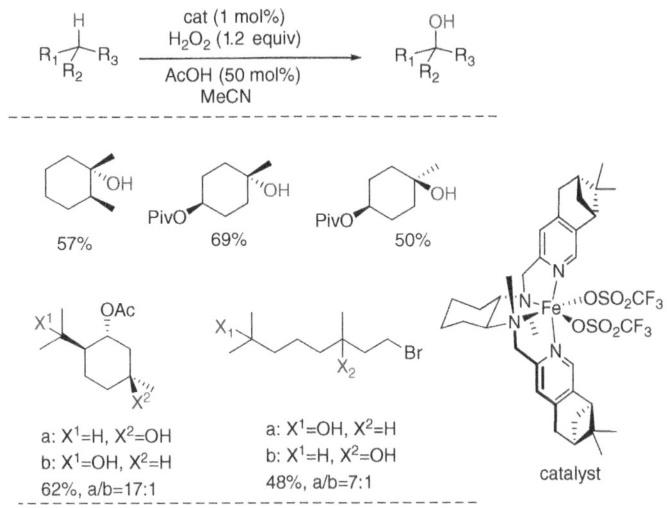

Scheme 1.34 Iron(II) catalyzed stereospecific C—H oxidation with H_2O_2.

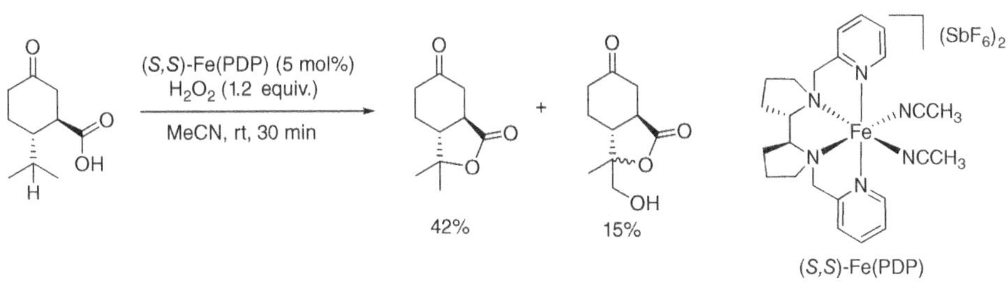

Scheme 1.35 Aliphatic C—H hydroxylations with H_2O_2 catalyzed by non-heme iron catalyst.

Under standard Fe(PDP)/H_2O_2 conditions, besides the anticipated butyrolactone products, unexpected hydroxylactone products derived from in situ generated olefins by the Fe(PDP) catalyst via desaturation were observed across a series of carboxylic acid substrates. Through the rearrangement of a taxane-based radical trap under Fe(PDP) oxidation, the authors provided the first direct evidence for the formation of a substrate radical using this class of stereoretentive hydroxylation catalysts.

Hitomi and co-workers demonstrated the first example of an iron(III)–monoamidate complex catalyzed efficient aliphatic C—H bond hydroxylation with H_2O_2 in 2012 (Scheme 1.36) [41]. The experimental result suggests that the reaction undergoes an iron(V) oxo mechanism. The catalytic system exhibited a high stereoretention rate and good regioselectivity. Previous examples mostly occupied acetic acid to increase both the selectivity and the efficiency. The acetic acid was supposed to facilitate the heterolytic scission of the O—O bond of Fe^{III}OOH species to form iron(V) oxo species, while in this case acetic acid was not needed.

In 2013, White's group reported a nonheme iron catalyst catalyzed aliphatic C—H oxidation over a range of topologically diverse substrates by hydrogen peroxide (Scheme 1.37) [42]. In most cases, the catalytic oxidation system gave a ketone in a good yield and a small amount of tertiary C—H bonded oxidation products (alcohol). However, the reaction could stop at the alcohol stage for some substrates.

In 2010, Mizuno's group reported the hydroxylation of alkanes through a bis(*m*-hydroxo) divanadium-substituted phosphotungstate/H_2O_2 system in high yields [43]. The bulky polyoxometalate framework of the catalyst showed an unusual selectivity to oxidize secondary rather than the weaker tertiary C—H bonds to give 2° alcohols in most cases. The oxidation system demonstrated mild conditions and good H_2O_2 efficiency ranging from 56 to 98%. Based on experimental observation, the authors proposed a catalytic cycle as shown in Scheme 1.38. First, the catalyst reacts with H_2O_2 to

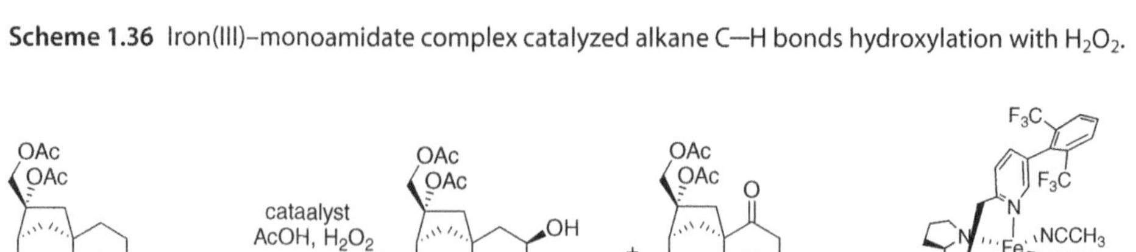

Scheme 1.36 Iron(III)–monoamidate complex catalyzed alkane C—H bonds hydroxylation with H_2O_2.

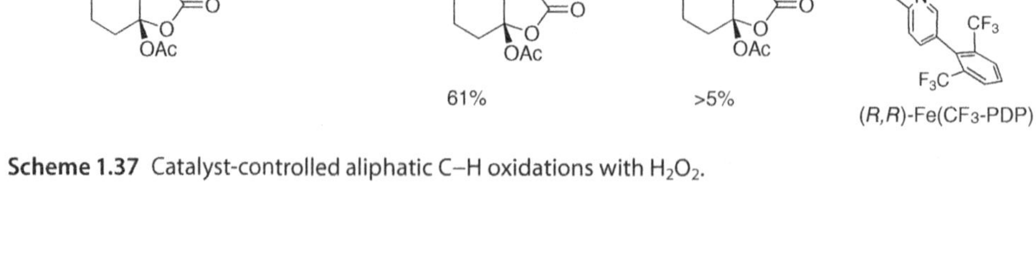

Scheme 1.37 Catalyst-controlled aliphatic C–H oxidations with H_2O_2.

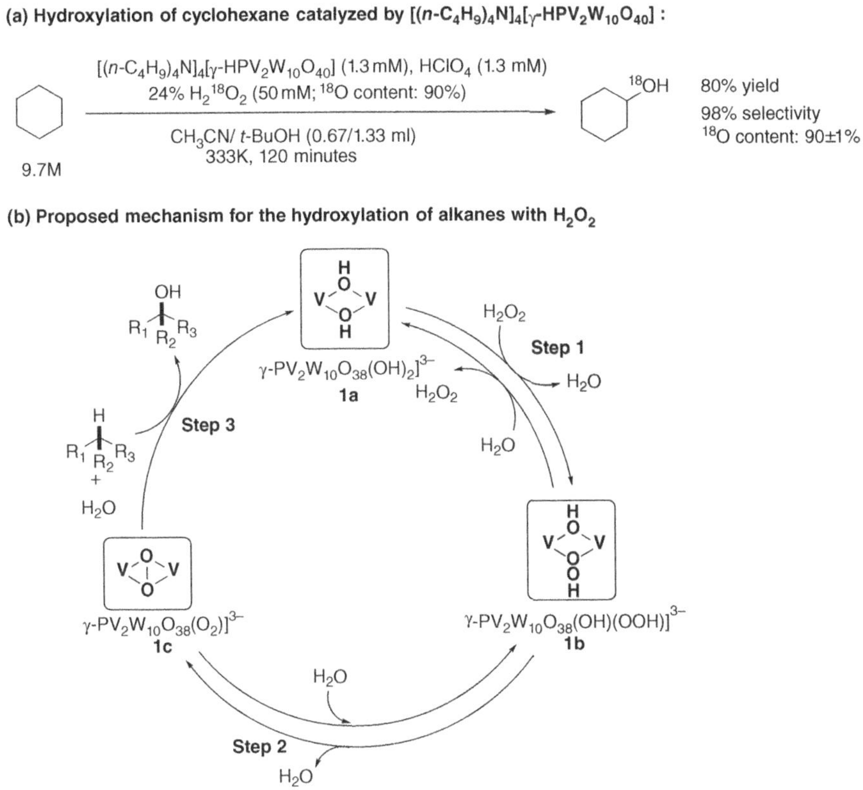

Scheme 1.38 Stereo- and regioselective hydroxylation of alkanes catalyzed by a bulky polyoxometalate.

generate $[\gamma\text{-}PV_2W_{10}O_{38}(OH)(OOH)]^{3-}$, which then goes through the dehydration to form the active oxidant. Eventually, that oxidant selectively oxidizes the alkane C—H bonds to furnish the hydroxylation product and the regenerated catalyst for the next turnover.

In addition to iron catalysts, copper and copper complexes have also been continuously studied for catalyzing the oxidation of alkanes by H_2O_2. In 2013, Pérez and co-workers reported that mononuclear Cu complexes bearing a trispyrazolylborate ligand could catalyze the oxidation of alkyl C—H bonds providing alcohols and/or ketones as the major products with small amounts of the alkenes in moderate yields (Scheme 1.39) [44]. Based on experimental data, the authors excluded the hydroxyl radical pathway. DFT studies supported the viability of the oxidation process through a copper-oxo active species.

In 2016, Siegler's group reported that Cu(I) salts, combined with commercially available ligand systems and H_2O_2, could catalyze the oxidation of C—H bonds with high efficiency (Scheme 1.40) [45]. This approach can be successfully applied to the oxidation of cyclohexane to produce cyclohexyl hydroperoxide as the main product and a 1:1 ratio of cyclohexanol and cyclohexanone at a roughly 50–60% total yield. Retention of the configuration was not observed in the tertiary C—H oxidation of *cis*-1,2-dimethylcyclohexane under these conditions, which excludes the mechanism of oxygen transfer from high-valent metal–oxo species. Combining other experimental evidence, the authors proposed that the mechanism involved C- and O-centered radical formation.

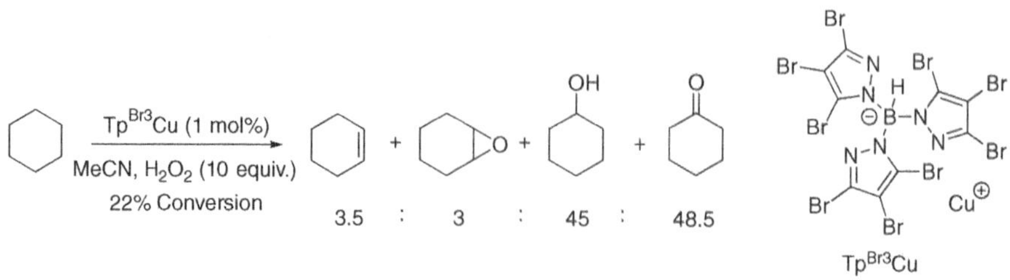

Scheme 1.39 Oxidation of alkanes with H_2O_2 catalyzed by a copper complex.

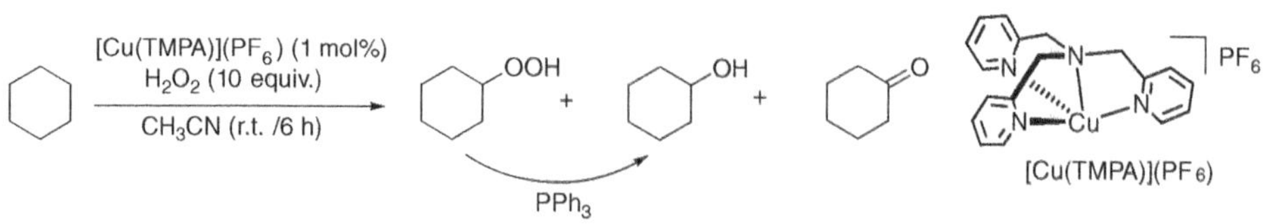

Scheme 1.40 Copper-catalyzed oxidation of alkanes with H_2O_2.

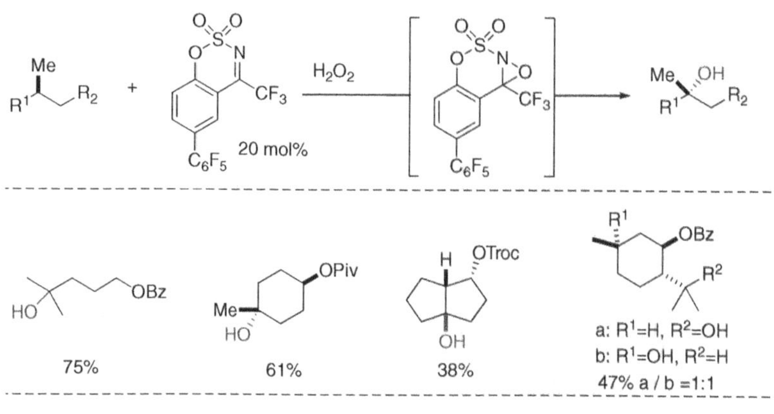

Scheme 1.41 C—H hydroxylation using a heterocyclic catalyst and aqueous H_2O_2.

Perfluorinated oxaziridines and dioxiranes have been known to be able to convert select alkanes to alcohols. However, the difficulties associated with the preparation or the inconvenience in using dilute solutions of these compounds prepared in advance have limited their use. Efforts have been made in the development of an analogous catalytic hydroxylation by using corresponding imine or ketone-based compounds to generate perfluorinated oxaziridines or dioxiranes in situ under oxidative conditions. As a result of continuous efforts seeking imine-based catalysts with application potential, in 2009, Du Bois' group described the catalytic reaction processes for selective C—H bond hydroxylation by using a benzoxathiazine catalyst and aqueous H_2O_2 (Scheme 1.41) [46]. The active oxidant was highly selective for tertiary C—H bonds over secondary C—H bonds.

In 2014, Hilinski's group introduced the first example of ketone-catalyzed oxidation of aliphatic C—H bonds to furnish the hydroxylation of tertiary C—H bonds using hydrogen peroxide as the terminal oxidant (Scheme 1.42) [47]. This catalytic hydroxylation reaction showed high selectivity for tertiary C—H bonds in trans-decalin and other

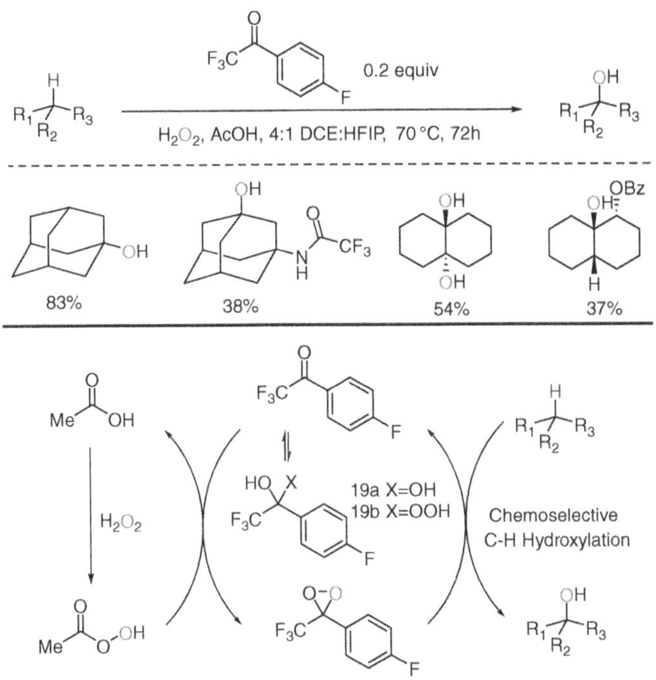

Scheme 1.42 Aliphatic C—H bonds hydroxylation using a ketone catalyst and aqueous H_2O_2.

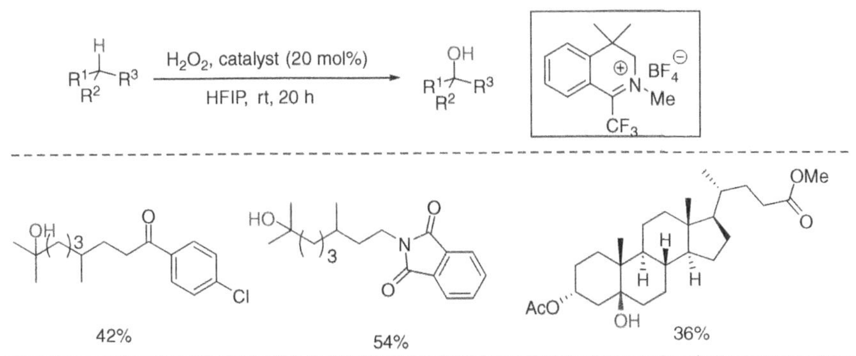

Scheme 1.43 Selective aliphatic C-H hydroxylation with H_2O_2 by iminium salt organocatalyst.

substrates. In contrast, the oxidation of trans-decalin by stoichiometric amount of TFDO gave a 2:1 ratio of 3°:2° oxidation products. The selectivity difference with traditional dioxirane oxidation may represent the advantage of this approach.

In 2016, Hilinski's group reported an example of aliphatic C—H hydroxylation by an iminium salt catalysis at room temperature using hydrogen peroxide as the terminal oxidant (Scheme 1.43) [48]. The reaction shares a similar catalytic mechanism with the previously mentioned benzoxathiazine catalysts.

In 2017, Takashi Ooi and co-workers reported an asymmetric α-hydroxylation of 3-substituted oxindoles under the catalysis of chiral 1,2,3-triazolium salts using

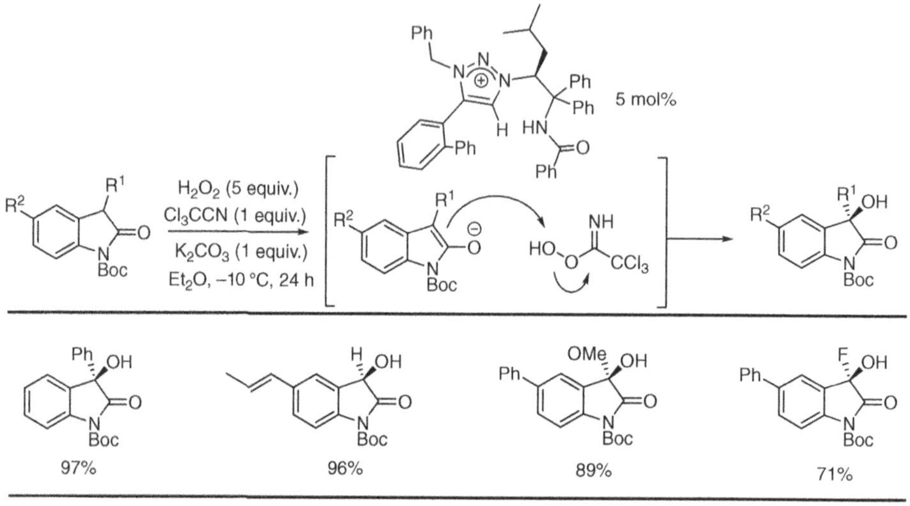

Scheme 1.44 Asymmetricα-hydroxylation of 3-substituted oxindoles with H_2O_2.

aqueous hydrogen peroxide as a terminal oxidant (Scheme 1.44) [49]. The combination of trichloroacetonitrile and H_2O_2 was crucial for the smooth transformation.

1.4 Photocatalytic Oxidation for Synthesis of Alcohols and Phenols

In 2010, Matsumura and co-workers proved that benzene could be oxidized by TiO_2 with UV light and oxygen to give the intermediate phenol (Scheme 1.45). However, the phenol was easily over-oxidized to carbon dioxide and water [50, 51]. In 2013, Sano's group further optimized the reaction conditions [52]. The layered silicate could be an excellent partner of a TiO_2 photocatalyst for efficient and selective green fine-chemical synthesis. The titanosilicate zeolites, a coexisting adsorbent composed of a silica framework and highly dispersed TiO_2, can promptly and selectively absorb phenol from a mixture solution, and efficiently prevent the over-oxidation of the phenol. Interestingly, in 2010 Ogawa's group found a benzene ring could be transformed into phenol with Au-nanoparticle-supported layered titanate as a visible-light-induced photocatalyst. The Au-nanoparticle selectively first oxidized benzene to phenol, thus obtaining a 62% yield and 96% selectivity for phenol formation under UV irradiation.

In 2015, Zhao and co-workers developed a visible-light-induced method for the hydroxytrifluoroethylation of styrenes under a molecular oxygen atmosphere in the

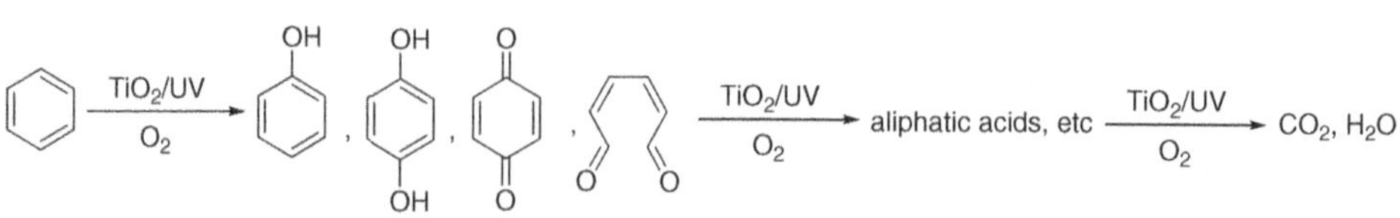

Scheme 1.45 Oxidation of Benzene on TiO_2 Photocatalysts in Aqueous Solutions.

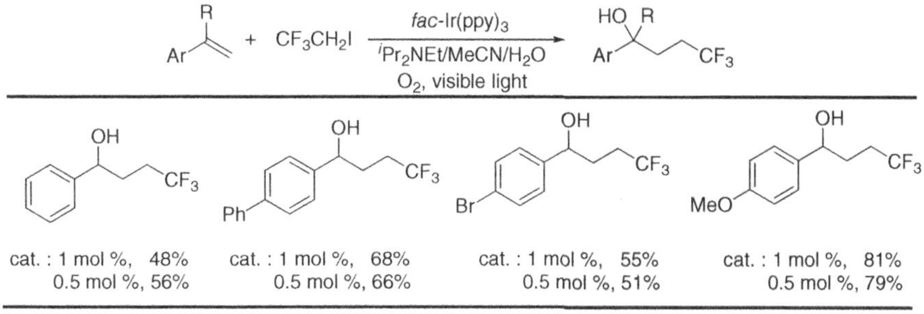

Scheme 1.46 Visible-light-induced hydroxytrifluoroethylation of styrenes under a molecular oxygen atmosphere.

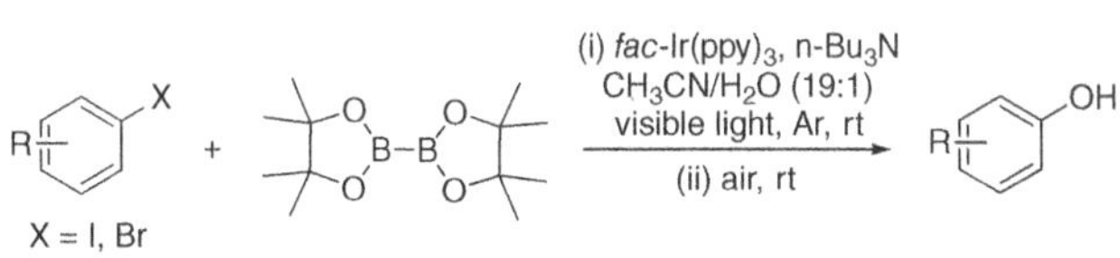

Scheme 1.47 Visible-light photoredox synthesis of phenols via a one-pot, two-step protocol.

presence of water (Scheme 1.46). The hydroxyl and trifluoroethyl groups were efficiently installed in one step, which is difficult to obtain by the conventional strategy [53]. The reaction was very compatible with various styrenes possessing electron-donating and halogen substituents at the aryl rings and also with substituents at the β-positions of double bonds. Importantly, the oxygen atom in the product originates from molecular oxygen in this radical reaction.

In 2016, Fu and co-workers illustrated a visible-light photoredox borylation of aryl haildes and subsequent aerobic oxidative hydroxylation at room temperature (Scheme 1.47) [54]. The methods show some advantages including simple equipment, mild conditions, easy operation, and wide substrate scope. In this reaction, commercially available aryl halides and bis(pinacolato)diboron as the starting materials, fac-Ir(ppy)$_3$ as the photocatalyst, and corresponding arylboronic esters and phenols were obtained in good yields. When the borylation was complete, the reaction was allowed to be exposed to air, and subsequent aerobic oxidation hydroxylation of arylboronic ester provided the corresponding phenols.

In 2016, Gao and co-workers reported a photo-organocatalytic enantioselective α-hydroxylation of β-dicarbonyl compounds by a series of Cinchona-derived *N*-oxide phase-transfer catalysts (Scheme 1.48) [55]. Moreover, the catalysts could be recycled and reused six times for such reactions with almost the original reactivity and enantioselectivity. Molecular oxygen was the sole oxidant in this reaction.

In 2016, Tung and co-workers came up with a blueprint for aromatic C—H functionalization via a combination of photocatalysis and cobalt catalysis (Scheme 1.49) [56]. They described the utility of this strategy for benzene hydroxylation as well as amination. As shown in Scheme 1.49, photocatalyst(PC$^+$) uses as the form of onium. When irradiated by UV light, the onium could generate excited state(s) of photocatalyst(PC^{+*}), which could then undergo single-electron transfer from benzene to produce benzene radical cation and photocatalyst radical(PC•). This may have then given an electron to

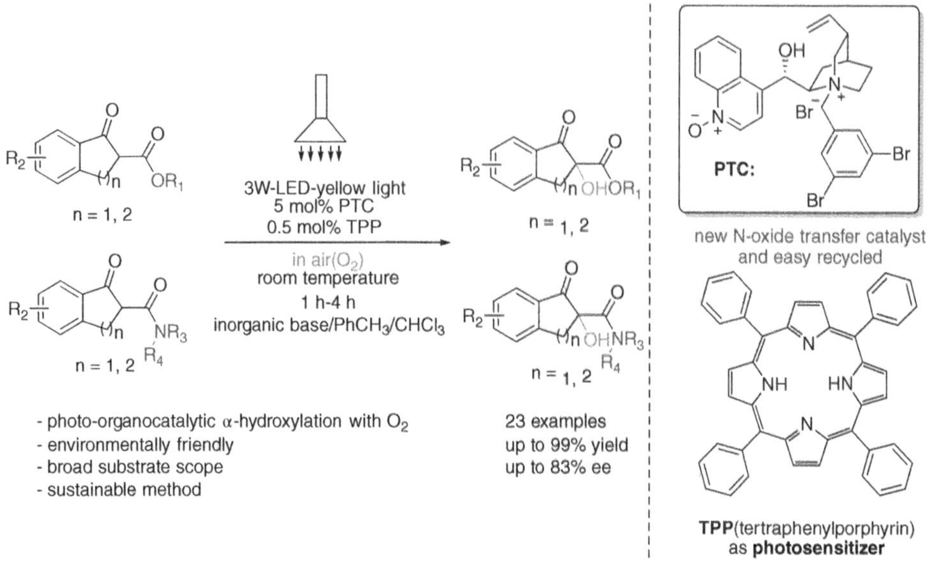

Scheme 1.48 Asymmetric α-hydroxylation of β-keto esters and β-keto amides by Cinchona-derived *N*-oxide phase-transfer catalysts.

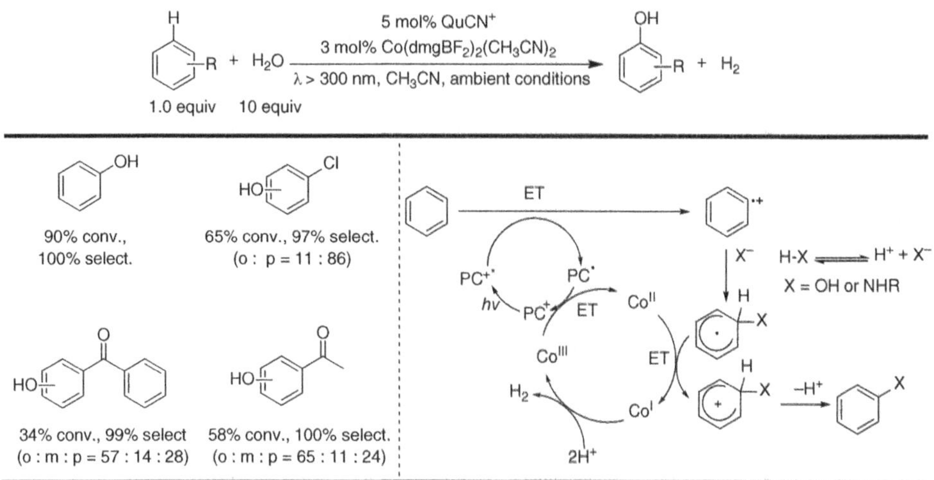

Scheme 1.49 Visible-light-induced hydroxytrifluoroethylation of styrenes under a molecular oxygen atmosphere.

the metal cocatalyst to produce Co^{II} and ground state photocatalyst(PC^{+}), completing the photocatalysis cycle. They used the dual catalyst system to produce phenol directly from benzene and water with evolution of hydrogen gas under unusually mild conditions in good yield and selectivity. These synthetic reactions show significant atom economy and step economy.

1.5 Enzymatic Oxidation and Electrochemical Oxidation for Synthesis of Alcohols

In 2011, Reetz and co-workers reported region- and stereoselectivity of P450-catalyzed hydroxylation of steroids (Scheme 1.50) [57]. According to the previous result, cytochrome P450 enzymes could enable C—H activation at non-activated positions. However, the simultaneous control of both region and stereoselectivity is problematic. In this work, starting from P450 BM(F87A) and using progesterone as a substrate, they demonstrated a series of mutagenesis could obtain 96–97% selectivity for either of the two regioisomers. The result showed the mutants could selectively oxidize them without performing an additional mutagenesis experiment. The shapes of the binding pockets of mutants have been evolved to enforce, in each case, a single active positional orientation of the substrate, setting the stage for the respective regio- and diastereoselective oxidation.

In 2014, Siegel's group demonstrated a protocol to generate phthaloyl peroxide in flow for the hydroxylation of arenes (Scheme 1.51) [58]. Improving on earlier batch procedures, the flow protocol they developed could directly yield phthaloyl peroxide in high purity (>95%) and be used to bypass the need to isolate and recrystallize phthaloyl peroxide. The flow protocol for the formation of phthaloyl peroxide could be combined with arene hydroxylation reactions and provided a method for the consumption of peroxide as it was generated to minimize the accumulation of large quantities of peroxide so as to make this protocol more efficient and environmentally friendly.

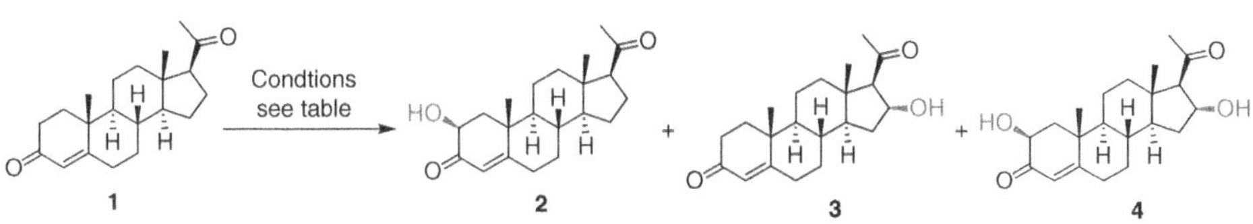

Mutant	Library	Mutations	2	3	4	Others*
Starting enzyme	-	F87A	18	82	0	0
KSA-1	Lib 330NNK	A330W/F87A	9	91	0	0
KSA-4	Lib A	R47Y/T49F/F87A†	23	63	12	2
KSA-5	Lib-B	V78L/A82F/F87A	26	51	12	12
KSA-7	Lib B	V78I/A82F/F87A	30	59	3	8
KSA-9	Lib B	V78V/A82N/F87A†	100	0	0	0
KSA-10	Lib B	V78C/A82G/F87A	13	81	4	3
KSA-14	Lib AB	R47Y/T49F/V78L/A82M/F87A‡	11	64	22	4
KSA-15	Lib AB	R47Y/T49F/V78L/A82M/F87A‡	8	75	14	4
KSA-16	Lib AB	R47Y/T49F/V78L/A82M/F87A‡	57	37	0	6

Conditions: 1 mM progesterone, 24 h, 25 °C, resting cells containing expressed P450 mutants. *Traces of side products were identified as 17a, 21-dihydroxyprogesterone and 21-hydroxyprogesterone by HPLC. †Wild-type amino acid was retained in at least one position; mutant KSA-9 contains a silent mutation V78V (GTA GTG). ‡Mutant KSA-4 was used as a parent and site B was addressed. Mean values are given, standard deviation for conversion is+5% and for selectivity+1%. Differences from 100% occur from rounding.

Scheme 1.50 Regio- and stereoselectivity of P450-catalyzed hydroxylation of steroids controlled by laboratory evolution.

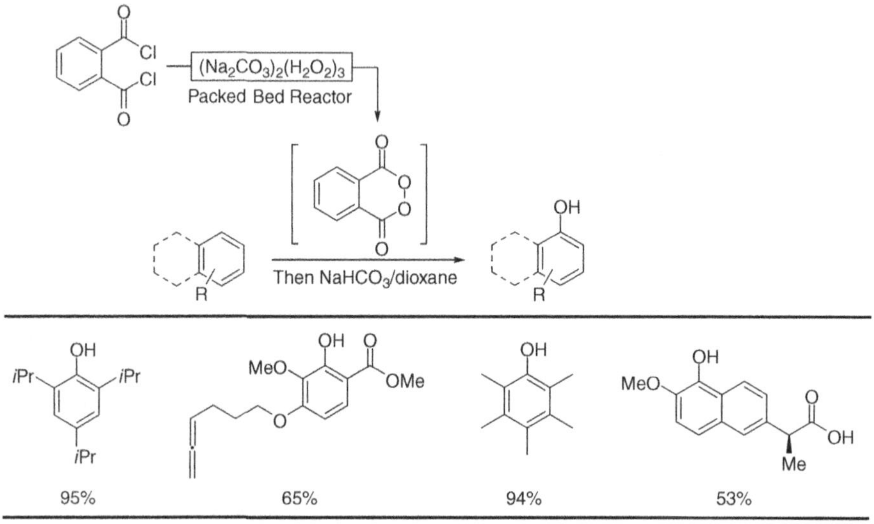

Scheme 1.51 The hydroxylation of arenes by generating phthaloyl peroxide in flow.

1.6 Conclusion and Perspectives

During the past few years, the green oxidative approach has emerged as an increasingly viable tool for the step-economical and sustainable functionalization and assembly for alcohols and phenols. The results summarized in this review highlight numerous important advances that have been made in the development of green oxidative reaction for synthesis of alcohols and phenols using molecular oxygen, air, and hydrogen peroxide as the oxidants. Furthermore, other green strategies such as the enzymatic reaction, photoredox reaction, and electrochemical oxidation were illustrated here. We have highlighted this progress in the areas of C—H hydroxylation, oxygenation of alkenes, dehydrogenative oxidation, and photoredox oxidation.

Despite momentous achievements, green oxidative synthesis of alcohols and phenols is still in an early stage of development, being used to install a hydroxyl group in some limited position and some simple scaffolds. We should recognize that only a small part of the transformation in these oxidative processes uses green oxidants as the terminal oxidant. However, these approaches provide environmentally friendly and step-economical methods. Many challenges remain to be addressed in the future development of green oxidative approaches, such as unactivated C(sp^3)-H hydroxylation, photocatalysis, and enzymatic reaction. A lot of effort is still required in order to extend in this field. It is believed that further research into new, highly active, and selective reaction development in the green oxidative synthesis of alcohols and phenols will continue to contribute toward the future progression of newly sustainable chemistry, which will, in turn, extend the utility of organic synthesis.

References

1 Zhang, Y.-H. and Yu, J.-Q. (2009). *J. Am. Chem. Soc.* 131: 14654–14655.
2 Liu, Q., Wu, P., Yang, Y. et al. (2012). *Angew. Chem. Int. Ed.* 51: 4666–4670, S4666/4661-S4666/4641.
3 Yan, Y., Feng, P., Zheng, Q.-Z. et al. (2013). *Angew. Chem. Int. Ed.* 52: 5827–5831.
4 Sun, S.-Z., Shang, M., Wang, H.-L. et al. (2015). *J. Org. Chem.* 80: 8843–8848.
5 Ozawa, J., Tashiro, M., Ni, J. et al. (2016). *Chem. Sci.* 7: 1904–1909.
6 Das, P., Saha, D., Saha, D., and Guin, J. (2016). *ACS Catal.* 6: 6050–6054.
7 Wendlandt, A.E., Suess, A.M., and Stahl, S.S. (2011). *Angew. Chem. Int. Ed.* 50: 11062–11087, S11062/11061-S11062/11062.
8 Garcia-Bosch, I., Company, A., Frisch, J.R. et al. (2010). *Angew. Chem. Int. Ed.* 49: 2406–2409, S2406/2401-S2406/2441.
9 Garcia-Bosch, I., Ribas, X., and Costas, M. (2012). *Chem. Eur. J.* 18: 2113–2122, S2113/2111-S2113/2118.
10 (a) Powers, D.C. and Ritter, T. (2009). *Nat. Chem.* 1: 302–309. (b) Powers, D.C., Xiao, D.Y., Geibel, M.A.L., and Ritter, T. (2010). *J. Am. Chem. Soc.* 132: 14530–14536.
11 Chuang, G.J., Wang, W., Lee, E., and Ritter, T. (2011). *J. Am. Chem. Soc.* 133: 1760–1762.
12 Yang, Y., Moinodeen, F., Chin, W. et al. (2012). *Org. Lett.* 14: 4762–4765.
13 Li, X., Jin, C., and Gu, L. (2015). *J. Org. Chem.* 80: 2443–2447.
14 Liang, Y.-F. and Jiao, N. (2014). *Angew. Chem. Int. Ed.* 53: 548–552.
15 Liang, Y.-F., Wu, K., Liu, Z. et al. (2015). *Sci. China Chem.* 58: 1334–1339.
16 Sim, S.-B.D., Wang, M., and Zhao, Y. (2015). *ACS Catal.* 5: 3609–3612.
17 Tsang, A.S.K., Kapat, A., and Schoenebeck, F. (2016). *J. Am. Chem. Soc.* 138: 518–526.
18 Schoenecker, B., Zheldakova, T., Liu, Y. et al. (2003). *Angew. Chem. Int. Ed.* 42: 3240–3244.
19 See, Y.Y., Herrmann, A.T., Aihara, Y., and Baran, P.S. (2015). *J. Am. Chem. Soc.* 137: 13776–13779.
20 Gupta, P., Diefenbach, M., Holthausen, M.C., and Foerster, M. (2017). *Chem. Eur. J.* 23: 1427–1435.
21 Izawa, Y., Pun, D., and Stahl, S.S. (2011). *Science (Washington, DC, U. S.)* 333: 209–213.
22 Izawa, Y., Zheng, C., and Stahl, S.S. (2013). *Angew. Chem. Int. Ed.* 52: 3672–3675.
23 Kikushima, K. and Nishina, Y. (2013). *RSC Adv.* 3: 20150–20156.
24 Rahaman, H., Ueda, M., Miyata, O., and Naito, T. (2009). *Org. Lett.* 11: 2651–2654.
25 Ueda, M., Miyabe, H., Kimura, T. et al. (2009). *Org. Lett.* 11: 4632–4635.
26 Taniguchi, T., Zaimoku, H., and Ishibashi, H. (2011). *Chem. Eur. J.* 17: 4307–4312.
27 Su, Y., Sun, X., Wu, G., and Jiao, N. (2013). *Angew. Chem. Int. Ed.* 52: 9808–9812.
28 Hashimoto, T., Hirose, D., and Taniguchi, T. (2014). *Angew. Chem. Int. Ed.* 53: 2730–2734.
29 Lu, Q., Liu, Z., Luo, Y. et al. (2015). *Org. Lett.* 17: 3402–3405.
30 Xi, H., Deng, B., Zong, Z. et al. (2015). *Org. Lett.* 17: 1180–1183.
31 Taniguchi, T., Sugiura, Y., Zaimoku, H., and Ishibashi, H. (2010). *Angew. Chem. Int. Ed.* 49: 10154–10157.
32 Yang, Y., Liu, Y., Jiang, Y. et al. (2015). *J. Org. Chem.* 80: 6639–6648.
33 Chen, X.-M., Ning, X.-S., and Kang, Y.-B. (2016). *Org. Lett.* 18: 5368–5371.

34 Makhlynets, O.V., Das, P., Taktak, S. et al. (2009). *Chem. Eur. J.* 15: 13171–13180, S13171/13171-S13171/13118.
35 Kamata, K., Yamaura, T., and Mizuno, N. (2012). *Angew. Chem. Int. Ed.* 51: 7275–7278, S7275/7271-S7275/7217.
36 Borah, P., Ma, X., Nguyen, K.T., and Zhao, Y. (2012). *Angew. Chem. Int. Ed.* 51: 7756–7761, S7756/7751-S7756/7757.
37 Raba, A., Cokoja, M., Herrmann, W.A., and Kuehn, F.E. (2014). *Chem. Commun.* 50: 11454–11457.
38 Morimoto, Y., Bunno, S., Fujieda, N. et al. (2015). *J. Am. Chem. Soc.* 137: 5867–5870.
39 Gomez, L., Garcia-Bosch, I., Company, A. et al. (2009). *Angew. Chem. Int. Ed.* 48: 5720–5723, S5720/5721-S5720/5725.
40 Bigi, M.A., Reed, S.A., and White, M.C. (2011). *Nat. Chem.* 3: 216–222.
41 Hitomi, Y., Arakawa, K., Funabiki, T., and Kodera, M. (2012). *Angew. Chem. Int. Ed.* 51: 3448–3452, S3448/3441-S3448/3419.
42 Gormisky, P.E. and White, M.C. (2013). *J. Am. Chem. Soc.* 135: 14052–14055.
43 Kamata, K., Yonehara, K., Nakagawa, Y. et al. (2010). *Nat. Chem.* 2: 478–483.
44 Conde, A., Vilella, L., Balcells, D. et al. (2013). *J. Am. Chem. Soc.* 135: 3887–3896.
45 Garcia-Bosch, I. and Siegler, M.A. (2016). *Angew. Chem. Int. Ed.* 55: 12873–12876.
46 Litvinas, N.D., Brodsky, B.H., and Du Bois, J. (2009). *Angew. Chem. Int. Ed.* 48: 4513–4516, S4513/4511-S4513/4512.
47 Pierce, C.J. and Hilinski, M.K. (2014). *Org. Lett.* 16: 6504–6507.
48 Wang, D., Shuler, W.G., Pierce, C.J., and Hilinski, M.K. (2016). *Org. Lett.* 18: 3826–3829.
49 Ohmatsu, K., Ando, Y., and Ooi, T. (2017). *Synlett* 28: 1291–1294.
50 Ide, Y., Matsuoka, M., and Ogawa, M. (2010). *J. Am. Chem. Soc.* 132: 16762–16764.
51 Bui, T.D., Kimura, A., Ikeda, S., and Matsumura, M. (2010). *J. Am. Chem. Soc.* 132: 8453–8458.
52 Ide, Y., Torii, M., and Sano, T. (2013). *J. Am. Chem. Soc.* 135: 11784–11786.
53 Li, L., Huang, M., Liu, C. et al. (2015). *Org. Lett.* 17: 4714–4717.
54 Jiang, M., Yang, H., and Fu, H. (2016). *Org. Lett.* 18: 5248–5251.
55 Wang, Y., Yin, H., Tang, X. et al. (2016). *J. Org. Chem.* 81: 7042–7050.
56 Zheng, Y.-W., Chen, B., Ye, P. et al. (2016). *J. Am. Chem. Soc.* 138: 10080–10083.
57 Kille, S., Zilly, F.E., Acevedo, J.P., and Reetz, M.T. (2011). *Nat. Chem.* 3: 738–743.
58 Eliasen, A.M., Thedford, R.P., Claussen, K.R. et al. (2014). *Org. Lett.* 16: 3628–3631.

2

Green Oxidative Synthesis of Aldehydes and Ketones

Yoshiharu Iwabuchi

Department of Organic Chemistry, Graduate School of Pharmaceutical Sciences, Tohoku University, Sendai, Japan

2.1 Introduction

Molecules equipped with a carbonyl group offer chemists an exceptional utility in organic chemistry, birthing the metaphor "carbonyl compounds are virtually the backbone of organic synthesis" [1]. Because of their indispensable roles, development of more selective, efficient, and economical methods capable of accessing carbonyl compounds have continuously been the central subject in chemical sciences. The oxidative approaches to synthesize carbonyl compounds are shown pictorially as the replacement of a functional group (or atom) on the substrate with a carbonyl group, in which most commonly oxidized are C_{sp3}—H, C_{sp2}=C_{sp2}, and C_{sp3}—O_{sp3} bonds (Scheme 2.1).

In the early days of synthetic organic chemistry, these transformations had mostly been conducted by using stoichiometric amounts of high valent metal oxidants such as chromate or permanganate. However, these traditional oxidants have been gradually discarded due to their economic and environmental costs in favor of green oxidants such as molecular oxygen and hydrogen peroxide [2]. For green oxidation (oxygenation) of the previously mentioned bonds to furnish the C_{sp2}=O_{sp2} bond, well-organized sequences of reactions have to be streamlined by linking rather high energy intermediates with molecular oxygen or hydrogen peroxide via kinetically acceptable transition states, where the intended products (carbonyl compounds) must be orthogonal to the oxidatively active species.

Today, aerobic oxidation has been widely executed in industries to produce feedstock chemicals from low-molecular-weight molecules, the practicality of which, however, relies critically on the inherent physical properties of substrates and products: they should easily be miscible and readily separable from each other by distillation. On increment of molecular size and complexity (numbers and kinds of functional group), the reactions necessarily require the use of appropriate solvents and formidable challenges are posed in terms of efficiency and selectivity where hydrogen peroxide offers alternative opportunities for green synthesis of carbonyl compounds.

This chapter showcases some modern oxidative methods capable of accessing carbonyl compounds using molecular oxygen or hydrogen peroxide as the terminal oxidants, in the order of increment of the innate reactivity of the substrates. As in many

Green Oxidation in Organic Synthesis, First Edition. Edited by Ning Jiao and Shannon S. Stahl.

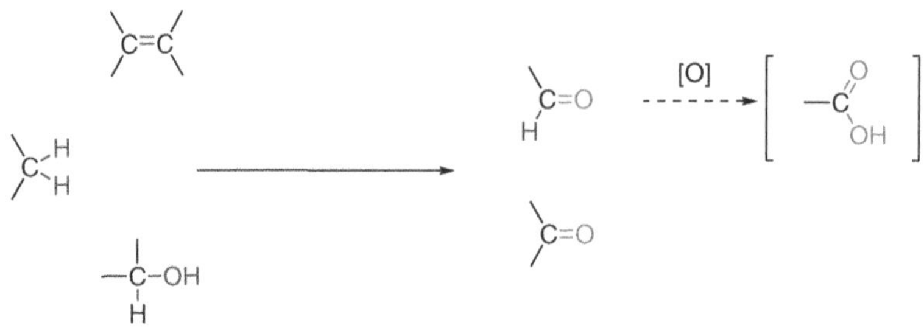

Scheme 2.1 The oxidative approach to synthesizing carbonyl compounds.

other oxidation reactions, the mechanisms of such aerobic transformations are far from being completely understood; several proposed mechanisms will be briefly discussed, showing that in some cases the pathways for the reaction seem to be catalyst dependent but several meaningful similarities can be found.

2.2 Overview of Oxidatively Active Species Employed for Green Synthesis of Carbonyl Compounds

The most stable form of dioxygen in the Earth's atmosphere is the spin triplet state (diradical), which is considered to quantum-chemically forbidden to react with closed-shell molecules, while it is ubiquitously involved in rancidity and autoxidation via a free radical mechanism [3].

A classical strategy for promoting aerobic oxidation is to activate substrates using a radical species, which is generated in situ from a suitable radical precursor, such as dialkyl peroxides, peroxy acids, or azo diesters (vide infra). Several classes of catalysts including transition metals, transition metal salts, transition metal oxides, and metal oxides have been identified that activate triplet oxygen in a chemical sense, from which various types of oxidatively active species are generated that may undergo C—H activation/oxygenation, oxidation of oxygenated intermediates (the oxidation state of which is same as that of alcohol), or decomposition of intermediates such as hydroperoxides (Scheme 2.2). For a catalytic oxidation reaction, the reduced form of the substrate-selective catalyst should be subsequently reoxidized by the stoichiometric oxidant, either molecular oxygen or hydrogen peroxide.

The reductive activation of dioxygen is the strategy commonly adopted in many metalloenzymes, in which NADH, NADPH, or flavins are used as popular reducing agents (Chapter 16). In this regard, use of H_2O_2, which is prepared from dioxygen via reduction, offers alternative opportunities for green oxidation since it is relatively cheap and easier to handle than molecular oxygen where the only side product after the oxidation is water. The standard reduction potential ($E°$ vs. SHE) of hydrogen peroxide itself is 1.776 V, showing only moderate reactivity toward most organic substrates, but it turns into various highly active oxidants on complexation with particular species, mostly transition metal ions.

Regarding transition metals, elements of platinum groups (Ru, Rh, Pd, Os, Ir, Pt) and Cu, Au, Ag are capable of activating dioxygen directly via reductive activation of dioxygen: [4] the other transition metals that have substantial affinities with oxygen and are

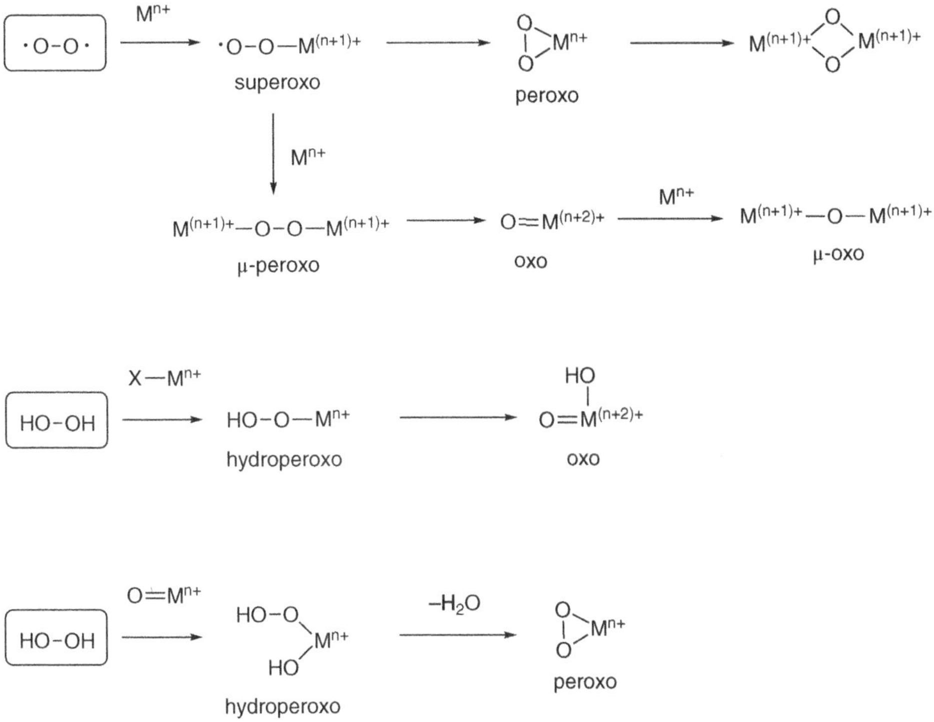

Scheme 2.2 Representative oxidatively active metal complexes derived from dioxygen and hydrogen peroxide.

oxidized on contacting with dioxygen to give the corresponding metal oxides. Fe, Mn, and Ru form higher valent metal-oxo species, and Os and Ru form metal-dioxo species that can be excellent active oxygen transfer intermediates for oxidation of alkenes, alkanes, and functionalized hydrocarbons (alcohols). Compounds based on d^0-metal, typically Ti^{4+}, V^{5+}, W^{6+}, Mo^{6+}, and Re^{7+} based compounds, can form metal-hydro (alkyl) peroxide species and side-on metal peroxides capable of oxygen transfer to nucleophilic substrate such as alkenes, sulfides, and amines. However, hydroperoxides in the presence of metals with higher oxidation potentials often lead to dismutation of the hydroperoxide and formation of highly reactive radicals such as hydroxy and alkoxy species, which often unselectively abstract hydrogen from C—H bonds leading to alkyl radicals. The major challenge in developing green oxidation methods using metal has been to inhibit the hemolytic cleavage of the peroxo O—O bond that produces a hydroxy radical that undergo non-selective C—H abstraction. An illustrative example is the Fenton reaction: the combination of an iron complex with peroxides is known to afford Haber–Weiss chemistry [5], generating hydroxyl (or alkoxyl) radicals that initiate radical chain autoxidation reactions (Scheme 2.3).

On irradiation with visible light in the presence of a suitable photosensitizer (organic dyes, such as methylene blue, tetraphenylprophyrin, or rose Bengal), triplet state oxygen is excited to the singlet state, which is 22.7 kcal mol^{-1} higher in energy than the triplet state. Singlet oxygen [6] exhibits distinct reactivity toward alkenes to exert the Shenck-ene reaction or [2 + 2]-cycloaddition, to give hydroperoxides,

$$Fe^{2+} + H_2O_2 \longrightarrow Fe^{3+} + HO^- + HO\cdot$$

$$Fe^{3+} + H_2O_2 \longrightarrow Fe^{2+} + H^+ + HOO\cdot$$

$$R{-}H + HO\cdot \longrightarrow R\cdot + H_2O$$

Scheme 2.3 Radical chain autoxidation reactions.

oxetanes, and endoperoxides, which are useful intermediates to convert to carbonyl compounds (Section 2.5.1.2).

On irradiation with ultraviolet light or subjection to electric discharge, molecular oxygen is broken apart to generate atomic oxygen, the reaction of which with a molecular oxygen gives rise to ozone, an allotrope of oxygen. Ozone is endowed with an extremely high oxidizing ability ($E° = 2/07$ [vs. SHE]), undergoing various oxidation of organic molecules in which ozonolysis plays an indispensable role for synthesizing carbonyl compounds from alkenes (Section 2.5.3.1).

2.3 Perspectives on the Structures of Catalytic Aerobic Oxidation Reactions

Under the reaction conditions that are applied to a partial and selective oxidation of organic molecules, most catalysts, such as metals, metal salts, and metal oxides, are innately solids, but a part of them would become soluble on complexation with suitable organic ligands, thereby offering adoption of various modes of catalytic reactions. Considering chemical requirements for a *partial* aerobic oxidation to proceed, the reactant(s), reagents, or catalyst(s) should be accommodated in a timely manner to satisfy their ideal orientations, for which the use of a suitable solvent plays a substantial role. When and where issues on chemo-, class-, diastereo-, or enantioselectivity become mandatory, most chemists first rely on homogeneous mixing of the regents to satisfy entropic and kinetic factors by setting the reaction temperature as low as possible. Research in academic fields and fine chemical industries directing toward controlling selectivity issues generally follow these disciplines in the homogeneous liquid phase.

On the other hands, when economic issues take priority, the rules of the game totally change: use of solvents is substantially avoided or refrained from as little as possible, and recycling of precious catalysts is a prerequisite. For separation of a certain chemical entity of interest from a mixture (reaction mixture), use of phase difference is the rule of thumb. Thus, development and discovery of efficient catalysts in innately solid forms is the primary concern in chemical industries. The phase difference makes them highly suitable for industrial implementation as processes can conveniently operate in continuous flow while eliminating the need to separate the product from the catalyst.

In the realm of solid catalysts, the wider the surface area available, the higher the catalytic performance, spurring the development of sciences and technology for designing the precise surface of finely divided materials or polymers. Since robustness is the essential factor asked for from a recyclable catalyst, designing and maintaining the catalytic function originating from finely tuned (often, finely elaborated) chemical structures on the surface poses significant challenges in current chemical sciences.

The first attempts of heterogenization would be made with polymeric materials as solid supports, but the greater part of the heterogenized catalysts are based on silica supports, since silica offers an excellent chemical and thermal stability, good accessibility, and porosity where organic moieties of catalysts can be robustly anchored to the surface to provide catalytic centers (or ligands) for metal-based catalysis [7]. These hybrid organic/inorganic catalysts can anchor the catalytic metal through covalent bonding or through simple adsorption. Note that significant issues have been posed, such as the accessibility of all active sites to reagents that renders heterogeneous catalysts often less efficient than homogeneous ones, and the leaching of metals from solid supports that again needs separation of traces of metal from the final product. Many of the considerations spur chemists to attach homogeneous catalysts onto a solid support for further application, selected successful examples of which will be illustrated in the following sections.

Nanoparticles are emerging as excellent sustainable alternatives to conventional solid supports, since they increase the exposed surface area of the active component of the catalyst, thus enhancing the contact between the reagents and the catalytic center, as happens in homogeneous catalysis. Nanocatalysis is currently considered as the frontier between homogeneous and heterogeneous catalysis that offers a novel sustainable alternative to conventional catalytic entities [8, 9].

2.4 Oxidation of Alkanes

Synthesis of carbonyl compounds by oxidation of alkanes necessarily requires selective oxy-functionalization of inactivated C_{sp3}—H bond in the initiation step, which is arguably one of the most challenging transformations in organic chemistry, because typical inactivated C_{sp3}—H bond has a high dissociation energy (90–100 kcal mol^{-1}) and low acidity ($pKa = 45–60$), and are ubiquitous in organic molecules. Most aerobic oxidation of alkanes is considered to follow the same initial mechanistic step, namely, the cleavage of a C—H bond via H atom transfer, which is the rate-determining step for the overall alkane oxidation.

Once a particular C_{sp3}—H bond is sufficiently activated to give rise to a reactive species in most typical case carbon radicals, which is readily captured by oxygen-atom transfer agent to form a C_{sp3}—O_{sp3} bond. The intermediate thus generated readily suffers further oxidation to furnish a $C_{sp2} = O_{sp2}$ bond. Conflictingly, the resulting carbonyl compounds are generally more reactive than the starting alkanes. Widely known chemical species that enable activation of C_{sp3}—H bond in alkanes or simple alkyl groups are radicals or highly electron-deficient species, such as oxometal and peroxometal derivatives of transition metals.

2.4.1 Autoxidation Approach

As mentioned in the previous section, dioxygen itself, existing in the spin triplet state (diradical), is not reactive enough to abstract a hydrogen from typical organic molecules with closed-shell electrons. However, when a trace of free radical (e.g. R'O •) is provided by some initiating event, energetically more favorable pathways become operative. The classical textbook pathway for the thermal autoxidation of alkanes features radical chain oxidation chemistry (Scheme 2.4).

Initiation

$$R'OOH \longrightarrow R'O\bullet + \bullet OH$$

Propagation

$$R'O\bullet / \bullet OH + RH \longrightarrow R'OH / H_2O + R\bullet$$

$$R\bullet + O_2 \longrightarrow ROO\bullet$$

$$ROO\bullet + RH \longrightarrow ROOH + R\bullet$$

Termination

$$ROO\bullet + ROO\bullet \longrightarrow ROH + R_{-\alpha H}{=}O + O_2$$

Scheme 2.4 Radical chain oxidation chemistry.

Table 2.1 Traditional radical initiating catalysts.

	R			
RO–OR	**H**	***t*-Bu**	**Ac**	**Bz**
BDE (kcal mol^{-1})	48	37	31	30

In general, selectivity of autoxidation primarily depends on innate reactivity of the substrates to generate a radical: the weaker the C—H bonding, the more likely it is to be attacked by a peroxy radical. The ease of generating carbon centered radical is aligned as tertiary C—H (D^o_{298}~ 95 kcal mol^{-1}) > secondary C—H (98 kcal mol^{-1}) > primary C—H (100 kcal mol^{-1}). The addition of dioxygen to a carbon radical proceeds in a diffusion-controlled manner to generate ROO•, which participates in a propagation reaction with RH to give R• together with ROOH.

Traditional radical initiating catalysts for autoxidation of hydrocarbons are shown in Table 2.1.

For autoxidation to meet a preparative purpose of carbonyl compounds, the substrates should easily be transformed into a particular ROOH, and the resulting ROOH should selectively be convertible into a particular carbonyl compound. The most celebrated example is the cumene oxidation process; the benzylic C—H bond (D^o_{298}~ 95 kcal mol^{-1}) is dominantly oxygenated to give cumene hydroperoxide (CHP), which is efficiently converted via Hock rearrangement to give phenol and acetone. As such, the autoxidation approach is inherently limited to substrates that undergo selective radical chemistry. It should be noted that aldehydes RC(=O)—H (D^o_{298}~ 87 kcal mol^{-1}) are prone to suffer over-oxidation to give the corresponding peracids or decarbonylative decomposition (Scheme 2.5).

2.4.1.1 Transition Metal-Catalyzed Autoxidation

Redox-active transition metals may affect autoxidation pathways in various ways, but most notably they have an influence both in initiating the radical chain propagation and decomposing intermediate alkyl hydroperoxide species to alkoxy and peralkoxy radicals.

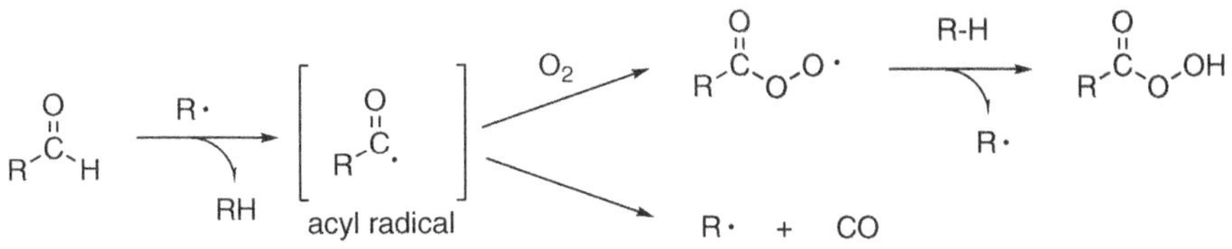

Scheme 2.5 Over-oxidation of aldehydes.

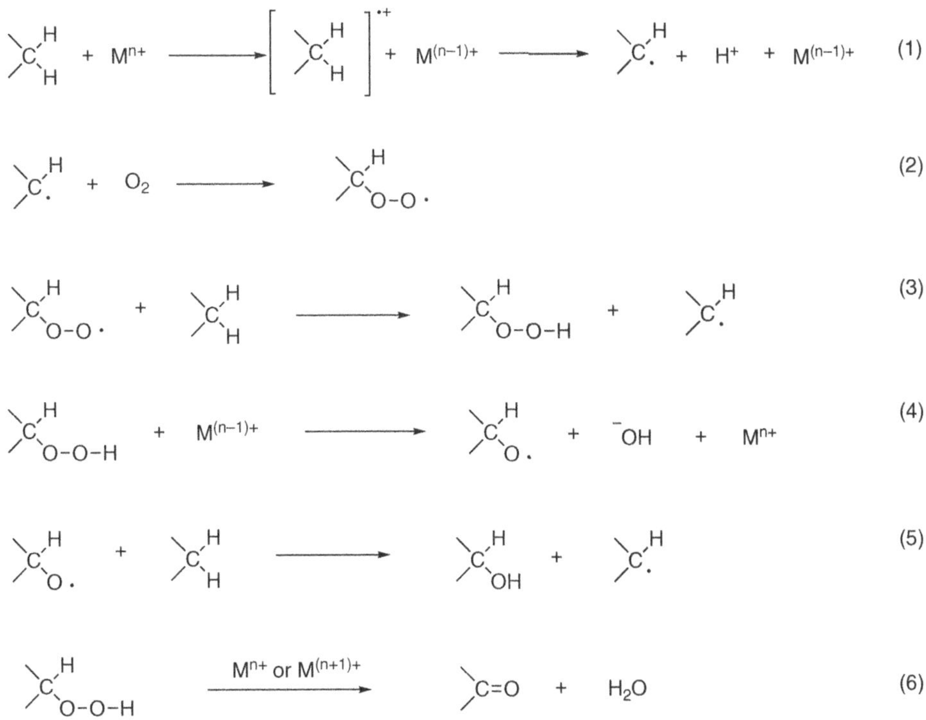

Scheme 2.6 Decomposition of alkyl peroxides.

Several transition metal salts that dissolve well in the liquid phase (e.g. $Co(OAc)_3$) have been used as catalysts for decomposition of alkyl peroxides (Scheme 2.6).

Despite the potential aid of the metal catalysts, autoxidation methods in preparation of carbonyl compounds suffer from inefficient and less selective natures. Although commercially successful, the industrial production of KA-oil, the feedstock chemical for Nylon-6 and Nylon-6,6 (producing roughly 10^6 tons per year) showcases the issues associated with autoxidation. The industrial production of KA-oil is typically carried out at 125–165 °C under 0.8–1.5 MPa air for 15–60 minutes with the aid of soluble Co^{2+} or Cr^{3+} salts as catalysts (less than 10 ppm) to give cyclohexanone together with cyclohexanol (KA-oil). The reaction is stopped around 10–12% conversion to preserve purity of KA-oils and unreacted cyclohexane is separated from the products and recycled. (The KA-oils thus obtained are subjected to the subsequent exhaustive oxidation to produce adipic acid using HNO_3 as a stoichiometric oxidant, by which cyclohexanol is oxidized to cyclohexane, see Scheme 2.7).

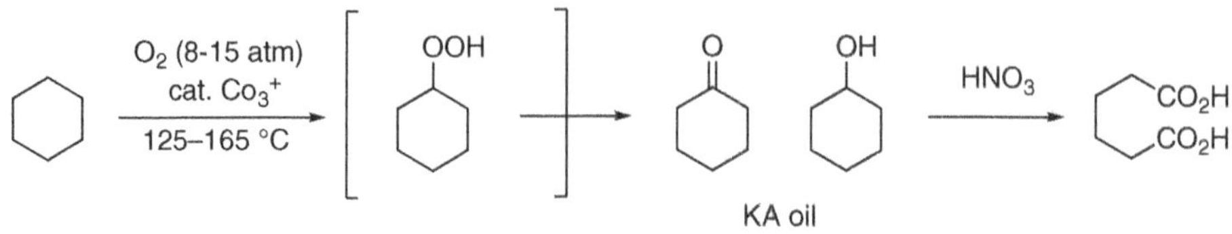

Scheme 2.7 Oxidation of cyclohexane.

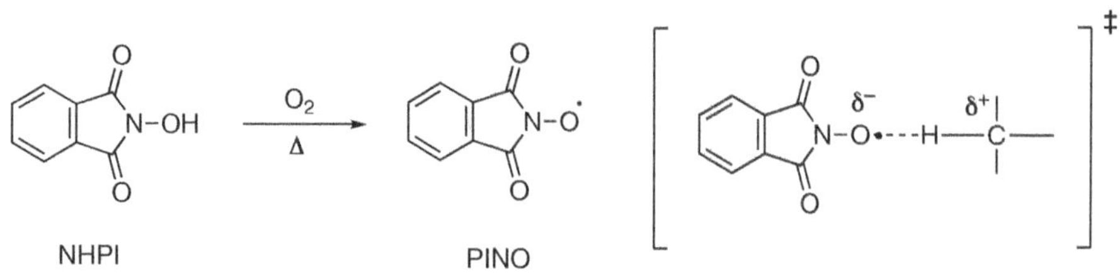

Scheme 2.8 The formation of the PINO radical.

2.4.1.2 Organocatalytic Autoxidation Approaches

Ishii and coworkers disclosed an innovative use of organocatalysts, namely, phthalimide *N*-oxyl (PINO) radical, which is generated in situ from *N*-hydroxyphthalimide (NHPI) and molecular oxygen, for aerobic oxidation of alkanes (Scheme 2.8) [10].

A variety of hydrocarbons were shown to undergo selective autoxidation under the influence of molecular oxygen to give oxygenated product such as alcohols, ketones, and carboxylic acids in good yields. Benzylic compounds are good substrates for the PINO catalyzed autoxidation to obtain the corresponding ketones, although benzylic methyl compounds are exhaustively oxidized to the corresponding benzoic acids (Scheme 2.9). As discussed in Section 2.5, secondary alcohols are converted to the corresponding ketones. Primary alcohols are converted to the corresponding carboxylic acids via the corresponding aldehydes in good yields. The higher reactivity of PINO than that of conventional alkyl peroxy radicals is due to the fact that PINO boosts the radical-chain length to increase the overall oxidation rate.

The innate instability issues of PINO toward self-decomposition into phthalimide, phthalic anhydride, and others, encourages chemists to design and synthesis of the second-generation catalysts. The relatively polar nature of NHPI and the congeners suffer low solubility in apolar mediums, spurring design and synthesis of less polar derivatives. Recent studies have shown that use of redox mediators significantly improves the catalytic efficiencies. The selected carbon-radical generating catalysts and mediators that improve the reaction performances are shown in Figure 2.1 [11].

It has been shown that the facile generation of nitroxyl radicals from hydroxylimides can significantly improve catalytic efficiencies. Extensive studies have showed that use of mediator give better results. To date, superior use of the following redox mediators such as, $Co(OAc)_2$, $Mn(OAc)_2$, anthraquinone, 1,4-diamino-2,3-dichloro-anthraquinone, xanthone, *o*-phenanthroline/Br_2, azobisisobutyronitrile (AIBN), V-70, benzoyl peroxide (BPO), cyclohexylbenzene-1-hydroperoxide (CHBHP), and dimethylglyoxime (DMG), have been documented (Figure 2.2).

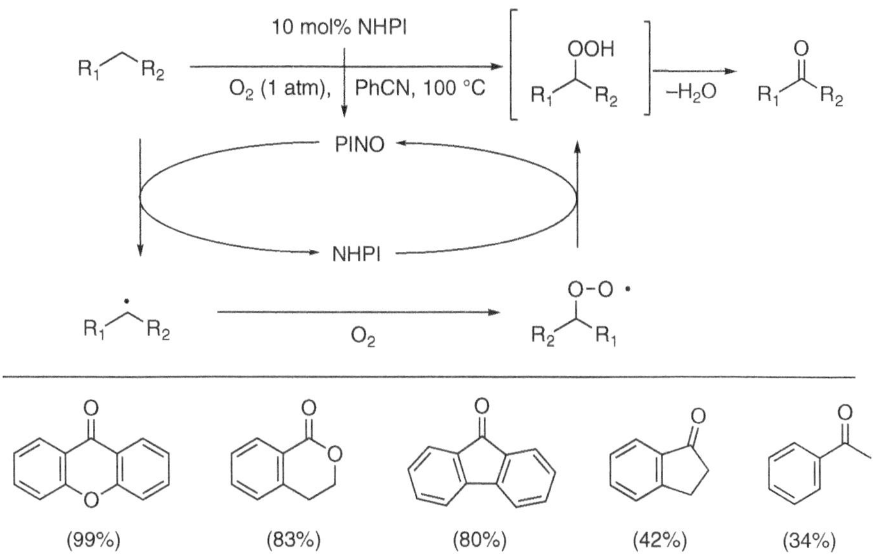

Scheme 2.9 Benzylic methyl oxidation.

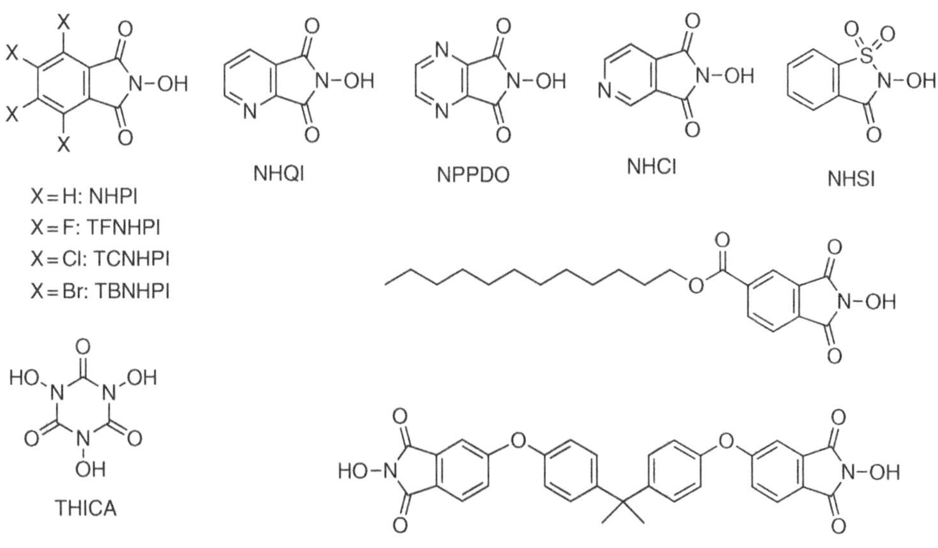

Figure 2.1 NHPI derivatives.

Regarding applicability of the nitroxyl radical, TEMPO and the related radicals find uses as catalysts for benzylic oxidation in the presence of $NaNO_2$ [12] (Scheme 2.10).

Recently, riboflavin tetraacetate was shown to function as a carbon-radical generating catalyst under UV irradiation that allows aerobic oxidation of benzylic C—H bonds in the presence of a catalytic amount of $Fe(CClO_4)_2$ (Scheme 2.11) [13].

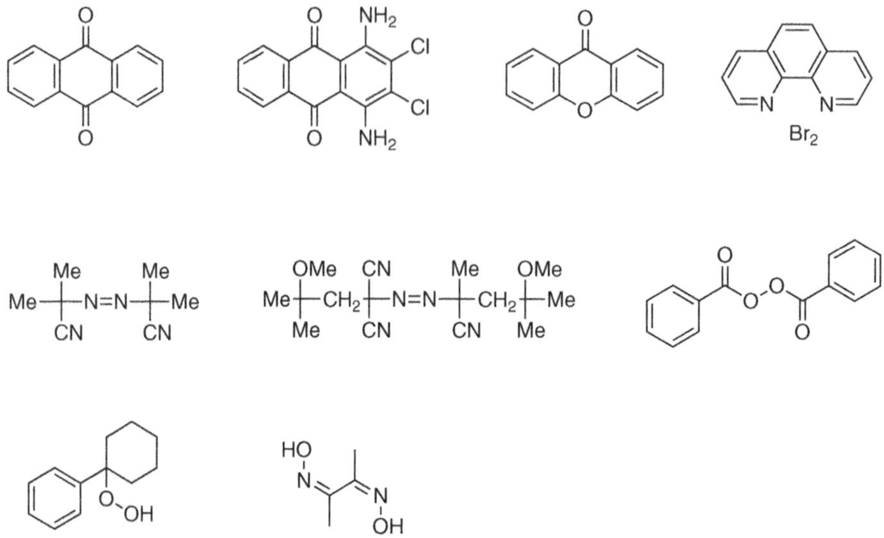

Figure 2.2 Redox mediators.

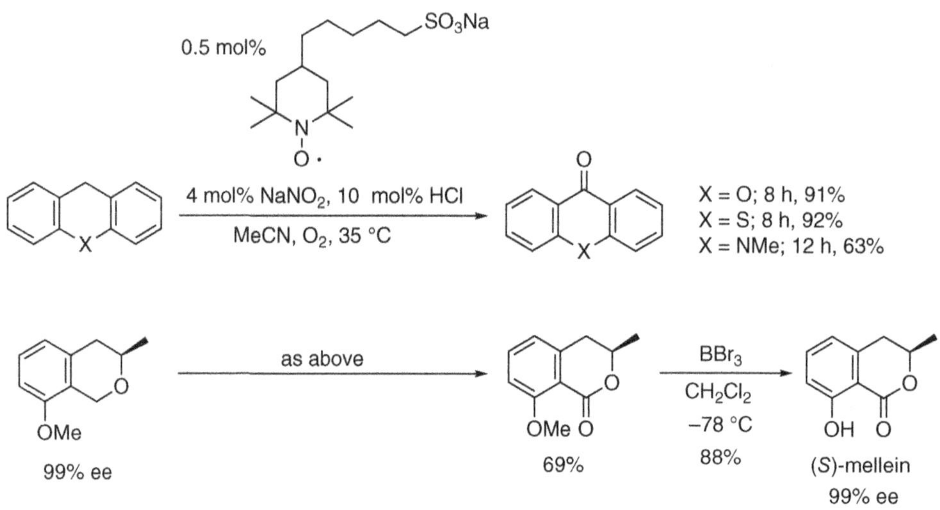

Scheme 2.10 Benzylic oxidation in the presence of $NaNO_2$.

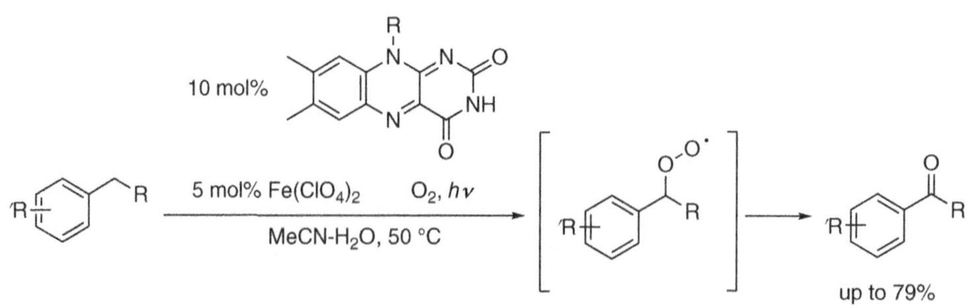

Scheme 2.11 Fe-catalyzed benzylic oxidation.

2.4.2 Metal Oxide-Catalyzed Aerobic Oxidation

Metal oxides have traditionally played an important role in organic synthesis due to their capacity for selective oxygen transfer to a wide variety of substrates under mild conditions. Participation by one or more M=O groups is a key mechanistic feature common to virtually all of these reactions. Metal oxides enjoy their versatile applications in catalytic oxidation of organic molecules by identifying cooperative systems to use bulk oxidants in various ways, as immobilized on the surface of solid supported metal oxides, or in chelating ligands or catalytic pocket of enzymes (see vide infra).

Certain metal oxides are capable of activating a molecular oxygen directly to generate active species that conduct C_{sp3}—H oxygenation of alkanes although markedly harsh conditions are required. On being attached on the surface of solid oxide supports, such as SiO_2, Al_2O_3, TiO_2, and CeO_2 via relatively simple procedures, metal oxides (supported metal oxides, Figure 2.3a) are endowed with metal economy and high thermal stability together with several functional advantages: the identity of the solid support chosen to immobilize a metal oxide exerts a profound influence on the catalytic activity as well as reaction selectivity (Figure 2.3) [7]. Mixed metal oxide catalysts offer further opportunities for controlling the catalytic activity and selectivity.

It is important to note that most of partial alkane oxidation using supported metal oxides may follow Mars–van Krevelen reactions kinetics,* which adopts different mechanistic features compared to that of autoxidation. The initial C—H cleavage of the alkane is considered to proceed via H atom transfer with concomitant one-electron reduction of the metal (Figure 2.3b), and the radical thus generated is trapped by the active surface of supported metal oxides in various ways to give the corresponding oxidized products. Supported metal oxides utilize lattice oxygen atoms to reoxidize the reduced metal centers while the gaseous O_2 reactant replenishes these lattice oxygen vacancies.

To date, supported metal oxide-catalyzed aerobic oxidation has been applied in commercial productions of several oxygenates with a higher oxidation level corresponding to carboxylic acids rather than production of carbonyl compounds (butane to maleic anhydride; *p*-xylene to *p*-phthalic acid). However, in the light of current production costs of petrochemical processes that use olefins, the use of alkanes as substrates presents an economic incentive due to the lower cost of alkanes compared to more oxidized compounds, much in part due to their current availability from natural gas. The current situations offer a valuable opportunity for academic and industrial researchers.

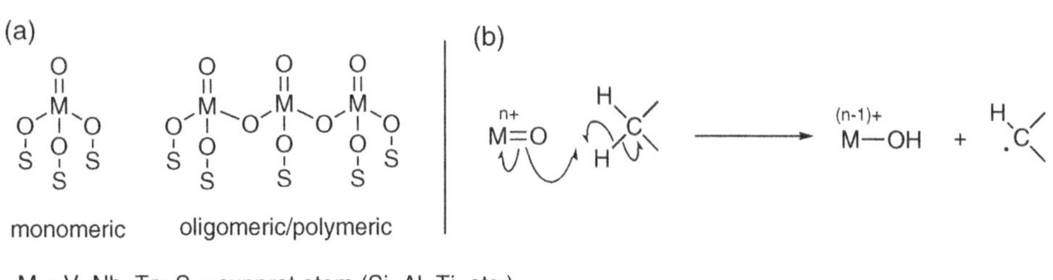

Figure 2.3 The solid supported metal oxides.

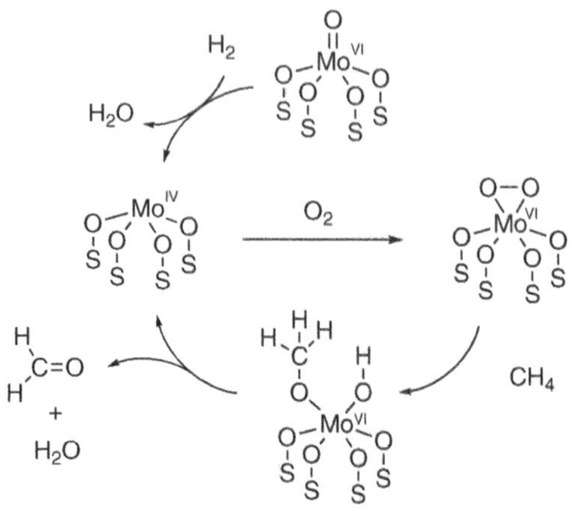

Scheme 2.12 Mechanistic studies on the $MoOx/SiO_2$.

Close mechanistic studies on the $MoOx/SiO_2$ catalyzed aerobic oxidation of CH_4 to formaldehyde indicated the in situ generation of metal peroxide species as the active center (Scheme 2.12) [14][1].

2.4.3 Bio-Inspired Approaches Using Metal Oxide Catalysis

Looking at this particularly challenging class of oxidation from another point of view, living systems cut a brilliant figure for aerobic oxidation of non-activated alkyl moiety. Thus, aerobic biological systems have evolved highly sophisticated metalloproteins and their coenzymes that realize efficient uses of molecular oxygen even in the presence of oxidation-sensitive functionality under physiological conditions, mostly ambient temperature and atmospheric air, to install to oxygen-functionalities onto non-activated alkyl groups, demonstrating an ultimate feature of catalytic oxidation using molecular oxygen.

There are two main types of pathway regarding enzymatic aerobic oxidation; (i) the monooxygenase pathway that catalyzes an oxygen-atom transfer from O_2 to the substrate via a metal center (Scheme 2.13a), and (ii) the oxidase pathway that catalyzes dehydrogenation using O_2 as an acceptor of electrons from a reduced-metal center (Scheme 2.13b).

Iron and copper ions are the metal ions of choice for many metallo-oxygenases, probably due to their abundance in the geosphere as well as versatile electronic properties. Typical examples are heme-containing enzymes, non-heme iron enzymes, and copper enzymes, in which oxygenation pertinent metal-oxygen species are shown in Scheme 2.14. It should be noted that the activation of dioxygen at the metallo-oxygenases generally entails its two-electron reduction to activate a molecular oxygen

1 Typically, olefins are produced through the cracking of oil fractions, and these olefins are used as feedstocks to produce oxygenates. In recent years, the use of natural gas, rather than petroleum, as a source of feedstock chemical has become advantageous; the proven reserves of natural gas have doubled over the past decade within the United States, thanks to part to the increase in extraction from unconventional sources such as those found in shale deposits. Classical processes that convert natural gas to olefins, such as the Fischer–Tropsch reaction, requires the conversion of light alkanes to syngas thorough steam reforming before ultimate transformation to olefins.

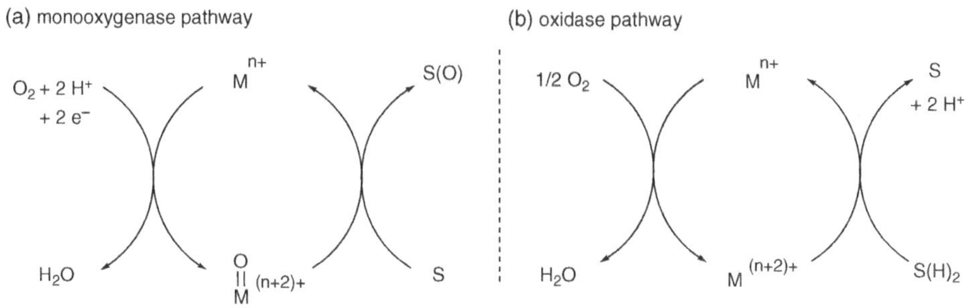

Scheme 2.13 Enzymatic aerobic oxidation pathway.

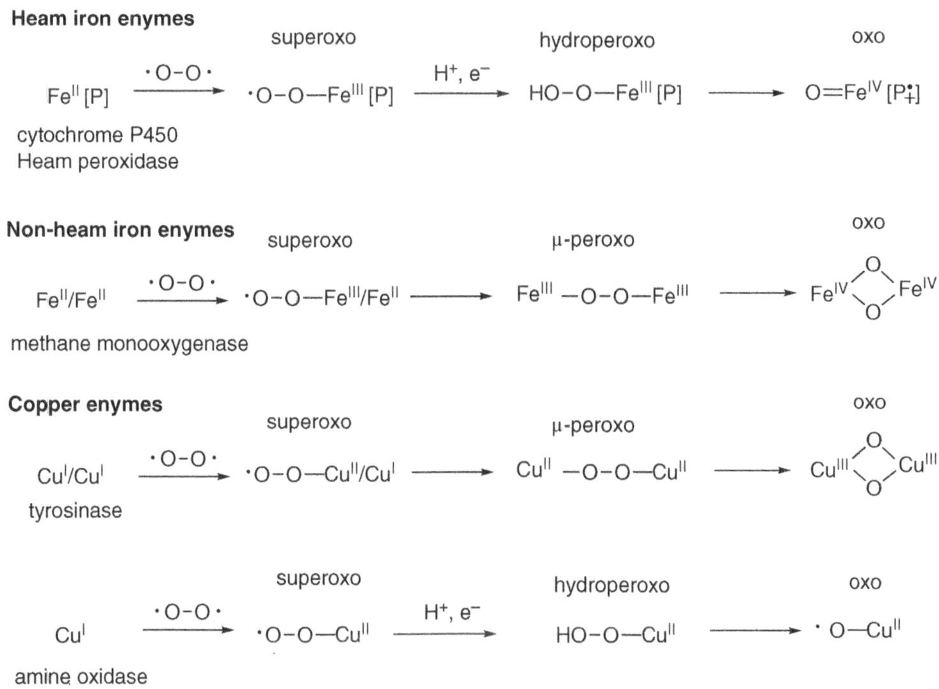

Scheme 2.14 Oxygenation pertinent metal-oxygen species.

via a stepwise single-electron-transfer process to furnish the peroxo state, which markedly facilitates a heterolytic cleavage of the O—O bond [15].

From a standpoint of synthesizing carbonyl compounds from alkanes, the corresponding alcohols are produced in the initial C—H oxygenation, therefore, the catalytic systems should oxidize the alcohols to give the corresponding carbonyl compounds via an oxidase mechanism. Effort focusing on development of oxidase-like catalyst will be discussed in Section 2.6.

The cytochrome P450 family of metalloenzymes containing a thiolate-ligated porphyrinic iron center generate extremely active iron spices, specifically, $Fe^{V}{=}O$ (or better described as $Fe^{IV}{=}O[P\bullet+]$), in their active site, inspiring chemists to model the chemistry for developing more efficient and greener oxidation (Scheme 2.15).

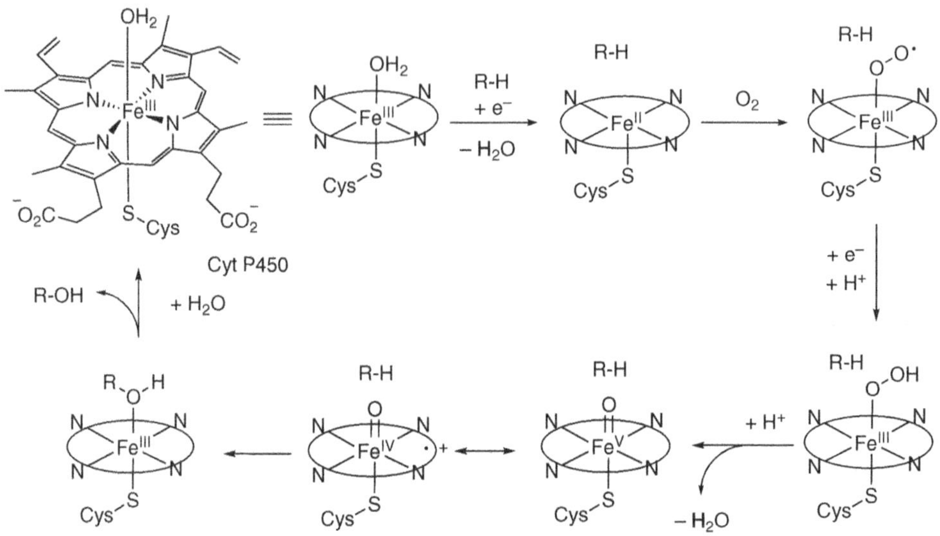

Scheme 2.15 Bioinspiration from cytochrome P450 for more efficient and greener oxidation.

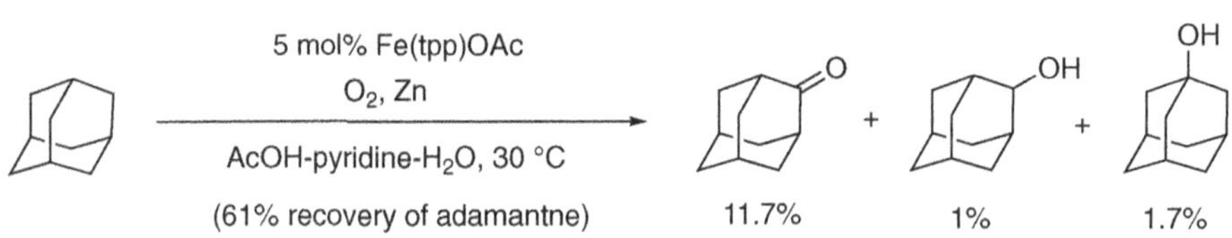

Scheme 2.16 Adamantine oxidation.

2.4.3.1 Reductive Activation of Dioxygen at Iron Center: Gif System

Based on the idea of an iron catalyst that is reduced by electron transfer and oxidized by molecular oxygen simultaneously in the source of proton (e.g. in Cyt P450s employ a reductant, such as NAD(P)H, for efficient generation of oxidatively active species via stepwise reduction of dioxygen to give M-OOH), Barton and coworkers implemented the Gif system [16], consisting of an iron catalyst, O_2, and a reductant such as Zn, Fe, or H_2S in AcOH-pyridine. The Gif system exhibited an unusual C—H oxygenation selectivity (2 °C—H > 3 °C—H > 1 °C—H) that is preferable for preparation of carbonyl compounds. The most efficient oxygenation was attained by a heterogeneous mixture of $Fe_{cat}/O_2/Zn$, affording a turnover number exceeding 2000 [17] (Scheme 2.16).

The Gif chemistry that oxidizes saturated hydrocarbons in the presence of rather easily oxidizable substances like H_2S, PPh_3, $P(OMe)_3$, PhSH, and especially PhSeH (Gif paradox [18]), spurred subsequent extensive studies to enrich our understanding of aerobic oxidation.

2.4.3.2 Transition Metal-Catalyzed Aerobic Oxidation of Alkanes Using Aldehydes as Co-Reductants

Much effort has been invested into development of oxidation catalyst based on metalloporphyrin motif, and several of which have been shown to catalyze epoxidation of alkenes, but applicable for oxidation for alkane oxidation is scarce. The susceptibility of

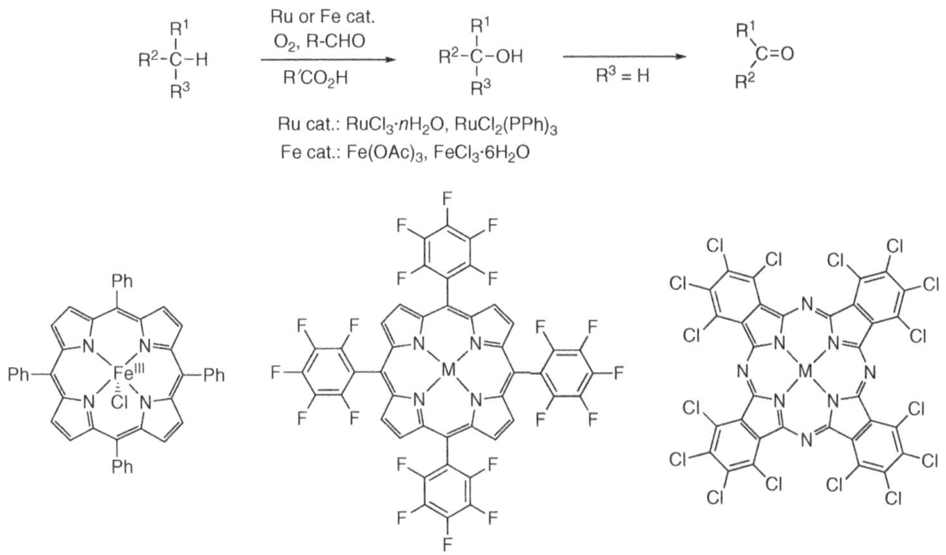

Scheme 2.17 Aerobic oxidation of alkanes.

porphyrin to oxidative self-degradation demands the use of an expensive oxidant like PhIO, hindering further development of this approach.

During systematic study on the cytochrome P-450 type oxidations, Murahashi and coworkers found a novel transition metal-catalyzed aerobic oxidation of alkanes in the presence of an aldehyde and catalytic amount of a carboxylic acid at room temperature (Scheme 2.17) [19]. The aldehyde was oxidized with molecular oxygen by the action of the transition metal to the corresponding peracid that subsequently react with transition metal complexes to afford oxometal species. The novel aerobic oxidation system was subsequently shown to exhibit markedly high turnover numbers using metalloporphyrins [M=Co^{II}, Mn^{III}(Cl), or Ru^{II}(CO)] bearing *meso*-pentafluorophenyl groups [20] or a chlorinated phthalocyanine iron (II) complex [21] as the catalyst.

2.4.3.3 Transition Metal-Catalyzed Activation of Hydrogen Peroxide for Oxygen Atom Transfer

Recently, significant progress has been made in the oxidation of hydrocarbons with hydrogen peroxide in the presence of vanadium, copper, iron, osmium, tungsten, manganese, molybdenum, or cobalt compounds and complexes with oxometal or peroxymetal species, to give the corresponding alkyl peroxides, from which the decompositions of the corresponding carbonyl compounds are obtained along with the corresponding alcohols [22]. In most cases, predictable reactivity has only been shown with simple hydrocarbons, such as cyclohexane, cyclooctane, *n*-heptane, and so on, and the initial products were the corresponding alkyl peroxides.

2.4.3.4 Non-Porphyrinic Iron Catalysts

Compared with porphyrin-type ligands, non-heme-type ligands offer relatively straightforward alkane modification for fine-tuning of the stereoelectronic features, leading to rapid development of green oxidation methods. In 1997, Que. and coworkers disclosed

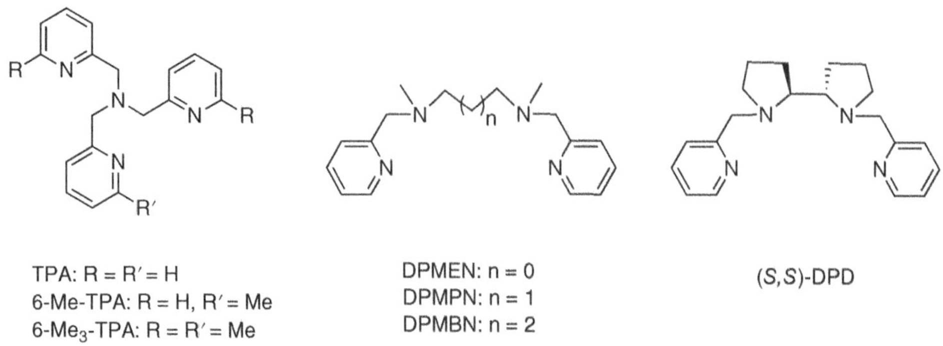

Figure 2.4 Design of tetradentate ligands.

the seminal paper on tripodal tetradentate ligand, namely, tris(2-pyridylmethyl)amine (TPA) [23]; $[Fe(TPS)(MeCN)_2]^{2+}$ was shown to catalyze stereospecific alkane hydroxylation using H_2O_2. The authors proposed the generation of a high-valent iron–oxo species $Fe^V(O)(OH)$ as catalytically active species, which was formed via a ferric hydroperoxo intermediate. The modular nature of TPS has spurred design of tetradentate ligands that exploit synthetic potentials of non-heme iron catalysts (Figure 2.4). Jacobsen and coworkers demonstrated the use of acetic acid as additive in iron Fe(II)-17 catalysis that converted alkenes into epoxides with a high yield.

Last decade witnessed the germination of catalytic C—H oxygenation of non-activated alkyl group that offer their practical and strategic uses. White and coworkers ingeniously applied the Jacobsen chemistry to selective hydroxylation of unactivated hydrocarbons using a closely related catalyst, Fe(PDP), which was capable of installing an oxygen functionality at the late-stage in the synthesis via regio-, diastereoselective C_{sp3}—H activation [24]. The catalytic system showed tertiary (3°) and secondary (2°) aliphatic C—H bonds can be differentiated on the basis of electronic, steric, and stereoelectronic factors that distinguish C—H bonds from one another within a molecule, thereby offering a predictable selectivity on the basis of the electronic and steric properties of the C—H bonds (substrate-controlled selectivity) (Scheme 2.18).

Close investigations on the mechanistic role of the acetic acid on Fe(DPD) catalysis suggest that acetic acid promotes formation of an iron (oxo)carboxylate via elimination of water by providing a critical proton to the H_2O_2 and that the iron(oxo)carboxylate thus generated abstracts a hydrogen from R-H to generate an FeOH intermediate together with a very short-lived carbon centered radical, which rebounds to FeOH to furnish the corresponding ROH and regenerates Fe(DPD) (Scheme 2.19).

Highly enantiomerically enriched ketones have been synthesized from the corresponding prochiral alkanes by enantioselective oxidation catalyzed by chiral manganese complexes (Scheme 2.20) [25].

2.5 Oxidation of Alkenes

The C_{sp2}=C_{sp2} bonding system endows molecules with salient electronic machineries to have interactions with electron-deficient chemicals, such as oxidants and coordinatively unsaturated transition metals, serving as a useful platform for conducting diverse

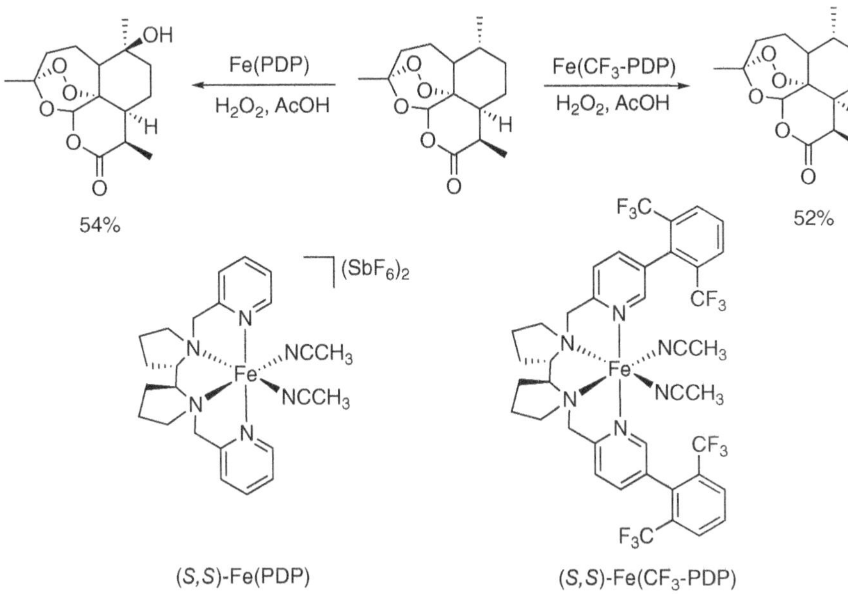

Scheme 2.18 Fe(PDP)-catalyzed selective C—H oxygenation of unactivated hydrocarbons.

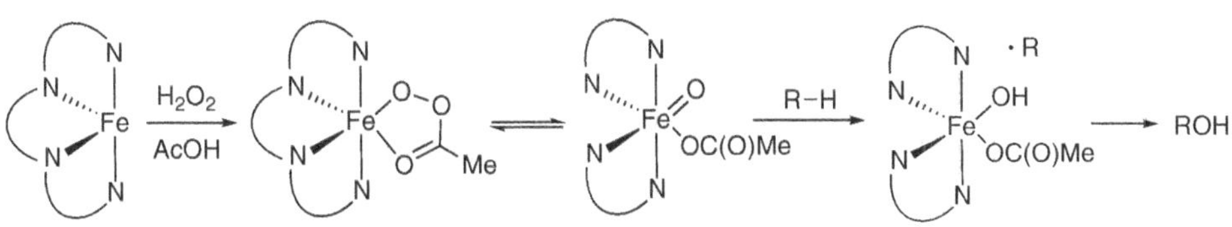

Scheme 2.19 Proposed mechanism.

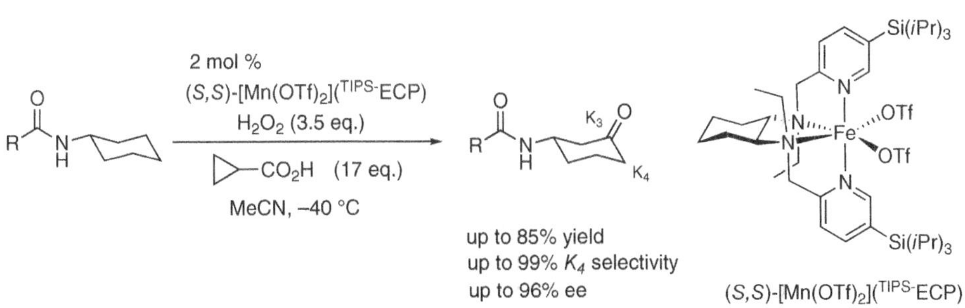

Scheme 2.20 Mn-catalyzed enantioselective oxidation.

ranges of functional group interconversions. Significantly, several distinct types of carbonyl compound can be synthesized from an alkene depending on the reagents and conditions applied, offering useful latitude for designing the synthesis of carbonyl compounds. Alkenes offer (i) oxygenation at the allylic position(s) to give α,β-unsaturated carbonyl compounds with or without accompanying allylic transposition; (ii) oxygenation at the double bond to give ketone or aldehyde with substrate- or reagent-controlled manners (Wacker oxidation in substrate-control or reagent-control/Markovnikov/

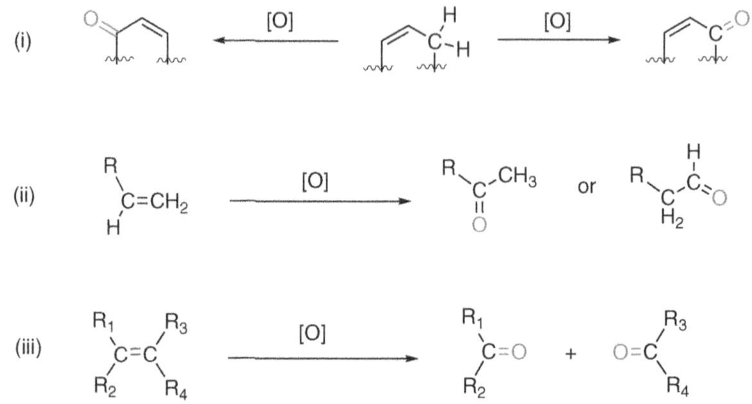

Scheme 2.21 Distinct types of carbonyl compound synthesized from alkenes.

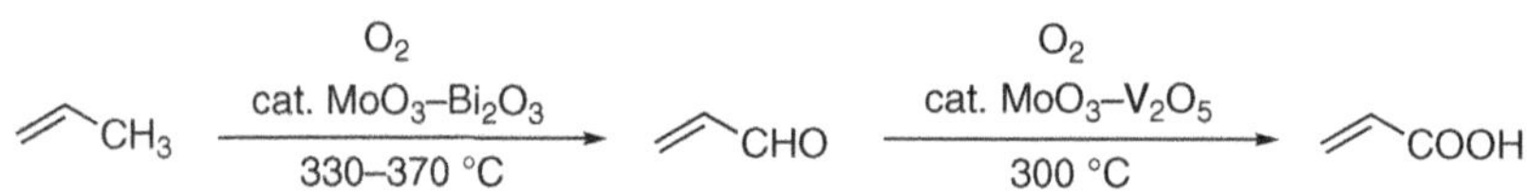

Scheme 2.22 Aerobic oxidation of propylene to acrolein.

anti-Markovnikov); and (iii) oxidative cleavage of the double bond to give ketones or aldehydes (Scheme 2.21).

Note that alkenes potentially undergo several other distinct types of synthetically important oxygenation reaction, such as epoxidation, dihydroxylation, and aminohydroxylation. As such, formation of undesired products along with the target carbonyl compounds due to competing side reactions, for example, alkene isomerization, epoxidation, dihydroxylation, is a common feature of many of the transformations discussed in this section. This section focuses on oxidative transformation of alkenes to give carbonyl compounds using O_2 as the terminal oxidant.

2.5.1 Oxygenation at the Allylic Position of Alkene

The selective oxygenation of alkenes at the periphery of the double bond offers attractive opportunities for obtaining synthetically versatile α,β-unsaturated carbonyl compounds. The most prototypical example is the industrial aerobic oxidation of propylene to acrolein [26] (Scheme 2.22).

Besides gas-phase aerobic oxidation in industrial applications, allylic oxygenation of alkenes had been classically effected by using stoichiometric amounts of chromium- or selenium-based reagents, and much effort has been mounted to conduct this type of oxidation under greener conditions.

Schematically, two distinct pathways can be drawn for the oxygenation of an alkene to give an enone; (i) direct oxygenation at the allylic position of alkene substrate to give the corresponding enone without migration of alkene moiety ("normal" allylic oxygenation, Scheme 2.23a), and the other type is an oxygenative allylic transposition that involves a migration of the alkene functionality to give regioisomeric enone (Scheme 2.23b).

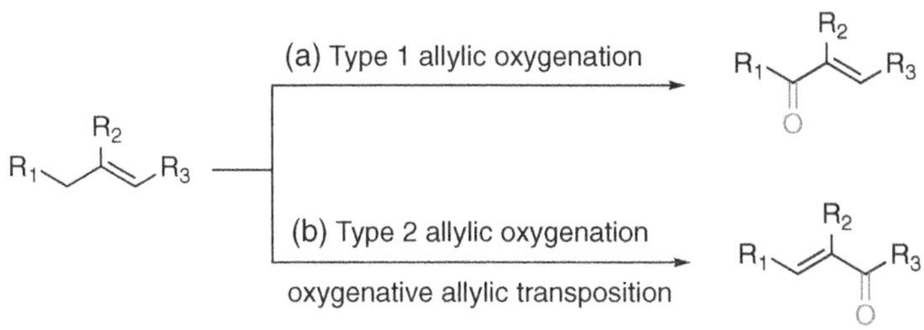

Scheme 2.23 Two distinct pathways on allylic oxygenation.

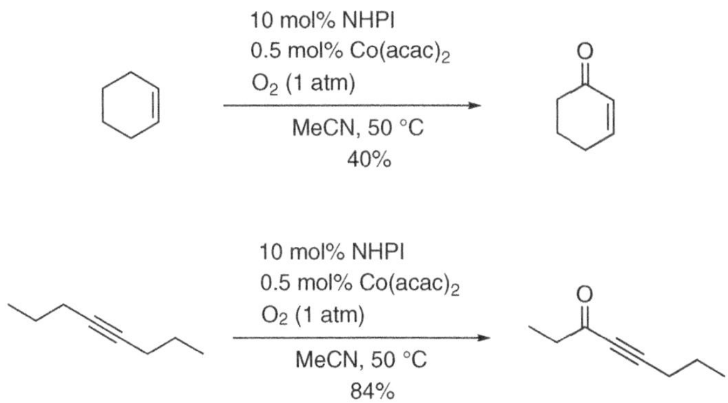

Scheme 2.24 Aerobic oxidation of cyclohexene and alkynes.

2.5.1.1 Type 1 Allylic Oxygenation

Most type 1 allylic oxygenations commence the reaction by abstracting an allylic hydrogen to give an allylic radical: compared to hydrogen abstraction from alkanes that occur at the allylic position (D^{o}_{298}~ 85 kcal mol^{-1}), which is more a facile process due to the resonance stabilization. The allylic radical thus generated can be readily captured by various oxygenating agents, in which the regioselectivity of the oxygenation follows the substrate-control. To date, considerable effort has been mounted for development of catalytic allylic oxygenation using transition metal, such as CrO_3, $Cr(CO)_6$, CuBr, CuI, MnO_2, $Pd(OH)_2$, $Fe(acac)_3$, $RuCl_3$, and $Rh_2(cap)_4$, confirming the superiority of *tert*-BuOOH or hypervalent iodine reagents to O_2 or H_2O_2 for obtaining the corresponding α,β-unsaturated carbonyl compounds [27].

On the other hand, the organocatalytic approach finds the better uses of O_2: aerobic oxidation of cyclohexene using PINO catalysis gave cyclohexenone in modest yields. Notably, propargylic positions of alkynes are efficiently oxygenated using O_2 and catalytic amounts of both NHPI and $Cu(acac)_2$ to give the corresponding ynones (Scheme 2.24).

The use of Ishii's catalytic aerobic oxidation system, NHPI/$Co(acac)_2$/O_2, was demonstrated in Toyota's total synthesis of serofendic acid A (Scheme 2.25) [28].

Recently, catalytic methods that use hydrogen peroxide have been reported employing limited numbers of simple alkenes. A few protocols have been reported that realized aerobic oxidation of cyclic alkenes to give cyclohexanone using iron complex, but the substrate applicability has not been explored. As such, the present survey indicates

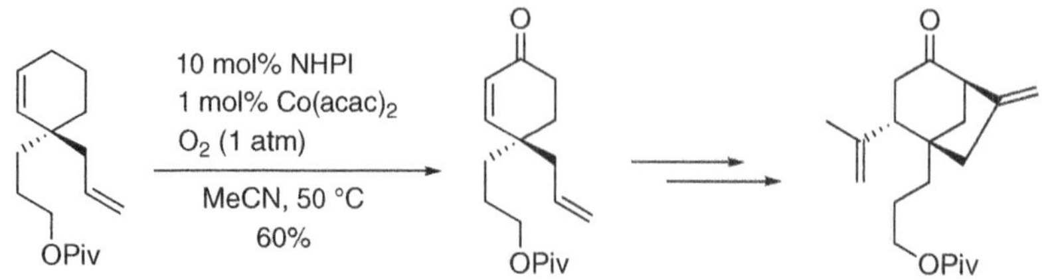

Scheme 2.25 Application in total synthesis of serofendic acid A.

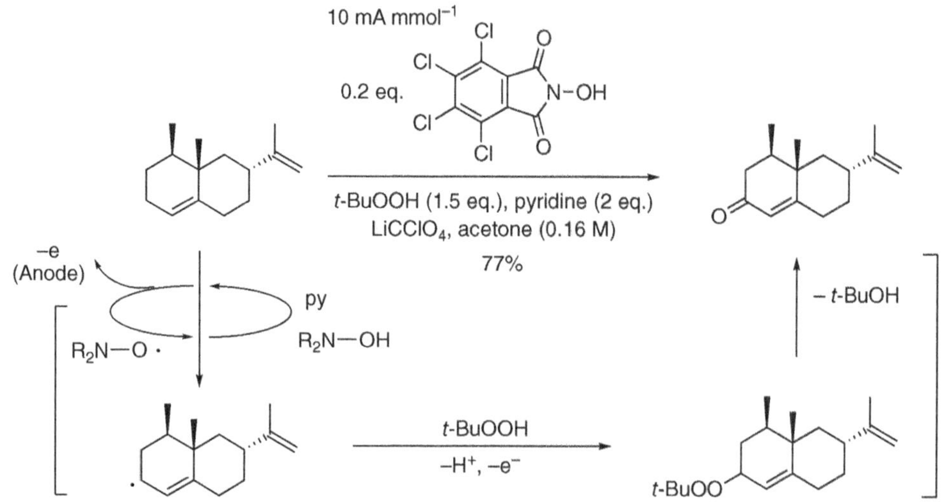

Scheme 2.26 Electrochemical allylic C—H oxidation.

paucity of reliable methods for aerobic oxidation that effect type 1 allylic oxygenation. Quite recently, Baran and coworkers disclosed a scalable and sustainable electrochemical allylic C—H oxidation using Cl_4NHPI as the redox catalyst and *tert*-BuOOH as the terminal oxidant [29], which will consolidate type 1 allylic oxygenation as a reliable strategy for synthesis of carbonyl compounds (Scheme 2.26).

2.5.1.2 Type 2 Allylic Oxygenation Mediated by $^1O_2$ (Schenck-ene Reaction)

Schenck and coworkers observed that singlet oxygen undergoes ene-like addition reaction to alkenes having an allylic hydrogen atom to give allylic hydroperoxides with migration of the double bond and that allylic hydroperoxides are easily converted to enones under mild conditions. Close experimental and theoretical studies indicate that a Schenck-ene reaction proceeds through a stepwise mechanism via a perepoxide-like intermediate (Scheme 2.27) [30].

Mihelich and Eickhoff developed a facile one-pot conversion of alkenes to α,β-unsaturated carbonyl compounds: [31] on photoirradiation with bubbling of O_2 in the presence of a catalytic amount of organic dye and a stoichiometric amount of Ac_2O and Et_3N, alkenes undergo a Schenck-ene reaction to give the corresponding allylic hydroperoxide, which concomitantly suffer acetylation followed by dehydroacetoxylation

Scheme 2.27 Schenck-ene reaction.

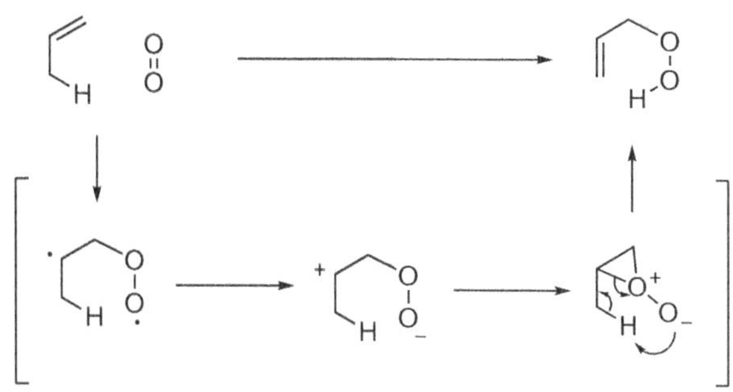

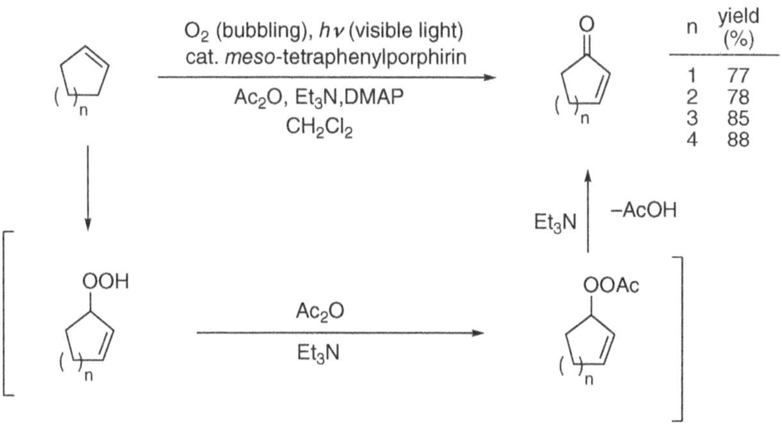

Scheme 2.28 One-pot conversion of alkenes to enones.

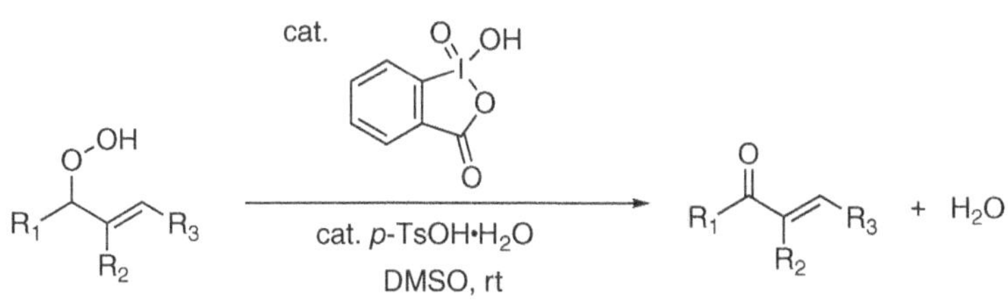

Scheme 2.29 Conversion of allylic hydroperoxides to enones.

to furnish the corresponding enones together with AcOH•NEt_3 as stoichiometric waste (Scheme 2.28).

Recently, IBX in combination with *p*-TsOH was found to catalyze efficient dehydration of allylic hydroperoxide to give the corresponding enone, which in combination with a singlet oxygen ene reaction enabled one-pot oxygenative allylic transposition of alkenes into enones with water as the only stoichiometric waste (Scheme 2.29) [32]. It is noteworthy that the IBX-catalyzed methods could produce distinct enones that were not obtainable by the traditional method (Scheme 2.30).

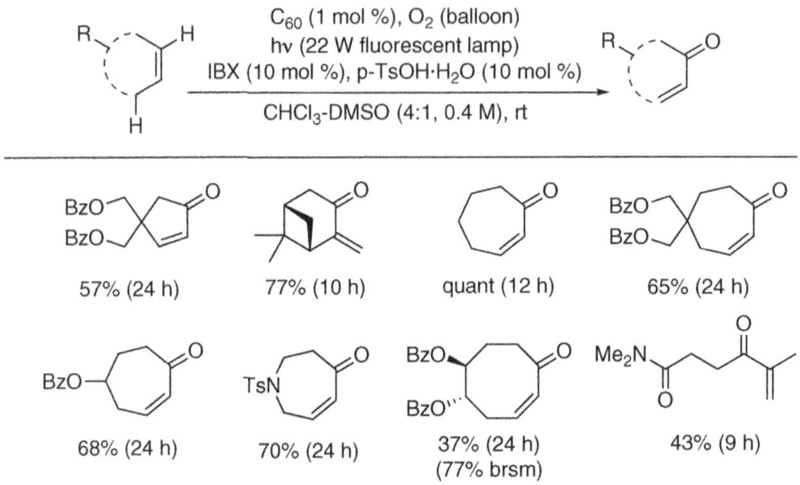

Scheme 2.30 Conversion of alkenes to enones.

$$C_2H_4 + PdCl_2 + H_2O \xrightarrow[-2\,Cl^-]{} CH_3CHO + 2\,HCl + Pd^0$$

$$Pd^0 + 2\,CuCl_2 \longrightarrow PdCl_2 + 2\,CuCl$$

$$2\,CuCl + 1/2\,O_2 + 2\,HCl \longrightarrow 2\,CuCl_2 + H_2O$$

Net: $$C_2H_4 + 1/2\,O_2 \xrightarrow[CuCl_2]{PdCl_2} CH_3CHO$$

Scheme 2.31 Synthesis of acetaldehyde from ethylene.

2.5.2 Oxygenation via Partial Cleavage of the C=C Double Bond

2.5.2.1 Wacker-Type Oxidation

On passing ethylene into an aqueous solution of palladium chloride, the generation of acetaldehyde was detected while metallic palladium precipitated. Since its publication in 1894 by F. C. Phillips [33], more than 60 years elapsed until the same phenomenon was rediscovered by researchers at Wacker-Chemie [34], which had culminated in the Wacker–Smidt process for manufacturing acetaldehyde from ethylene, in which re-oxidation of Pd^0 with $CuCl_2$ realizes dioxygen coupled turnover (Scheme 2.31).

To be commercially successful, substantial concentrations of copper salts (around 1 M) and chlorides (around 2 M) are added to achieve favorable redox potentials of the Pd^{II}/Pd^0 and Cu^{II}/Cu^I couples and to solubilize CuI as $CuICl_2$, HCl is required to circumvent clustering of transient atomic palladium. Low-boiling point acetaldehyde is removed continuously by distillation.

Much efforts have been mounted to expand the use of one-pot oxidation of alkenes to the corresponding carbonyl compounds with catalytic amount of Pd^{II}.

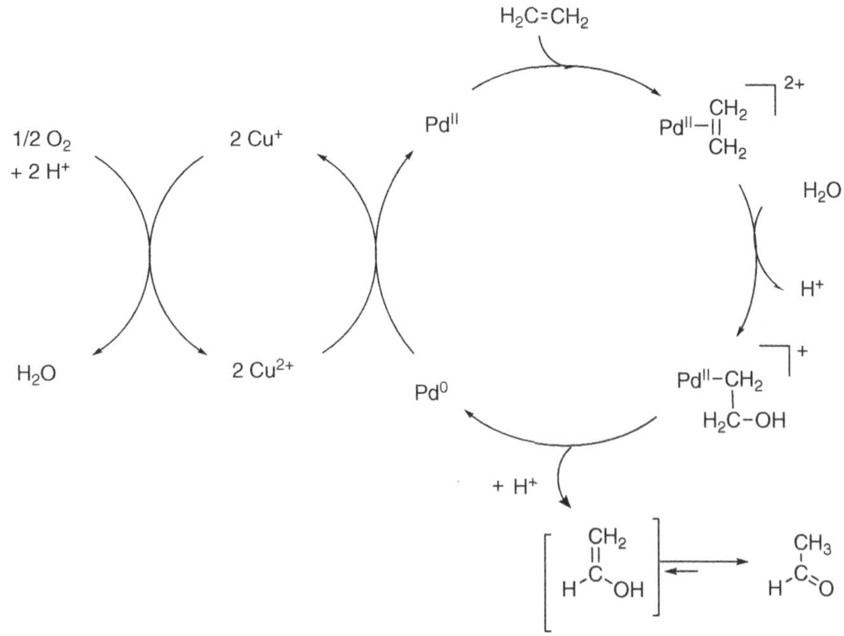

Scheme 2.32 The mechanism of Wacker oxidation.

The oxidation of higher alkenes demands use of appropriate solvent for mixing both alkenes and water, for which the superior use of DMF spurred extensive expansion of the synthetic use by Tsuji and coworkers that compatible that consequently allowed many functional groups such as ester, sulfonate, acetal, and MOM ether. Today, the one-pot oxidation of alkenes to the corresponding ketones with catalytic amounts of Pd^{II} is referred to as Wacker–Tsuji oxidation [35]. A widely accepted mechanism, Wacker-oxidation, is shown in Scheme 2.32, where Pd^{II} promotes hydroxypalladation of a coordinated olefin, according to Markovnikov's rule, followed by β-hydride elimination (Scheme 2.32) [36].

In accordance with the mechanism, classical Wacker–Tsuji oxidations of nonbiased terminal alkenes predominantly afford methyl ketones. But the oxidation has proven to be sensitive and influenced by substrates: the presence of a proximal heteroatom at the allylic or homoallylic position directs the addition of water in an anti-Markovnikov fashion to give the corresponding aldehydes, resulting in a mixture of ketone and aldehyde [37].

The additional use of organic solvent necessarily attenuates the redox communication between Pd^{II}/Pd and Cu^{II}/Cu^{I} couples where dioxygen had not always been useful as the stoichiometric oxidant. Furthermore, in palladium-catalyzed aerobic oxidations, the precipitation of palladium metal (palladium black) from soluble palladium species (Pd—H, Pd^0) often competes kinetically with reoxidation by O_2 [38]. To circumvent this problem one can employ ligands that stabilize Pd^0 and that are stable under oxidative conditions, spurring effort for identifying efficient reagents and conditions that help swift regeneration of Pd^{II} from Pd^0 before it undergoes intractable aggregation. Numbers of reliable variants have been developed [39, 40].

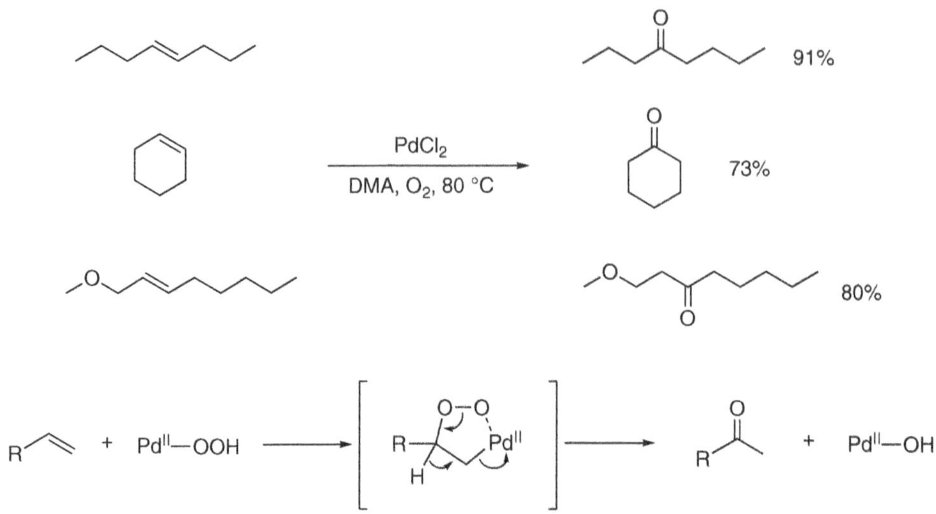

Scheme 2.33 Selective oxidation of internal alkenes.

2.5.2.2 Overcoming the Oxidant Problem: Use of O_2 as the Solo Oxidant

Keneda and coworkers disclosed that $PdCl_2$ in combination with DMA as the solvent established an extremely simple and highly efficient catalytic system for the Wacker–Tsuji oxidation of terminal alkenes [41]. From the study of CV and kinetics, DMA was shown to act as the most efficient solvent for promoting the reoxidation of the Pd(0) species by O_2. The extensive application of this chemistry had formulated a common understanding that the Wacker–Tsuji oxidation would operate best in the conversion of terminal alkenes into methyl ketones. The oxidation of internal alkenes is very slow: clean selective oxidation of the terminal double bond in the presence of an internal alkene is possible under normal conditions (Scheme 2.33).

Some anomalous examples have been accumulated: in the presence of proximal functional groups, regioselectivity of oxidation can be difficult to rationally predict. Apart from these irregular cases, the innate Markovnikov selectivity of the Wacker–Tsuji oxidation poses a challenge to alter the selectivity.

2.5.2.3 The Anti-Markovnikov Wacker-Type Oxidation

It has been shown that some stereoelectronically biased alkenes circumvent Markovnikov's rule and produce aldehydes. Allyl alcohols (protected) are notorious substrates giving an intractable mixture of aldehyde and ketones. Grubbs and coworkers have made a breakthrough in Wacker-type oxidation of unbiased alkenes affording anti-Markovnikov aldehyde products (Scheme 2.34) [42].

A careful ^{18}O-labeling experiment suggested an unusual mechanistic manifold, potentially involving the metal-catalyzed attack at the terminal position of the alkene by NO_2 radical.

The applicability of the nitrite-modified Wacker–Tsuji reaction to a sterically hindered system has been demonstrated, consolidating the utility for synthesizing aldehydes from alkenes (Scheme 2.35) [43].

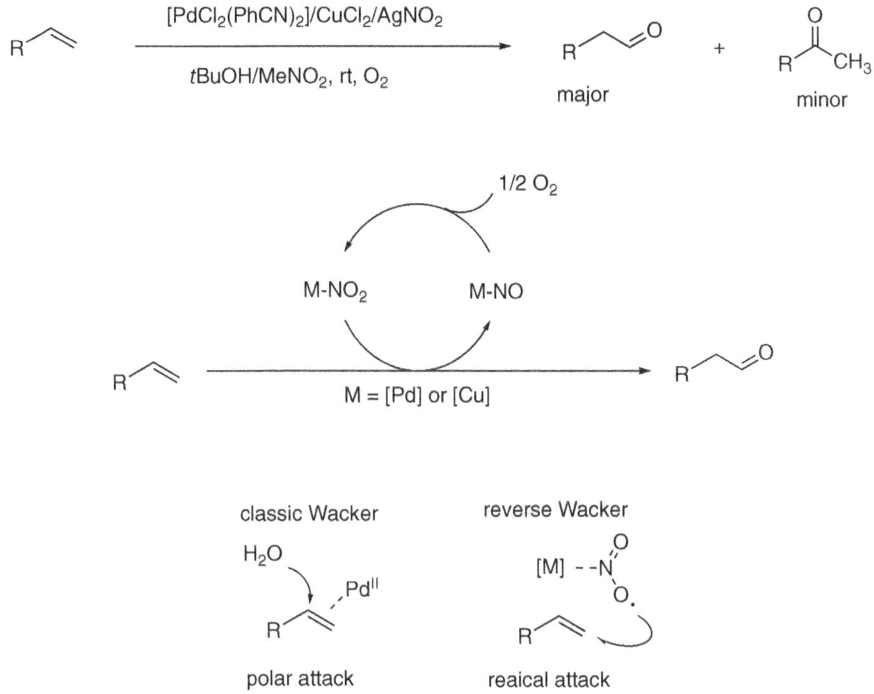

Scheme 2.34 Selective oxidation of terminal alkenes.

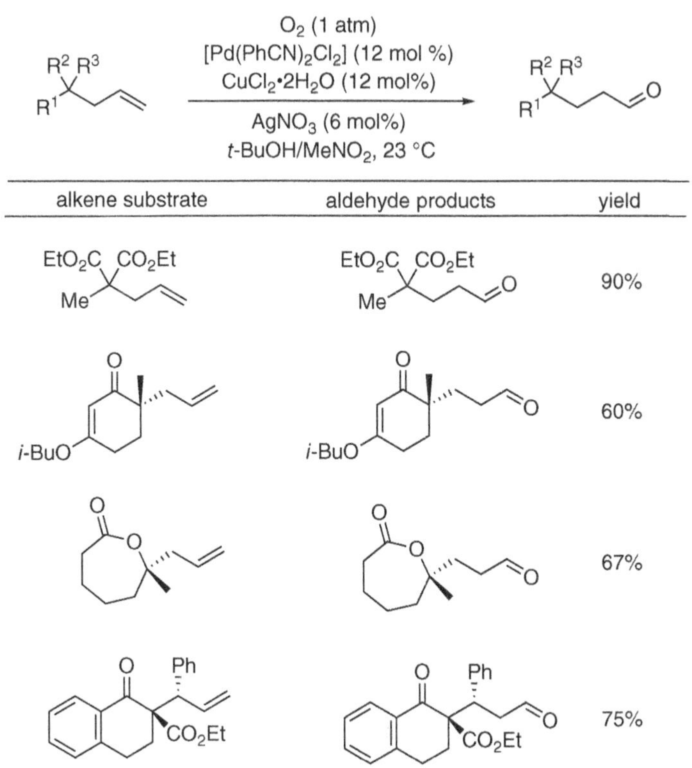

Scheme 2.35 Synthesis of aldehydes from alkenes.

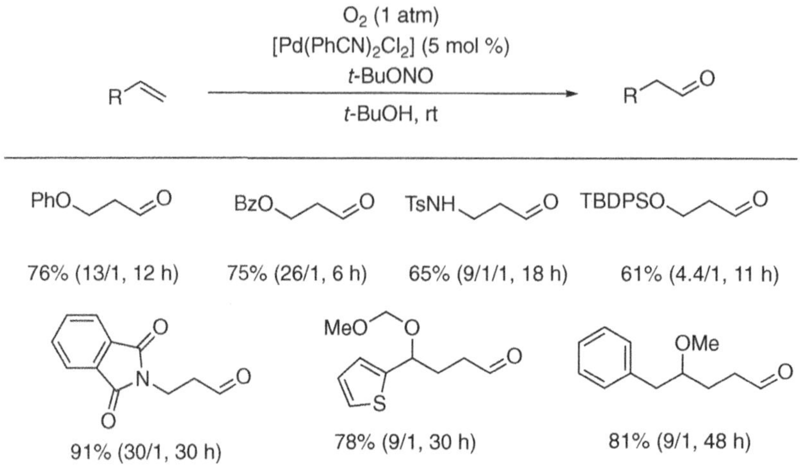

Scheme 2.36 Aldehyde-selective aerobic Wacker–Tsuji oxidation.

Recently, a highly aldehyde-selective aerobic Wacker–Tsuji oxidation at room temperature under copper- and silver-free conditions has been developed that features the use of *t*-BuONO as a simple organic redox catalyst [44] (Scheme 2.36).

2.5.2.4 Wacker-Type Oxidation

In 2017, Han and coworkers disclosed a practical and general iron-catalyzed Wacker-type oxidation of olefins to ketones using ambient air as the sole oxidant. The mild oxidation conditions allow an exceptional functional-group tolerance, which has not been demonstrated for any other Wacker-type reaction to date. The inexpensive and nontoxic reagents (iron(II) chloride, polymethylhydrosiloxane (PMHS), and air) have been shown to be applicable to late-stage oxidation of complex natural-product-derived and polyfunctionalized molecules (Scheme 2.37) [45].

2.5.2.5 Heterogeneous Catalysts for Wacker Reactions

Beside the great performances of Pd-catalyzed reactions, the use of soluble Pd catalysts poses significant problems, including the toxicity caused by residual Pd and high costs attributed to the scarcity of the metal, thus hindering their industrial applications [46]. To overcome these drawbacks, heterogeneous Pd catalysts have been investigated [47]. Heterogeneous catalytic processes usually employ biphasic systems or immobilization on supports. For biphasic systems, separable ionic liquids [48] or phase-transfer methods have been used [49]. In the case of immobilization, materials such as carbon nanotubes, graphene, high-surface-area solids (silicates, zeolites, alumina, etc.), polymers, metal oxides [50], and nanocrystals derived from Fe_3O_4 [51] have been reported as metal supports.

2.5.3 Oxygenation via Full Cleavage of the C=C Double Bond

2.5.3.1 Ozonolysis

Ozone is an allotrope of oxygen, formed from dioxygen by the action of ultraviolet light or electrical discharges, and decays into dioxygen ($t1/2$~ a few days at ambient

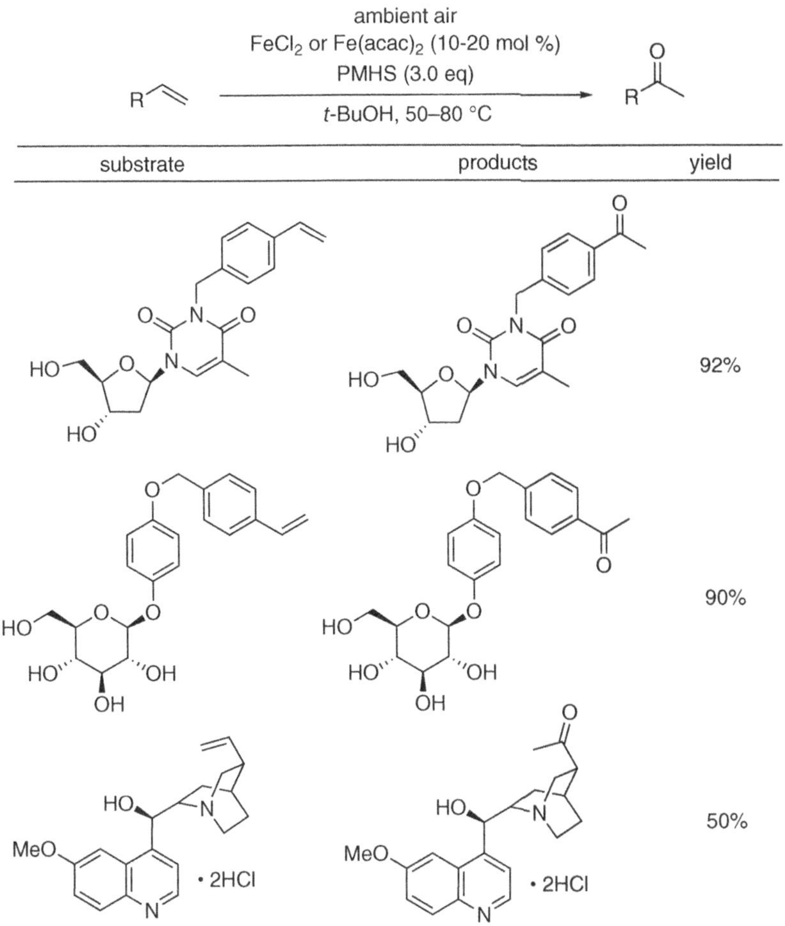

Scheme 2.37 Fe-catalyzed Wacker-type oxidation of olefins to ketones.

atmosphere), being present in very low concentrations throughout the Earth's atmosphere. Since Schönbein first studied ozonolysis in 1855 [52], this reaction has been applied to many synthetic transformations [53]. The extremely high oxidizing ability ($E°$ = 2.07 [vs. SHE]) and peculiarly bent structure of ozone finds the C=C double bond in alkenes a compatible partner to undergo a 1,3-dipolar cycloaddition, giving rise to molozonides (or primary ozonide), which is mostly transformed to ozonide via fragmentation and recombination in a less polar solvent, but, it is revealed that the reaction would proceed in different ways to give carbonyl oxide, reductive treatment of which affords aldehyde and/or ketones depending on substitution patterns of the alkene (Scheme 2.38).

The classical but versatile uses of ozolysis in synthesizing carbonyl compounds are shown in Scheme 2.39.

The primary concern with ozonolysis chemistry rests on safety issues: ozone itself is highly toxic, and low-molecular-weight ozonides and peroxides generated present an explosive hazard on concentration during the workup. Notwithstanding these issues,

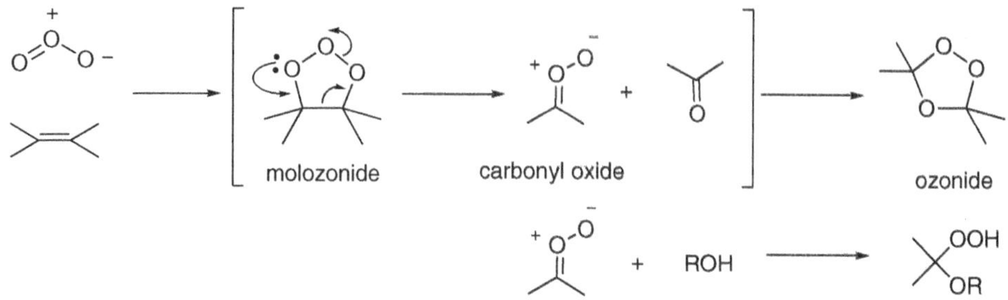

Scheme 2.38 1,3-dipolar cycloaddition of ozone and alkene.

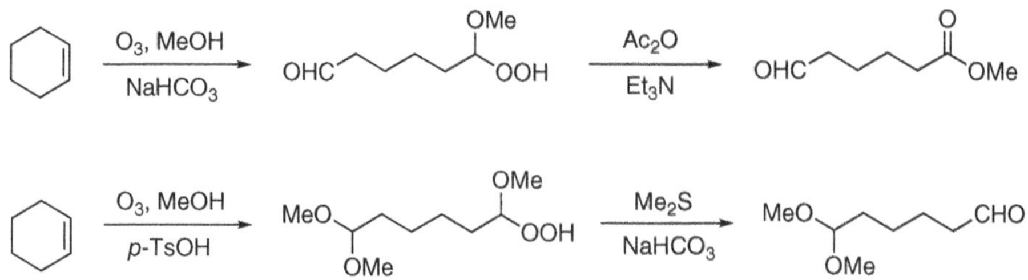

Scheme 2.39 Synthesis of carbonyl compounds from ozolysis.

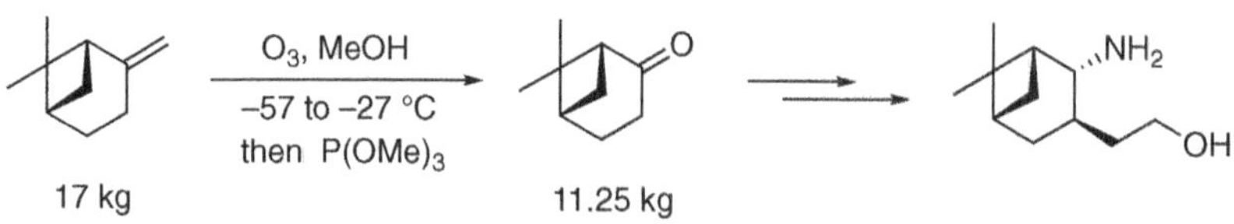

Scheme 2.40 Batch-scale ozonolysis reactions.

ozone is considered to be "green" oxidant because it only produces oxygen as a side product after the reaction, rendering ozonolysis a method of choice for obtaining carbonyl compounds in academia and industry [54]. In practice, ozone is generated from oxygen by and used as a mixture with oxygen: typically, a stream of oxygen containing 2–10% of ozone is passed into a solution or suspension of the compound in a suitable solvent. Ozonolyses are highly exothermic and are typically performed in organic solvents at low temperature [55]. The risk of fire or explosion in large-scale reactions has been ameliorated via reduction of head space oxygen content, or reaction in ionic liquids or supercritical CO_2.

Batch-scale ozonolysis reactions have been reported, including the reactions illustrated in Scheme 2.40.

A spectacular use of ozonolysis is exemplified in the multi-kilogram manufacturing process of the Halaven® C27-C35 fragment (Scheme 2.41) [56].

The indispensable use of ozonolysis encourages development of several alternatives to conventional batch reactions that offer a safer (greener) conduction of it based on the use of flow-based technologies to facilitate reagent mixing and temperature control [57]. Scientists at Lonza have developed large scale ozonolysis for the production of an insecticide key intermediate [58]. A 450 L loop reactor was used for the conversion of

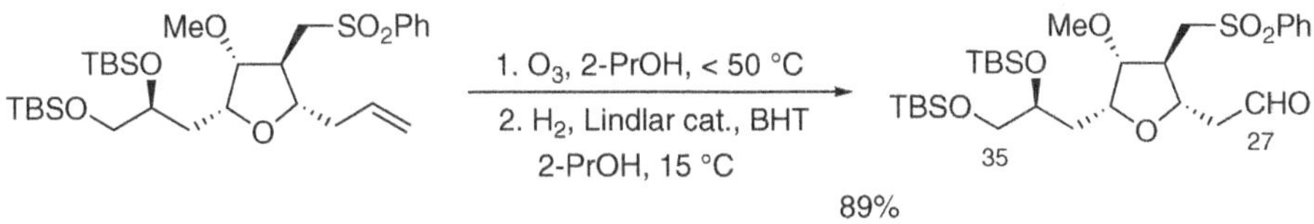

Scheme 2.41 Multi-kilogram-scale ozonolysis reactions.

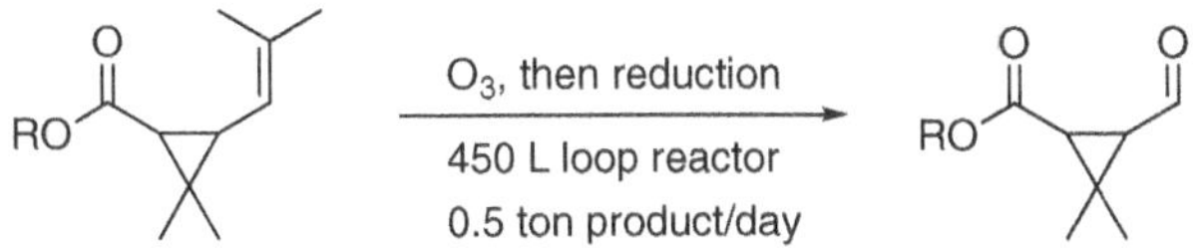

Scheme 2.42 The reaction of chrysanthemic acid ester with ozone.

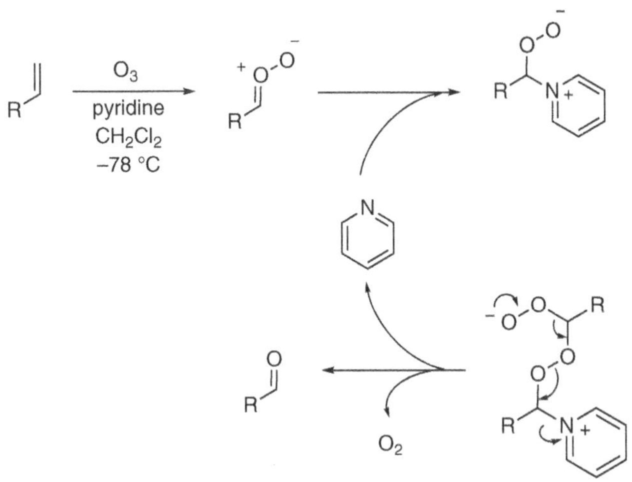

Scheme 2.43 Reductive ozonolysis.

chrysanthemic acid ester with ozone (Scheme 2.42) [59]. The process was optimized to produce 0.5 t per day in continuous mode. Approaches based on novel chemical engineering technology have been reviewed.

Although organic chemistry textbooks describe that carbonyl compounds could be obtained by reductive treatment of ozonide intermediates, Dussault and coworkers disclosed the innovative methods coined "reductive ozonolysis" that offer elimination of reducing agents [60]. The novel protocols based on unprecedented trapping and fragmentation of short-lived carbonyl oxide intermediates, have been realized by several nucleophilic reagents, such as amine-*N*-oxides, water, and pyridine. Carbonyl oxides, avoiding potential hazards associated with generation and isolation of ozonides or other peroxide products (Scheme 2.43).

The reductive ozonolysis protocol allows the in situ generation of aldehydes or ketones via ozonolytic cleavage of alkenes in tandem with a number of C—C-bond forming reactions, including 1,2-addition of Grignard reagents, Wittig or Horner–Wadsworth–Emmons reactions, and direct aldol reaction of lithium enolates, offering a rapid and high-yielding alternative to traditional two-step protocols (Scheme 2.44) [61].

2.5.3.2 Organocatalytic Aerobic Oxidative Cleavage of C=C Double Bond

The toxicity and potential risk of ozone and other oxidants have encouraged the search for more sustainable reagents. Besides ozone, molecular oxygen has been shown to have practical use in the oxygenative double bond cleave of a particular class of alkenes (that do not have an allylic hydrogen).

Jiao and coworkers cleverly exploited the electrophilic property of PINO radical, demonstrating another use of carbon radical generating catalyst via addition to alkenes. NHPI in DMA at 1 atm of O_2 and 80 °C disclosed novel oxidative transformation. The authors proposed a mechanism where formation of a PINO radical and electrophilic addition to the alkene generated a carbon radical that could be trapped by molecular oxygen: 1,2-dioxetane formation would release the PINO radical and the thermal cleavage of the former would provide the ketone (Scheme 2.45) [62].

The same authors reported a TEMPO-catalyzed one-pot oxidative cleavage of/nitrogenation in the presence of $TMSN_3$, leading to a keto nitrile from internal alkenes (Scheme 2.46) [63].

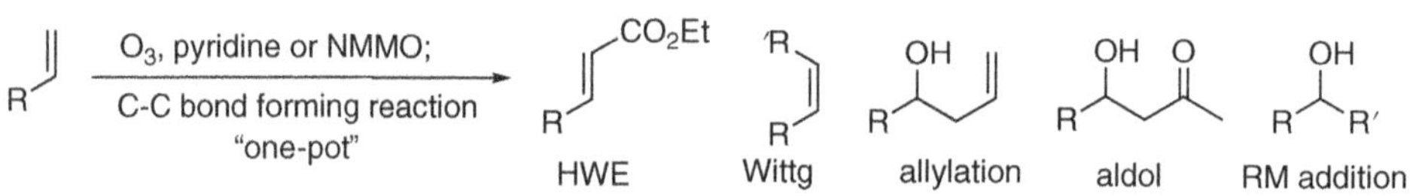

Scheme 2.44 Tandem application of C-C bond forming reactions with reductive ozonolysis.

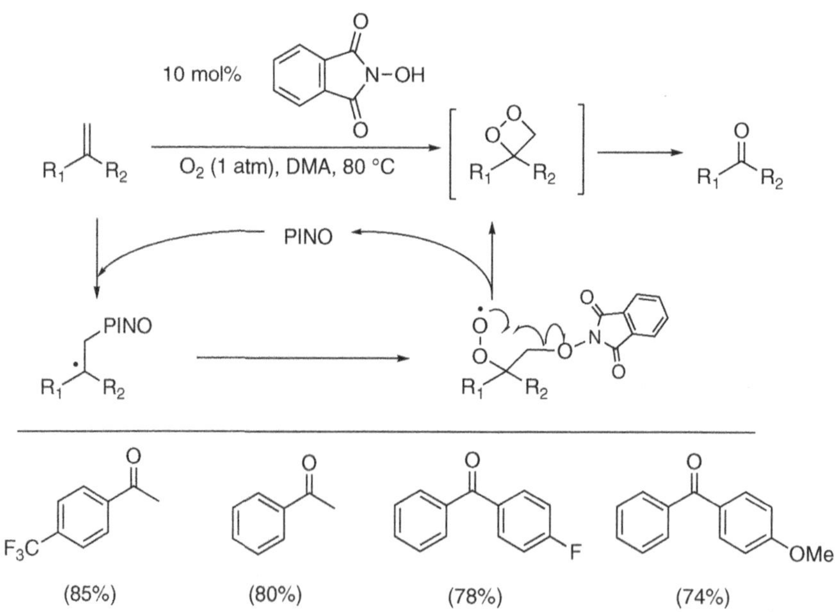

Scheme 2.45 NHPI-catalyzed oxidation of alkenes.

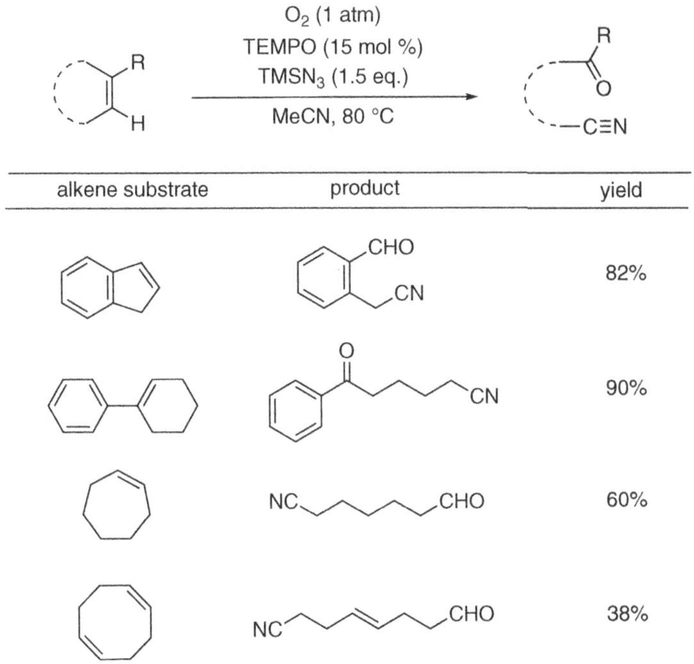

Scheme 2.46 TEMPO-catalyzed oxidative cleavage C=C bonds.

Wang and co-workers have developed another non-metal catalyzed method for the aerobic oxidative cleavage of the C=C double bond under visible light at room temperature. In seeking a photoinitiated radical that could reversibly add to the C=C bond to undergo dioxetane formation, they identified *bis*(4-methoxyphenyl) disulfide as an enabling photocatalyst, and typical monosubstituted as well as 1,1-and 1,2-disubstituted aromatic alkenes could be converted into the corresponding ketones and aldehydes. Experimental and computational studies suggested that a disulfide-olefin charge transfer complex is possibly responsible for the unconventional dissociation of the S—S bond under visible light (Scheme 2.47) [64].

2.6 Oxidation of Alcohols

Among methods for synthesizing carbonyl compounds, oxidation of alcohols offers chemists the most straightforward entry, making this particular oxidation a popular strategy and continuously enriching the chemist's toolbox to meet the ever-growing requirements in organic synthesis. However, traditional textbook oxidants, such as Cr(VI) or Mn(VII) derivatives, are notorious for their toxicity, and even more renowned reagents, such as Dess–Martin periodinane, or its precursor IBX, the Swern reagent, and TPAP/NMO system, still have poor atom efficiency, actually being avoided by the pharmaceutical industry. Despite these facts, the conventional methods offering reasonably mild reaction conditions (below room temperature), simple operation (just mixing), and promising results render them a method of choice in laboratory synthesis, still

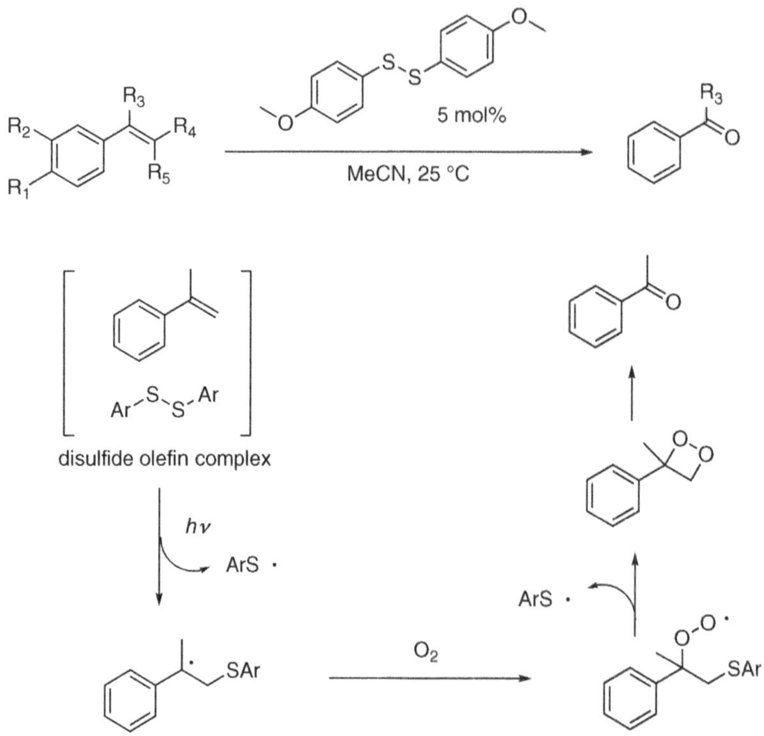

Scheme 2.47 Aerobic oxidative cleavage of the C=C bonds by bis(4-methoxyphenyl)disulfide.

causing a divide between large-scale preparative chemistry. In this context, development of truly practical, scalable, green sustainable methods to realize the oxidation of alcohols to the corresponding carbonyl compounds is of great importance in current organic chemistry. To this end, much effort has been devoted to search for competent mediators that link oxidatively competent catalysts and environmentally benign bulk oxidants.

2.6.1 Platinum-Group-Metal-Catalyzed Aerobic Oxidation

Before the concept of green chemistry was introduced, noble metal compounds had already shown aerobic alcohol oxidations. The first observation of catalytic aerobic oxidation could be attributed to Davy, who showed in 1820 that ethanol is oxidized to acetic acid in the presence of platinum [65]. In 1855, Strecker reported that a platinum oxide catalyzed oxidation of cinnamyl alcohol to cinnamaldehyde in the presence of dioxygen (air) [66]. Exhaustive investigations were conducted by Wieland to demonstrate that different simple alcohols yielded the corresponding aldehydes in dilute aqueous solution with finely divided platinum and oxygen [67]. Wieland designated theses reactions as dehydrogenation in which the platinum activates the hydrogen of the alcohol, establishing the basis of heterogeneous catalytic oxidations. Today, various bi- and multi-metallic catalysts, which can be prepared by simultaneous deposition and reduction of the metal precursors onto a supported catalyst, are applied that are more

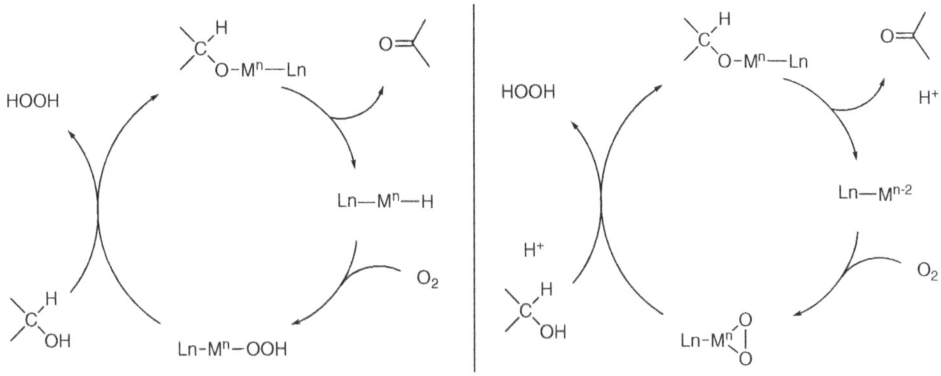

Scheme 2.48 Aerobic oxidation of alcohols by low-valent transition metal.

active and selective than monometallic catalysts. The most commonly used catalysts consist of Pt or Pd as active components and Bi or Pb as promoters. The oxidations are commonly carried out at 330–370 K under ambient pressure of air or oxygen. An advantage of the method is that the catalysts are active even in water. Dioxygen is considered to serve as an acceptor for the activated hydrogen in this particular system to give a water and thus remove it from the equilibrium. The catalytic dehydrogenation systems that just release hydrogen (acceptor-free oxidation) will not be discussed [68, 69].

2.6.2 Palladium(II)-Catalyzed Aerobic Oxidation

1n 1828, Berzelius observed that upon reflux of a wet ethanolic solution of K_2PdCl_4 most of the palladium was reduced [70]. The reaction is now rationalized that alcohol is oxidized to carbonyl compounds via the β-hydride elimination of a metal alkoxide intermediate. This method has often been employed for the synthesis of Group VIII metal hydride from Pt(II), Pd(II), and Ru(III) and Rh(III) salts. Simplified mechanisms for aerobic oxidation of alcohols by low-valent transition metal ions is shown in Scheme 2.48.

In the catalytic cycle, a hydridometal species, formed by β-hydride elimination from an alkoxymetal intermediate, is reoxidized by dioxygen, presumably via insertion of dioxygen into the M-H bond with the formation of H_2O_2. Alternatively, an alkoxymetal species can decompose to a proto and the reduced form of the catalyst, either directly or via the intermediacy of a hydridometal intermediate.

The methods received little attention as preparation for carbonyl compounds until Lloyd closely investigated homogeneous oxidation of alcohols with palladium (II) using copper(II) as a cocatalyst at 70–120 °C and 3 atm O_2 pressure. In 1977, Blackburn and Schwartz first disclosed that $PdCl_2$ underwent aerobic oxidation of secondary alcohols via direct dioxygen-coupled turnover in the presence of NaOAc [71]. Since then, numerous Pd catalyst systems have been reported for aerobic alcohol oxidation.

The first synthetically useful system, consisting of $Pd(OAc)_2$ in combination of $NaHCO_3$ as a base in DMSO, which enabled aerobic oxidation of primary and secondary allylic and benzylic alcohols to the corresponding carbonyl compounds, was reported in 1998 by Peterson and Larock [72]. Uemura and coworkers reported an improved procedure using $Pd(OAc)_2$ in combination with pyridine that allowed oxidation of primary and

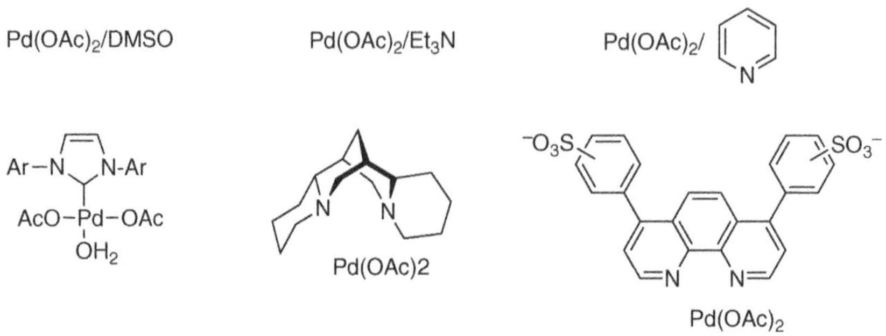

Scheme 2.49 Several palladium-catalyzed aerobic oxidation systems.

secondary aliphatic alcohols in addition to allylic and benzylic ones [73]. Sheldon and coworkers disclosed that a water-soluble palladium(II) complex of sulfonated bathophenantroline allowed a highly efficient aerobic oxidation in a two-phase aqueous-organic medium, in which no organic solvent was required except for solid alcohols.

The formation of Pd black is one of the main problems with using homogeneous Pd(II)-catalysts (see Section 2.4.2). Tsuji and coworkers used substituted pyridines as ligands to prevent formation of Pd black, allowing oxidations using low catalyst loading [74]. Sigman and coworkers developed several novel Pd(II)-catalysts, from which $Pd(OAc)_2$/TEA system was identified as the most convenient. Impacts of ligands on palladium-catalyzed aerobic oxidation reactions have been reviewed (Scheme 2.49) [75].

2.6.3 Cu-Catalyzed Aerobic Alcohol Oxidation

Copper(II) salts play salient roles as one-electron oxidants in organic chemistry, and the resultant Cu(I) species have shown to be reoxidized to Cu(II) with O_2. This prominent reactivity renders Cu as indispensable at the heart of many aerobic oxidative transformations [76]. A wide range of oxygenase and oxidase enzymes construct their active centers using Cu, which has spurred extensive studies of the fundamental reactivity of Cu^I complexes with O_2 [77]. In addition to the academic subjects of bioinorganic chemistry, Cu-catalyzed aerobic oxidation reactions have been applied in several important industrial processes, many of which are discussed in the previous section. In the history of the development of alcohol oxidation processes, a class of stable nitroxyl radicals has been playing salient roles. In 1984, Semmelhack et al. disclosed the first catalytic aerobic alcohol oxidation using TEMPO with CuCl in DMF [78], although the substrate scope was limited for activated alcohols such as benzylic and allylic alcohols. After their seminal report, significant progress in terms of catalytic efficiency and substrate applicability has been achieved by Sheldon and coworkers, who showed that $CuBr_2$ and TEMPO in the presence of 2,2-bpy as a ligand for Cu led to the oxidation of several primary alcohols with no overoxidation of carboxylic acids [79]. The authors postulated a copper mediated dehydrogenation mechanism, in which TEMPO acts as a hydrogen acceptor and is able to regenerate the active Cu(II) species. Kumpulainen and Koskinen reported a catalyst composed of CuII/bpy/TEMPO with DBU and/or *N*-methylimidazole (NMI) that exhibited excellent reactivity toward unactivated aliphatic alcohols at

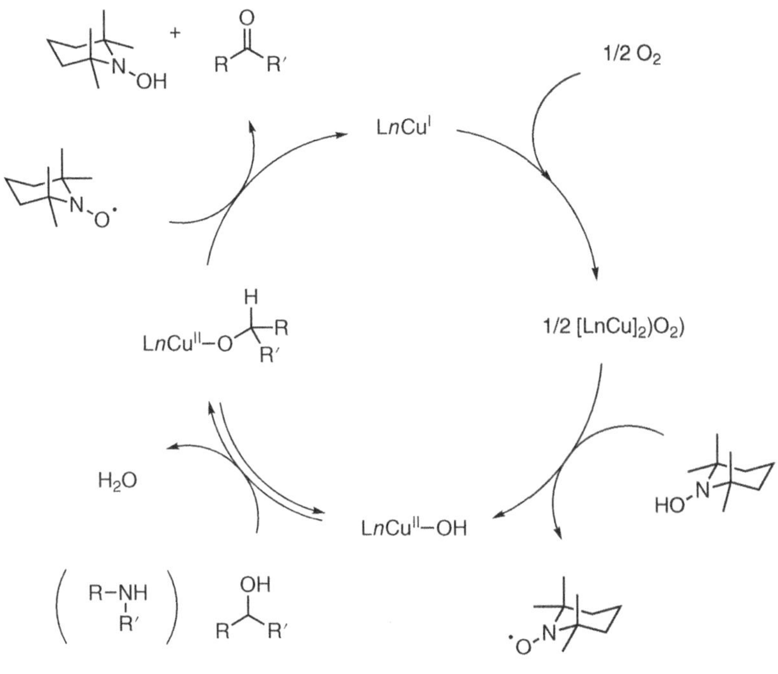

Scheme 2.50 Mechanism for Cu-catalyzed aerobic alcohol oxidation.

room temperature, expanding the scope of TEMPO/Cu-catalyzed aerobic oxidation (Scheme 2.50) [80].

In 2011, Hoover and Stahl reported a CuOTf/bpy/TEMPO/NMI catalyst system that was effective in the oxidation of benzylic, allylic, propargylic, and aliphatic alcohols under ambient air (Scheme 2.2). A significant improvement in catalytic activity was associated not only with the use of a copper(I) rather than a copper(II), but with non-coordinating anions (e.g. CuOTf). The mild reaction conditions were compatible with numerous functional groups, including aryl halides, anilines, nitrogen and sulfur heterocycles, and sulfides. This method is compatible with base-sensitive substrates thanks to the lack of stoichiometric or strongly basic additives: *N*-Boc-prolinol was oxidized to the aldehyde without epimerization. Although Cu/TEMPO catalyst systems suffered from low efficiency in the oxidation of secondary alcohols, the chemoselectivity for primary over secondary alcohols enabled class-selective oxidation of diols that contained both primary and secondary alcohols. Reactions of 1,5-diols led to efficient lactonization in high yields (Scheme 2.51) [81].

The observations noted above prompted Steves and Stahl to examine nitroxyl derivatives other than TEMPO. Much faster reaction rates were observed with less-hindered class of nitroxyls ABNO, keto-ABNO, and AZADO [82]. The Cu/ABNO catalyst system exhibited nearly identical oxidation rates with essentially all classes of alcohols, including 1° and 2° benzylic and 1° and 2° aliphatic alcohols (Scheme 2.52).

A recent experimental and computational study of the mechanism of Cu/nitroxyl catalyzed alcohol oxidations indicated that the oxidation proceeds via an Oppenauer-type six-membered-ring transition state [83].

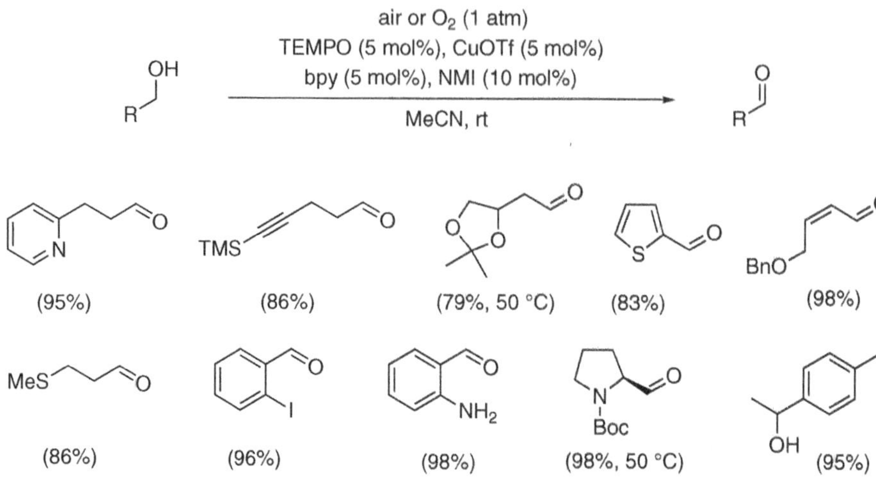

Scheme 2.51 Stahl's Cu-catalyzed aerobic alcohol oxidation system.

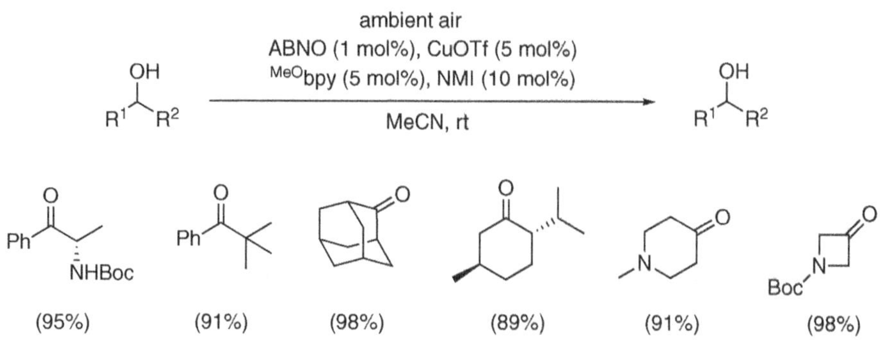

Scheme 2.52 Cu-catalyzed aerobic alcohol oxidation.

Iwabuchi and co-workers subsequently reported a complementary catalyst system employing AZADO as a nitroxyl in a study that focused on the oxidation of unprotected amino alcohols [84]. Such reactions are typically challenging due to competitive oxidation of the alcohol and amine groups and/or unfavorable interaction of the electron-rich amine with the oxidant. Synthetic methods to access carbonyl products bearing unprotected amines typically require the protection of the amino group. The Cu/AZADO catalytic method was shown to be superior to several conventional oxidation methods (e.g. with pyridinium chlorochromate, Swern, Dess–Martin periodinane, and tetrapropylammonium perruthenate oxidants) (Scheme 2.53).

The AZADO/Cu catalysis has shown to be applicable to chemoselective aerobic oxidation of alcohols containing electron-rich divalent sulfur functionalities (e.g. 1,3-dithianes and sulfides) into the corresponding carbonyl compounds (Scheme 2.54) [85].

The advantage of the Cu/AZADO catalyst system relative to traditional oxidation methods was showcased in synthetic routes to two small alkaloid targets, (−)-mesembrine and myosmine (Scheme 2.55).

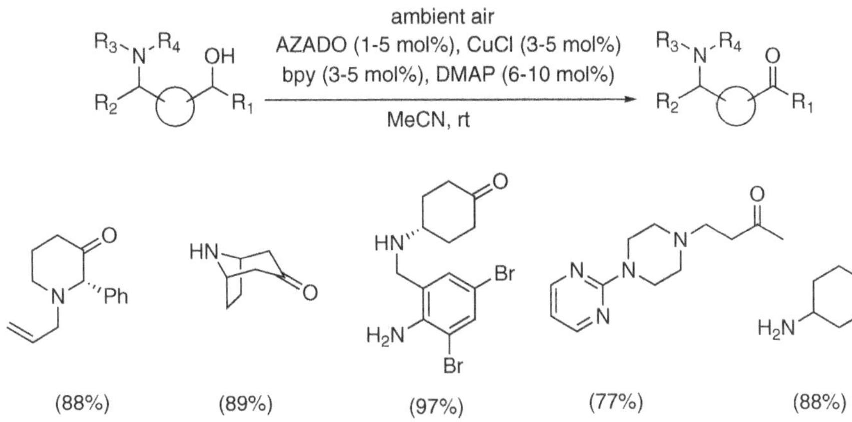

Scheme 2.53 Employing AZADO in Cu-catalyzed aerobic alcohol oxidation.

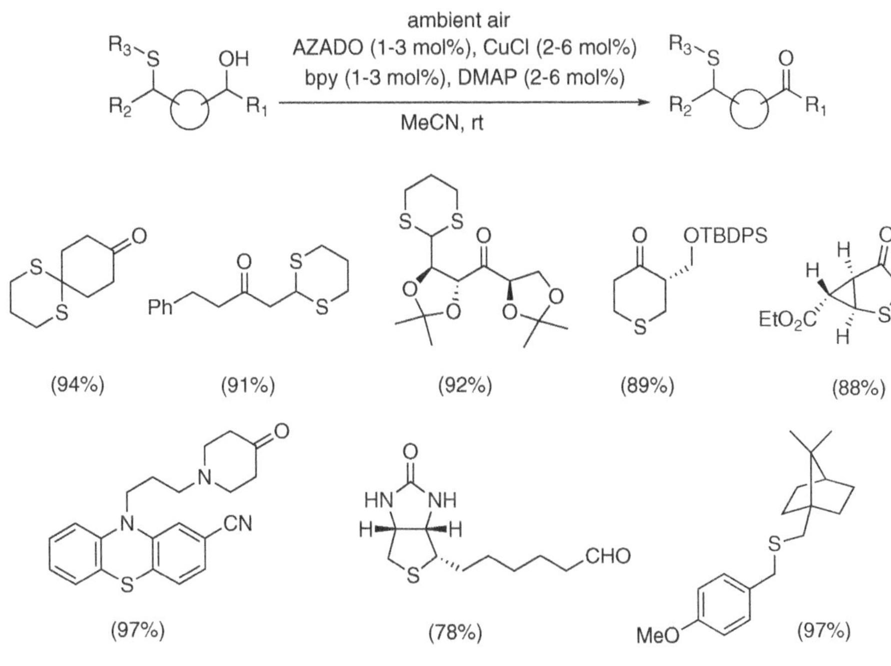

Scheme 2.54 Aerobic oxidation of S-containing alcohols.

2.6.4 Transition Metal-Catalyzed Alcohol Oxidation with H_2O_2

Many acidic metal oxides such as MnO_3, WO_3, V_2O_5 catalyze the reaction of H_2O_2 through the formation of inorganic peroxides, most of which likely undergo heterolysis at the O—O bond in the oxidation of alcohols. Metal-catalyzed oxidation of alcohols with H_2O_2 can be conveniently divided into two categories, involving peoxyometal and oxyometal species as the oxidatively active entities. In the peroxometal pathway, the metal ion remains in the same oxidation state throughout the catalytic cycle and no

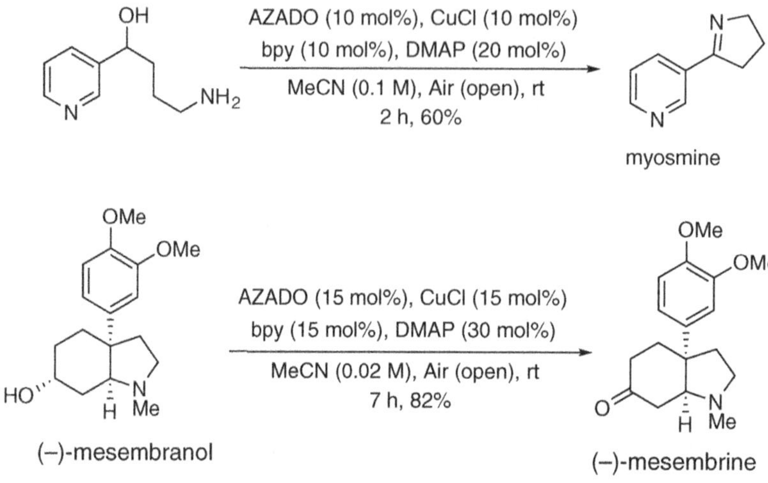

Scheme 2.55 Application in total synthesis.

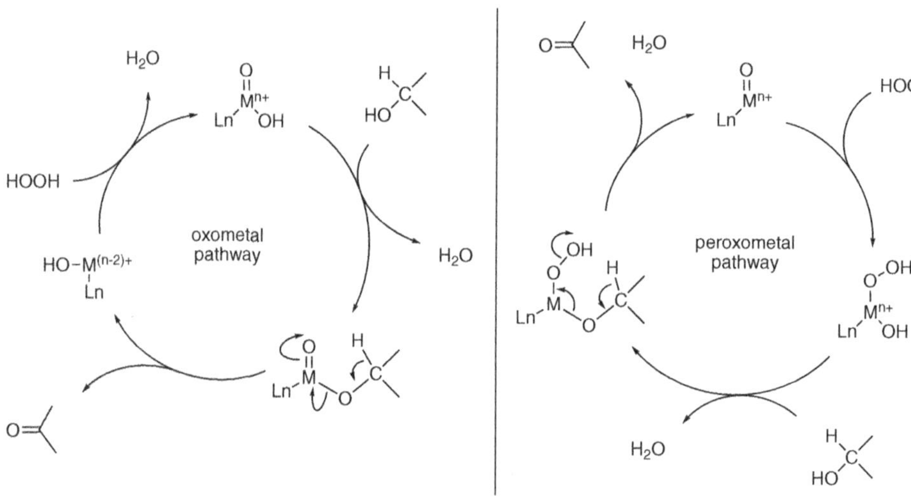

Scheme 2.56 Mechanism of transition metal-catalyzed alcohol oxidation with H_2O_2.

stoichiometric oxidation is observed in the absence of the peroxide. In contrast, the oxometal pathway involves a two-electron change in oxidation state of the metal ion and a stoichiometric oxidation is observed, with the oxidized form of the catalyst, in the absence of H_2O_2 (Scheme 2.56).

Peroxometal pathways are typically operative in early transition metal ions with a d^0 configuration, for example, Mo^{6+}, W^{6+}, Ti^{4+}, and Re^{7+}. Oxometal pathways are characteristic of late transition metals and first row transition metals, for example Cr^{6+}, Mn^{5+}, Os^{8+}, Ru^{6+}, Ru^{8+}, which are strong oxidants.

Bortolini and coworkers disclosed the first tungstate catalyzed oxidation of alcohols with aqueous hydrogen peroxide, in a biphasic system composed of water and 1,2-dichloroethane using a tetraalkylammonium salt as a phase transfer agent [86]. Noyori and coworkers optimized this methodology to establish an extremely effective, halide- and organic solvent-free system using 1.1 equiv. of 30% H_2O_2 at 90 °C and the lipophilic methyltrioctylammonium bisulfate, [CH3(n- C8H17)3N] + HSO4−, as a phase transfer agent [87]. Substrate catalyst ratios as high as 400 000 were realized, affording turnover numbers up to 180 000. A variety of secondary alcohols, including unsaturated alcohols, afforded the corresponding ketones in high yields. Primary alcohols gave the corresponding carboxylic acid in high yields via further oxidation of the hydrate of the intermediate aldehyde. The catalytic system is shown to be applicable to oxidations of alkenes to epoxides as well as sulfides to sulfoxides [88]. In an attempt to expand the application of nonheme system, manganese complex Mn(S-PMB)(OTf)2 to the oxidation of a series of secondary alcohols. Various alcohols were oxidized to ketones with good to excellent yields, in which maximum turnover of 4700 was observed [86].

2.6.5 Heterogeneous Catalysts for Alcohol Oxidation

Application of solid catalysts for the gas- or vapor- phase oxidation of simple, small-chain alcohols to the corresponding carbonyl compounds is well established [87] (Section 2.6.1), where volatility and thermal stability of reactant and products shape the structures of reaction processes. Solid catalysts that sustain their catalytic activity, even in the liquid phase under mild conditions, have a much broader application range [89] in which preventing of leaching out of the active species becomes a major challenge [90].

The reactants and particularly the carboxylic acid-type (by)products are frequently excellent chelating agents and accelerate dissolution and deactivation of the catalysts. Indeed, especially in industrial chemistry, heterogeneous catalytic systems are preferred over homogeneous ones due to easier recyclability. However, they usually suffer from low catalytic activity relative to their homogeneous counterparts. Much effort has been made to overcome the difficulties encountered with heterogeneous catalysis, because reduction of environmental loading due to easy separation and reuse of the catalyst could result [91]. These catalysts can be applied to continuous-flow systems composing tandem reactions that do not require workup or much purification of the intermediate, thereby offering efficient use of unstable intermediates. Developments of multifunctional heterogeneous metal nanocluster catalysts for use in tandem reactions based on aerobic oxidation reactions as key processes has been reviewed [92].

References

1 Seebach, D., Weidmann, B., and Wilder, L. (1983). Titanium and zirconium derivatives in organic synthesis. In: *Modern Synthetic Methods* (ed. R. Scheffold), 324. Frankfurt: Otto Salle Verlag.

2 Sheldon, R.A. (2017). Metrics of green chemistry and sustainability: past, present, and future. *ACS Sustain. Chem. Eng.* 6: 32–48.

3 Hermans, I. (2016). Overview of radical chain oxidation chemistry. In: *Liquid Phase Aerobic Oxidation Catalysis* (ed. S.S. Stahl and P.L.E. Alsters), 3–14. Weinheim: Wiley-VCH Verlag.

4 Punniyamurthy, T., Velusamy, S., and Iqbal, J. (2005). Recent advances in transition metal-catalyzed oxidation of organic substrates with molecular oxygen. *Chem. Rev.* 105: 2323–2363.

5 Walling, C. (1975). Fenton's reagent revisited. *Acc. Chem. Res.* 8: 125–131.
6 Hang, J., Ghorai, P., Finkenstaedt-Quinn, S.A. et al. (2012). Generation of singlet oxygen from fragmentation of monoactivated 1,1-dihydroperoxides. *J. Org. Chem.* 77: 1233–1243.
7 Grant, J.T., Venegas, J.M., McDermott, W.P., and Hermans, I. (2018). Aerobic oxidations of light alkanes over solid metal oxide catalysts. *Chem. Rev.* 118: 2769–2815.
8 Cao, S., Tao, F.F., Tang, Y. et al. (2016). Size- and shape-dependent catalytic performances of oxidation and reduction reactions on nanocatalysts. *Chem. Soc. Rev.* 45: 4747–4765.
9 Chng, L.L., Erathodiyil, N., and Ying, J.Y. (2013). Nanostructured catalysts for organic transformations. *Acc. Chem. Res.* 46: 1825–1837.
10 Ishii, Y.S. (2001). Satoshi; Iwahama, Takahiro, innovation of hydrocarbon oxidation with molecular oxygen and related reactions. *Adv. Synth. Catal.* 343: 343–427.
11 Chen, K., Zhang, P., Wang, Y., and Li, H. (2014). Metal-free allylic/benzylic oxidation strategies with molecular oxygen: recent advances and future prospects. *Green Chem.* 16: 2344–2374.
12 Zhang, Z., Gao, Y., Liu, Y. et al. (2015). Organocatalytic aerobic oxidation of Benzylic sp(3) C-H bonds of ethers and alkylarenes promoted by a recyclable TEMPO catalyst. *Org. Lett.* 17: 5492–5495.
13 Lesieur, M., Genicot, C., and Pasau, P. (2018). Development of a flow photochemical aerobic oxidation of benzylic C-H bonds. *Org. Lett.* 20: 1987–1990.
14 Schwarz, H., González-Navarrete, P., Li, J. et al. (2016). Unexpected mechanistic variants in the thermal gas-phase activation of methane. *Organometallics* 36: 8–17.
15 Gamenara, G., Seoane, G., Saenz Méndez, P., and Domínguez de María, P. (2012). *Redox Biocatalysis: Fundamentals and Applications*. Hoboken, NJ: Wiley.
16 Barton, D.H.R. and Dario, D. (1992). The selective functionalization of saturated hydrocarbons: Gif chemistry. *Acc. Chem. Res.* 25: 504–512.
17 Barton, D.H.R., Boivin, J., Gastiger, M. et al. (1986). Functionalization of saturated hydrocarbons. Part 4. The gif system for selective oxidation using molecular oxygen. *J. Chem. Soc. Perkin Trans.* 1: 947–955.
18 Barton, D.H.R. (1988). Gif chemistry: the present situation. *Tetrahedron* 54: 5805–5817.
19 Murahashi, S., Oda, Y., and Naota, T. (1992). Iron- and ruthenium-catalyzed oxidations of alkanes with molecular oxygen in the presence of aldehydes and acids. *J. Am. Chem. Soc.* 114: 7913–7914.
20 Murahashi, S.N., Naota, T., and Komiya, N. (1995). Metalloporphyrin-catalyzed osidaion of alkanes with molecular oxygen in the presence of acetaldehyde. *Tetrahedron Lett.* 36: 8059–8062.
21 Murahashi, S.-I., Zhou, X.-G., and Komiya, N. (2003). Chlorinated phthalocyanine iron(II) complex catalyzed oxidation of alkanes and alkenes with molecular oxygen in the presence of acetaldehyde. *Syn. Lett.* 321–324.
22 Wojtowicz-Mlochowska, H. (2017). Synthetic utility of metal catalyzed hydrogen peroxide oxidation of C-H, C-C and C=C bonds in alkanes, arenes and alkenes: recent advances. *ARKIVOC* 2: 12–58.
23 Kim, C., Kui, C., Kim, J., and Que, L. (1997). Stereospecific alkane hydroxylation with H2O2 catalyzed by an iron(II)-Tris(2-pyridylmethyl)amine complex. *J. Am. Chem. Soc.* 119: 5964–5965.
24 Howell, J.M., Feng, K., Clark, J.R. et al. (2015). Remote oxidation of aliphatic C-H bonds in nitrogen-containing molecules. *J. Am. Chem. Soc.* 137: 14590–14593.

25 Milan, M., Bietti, M., and Costas, M. (2017). Highly enantioselective oxidation of nonactivated aliphatic C-H bonds with hydrogen peroxide catalyzed by manganese complexes. *ACS Cent. Sci.* 3: 196–204.

26 Liu, L., Ye, X.P., and Bozell, J.J. (2012). A comparative review of petroleum-based and bio-based acrolein production. *ChemSusChem* 5: 1162–1180.

27 Maison, W. and Weidmann, V. (2013). Allylic oxidations of olefins to enones. *Synthesis* 45: 2201–2221.

28 Toyota, M.A., Asano, T., and Ihara, M. (2005). Total synthesis of serofendic acids A and B employing tin-free homoallyl-homoallyl radical rearrangement. *Org. Lett.* 7: 3929–3932.

29 Horn, E.J., Rosen, B.R., Chen, Y. et al. (2016). Scalable and sustainable electrochemical allylic C-H oxidation. *Nature* 533: 77–81.

30 Alberti, M.N. and Orfanopoulos, M. (2010). Unraveling the mechanism of the singlet oxygen ene reaction: recent computational and experimental approaches. *Chem. Eur. J.* 16: 9414–9421.

31 Mihelich, E.D. and Eickhoff, D.J. (1983). A one-pot conversion of olefins to α,β-unsaturated carbonyl compounds. An easy synthesis of 2-cyclopentenone and related compounds. *J. Org. Chem.* 48: 4135–4137.

32 Kuga, T., Sasano, Y., and Iwabuchi, Y. (2018). IBX as a catalyst for dehydration of hydroperoxides: green entry to α,β-unsaturated ketones via oxygenative allylic transposition. *Chem. Commun.* 54: 798–801.

33 Phillips, F.C. (1894). Note on a form of silver obtained in the reduction of the sulphide by hydrogen. *J. Am. Chem. Soc.* 16: 700–703.

34 Jira, R. (2009). Acetaldehyde from ethylene–a retrospective on the discovery of the Wacker process. *Angew. Chem. Int. Ed. Eng.* 48: 9034–9037.

35 Tsuji, J. (1984). Synthetic applications of the palladium-catalyzed oxidation of olefins to ketones. *Synthesis* 369–384.

36 (a) Keith, J.A. and Henry, P.M. (2009). The mechanism of the Wacker reaction: a tale of two hydroxypalladations. *Angew. Chem. Int. Ed. Eng.* 48: 9038–9049. (b) Kocovsky, P. and Backvall, J.E. (2015). The syn/anti-dichotomy in the palladium-catalyzed addition of nucleophiles to alkenes. *Chemistry* 21: 36–56.

37 Dong, J.J., Browne, W.R., and Feringa, B.L. (2015). Palladium-catalyzed anti-Markovnikov oxidation of terminal alkenes. *Angew. Chem. Int. Ed. Eng.* 54: 734–744.

38 Stahl, S.S. (2004). Palladium oxidase catalysis. Selective oxidation of organic chemicals by direct dioxygen-coupled turnover. *Angew. Chem. Int. Ed.* 43: 3400–3420.

39 Baiju, T.V., Gravel, E., Doris, E., and Namboothiri, I.N.N. (2016). Recent developments in Tsuji–Wacker oxidation. *Tetrahedron Lett.* 57: 3993–4000.

40 ten Brink, G.-J.A., Isabel, W.C.E., Papadogianakis, G., and Sheldon, R.A. (1998). Catalytic conversions in water. Part 10. Aerobic oxidation of terminal olefins to methyl ketones catalyzed by water soluble palladium complexes. *Chem. Commun.* 2359–2360.

41 (a) Mitsudome, T., Yoshida, S., Mizugaki, T. et al. (2013). Highly atom-efficient oxidation of electron-deficient internal olefins to ketones using a palladium catalyst. *Angew. Chem. Int. Ed. Eng.* 52: 5961–5964. (b) Mitsudome, T., Umetani, T., Nosaka, N. et al. (2006). Convenient and efficient Pd-catalyzed regioselective oxyfunctionalization of terminal olefins by using molecular oxygen as sole reoxidant. *Angew. Chem. Int. Ed. Eng.* 45: 481–485.(c) Mitsudome, T., Mizumoto, K., Mizugaki, T. et al. (2010).

Wacker-type oxidation of internal olefins using a PdCl2/N,N-dimethylacetamide catalyst system under copper-free reaction conditions. *Angew. Chem. Int. Ed. Eng.* 49: 1238–1240.

42 (a) Wickens, Z.K., Skakuj, K., Morandi, B., and Grubbs, R.H. (2014). Catalyst-controlled Wacker-type oxidation: facile access to functionalized aldehydes. *J. Am. Chem. Soc.* 136: 890–893. (b) Wickens, Z.K., Morandi, B., and Grubbs, R.H. (2013). Aldehyde-selective Wacker-type oxidation of unbiased alkenes enabled by a nitrite co-catalyst. *Angew. Chem. Int. Ed. Eng.* 52: 11257–11260.

43 Kim, K.E., Li, J., Grubbs, R.H., and Stoltz, B.M. (2016). Catalytic anti-Markovnikov transformations of hindered terminal alkenes enabled by aldehyde-selective Wacker-type oxidation. *J. Am. Chem. Soc.* 138: 13179–13182.

44 Ning, X.S., Wang, M.M., Yao, C.Z. et al. (2016). Tert-butyl nitrite: organic redox cocatalyst for aerobic aldehyde-selective Wacker–Tsuji oxidation. *Org. Lett.* 18: 2700–2703.

45 Liu, B., Jin, F., Wang, T. et al. (2017). Wacker-type oxidation using an iron catalyst and ambient air: application to late-stage oxidation of complex molecules. *Angew. Chem. Int. Ed. Eng.* 56: 12712–12717.

46 Astruc, D. (2007). Palladium nanoparticles as efficient green homogeneous and heterogeneous carbon-carbon coupling precatalysts: a unifying view. *Inorg. Chem.* 46: 1884.

47 Lamblin, M., Nassar-Hardy, L., Hierso, J.-C. et al. (2010). Recyclable heterogeneous palladium catalysts in pure water: sustainable developments in Suzuki, Heck, Sonogashira and Tsuji-Trost reactions. *Adv. Synth. Catal.* 352: 33–79.

48 Zhang, Y., Quek, X.-Y., Wu, L. et al. (2013). Palladium nanoparticles entrapped in polymeric ionic liquid microgels as recyclable hydrogenation catalysts. *J. Mol. Catal. A Chem.* 379: 53–58.

49 Lee, D., Lee, H., Kim, S. et al. (2006). A novel chemoentrapment approach for supportless recycling of a catalyst: catalytic asymmetric dihydroxylation. *Adv. Synth. Catal.* 348: 1021–1024.

50 Akai, Y., Yamamoto, T., Nagata, Y. et al. (2012). Enhanced catalyst activity and enantioselectivity with chirality-switchable polymer ligand PQXphos in Pd-catalyzed asymmetric silaborative cleavage of meso-methylenecyclopropanes. *J. Am. Chem. Soc.* 134: 11092–11095.

51 Byun, S., Chung, J., Kwon, J., and Moon Kim, B. (2015). Mechanistic studies of magnetically recyclable Pd - Fe3O4 heterodimeric nanocrystal-catalyzed organic reactions. *Chem. Asian J.* 10: 982–988.

52 Criegee, R. (1975). Mechanism of ozonolysis. *Angew. Chem. Int. Ed. Eng.* 14: 765–771.

53 Fisher, T.J. and Dussault, P.H. (2017). Alkene ozonolysis. *Tetrahedron* 73: 4233–4258.

54 Van Ornum, S.G.C., Robin, M., and Pariza, R. (2006). Ozonolysis applications in drug synthesis. *Chem. Rev.* 106: 2990–3001.

55 Ragan, J.A., am Ende, D.J., Brenek, S.J. et al. (2003). Safe execution of a large-scale ozonolysis: preparation of the bisulfite adduct of 2-hydroxyindan-2-carboxaldehyde and its utility in a reductive amination. *Org. Process. Res. Dev.* 7: 155–160.

56 Chase, C., Austad, B., Benayoud, F. et al. (2013). Process development of Halaven®: synthesis of the C14-C35 fragment via iterative Nozaki-Hiyama-Kishi reaction-Williamson ether cyclization. *Synlett* 24: 327–332.

57 Roydhouse, M.D., Ghaini, A., Constantinou, A. et al. (2011). Ozonolysis in flow using capillary reactors. *Org. Process Res. Dev.* 15: 989–996.

58 Roberge, D. (2012). Lonza – hazardous flow chemistry for streamlined large scale synthesis. *Green Processes. Synth.* 1: 129–130.
59 Nobis, M.R. and Dominique, M. (2011). Mastering ozonolysis: production from laboratory to ton scale in continuous flow. *Chim. Oggi* 29: 56–58.
60 Willand-Charnley, R., Fisher, T.J., Johnson, B.M., and Dussault, P.H. (2012). Pyridine is an organocatalyst for the reductive ozonolysis of alkenes. *Org. Lett.* 14: 2242–2245.
61 Willand-Charnley, R. and Dussault, P.H. (2013). Tandem application of C-C bond-forming reactions with reductive ozonolysis. *J. Org. Chem.* 78: 42–47.
62 Lin, R., Feng, C., and Jiao, N. (2012). Metal-free, NHPI catalyzed oxidative cleavage of C-C double bond using molecular oxygen as oxidant. *Org. Lett.* 14: 4158–4161.
63 Wang, T. and Jiao, N. (2013). TEMPO-catalyzed aerobic oxygenation and nitrogenation of olefins via C horizontal line C double-bond cleavage. *J. Am. Chem. Soc.* 135: 11692–11695.
64 Deng, Y., Wei, X.J., Wang, H. et al. (2017). Disulfide-catalyzed visible-light-mediated oxidative cleavage of C=C bonds and evidence of an olefin-disulfide charge-transfer complex. *Angew. Chem. Int. Ed. Eng.* 56: 832–836.
65 Sermon, P.A. (1979). *Chem. Ind.* 681–690.
66 Strecker, A. (1855). Künstliche Darstellung des Zimmtöles. *Liebigs Ann. Chem.* 93: 370–371.
67 Wieland, H. (1921). *Ber. Dtsch. Chem. Ges.* 54: 2353.
68 Kim, W.-H.P., Park, I.S., and Park, J. (2006). Acceptor-free alcohol dehydrogenation by recyclable ruthenium catalyst. *Org. Lett.* 8: 2543–2545.
69 Fujita, K.-i., Yoshida, T., Imori, Y., and Yamaguchi, R. (2011). Dehydrogenative oxidation of primary and secondary alcohols catalyzed by a Cp*Ir complex having a functional C,N-chelate ligand. *Org. Lett.* 13: 2278–2281.
70 Berzelius, J.J. (1828). *Ann. Phys. Chem.* 13: 435–488.
71 Blackburn, T.F. and Schwartz, J. (1977). Homogeneous catalytic oxidation of secondary alcohols to ketones by molecular oxygen under mild conditions. *J. Chem. Commun.* 157–158.
72 Peterson, K.P.L. and Richard, C. (1998). Palladium-catalyzed oxidation of primary and secondary allylic and benzylic alcohols. *J. Org. Chem.* 63: 3185–3189.
73 Nishimura, T., Onoue, T., Ohe, K., and Uemura, S. (1999). Palladium(II)-catalyzed oxidation of alcohols to aldehydes and ketones by molecular oxygen. *J. Org. Chem.* 64: 6750–6755.
74 Iwasawa, T., Tokunaga, M., Obora, Y., and Tsuji, Y. (2004). Homogeneous palladium catalyst suppressing Pd black formation in air oxidation of alcohols. *J. Am. Chem. Soc.* 6554–6555.
75 Wang, D., Weinstein, A.B., White, P.B., and Stahl, S.S. (2018). Ligand-promoted palladium-catalyzed aerobic oxidation reactions. *Chem. Rev.* 118: 2636–2679.
76 (a) McCann, S.D. and Stahl, S.S. (2015). Copper-catalyzed aerobic oxidations of organic molecules: pathways for two-electron oxidation with a four-electron oxidant and a one-electron redox-active catalyst. *Acc. Chem. Res.* 48: 1756–1766. (b) Allen, S.E., Walvoord, R.R., Padilla-Salinas, R., and Kozlowski, M.C. (2013). Aerobic copper-catalyzed organic reactions. *Chem. Rev.* 113: 6234–6458.
77 (a) Mirica, L.M., Ottenwaelder, X., and Stack, T.D.P. (2004). Structure and spectroscopy of copper-Dioxygen complexes. *Chem. Rev.* 104: 1013–1045. (b) Lewis, E.A.T. and William, B. (2004). Reactivity of dioxygen-copper systems. *Chem. Rev.* 104: 1047–1076.

78 Semmelhack, M.F.S., Christopher, R., Cortes, D.A., and Chou, C.S. (1984). Oxidation of alcohols to aldehydes with oxygen and cupric ion, mediated by nitrosonium ion. *J. Am. Chem. Soc.* 106: 3374–3376.

79 (a) Gamez, P.A., Isabel, W.C.E., Reedijk, J., and Sheldon, R.A. (2003). Copper(II)-catalyzed aerobic oxidation of primary alcohols to aldehydes. *Chem. Commun.* 2414–2415. (b) Gamez, P.A., Isabel, W.C.E., Sheldon, R.A., and Reedijk, J. (2004). Room temperature aerobic copper-catalyzed selective oxidation of primary alcohols to aldehydes. *Adv. Synth. Catal.* 347: 805–811.

80 Kumpulainen, E.T.T.K. and Ari, M.P. (2009). Catalytic activity dependency on catalyst components in aerobic copper-TEMPO oxidation. *Chem. Eur. J.* 15: 10901–10911.

81 Hoover, J.M.S. and Shannon, S. (2011). Highly practical copper(I)/TEMPO catalyst system for chemoselective aerobic oxidation of primary alcohols. *J. Am. Chem. Soc.* 133: 16901–16910.

82 Steves, J.E. and Stahl, S.S. (2013). Copper(I)/ABNO-catalyzed aerobic alcohol oxidation: alleviating steric and electronic constraints of Cu/TEMPO catalyst systems. *J. Am. Chem. Soc.* 135: 15742–15745.

83 Ryland, B.L., McCann, S.D., Brunold, T.C., and Stahl, S.S. (2014). Mechanism of alcohol oxidation mediated by copper(II) and nitroxyl radicals. *J. Am. Chem. Soc.* 136: 12166–12173.

84 Sasano, Y., Nagasawa, S., Yamazaki, M. et al. (2014). Highly chemoselective aerobic oxidation of amino alcohols into amino carbonyl compounds. *Angew. Chem. Int. Ed. Eng.* 53: 3236–3240.

85 Sasano, Y., Kogure, N., Nagasawa, S. et al. (2018). 2-Azaadamantane N-oxyl (AZADO)/Cu catalysis enables chemoselective aerobic oxidation of alcohols containing electron-rich divalent sulfur functionalities. *Org. Lett.* 20: 6104–6107.

86 Shen, D., Miao, C., Xu, D. et al. (2015). Highly efficient oxidation of secondary alcohols to ketones catalyzed by manganese complexes of N4 ligands with H2O2. *Org. Lett.* 17: 54–57.

87 Hoelderich, W.F. (2000). Environmentally benign manufacturing of fine and intermediate chemicals. *Catal. Today* 62: 115–130.

88 Bortolini, O., Conte, V., Di Furia, F., and Modena, G. (1986). Metal catalysis in oxidation by peroxides. Part 25. Molybdenum- and tungsten-catalyzed oxidations of alcohols by diluted hydrogen peroxide under phase-transfer conditions. *J. Org. Chem.* 51: 2661–2663.

89 Sheldon, R.A., Arends, I.W., and Dijksman, A. (2000). New developments in catalytic alcohol oxidations for fine chemicals synthesis. *Catal. Today* 57: 157.

90 Sheldon, R.A., Wallau, M., Arends, I.W.C.E., and Schuchardt, U. (1998). *Acc. Chem. Res.* 31: 485–493.

91 Matsumoto, T., Ueno, M., Wang, N., and Kobayashi, S. (2008). Recent advances in immobilized metal catalysts for environmentally benign oxidation of alcohols. *Chem. Asian J.* 3: 196–214.

92 Miyamura, H. and Kobayashi, S. (2014). Tandem oxidative processes catalyzed by polymer-incarcerated multimetallic nanoclusters with molecular oxygen. *Acc. Chem. Res.* 47: 1054–1066.

3

Green Oxidative Synthesis of Ethers, Esters, and Organic Halides

Lu Ouyang, Wanqing Wu, and Huanfeng Jiang

Key Laboratory of Functional Molecular Engineering of Guangdong Province, School of Chemistry and Chemical Engineering, South China University of Technology, Guangzhou, China

3.1 Introduction

With the advent of the words "environmentally friendly," "sustainable," and "green," high efficiency and low consumption, non-toxic, harmless chemical transformation is an irresistible trend in today's chemistry [1]. The oxidation reaction is the most important synthetic transformation for the synthesis of both functional group manipulation and heteroatom incorporation biologically relevant compounds in organic chemistry [2]. Commonly, the traditional oxidation reaction employs stoichiometric oxidizing reagents, which can be expensive, hazardous, or toxic. Meanwhile, a toxic solvent is usually used to dissolve the oxidizing reagent and a large amount of inorganic waste is produced after workup. With the development of "green chemistry," the chemistry community faces enormous challenges to develop newly discovered and significantly improved oxidation processes. Therefore, the development of green oxidation reactions – economical and environmentally concerned oxidation processes with versatile, inexpensive, and environmentally friendly oxidants – has emerged as required [3].

Ethers are ubiquitous in many natural and synthetic substances, particularly in agrochemicals and bioactive molecules targeted as pharmaceuticals. Compounds that substitute hydrocarbyl for the two hydrogens in water are called ethers, the carbon–oxygen–carbon bond is called the ether bond, and is also the functional group of ether. The atom oxygen in ether is sp^3 hybridized; two pairs of lone pair electrons exist in two different sp^3 hybrid orbitals, respectively. Another two sp^3 hybrid orbitals combine with the sp^3 hybrid orbitals in alkyl carbon to form a σ bond. So, the bonding style of ether leads to a unique chemical property. Esters are among the most important functional groups and are found widely in commodity and fine chemicals, pharmaceuticals and agrochemicals, natural products, and polymers. Traditional methods of esterification are accomplished with carboxylic acid derivatives with alcohols through a dehydration-condensation reaction. Thus, a carbonyl and a C—O bond exist in the structure of ester. Organic halides refer to a kind of carbohydrate equipped with C—X bonds (X refers to fluorine, chlorine, bromine, or iodine), which is a versatile building block in organic

Green Oxidation in Organic Synthesis, First Edition. Edited by Ning Jiao and Shannon S. Stahl.

synthesis and plays an important role in science research and industry production. Simultaneously, the existence of halogens in organic compounds will significantly change many of their properties.

Considering the great importance and difference in structures and their properties, a variety of approaches have been used to synthesize these chemical compounds. In this chapter, we will focus on protocols that are much greener and more sustainable for the construction of ethers, esters, and organic halides through oxidation reactions.

3.2 Green Oxidative Synthesis of Ethers

The development of synthetic approaches toward ethers is highly desirable due to their significant importance. Up to now, a large number of methods have been reported for the synthesis of ethers involving metal-free or transition metal-catalyzed reactions (Scheme 3.1), such as acid-catalyzed condensation of alcohols, Williamson ether synthesis [4], the Ullmann reaction [5], and the Buchwald–Hartwig reaction [6]. Although significant advancements have been made in this area, there are still some disadvantages of these methods, such as long reaction time, high reaction temperature, excess amounts of metal salts, and the production of metal halides as by-products, which goes against the trend of modern chemistry. For green and sustainable chemistry, oxidation reactions with green oxidants are considered environmentally friendly methods for the construction of C—O ether bonds, and also offer attractive academic and industrial prospects. Recent advances in oxidation reactions for the synthesis of ethers with green oxidants via intermolecular/intramolecular etherification and alkoxylation will be described in detail.

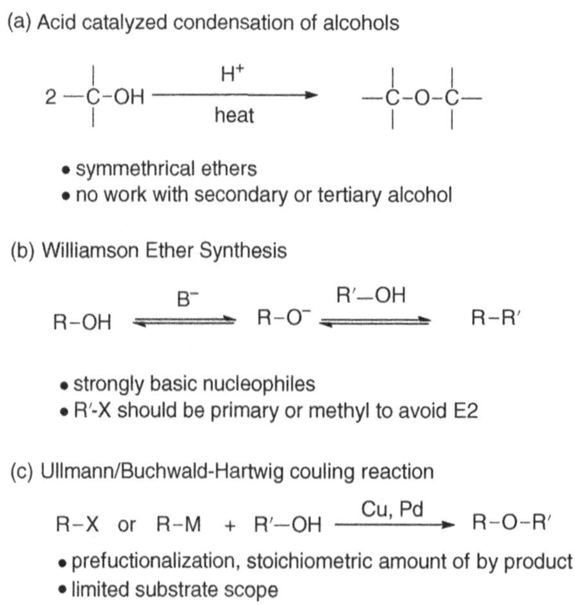

Scheme 3.1 Traditional methods for the synthesis of ethers.

3.2.1 Intramolecular Cycloetherification

Over the past few years, transition-metal-catalyzed C—H functionalization has emerged as a valuable and versatile tool in organic synthesis [7]. Intramolecular C—H functionalization, which offers a straightforward route to heterocyclic compounds, has recently attracted extensive attention from synthetic chemists [8]. Despite great achievements that have been made in this area, the development of intramolecular C—O bond formation for the direct functionalization of the C—H bond to form complex organic structures remains a tremendous challenge.

In 1995, Andersson and co-workers reported an efficient Pd-catalyzed intramolecular oxidative cyclization of acyclic enols (Scheme 3.2) [9]. A variety of cyclic ethers have been synthesized through this method with high stereo- and regioselectivity and yields under mild conditions. More importantly, this reaction proceeds without the use of environmentally unfriendly oxidants such as 1,4-benquinone, Cu(II), DDQ, $PhI(OAc)_2$, or $K_2S_2O_8$, which was performed in DMSO under an atmosphere of oxygen.

Hydroxyl groups can be used not only as internal nucleophiles, but also as weak coordinated directing groups. More recently, the first example of intramolecular metal-catalyzed C—O formation via aliphatic alcohol-directed C—H activation was discovered by Yu's group [10]. In contrast, a similar strategy to the formation of C—O bond via phenol-directed C—H activation is ineffectual, because phenols are not stable enough to strong oxidants and are prone to undergo homocoupling [11]. Thus, research on phenol-directed C—H activation/C—O cyclization is still rarely reported. Recently, Liu's group have reported a practical and efficient Pd(II)-catalyzed phenol-directed C—H activation/C—O cyclization using air as the oxidant (Scheme 3.3) [12]. Moreover, the present method provides a new strategy to the synthesis of substituted dibenzofurans, which is a significant structural motif in many biologically active compounds. Later, Zhu and co-workers developed a Cu-catalyzed aerobic cycloetherification of 2-arylphenols (Scheme 3.4) [13]. A strong *para*-electron-withdrawing group on the phenol is indispensable, thus the substrate scope of this copper-catalyzed aerobic C(sp^2)—H activation is limited.

Scheme 3.2 Pd-catalyzed cycloetherification of acyclic enols.

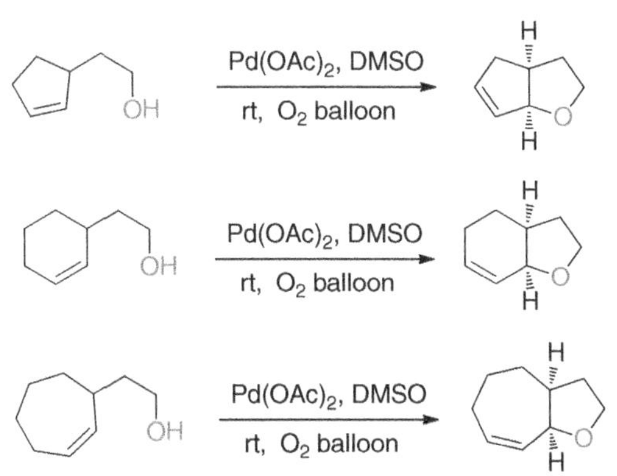

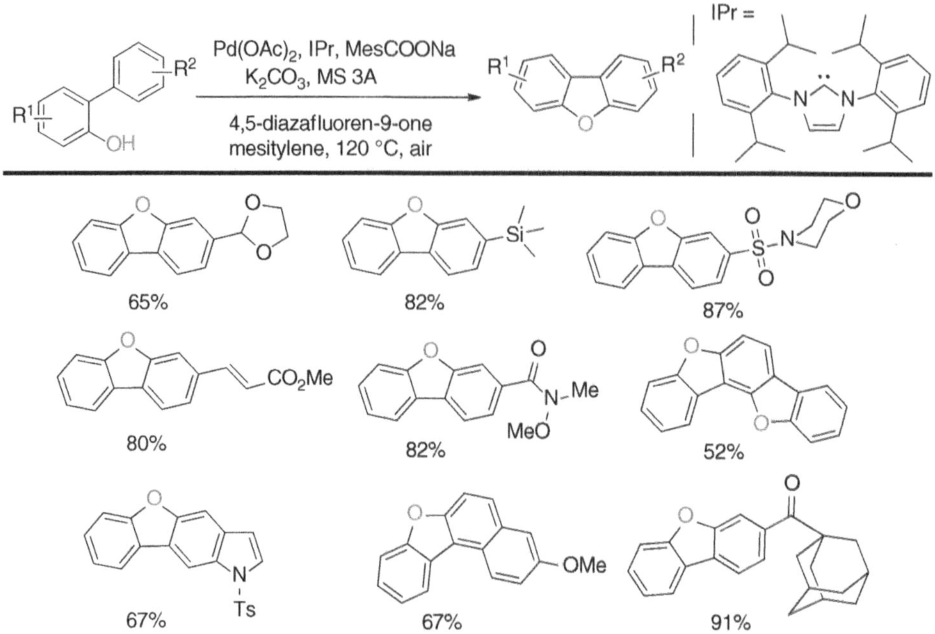

Scheme 3.3 Pd-catalyzed phenol-directed C—H activation/C—O cyclization of *o*-arylphenols.

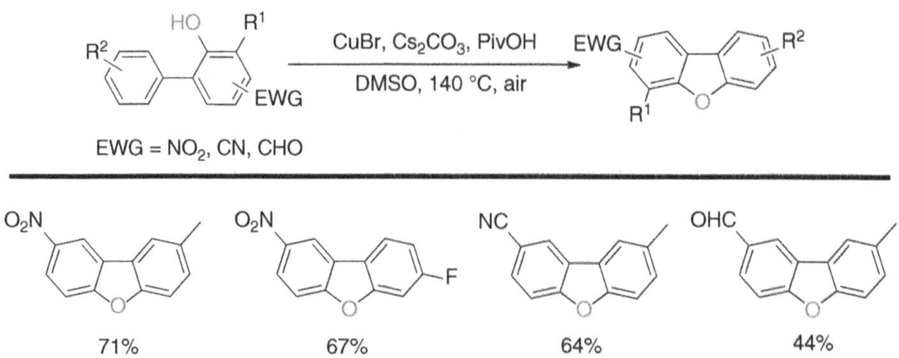

Scheme 3.4 Cu-catalyzed oxidative C(sp^2)-H cycloetherification of *o*-arylphenols.

3.2.2 Intermolecular Etherification

3.2.2.1 Auxiliary-Assisted Alkoxylation or Phenoxylation via Dehydrogenative Cross-Coupling

Dehydrogenative cross-coupling represents the most effective strategy to introduce carbon- or heteroatoms into organic compounds [14]. This new strategy merits attention due to the substrates not requiring prefunctionalization, and high atom economy and efficiency. In view of these advantages, an impressive number of efficient dehydrogenative cross-coupling under oxidative conditions have been extensively investigated by chemical scientists [15]. In 2013, Gooßen and co-workers discovered a bimetallic

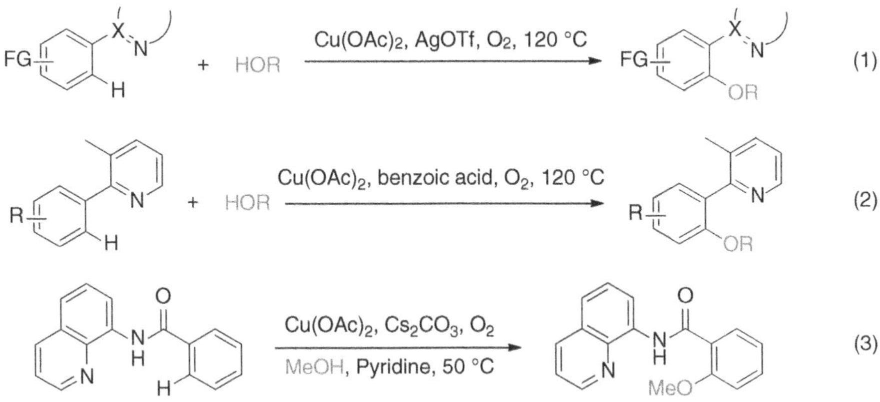

Scheme 3.5 Copper-catalyzed auxiliary-assisted aromatic C—H alkoxylation.

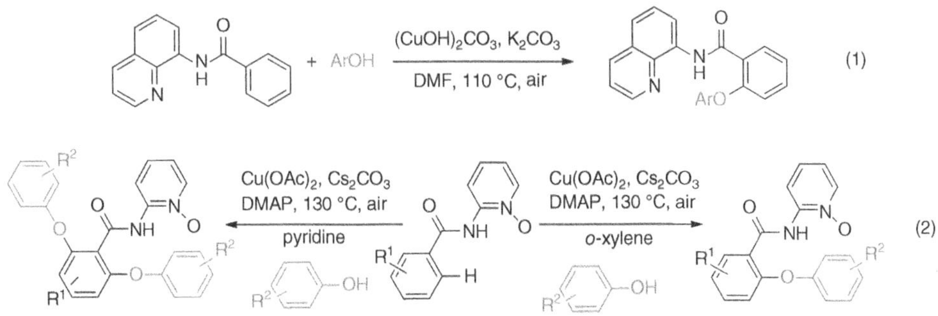

Scheme 3.6 Copper-catalyzed auxiliary-assisted aromatic C—H phenoxylation.

copper/silver-catalyzed dehydrogenative cross-coupling reaction of *o*-C(sp^2)—H bonds of arenes and alcohols (Scheme 3.5(1)) [16]. The directing group on arenes such as pyridine or pyrimidine is essential to the reaction. And this reaction provides an alternative route to synthesize aryl ethers. After that, another copper-catalyzed dehydroalkoxylation of arenes with alcohols under aerobic conditions has been described by Shen et al. (Scheme 3.5(2)) [17]. The use of catalytic amounts of benzoic acid improved this reaction significantly. Meanwhile, in 2013, a novel Cu-mediated direct C–H methoxylation of *N*-(8-quinolinyl)benzamide using molecular oxygen (O_2) as the sole oxidant has also been developed (Scheme 3.5(3)) [18]. Experimental studies indicated a single-electron-transfer mechanism for the C—H methoxylation.

Biaryl ethers are a kind of important structure motif in many natural products and biologically active compounds [19]. Although there have been many studies on the synthesis of biaryl ethers over the past few years [20], it is still necessary to develop a greener and more convenient method to obtain them. Recently, Daugulis' group discovered an auxiliary-assisted phenoxylation, which was performed with the catalyst of $(CuOH)_2CO_3$, phenol, or alcohol as a coupling partner, and K_2CO_3 or tetramethylguanidine as a base under aerobic conditions (Scheme 3.6(1)) [21]. This reaction shows good functional group tolerance and provides an efficient and convenient approach for the synthesis of diverse aryloxylated benzoic acids. Then, Song's group developed a new Cu-mediated mono- and diaryloxylation of 2-aminopyridine 1-oxide in 2014 (Scheme 3.6(2)) [22]. This strategy used a removable *N,O*-bidentate as the directing group, providing an efficient and straightforward method for the construction of various mono- and diaryloxylated benzoic acids.

3.2.2.2 Decarboxylative Etherification

Transition metal-catalyzed decarboxylative coupling reactions have emerged over recent years, and have been used as a powerful strategy to generate a variety of allylated and benzylated compounds [23]. In these catalytic transformations, organometallic species are generated via extrusion of CO_2 from metal carboxylates, which react with other organic compounds to form C—C or C—heteroatom bonds [24]. Therefore, they have attracted great interest from chemists for the development of new synthetic methods using this protocol, although decarboxylative etherifications remain challenging. In 2012, the first example of etherification of aromatic carboxylic acids via decarboxylative alkoxylation was reported by Gooßen and co-workers (Scheme 3.7) [25]. The reaction employs silver carbonate as the decarboxylation catalyst and copper acetate as the cross-coupling catalyst under aerobic conditions. This strategy provided a valuable and interesting synthetic entry to aryl ethers with widely available carboxylic acids as starting materials.

The following year, they discovered a new decarboxylative alkoxylation of aromatic carboxylic acids using trialkylborate as the precursor of alkoxyl (Scheme 3.8) [26], which presented a rare example of decarboxylative alkoxylation.

3.2.2.3 Chan–Evans–Lam-Type Etherification

The Chan–Evans–Lam coupling reaction. which was discovered in the 1990s, is a cross-coupling reaction between aryl boronic acid and alcohol or amine to form the corresponding secondary aryl amines or aryl ethers [27]. The transformation is catalyzed in the presence of copper(II) species and conducted with the feature of mild reaction conditions, functional group tolerance, and open-flask chemistry. Chan–Evans–Lam coupling has several benefits over the C—C, C—N, or C—O coupling reactions, and has seen a number of applications in total synthesis [28]. The proposed mechanism is shown in Scheme 3.9. The catalytic cycle is characterized by (i) ligand exchange, (ii) transmetallation, (iii) reductive elimination, and (iv) oxidation.

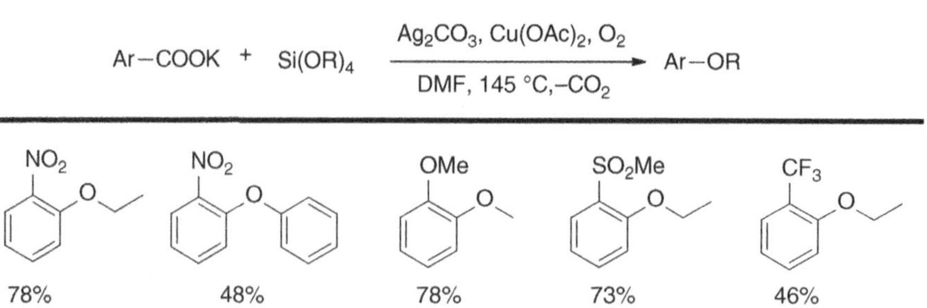

Scheme 3.7 Decarboxylative etherification of aromatic carboxylic acids with $Si(OR)_4$.

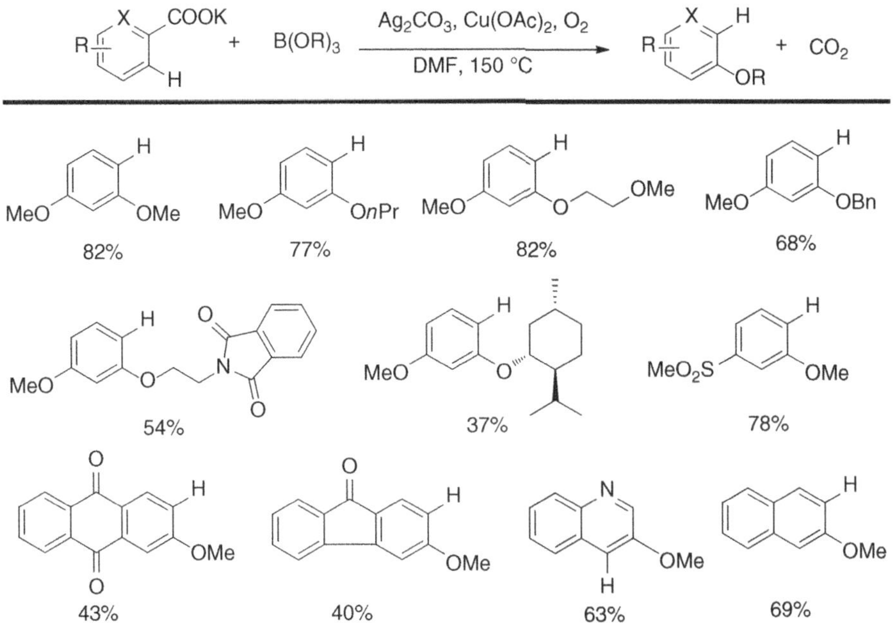

Scheme 3.8 Decarboxylative etherification of aromatic carboxylic acids with $B(OR)_3$.

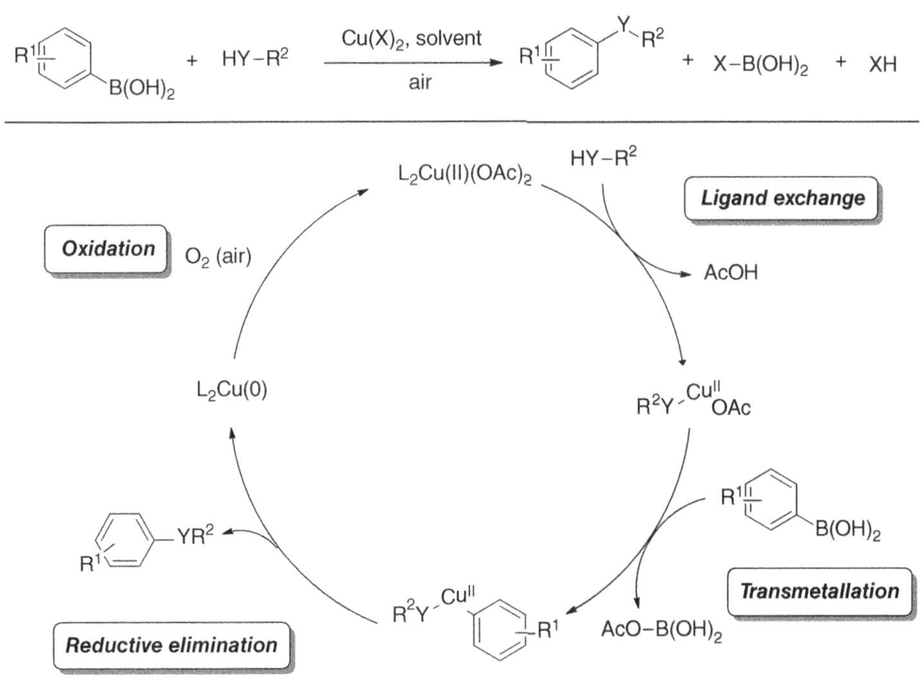

Scheme 3.9 The Chan–Evans–Lam coupling reaction and proposed mechanism.

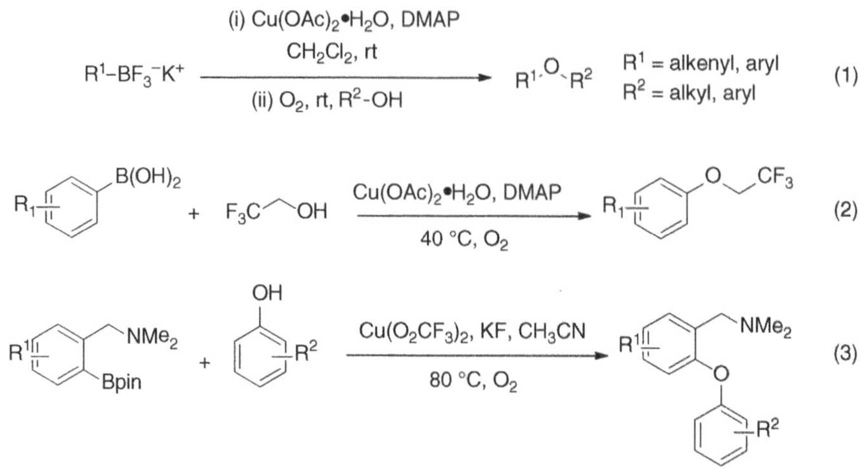

Scheme 3.10 Cu-catalyzed Chan–Evans–Lam-type etherification.

In 2003, Batey's group developed a copper-catalyzed Chan–Evans–Lam-type etherification of potassium trifluoroborate salts (Scheme 3.10(1)) [29]. A variety of aliphatic primary and secondary alcohols and phenols were employed to give alkyl-aryl and alkyl-vinyl ethers with high functional group tolerance. In addition, a trifluoroethoxylation of boronic acids was described by Wu's group in 2015 (Scheme 3.10(2)) [30], which was catalyzed by $Cu(OAc)_2$ via Chan–Evans–Lam-type etherification reactions. Diverse trifluoroethyl aryl ethers could be obtained in moderate to good yields under mild conditions. Moreover, an *ortho*-benzylic amine assisted Cu-catalyzed Chan–Evans–Lam etherification from boronate esters, which was achieved by Clark and co-workers in 2016 (Scheme 3.10(3)) [31]. This strategy provided an efficient and convenient tool to construct a variety of substituted aryl ethers. The existence of benzylic amine was essential and inhibited the hydrolysis of boronate ester.

3.2.2.4 Nucleopalladation Triggered Etherification Reactions

The first example of nucleopalladation was reported by researchers at Wacker Chemie in 1959 [32]. This method applied Pd sources as catalyst, O_2 or air as the oxidant, and H_2O as the nucleophile, which transformed ethylene to acetaldehyde. Nowadays, nucleopalladation of alkenes and alkynes has been identified as a highly important route to their difunctionalization [33]. In 1999, we first reported our investigation on the Pd^{II}-catalyzed dialkoxylation of terminal olefins with electron-withdrawing groups in supercritical CO_2 with high conversion and selectivity at low temperature (Scheme 3.11) [34].

In 2005, Sigman and co-workers presented a novel Pd^{II}-catalyzed dialkoxylation of styrene, which contained an *o*-phenol. The existence of *o*-phenol could efficiently prevent β-hydride elimination (Scheme 3.12) [35]. Mechanism investigations indicated that the nucleopalladation occurred at the β-carbon of the styrene, followed by the subsequent nucleophile attack with MeOH to give the desired products.

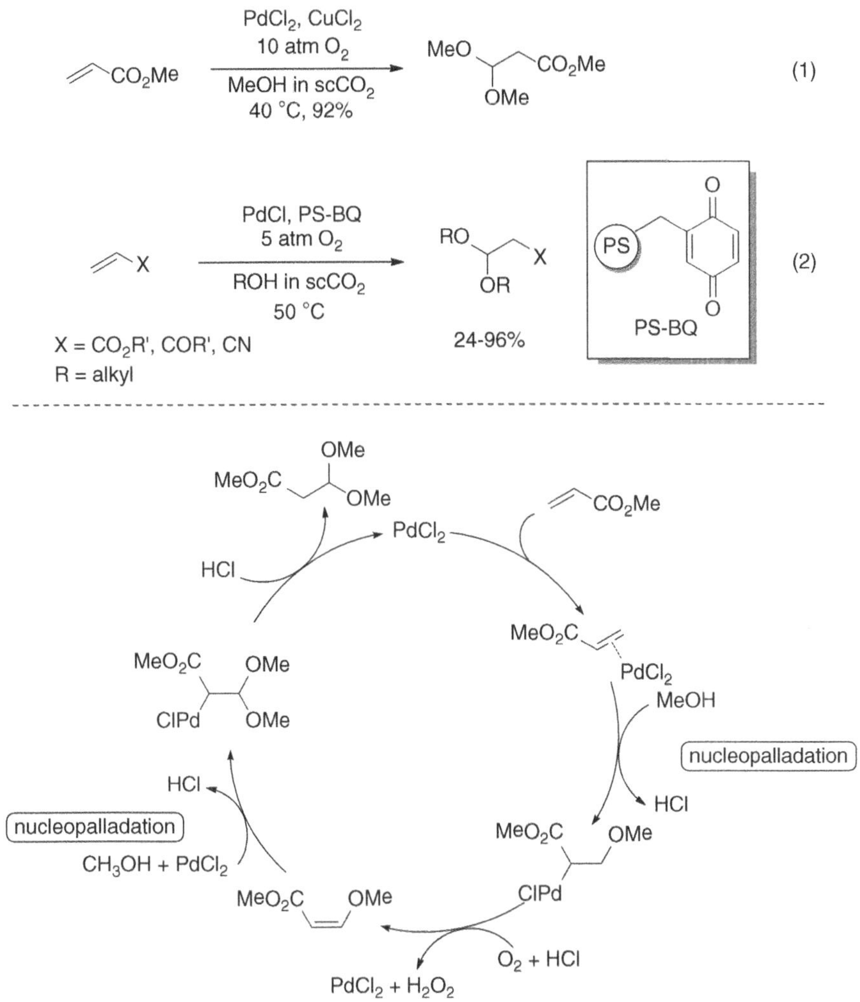

Scheme 3.11 Pd^{II}-catalyzed dialkoxylation of terminal olefins with electron-withdrawing groups and proposed mechanism.

3.2.2.5 Other Etherification Reactions

As we know, the overwhelming majority of aryl etherification has been conducted with aryl precursors such as phenol, although there have been a few reports of aryl etherification with non-aromatic precursors. Recently, a novel and attractive Cu-mediated oxidative condensation of alcohols and 2-cyclohexenones was reported by Li and co-workers (Scheme 3.13). Diverse aromatic ethers were achieved from non-aromatic precursors, 2-cyclohexenones [36]. This catalytic system was performed with O_2 as the terminal oxidant and eliminated water as the only by-product, which is in agreement with the principle of green chemistry. This catalytic system was performed with O_2 as the terminal oxidant and water as the only by-product, which has been shown to conform with the principle of green chemistry.

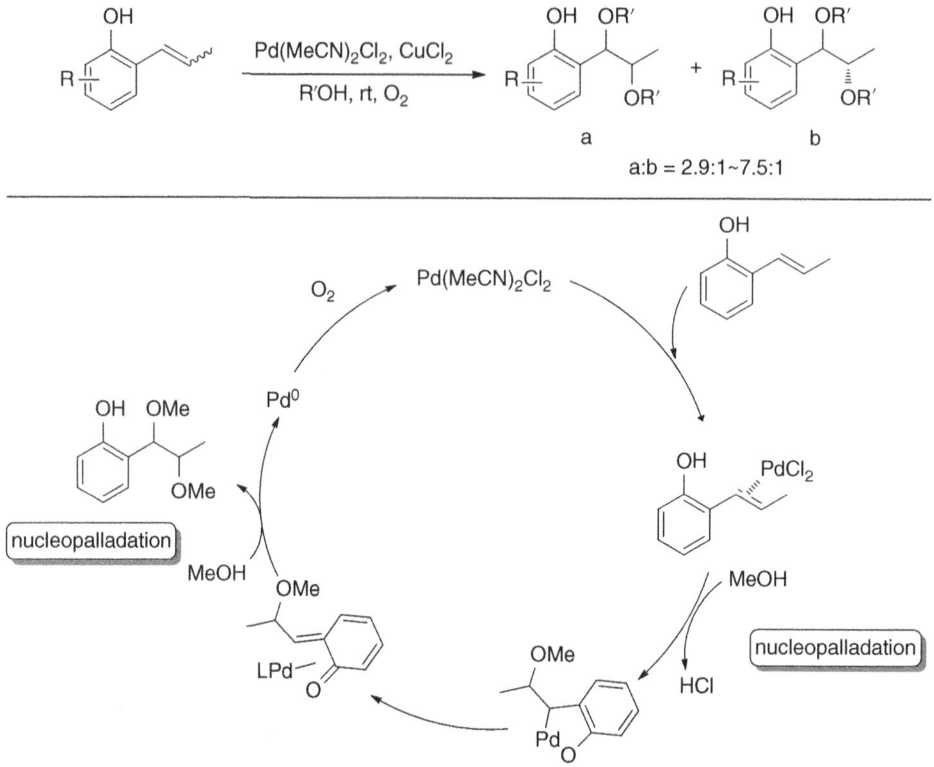

Scheme 3.12 Pd^{II}-catalyzed aerobic dialkoxylation of *o*-hydroxyl styrene and proposed mechanism.

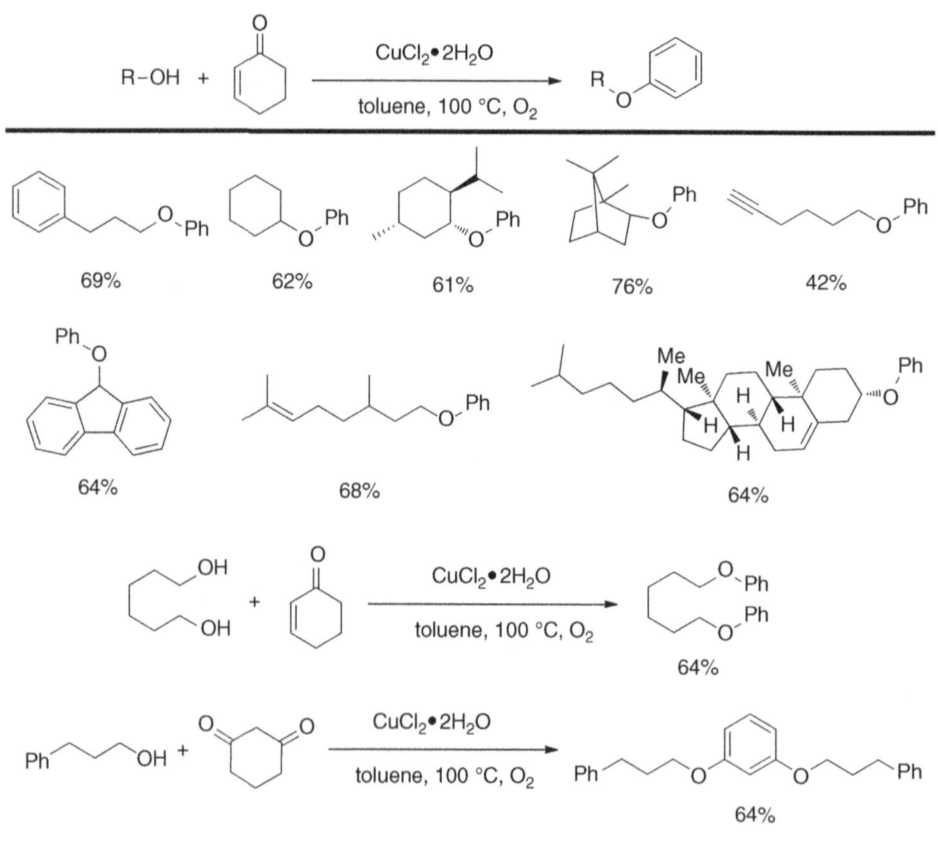

Scheme 3.13 Cu-mediated oxidative condensation and oxidative aromatization of alcohols.

3.3 Green Oxidative Synthesis of Esters

In chemistry, an ester is a chemical compound derived from an acid (organic or inorganic) in which at least one hydroxyl group is replaced by an alkoxy group. Esters contain a carbonyl center, which gives rise to 120 °C C═O and C—O angles. Esters are common in organic chemistry and biological materials, and often have a characteristic pleasant, fruity odor [37]. Moreover, it is also a very important synthon and it has the potential for further transformation [38]. Because of their great importance, a rapidly growing number of reactions have been reported for synthetic esters within recent years. However, more environmentally friendly and energy-efficient synthetic methods are still in great demand.

3.3.1 Oxidation of Alcohols and Aldehydes to Esters

The transformation of alcohols and aldehydes to esters is a fundamental reaction in organic synthesis via one- or two-step oxidation [39]. Over the past few decades, several successful protocols have been discovered using different oxidants in the presence of a metal catalyst or metal-free condition [40]. Undoubtedly, transition metal-catalyzed aerobic oxidations of alcohols and aldehydes for the construction of esters represent a step forward for processes. Therefore, chemists have attracted considerable attention with this interesting process. In 2011, Beller and co-workers developed a novel Pd-catalyzed oxidative esterification of benzylic and aliphatic alcohols in high selectivity (Scheme 3.14(1)) [41]. This process proceeded with air as the oxidant and water as the only side-product. In the same year, Lei's group also reported a Pd-catalyzed oxidative esterification of alcohols under a similar condition (Scheme 3.14(2)) [42]. Later in 2014, a highly efficient and versatile Pd/C-catalyzed microwave-assisted oxidative esterification of alcohols and aldehydes was reported by Cravotto's group (Scheme 3.14(3)) [43]. This environmentally friendly process did not require a ligand and was conducted in almost quantitative yields under an oxygen atmosphere. A variety of benzoate esters could be obtained in high convention and selectivity.

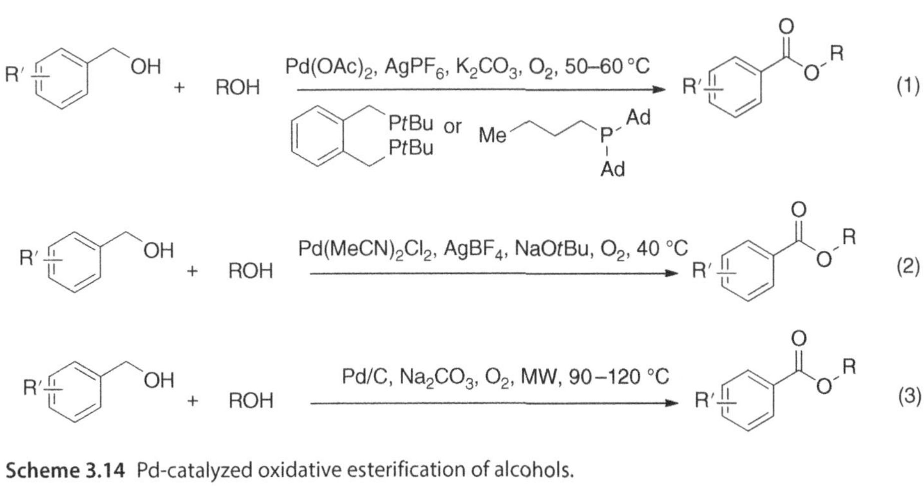

Scheme 3.14 Pd-catalyzed oxidative esterification of alcohols.

Traditionally, esterification of alcohols or aldehydes has been well developed with the assistance of transition metal catalysts. However, studies about metal-free oxidative esterification are rarely reported. The development of environmentally benign oxidative esterification is still in high demand. Recently, Adimurthy and co-workers reported a transition metal-free oxidative esterification of benzylic alcohols with a catalytic amount of HBr-H_2O_2 in aqueous medium (Scheme 3.15(1)) [44]. The reactive species BrOH was generated in situ from HBr-H_2O_2 for the oxidative esterification. The electronegativity of substituents did not affect this transformation under mild conditions. Simultaneously, in 2017, Lei's group disclosed a mild and simple photoredox-catalyzed selective oxidative esterification from two different alcohols using O_2 as the sole oxidant (Scheme 3.15(2)) [45]. A variety of esters could be received through this catalytic system, especially for long chain and secondary esters.

A novel *N*-heterocyclic carbine-catalyzed esterification of aromatic aldehydes under ambient conditions was described by Sudalai's group in the presence of catalytic amount of 1,8-diazabicyclo[5.4.0]undec-7-ene (DBU) (Scheme 3.16) [46]. This efficient and easy operation process proceeded smoothly to give corresponding aromatic esters in high yields. Then, a riboflavin tetraacetate-catalyzed visible-light-driven aerobic esterification of benzaldehydes was also reported using air as a terminal oxidant (Scheme 3.17) [47]. H_2O_2 as a green oxidant was also applied in oxidative esterification reactions, which generated water as the only by-product.

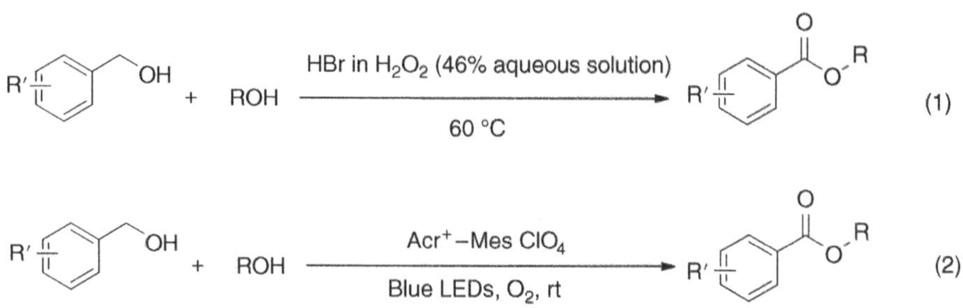

Scheme 3.15 Transition metal-catalyzed oxidative esterification of alcohols.

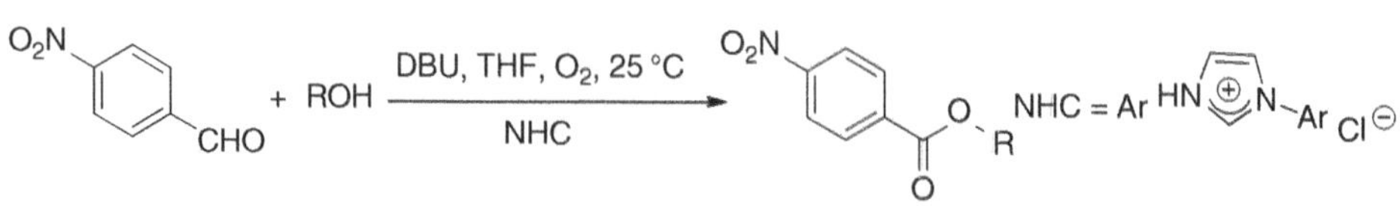

Scheme 3.16 *N*-Heterocyclic carbine-catalyzed esterification of aromatic aldehydes.

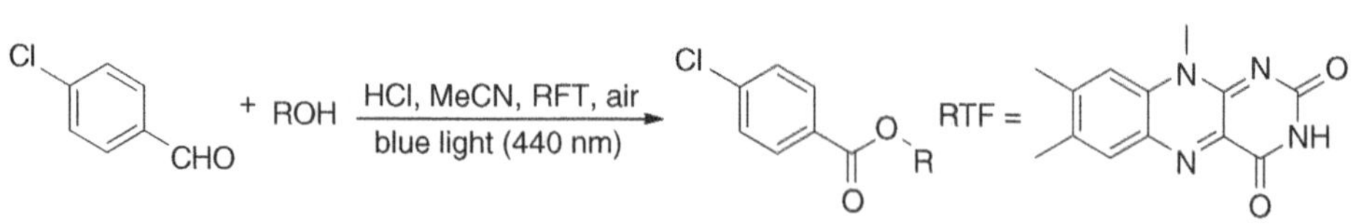

Scheme 3.17 Visible-light-driven aerobic esterification of benzaldehydes.

A V_2O_5-H_2O_2-mediated oxidative esterification of aldehydes was discovered by Patel's group (Scheme 3.18(1)) [48]. This new strategy salient merits attention due to the mild reaction conditions, short reaction time, high efficiency, cost-effectiveness, and facile isolation. After that, Wu reported a general and efficient Zn-catalyzed oxidative esterification of aldehydes (Scheme 3.18(2)) [49]. The reaction underwent with 10 mol% of $ZnBr_2$ and 4 equiv. of H_2O_2 with high functional group tolerance.

3.3.2 Oxidative Carbonylation to Esters

Transition-metal-catalyzed carbonylation has become a powerful tool to produce many carbonyl compounds such as esters, amides, and ketones [50]. This type of reaction is of significant value both in academic and industrial laboratories, which has attracted much interest and developed rapidly by synthetic chemists over the past few years. Carbon monoxide, as the most important C1 building block in organic synthesis, is valuable due to its easily available nature, cost-efficiency, and atom economy [51], thus the caronylation process is a fundamental and promising transformation. Traditional carbonylation uses R-X as starting materials [52], which act as electrophiles and normally need to be prepared in advance from the corresponding nucleophiles (R—H). Obviously, it is not economically efficient, so the need for the development of a more straightforward method for carbonylation has been recognized around the chemical community, and oxidative carbonylation has emerged and been rapidly developed [53]. In contrast with classical carbonylation, oxidative carbonylation proceeds with two different nucleophiles in the presence of oxidation, which reduces the cost, makes the subtracts more abundant, and enriches the diversity of carbonylation (Scheme 3.19). Oxidizing reagents including organic compounds and inorganic salts, such as benzoquinone (BQ), copper salts, and silver salts, are usually applied in oxidative carbonylation. However, the most popular oxidant is oxygen or air, which is environmentally sound and efficient at producing many versatile carbonyl derivatives. The process of oxidative carbonylation is shown in Scheme 3.20.

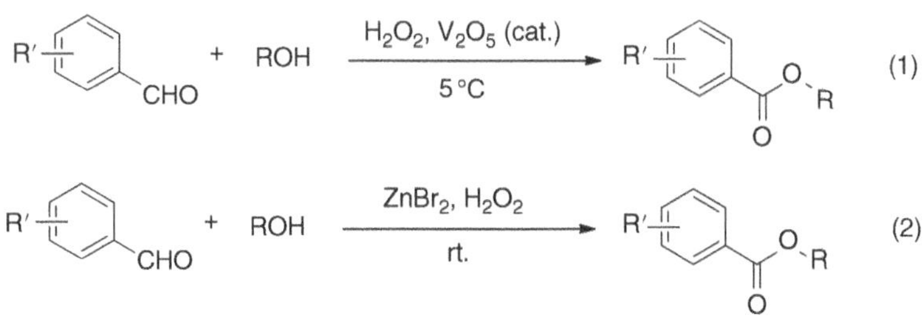

Scheme 3.18 Oxidative esterification of benzaldehydes by H_2O_2.

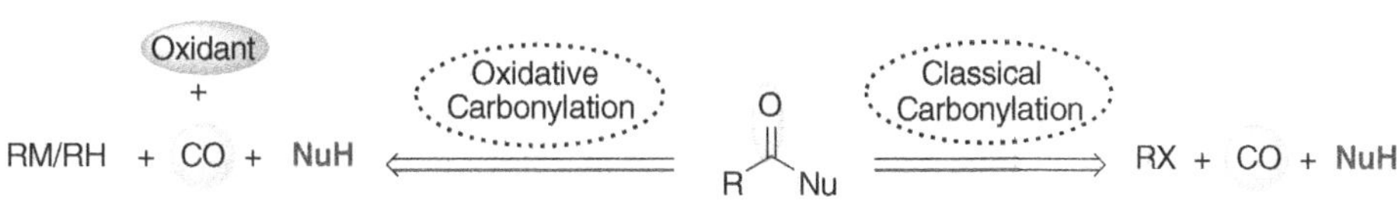

Scheme 3.19 Comparison of the classical carbonylation and oxidative carbonylation.

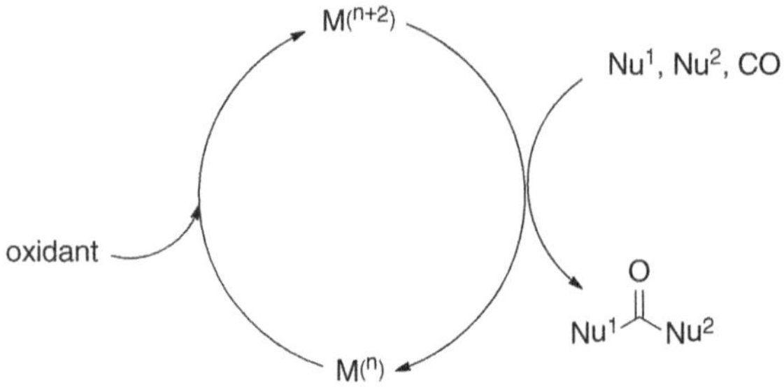

Scheme 3.20 General process for the oxidative carbonylation.

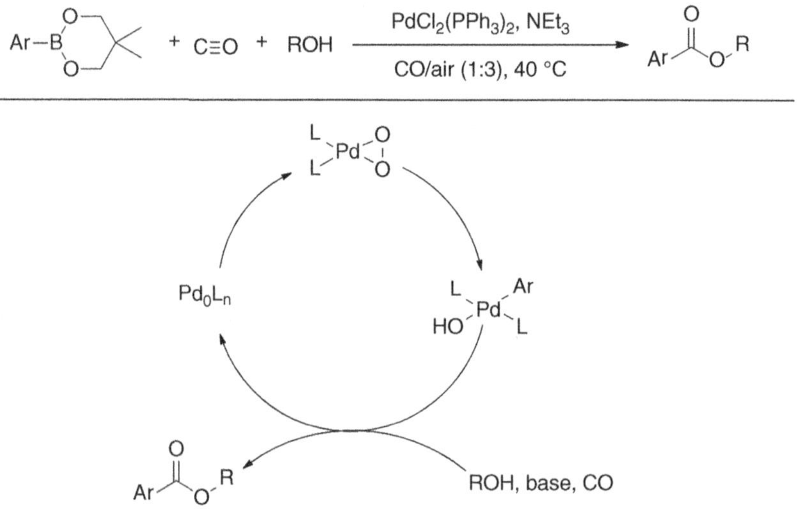

Scheme 3.21 Pd-catalyzed aerobic oxidative carbonylation of arylboronate esters and proposed mechanism.

Arylboronic acid derivatives represent one of the most important compounds and have been widely applied in chemical syntheses, since they are extremely stable in conditions of air, moisture, and light, and are compatible with a broad range of common functional groups [54]. Recently, arylboronic acid derivatives employed as nucleophiles were reported by Hou's and Iwasawa's groups for the metal-catalyzed activation of CO_2 [55]. In 2010, Lei and co-workers described a novel Pd-catalyzed oxidative carbonylation of arylboronate esters (Scheme 3.21) [56]. The reaction proceeded under balloon pressure of CO/O_2 at temperature of 40–50 °C. $PdCl_2(PPh_3)_2$ was used as the catalyst and the corresponding products were given efficiently.

α,β-Unsaturated esters are not only important chemical feedstocks but also useful synthetic intermediates. In 2014, Lei and co-workers also developed an interesting method for the synthesis of α,β-unsaturated esters by using a palladium-catalyzed aerobic oxidative carbonylative esterification (Scheme 3.22) [57]. The reaction was performed with stoichiometric amounts of the alcohol under base- and ligand-free conditions. The CO/olefinic system was employed as the acylating reagent with O_2 as the terminal oxidant.

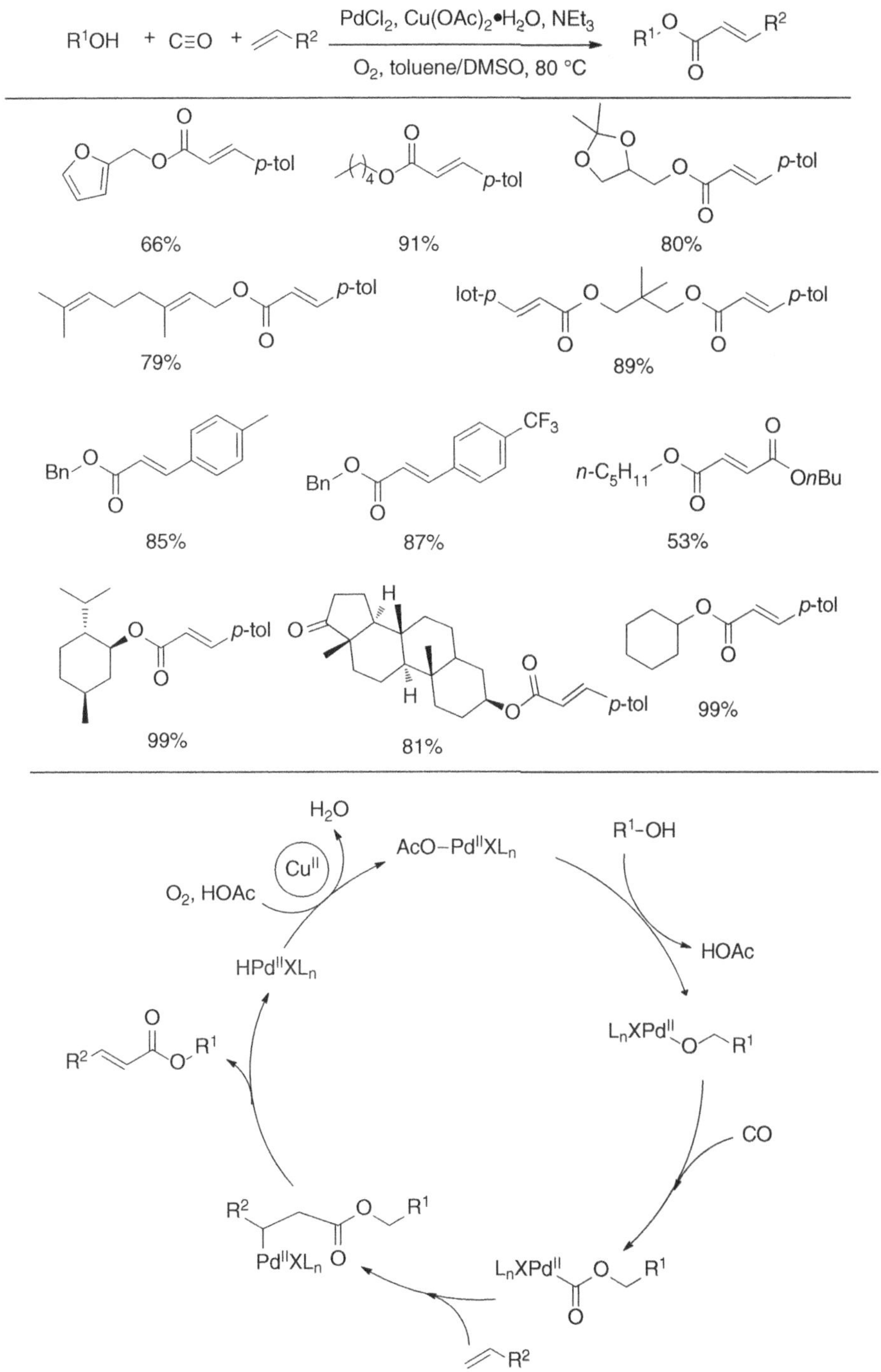

Scheme 3.22 Pd-catalyzed oxidative carbonylative esterification for the construction of *α,β*-unsaturated esters.

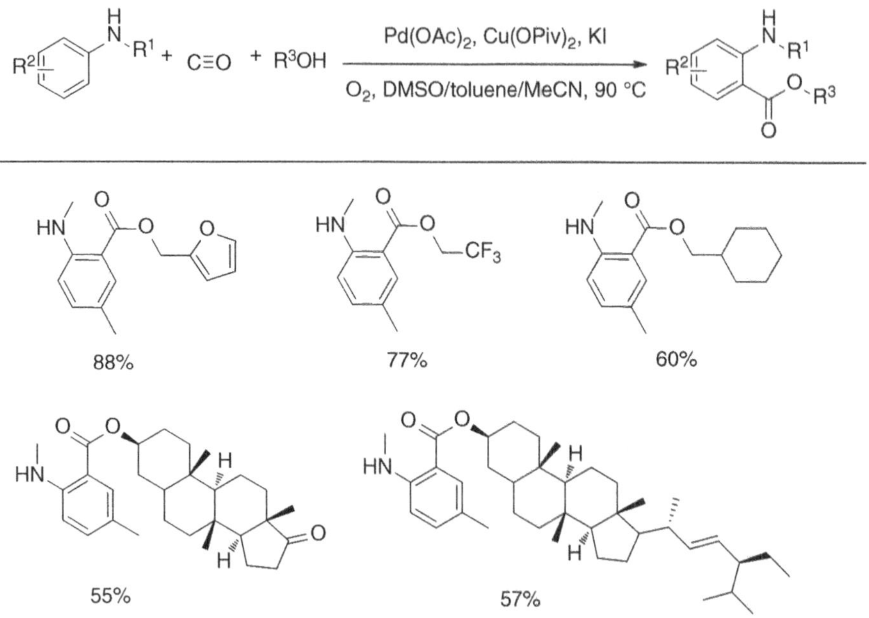

Scheme 3.23 Palladium/copper-catalyzed aerobic oxidative C—H carbonylation.

Simultaneously, they developed a method for the esterification of *N*-substituted anilines and alcohols via palladium/copper-catalyzed aerobic oxidative carbonylation of the *o*-C-H of amino group (Scheme 3.23) [58]. Notably, this method provided a convenient entry to construct valuable *o*-aminobenzoates, which was tolerated with primary and secondary alcohols, especially natural alcohols.

3.3.3 Acyloxylation

3.3.3.1 Acetoxylation

Transition metal-catalyzed oxidative acetoxylation reactions represent one of the most efficient and available methods to introduce an acetoxyl in carbon-based molecules, especially for Pd catalysis [59]. Methods for selective acetoxylation have expanded significantly over the past decades, however, a stoichiometric oxidant such as Cu^{II}, Ag^{I} salts, $PhI(OAc)_2$, or benzoquinone is necessary. Also, Pd-catalyzed oxidative acetoxylation is rarely compatible with O_2. It would be a challenge to realize oxidative acetoxylation using O_2 as the terminal oxidant. As our interest continues in palladium-catalyzed oxidation reactions, we have developed a Pd-catalyzed acetoxylation of sp^3 C—H bond of 2-substituted pyridines and pyrazines (Scheme 3.24) [60]. This reaction proceeds with O_2 as the oxidant and affords a α-oxygenated product without hazardous by-products.

Then, Stahl and co-workers demonstrated a Pd-catalyzed oxidation acetoxylation of a unactivated sp^3 C—H bond (Scheme 3.25) [61]. The optimized conditions for this transformation involved the use of 5 mol% $Pd(OAc)_2$, 25–100 mol% $NaNO_3$, in $AcOH/Ac_2O$ at 100–110 °C under 1 atm O_2 or air. In 2015, they also developed an efficient and high selectivity palladium-catalyzed aerobic acetoxylation of benzene with NO_x sources

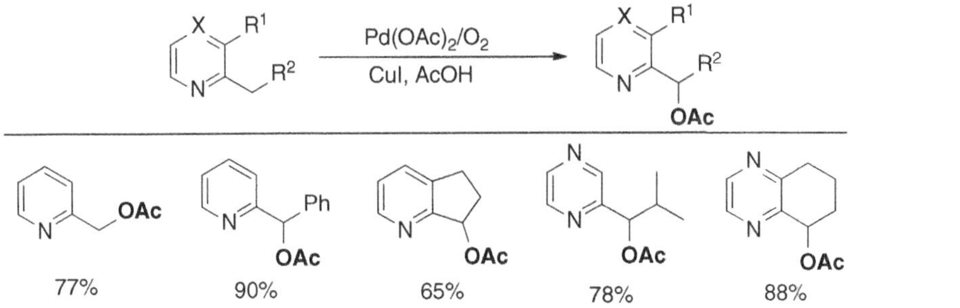

Scheme 3.24 Pd-catalyzed acetoxylation of *sp*^{3} C—H bond of 2-substituted pyridines and pyrazines.

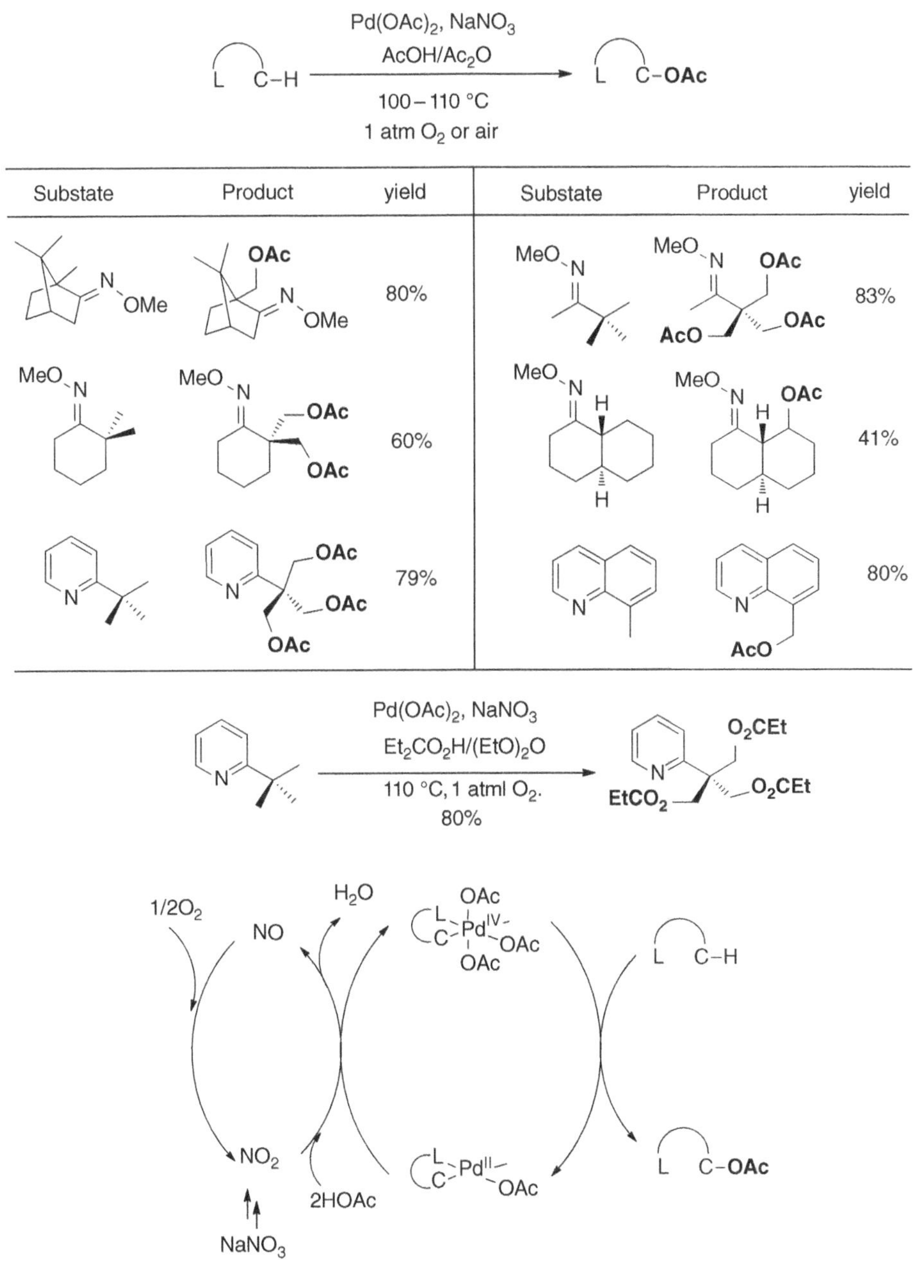

Scheme 3.25 Pd-catalyzed oxidative acetoxylation of unactivated *sp*^{3} C—H bond and proposed mechanism.

serving as effective redox mediators with O_2 as the oxidant (Scheme 3.26) [62]. A variety of phenyl acetates have been achieved through this transformation.

Simultaneously, Stahl et al. discovered a ligand assisted Pd-catalyzed allylic C—H acetoxylation with O_2 as the stoichiometric oxidant (Scheme 3.27) [63], which was more advanced than previous reports using BQ as the oxidant. Mechanistic studies suggested

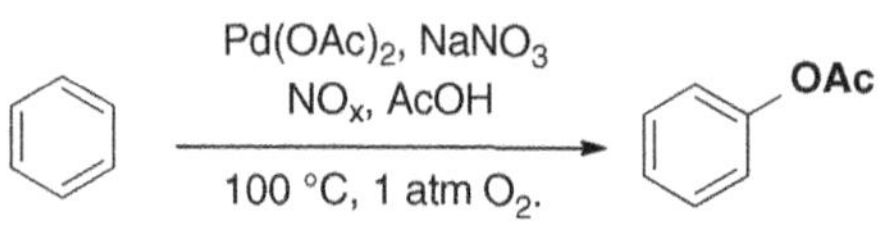

Scheme 3.26 Pd-catalyzed aerobic acetoxylation of benzene.

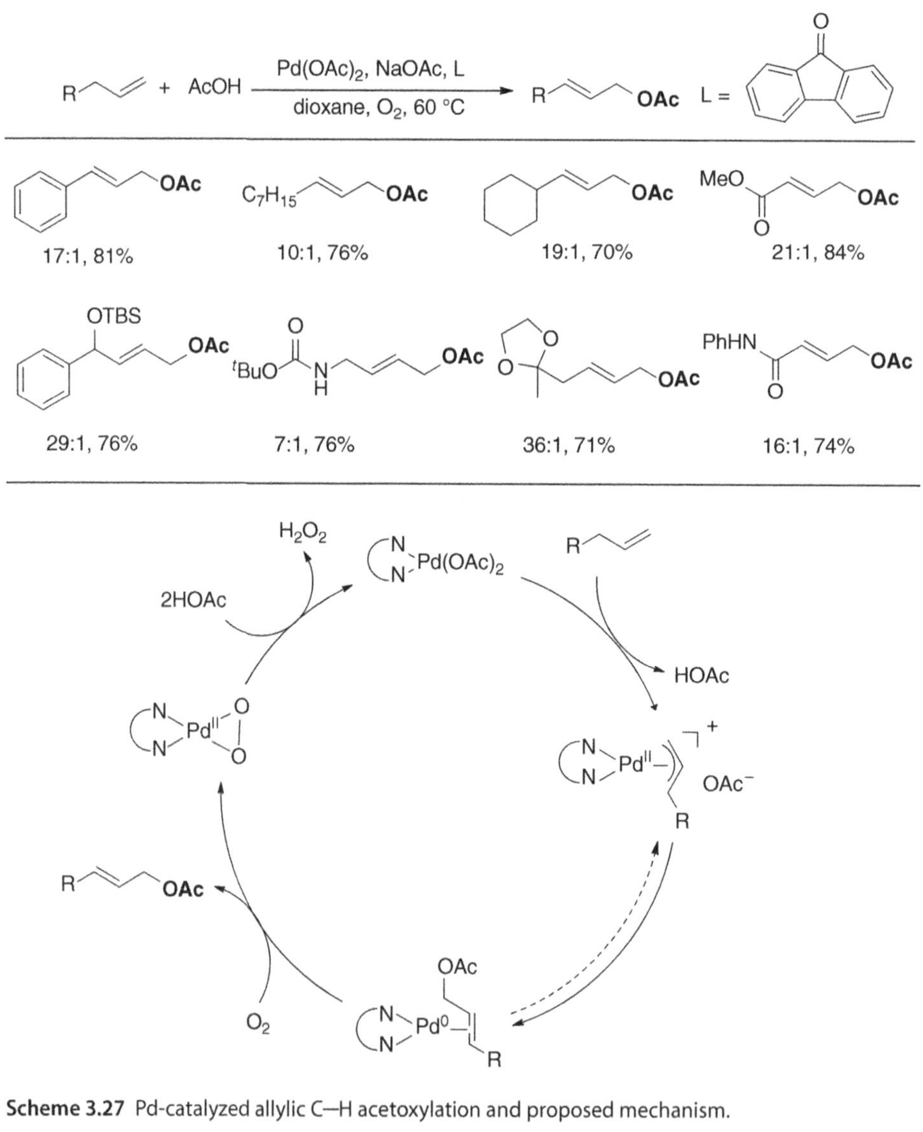

Scheme 3.27 Pd-catalyzed allylic C—H acetoxylation and proposed mechanism.

that ligands could facilitate the reductive elimination from π-allyl-Pd^{II}-intermediate to form C—O bond and thereby avoid the utilization of BQ.

In addition to mono-acetoxylation, Pd-catalyzed oxidative diacetoxylation was also well developed for difunctionalization in recent years. It is worth noting that the first example of Pd-catalyzed diacetoxylation of alkenes using oxygen as the sole oxidant was reported by our group (Scheme 3.28) [64]. A range of 1,2-diacetates were synthesized through this green and sustainable, attractive, and straightforward method in high stereoselectivity. Interestingly, the final products could be also easily transformed to *cis*-vicinal diols. Since then, Grubbs and co-workers have also provided a non-toxic and environmentally benign strategy for diacetoxylation of traditional alkenes (Scheme 3.29) [65]. The existence of a nitrite co-catalyst and molecular oxygen at normal pressure could facilitate the elimination of alkyl-Pd species for the construction of a C—O bond.

In recent years, acetoxypalladation of alkynes has been a widely used method for the formation of both carbon–carbon and carbon–oxygen bonds in a single step.

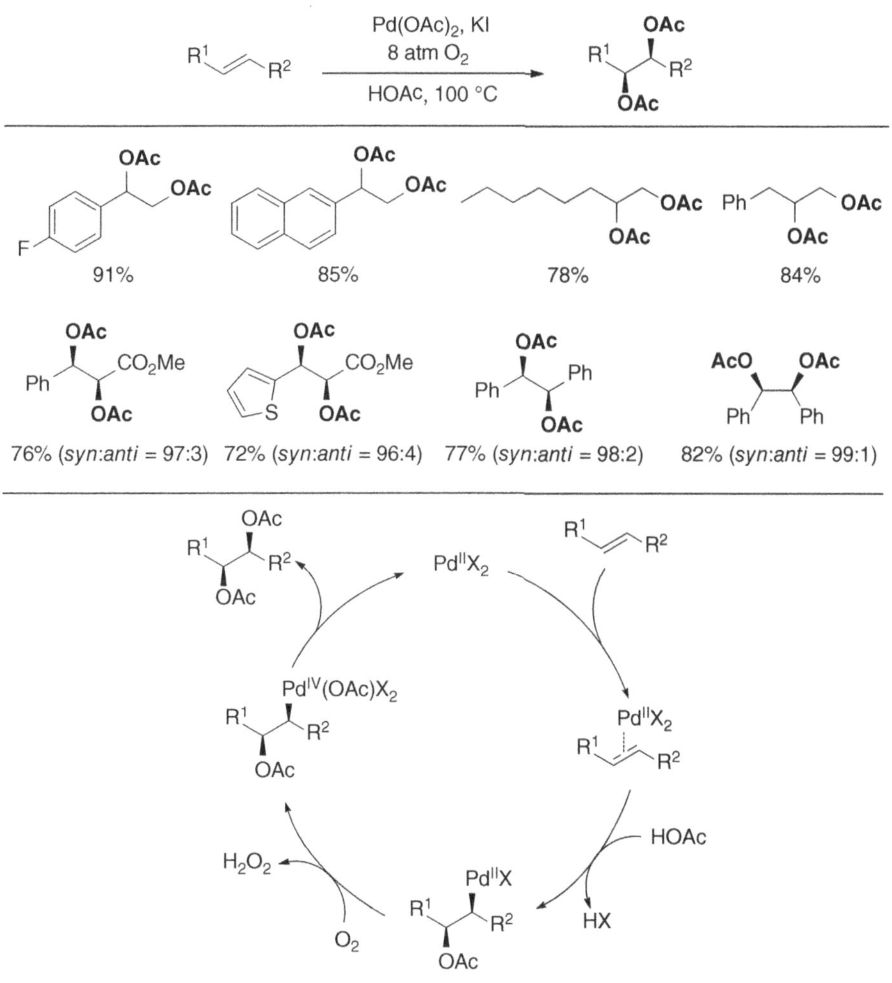

Scheme 3.28 Pd-catalyzed diacetoxylation of alkenes and proposed mechanism.

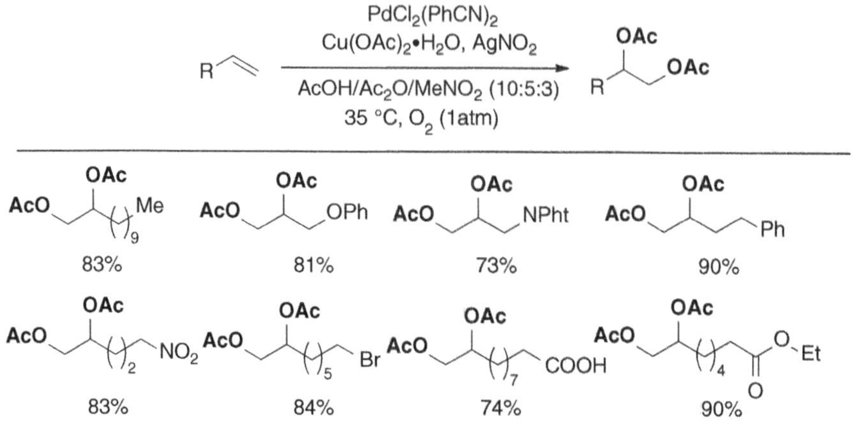

Scheme 3.29 Pd-catalyzed diacetoxylation of alkenes under O_2 atmosphere.

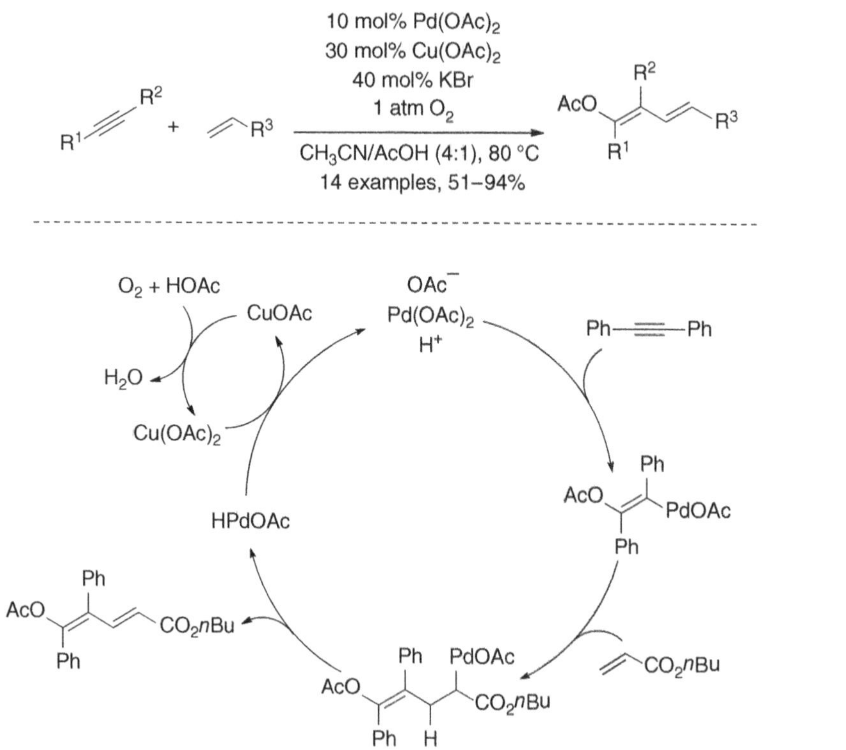

Scheme 3.30 Acetoxypalladation of alkynes to generate 1-acetoxy-1,3-dienes and proposed mechanism.

We discovered the first example of acetoxypalladation of alkynes to generate 1-acetoxy-1,3-dienes with dioxygen as the terminal oxidant in high regio- and stereoselectivity with high atom efficiency (Scheme 3.30) [66]. This strategy proceeded through the tandem process of acetoxypalladation/Heck cross-coupling/β-H elimination.

Difunctionalization of alkenes, which involve the Pd(0/II) catalytic cycles, have been studied in depth in synthetic organic chemistry, such as aminoarylation, oxyarylations, and aminocarbonylation [67]. Some limitations still exist such as the inevitably rapid β-hydride elimination of Pd^{II}-intermediate. Different from the Pd^{II}-intermediate, Pd^{IV} intermediates undergo reductive elimination efficiently to provide the corresponding alkyl-X bond (X=O, halogen) due to the electron-deficiency of palladium center [68]. Therefore, the strong oxidants are indispensable to generate high-valent palladium species. The wildly used strong oxidants are $PhI(OAc)_2$ and NFSI, but both of them are environmentally harmful and expensive. Recently, a novel Pd-catalyzed aminoacetoxylation of unactivated alkenes was discovered by Liu's group (Scheme 3.31) [69]. This protocol utilized H_2O_2 as the sole oxidant, which is available and environmentally

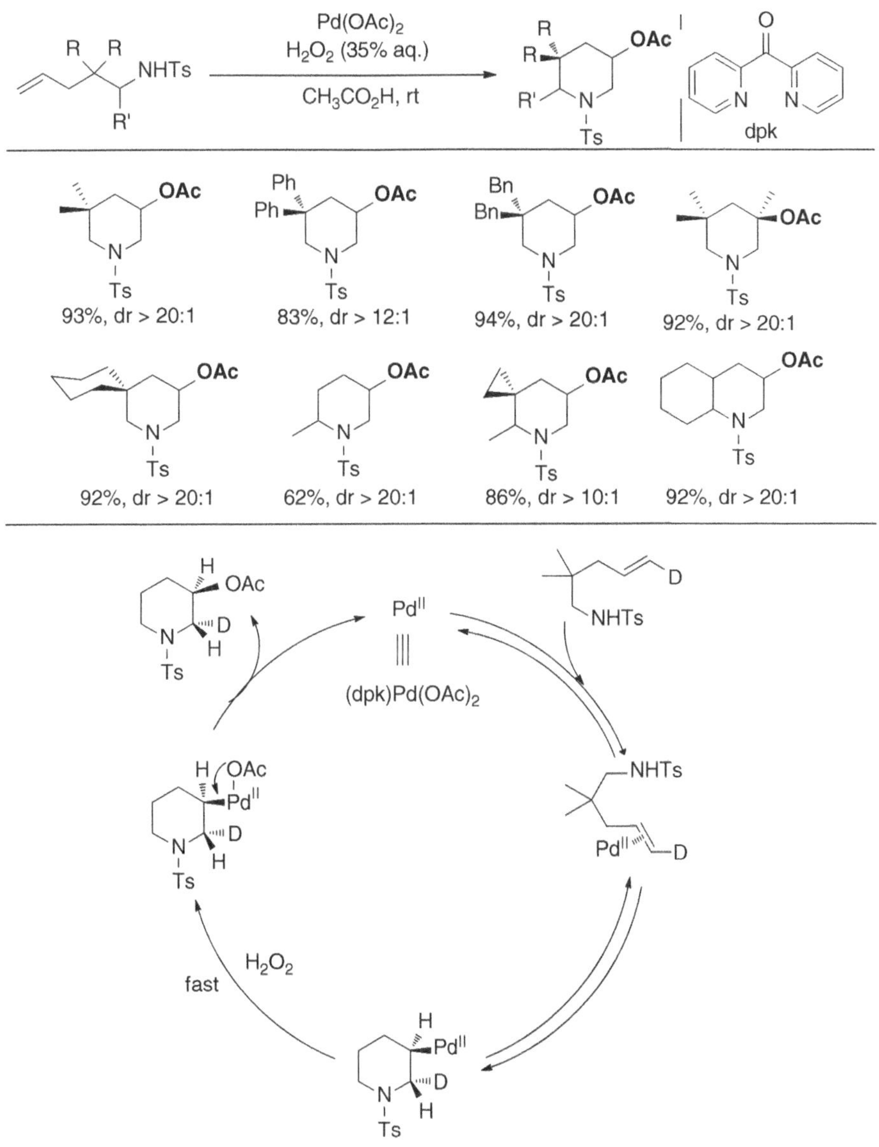

Scheme 3.31 Pd-catalyzed aminoacetoxylation of unactivated alkenes using H_2O_2 as oxidant.

friendly. Various 3-acetoxylated piperidines were synthesized in high yields with excellent regio- and diastereoselectivity in the presence of a di(2-pyridyl) ketone (dpk) ligand, which promoted the formation of a C—OAc bond from C—Pd^{II}. Then, Stoltz's group also reported an aminoacetoxylation of unactivated alkenes via Pd(II/IV) catalytic cycles under aerobic conditions (Scheme 3.32). The catalytic NO_x species played a role of electron transfer mediator to access the Pd^{IV} intermediates [70].

3.3.3.2 Acyloxylation of Benzyl

Direct C—H acyloxylation has made noticeable progress in recent decades for installing acyloxyl in carbon-bearing compounds. Until recently, however, the synthetic utility of the reaction had largely been limited to acetoxylation with environmentally harmful strong oxidants, or had to rely on the use of directing groups that restrict its development and application [71]. To address this problem, Zhang and co-workers developed a facile and efficient Pd-catalyzed actoxylation of toluene using 1 atm O_2 as the oxidant (Scheme 3.33) [72]. A variety of carboxylic acids could participate in the reaction to form desired benzyl esters in high yields and with good atom economy.

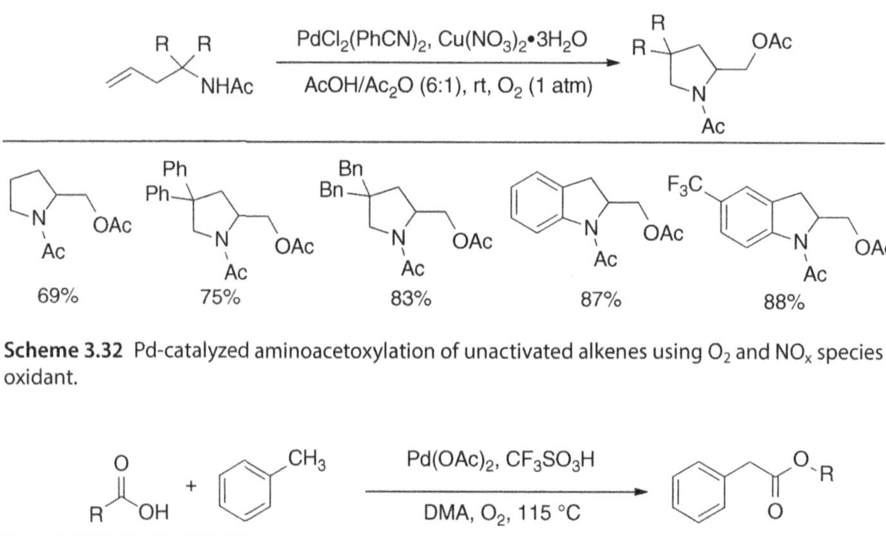

Scheme 3.32 Pd-catalyzed aminoacetoxylation of unactivated alkenes using O_2 and NO_x species as oxidant.

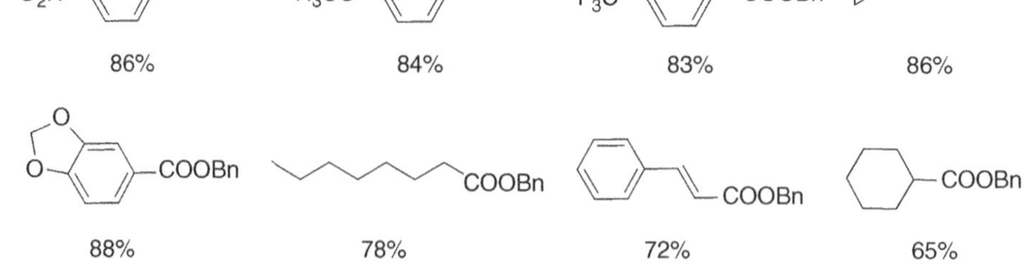

Scheme 3.33 Palladium-catalyzed benzylation of carboxylic acids.

3.3.4 Carboesterification of Alkenes and Alkynes

A strategy for difunctionalization of alkenes or alkynes has attracted broad interest for converting simple alkenes or alkynes into diverse valuable synthetic products. Among the transformations, palladium-catalyzed difunctionalization presented the most efficient strategy and has achieved great advancements such as aminohalogenation, carboamination, carboesterification, carboetherification, and diacetoxylation [73]. Recently, our group discovered a new copper-catalyzed oxidative [3 + 2] cycloaddition of alkenes for the synthesis of *γ*-lactones (Scheme 3.34) [74], which has been identified as an essential moiety in the natural products and biological fields. This process uses an alkene as the substrate that is simple and readily available, $Cu(OTf)_2$ as the catalyst, LiBr as the additive, Ac_2O as the sources of acetoxyl, O_2 as the sole oxidant for the formation of C(sp^3)-C(sp^3), and C—O bonds through carboesterification with a broad substrate scope. After that, a convenient carboesterification method to construct α-methylene-γ-lactone rings by using either alkynoates or alkynamides as substrates together with alkenes also has been reported by our group (Scheme 3.35) [75].

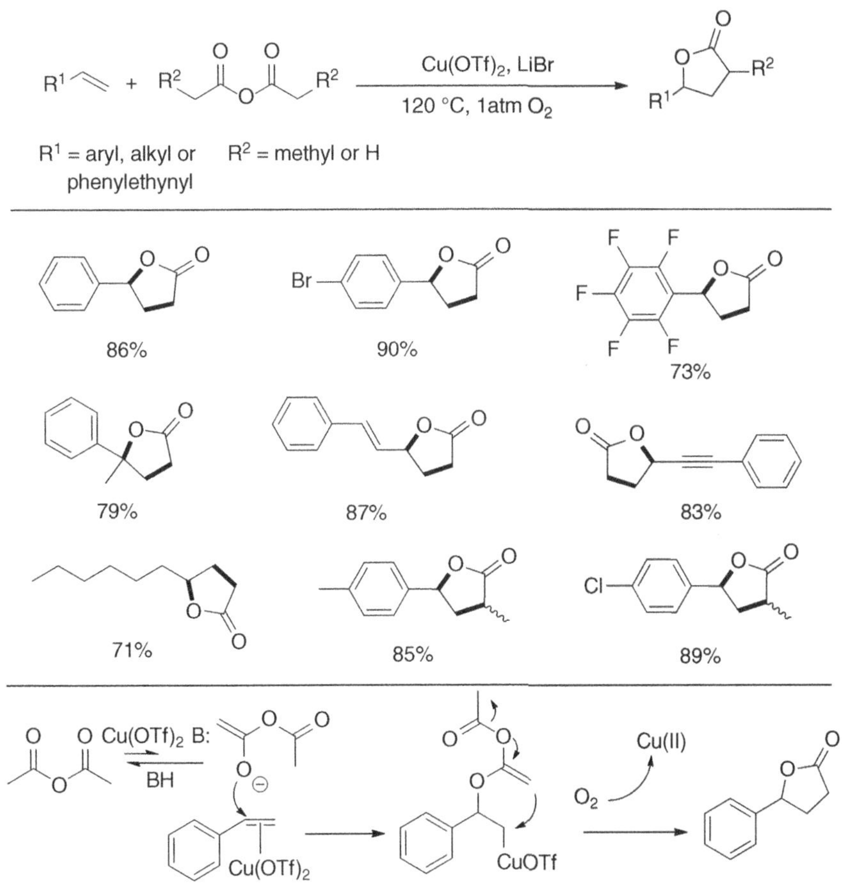

Scheme 3.34 Copper-catalyzed carboesterification of alkenes.

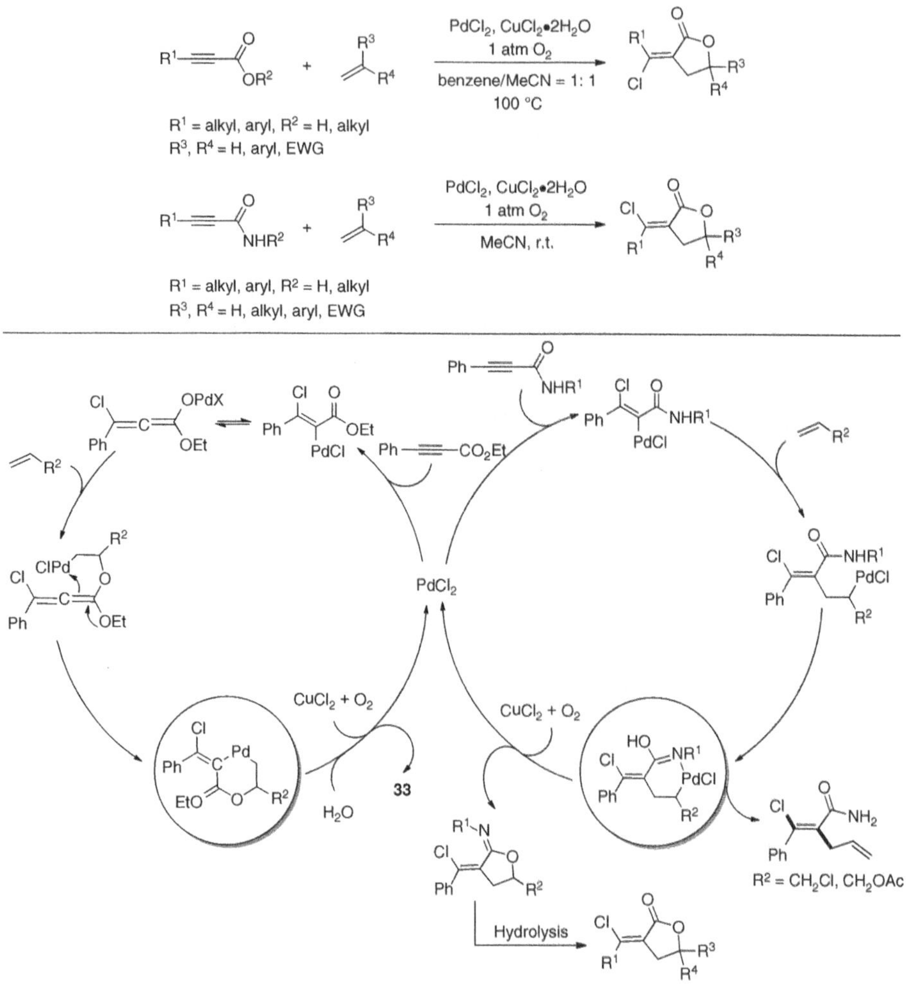

Scheme 3.35 Palladium-catalyzed intermolecular carboesterification of alkenes with alkynoates or alkynamides.

The functionalization of an allylic C—H bond has been a practical and essential method for the functionalization of allylic at the positions of C1, C2, and C3, which has attracted much attention and has been a convenient method for the transformation of simple alkenes into highly functional products [76]. Recently, our group discovered a strategy for the construction of aryl α,β-unsaturated esters through palladium-catalyzed aerobic oxidative double allylic C—H oxygenation with the catalytic system of $Pd(OAc)_2$/DMA under 1 atm O_2 (Scheme 3.36) [77]. Mechanistic studies indicated that this reaction undergoes allylic C—H activation to form a C—O bond smoothly, followed by the nucleophilic attack of another alcohol, hydrolysis, and finally oxidation–dehydrogenation to give the desired products.

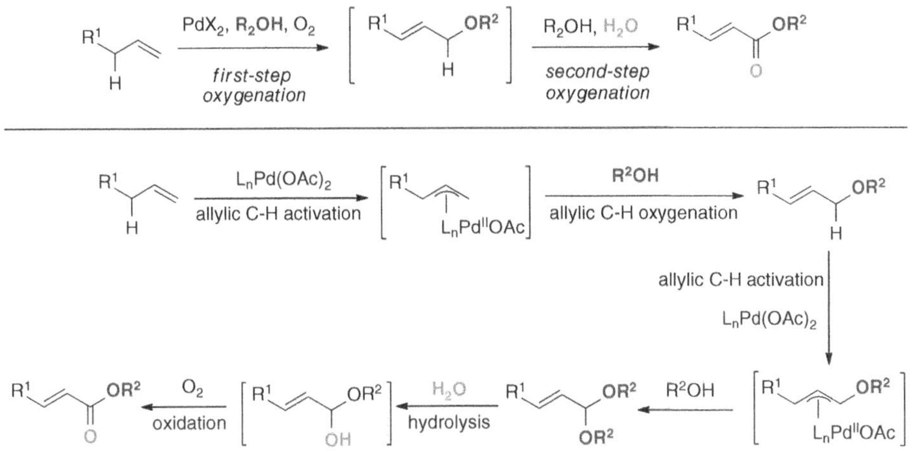

Scheme 3.36 Palladium-catalyzed oxidative allylic C—H for the synthesis of α,β-unsaturated esters.

3.3.5 Esterification via Oxidative Cleavage of C—C Bonds

3.3.5.1 Esterifications of Arylacetonitriles via Oxidative Cleavage of C—CN Bonds

Direct cleavage of a C—CN bond has been explored extensively over the past years via transition metal-catalyzed such as Pd, Ni, Rh, and Cu in modern synthetic chemistry, even through the high bond dissociation energy of C—CN bond [78]. Recently, a novel Fe-catalyzed C—CN bond cleavage reaction in arylacetonitriles affording aryl carboxylic esters has been demonstrated by Song's group (Scheme 3.37) [79]. This strategy proceeded with readily accessible starting materials, abundant, and an environmentally friendly oxidant in good to excellent yields with high functional group tolerance.

3.3.5.2 Esterification of Simple Ketones via Oxidative Cleavage of C(CO)–C(Alkyl) Bonds

C—C bond cleavage has emerged as an active research topic in organic synthesis and has broad applications in both academic research and industrial applications [80]. In 2016, Song and co-workers reported a Cu^{II}-catalyzed oxidative C(CO)-C(alkyl) cleavage of aryl alkyl ketones with environmentally friendly O_2 as the oxidant (Scheme 3.38) [81]. Various esters were synthesized through this efficient and facile strategy. Preliminary mechanistic investigations suggested an interesting reaction sequence, involving α-oxygenation/esterification/nucleophilic addition/C—C bond cleavage with the release of CO_2 as the by-product.

3.3.5.3 Esterification of Alkynes via Oxidative Cleavage of C—C Triple Bonds

Though great progress has been made in the cleavage of the C—C bond, the cleavage of C—C triple bond is still a challenge in modern synthetic organic chemistry [82]. In 2008, we described a novel palladium-catalyzed cleavage of C—C triple bonds to synthesis a carboxylic ester with molecular oxygen (Scheme 3.39). The Lewis acid of $ZnCl_2{\cdot}H_2O$ was necessary and promoted this transformation efficiently [83].

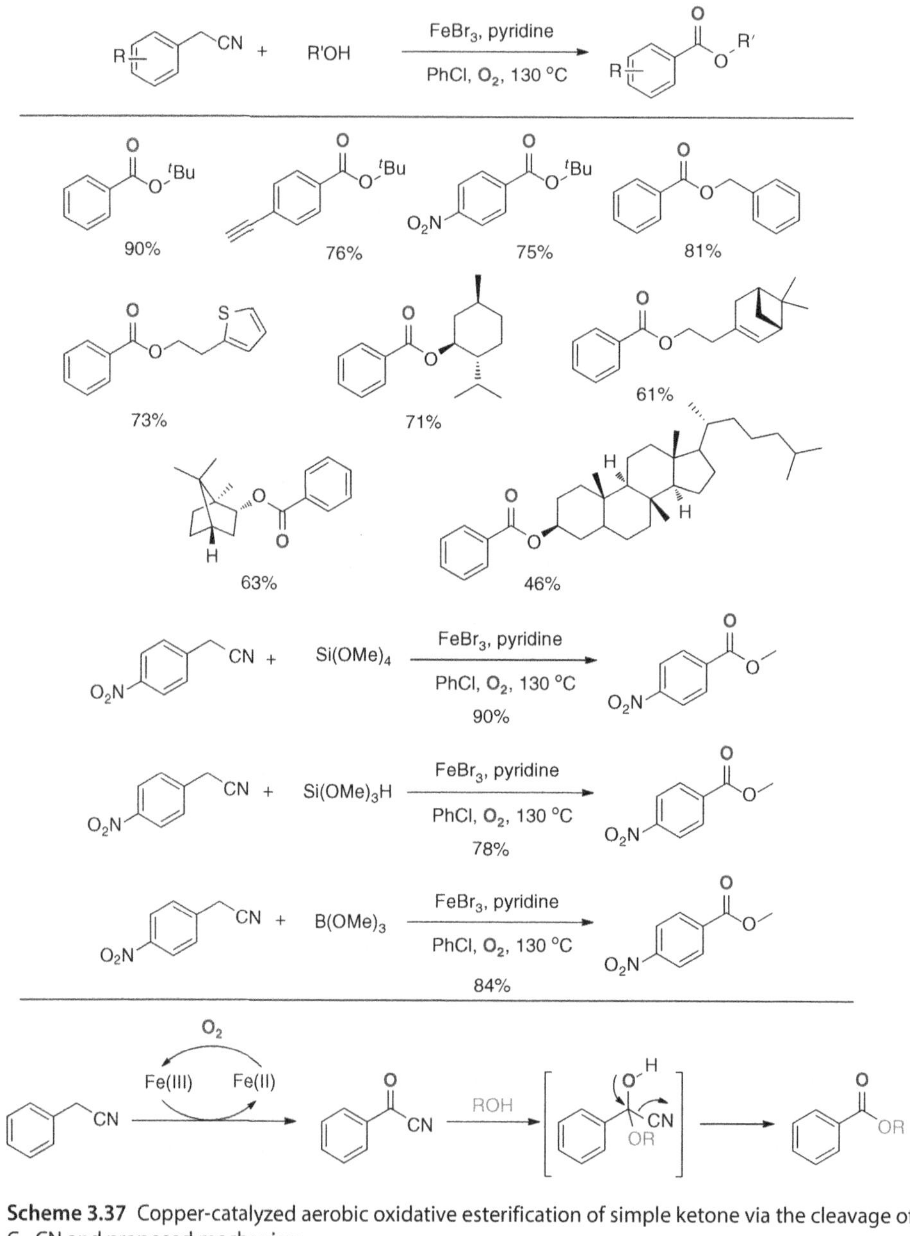

Scheme 3.37 Copper-catalyzed aerobic oxidative esterification of simple ketone via the cleavage of C—CN and proposed mechanism.

3.4 Green Oxidative Synthesis of Organic Halides

Organo-halogen compounds form a class of compounds that are essential in organic chemistry, and are closely related with our life [84]. The synthesis and further transformation of organo-halogen compounds are key industrial process. However, traditional preparation methods of organo-halogen compounds have several drawbacks, such as

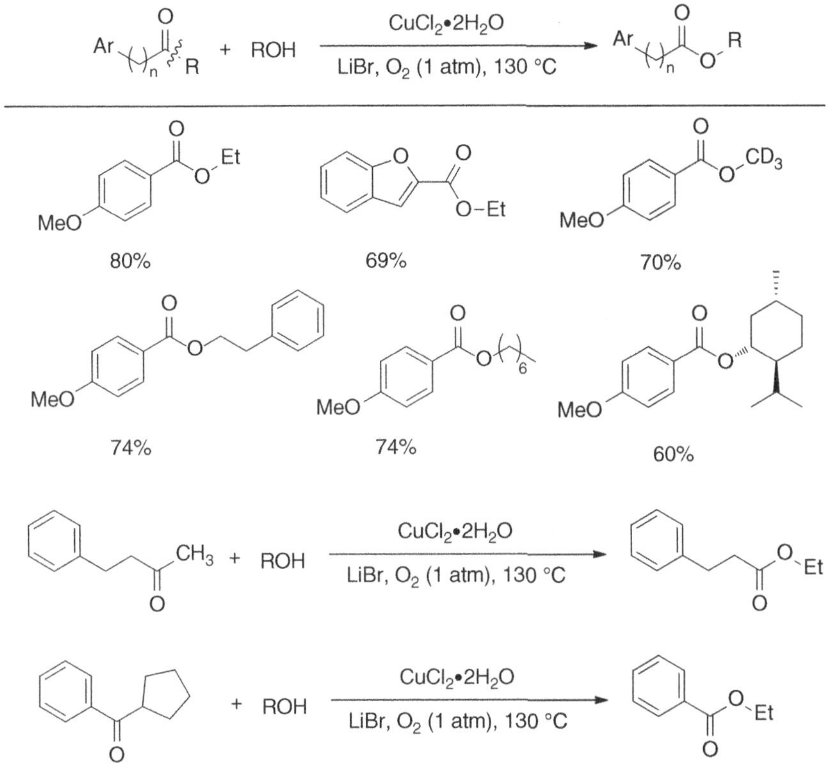

Scheme 3.38 Cu-catalyzed oxidative C(CO)-C(alkyl) cleavage for the synthesis of esters.

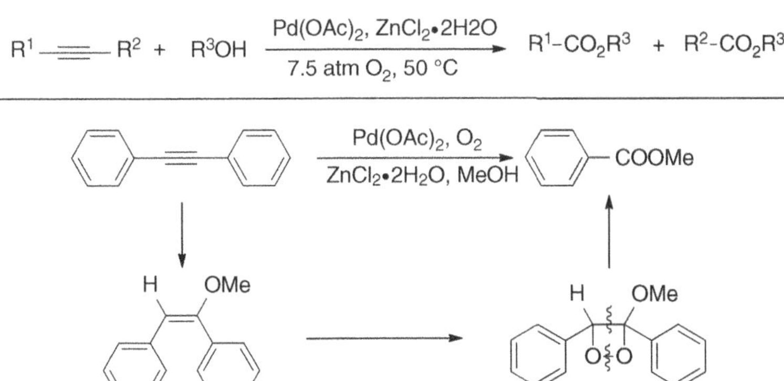

Scheme 3.39 Pd-catalyzed oxidative triple bond cleavage for the synthesis of esters and possible pathway.

the use of pollutant and toxic elemental halogens (X_2) and high energy consumption. Thus, oxidative halogenation has been developed employing halogen salts as halogen sources, and most work has been done in this field [85]. From the green chemistry point of view, dioxygen and hydrogen peroxide might be the most abundant and environmentally friendly oxidants with water as the only oxidation by-product.

3.4.1 Catalyzed/Noncatalyzed Oxidative Halogenation with Hydrogen Peroxide

In the 1920s, Bray and Livingston discovered that bromide in the acid solution of hydrogen peroxide could be oxidized to bromine and they found bromine also could transform to bromide in the presence of hydrogen peroxide [86]. Oxidative halogenation of this type has found significant development in the preparation of organo-halogen compounds, which avoid the use of halogens (X_2). It is a safe, low-cost, and green process without using halogens (X_2) directly with 100% halogen atom economy.

3.4.1.1 Chlorination

Conventional methods for chlorination typically involve stoichiometric chlorine, which utilized only half of the reagent and other half forming corrosive HCl was wasted. Therefore, methods involving HCl or a chloride salt in the presence of an oxidant are more attractive for many chemists. In general, it is difficult to achieve selective chlorination, mainly because of the high oxidation potential of HCl. Therefore, oxidative chlorination usually needs an excess of HCl and high temperatures. Chandalia and co-workers reported the oxidative mono- or dichlorination of aromatic compounds by using HCl-H_2O_2 in 1999 (Scheme 3.40), which has taken a backward step in chlorination of aromatic compounds [87]. After that, many research groups began their studies on oxidative chlorination with HCl or a chloride salt by using H_2O_2 as the oxidant (see the references for details) [88].

3.4.1.2 Bromination

Oxidative bromination by using HBr—H_2O_2 is easier and faster than that of HCl, since the oxidation potential of HBr is lower than HCl, which means that the oxidative bromination can be carried out under milder conditions and with a smaller excess of reagents. Raghavan et al. reported oxidative bromination by using HBr or ammonium

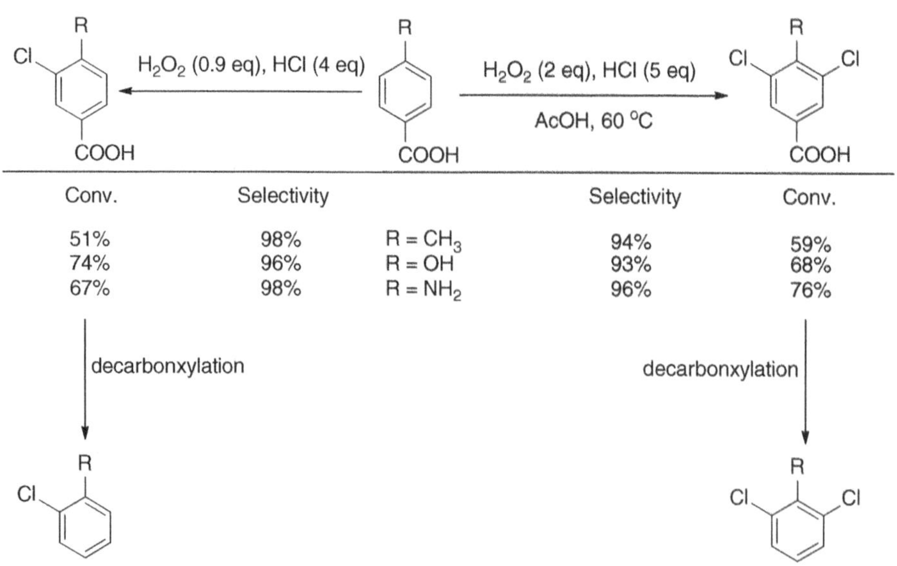

Scheme 3.40 H_2O_2 oxidative chlorination.

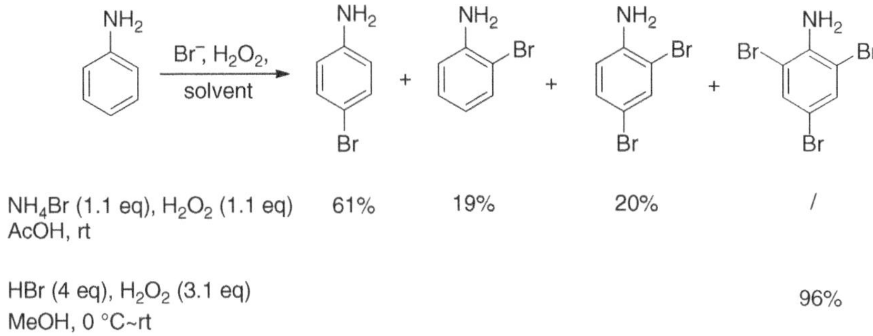

Scheme 3.41 H_2O_2 oxidative bromination.

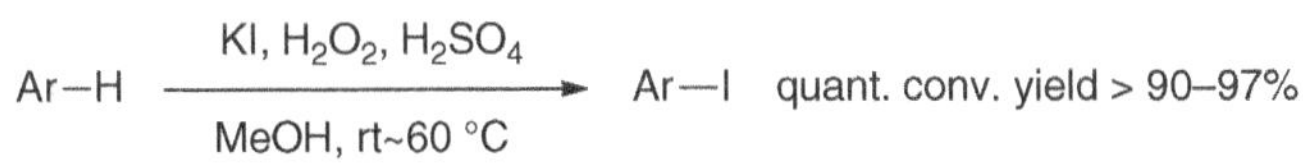

Scheme 3.42 Cu-catalyzed pyridine-directed halogenation of aryl C—H bonds.

bromide as the bromine source and H_2O_2 as the terminal oxidant (Scheme 3.41) [89]. Different bromated aromatic compounds are synthesized through this simple, efficient, environmentally safe, and economical method. More examples of oxidative bromination by using HBr/H_2O_2 or bromide salt/H_2O_2 have been listed in the references [90].

3.4.1.3 Iodination

In contrast with other halogens, iodine has some unique properties; for example, iodide (I^-) has the lowest oxidation potential among all the halides so it is more easily oxidized than others. Meanwhile, HI is unstable in air and light, and iodine is the least reactive halogen, so the activation of the iodination reaction is particularly necessary. A novel and simple, eco-friendly, and inexpensive oxidative iodination of arenes was reported by Iskra's group in 2004 (Scheme 3.42) [91]. This process proceeded by applying 30% aqueous H_2O_2 as the oxidant and KI as the source of iodine in MeOH in the presence of a strong acid. Various substituted arenes such as anisoles, phenols and anilines, mesitylene, and uracil were tolerated, and the desired aryl iodides were obtained in high selectivity and effectivity with good yields. Other oxidative iodinations with H_2O_2 as the oxidant have been listed in the references (Scheme 3.43) [92].

3.4.1.4 Fluorination

The high oxidation potential of fluorine makes oxidative fluorination quite hard even with the help of catalysts. Few examples have been achieved for oxidative fluorination, however, all of them proceeded under harsh conditions and with expensive and environmentally unfriendly strong oxidants [93].

3.4.2 Aerobic Oxidative Halogenation

O_2 as a four-electro oxidant, is an abundant, least expensive, environment being oxidant, which has gained widespread attention in academy and industry [94]. Despite the merits of dioxygen, there are still some shortcomings such as the low oxidation

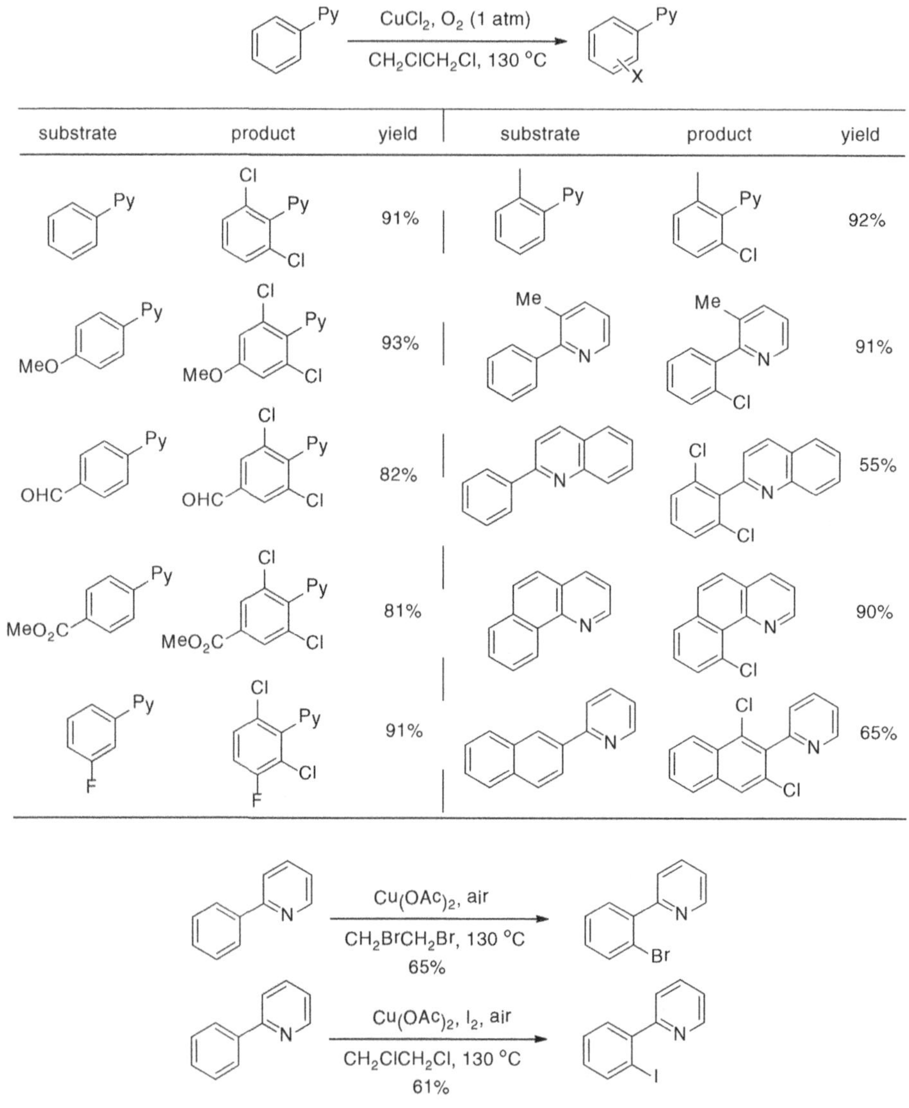

Scheme 3.43 H_2O_2 oxidative iodination.

capability. Therefore, in the presence of a metal catalyst, such as copper, palladium, platinum, ruthenium, rhodium, or iridium, it would effectively promote the oxidative reaction, and various organic molecules could be generated through metal-catalyzed aerobic oxidative reactions [95]. Recently, great progress has been achieved in metal-catalyzed aerobic oxidative halogenation reactions, and a series of organic halides have been synthesized through this strategy.

3.4.2.1 Oxidative Halogenation of Arenes or Alkenes

Transition-metal-catalyzed C—H halogenation has been the focus of growing attention and has witnessed substantial progress in recent years. In 2005, Yu and co-workers

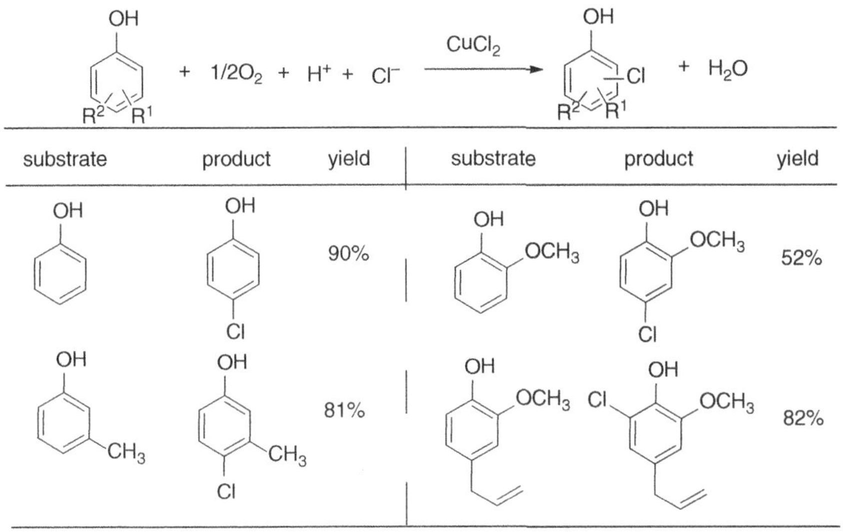

Scheme 3.44 Cu-catalyzed oxidative chlorination of phenols.

reported a Cu(II)-catalyzed chlorination of aryl C—H bonds (Scheme 3.42) [96]. Different kinds of chlorinated product have been generated through this protocol by using inexpensive $CuCl_2$ as the catalyst and O_2 as an oxidant. When substituents are in the *ortho* position of pyridine, a monochlorinated product is the mainstay, thus demonstrating the impact of steric hindrances in the N atom that prevent further chlorination. Simultaneously, bromated and iodinated products could be also formed via this process.

Meanwhile, Gusevskaya et al. also discovered a Cu-catalyzed oxidative chlorination of phenols in high yields (Scheme 3.44) [97]. Various solvents, even water, could promote the reaction smoothly. Chloride ions served as the halogenating reagent and O_2 as the final oxidant in high regioselectivity and fast reaction rate for the synthesis of chlorophenols, which have potential applications in pharmaceutical, agricultural, and dye industries. In 2009, a novel regioselective Cu-catalyzed chlorination and bromination of electron-rich arenes was also described by Stahl and co-workers under mild conditions (Scheme 3.45) [98]. Preliminary studies indicated that bromination and chlorination reactions occur with different mechanisms.

In addition to oxidative halogenation of arenes, direct oxidative halogenation of alkenes represents a versatile strategy in constructing the Csp^2—X bond rapidly and has attracted attention. Over the past few years, we set out to study the oxidative halogenation and discovered a highly regio- and stereoselective Pd-catalyzed dehydrogenative aminohalogenation of alkenes for the first time (Scheme 3.46) [99]. A wide range of brominated enamines was assembled through this novel and efficient process by using O_2 as the final oxidant, which was of great significance due to the versatile reactivity for further transformations. We then successfully developed a solvent-controlled regioselective bromination via Pd catalysis under 1 atm O_2 (Scheme 3.47) [100]. The presence of H_2O_2 is crucial and acts as co-oxidant.

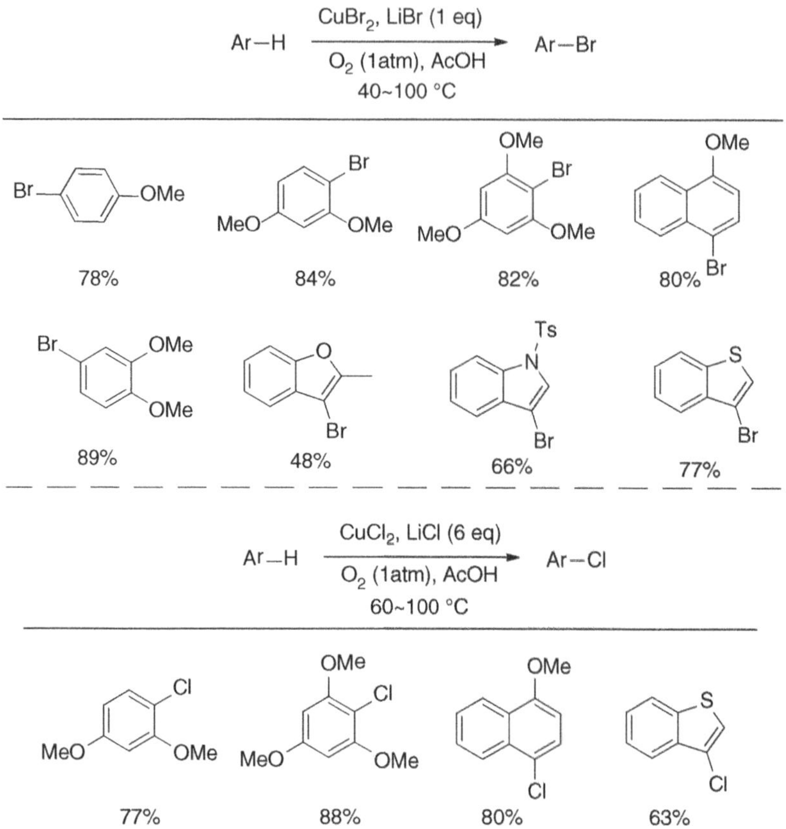

Scheme 3.45 Cu-catalyzed chlorination and bromination of electron-rich arenes.

3.4.2.2 Oxidative Halogenation of Alkynes-Halopalladation Reaction

At present, halopalladation reaction is receiving considerable attention and has proven to be a versatile and powerful strategy for the construction of carbon–carbon and carbon-halide bonds in one step with high efficiency and atom economy [101]. The initial studies in this area were reported by Lu and co-workers [102], and this has now been developed rapidly and widely used in the synthesis of highly functionalized haloalkenes. Recently, our group developed a Pd-catalyzed tandem bromoalkylation reaction of alkynoates and allylic alcohols (Scheme 3.48) [103]. Different from previous reports, this process used $CuBr_2$ as the sources of halide ions. A wide range of δ-bromo-γ,δ-unsaturated carbonyls could be assembled through this efficient and atom economic process with high regio- and stereoselectivity.

Next, we also discovered a novel and efficient palladium-catalyzed intermolecular aerobic oxidative annulation of 2-ethynylanilines with isocyanides for the first time (Scheme 3.49) [104]. Various 4-halo-2-aminoquinolines were constructed with high functional group tolerance using molecular oxygen as the sole oxidant. The final product could also be further converted to 6*H*-indolo[2,3-b]quinoline.

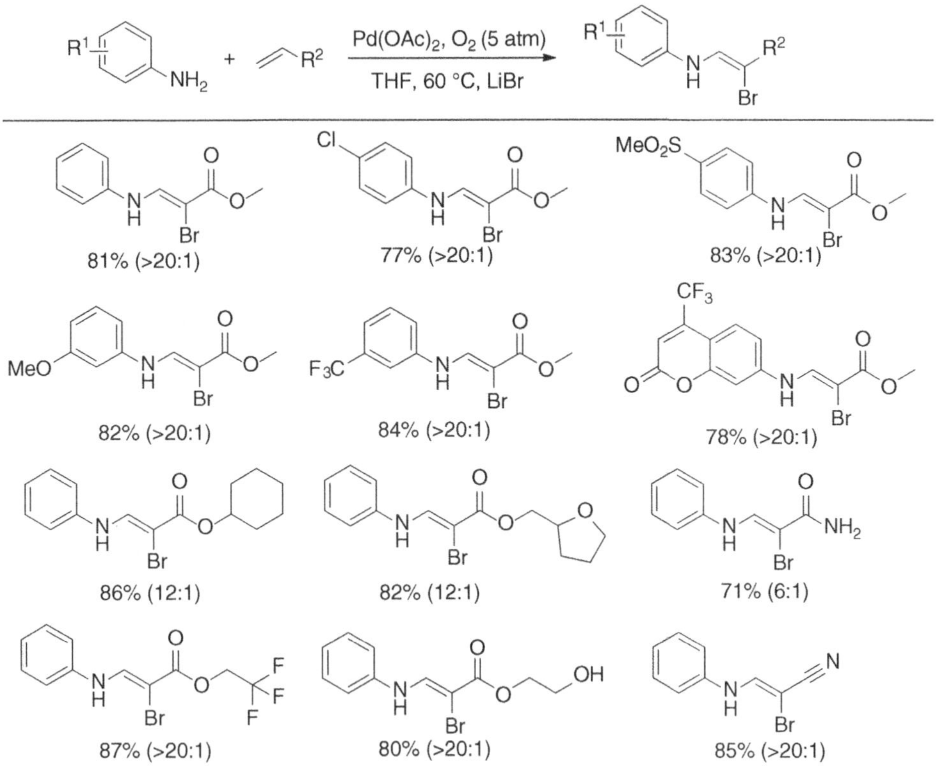

Scheme 3.46 Pd-catalyzed dehydrogenative aminohalogenation of alkenes.

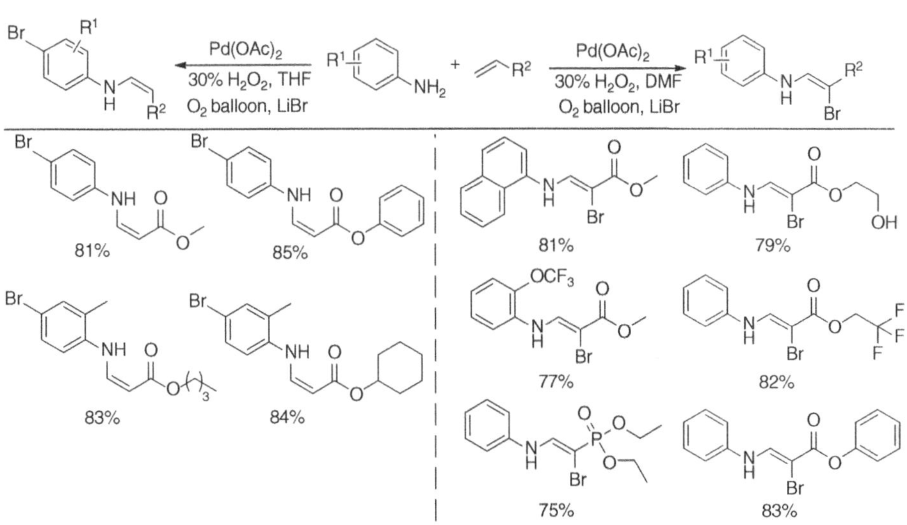

Scheme 3.47 Pd-catalyzed solvent-controlled regioselective bromination.

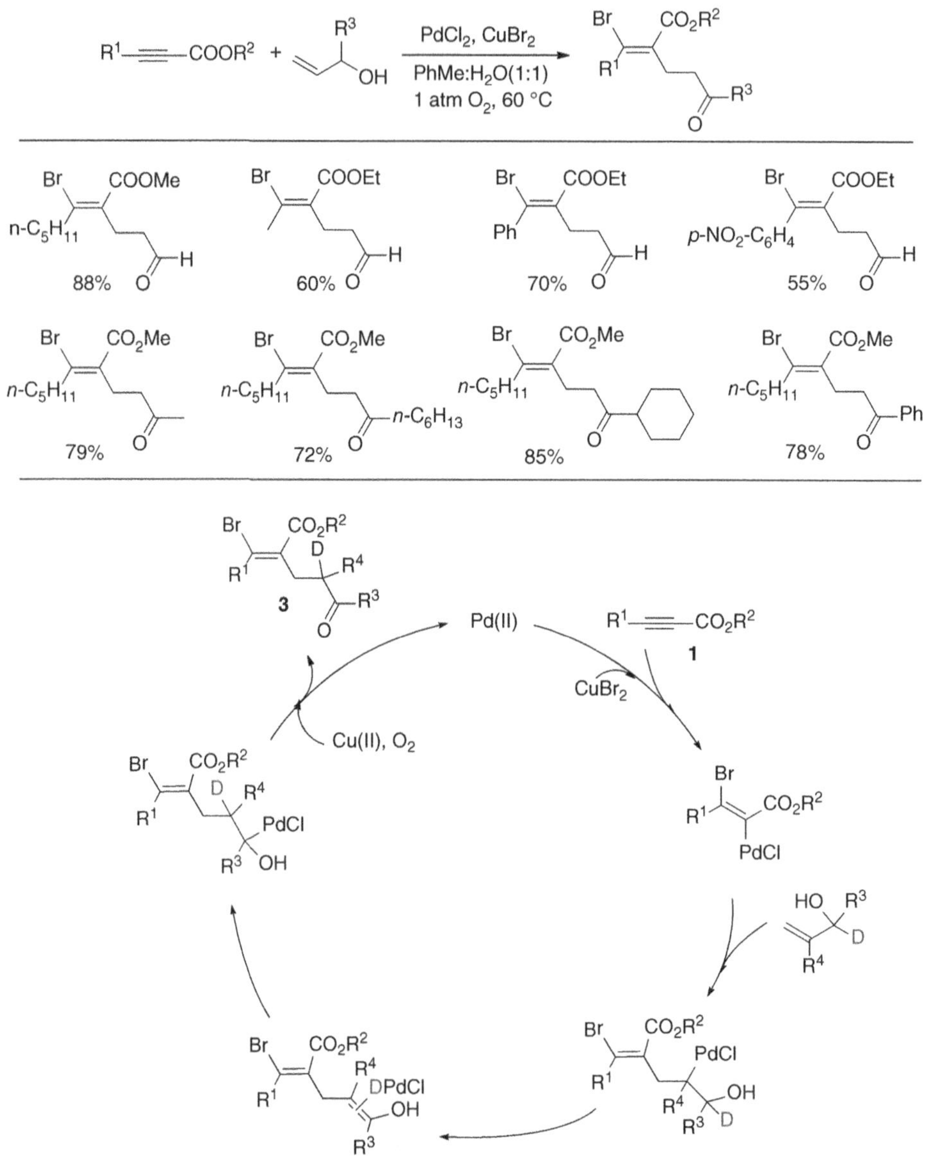

Scheme 3.48 Pd-catalyzed bromoalkylation tandem reaction of alkynoates and allylic alcohols.

More importantly, we discovered the first example of chloropalladation-initiated intermolecular asymmetric carboesterification of alkenes with O_2 as the terminal oxidant (Scheme 3.50) [105], which provided a new strategy for the synthesis of α-methylene-γ-lactones in good yields and excellent enantioselectivities at room temperature.

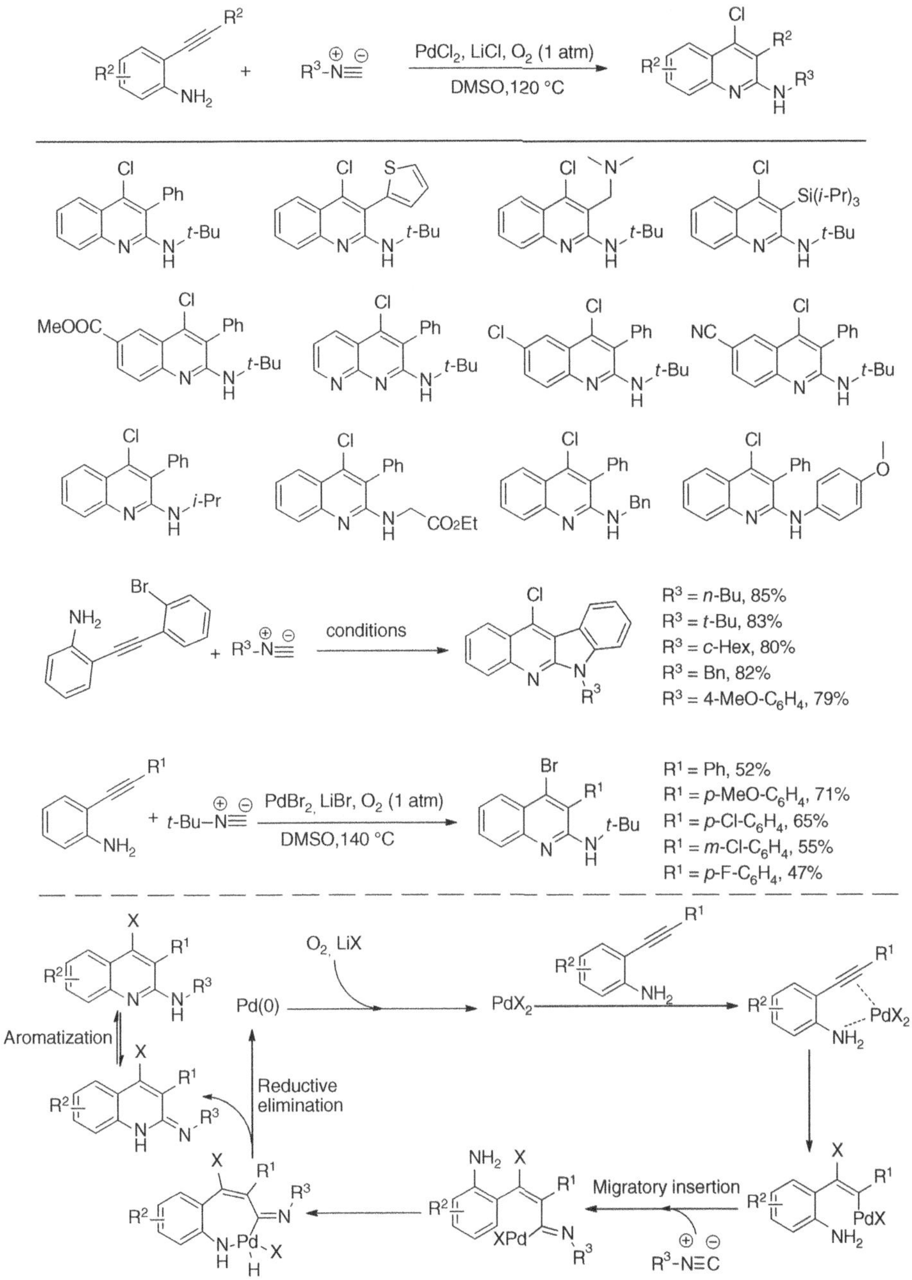

Scheme 3.49 Palladium-catalyzed oxidative cyclization of 2-ethynylanilines with isocyanides for the synthesis of 4-halo-2-aminoquinolines.

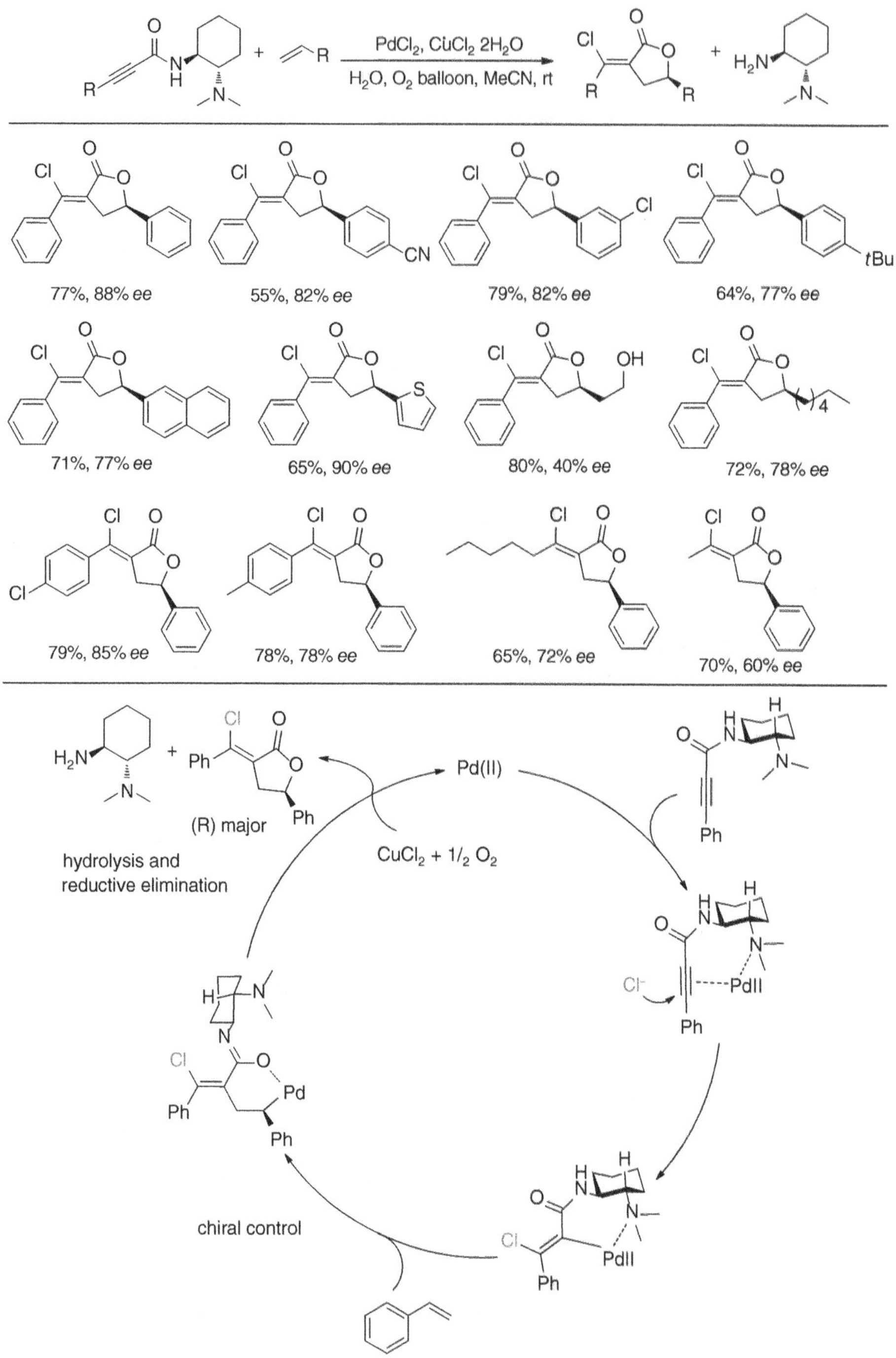

Scheme 3.50 Chloropalladation-initiated intermolecular asymmetric carboesterification of alkenes for the construction of halo-α-methylene-γ-lactones.

3.5 Conclusions and Outlook

In the past few decades, research into the synthesis of ethers, esters, and organic halides has become a vibrant area of study, due to the great importance and ubiquity of them in chemical industry. In the preceding chapters, we have clearly demonstrated the development in the area of green oxidative synthesis of ethers, esters, and organic halides. The methods described have provided an avenue for the growth of etherification, esterification, and halogenation using green oxidants such as hydrogen peroxide and O_2. Many of these reactions catalyzed by transition metals were developed with mechanistic studies, which illustrate the details of oxidation process in the catalytic system. More importantly, the utilization of green oxidants with no poisonous and harmful by-products makes the synthesis more attractive, highly desirable, economically viable, and environmentally acceptable.

The future seems bright, but there are still many challenges to be addressed. Although green oxidative synthesis of ethers, esters, and organic halides has made adequate progress, there are still some transformations that proceed with environmentally unfriendly oxidants. So, the development of new reactions with increased reactivity that are compatible with green oxidants as an environmentally benign oxidants will have important implications. Moreover, the asymmetric catalysis and mechanistic investigations will also be an important area for development in the future.

References

1 (a) Cernansky, R. (2015). *Nature* 519: 379. (b) Sanderson, K. (2011). *Nature* 469: 18. (c) Poliakoff, M. and Licence, P. (2007). *Nature* 450: 810. (d) Clark, J.H. (1999). *Green Chem.* 1: 1.

2 Bäckvall, J.-E. (2004). *Modern Oxidation Methods*. Weinheim: VCH-Wiley.

3 (a) Sheldon, R.A., Arends, I.W.C.E., ten Brink, G.J., and Dijksman, A. (2002). *Acc. Chem. Res.* 35: 774. (b) Punniyamurthy, T., Velusamy, S., and Iqbal, J. (2005). *Chem. Rev.* 105: 2329. (c) Wu, W. and Jiang, H. (2012). *Acc. Chem. Res.* 45: 1736. (d) Ouyang, L. and Wu, W. (2017). *Curr. Opin. Green Sustain. Chem.* 7: 46. (e) Boisvert, L. and Goldberg, K. (2012). *Acc. Chem. Res.* 45: 899.

4 (a) Dermer, O.C. (1934). *Chem. Rev.* 14: 385. (b) Williamson, A.W. (1852). *J. Chem. Soc.* 4: 229.

5 (a) Kunz, K., Scholz, U., and Ganzer, D. (2003). *Synlett* 2428. (b) Ley, S.V. and Thomas, A.W. (2003). *Angew. Chem. Int. Ed.* 42: 5400.

6 (a) Hartwig, J.F. (1998). *Acc. Chem. Res.* 31: 852. (b) Hartwig, J.F. (1998). *Angew. Chem. Int. Ed.* 37: 2046. (c) Enthaler, S. and Company, A. (2011). *Chem. Soc. Rev.* 40: 4912. (d) Wu, X., Fors, B.P., and Buchwald, S.L. (2011). *Angew. Chem. Int. Ed.* 50: 9943. (e) Altman, R.A., Shafir, A., Choi, A. et al. (2008). *J. Org. Chem.* 73: 284. (f) Wolter, M., Nordmann, G., Job, G.E., and Buchwald, S.L. (2002). *Org. Lett.* 4: 973.

7 (a) Wencel-Delord, J., Dröge, T., Liu, F., and Glorius, F. (2011). *Chem. Soc. Rev.* 40: 4740. (b) Newhouse, T. and Baran, P.S. (2011). *Angew. Chem. Int. Ed.* 50: 3362. (c) Lyons, T.W. and Sanford, M.S. (2010). *Chem. Rev.* 110: 1147. (d) Daugulis, O., Do, H.-Q., and Shabashov, D. (2009). *Acc. Chem. Res.* 42: 1074. (e) Ackermann, L., Vicente, R., and Kapdi, A.R. (2009). *Angew. Chem. Int. Ed.* 48: 9792. (f) Sun, C.-L., Li, B.-J., and Shi, Z.-J. (2011). *Chem. Rev.* 111: 1293.

8 (a) Tang, L., Pang, Y., Yan, Q. et al. (2011). *J. Org. Chem.* 76: 2744. (b) Uyanik, M., Okamoto, H., Yasui, T., and Ishihara, K. (2010). *Science* 328: 1376. (c) Uyanik, M., Suzuki, D., Yasui, T., and Ishihara, K. (2011). *Angew. Chem. Int. Ed.* 50: 5331.
9 Rönn, M., Bäckvall, J.-E., and Andersson, D.G. (1995). *Tetrahedron Lett.* 36: 7749.
10 Wang, X., Lu, Y., Dai, H.-X., and Yu, J.-Q. (2010). *J. Am. Chem. Soc.* 132: 12203.
11 (a) Smrčina, M., Poláková, J., Vyskoćil, S., and Kočovský, P. (1993). *J. Org. Chem.* 58: 4534. (b) Li, X., Hewgley, B., Mulrooney, C.A. et al. (2003). *J. Org. Chem.* 68: 5500.
12 Xiao, B., Gong, T.-J., Liu, Z.-J. et al. (2011). *J. Am. Chem. Soc.* 133: 9250.
13 Zhao, J., Wang, Y., He, Y. et al. (2012). *Org. Lett.* 14: 1078.
14 (a) Kuhl, N., Hopkinson, M.N., Wencel Delord, J., and Glorius, F. (2012). *Angew. Chem. Int. Ed.* 51: 10236. (b) Yeung, C.S. and Dong, V.M. (2011). *Chem. Rev.* 111: 1215. (c) Ackermann, L. (2011). *Chem. Rev.* 111: 1315. (d) Li, C.-J. (2009). *Acc. Chem. Res.* 42: 335.
15 (a) Girard, S.A., Knauber, T., and Li, C.-J. (2014). *Angew. Chem. Int. Ed.* 53: 74. (b) Tran, L.D., Roane, J., and Daugulis, O. (2013). *Angew. Chem. Int. Ed.* 52: 6043. (c) Pan, J., Su, M., and Buchwald, S.L. (2011). *Angew. Chem. Int. Ed.* 50: 8647. (d) Mei, T.-S., Wang, X., and Yu, J.-Q. (2009). *J. Am. Chem. Soc.* 131: 10806. (e) Tsang, W.C.P., Zheng, N., and Buchwald, S.L. (2005). *J. Am. Chem. Soc.* 127: 14560. (f) Wang, T., Zhou, W., Yin, H. et al. (2012). *Angew. Chem. Int. Ed.* 51: 10823. (g) Wang, X., Lu, Y., Dai, H.-X., and Yu, J.-Q. (2010). *J. Am. Chem. Soc.* 132: 12203.
16 Bhadra, S., atheis, C.M., Katayev, D., and Gooßen, L.J. (2013). *Angew. Chem. Int. Ed.* 52: 9279.
17 Lu, W., Xu, H., and Shen, Z. (2017). *Org. Biomol. Chem.* 15: 1261.
18 Suess, A.M., Ertem, M.Z., Cramer, C.J., and Stahl, S.S. (2013). *J. Am. Chem. Soc.* 135: 9797.
19 (a) Evano, G., Blanchard, N., and Toumi, M. (2008). *Chem. Rev.* 108: 3054. (b) Wang, C., Piel, I., and Glorius, F. (2009). *J. Am. Chem. Soc.* 131: 4194. (c) Campeau, L.-C., Parisien, M., Jean, A., and Fagnou, K. (2006). *J. Am. Chem. Soc.* 128: 581. (d) Zhang, H., Shi, R., Gan, P. et al. (2012). *Angew. Chem. Int. Ed.* 51: 5204.
20 (a) Ley, S.V. and Thomas, A.W. (2003). *Angew. Chem. Int. Ed.* 42: 5400. (b) Beletskaya, I.P. and Cheprakov, A.V. (2004). *Coord. Chem. Rev.* 248: 2337. (c) Ma, D. and Cai, Q. (2008). *Acc. Chem. Res.* 41: 1450. (d) Monnier, F. and Taillefer, M. (2009). *Angew. Chem. Int. Ed.* 48: 6954.
21 Roane, J. and Daugulis, O. (2013). *Org. Lett.* 15: 5842.
22 Hao, X.-Q., Chen, L.-J., Ren, B. et al. (2014). *Org. Lett.* 16: 1104.
23 (a) Tsuji, J. (2004). *Proc. Jpn. Acad. Ser. B* 80: 349. (b) Tunge, J.A. and Burger, E.C. (2005). *Eur. J. Org. Chem.* 1715. (c) You, S.-L. and Dai, L.-X. (2006). *Angew. Chem. Int. Ed.* 45: 5246. (d) Mohr, J.T. and Stoltz, B.M. (2007). *Chem. Asian J.* 2: 1476. (e) Mohr, J.T., Ebner, D.C., and Stoltz, B.M. (2007). *Org. Biomol. Chem.* 5: 3571. (f) Austeri, M., Buron, F., Constant, S. et al. (2008). *Pure Appl. Chem.* 80: 967. (g) Weaver, J.D., Reciolll, A., Grenning, A.J., and Tunge, J.A. (2011). *Chem. Rev.* 111: 1846. (h) Hong, A.Y. and Stoltz, B.M. (2013). *Eur. J. Org. Chem.* 2745. (i) Huang, L., Qi, J., Wu, X. et al. (2013). *Org. Lett.* 15: 1130.
24 Rodríguez, N. and Gooßen, L.J. (2011). *Chem. Soc. Rev.* 40: 5030.
25 Bhadra, S., Dzik, W.I., and Gooßen, L.J. (2012). *J. Am. Chem. Soc.* 134: 9938.
26 Bhadra, S., Dzik, W.I., and Gooßen, L.J. (2013). *Angew. Chem. Int. Ed.* 52: 2959.
27 (a) Chan, D., Monaco, K., Wang, R., and Winter, M. (1998). *Tetrahedron Lett.* 39: 2933. (b) Evans, D., Katz, J., and West, T. (1998). *Tetrahedron Lett.* 39: 2937. (c) Lam, P., Clark, C., Saubern, S. et al. (1998). *Tetrahedron Lett.* 39: 2941.

28 (a) Ley, S.V. and Thomas, A.W. (2003). *Angew. Chem. Int. Ed.* 42: 5400. (b) Shafir, A., Lichtor, P.A., and Buchwald, S.L. (2007). *J. Am. Chem. Soc.* 129: 3490. (c) Altman, R.A., Shafir, A., Choi, A. et al. (2008). *J. Org. Chem.* 73: 284. (d) Naidu, A.B., Jaseer, E.A., and Sekar, G. (2009). *J. Org. Chem.* 74: 3675.
29 Quach, T.D. and Batey, R.A. (2003). *Org. Lett.* 5: 1381.
30 Wang, R., Wang, L., Zhang, K. et al. (2015). *Tetrahedron Lett.* 56: 4815.
31 Marcum, J.S., McGarry, K.A., Ferber, C.J., and Clark, T.B. (2016). *J. Org. Chem.* 81: 7963.
32 (a) Smidt, J., Hafner, W., Jira, R. et al. (1959). *Angew. Chem.* 71: 176. (b) Smidt, J., Hafner, W., Jira, R. et al. (1962). *Angew. Chem. Int. Ed.* 1: 80. (c) Jira, R. (2009). *Angew. Chem. Int. Ed.* 48: 9034.
33 For recent reviews of the Wacker process, see: (a) Takacs, J.M. and Jiang, X. (2003). *Curr. Org. Chem.* 7: 369. (b) Cornell, C.N. and Sigman, M.S. (2007). *Inorg. Chem.* 46: 1903. (c) Muzart, J. (2007). *Tetrahedron* 63: 7505. (d) Zeni, G. and Larock, R.C. (2006). *Chem. Rev.* 106: 4644. (e) Patil, N.P. and Yamamoto, Y. (2008). *Chem. Rev.* 108: 3395. (f) McDonald, R.I., Liu, G., and Stahl, S.S. (2011). *Chem. Rev.* 111: 2981.
34 Jia, L., Jiang, H., and Li, J. (1999). *Chem. Commun.* 985.
35 Wang, Z., Jiang, H., Qi, C. et al. (2005). *Green Chem.* 7: 582.
36 Simon, M.-O., Girard, S.A., and Li, C.-J. (2012). *Angew. Chem. Int. Ed.* 51: 7537.
37 (a) Otera, J. (2003). *Esterification: Methods, Reactions, and Applications.* Weinheim: Wiley-VCH. (b) Otera, J. (1993). *Chem. Rev.* 93: 1449. (c) Otera, J. (2001). *Angew. Chem. Int. Ed.* 40: 2044.
38 Larock, R.C. (1989). *Comprehensive Organic Transformations: A Guide to Functional Group Preparation.* New York: VCH.
39 (a) Sheldon, R.A. and Kochi, J.K. (1994). *Metal-Catalyzed Oxidations of Organic Compounds.* New York: Academic Press. (b) Hudlucky, M. (1990). *Oxidations in Organic Chemistry.* Washington, DC: ACS Monograph. (c) Procter, G. (1991). *Comprehensive Organic Synthesis*, vol. 7 (ed. S.V. Ley), 305. Oxford: Pergamon.
40 (a) Jagadeesh, R.V., Junge, H., Pohl, M.M. et al. (2013). *J. Am. Chem. Soc.* 135: 10776. (b) Karimia, B. and Esfahani, F.K. (2012). *Adv. Synth. Catal.* 354: 1319. (c) Chen, L., Hu, J., Sankar Mal, S. et al. (2009). *Chem. Eur. J.* 15: 7490. (d) Yamaguchi, K. and Mizuno, N. (2002). *Angew. Chem., Int. Ed.* 41: 4538. (e) Mori, K., Hara, T., Mizugaki, T. et al. (2004). *J. Am. Chem. Soc.* 126: 10657. (f) Ekoue-Kovi, K. and Wolf, C. (2008). *Chem. Eur. J.* 14: 6302. (g) Rajabi, F., Aracon, R.A.D., and Luque, R. (2015). *Catal. Commun.* 59: 101. (h) Marsden, C., Taarning, E., Hansen, D. et al. (2008). *Green Chem.* 10: 168. (i) Gopinath, R., Barkakaty, B., Talukdar, B., and Patel, B.K. (2003). *J. Org. Chem.* 68: 2944. (j) Travis, B.R., Sivakumar, M., Hollist, G.O., and Borhan, B. (2003). *Org. Lett.* 5: 1031.
41 Gowrisankar, S., Neumann, H., and Beller, M. (2011). *Angew. Chem. Int. Ed.* 50: 5139.
42 Liu, C., Wang, J., Meng, L. et al. (2011). *Angew. Chem. Int. Ed.* 50: 5144.
43 Caporaso, M., Cravotto, G., Georgakopoclos, S. et al. (2014). *Beilstein J. Org. Chem.* 10: 1454.
44 Samanta, S., Pappula, V., Dinda, M., and Adimurthy, S. (2014). *Org. Biomol. Chem.* 12: 9453.
45 Yi, H., Hu, X., Bian, C., and Lei, A. (2017). *ChemSusChem* 10: 79.
46 Kiran, I.N.C., Lalwani, K., and Sudalai, A. (2013). *RSC Adv.* 3: 1695.
47 Mhldorf, B. and Wolf, R. (2017). *ChemCatChem* 9: 1.
48 Gopinath, R. and Patel, B.K. (2000). *Org. Lett.* 2: 577.
49 Wu, X.-F. (2012). *Tetrahedron Lett.* 53: 3397.

50 (a) Díaz, D.J., Darko, A.K., and McElwee-White, L. (2007). *Eur. J. Org. Chem.* 4453. (b) Xiao, X.-F., Neumann, H., and Beller, M. (2011). *Chem. Soc. Rev.* 40: 4986. (c) Luo, Y., Pan, X., Yu, X., and Wu, J. (2014). *Chem. Soc. Rev.* 43: 834. (d) Wu, X.-F., Fang, X., Wu, L. et al. (2014). *Acc. Chem. Res.* 47: 1041.

51 (a) Skoda-Földes, R. and Kollár, L. (2002). *Curr. Org. Chem.* 6: 1097. (b) Tsuji, J. (1995). *Palladium Reagents and Catalysts: Innovations in Organic Synthesis.* Chichester: Wiley. (c) Colquhoun, H.M., Thompson, D.J., and Twigg, M.V. (1991). *Carbonylation, Direct Synthesis of Carbonyl Compounds.* New York: Plenum Press. (d) Trzeciak, A.M. and Ziolkowski, J.J. (2005). *Coord. Chem. Rev.* 249: 2308. (e) Morimoto, T. and Kakiuchi, K. (2004). *Angew. Chem. Int. Ed.* 43: 5580. (f) Brennführer, A., Neumann, H., and Beller, M. (2009). *Angew. Chem. Int. Ed.* 48: 4114. (g) Grigg, R. and Mutton, S.P. (2010). *Tetrahedron* 66: 5515.

52 (a) Heck, R.F. and Breslow, D.S. (1963). *J. Am. Chem. Soc.* 85: 2779. (b) Schoenberg, A. and Heck, R.F. (1974). *J. Org. Chem.* 39: 3327. (c) Beller, M., Cornils, B., Frohning, C.D., and Kohlpaintner, C.W. (1995). *J. Mol. Catal. A* 104: 17.

53 (a) Gabriele, B., Salerno, G., Costa, M., and Chiusoli, G.P. (2004). *Curr. Org. Chem.* 8: 919. (b) Gabriele, B., Salerno, G., and Costa, M. (2006). *Top. Organomet. Chem.* 18: 239. (c) Liu, Q., Zhang, H., and Lei, A. (2011). *Angew. Chem. Int. Ed.* 50: 10788. (d) Xiao, X.-F., Neumann, H., and Beller, M. (2013). *ChemSusChem* 6: 229. (e) Macho, V. and Harastiak, M. (1994). *J. Mol. Catal* 91: L155.

54 (a) Suzuki, A. and Synth, J. (2005). *Org. Chem. Jpn.* 63: 312. (b) Hall, D.G. (ed.) (2005). Boronic acids. In: *Preparation and Applications in Organic Synthesis and Medicine.* Weinheim: Wiley-VCH.

55 (a) Ohishi, T., Nishiura, M., and Hou, Z. (2008). *Angew. Chem. Int. Ed.* 47: 5792. (b) Ukai, K., Aoki, M., Takaya, J., and Iwasawa, N. (2006). *J. Am. Chem. Soc.* 128: 8706.

56 Liu, Q., Li, G., He, J. et al. (2010). *Angew. Chem. Int. Ed.* 49: 3371.

57 Wang, L., Wang, Y., Liu, C., and Lei, A. (2014). *Angew. Chem. Int. Ed.* 53: 5657.

58 Li, W., Duan, Z., Jiang, R., and Lei, A. (2015). *Org. Lett.* 17: 1397.

59 (a) Zhao, F., Chen, Z., Ma, X. et al. (2017). *Tetrahedron Lett.* 58: 614. (b) Rao, W.-H., Yin, Y.-S., and Shi, B.-G. (2015). *Org. Lett.* 17: 3758. (c) Gary, J.B., Cook, A.K., and Sanford, M.S. (2013). *ACS Catal.* 3: 700. (d) Dick, A.R., Hull, K.L., and Sanford, M.S. (2004). *J. Am. Chem. Soc.* 126: 2300. (e) Desai, L.V., Hull, K.L., and Sanford, M.S. (2004). *J. Am. Chem. Soc.* 126: 9542. (f) Giri, R., Liang, J., Lei, J.G. et al. (2005). *Angew. Chem. Int. Ed.* 44: 7420. (g) Vermeulen, N.A., Delcamp, J.H., and White, M.C. (2010). *J. Am. Chem. Soc.* 132: 11323. (h) Yin, G., Wu, Y., and Liu, G. (2010). *J. Am. Chem. Soc.* 132: 11978.

60 Jiang, H., Chen, H., Wang, A., and Liu, X. (2010). *Chem. Commun.* 46: 7259.

61 Stowers, K.J., Kubota, A., and Sanford, M.S. (2012). *Chem. Sci.* 3: 3192.

62 Zultanski, S.L. and Stahl, S.S. (2015). *J. Organomet. Chem.* 793: 263.

63 Campbell, A.N., White, D.B., Guzei, I.A., and Stahl, S.S. (2010). *J. Am. Chem. Soc.* 132: 15116.

64 Wang, A., Jiang, H., and Chen, H. (2009). *J. Am. Chem. Soc.* 131: 3846.

65 Wickens, Z.K., Guzmán, P.E., and Grubbs, R.H. (2015). *Angew. Chem. Int. Ed.* 54: 236.

66 Zhou, P., Jiang, H., Huang, L., and Li, X. (2011). *Chem. Commun.* 47: 1003.

67 (a) Bäckvall, J.E. (1983). *Acc. Chem. Res.* 16: 335–342. (b) Schultz, D.M. and Wolfe, J.P. (2012). *Synthesis* 44: 351. (c) Nigishi, E.J. (2002). *Handbook of Organopalladium Chemistry for Organic Synthesis*, vol. Vol. 2. Weinheim: Wiley-VCH.

68 (a) Alexanian, E.J., Lee, C., and Sorensen, E.J. (2005). *J. Am. Chem. Soc.* 127: 7690. (b) Yin, G., Mu, X., and Liu, G. (2016). *Acc. Chem. Res.* 49: 2413. (c) Muñiz, K. (2009). *Angew. Chem. Int. Ed.* 48: 9412. (d) McDonald, R.I., Liu, G., and Stahl, S.S. (2011). *Chem. Rev.* 111: 2981. (e) Martínez, C. and Muñiz, K. (2012). *Angew. Chem. Int. Ed.* 51: 7031. (f) Kalyani, D. and Sanford, M.S. (2008). *J. Am. Chem. Soc.* 130: 2150. (g) Rosewall, C.F., Sibbald, P.A., Liskin, D.V., and Michael, F.E. (2009). *J. Am. Chem. Soc.* 131: 9488. (h) Ouyang, L., Li, J., Zheng, J. et al. (2017). *Angew. Chem. Int. Ed.* 56: https://doi.org/10.1002/anie.201709285.

69 Zhu, H., Chen, P., and Liu, G. (2015). *Org. Lett.* 17: 1485.

70 Li, J., Grubbs, R.H., and Stoltz, B.M. (2016). *Org. Lett.* 18: 5449.

71 (a) Desai, L.V., Malik, H.A., and Sanford, M.S. (2006). *Org. Lett.* 8: 1141. (b) Reddy, B.V.R., Reddy, L.R., and Corey, E.J. (2006). *Org. Lett.* 8: 3391. (c) Wang, G.-W., Yuan, T.-T., and Wu, X.-L. (2008). *J. Org. Chem.* 73: 4717. (d) Vickers, C.J., Mei, T.-S., and Yu, J.-Q. (2010). *Org. Lett.* 12: 2511. (e) Ye, Z., Wang, W., Luo, F. et al. (2009). *Org. Lett.* 11: 3974. (f) Hua, C.-J., Zhang, X.-H., Ding, Q.-P. et al. (2012). *Tetrahedron Lett.* 53: 2465. (g) Dick, A.R., Kampf, J.W., and Sanford, M.S. (2005). *J. Am. Chem. Soc.* 127: 12790. (h) Racowski, J.M., Dick, A.R., and Sanford, M.S. (2009). *J. Am. Chem. Soc.* 131: 10974.

72 Liu, H., Shi, G., Pan, S. et al. (2013). *Org. Lett.* 15: 4098.

73 (a) Ji, X., Huang, H., Wu, W., and Jiang, H. (2013). *J. Am. Chem. Soc.* 135: 5286. (b) Ji, X., Huang, H., Xiong, W. et al. (2014). *J. Org. Chem.* 79: 7005. (c) Zhang, G., Cui, L., Wang, Y., and Zhang, L. (2010). *J. Am. Chem. Soc.* 132: 1474. (d) Houlden, C.E., Bailey, C.D., Ford, J.G. et al. (2008). *J. Am. Chem. Soc.* 130: 10066. (e) Li, J., Hu, W., Li, C. et al. (2017). *Org. Chem. Front.* 4: 373. (f) Li, J., Yang, W., Yang, S. et al. (2014). *Angew. Chem. Int. Ed.* 53: 7219. (g) Li, J., Yang, S., Wu, W., and Jiang, H. (2014). *Chem. Commun.* 50: 1381.

74 Huang, L., Jiang, H., Qi, C., and Liu, X. (2010). *J. Am. Chem. Soc.* 132: 17652.

75 Huang, L., Wang, Q., Liu, X., and Jiang, H. (2012). *Angew. Chem. Int. Ed.* 51: 5696.

76 (a) Li, H., Li, B.-J., and Shi, Z.-J. (2011). *Catal. Sci. Technol.* 1: 191. (b) Jensen, T. and Fristrup, P. (2009). *Chem. Eur. J.* 15: 9632. (c) Trost, B.M. (1980). *Acc. Chem. Res.* 13: 385. (d) Xu, J., Fu, Y., Luo, D.-F. et al. (2011). *J. Am. Chem. Soc.* 133: 15300. (e) Parsons, A.T. and Buchwald, S.L. (2011). *Angew. Chem., Int. Ed.* 50: 9120. (f) Yang, B., Zhang, H.-Y., and Yang, S.-D. (2015). *Org. Biomol. Chem.* 13: 3561.

77 Yang, H. Chen, W., Li, J., Li, C. et al. (2015). *Chem. Commun.* 51: 9575.

78 (a) Tobisu, M. and Chatani, N. (2008). *Chem. Soc. Rev.* 37: 300. (b) Souillart, L. and Cramer, N. (2015). *Chem. Rev.* 115: 9410. (c) Nakao, Y., Oda, S., and Hiyama, T. (2004). *J. Am. Chem. Soc.* 126: 13904. (d) McCormick, M.M., Duong, H.A., Zuo, G., and Louie, J. (2005). *J. Am. Chem. Soc.* 127: 5030. (e) Miller, J.A. and Dankwardt, J.W. (2003). *Tetrahedron Lett.* 44: 1907.

79 Kong, W., Li, B., Xu, X., and Song, Q. (2016). *J. Org. Chem.* 81: 8436.

80 (a) Gooßen, L.J., Rodríguez, N., and Gooßen, K. (2008). *Angew. Chem. Int. Ed.* 47: 3100. (b) He, C., Guo, S., Huang, L., and Lei, A. (2010). *J. Am. Chem. Soc.* 132: 8273. (c) Horino, Y. (2007). *Angew. Chem. Int. Ed.* 46: 2144. (d) Nájera, C. and Sansano, J.M. (2009). *Angew. Chem. Int. Ed.* 48: 2452. (e) Rubin, M., Rubina, M., and Gevorgyan, V. (2007). *Chem. Rev.* 107: 3117. (f) Chem, F., Wang, T., and Jiao, N. (2014). *Chem. Rev.* 114: 8613.

81 Ma, R., He, L.-N., Liu, A.-H., and Song, Q.-W. (2016). *Chem. Commun.* 52: 2145.

82 (a) Bunz, U.H.F. and Kloppenburg, L. (1999). *Angew. Chem. Int. Ed.* 38: 478. (b) Liu, Y., Song, F., and Guo, S. (2006). *J. Am. Chem. Soc.* 128: 11332. (c) Jennings, P.W. and

Johnson, L.L. (1994). *Chem. Rev.* 94: 2241. (d) Rybtchinski, B. and Milstein, D. (1999). *Angew. Chem. Int. Ed.* 38: 870. (e) Bunz, U.H.F. (2001). *Acc. Chem. Res.* 34: 998. (f) Fürstner, A., Mathes, C., and Lehmann, C.W. (2001). *Chem. Eur. J.* 7: 5299. (g) Cairns, G.A., Carr, N., Green, M., and Mahon, M.F. (1996). *Chem. Commun.* 2431. (h) Hayashi, N., Ho, D.M., and Pascal, R.A. (2000). *Tetrahedron Lett.* 41: 4261.

83 Wang, A. and Jiang, H. (2008). *J. Am. Chem. Soc.* 130: 5030.

84 (a) Dagani, M.J., Barda, H.J., Benya, T.J., and Sanders, D.C. (2002). *Ullmann's Encyclopedia of Industrial Chemistry: Bromine Compounds.* Weinheim: Wiley-VCH. (b) Fauvarque, J. (1996). *Pure Appl. Chem.* 68: 1713. (c) Stavber, S., Jereb, M., and Zupan, M. (2008). *Synthesis* 1487. (d) Kolvari1, E., Koukabi1, N., Khoramabadi-zad, A. et al. (2013). *Curr. Org. Synth.* 10: 837.

85 (a) Podgoršek, A., Zupan, M., and Iskra, J. (2009). *Angew. Chem. Int. Ed.* 48: 8424. (b) Barhate, N.B., Gajare, A.S., Wakharkar, R.D., and Bedekar, A.V. (1999). *Tetrahedron* 55: 11127.

86 (a) Bray, W.C. and Livingston, R.S. (1923). *J. Am. Chem. Soc.* 45: 1251. (b) Bray, W.C. and Livingston, R.S. (1928). *J. Am. Chem. Soc.* 50: 1654. (c) Leulier, A. (1924). *Bull. Chem. Soc. Fr.* 35: 1325. (d) Speyer, E. and Rosenfeld, H. (1925). *Ber. Dtsch. Chem. Ges. A* 58: 1110. (e) Maass, O. and Hiebert, P.G. (1924). *J. Am. Chem. Soc.* 46: 290. (f) Lengyel, I., Epstein, I.R., and Kustin, K. (1993). *Inorg. Chem.* 32: 5880.

87 Vyas, P.V., Bhatt, A.K., Ramachandraiah, G., and Bedekar, A.V. (2003). *Tetrahedron Lett.* 44: 4085.

88 (a) Mukhopadhyay, S. and Chandalia, S.B. (1999). *Org. Process Res. Dev.* 3: 10. (b) Mukhopadhyay, S., Chandnani, K.H., and Chandalia, S.B. (1999). *Org. Process Res. Dev.* 3: 196. (c) Terent'ev, A.O., Krylov, I.B., Ogibin, Y.N., and Nikishin, G.I. (2006). *Synthesis* 3819. (d) Terent'ev, A.O., Khodykin, S.V., Troitskii, N.A. et al. (2004). *Synthesis* 2845. (e) Suryakiran, N., Srinivasulu, M., and Venkateswarlu, Y. (2007). *J. Sulfur Chem.* 28: 345. (f) Suryakiran, N., Prabhakar, P., Srikanth Reddy, T. et al. (2007). *Tetrahedron Lett.* 48: 877.

89 Mukhopadhyay, S., Ananthakrishnan, S., and Chandalia, S.B. (1999). *Org. Process Res. Dev.* 3: 451.

90 (a) Krishna Mohan, K.V.V., Narender, N., Srinivasu, P. et al. (2004). *Synth. Commun.* 34: 2143. (b) Lazarev, I.M. and Nedolya, N.A. (2000). *Russ. J. Org. Chem.* 36: 1758. (c) Kirihara, M., Ogawa, S., Noguchi, T. et al. (2006). *Synlett* 2287. (e) Terent'ev, A.O., Khodykin, S.V., Krylov, I.B. et al. (2006). *Synthesis* 1087. (f) Tillu, V.H., Shinde, P.D., Bedekar, A.V., and Wakharkar, R.D. (2003). *Synth. Commun.* 33: 1399. (g) Mestres, R. and Palenzuela, J. (2002). *Green Chem.* 4: 314. (h) Khan, A.T., Parvin, T., Choudhury, L.H., and Ghosh, S. (2007). *Tetrahedron Lett.* 48: 2271. (i) Moriuchi, T., Yamaguchi, M., Kikushima, K., and Hirao, T. (2007). *Tetrahedron Lett.* 48: 2667. (j) Sinha, J., Layek, S., Mandal, G.C., and Bhattacharjee, M. (2001). *Chem. Commun.* 1916. (k) Sels, B.F., De Vos, D.E., and Jacobs, P.A. (2001). *J. Am. Chem. Soc.* 123: 8350.

91 Iskra, J., Stavber, S., and Zupan, M. (2004). *Synthesis* 1869.

92 (a) Prebila, R. and Stavber, S. (2014). *Adv. Synth. Catal.* 356: 1266. (b) Hosseini, A., Khalilzadeh, M.A., Keipour, H., and Tajbakhsh, M. (2012). *Synth. Commun.* 42: 2407. (c) Higgs, D.E., Nelen, M.I., and Detty, M.R. (2001). *Org. Lett.* 3: 349. (d) Sels, B., De Vos, D., Buntinx, M. et al. (1999). *Nature* 400: 855. (e) Choudary, B.M., Someshwar, T., Reddy, C.V. et al. (2003). *App. Catal. A. Gen.* 251: 397. (f) Drake, M.D., Bateman, M.A., and Detty, M.R. (2003). *Organometallics* 22: 4158. (g) Narender, N., Mohan, K.V.V.K.,

Reddy, R.V. et al. (2003). *J. Mol. Catal. A Chem.* 192: 73. (h) Bora, U., Bose, G., Chaudhuri, M.K. et al. (2000). *Org. Lett.* 2: 247. (i) Ben Daniel, R., de Visser, S.P., Shaik, S., and Neumann, R. (2003). *J. Am. Chem. Soc.* 125: 12116. (j) Vyas, P.V., Bhatt, A.K., Ramachandraiah, G., and Bedekar, A.V. (2003). *Tetrahedron Lett.* 44: 4085. (k) Barhate, N.B., Gajare, A.S., Wakharkar, R.D., and Bedekar, A.V. (1999). *Tetrahedron* 55: 11127.

93 (a) Liu, W. and Groves, J.T. (2013). *Angew. Chem. Int. Ed.* 52: 6024. (b) Gamache, R.F., Waldmann, C., and Murphy, J.M. (2016). *Org. Lett.* 18: 4522. (c) Truong, T., Klimovica, K., and Daugulis, O. (2013). *J. Am. Chem. Soc.* 135: 9342. (d) Fier, P.S. and Hartwig, J.F. (2012). *J. Am. Chem. Soc.* 134: 10795. (e) Jiang, X., Zhang, F., Yang, J. et al. (2017). *Adv. Synth. Catal.* 359: 853.

94 Cavani, F. and Teles, J.H. (2009). *ChemSusChem* 2: 508.

95 (a) Punniyamurthy, T., Velusamy, S., and Iqbal, J. (2005). *Chem. Rev.* 105: 2329. (b) Boisvert, L. and Goldberg, K.I. (2012). *Acc. Chem. Res.* 45: 899. (c) McCann, S.D. and Stahl, S.S. (2015). *Acc. Chem. Res.* 48: 1756. (d) Campbell, A.N. and Stahl, S.S. (2012). *Acc. Chem. Res.* 45: 851. (e) Scheuermann, M.L. and Goldberg, K.I. (2014). *Chem. Eur. J.* 20: 14556.

96 Chen, X., Hao, X.-S., Goodhue, C.E., and Yu, J.-Q. (2006). *J. Am. Chem. Soc.* 128: 6790.

97 Menini, L. and Gusevskaya, E.V. (2006). *Chem. Commun.* 209.

98 Yang, L., Lub, Z., and Stahl, S.S. (2009). *Chem. Commun.* 6460.

99 Ji, X., Huang, H., Wu, W., and Jiang, H. (2013). *J. Am. Chem. Soc.* 135: 5286.

100 Ji, X., Huang, H., Xiong, W. et al. (2014). *J. Org. Chem.* 79: 7005.

101 (a) Kaneda, K., Kawamoto, F., Fujiwara, Y. et al. (1974). *Tetrahedron Lett.* 15: 1067. (b) Kaneda, K., Uchiyama, T., Fujiwara, Y. et al. (1979). *J. Org. Chem.* 44: 55. (c) Thadani, A.N. and Rawal, V.H. (2002). *Org. Lett.* 4: 4317. (d) Huang, J., Zhou, L., and Jiang, H. (2006). *Angew. Chem. Int. Ed.* 45: 1945. (e) Jiang, H., Liu, X., and Zhou, L. (2008). *Chem. Eur. J.* 14: 11305. (f) Jiang, H., Qiao, C., and Liu, W. (2010). *Chem. Eur. J.* 16: 10968. (g) Hoye, T.R. and Wang, J. (2005). *J. Am. Chem. Soc.* 127: 6950. (h) Huang, X.-C., Wang, F., Liang, Y., and Li, J.-H. (2009). *Org. Lett.* 11: 1139. (i) Ye, S.Q., Gao, K., Zhou, H. et al. (2009). *Chem. Commun.* 5406.

102 (a) Ma, S. and Lu, X. (1990). *J. Chem. Soc. Chem. Commun.* 733. (b) Ma, S. and Lu, X. (1991). *J. Org. Chem.* 56: 5120. (c) Ma, S., Zhu, G., and Lu, X. (1993). *J. Org. Chem.* 58: 3692. (d) Zhu, G. and Lu, X. (1995). *Organometallics* 14: 4899. (e) Wang, Z. and Lu, X. (1996). *J. Org. Chem.* 61: 2254. (f) Ji, J., Wang, Z., and Lu, X. (1996). *Organometallics* 15: 2821. (g) Xie, X., Lu, X., Liu, Y., and Xu, W. (2001). *J. Org. Chem.* 66: 6545. (h) Shen, Z. and Lu, X. (2006). *Tetrahedron* 62: 10896. (i) Song, J., Shen, Q., Xua, F., and Lu, X. (2007). *Tetrahedron* 63: 5148. (j) Zhou, F., Han, X., and Lu, X. (2011). *J. Org. Chem.* 76: 1491.

103 Wen, Y., Huang, L., Jiang, H., and Chen, H. (2012). *J. Org. Chem.* 77: 2029.

104 Liu, B., Gao, H., Yu, Y. et al. (2013). *J. Org. Chem.* 78: 10319.

105 Zhang, Z., Wu, W., Liao, J. et al. (2015). *Chem. Eur. J.* 21: 6708.

4

Green Oxidative Synthesis of Epoxides

Miquel Costas

Facultat de Ciències, Departament de Química I Institut de Química Computacional i Catàlisi, Universitat de Girona, Girona, Spain

4.1 Introduction

Epoxides are strained three-membered heterocycles that have a high interest in organic synthesis because they exhibit a very versatile reactivity (Scheme 4.1). Most remarkably, they can be ring-opened by a number of nucleophiles with high control over regio- and stereochemistry [1]. This aspect makes epoxides convenient starting points for the preparation of a number of 1,2-difunctionalized products.

Epoxides are most commonly prepared by three different methods (Scheme 4.2).

The most common method is the direct epoxidation of olefins, by reaction with an oxidant, most commonly a peroxide or peracid (Scheme 4.2a), but also via a metal-catalyzed oxidation reaction with a number of oxidants [2–4]. Epoxides can be also prepared by ring closing of halohydrins on reaction with a base (Scheme 4.2b). Halohydrins are initially formed by reaction of an olefin and hypochlorous acid. Alternatively, the hypochlorous acid can be prepared *in situ* by reaction of Cl_2 and water. This method is used in the large-scale production of propene oxide and in the production of epichlorohydrin [5–7]. Finally, epoxides can be also synthetized by reaction of aldehydes or ketones with ylides (Scheme 4.2c) [8].

Epoxidation of olefins is a well-established and particularly useful reaction because olefins are largely available starting materials. However, the vast majority of current methods are far from satisfactory from a number of perspectives. In the first place, some epoxidizing agents exhibit poor atom economy in their reactions, and delivery of the oxygen atom occurs at the expense of the production of a large number of byproducts. In addition, because of the rich reactivity of epoxides, chemoselectivity remains a critical aspect in epoxidations with aggressive oxidants and acidic conditions. Paradigmatically, the epoxidation performed with peracids, known as the Prilezhaev reaction, is widely employed in organic synthesis [9]. For example, epoxidations with electron poor peracids, such as *m*-chloro perbenzoic acid (MCPBA), are reliable reactions, with a wide substrate scope. However, they produce the corresponding carboxylic acid as a by-product. Peracetic acid is also employed in large-scale epoxidations [2].

Green Oxidation in Organic Synthesis, First Edition. Edited by Ning Jiao and Shannon S. Stahl.

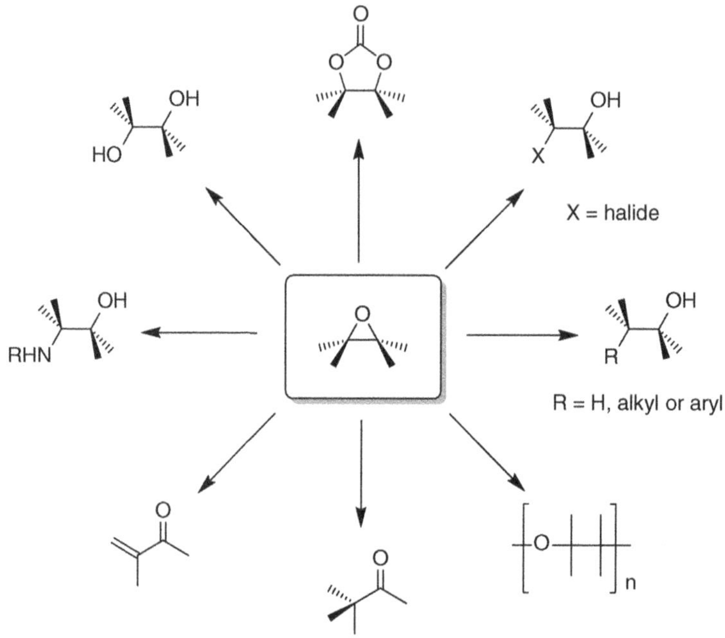

Scheme 4.1 Representative examples of the reactivity of epoxides.

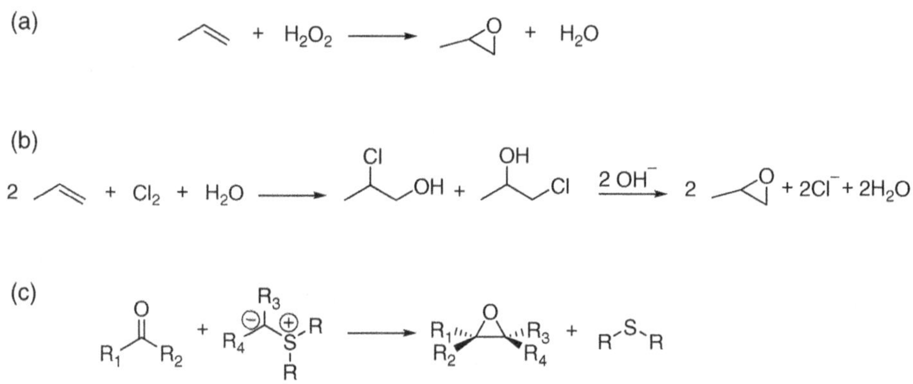

Scheme 4.2 General methods for preparation of epoxides.

Epoxidation methodologies based on O_2 and H_2O_2 as oxidants are particularly appealing because of the low cost, lack of toxicity, and high atom economy of these oxidants [10–14]. The only by-product originating from these is water. Thus, these oxidants have the potential to be developed into large-scale green processes. However, they need to be activated with the use of catalysts, and the rich redox chemistry of dioxygen and its reduced forms makes catalyst design a non-trivial issue. On the other hand, because of sustainability considerations, catalysts based on earth abundant metals are especially suitable. Of these, the group of biologically relevant metals (Mn, Fe, Co, Ni, Cu, and Zn) are probably the most interesting and promising because of availability and limited toxicity considerations.

4.1.1 Scope of the Chapter

Based in these considerations, the current chapter focuses on epoxidation methodologies that employ O_2 and H_2O_2 as oxidants and base metals as catalysts. These methods have also concentrated most of the research effort devoted to this reaction in the last decade. Few examples describing the use of peracids will be also discussed since these systems historically constitute a starting point from where methods based on H_2O_2 have eventually evolved. The chapter also mainly focuses on the description of systems that operate under homogeneous conditions, while few especially remarkable heterogeneous systems are only briefly mentioned. Methodologies based in organocatalysts are very powerful for asymmetric epoxidation and can constitute by themselves a complete chapter, but they usually require relatively high catalyst loadings and long reaction times. The reader is directed toward excellent recent reviews on this specific topic [15, 16].

Assessment of an epoxidation method requires the study of its performance against different olefins. It is well-known that the reactivity of the olefins and the stability of the resulting epoxides are highly dependent on the structure of the epoxide and olefin. For example, terminal olefins are notoriously more challenging than cyclic olefins and aromatics. On the other hand, aromatic epoxides are highly sensitive to acidic conditions while linear aliphatic epoxides are quite stable. Because of that, whenever the data is available, we have tried to include different olefins in the discussion of the different methods, with the aim to provide a balanced view of their strengths and limitations.

4.2 Metal-Catalyzed Epoxidation with O_2

The use of O_2 as an oxidant in oxidation reactions with synthetic value is a standing problem [10]. O_2 can be seen as a four-electron oxidant, forming two molecules of water. This poses serious difficulties in accommodation with most of the oxidation reactions of organic functionalities, which are two-electron oxidation processes. Olefin epoxidation can be seen as a paradigmatic case. The combination of O_2 and redox active metals can readily initiate radical chain reactions, forming peroxyl radicals eventually leading to epoxides. However, these reactions often lead to mixtures of products besides the epoxide, and rarely provide the epoxide in good product yields and selectivity. A few interesting examples that overcome these difficulties are described in the following text.

4.2.1 Metal-Catalyzed Epoxidations with O_2

4.2.2 Epoxidations with O_2 in the Absence of a Cosubstrate

Heterogeneous silver catalyzed epoxidation is the most notable example of catalytic epoxidation employing O_2 as the single oxidant. The reaction is technologically very important because it is used in the large-scale epoxidation of ethylene [17, 18]. The reaction is presumed to take place over Ag surfaces via initial lysis of the O—O bond. Extension toward other substrates remains a challenge because allylic C—H bonds are also susceptible to oxidation. Furthermore, selective monooxygenation is very difficult to achieve with other heterogeneous metal catalysts, which tend to favor full oxidation to form CO_2. Coinage metals are a notable exception. Copper and silver surfaces are explored in the quest for more active and selective catalysts. Notably, Cu(111) exhibits a very high selectivity toward epoxidation, but it is oxidized to copper oxides that show reduced activity and selectivity [19].

Nanocrystalline gold catalysts supported on carbon (Au/C) oxidize alkenes with air under relatively mild conditions (80 °C) providing the epoxide as the major product, with moderate to good (50–80%) selectivities. A peroxide initiator (H_2O_2 or tertbutyl hydroperoxide, TBHP) is required in these reactions, and epoxidation proceeds with stereoscrambling when *cis*-alkenes are employed as substrates. For example, *cis*-stilbene is epoxidized to the corresponding *trans*-stilbene oxide [20]. Interestingly, the use of very small gold entities (~1.4 nm) derived from 55-atom gold clusters and supported on inert materials are efficient and robust catalysts for the selective oxidation of styrene by dioxygen. In these reactions, the epoxide is obtained with modest selectivity (up to 27%), but they are remarkable because no initiator is necessary. It is proposed that O_2 activation at the gold clusters takes place by splitting the O—O bond [21].

4.2.3 Epoxidations with O_2 in the Presence of a Cosubstrate

Use of a cosubstrate that acts as a $2e^-$ reducing agent can lead to a high yield and highly selective epoxidation reactions, with interest for synthetic organic chemistry. However, these reactions have the drawback that the oxidized cosubstrate, produced in stoichiometric amounts, is the necessary by-product of the reaction.

Several metal complexes catalyze the aerobic epoxidation of olefins in the presence of aldehydes. This process is known as the Mukaiyama reaction [22]. Nickel complexes with 1,3-diketone ligands are particularly efficient, and both substrate conversions and selectivities toward the epoxide are sensitive to the nature of the aldehyde, which is converted to the corresponding carboxylic acid. Aldehydes containing a secondary or a tertiary carbon next to the carbonyl moiety provide the best results. The substrate scope of the reaction covers both aliphatic and aromatic olefins, and is illustrated in Scheme 4.3 [23]. The method is mild, reactions are performed at room temperature under an atmosphere of O_2, and epoxides are obtained with high selectivity. An important limitation of the method is that epoxidation is not stereospecific; *cis*- and *trans*-olefins are oxidized to a mixture of *cis*- and *trans*-epoxide isomers.

The mechanism of Mukaiyama reactions may be complex and dependent on specific reaction conditions and catalysts, but the dominant epoxidizing agents is most likely an acylperoxy radical formed by the autoxidation of the aldehyde. In addition, peroxide carboxylic acids may also be formed in the radical chain, and can also contribute to the epoxidation reaction. In this scenario, the role of the catalyst appears to be the initiation of the radical chain, but it has also been proposed that it can stabilize the acylperoxy radical [24, 25].

Mukaiyama reactions are robust, operationally simple, and a number of metal catalysts appear to be competent. Different strategies have been pursued for finding catalysts that are highly active or that present interesting properties from the viewpoint of sustainability. A comprehensive review is beyond the scope of this chapter but representative examples are detailed in the following.

The combination of imidazolium salts containing carboxylic acid tails (Imace-R, Scheme 4.4a) [26], and copper salts results in the formation of copper complexes that

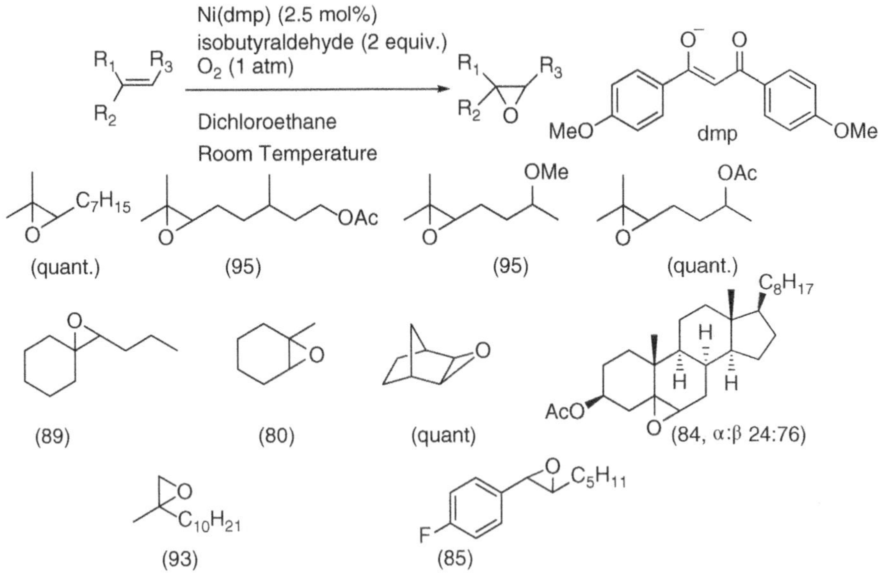

Scheme 4.3 Representative example of a Mukaiyama epoxidation reaction [23].

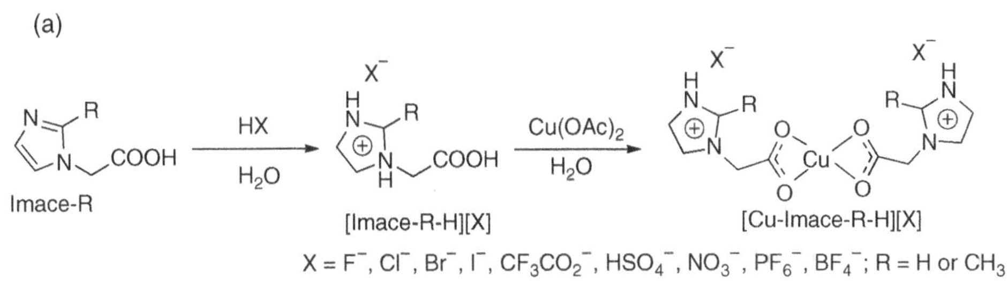

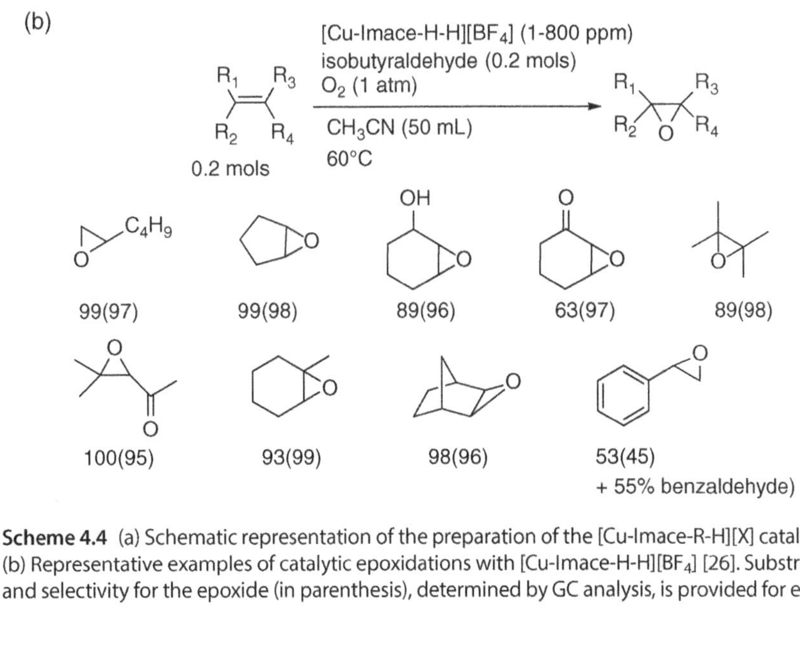

Scheme 4.4 (a) Schematic representation of the preparation of the [Cu-Imace-R-H][X] catalysts. (b) Representative examples of catalytic epoxidations with [Cu-Imace-H-H][BF_4] [26]. Substrate conversion and selectivity for the epoxide (in parenthesis), determined by GC analysis, is provided for each substrate.

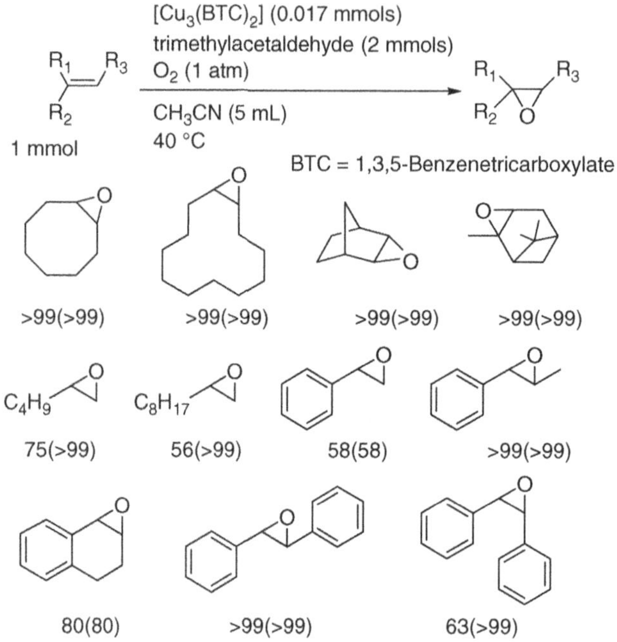

Scheme 4.5 Representative example of the use of a MOF in Mukaiyama type epoxidations [28]. Product yields and selectivities in parenthesis determined by GCMS analysis.

show excellent activity in the catalytic epoxidation of non-aromatic olefins. Catalytic activity appears to be influenced by the anion, non-coordinating BF_4^-, the most active of the series of catalysts (Scheme 4.4a). These copper complexes are extraordinarily active and can operate at ppm levels, with low temperatures and pressures. Because of their ionic character, they are quite insoluble in the organic media, and can be easily filtered and recycled at the end of the reactions. Recycled catalysts (up to six times) retain the original activity in terms of both efficiency and selectivity. In terms of substrate scope, this system works especially well for aliphatic olefins, irrespective of their substitution pattern (Scheme 4.4b). However, aromatic olefins are oxidized in moderate yields and selectivities, forming substantial amounts of aldehyde products resulting from olefin C—C cleavage.

An interesting alternative approach is the use of heterogeneous catalysts prepared by immobilizing homogeneous catalysts in heterogeneous supports [27]. Along this line, a very promising strategy is the use of metal organic frameworks (MOFs) as catalysts [28–31]. These structures tend to be highly robust and are characterized by the molecular well-defined, homogenous distribution of the metal centers, and by exhibiting high surface area and porosity. A representative example of the use of MOFs as epoxidation catalysts is shown in Scheme 4.5 [28].

Related to Mukaiyama reactions is a biologically inspired approach that employs a β-ketoester as a cosubstrate. The combination of $FeCl_3$ and imidazole ligands, assisted by a β-ketoester, activates O_2 from air, catalyzing the selective epoxidation of olefins at room temperature in synthetically useful yields (Scheme 4.6) [32]. Activation of O_2 is presumed to occur via a synergistic reaction with the iron center and the ketoester cosubstrate, which is converted to the corresponding α-hydroxylated ketoester, forming the iron-based epoxidizing species. Competitive four-electron-oxidation of the ketoester is also observed as a side-reaction.

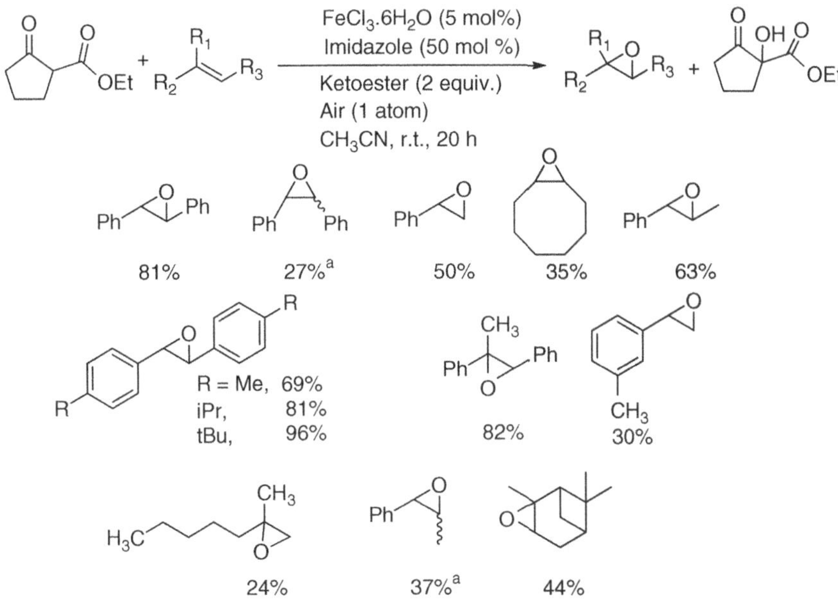

Scheme 4.6 Iron-catalyzed epoxidation with air assisted by a β-ketoester [32]. Epoxide yields determined by GC analysis. [a]The corresponding *cis*-stilbene or *cis*-methyl styrene oxide could not be observed. Instead *trans*-stilbene oxide and *trans*-methyl oxide formation was detected.

The system shows a limited substrate scope. *Trans*-stilbene derivatives are epoxidized in high yields and excellent chemoselectivity. Aliphatic olefins were also catalytically epoxidized but in general lower yields and selectivities. Importantly, *cis*-aromatic olefins such as *cis*-stilbene and *cis*-β-methylstyrene were epoxidized to the corresponding *trans*-epoxide. This indicates that the oxygen transfer does not proceed in a concerted manner and a long-lived radical or carbocationic intermediate is created before the second C—O epoxide bond is formed. Oxidation of α-pinene showed minimum amounts of allylic oxidation products, and instead the main product was α-pinene oxide, strongly arguing against a mechanism occurring via free diffusing radicals.

Enantioselective oxidation with the Mukaiyama type of system has been explored with limited success, presumably because the radical nature of the intermediates easily erodes if it is not incompatible with stereoselectivity. Enantioselective epoxidation of activated aromatic olefins can be done with manganese-salen [33] and structurally related complexes [34]. Pivalic aldehyde is the optimum cosubstrate and N-alkyl imidazoles are necessary co-ligands for reactions to proceed with moderate yields and moderate to good enantiomeric excesses (*ees*). The substrate scope of these reactions is very narrow, and epoxidations occur with stereoscrambling. An interesting approach is the use of fluorous biphasic systems; the concentration of O_2 in the fluorinated solvent is very high and the use of fluorinated Mn-salen catalysts soluble in these systems results in the epoxidation of indene in high yields and *ees* (Scheme 4.7). Unfortunately, the scope of the system appears to be very narrow [35].

Enantioselective epoxidation of styrenes with a camphor-derived mononuclear ferric catalyst, under Mukaiyama's conditions has been described (Scheme 4.8) [36]. Moderate

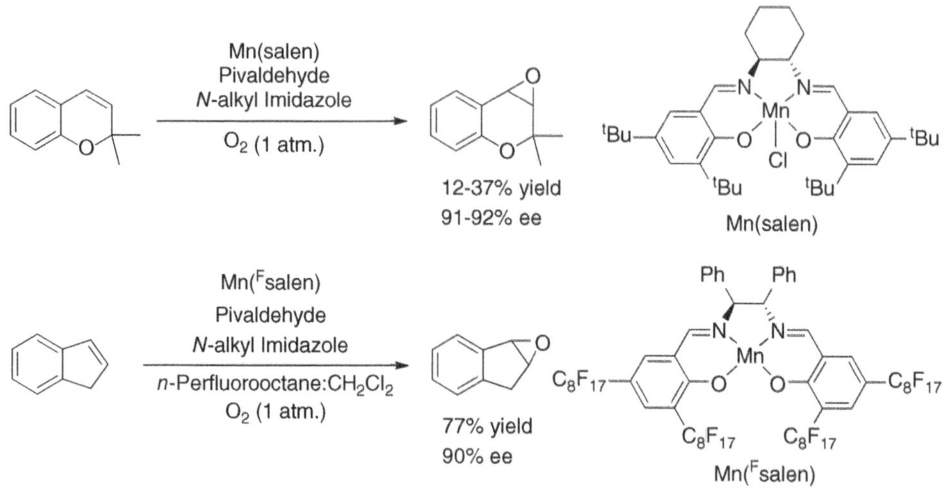

Scheme 4.7 Representative example of a Mukaiyama epoxidation reaction with a Mn-salen complex [33], and a perfluorinated Mn-salen complex operating in a fluorous biphasic system [35].

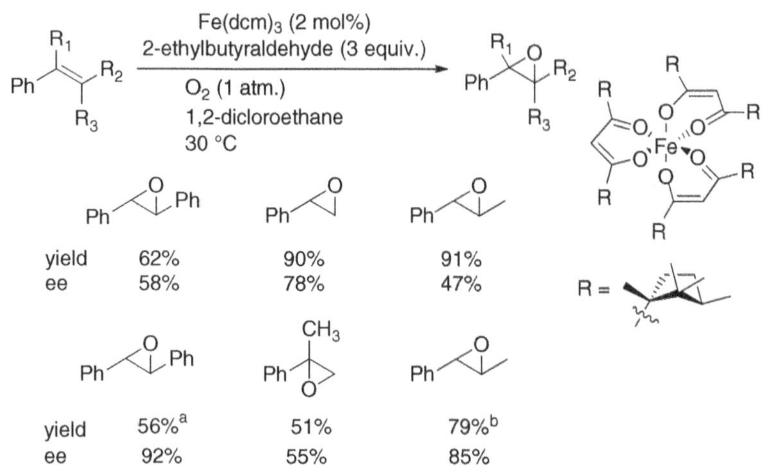

Scheme 4.8 Aerobic iron-catalyzed enantioselective epoxidation of styrene substrates under Mukaiyama conditions [36]. [a]From *cis*-stilbene. [b]From *cis*-β-methyl-styrene.

to good product yields and enantioselectivities were obtained for this class of substrates for which high enantioselectivity is difficult to accomplish.

4.3 Metal-Catalyzed Epoxidation with H_2O_2

Hydrogen peroxide is a convenient oxidant from an economical and environmental point of view. It is relatively cheap, and its only by-product is water. Safe manipulation of commercially available solutions is usually operationally simpler than reactions with gases such as O_2. Finally, a number of metals can activate this oxidant, leading to powerful oxidation species that can engage in olefin epoxidation [13, 37]. Despite all these a priori positive aspects, use of H_2O_2 in an efficient manner, producing selective oxidation reactions, is not straightforward because this peroxide has a rich chemistry when combined when redox active metal species. The following lines will focus on catalytic methodologies that provide reactions of interest from a synthetic perspective.

4.3.1 Heterogeneous Systems and Polyoxometallate Catalyzed Epoxidations

Heterogeneous systems based in zeolites and hydrotalcites constitute robust catalysts, suitable for large-scale epoxidation with H_2O_2. A titanium(IV)-silicalite zeolite catalyst (TS-1) is effective because the silicalite is hydrophobic [38, 39], in contrast to more traditional Ti(IV)-SiO_2 catalysts that only operate with organic peroxides and become rapidly inhibited under aqueous conditions. This catalyst has a limited substrate scope. It operates only with small olefins, and linear olefins that can fit the porous of the matrix. Zeolites having larger cavities have been designed that catalyze epoxidation of larger olefins [40, 41]. A major limitation of zeolites is that their acidic character commonly limits their utility to simple aliphatic epoxides that are not acid sensitive.

Hydrotalcites are double-layered hydroxides of aluminum and magnesium that activate hydrogen peroxide and epoxidize a larger variety of substrates [42]. $Mg_{10}Al_2(OH)_{24}CO_3$ is a prototypical example. Their basic character promotes nucleophilic epoxidations of electron-deficient olefins [43], but in combination with peroxide activating agents, such as amides or nitriles, they can also participate in electrophilic epoxidations [44, 45].

Chemically related to these solid metal-oxides (hydroxides) are polyoxometallates, which are discrete polymetallic structures with anionic character. They are highly robust and resistant to oxidation and hydrolytic conditions [46]. Depending on their cationic counterion, they can be made soluble in different solvents, converting them into homogeneous catalysts. Most interestingly, their acidic and redox properties but also their topology can be controlled at molecular level. The combination of these elements can be used in catalyst design, translating into highly effective and selective systems. They have been extensively studied as oxidation catalysts. An extensive overview falls beyond the scope of this chapter but representative examples are described in the following lines.

Polyoxometallates employed as epoxidation catalysts can be divided into two groups on the basis of their structure and chemical composition [47]; (i) catalysts based in peroxotungstate and peroxomolybdate species, and (ii) transition metal-substituted polyoxometallates.

4.3.1.1 Catalysts Based in Peroxotungstate and Peroxomolybdate Species

One of the pioneering and most extensively explored examples is Venturello's catalyst. Composed of a mixture of phosphate and tungstate in the presence of a tetraalkylammonium salt as a phase-transfer agent, this system catalyzes epoxidations with H_2O_2 in a two phase dichloroethane/water mixture [48, 49].

Venturello's catalyst has been further elaborated into a more attractive system that operates under halogenated solvent free conditions, toluene, water, or simply water media (Scheme 4.9). It basically consists of a combination of Na_2WO_4, (aminomethyl)

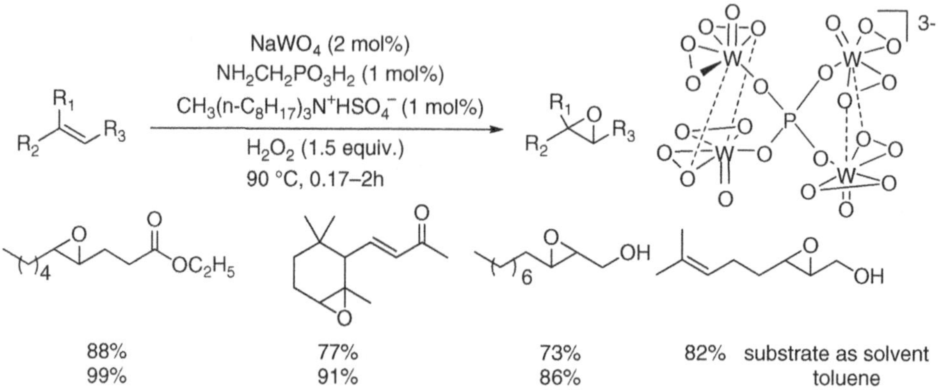

Scheme 4.9 Catalytic epoxidations with Venturello's catalyst improved by Noyori, and molecular structure of the peroxotungstate anion [12].

phosphonic acid, and $CH_3(n\text{-}C_8H_{17})_3N^+HSO_4^-$, and catalyzes epoxidation of simple olefins with a high yield [12, 50]. This is a biphasic system and toluene may be used as a convenient cosolvent. By using 1.5 equiv. of 30% H_2O_2 and 1–2 mol% of the catalyst, a wide range of olefins are epoxidized with excellent yields and selectivities, and the reaction can be easily scaled up. The temperature of the reaction depends on the reactivity of the substrate and goes from 5 °C for the most reactive cyclic substrates to 90 °C for the more inert monosubstituted terminal aliphatic olefins. Interestingly, the epoxidation of 1,2-disubstituted olefins proceeds with retention of the configuration. The active species in these reactions has been identified as the peroxotungstate anion $[PO_4(WO(O_2)_2)_4]^{3-}$ although some other peroxotungstate species may also be contributing.

An interesting reaction controlled phase-transfer behavior has been described for $[\pi\text{-}C_5H_5NC_{16}H_{33}]_3[PO_4(WO_3)_4]$ [51]. This salt is insoluble but it converts into a soluble peroxo complex, $[\pi\text{-}C_5H_5NC_{16}H_{33}]_3\{PO_4[WO_2(O_2)]_4\}$ in the presence of H_2O_2 in organic solvents, which engages in selective epoxidation reactions. When the peroxide is consumed, the salt returns to the insoluble form, permitting its easy separation and recycling. The system could be applied to the epoxidation of light alkenes such as propene, linear terminal alkenes, internal and cyclic aliphatic alkenes, styrene, and allyl chloride. The system can be also coupled with the 2-ethylanthraquinone/2-ethylanthrahydroquinone redox process that generates H_2O_2 *in situ* from O_2 and its use is exemplified in the waste-free epoxidation of propene.

The dinuclear peroxotungstate $K_2\{[WO(O_2)_2(H_2O)]_2(\mu\text{-}O)\}\cdot 2H_2O$ catalyzes the selective epoxidation of allylic alcohols in water at 305 K with an excellent efficiency in the consumption of H_2O_2 (Scheme 4.10) [52]. The reactions are stereospecific and chemoselective; the hydroxyl group remains unaltered and the epoxide is not opened despite the reactions being conducted in water. Notably, the catalyst can be easily recycled by extraction and retains its activity.

The XO_4^{n-} groups tune the reactivity of peroxotungstate units by increasing the Lewis acidity of the active tungsten sites. The catalytic activity of $\{XO_4[WO(O_2)_2]_2\}^{2-}$ decreases in the sequence Se > S > As > P > Si, in line with that of the pK_a values of H_nXO_4. Consequently, the W atoms in $\{SeO_4[WO(O_2)_2]_2\}^{2-}$ exhibit the highest electrophilicity, resulting in the highest activity for alkene epoxidation. With $TBA_2\{SeO_4[WO(O_2)_2]_2\}$,

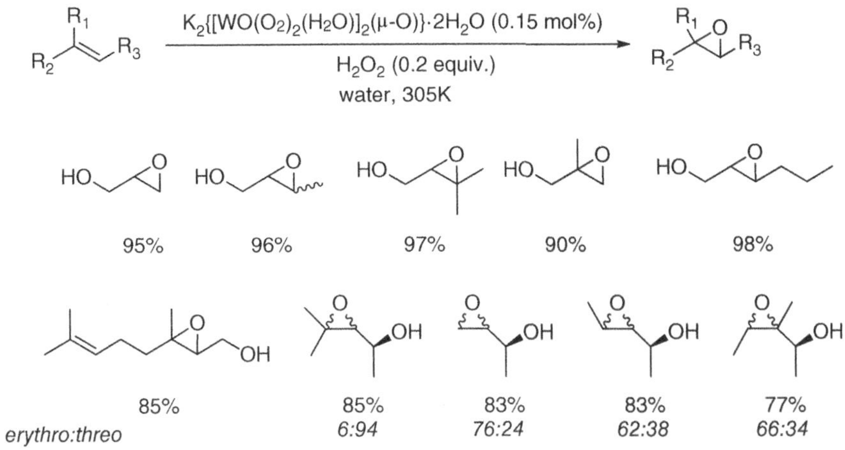

Scheme 4.10 Catalytic epoxidations of allylic alcohols with the dinuclear peroxotungstate catalyst $K_2\{[WO(O_2)_2(H_2O)]_2(\mu\text{-}O)\}\cdot 2H_2O$ [52].

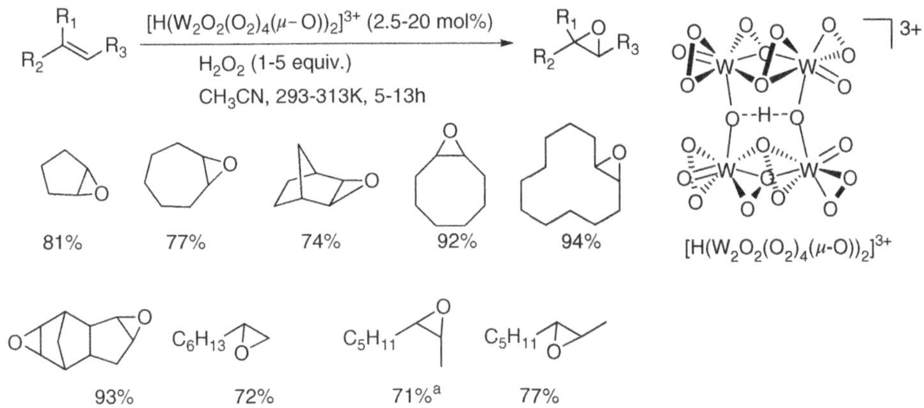

Scheme 4.11 Catalytic epoxidations with the tetranuclear peroxotungstate complex $TPA_3[H(W_2O_2(O_2)_4(\mu\text{-}O))_2]$ [54].

diverse homoallylic and allylic alcohols are epoxidized to the corresponding epoxides with very large TOF (up to 150 per hour) [53].

The tetranuclear peroxotungstate complex $TPA_3[H(W_2O_2(O_2)_4(\mu\text{-}O))_2]$, the TPA = *tetra-n*-propyl ammonium ion, prepared by protonation of the dinuclear peroxotungstate complex $[(WO(O_2)_2)_2(\mu\text{-}O)]^{2-}$, is an extraordinarily active system [54]. Using equimolar or slightly excess of hydrogen peroxide, the catalyst provides excellent yields and selectivities for epoxides in the oxidation of internal and cyclic alkenes (Scheme 4.11). Epoxidation of *cis*- and *trans*-alkenes proceeds with retention of the configuration. Furthermore, its activity in terms of reaction rates exceeds that attained with Venturello's catalyst and related peroxotungstate systems containing XO_4^{n-} (X = Se, S, As, P, and Si) ligands. Kinetic studies suggest that the tetranuclear complex is indeed the oxygen atom transfer agent, which constitutes the rate-determining step of the reaction.

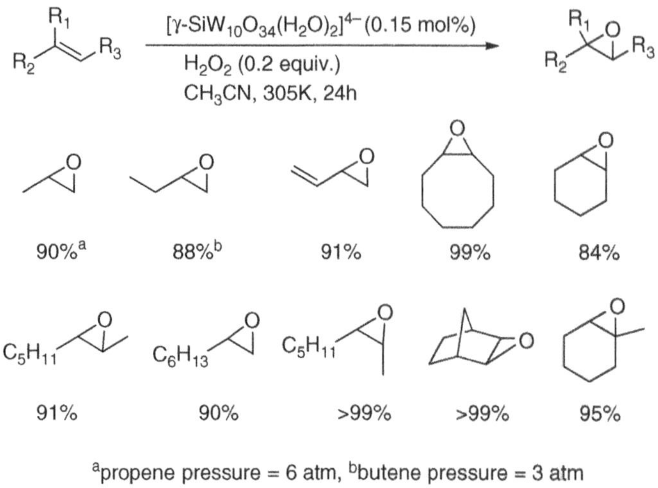

Scheme 4.12 Catalytic epoxidations with silicotungstate compound $[\gamma\text{-}SiW_{10}O_{34}(H_2O)_2]^{4-}$ [55].

The divacant Kegging type dianion $[\gamma\text{-}SiW_{10}O_{36}]^{8-}$ has become an extraordinary structural motif for the preparation of highly active oxidation catalysts. Extraordinarily efficient and selective epoxidation of olefins with respect to H_2O_2 is accomplished with the silicotungstate compound $[\gamma\text{-}SiW_{10}O_{34}(H_2O)_2]^{4-}$ (Scheme 4.12), which was prepared by protonation of the divacant, lacunary, Keggin-type polyoxometallate $[\gamma\text{-}SiW_{10}O_{36}]^{8-}$ [55]. Under conditions of excess of substrate (0.2 equiv. H_2O_2 per equiv. substrate) this complex epoxidizes a wide array of aliphatic olefins, including light alkenes such as propene and butene at 305 K, providing the epoxides in excellent yields (84 to >99% with respect to H_2O_2 employed) and outstanding product selectivities (>99%). Efficiency in the use of H_2O_2 is also extraordinarily high (>99%).

4.3.1.2 Catalysts Based in Transition Metal-Substituted Polyoxometallates

To improve the catalytic performances of polyoxometallate catalysts, secondary transition metals can be introduced into the skeleton of vacant polyoxometallates. Ti substituted POMs have been studied in a lot of detail because they can be considered small, molecular analogs of TS-1 [56, 57].

Disubstituted Keggin-type POMs have also been extensively studied as oxidation catalysts. Diiron substituted $\{\gamma\text{-}SiW_{10}[Fe(OH_2)]_2O_{38}\}^{6-}$ ($\gamma\text{-}Fe_2SiW_{10}$) is an excellent epoxidation catalyst that outperforms the activity of dimanganese- and dicopper-substituted, and also mono- and triiron-substituted structural analogs [58].

A particularly effective catalyst is $[\gamma\text{-}1,2\text{-}H_2SiV_2W_{10}O_{40}]^{4-}$, which contains a V(O)-(μ-OH)$_2$-V(O) core, and catalyzes the epoxidation of a wide arrange of substrates with one equivalent of hydrogen peroxide (Scheme 4.13) [59]. Aliphatic substrates including light alkenes are epoxidized with excellent yields and selectivities. Styrene is also selectively epoxidized without significant formation of side products. One of the most remarkable aspects of this system is the unique selectivity exhibited by the catalyst with respect to the degree and type of substitution of the olefin. Epoxidation of *cis*-substituted alkenes is strongly favored over the *trans*-isomers. For example, *cis*- and *trans*-2-octene are epoxidized in a stereoretentive manner to the corresponding epoxides in 90 and 6% yields, respectively. Epoxidation of three-substituted cyclohexenes occurs in a highly

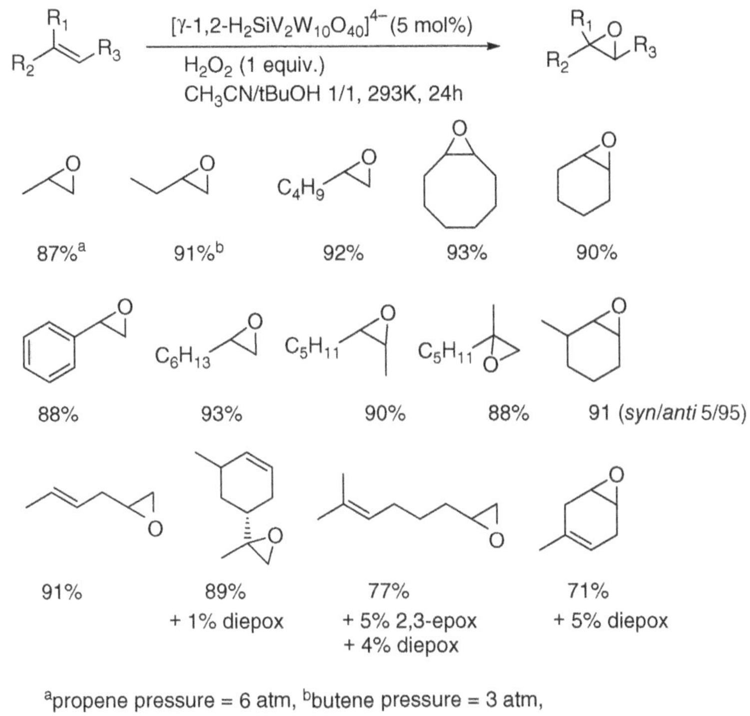

Scheme 4.13 Catalytic epoxidations with divanadium polyoxometallate $[\gamma\text{-}1,2\text{-}H_2SiV_2W_{10}O_{40}]^{4-}$ [59].

diastereoselective manner, providing the epoxide resulting from oxidation at the more accessible olefinic site, with the oxirane ring *trans* to the substituents. The less nucleophilic but more accessible olefinic site in non-conjugated dienes such as *trans*-1,4-hexadiene, (R)-(+)-limonene, 7-methyl-1,6-cyclohexadiene is epoxidized in high yields and a highly regioselective manner. All of these observations indicate that epoxidation selectivity is strongly influenced by the steric properties of the olefin. Most likely, the active species are embedded in the polyoxometallate framework and sterically demanding olefins cannot approach them.

Phosphorous monosubstituted $[\gamma\text{-}V_2(\mu\text{-}OH)_2\text{-}PW_{10}O_{38}]^{3-}$ ($\gamma\text{-}V_2PW_{10}$) is even more active and permits efficient epoxidation of electron poor olefins [60].

The sandwich types of POM with the general formula $[ZnWM_2(ZnW_9O_{34})_2]^{x-}$ ($M = Mn^{II}, Ru^{III}, Fe^{III}, Pd^{II}, Pt^{II}, Zn^{II}$; $x = 10–12$) are also very active epoxidation catalysts [61]. $Na_{12}[WZn_3(ZnW_9O_{34})_2]$ ($Na_{12}[WZn_3\text{-}(ZnW_9)_2]$ is a particularly interesting system because it is easily and readily assembled in water by combining simple bench stable reagents, and the complex can be used after *in situ* formation, without isolation and purification [62, 63]. Furthermore, the water phase containing the catalyst can be reused by extraction with an organic phase and reused.

4.3.2 Manganese-Catalyzed Epoxidations

Manganese holds a prominent role in metal-catalyzed epoxidation reactions. Simple manganese salts can exhibit good epoxidation activity, and the use of elaborated chiral ligands can lead to highly stereoselective processes [64, 65].

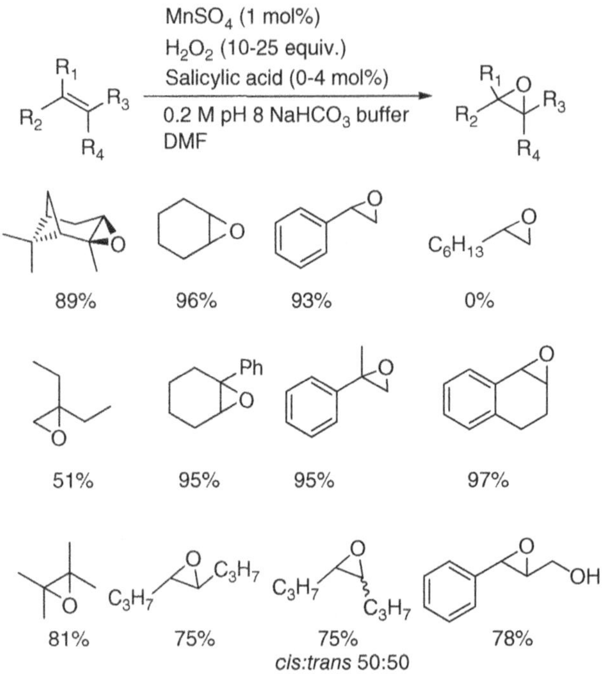

Scheme 4.14 Catalytic epoxidations with manganese salts in bicarbonate buffered solutions [68].

One of the simplest epoxidizing methods is the use of manganese(II) salts in combination with hydrogen peroxide and catalytic amounts of bicarbonate buffer in either DMF or tBuOH. Under these conditions, peroxymonocarbonate (HCO_4^-) is formed [66]. This anion is a mild electrophilic epoxidizing agent [67], but several metal ions react with it to produce a more powerful oxidant. Manganese salts ($MnSO_4$ and $Mn(OAc)_3$) are particularly efficient in this reaction. When reactions are performed in DMF, addition of different carboxylic acids exerts a beneficial effect, accelerating the reaction and limiting the non-productive use of H_2O_2 [68, 69]. Salicylic acid was identified as the optimal choice. In tBuOH, the best results were obtained with the addition of sodium acetate. The method shows a relatively wide substrate scope (Scheme 4.14) that includes aryl substituted alkenes, cyclic aliphatic alkenes, and tris- and tetraalkylsubstituted olefins. Dialkyl substituted olefins show more modest results and alkyl monosubstituted terminal olefins are unreactive. Sensitive functional groups such as primary allylic alcohols and aldehydes are well tolerated. Furthermore, the mild conditions of the method permit epoxidation of acid sensitive substrates without subsequent ring opening. On the other hand, the major limitations of the method are the relatively large excess of hydrogen peroxide required for optimum activity (generally 5–25 equiv.) and the lack of stereospecificify of the reactions.

Mono and dinuclear manganese complexes with the tridentate ligand 1,4,7-trimethyl-1, 4,7-triazacyclononane (tmtacn) are very powerful epoxidation catalysts using hydrogen peroxide as an oxidant [70–74]. Initial reports made use of a large excess of hydrogen peroxide, because peroxide disproportionation competed efficiently with oxidation of the olefin. Use of acetone as a solvent, and performing the reaction at 0–5 °C substantially decreases the catalase reaction, and permits the epoxidation of aliphatic alkenes

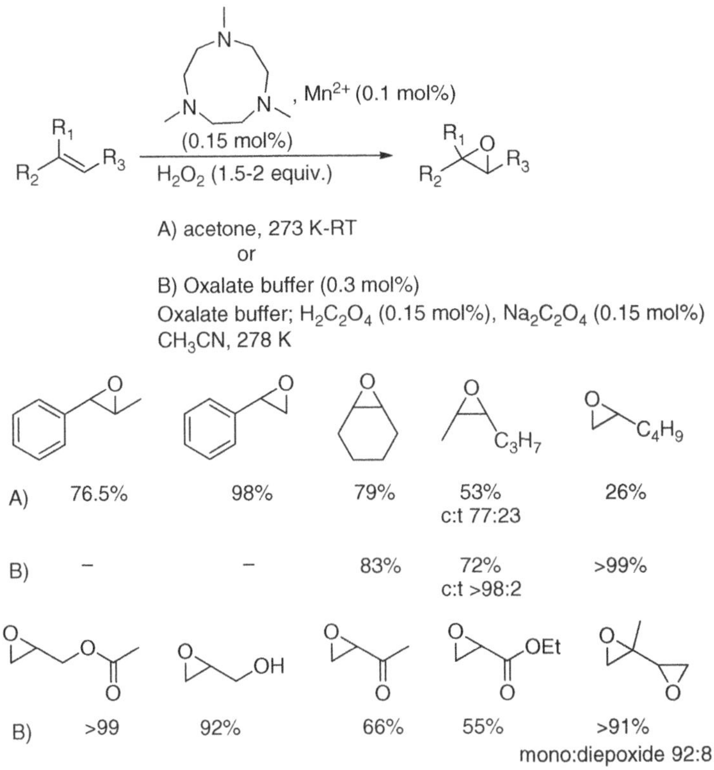

Scheme 4.15 (a) tmtacn-Mn catalyzed epoxidations in acetone [75]. (b) In the presence of an oxalate buffer [76].

and styrenes with excellent yields and chemoselectivities (Scheme 4.15). However, use of acetone is discouraged for safety reasons since the combination of this solvent with hydrogen peroxide produces explosive organic peroxides. In addition, under these conditions, the reaction is not stereospecific. For example, epoxidation of *cis*-3-heptene yields a 77:23 mixture of the *cis*- and *trans*-epoxides, while the corresponding *trans*-alkene isomer yields a 9:91 *cis*:*trans* isomer ratio [75].

Alternatively, the use of different additives limits peroxide disproportionation, leading to highly efficient oxidation reactions [77]. For example, use of an oxalate buffer leads to a very efficient system that employs a slight excess of peroxide to epoxidize *cis*- and *trans*-aliphatic alkenes stereospecifically (Scheme 4.15b). Also, terminal monosubstituted aliphatic alkenes and electron-deficient olefins are also epoxidized with excellent yields and selectivities. Dienes are also rapidly and selectively converted into the corresponding diepoxides. Sensitive groups such as alcohols and ketones are tolerated [76]. Ascorbic acid also proves to exert similar positive effects in the efficient and selective use of H_2O_2 by this system [78].

The high activity of the Mn-tmtacn catalyst has prompted its link to heterogeneous supports, leading to highly active and recyclable systems. This has been initially accomplished via modification of the triazacyclononane ligand, enabling its covalent attachment to the support, but a more recent strategy is the attachment of the

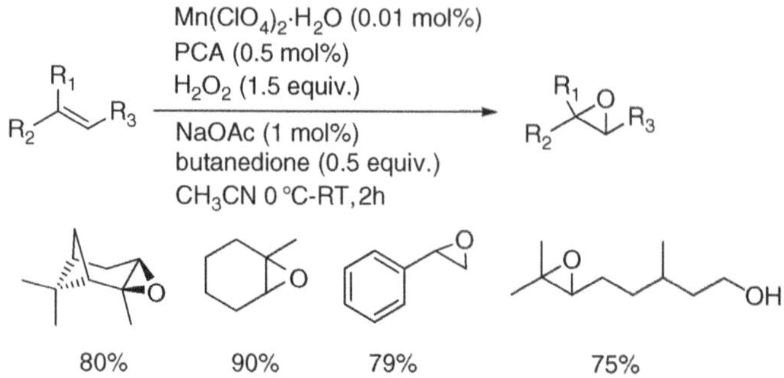

Scheme 4.16 Representative examples of PCA-Mn catalyzed epoxidations [82].

manganese catalyst to the support via metal coordination bonds with simple carboxylate anions installed in the support, without need to modify the tacn ligand [73, 79–81].

Combination of manganese salts with picolinic acid (PCA) also produces an extraordinarily active epoxidation catalyst (Scheme 4.16) [82–84]. For example, a simple, high yielding catalytic method for the selective epoxidation of alkenes using near-stoichiometric H_2O_2 under ambient conditions consists of a Mn(II) salt (<0.01 mol%), PCA (<0.5 mol%), and substoichiometric amounts of butanedione. The system yields high TON (up to 300 000) and TOF (up to $40\,s^{-1}$) for a wide range of substrates with good to excellent selectivity, remarkable functional group tolerance, and a wide solvent scope. Spectroscopic monitoring of the reaction reveals the initial formation of 3-hydroperoxy-3-hydroxybutan-2-one from butanedione, and H_2O_2, which appears to be the oxidant activated by the manganese catalyst [82].

The mononuclear Mn(III) complex $Mn(Q)_3$, where Q3 is a bidentate isoquinolinato ligand (Scheme 4.17) is also an excellent epoxidation catalyst with H_2O_2 as an oxidant [85]. Use of an acetic acid/ammonium acetate buffer is required for efficient use of H_2O_2. Aromatic and aliphatic olefins are epoxidized in good yields, under mild experimental conditions, employing a slight excess of H_2O_2.

The manganese(V) nitride complex $[Mn(N)(CN)_4](PPh_4)_2$ (1 mol%) catalyzes the oxidation of a wide range of alkenes using nearly stoichiometric (1–1.05 equiv.) amounts of H_2O_2 [86]. The complex exhibits a wide substrate scope covering aliphatic and aromatic olefins (Scheme 4.18). Reactions are stereospecific and take place under mild experimental conditions (room temperature). The use of acetic acid accelerates the reaction but it was not necessary for efficient catalysis. The complex is proposed to act as a Lewis acid facilitating O-atom transfer from H_2O_2 to the alkene.

Manganese complexes with tetradentate N-based ligands show excellent catalytic activity in alkene epoxidation. The prototypical example is $[Mn(OTf)_2(mcp)]$ (OTf = trifluoromethanesulfonate anion, Scheme 4.19), which was first described as a very active epoxidation catalyst in combination with peracetic acid [88]. A second complex that also shows very high catalytic epoxidation activity under analogous conditions is $[Mn(OTf)_2(^{Me,H}Pytacn)]$ (Scheme 4.19) [89]. Both complexes show excellent epoxidation activity with H_2O_2 as the oxidant when reactions are performed in the presence of carboxylic acids, while peroxide decomposition takes place in the absence of the acid [87]. In the presence of acetic acid (14 equiv.), a small catalyst loading (0.1 mol%) and a

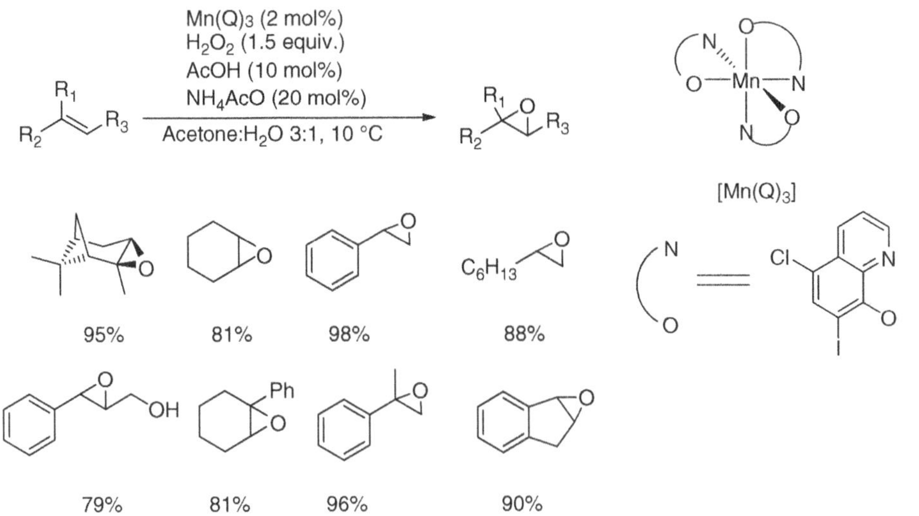

Scheme 4.17 Representative examples of $Mn(Q)_3$-catalyzed epoxidations. Epoxide yields determined by GC [85].

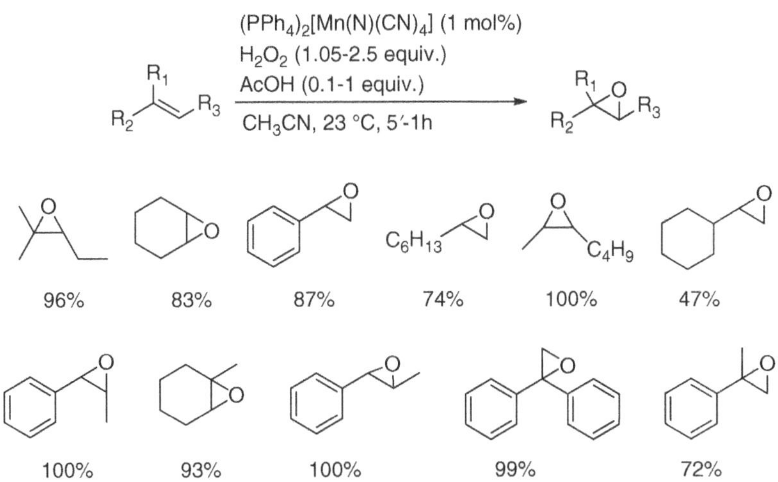

Scheme 4.18 Catalytic epoxidations with the $[Mn(N)(CN)_4](PPh_4)_2$ catalyst [86].

slight excess of H_2O_2 (1.2 equiv.) suffice to provided excellent conversions and yields of the epoxide products (80–100%), comparable to those obtained with peracetic acid. Reactions proceed very fast (in minutes) at 0 °C, and the peroxide should be added via syringe pump because of its very fast reaction with the catalyst. These systems show an excellent substrate scope (Scheme 4.19) covering aliphatic olefins with all types of substitution pattern, cyclic aliphatic olefins, aromatic olefins, and electron-deficient olefins. Reactions occur with retention of configuration, and even the highly sensitive *cis*-stilbene is epoxidized with minimum epimerization.

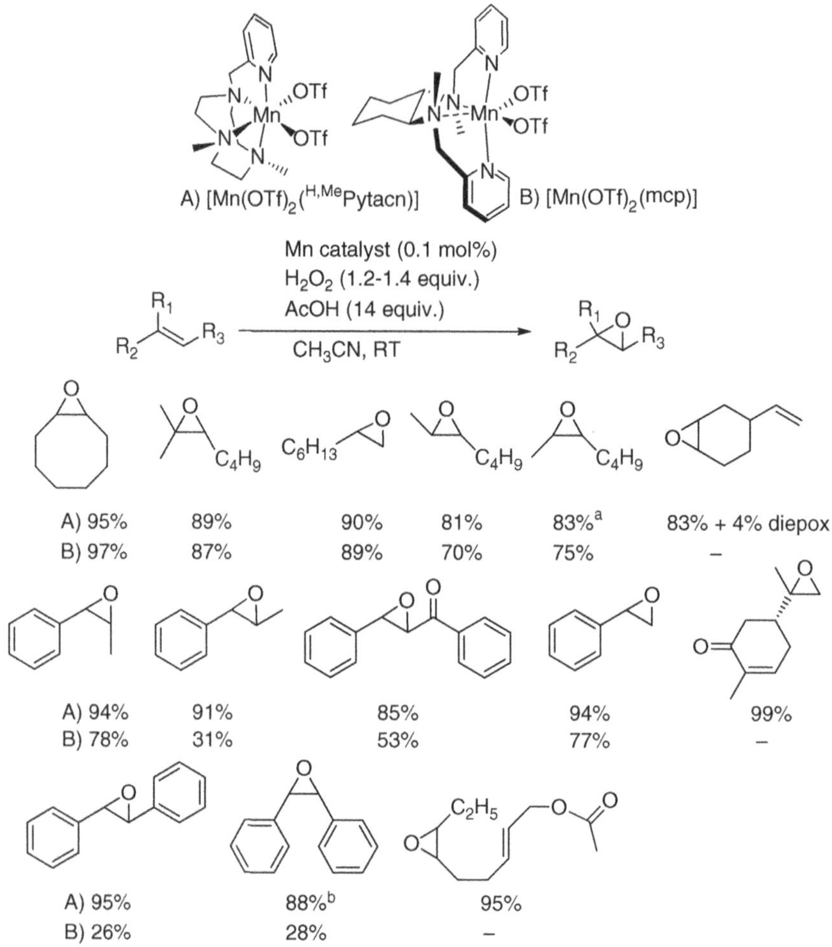

Scheme 4.19 Catalytic epoxidations with H_2O_2 and manganese complexes bearing tetradentate ligands. GC yields of epoxides. [a]98% *cis*. [b]90% *cis* [87].

4.3.3 Asymmetric Epoxidations with Manganese Catalysts

Manganese catalyzed asymmetric epoxidation of alkenes was pioneered by Jacobsen et al. [90] and Katsuki et al. [91] in the 1990s by employing Mn-salen complexes. Salen systems commonly operate with oxidants such as sodium hypochlorite, iodosyl arenes, or peracids, while H_2O_2 leads to rapid catalyst degradation and reduced yields and enantioselectivities, presumably because of homolytic cleavage of the O—O bond [92]. Strategies have been explored to combine the use of this convenient oxidant with the high enantioselectivity provided by this type of catalyst. The addition of imidazoles [93, 94] and carboxylates [95] enables the use of H_2O_2 as the oxidant. Presumably, these molecules bind at one of the axial positions of the Mn-salen complex and assist the heterolytic cleavage of the O—O bond of bound H_2O_2, providing yields and enantioselectivities comparable to those obtained with other oxidants. A more elaborate

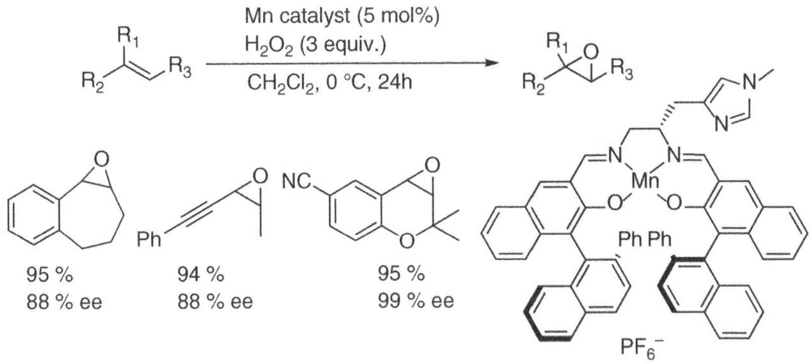

Scheme 4.20 Modified salen complex that employs aqueous H_2O_2 in asymmetric epoxidations [96].

strategy entails the use of catalysts where an imidazole ring is covalently attached to the ligand (Scheme 4.20) [96–98].

Manganese complexes containing tetradentate aminopyridine ligands are more robust than salen and show excellent catalytic activity in asymmetric epoxidation of different types of olefins employing hydrogen peroxide as the terminal oxidant. This type of complex was first explored in asymmetric epoxidation using peracetic acid as the oxidant [88, 99–101]. As shown earlier for [Mn(OTf)$_2$(mcp)] [87], the addition of a carboxylic acid (initially acetic acid) is key for promoting a productive use of the peroxide, avoiding peroxide disproportionation reactions [102]. [Mn(OTf)$_2$(mcp)] yields modest stereoselectivities, but several variations in the basic structure of the catalyst have led to improved systems. These changes include derivatization of the benzylic positions with bulky arene groups [100], or even with simple methyl groups [103], replacement of the aliphatic diamine backbone by other chiral diamines such as chiral pyrrolidines [104] and bipyrrolidines [102, 105–108], and replacement of the pyridines by N-alkyl benzylimidazoles [104, 109, 110]. Interestingly, the nature of the carboxylic acid also impacts on the enantioselectivity of the reactions; for example, replacement of acetic acid by branched alkyl carboxylic acids such as 2-ethyl-hexanoic acid [105, 106] or pivalic acid [106] has permitted important improvements in the levels of enantioselectivity. These systems perform extraordinarily well in the epoxidation of aromatic enones and chromenes (Scheme 4.21). Remarkably, installing electron-donating groups in the pyridines of the ligands consistently led to highly enantioselective catalysts (Scheme 4.22) [106, 111, 112].

Highly enantioselective epoxidation of alkenyl amides has recently been described with aminopyridine manganese catalysts that contain a rigid benzene fused 1,1′-bisindoline diamine backbone, and electron-rich pyridines (Scheme 4.23). High yield and highly enantioselective epoxidation of this class of substrates yielded epoxides that are precursors that can be further elaborated to yield pharmaceuticals and natural products [108].

Quite recently, a novel family of manganese catalysts bearing chiral ligand scaffolds inspired in the structure of porphyrins have been described as highly enantioselective epoxidation catalysts employing H_2O_2 as an oxidant. The systems are based in tetradentate ligands that combine an aromatic 1,2-diamine site, and two oxazoline moieties as

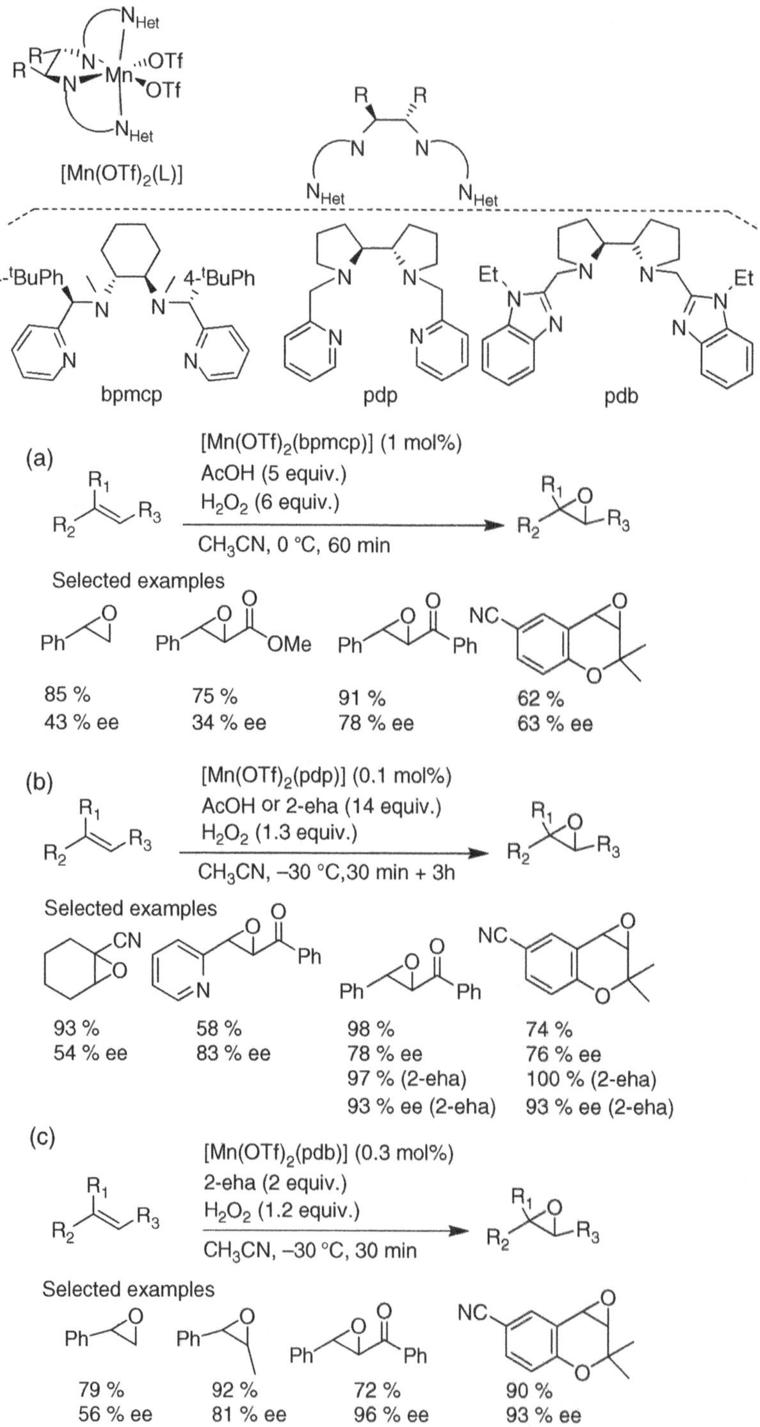

Scheme 4.21 Representative examples of asymmetric epoxidations with H_2O_2 and manganese complexes bearing tetradentate aminopyridine ligands. (a) Ref. [100], (b) Ref. [105], (c) Ref. [110].

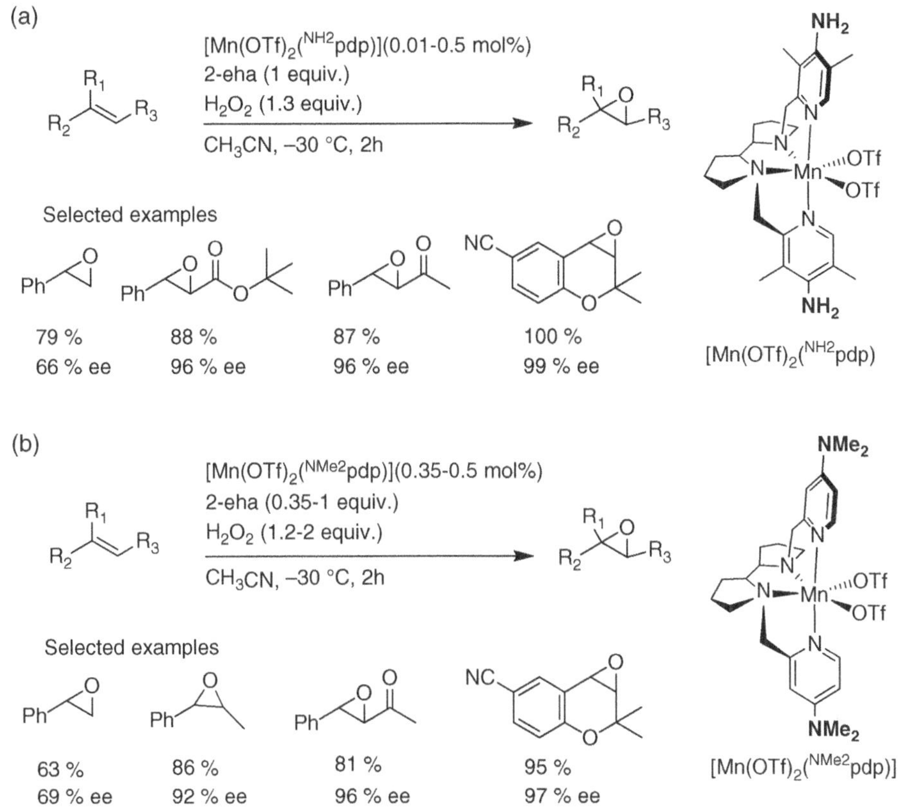

Scheme 4.22 Representative examples of asymmetric epoxidations with H_2O_2 and manganese complexes bearing electron-rich tetradentate aminopyridine ligands. (a) Ref. [111], (b) Ref. [106].

the chiral source [113, 114]. The corresponding manganese catalysts were generated *in situ* by mixing the ligand and $Mn(OTf)_2$ just before initiating the epoxidation reaction. One of the main advantages of this system is the easily tunable ligand, based in oxazolines. Upon reaction with hydrogen peroxide, in the presence of a carboxylic acid, the system provides excellent yields and enantioselectivities for electron-deficient chromenes, *trans*-stilbenes, indene, and 1,2-dihydronaphtalene (up to 99% *ee*, Scheme 4.24).

4.3.4 Iron-Catalyzed Epoxidation

Iron is an interesting metal for sustainable catalysis because of its availability, low price, and lack of toxicity [115–117]. This metal is thus an ideal choice for development of catalytic epoxidation reactions [117]. Interestingly, iron is found at the active site of a number of metalloenzymes that catalyze oxidation reactions, including epoxidations. Epoxidation by the nonheme alfa-ketoglutarate dependent oxygenase AsqJ is involved in quinolone alkaloid biosynthesis [118], and P450 and peroxidases are well-known to catalyze olefin epoxidations [119, 120]. Based on these considerations, iron-catalyzed epoxidations with green oxidants such as O_2 and H_2O_2 are potentially very attractive reactions, and their development has been actively investigated [121–123].

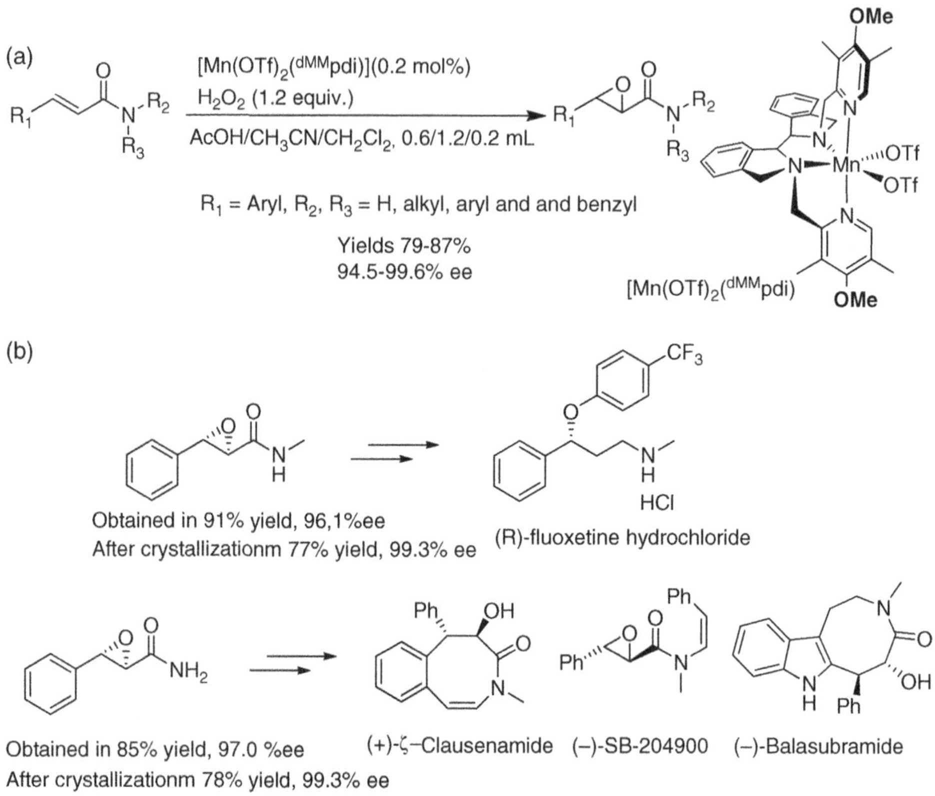

Scheme 4.23 Asymmetric epoxidation of alkenyl amides (top), and elaboration of the epoxides into products of interest in pharmaceuticals and natural products (bottom) [108].

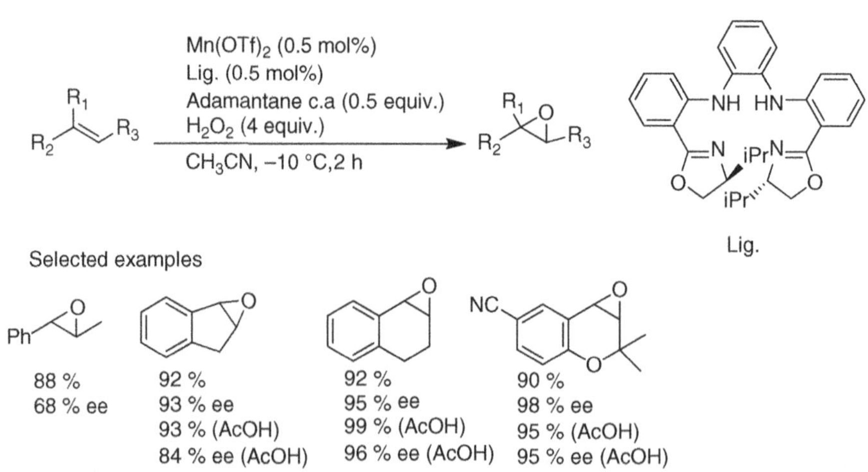

Scheme 4.24 Asymmetric epoxidation of aromatic olefins with H_2O_2 and adamantane carboxylic acid (c.a.) with an oxazoline based manganese catalyst [113].

Two basic strategies have been pursued in the development of iron catalyze epoxidations. A simple, straightforward, and convenient strategy relies on the assembly of the catalyst *in situ* by combining carboxylic acids (alkyl carboxylic acids and PCAs), different types of amines (generally heterocyclic), and iron salts. A second strategy has focused on studying well-defined coordination compounds as catalysts. From this approach, selected iron complexes with tetradentate ligands combining aliphatic and aromatic heterocyclic amines have been discovered to be powerful epoxidation catalysts.

4.3.5 Epoxidations with Iron Catalysts Prepared *In Situ*

An environmentally benign and practical epoxidation method entails the use of $FeCl_3{\cdot}6H_2O$ in combination with commercially available pyridine-2,6-dicarboxylic acid and pyrrolidine (other bases also provide good activity), showing excellent reactivity and selectivity toward aromatic olefins and moderate reactivity toward 1,3-cyclooctadiene utilizing H_2O_2 as the terminal oxidant. The neutral conditions of the reaction media permit the epoxidation of acid sensitive epoxides such as α-methyl styrene. However, epoxidation of *cis*-β-methyl styrene provides a mixture of *cis* and *trans* epoxides (~2:1), showing that formation of the two C—O bonds is not concerted [124] (Scheme 4.25).

Replacement of pyrrolidine by formamidines results in a method that furnishes improved yields in the epoxidation of aromatic alkenes and expands the scope to 1,3-dienes [125]. Use of benzylamines instead of pyrrolidine extends the substrate scope of the method to a variety of aliphatic olefins (Scheme 4.26) [126].

A variation of this method consists in using $FeCl_3{\cdot}6H_2O$ and imidazoles (Scheme 4.27). Under these conditions, dipicolinic acid and pyrrolidine are not needed [127]. Epoxidation of aromatic olefins proceeds with moderate to good yields and trisubstituted aromatic olefins are epoxidized with improved yields with respect to the dipicolinic acid method.

Extensive screening of imidazole derivatives leads to the identification of aryl substituted imidazoles as particularly suitable ligands. Furthermore, the free 2-position of the imidazole ligand is a key element for catalytic activity, as substitution leads to a dramatic depletion of yield and conversion [128].

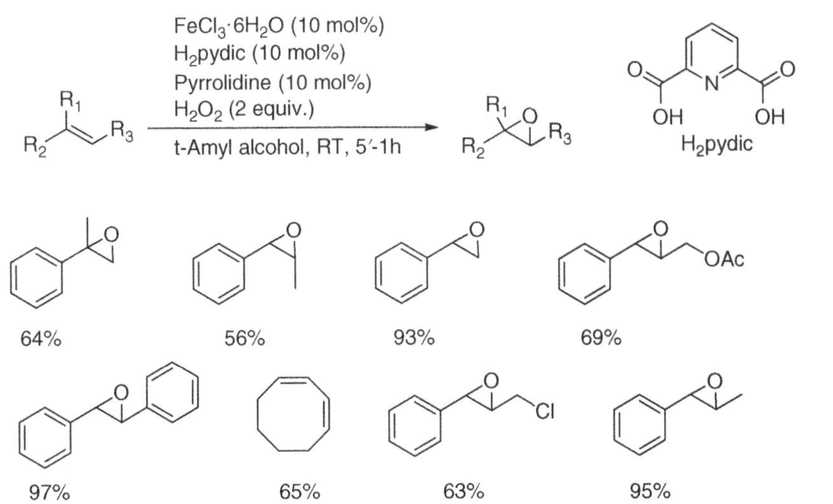

Scheme 4.25 Catalytic epoxidation using $FeCl_3$-$6H_2O$, dipicolinic acid and pyrrolidine [124].

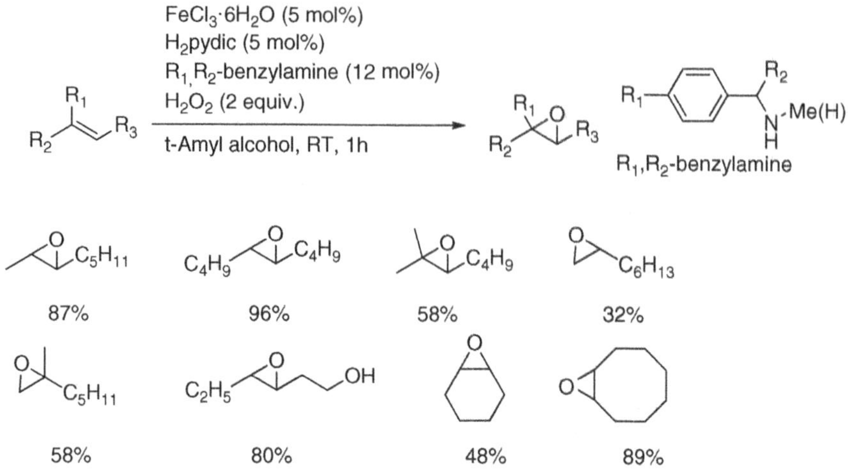

Scheme 4.26 Catalytic epoxidation using $FeCl_3$-$6H_2O$, dipicolinic acid and benzylamines [126].

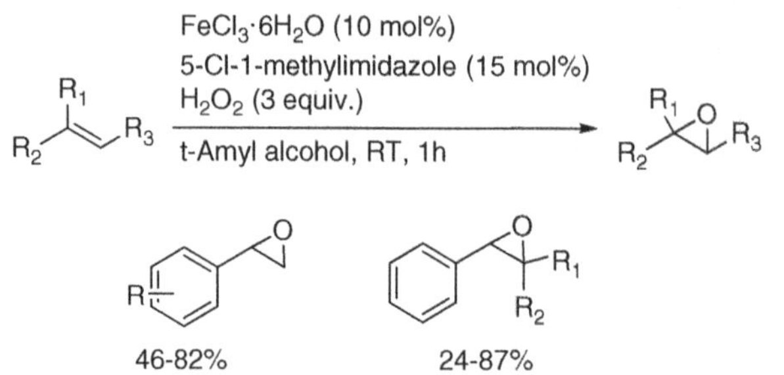

Scheme 4.27 Catalytic epoxidations with $FeCl_3$-$6H_2O$ and imidazole ligands [127].

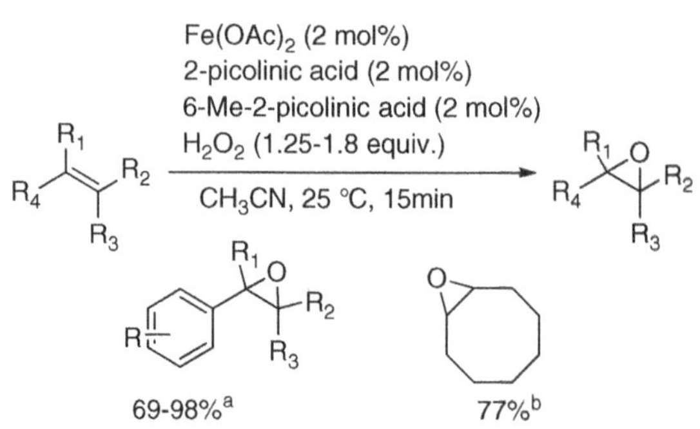

Scheme 4.28 Catalytic epoxidations with $Fe(OAc)_2$ and picolinic acids [129].

Related to the previous findings, combination of ferrous acetate, 2-picolinic acid and 6-methyl-2-picolinic acid in acetonitrile results in a mononuclear complex that quickly (15 minutes) catalyzes the epoxidation of styrenes at 25 °C, employing a slight excess (1.25–1.8 equiv.) of 35% aqueous hydrogen peroxide solution [129]. Excellent yields and selectivities are obtained (Scheme 4.28). Most remarkable, oxidation of *cis*-β-methyl styrene is stereospecific, evidencing important mechanistic differences with respect to reactions with dipicolinic acid.

4.3.6 Epoxidation with Well-Defined Iron Coordination Complexes

Reaction of iron species with peroxides is very complex and highly dependent on reaction conditions and the nature of the specific nature of the iron species. The use of well-defined coordination complexes has shown that small modifications in the coordination sphere of the iron complexes can lead to dramatic changes in the reactivity, and Fenton type reactions may dominate the chemistry [130]. A number of iron complexes produce epoxides on reaction with hydrogen peroxide, but the selectivities and yields of most of these reactions are modest. Nevertheless, few of them turn out to be excellent epoxidation catalysts.

Some of the most active catalyst are mononuclear iron complexes with tetradentate aminopyridine ligands. The prototypical case is $[Fe(mep)(CH_3CN)_2](SbF_6)_2$ (Scheme 4.29), which shows excellent activity in the epoxidation of aliphatic olefins. High epoxide yields (60–90%) were obtained in short reaction times by using 1.5 equiv. of H_2O_2 and 30 mol% of acetic acid was identified as a key additive to ensure high product yields [130, 131]. Reactions are stereospecific. The most obvious limitation of the system is that it cannot operate in aromatic substrates, that are rapidly oxidatively degraded.

The triflate complex $[Fe(OTf)_2(mep)]$ provides identical results in these simple aliphatic olefins and has also been used in the epoxidation of steroidal substrates (Scheme 4.30) [132, 133].

Modification of the ethylendiamine backbone of the mep ligand by more rigid cyclic diamines such as cyclohexandiamine [134], bipyrrolidine [135], or bipyperidine [136] results in iron catalysts with excellent epoxidation activity and improved stability. These systems have been explored in asymmetric epoxidations (vide infra).

Helmet-phthalocyanine iron complexes are also excellent epoxidation catalysts using H_2O_2 as the terminal oxidant (Scheme 4.31) [137]. Aromatic and aliphatic olefins, including terminal monosubstituted alkenes, are epoxidized with very good yields and selectivities in short reaction times (<1 hour) at 25 °C using 1 mol% catalyst loading and 2 equiv. of H_2O_2.

4.3.7 Asymmetric Epoxidations with Iron Catalysts

Because of the chemical versatility of epoxides, asymmetric epoxidation represents a very interesting reaction to create chiral building blocks suitable for multiple transformations. Asymmetric epoxidation with iron catalysts and hydrogen peroxide as the

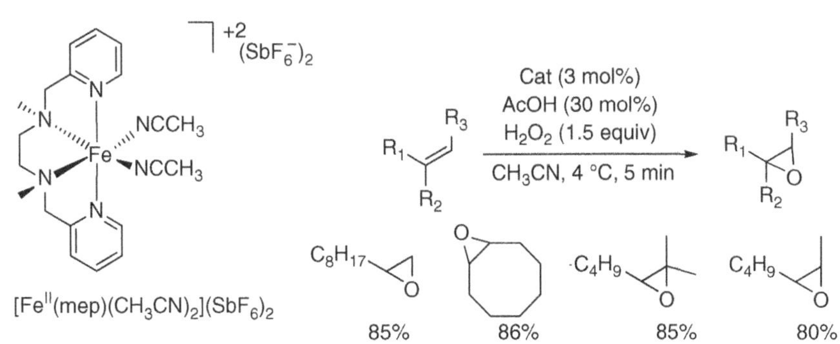

Scheme 4.29 Catalytic epoxidation with the $[Fe(mep)(CH_3CN)_2](SbF_6)_2$ complex [131].

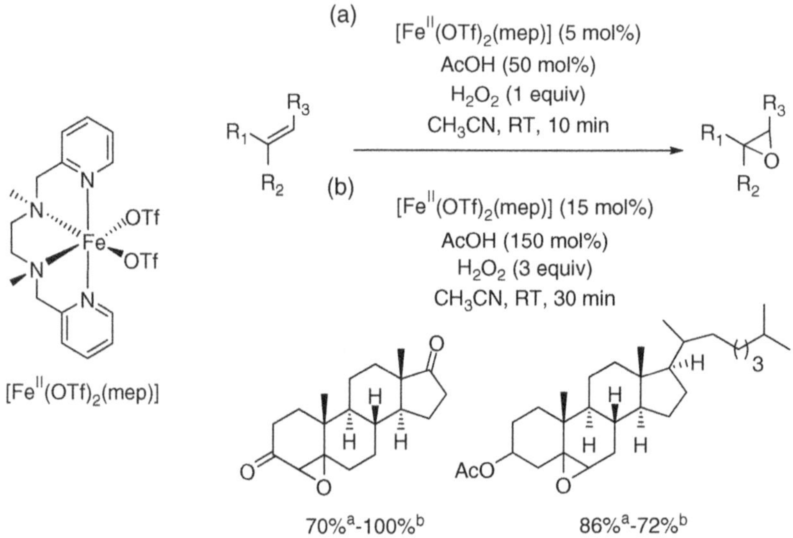

Scheme 4.30 Catalytic epoxidations with the [Fe^{II}(OTf)$_2$(mep)] complex [132, 133].

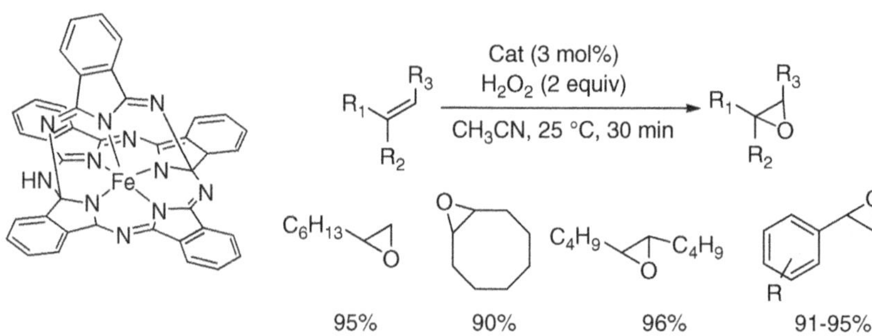

Scheme 4.31 Catalytic epoxidations with a helmet-phthalocyanine iron complex [137].

oxidant has been actively explored during the last decade [11, 122, 138, 139]. The substrate scope still remains relatively narrow; aromatic enones and chromenes are epoxidized with excellent *ee*s and yields, and epoxidation of difficult classes of substrates has been accomplished in several cases. However, simple styrenes and aliphatic olefins remain a challenge.

A simple method for asymmetric epoxidation of aromatic substrates involves the use of $FeCl_3{\cdot}6H_2O$, pyridine-2,6-dicarboxylic acid, and chiral benzylamines as ligands. The substrate scope is relatively narrow, reaching good yields and best *ee*s (up to 97%) in the epoxidation of *trans*-stilbene substrates (Scheme 4.32) [140, 141].

Arguably, the most versatile type of iron catalysts described so far is based on iron complexes with linear tetradentate ligands containing two aromatic and two aliphatic N donors. The basic structure, and representative variations therefrom, are shown in Scheme 4.33 [105, 142, 143]. The most commonly explored catalysts contain pyridines as aromatic donors [105, 135, 142, 144], but catalysts containing benzylimidazole rings

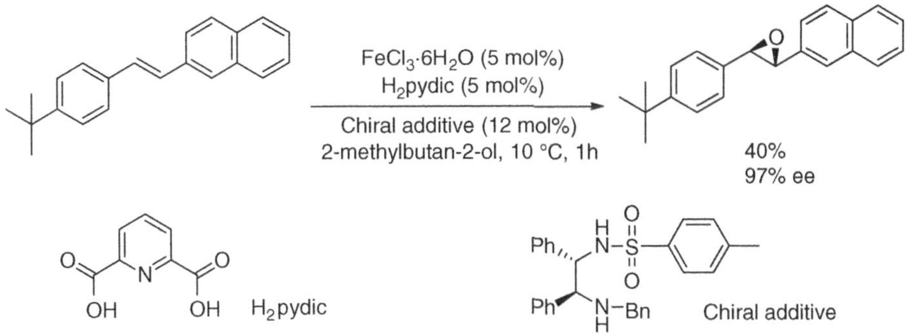

Scheme 4.32 *In situ* generated asymmetric catalyst based in $FeCl_3$/H_2pydic/chiral diamine [86, 88].

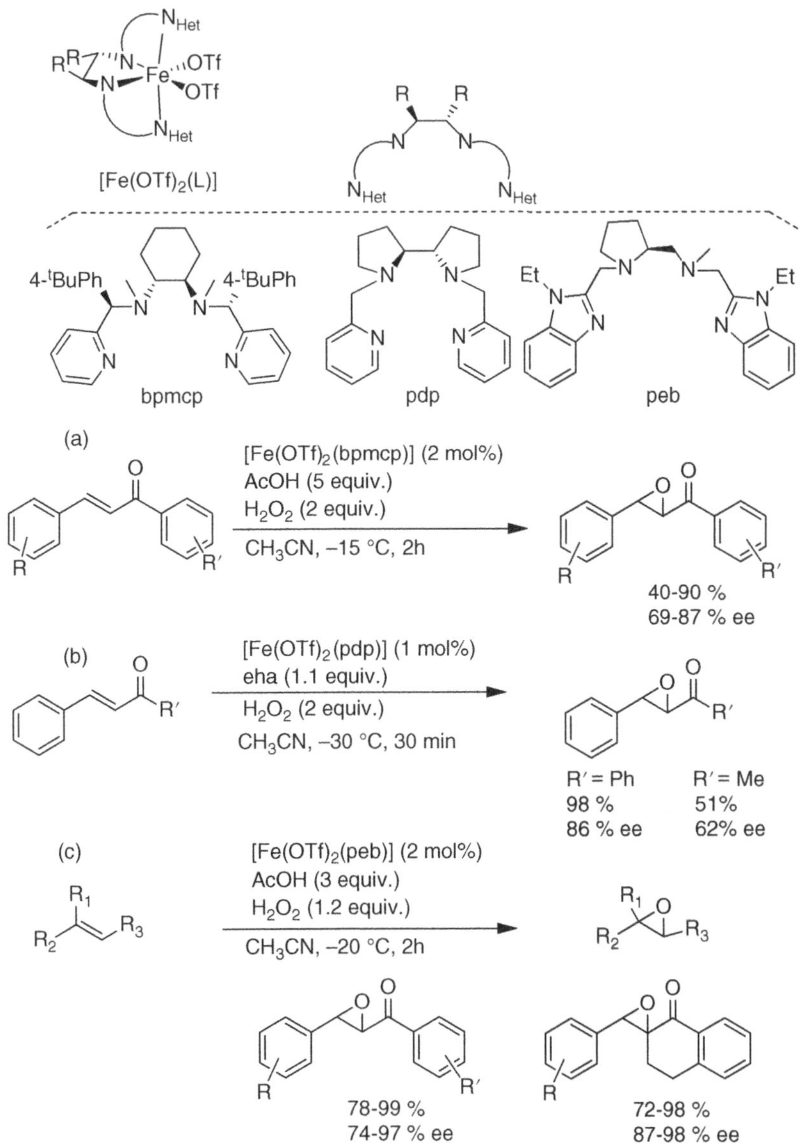

Scheme 4.33 Representative examples of asymmetric epoxidations with iron catalysts bearing tetradentate chiral ligands. (a) Ref. [142], (b) Ref. [105], and (c) Ref. [143].

have been also studied and provide excellent yields and enantioselectivities in the epoxidation of aromatic enones [109, 143]. The nature of the aliphatic diamine has been also studied as key element in defining the enantioselectivity of the catalysts, and, so far, 2,2′-bipyrrolidine appears to be the optimum choice. As shown in the case of non-chiral epoxidation reactions [131, 145], carboxylic acids are key for efficient activation of H_2O_2, and can modulate the enantioselectivity of the reactions, presumably acting as ligands [105, 135, 144]. Use of electron-rich pyridines enables efficient activation of H_2O_2 by the metal center, requiring only catalytic amounts of carboxylic acids [135, 144]. The catalyst [Fe(CF_3SO_3)(^{NMe2}PDP)] (Scheme 4.34) has been explored in combination with different carboxylic acids, expanding its substrate scope. A wide range of aromatic enones and chromenes are epoxidized with excellent yields and enantioselectivities employing 2-ethyl hexanoic acid (eha) or S-ibuprofen as cocatalyst [135]. Use of protected amino acids [146] and even small peptides [147] expands the scope to 1,1-disubstituted styrenes, which are recognized as particularly difficult substrates for asymmetric epoxidation.

Elaboration of the structure of the catalysts by combining a benzylimidazole and a bulky pyridine has led to catalysts that show excellent enantioselectivities in the epoxidation of cyclic enones and cyclohexene ketones (Scheme 4.35) [148].

An interesting aspect of iron catalysts is that they generate electrophilic oxidants in sharp contrast with the activation of H_2O_2 with basic organocatalysts that generates nucleophilic oxidants. This aspect can be used to oxidize dienes with a selectivity orthogonal to that obtained with organocatalysts (Scheme 4.36) [144].

4.4 Future Perspectives

Epoxidation methodologies that can be considered satisfactory from a sustainable perspective are currently available for most classes of olefins. Hydrogen peroxide and O_2 can be reliably and efficiently used as green oxidants with catalysts based in first row transition metal ions. However, important problems remain. In the first place, catalytic epoxidations with O_2 mostly require the use of a cosubstrate. Methods that could employ only dioxygen will represent an extraordinary step forward, but major breakthroughs on the discovery of O_2 activation reactions and catalysts that engage in such processes are needed. Hydrogen peroxide currently represents a better alternative, and nowadays catalysts exist that permit their use in an extraordinarily efficient manner, limiting or avoiding peroxide disproportionation reactions. Highly active transition metal complexes and polyoxometallates are already known, but it is likely that their activity could be further improved by careful catalyst design. Asymmetric epoxidation constitutes a very powerful reaction in organic synthesis, but green methods for enantioselective epoxidation are far from satisfactory. Organocatalytic methods that employ hydrogen peroxide as the oxidant are extraordinarily powerful in terms of chemo-, regio-, and stereoselectivity, but most organocatalysts are slow and provide quite limited turnover numbers. Asymmetric epoxidations with hydrogen peroxide and metal catalysts based on first row transition metal complexes constitute a potentially more satisfactory option [14], but the substrate scope exhibited by these catalysts is still quite limited. Besides the manganese and iron-based systems described in the current chapter, particularly promising and interesting are catalysts based on scandium [149, 150] and in titanium [151–153].

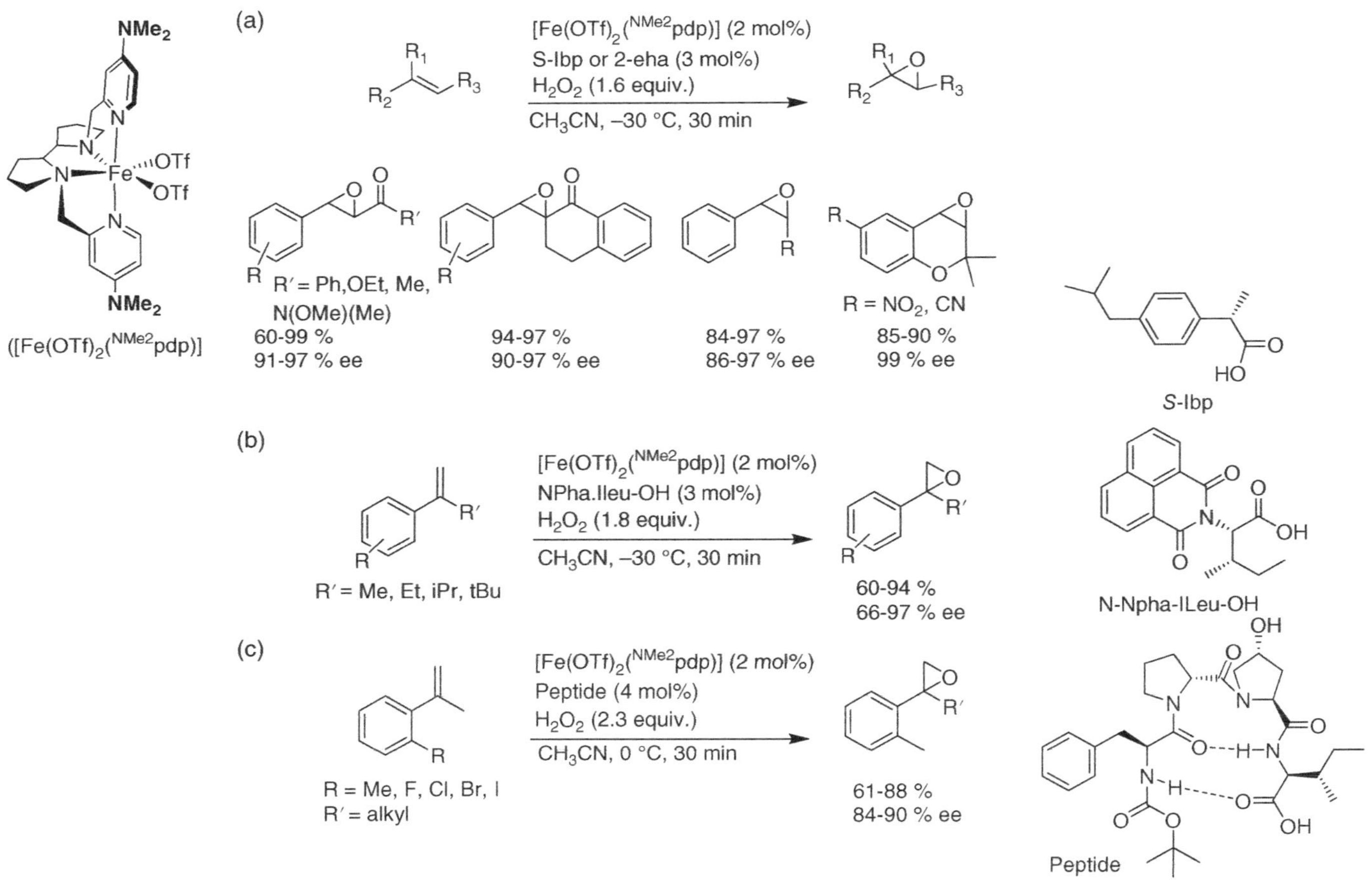

Scheme 4.34 Catalytic epoxidations with the electron-rich $[Fe(OTf)_2(^{NMe2}PDP)]$ catalyst. (a) Ref. [135], (b) Ref. [146], and (c) Ref. [147].

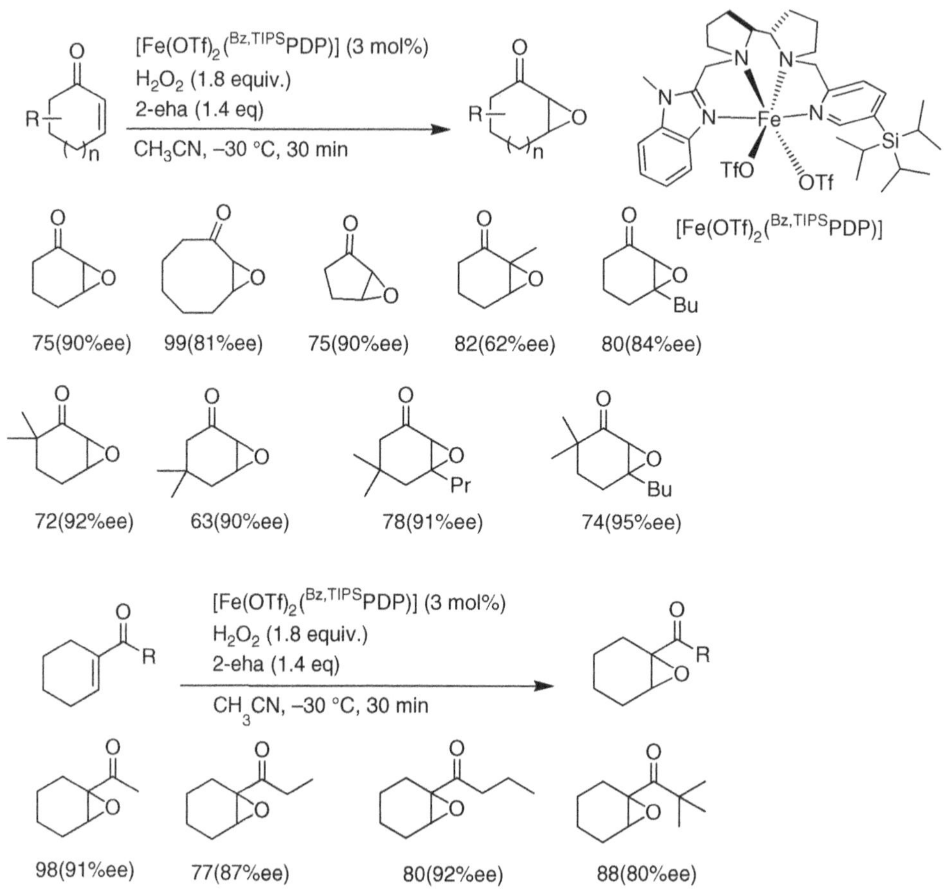

Scheme 4.35 Asymmetric epoxidation of cyclic enones and cyclohexene ketones [148].

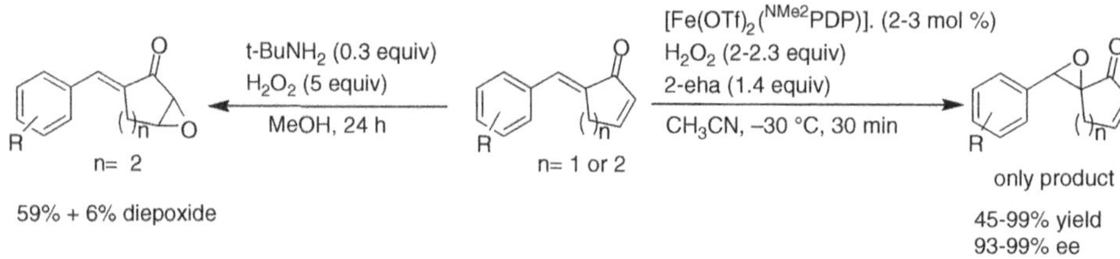

Scheme 4.36 Orthogonal regioselectivity in epoxidation of dienes with iron catalysts and basic H_2O_2 [148].

The latter provide high enantioselectivities in the epoxidation of recognizable difficult substrates such as terminal aliphatic olefins. Unfortunately, a highly undesirable media such as chlorinated solvents appear to be required for these titanium catalysts. Careful design of catalysts based on the biologically relevant transition metals (Fe, Mn, Co, Ni, Cu, or Zn) is likely to provide satisfactory solutions in the future.

References

1 Bäckvall, J.-E. (2004). *Modern Oxidation Methods*. Weinheim: Wiley-VCH.
2 Swern, D. (1953). *Organic Reactions* 7: 378.
3 Joergensen, K.A. (1989). *Chemical Reviews* 89: 431.
4 Hauser, S.A., Cokoja, M., and Kuhn, F.E. (2013). *Catalysis Science & Technology* 3: 552.
5 Cavani, F. and Teles, J.H. (2009). *ChemSusChem* 2: 508.
6 Nijhuis, T.A., Makkee, M., Moulijn, J.A., and Weckhuysen, B.M. (2006). *Industrial & Engineering Chemistry Research* 45: 3447.
7 Krähling, L., Krey, J., Jakobson, G. et al. (2000). *Ullmann's Encyclopedia of Industrial Chemistry*. Weinheim: Wiley-VCH Verlag GmbH & Co. KGaA.
8 Li, A.-H., Dai, L.-X., and Aggarwal, V.K. (1997). *Chemical Reviews* 97: 2341.
9 Wang, Z. (2010). *Comprehensive Organic Name Reactions and Reagents*. Wiley.
10 Punniyamurthy, T., Velusamy, S., and Iqbal, J. (2005). *Chemical Reviews* 105: 2329.
11 Bryliakov, K.P. (2017). *Chemical Reviews* 117: 11406.
12 Noyori, R., Aoki, M., and Sato, K. (2003). *Chemical Communications* 16: 1977.
13 Lane, B.S. and Burgess, K. (2003). *Chemical Reviews* 103: 2457.
14 Wang, C. and Yamamoto, H. (2015). *Chemistry – An Asian Journal* 10: 2056.
15 Zhu, Y., Wang, Q., Cornwall, R.G., and Shi, Y. (2014). *Chemical Reviews* 114: 8199.
16 Davis, R.L., Stiller, J., Naicker, T. et al. (2014). *Angewandte Chemie, International Edition* 53: 7406.
17 Serafin, J.G., Liu, A.C., and Seyedmonir, S.R. (1998). *Journal of Molecular Catalysis A: Chemical* 131: 157.
18 Montemore, M.M., van Spronsen, M.A., Madix, R.J., and Friend, C.M. (2017). *Chemical Reviews* 118: 2816.
19 Santra, A.K., Cowell, J.J., and Lambert, R.M. (2000). *Catalysis Letters* 67: 87.
20 Hughes, M.D., Xu, Y.-J., Jenkins, P. et al. (2005). *Nature* 437: 1132.
21 Turner, M., Golovko, V.B., Vaughan, O.P.H. et al. (2008). *Nature* 454: 981.
22 Teruaki, M. and Tohru, Y. (1995). *Bulletin of the Chemical Society of Japan* 68: 17.
23 Yamada, T., Takai, T., Rhode, O., and Mukaiyama, T. (1991). *Bulletin of the Chemical Society of Japan* 64: 2109.
24 Wentzel, B.B., Alsters, P.L., Feiters, M.C., and Nolte, R.J.M. (2004). *The Journal of Organic Chemistry* 69: 3453.
25 Nam, W., Kim, H.J., Kim, S.H. et al. (1996). *Inorganic Chemistry* 35: 1045.
26 Yang, G., Du, H., Liu, J. et al. (2017). *Green Chemistry* 19: 675.
27 Opre, Z., Mallat, T., and Baiker, A. (2007). *Journal of Catalysis* 245: 482.
28 Qi, Y., Luan, Y., Yu, J. et al. (2015). *Chemistry – A European Journal* 21: 1589.
29 Beier, M.J., Kleist, W., Wharmby, M.T. et al. (2012). *Chemistry – A European Journal* 18: 887.
30 Dhakshinamoorthy, A., Alvaro, M., and Garcia, H. (2012). *Journal of Catalysis* 289: 259.
31 Dhakshinamoorthy, A., Alvaro, M., and Garcia, H. (2011). *ACS Catalysis* 1: 836.
32 Schröder, K., Join, B., Amali, A.J. et al. (2011). *Angewandte Chemie, International Edition* 50: 1425.
33 Yamada, T., Imagawa, K., Nagata, T., and Mukaiyama, T. (1994). *Bulletin of the Chemical Society of Japan* 67: 2248.
34 Nagata, T., Imagawa, K., Yamada, T., and Mukaiyama, T. (1995). *Bulletin of the Chemical Society of Japan* 68: 1455.

35 Pozzi, G. and Cinato, F. (1998). *Chemical Communications* 877.
36 Cheng, Q.F., Xu, X.Y., Ma, W.X. et al. (2005). *Chinese Chemical Letters* 16: 1467.
37 Grigoropoulou, G., Clark, J.H., and Elings, J.A. (2003). *Green Chemistry* 5: 1.
38 Notari, B. (1993). *Catalysis Today* 18: 163.
39 Clerici, M.G. and Ingallina, P. (1993). *Journal of Catalysis* 140: 71.
40 Wu, P. and Tatsumi, T. (2001). *Chemical Communications* 897.
41 Wu, H., Wang, L., Zhang, H. et al. (2006). *Green Chemistry* 8: 78.
42 Nagendrappa, G. (2011). *Applied Clay Science* 53: 106.
43 Honma, T., Nakajo, M., Mizugaki, T. et al. (2002). *Tetrahedron Letters* 43: 6229.
44 Yamaguchi, K., Ebitani, K., and Kaneda, K. (1999). *The Journal of Organic Chemistry* 64: 2966.
45 Ueno, S., Yoshida, K., Ebitani, K., and Kaneda, K. (1998). *Chemical Communications* 295.
46 Wang, S.-S. and Yang, G.-Y. (2015). *Chemical Reviews* 115: 4893.
47 Mizuno, N., Yamaguchi, K., and Kamata, K. (2005). *Coordination Chemistry Reviews* 249: 1944.
48 Venturello, C., Alneri, E., and Ricci, M. (1983). *The Journal of Organic Chemistry* 48: 3831.
49 Venturello, C., D'Aloisio, R., Bart, J.C.J., and Ricci, M. (1985). *Journal of Molecular Catalysis* 32: 107.
50 Sato, K., Aoki, M., Ogawa, M. et al. (1996). *The Journal of Organic Chemistry* 61: 8310.
51 Zuwei, X., Ning, Z., Yu, S., and Kunlan, L. (2001). *Science* 292: 1139.
52 Kamata, K., Yamaguchi, K., Hikichi, S., and Mizuno, N. (2003). *Advanced Synthesis & Catalysis* 345: 1193.
53 Kamata, K., Hirano, T., Kuzuya, S., and Mizuno, N. (2009). *Journal of the American Chemical Society* 131: 6997.
54 Ishimoto, R., Kamata, K., and Mizuno, N. (2012). *Angewandte Chemie, International Edition* 51: 4662.
55 Kamata, K., Yonehara, K., Sumida, Y. et al. (2003). *Science* 300: 964.
56 Kato, C.N., Negishi, S., Yoshida, K. et al. (2005). *Applied Catalysis A: General* 292: 97.
57 Kholdeeva, O.A. (2006). *Topics in Catalysis* 40: 229.
58 Mizuno, N., Nozaki, C., Kiyoto, I., and Misono, M. (1999). *Journal of Catalysis* 182: 285.
59 Nakagawa, Y., Kamata, K., Kotani, M. et al. (2005). *Angewandte Chemie, International Edition* 44: 5136.
60 Kamata, K., Sugahara, K., Yonehara, K. et al. (2011). *Chemistry – A European Journal* 17: 7549.
61 Adam, W., Alsters, P.L., Neumann, R. et al. (2003). *The Journal of Organic Chemistry* 68: 1721.
62 Witte, P.T., Alsters, P.L., Jary, W. et al. (2004). *Organic Process Research & Development* 8: 524.
63 Sloboda-Rozner, D., Alsters, P.L., and Neumann, R.J. (2003). *Journal of the American Chemical Society* 125: 5280.
64 Krishnan, K.K., Thomas, A.M., Sindhu, K.S., and Anilkumar, G. (2016). *Tetrahedron* 72: 1.
65 Saisaha, P., de Boer, J., and Browne, W. (2013). *Chemical Society Reviews* 42: 2059.
66 Richardson, D.E., Yao, H., Frank, K.M., and Bennett, D.A. (2000). *Journal of the American Chemical Society* 122: 1729.

67 Yao, H. and Richardson, D.E. (2000). *Journal of the American Chemical Society* 122: 3220.
68 Lane, B.S., Vogt, M., DeRose, V.J., and Burgess, K. (2002). *Journal of the American Chemical Society* 124: 11946.
69 Lane, B.S. and Burgess, K. (2001). *Journal of the American Chemical Society* 123: 2933.
70 Hage, R., Iburg, J.E., Kerschner, J. et al. (1994). *Nature* 369: 637.
71 de Boer, J.W., Browne, W.R., Brinksma, J. et al. (2007). *Inorganic Chemistry* 46: 6353.
72 Boer, J.W.d., Brinksma, J., Browne, W.R. et al. (2005). *Journal of the American Chemical Society* 127: 7990.
73 De Vos, D.E., de Wildeman, S., Sels, B.F. et al. (1999). *Angewandte Chemie, International Edition* 38: 980.
74 Vos, D.E.D., Meinershagen, J.L., and Bein, T. (1996). *Angewandte Chemie, International Edition in English* 35: 2211.
75 Vos, D.E.D. and Bein, T. (1996). *Chemical Communications* 917.
76 De Vos, D.E., Sels, B.F., Reynaers, M. et al. (1998). *Tetrahedron Letters* 39: 3221.
77 de Boer, J.W., Alsters, P.L., Meetsma, A. et al. (2008). *Dalton Transactions* 6283.
78 Berkessel, A. and Sklorz, C.A. (1999). *Tetrahedron Letters* 40: 7965.
79 Schoenfeldt, N.J., Ni, Z.J., Korinda, A.W. et al. (2011). *Journal of the American Chemical Society* 133: 18684.
80 Schoenfeldt, N.J. and Notestein, J.M. (2011). *ACS Catalysis* 1: 1691.
81 Ignacio-de Leon, P.A.A., Contreras, C.A., Thornburg, N.E. et al. (2016). *Applied Catalysis A: General* 511: 78.
82 Dong, J.J., Saisaha, P., Meinds, T.G. et al. (2012). *ACS Catalysis* 2: 1087.
83 Saisaha, P., Dong, J.J., Meinds, T.G. et al. (2016). *ACS Catalysis* 6: 3486.
84 Moretti, R.A., Du Bois, J., and Stack, T.D.P. (2016). *Organic Letters* 18: 2528.
85 Zhong, S., Fu, Z., Tan, Y. et al. (2008). *Advanced Synthesis and Catalysis* 350: 802.
86 Kwong, H.-K., Lo, P.-K., Lau, K.-C., and Lau, T.-C. (2011). *Chemical Communications* 47: 4273.
87 Garcia-Bosch, I., Ribas, X., and Costas, M. (2009). *Advanced Synthesis and Catalysis* 351: 348.
88 Murphy, A., Dubois, G., and Stack, T.D.P. (2003). *Journal of the American Chemical Society* 125: 5250.
89 Garcia-Bosch, I., Company, A., Fontrodona, X. et al. (2008). *Organic Letters* 10: 2095.
90 Zhang, W., Loebach, J.L., Wilson, S.R., and Jacobsen, E.N. (1990). *Journal of the American Chemical Society* 112: 2801.
91 Katsuki, T. (1995). *Coordination Chemistry Reviews* 140: 189.
92 Matsumoto, K. and Katsuki, T. (2010). *Catalytic Asymmetric Synthesis*, 839. Wiley.
93 Irie, R., Hosoya, N., and Katsuki, T. (1994). *Synlett* 1994: 255.
94 Pietikainen, P. (1994). *Tetrahedron Letters* 35: 941.
95 Pietikainen, P. (1998). *Tetrahedron* 54: 4319.
96 Shitama, H. and Katsuki, T. (2006). *Tetrahedron Letters* 47: 3203.
97 Schwenkreis, T. and Berkessel, A. (1993). *Tetrahedron Letters* 34: 4785.
98 Berkessel, A., Frauenkron, M., Schwenkreis, T. et al. (1996). *Journal of Molecular Catalysis A: Chemical* 113: 321.
99 Gómez, L., Garcia-Bosch, I., Company, A. et al. (2007). *Dalton Transactions* 5539.
100 Wu, M., Wang, B., Wang, S. et al. (2009). *Organic Letters* 11: 3622.
101 Murphy, A., Pace, A., and Stack, T.D.P. (2004). *Organic Letters* 6: 3119.

102 Ottenbacher, R.V., Bryliakov, K.P., and Talsi, E.P. (2011). *Advanced Synthesis and Catalysis* 353: 885.
103 Maity, N.C., Kumar Bera, P., Ghosh, D. et al. (2014). *Catalysis Science & Technology* 4: 208.
104 Wang, B., Miao, C., Wang, S. et al. (2012). *Chemistry, A European Journal* 18: 6750.
105 Lyakin, O.Y., Ottenbacher, R.V., Bryliakov, K.P., and Talsi, E.P. (2012). *ACS Catalysis* 2: 1196.
106 Cussó, O., Garcia-Bosch, I., Font, D. et al. (2013). *Organic Letters* 15: 6158.
107 Garcia-Bosch, I., Gómez, L., Polo, A. et al. (2012). *Advanced Synthesis & Catalysis* 354: 65.
108 Chen, X.N., Gao, B., Su, Y.J., and Huang, H.M. (2017). *Advanced Synthesis & Catalysis* 359: 2535.
109 Wang, X., Miao, C., Wang, S. et al. (2013). *ChemCatChem* 5: 2489.
110 Shen, D., Miao, C., Wang, S. et al. (2014). *European Journal of Inorganic Chemistry* 2014: 5777.
111 Ottenbacher, R.V., Samsonenko, D.G., Talsi, E.P., and Bryliakov, K.P. (2014). *ACS Catalysis* 4: 1599.
112 Shen, D., Qiu, B., Xu, D. et al. (2016). *Organic Letters* 18: 372.
113 Dai, W., Li, J., Li, G. et al. (2013). *Organic Letters* 15: 4138.
114 Dai, W., Shang, S., Chen, B. et al. (2014). *The Journal of Organic Chemistry* 79: 6688.
115 Enthaler, S., Junge, K., and Beller, M. (2008). *Angewandte Chemie, International Edition* 47: 3317.
116 Furstner, A. (2016). *ACS Central Science* 2: 778.
117 Bauer, I. and Knolker, H.J. (2015). *Chemical Reviews* 115: 3170.
118 Ishikawa, N., Tanaka, H., Koyama, F. et al. (2014). *Angewandte Chemie, International Edition* 53: 12880.
119 Meunier, B., de Visser, S.P., and Shaik, S. (2004). *Chemical Reviews* 104: 3947.
120 van Rantwijk, F. and Sheldon, R.A. (2000). *Current Opinion in Biotechnology* 11: 554.
121 Fingerhut, A., Serdyuk, O.V., and Tsogoeva, S.B. (2015). *Green Chemistry* 17: 2042.
122 Gelalcha, F.G. (2014). *Advanced Synthesis & Catalysis* 356: 261.
123 Olivo, G., Cusso, O., and Costas, M. (2016). *Chemistry – An Asian Journal* 11: 3148.
124 Anilkumar, G., Bitterlich, B., Gelalcha, F.G. et al. (2007). *Chemical Communications* 3: 289.
125 Schroeder, K., Enthaler, S., Join, B. et al. (2010). *Advanced Synthesis and Catalysis* 352: 1771.
126 Bitterlich, B., Schroeder, K., Tse, M.K., and Beller, M. (2008). *European Journal of Organic Chemistry* 4867.
127 Schröder, K., Tong, X., Bitterlich, B. et al. (2007). *Tetrahedron Letters* 6339.
128 Schroeder, K., Enthaler, S., Bitterlich, B. et al. (2009). *Chemistry, A European Journal* 15: 5471.
129 Chishiro, T., Kon, Y., Nakashima, T. et al. (2014). *Advanced Synthesis & Catalysis* 356: 623.
130 Chen, K., Costas, M., Kim, J. et al. (2002). *Journal of the American Chemical Society* 124: 3026.
131 White, M.C., Doyle, A.G., Jacobsen, E.N. et al. (2001). *Journal of the American Chemical Society* 2: 7194.
132 Clemente-Tejeda, D., López-Moreno, A., and Bermejo, F.A. (2013). *Tetrahedron* 69: 2977.

133 Clemente-Tejeda, D., López-Moreno, A., and Bermejo, F.A. (2012). *Tetrahedron* 68: 9249.

134 Costas, M. and Que, L. Jr. (2002). *Angewandte Chemie, International Edition* 12: 2179.

135 Cussó, O., Garcia-Bosch, I., Ribas, X. et al. (2013). *Journal of the American Chemical Society* 135: 14871.

136 Mikhalyova, E.a., Makhlynets, O.V., Palluccio, T.D. et al. (2012). *Chemical Communications* 48: 687.

137 Skobelev, I.Y., Kudrik, E.V., Zalomaeva, O.V. et al. (2013). *Chemical Communications* 49: 5577.

138 Cusso, O., Ribas, X., and Costas, M. (2015). *Chemical Communications* 51: 14285.

139 Gopalaiah, K. (2013). *Chemical Reviews* 113: 3248.

140 Gelalcha, F.G., Anilkumar, G., Tse, M.K. et al. (2008). *Chemistry, A European Journal* 14: 7687.

141 Gelalcha, F.G., Bitterlich, B., Anilkumar, G. et al. (2007). *Angewandte Chemie, International Edition* 46: 7293.

142 Wu, M., Miao, C.-X., Wang, S. et al. (2011). *Advanced Synthesis and Catalysis* 353: 3014.

143 Wang, B., Wang, S., Xia, C., and Sun, W. (2012). *Chemistry – A European Journal* 18: 7332.

144 Wang, W.F., Sun, Q.S., Xu, D.Q. et al. (2017). *ChemCatChem* 9: 420.

145 Mas-Balleste, R. and Que, L. (2007). *Journal of the American Chemical Society* 129: 15964.

146 Cussó, O., Ribas, X., Lloret-Fillol, J., and Costas, M. (2015). *Angewandte Chemie, International Edition* 54: 2729.

147 Cusso, O., Giuliano, M.W., Ribas, X. et al. (2017). *Chemical Science* 8: 3660.

148 Cusso, O., Cianfanelli, M., Ribas, X. et al. (2016). *Journal of the American Chemical Society* 138: 2732.

149 Chu, Y., Liu, X., Li, W. et al. (2012). *Chemical Science* 3: 1996.

150 Chu, Y., Hao, X., Lin, L. et al. (2014). *Advanced Synthesis & Catalysis* 356: 2214.

151 Matsumoto, K., Sawada, Y., Saito, B. et al. (2005). *Angewandte Chemie, International Edition* 44: 4935.

152 Berkessel, A., Günther, T., Wang, Q., and Neudörfl, J.-M. (2013). *Angewandte Chemie, International Edition* 52: 8467.

153 Wang, Q., Neudörfl, J.-M., and Berkessel, A. (2015). *Chemistry – A European Journal* 21: 247.

5

Green Oxidative Synthesis of Carboxylic Acids

Mingxin Liu and Chao-Jun Li

Department of Chemistry and FRQNT Centre for Green Chemistry and Catalysis, McGill University, Montreal, Quebec, Canada

Carboxylic acid (or carboxylate derivative) is one of the most important functional groups concerning all life forms. Proteins, as the main component of life, are constructed via carboxylic acid derivatives. Furthermore, carboxylate serves as an extremely important functional group in pharmaceuticals to bring crucial water-solubility to drug molecules. As the carbon in carboxylic acid is in its high oxidation state, methods to produce carboxylic acids are mainly based on oxidative methods. Herein, we will briefly summarize oxidative methods to produce carboxylic acids, with a particular focus on recent advances using "greener" oxidants.

5.1 Oxidation of Aldehyde

The pursuit of efficient and widely applicable greener oxidation of aldehyde has been challenging for the chemistry community. Although the direct use of the oxygen in air as an oxidant for large-scale aldehyde oxidation is widely applied in industry [1], usually catalyzed by simple transition-metal salts, such a protocol is only compatible with large-scale production of a specific carboxylic acid in the gas phase. It would be impractical, especially for relatively small-scale synthesis such as pharmaceuticals, to widely adapt the industry protocol. As a matter of fact, it is highly desirable for the development of widely applicable aldehyde oxidation using greener oxidants to install a highly useful carboxylate functional group in molecules [2].

5.1.1 Historic Methods and Challenges for Aldehyde Oxidation

Although the autoxidation of aldehyde in air has long been known about and can be significant, high yields with broad substrate scope of the corresponding carboxylic acids are very hard to achieve, as mixtures of products have often been obtained with autoxidation and with other common oxidants. In fact, oxidation of aldehydes poses a highly challenging task for chemists. Kinetically, the interaction between electrophilic aldehyde and also-electrophilic oxidant is difficult (Figure 5.1, left). Furthermore, the

Green Oxidation in Organic Synthesis, First Edition. Edited by Ning Jiao and Shannon S. Stahl.

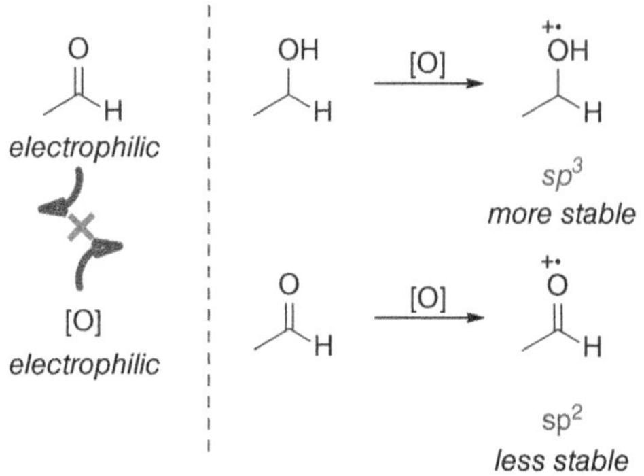

Figure 5.1 Challenges for aldehyde oxidation.

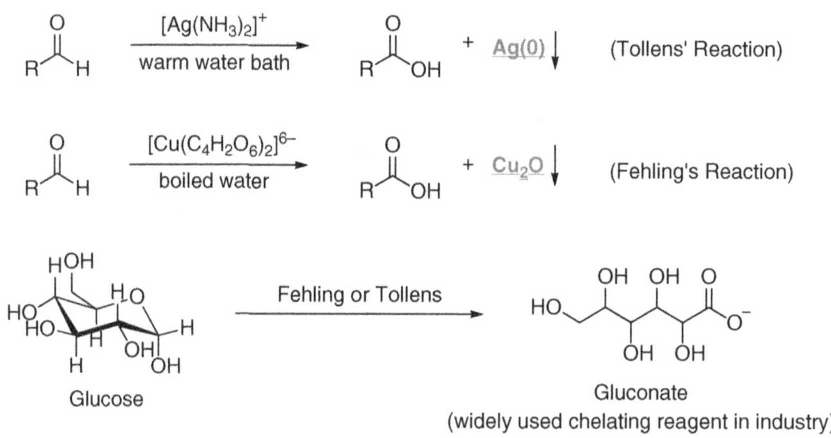

Figure 5.2 Fehling's and Tollens' reagents in aldehyde oxidation.

well-established single electron transfer (SET) mechanism for alcohol oxidations [3] is impractical for aldehydes, as the sp^2-hybridized radical-cation intermediate generated from aldehyde is much more unstable than the sp^3-intermediate from alcohol (Figure 5.1, right).

Historically, the most important aldehyde oxidation methods use Fehling's [4] or Tollens' oxidation [5]. Using metallic oxidants such as Cu(II) or Ag(I), remarkably efficient aldehyde oxidation can readily be achieved in a very short reaction time. Both reactions use water as the sole solvent and are performed under mild reaction temperatures (a warm water bath). Almost every type of aldehyde (aromatic, aliphatic, unsaturated, and even aldoses such as glucose and fructose) are efficiently oxidized into the corresponding carboxylic acid (aldonic acid). Such versatility has made Fehling's and Tollens' oxidation among the most important aldehyde oxidation methods in history, especially for aldehyde titration-analysis and the production of aldonic acid, which are extremely difficult to achieve using other methods (Figure 5.2). The mechanism for those reactions is still controversial, even today [6]. One hypothesis suggests that the aldehyde was first hydrated into the corresponding gem-diol and deprotonated in basic aqueous solution; then the Ag(I)/Cu(II) coordinated to the deprotonated gem-diol and extract one electron via SET, generating the gem-diol radical, which was further extracted one electron to give the carboxylic acid. Another hypothesis involves the formation of Ag(I)/Cu(II)–OH intermediate. The aldehyde was activated by the metal followed by the –OH nucleophilic attack on the activated carbonyl. The generation of the tetrahedral intermediate was followed by β-H elimination. Since Ag(I)/Cu(II)-H is not stable without an additional ligand, the metal hydrides collapse into lower valent metal species and proton.

Another important example of classical aldehyde oxidation is Jones' oxidation [7]. Although mainly used for alcohol oxidation, the Jones' reagent has made a significant contribution in the oxidation of aldehyde due to good functional tolerance and selectivity. The mechanism of the Jones' oxidation involves the initial formation of alcohol perchromate ester (general formula $CrO_3(OCH_2R)^-$). The ester then collapses with Cr(VI) obtaining two electrons to become Cr(IV) and releasing the carboxylic acid. Even today, the Jones' oxidation of aldehyde has broad applications. For example, in Hao's very recent total synthesis of Perforanoid A [8], Jones' reagent played a key role in selectively oxidizing specific acetal into the corresponding ester (Figure 5.3).

Although achieving great utility and efficiency, the stoichiometric metallic wastes generated in previously mentioned methods are particularly hazardous and difficult to process. As more and more restrictions were added regarding metal waste generation, the development of new aldehyde oxidation method without generation of stoichiometric metallic waste is eagerly demanded by academia and industry. In 1973, Lindgren first demonstrated the use of chlorite (ClO_2^-) as an oxidant toward oxidation of aldehyde [9]. Though a very limited range of substrates were obtained, the scope includes both aliphatic and aromatic aldehydes. Later in 1981, Pinnick demonstrated the oxidation of unsaturated aldehyde in a similar system [10]. This system was later recognized as the Pinnick oxidation. As one of the most widely applied aldehyde oxidation methods, even today, many products in chemical industry, especially pharmaceuticals, have been synthesized via this method (Figure 5.4) [11]. Despite the great utility achieved by Pinnick oxidation, certain problems still remain. Since the oxidation generates stoichiometric amounts of sodium hypochlorite (HOCl), which consumes the ClO_2^- oxidant and inhibits the desired transformation, a scavenger was generally required to eliminate HOCl. The most commonly used scavenger was 2-methyl-2-butene, which

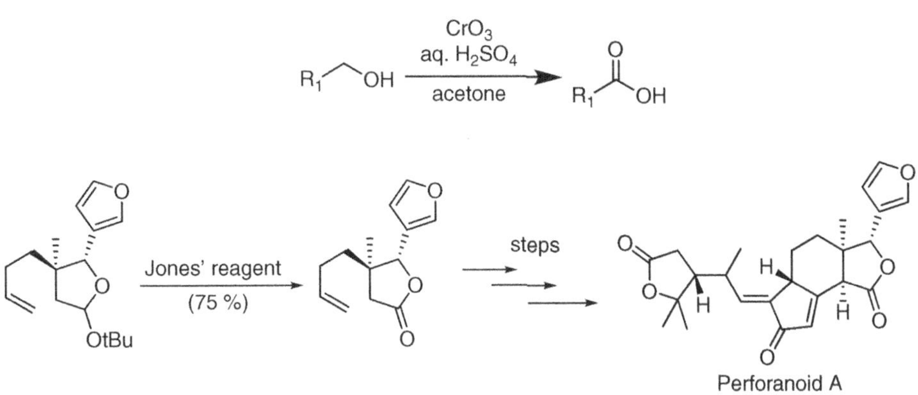

Figure 5.3 The Jones' reagent and its application in aldehyde oxidation.

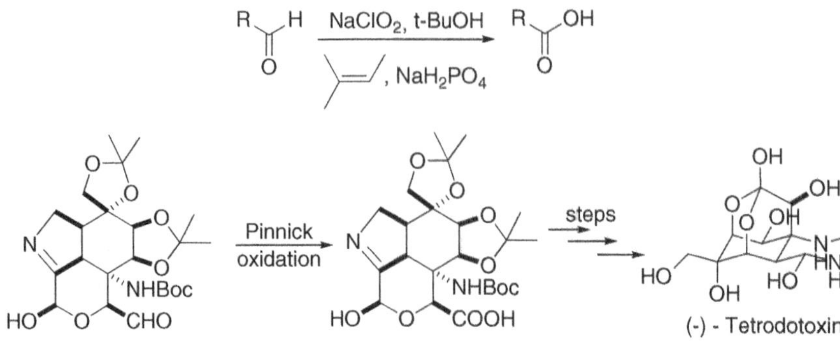

Figure 5.4 Pinnick oxidation and its application in pharmaceutical synthesis.

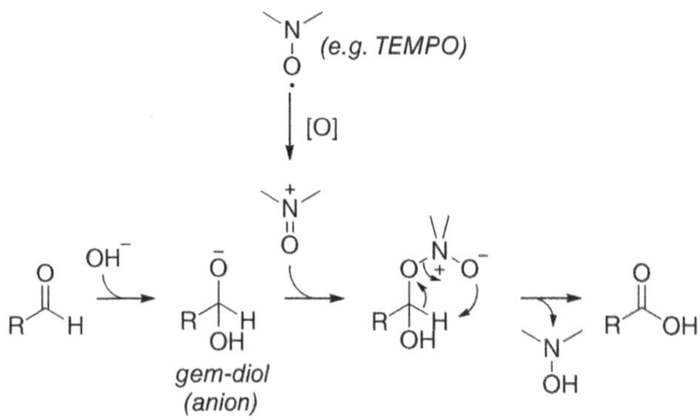

Figure 5.5 A typical mechanism for aldehyde oxidation using persistent radical system.

was added stoichiometrically in almost every practice. Furthermore, some aldehydes show unsatisfying reactivity. For example, aliphatic conjugated C=C, unprotected amine/pyrrole and thioether contained in aldehydes are easily oxidized in the Pinnick oxidation conditions.

As early as 1979, Evans and co-workers have studied the effect of anions toward aldehyde oxidation using electrochemistry with different electrodes [12]. The results showed that the addition of OH^- to the aldehyde can dramatically boost its tendency toward oxidation. The generated gem-diol has a great oxidation potential shift (can be as great as 3.6 V), indicating the potential to conduct aldehyde oxidation using this alternative approach. The persistent radical (TEMPO, bicyclic-nitroxyl, etc.) /NO_x-catalyzed method, which has been used for alcohol oxidation [13], has been directly applied to the oxidation of gem-diol of the corresponding aldehyde (Figure 5.5) [14]. However, with those persistent radicals, which are generally volatile and reactive, there was poor functional tolerance.

5.1.2 Enzymatic Oxidation of Aldehyde

Despite the challenges and difficulties in developing aldehyde oxidation method, such a process also widely exists in nature and the human body [15]. Those biochemical oxidations of aldehyde *in vivo* are generally catalyzed by a well-engineered enzyme from nature using water as the sole solvent at a mild temperature (in most cases at 37 °C). Dissolved oxygen is either directly used from water, or consumed through biochemical oxidative cascade. Although most natural enzymes exhibit specificity toward the substrate, many functionalized enzymes have been isolated and investigated as catalysts for the oxidation of different aldehydes. With the human dehydrogenase family alone in our liver cell, the combined substrate scope has already covered almost all common aldehydes, including aliphatic, aromatic, and unsaturated aldehydes [16]. Those processes are well adapted in some of the specific synthesizing processes. However, it is usually difficult for many industries to carry out *in vivo* or biomimetic reactions due to the lack of equipment, and so on. Furthermore, the enzyme specificity usually demands the industry to obtain a large number of those specific enzymes, and many of them require different storage and reaction conditions.

5.1.3 New Oxidants for Catalytic Aldehyde Oxidation

Generally, due to the difficulty described in Section 5.1.1, up till now, methods that can achieve efficient and widely applicable oxidation of aldehydes are very limited. In attempt to solve these problems, at first, chemists were looking forward to alternative oxidants that does not require or generate undesirable materials. Those oxidants generally consist of high-valent or electron-deficient center, and anionic oxygen to increase the affinity toward the already-electrophilic carbonyl (Figure 5.6). The mechanism of those aldehyde oxidations is generally initiated by anionic oxygen attack on the carbonyl to generate the tetrahedral intermediate; then, the tetrahedron collapses and the electron is extracted by the electron-deficient oxidant to generate the desired product.

Among those catalysts, potassium hydropersulfate (oxone) showed great potential for its structural stability and cost-effectiveness. In 2003, Borhan systematically studied the use of oxone toward oxidation of aldehydes (Scheme 5.1) [17]. The oxidation showed good efficiency and functional group tolerance and proceeded under mild reaction conditions (50 °C, DMF, 3 h). Furthermore, the oxidation can also afford the desired ester by switching solvent to the corresponding alcohol. However, the oxone oxidation is not effective for unsaturated aldehyde and hydroxyl/alkoxyl containing substrates. Bulky *tert*-pentanal and even less bulky *iso*-butanal were not reactive either.

Another effort toward the development of alternative oxidant was introduced in 2005. While pyridinium chloroperchromate (PCC) was developed as a highly efficient alcohol oxidation reagent, Hunsen developed the combined use of catalytic PCC and stoichiometric periodate as efficient oxidation of aldehyde [18], which generates the active oxidant chlorochromatoperiodate *in situ* (Scheme 5.2). A satisfying substrate

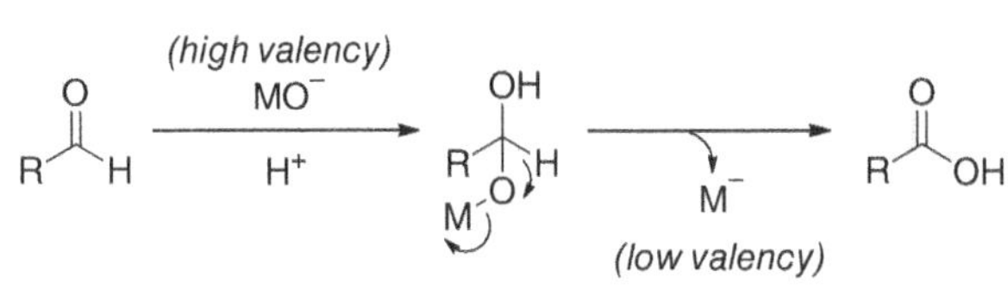

Figure 5.6 Typical mechanism for previous aldehyde oxidation.

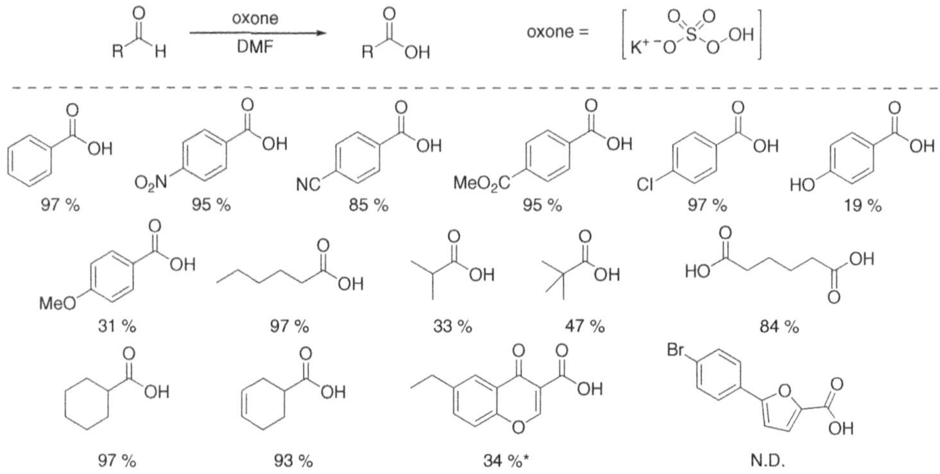

Scheme 5.1 Oxidation of aldehydes using oxone.

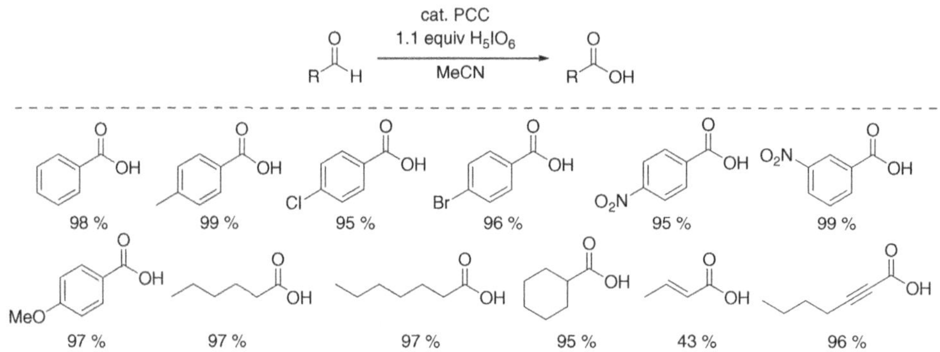

Scheme 5.2 PCC-catalyzed aldehyde oxidation using periodic acid.

scope was achieved. However, excessive amount of periodate oxidant was required, which potentially resulted in side reactions such as C—C cleavage [19].

To circumvent the undesired side-reaction caused by excessive oxidant, one solution is to use gaseous oxidant that can be introduced slowly in a controlled manner. As ozonolysis with an alkene into the corresponding aldehyde/ketone is well established [20], it is natural to assume that ozone can also oxidize aldehyde into the corresponding carboxylic acid. In 2008, Johnson summarized the ozone oxidation of aldehyde attempts [21]. While achieving significant results, however, poor functional group tolerance was obtained due to the harsh oxidation by ozone. In 2016, nitrous oxide was examined as a milder alternative of ozone to oxidize aldehyde [22]. It was demonstrated that nitrous oxide oxidation of cyclohexanecarboxyaldehyde was very efficient under atmospheric pressure and room temperature (Scheme 5.3. Although a decarbonylation side-reaction was also observed, the use of nitrous oxide still shows interesting potential.

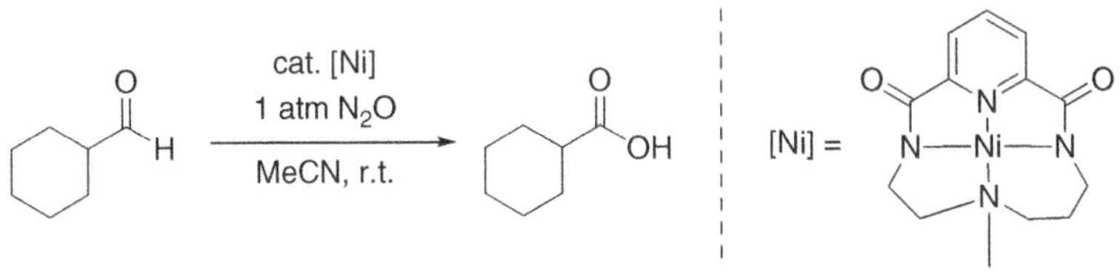

Scheme 5.3 Nitrous oxide oxidation of aldehyde.

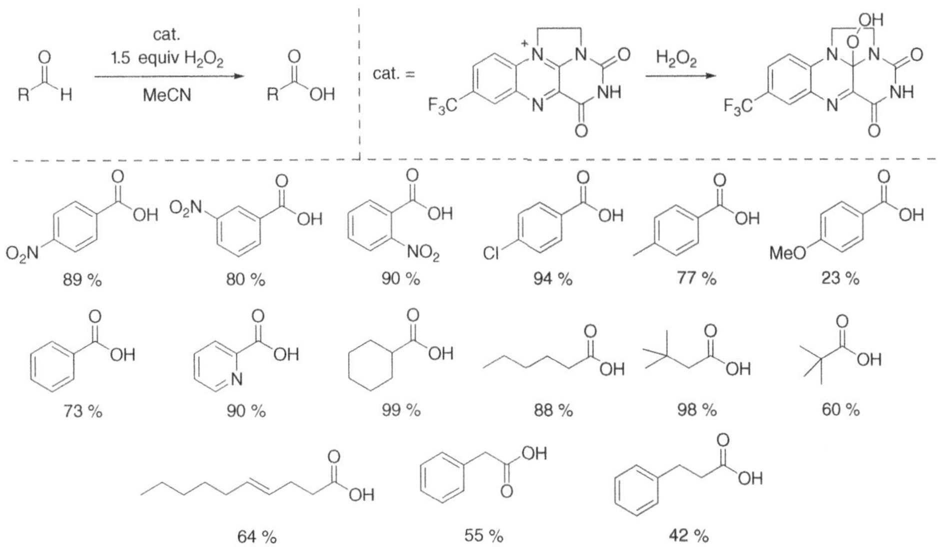

Scheme 5.4 Flavin-catalyzed biomimetic oxidation of aldehydes.

Another possible solution is the use of a milder oxidant in aldehyde oxidation. Peroxide (OOH^-/O_2^{2-}) is more desirable for its cost-effectiveness and high atom economy. Inspired by the mechanism of flavin bioluminescence, which undergoes catalytic nucleophilic addition of peroxide to the fatty aldehyde followed by the collapse of tetrahedral intermediate to eject fatty acid and a photon, a flavin-catalyzed biomimetic oxidation of aldehyde was reported by Carbery using hydrogen peroxide in 2012 [23]. An impressive substrate scope was obtained, and even bulky *tert*-pentanal gave a satisfying yield (60%). However, unsaturated aldehydes were not reactive (Scheme 5.4).

In 2017, Hlavac noticed that hydrazine, under certain circumstances, can oxidize aldehyde into the corresponding carboxylic acid by cleavage of a C—N bond and the ejection of nitrogen gas [24]. He later further developed the system using resin linker and presynthesized hydrazone [25]. A series of aromatic aldehydes successfully underwent the desired transformation (Scheme 5.5). Despite the scope of substrates was limited to aromatic aldehydes, this approach still shows interesting potential by (i) avoiding the use of transition-metal catalyst and (ii) using resin support that can be easily cleaved to facilitate the purification of product.

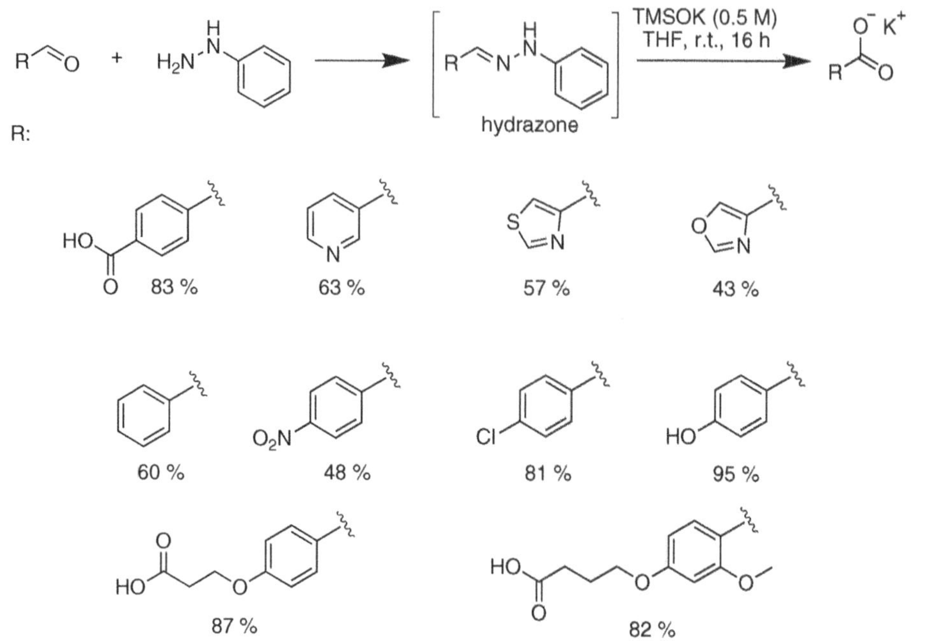

Scheme 5.5 Catalytic oxidation of hydrazones.

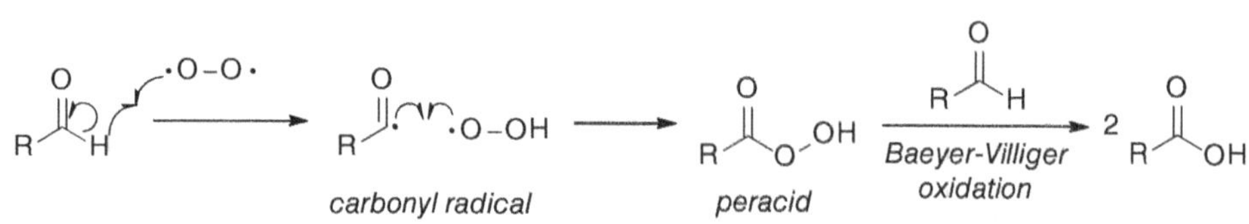

Figure 5.7 Mechanism of autoxidation of aldehyde.

5.1.4 Catalyzed Aerobic Oxidation of Aldehyde

Aerobic oxidation of aldehyde toward carboxylic acid is probably one of the most exciting approaches, as no hazardous oxidant is used and the highest atom economy can be achieved. Despite the promising outcome, this area of research has received very limited attention. The autoxidation of aldehyde by aerobic oxygen is generally significant, however, giving incomplete conversion and mixture of products as well as having a limited substrate scope. The mechanism for the autoxidation has been thoroughly studied (Figure 5.7) [26]. The process involves hydrogen extraction by oxygen radical, generating a peracid, which oxidizes another aldehyde, similar to the Baeyer–Villiger mechanism, to generate two carboxylic acids. Both the carbonyl radical and peracid are very reactive and cause many side reactions, which potentially contribute to the low efficiency of autoxidation and many attempts at the catalytic aerobic oxidation of aldehyde.

In 2009, Tada reported the synthesis of a SiO_2 supported Ru-*p*-cymene catalyst for aerobic oxidation of aldehyde [27]. A labile *p*-cymene ligand was designed to generate an empty coordination site for oxygen (Figure 5.8). The catalyst shows good recyclability

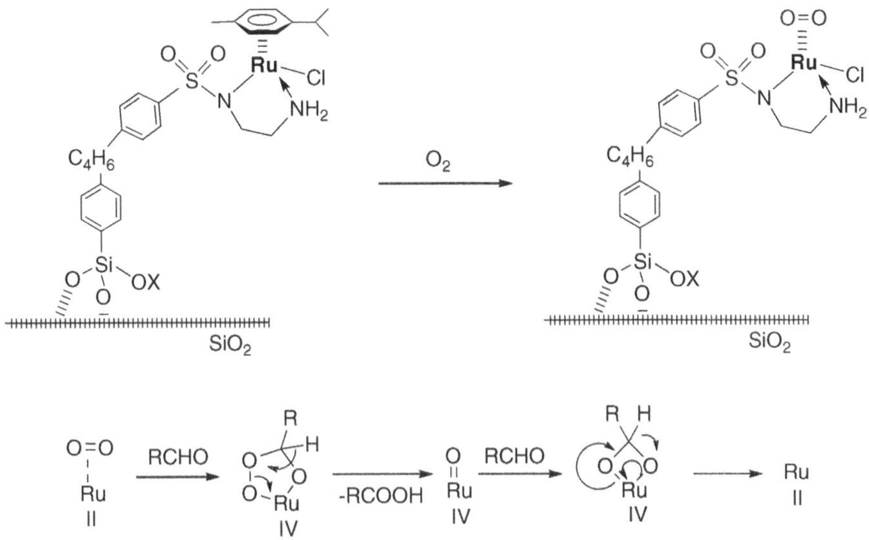

Figure 5.8 Ru-catalyzed aerobic oxidation of aldehyde.

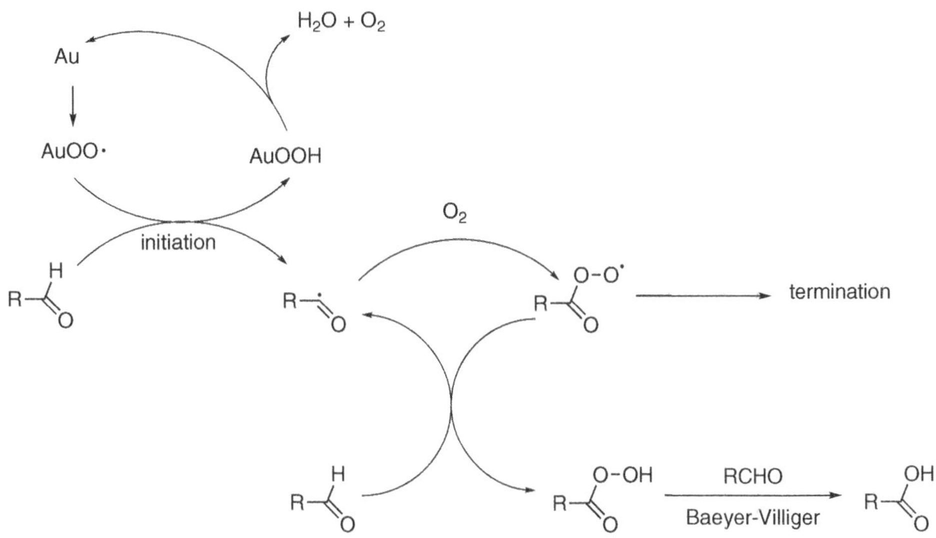

Figure 5.9 Mechanism of AuNP-catalyzed aerobic oxidation of aldehydes.

and ability to catalyze the aerobic oxidation of a variety of substrates. However, the catalyst is also shown to catalyze epoxidation of alkene, and thus greatly limits the substrate scope of this method.

In 2009, Chechik and co-workers reported the aerobic oxidation of aldehyde using simple oxygen and gold-nanoparticles as the catalyst [28]. Although significant, those oxidations did not avoid the generation of previously mentioned unstable sp^2-hybridized carbonyl radical and the peracid (Figure 5.9), which resulted in low reaction efficiency and limited scope of substrates.

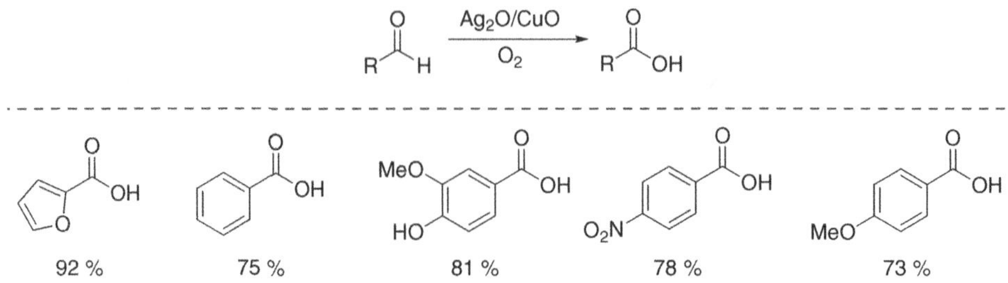

Scheme 5.6 Ag_2O/CuO-catalyzed aldehyde aerobic oxidation.

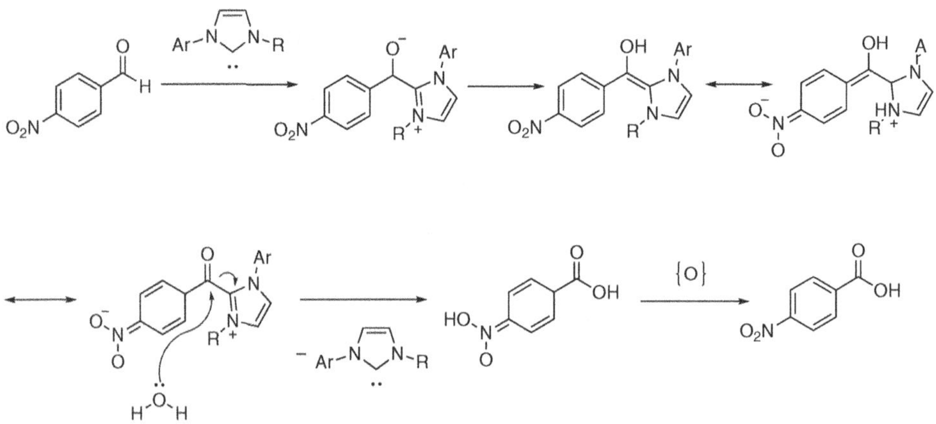

Figure 5.10 NHC-catalyzed aerobic oxidation of aromatic aldehydes.

In 2008, aerobic oxidation of aldehyde using a heterogeneous Ag_2O/CuO catalyst in water was reported [29]. A few examples of substrate were reported to be efficiently oxidized (Scheme 5.6). However, in addition to the requirement of high catalyst loading, limited substrate scope and side reactions were also obtained.

In 2009, N-heterocyclic carbene (NHC) was developed as aerobic oxidation organocatalyst [30]. Significantly different from the previous aerobic oxidation mechanism, the oxidation proceeds via the nucleophilic attack of the carbene to generate the Breslow intermediate, which was further hydrolyzed to give the carboxylic group (Figure 5.10). Oxygen was then involved to restore aromaticity of the system. However, the activation of oxygen was driven by the restoring of aromaticity, and therefore the substrate was limited to aromatic aldehyde. Furthermore, this method cannot evade the requirement of organic solvents, which have to be kept in an air-free and anhydrous environment.

Inspired by the biological aldehyde oxidation described in Section 5.1.2 and the classic Tollens' oxidation, in 2015, Li and co-workers reported a highly efficient silver(I)-catalyzed aerobic oxidation of aldehyde in water [31]. With the use of a common ligand, 2,2′-bipyridyl, the initially developed oxidation uses silver(I) hexafluorophosphate as catalyst. Under moderately basic conditions and sodium fluoride additive in water, a series of simple aldehydes was successfully oxidized into the corresponding carboxylic acids (Scheme 5.7).

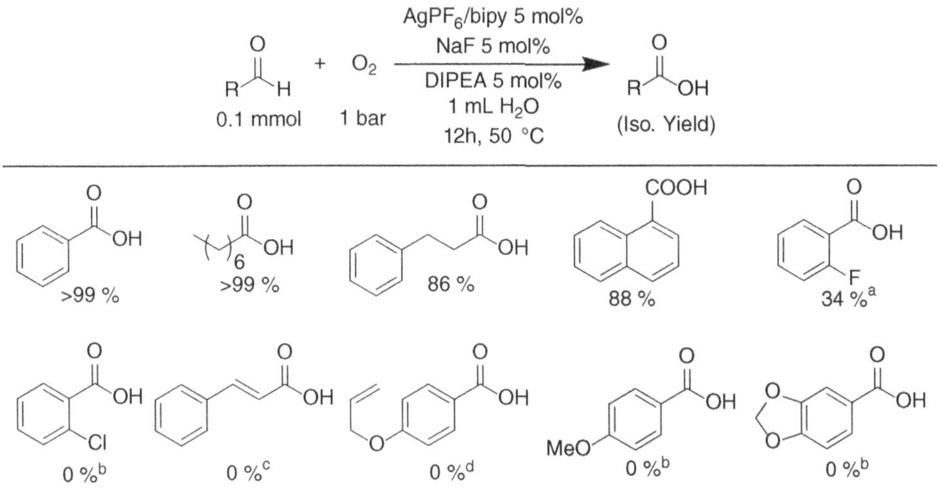

a 60 % of starting material was converted
b Starting material all recovered
c Starting material all converted
d 71 % of starting material was converted

Scheme 5.7 The first protocol of Li's silver-catalyzed aerobic oxidation.

However, more functionalized aldehydes, especially those containing C—C double bonds and heteroatoms, did not give satisfying results. To address this problem, it was later found that NHC ligand was extremely efficient at facilitating the silver(I)-catalyzed aerobic oxidation of aldehydes. With the new conditions, over 50 examples of different aldehydes (Scheme 5.8), including a natural product aldehyde, was readily oxidized into their corresponding carboxylic acid (even on a large scale) very efficiently, under ambient temperature and atmospheric oxygen. Another highlight of this work is that the work-ups only contain simple liquid-liquid extraction. Chromatography was generally not necessary to obtain analytical-grade products.

Despite the demonstrated efficiency of Li's silver(I)-NHC catalyst system, expensive silver was still required. Historically, since the Fehling's oxidation was very similar to and almost as equally important as the Tollens' oxidation, in 2016 Li and co-workers reported a copper-catalyzed aerobic oxidation of aldehydes in water using NHC as ligand [32]. With very similar conditions compared to their previous work, the copper-NHC catalyst also showed exceptional oxidation efficiency toward a variety of aldehydes to provide their corresponding carboxylic acids (Scheme 5.9). The simple and facile work-up also demonstrated without the need of chromatography. Mechanistic studies suggested a copper-hydride being the crucial intermediate, which was generated from β-hydride elimination of the aldehyde intermediate (Figure 5.11).

Thus, in the effort of making the aldehyde oxidation even "greener," Li and co-workers have shown the feasibility of using copper to substitute silver catalyst. Furthermore, the ideal catalyst in this scenario should be iron. Indeed, later in 2016, Ma and co-workers successfully demonstrated that iron-catalyzed aerobic oxidation of aldehyde can also be highly efficient under ambient conditions [33]. Abundant iron(III) nitrate

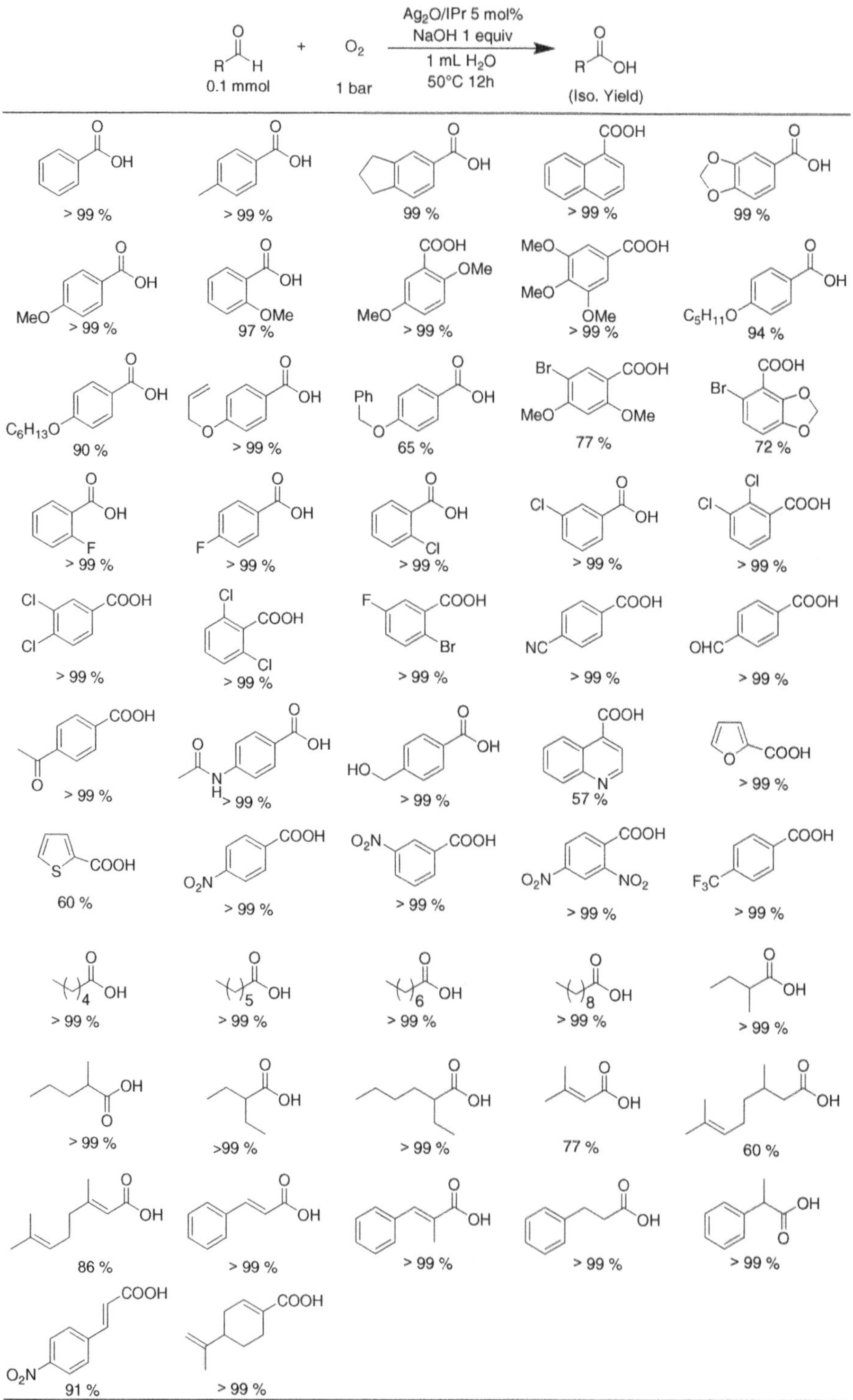

Scheme 5.8 Improved conditions for Li's silver(I)-catalyzed aerobic aldehyde oxidation.

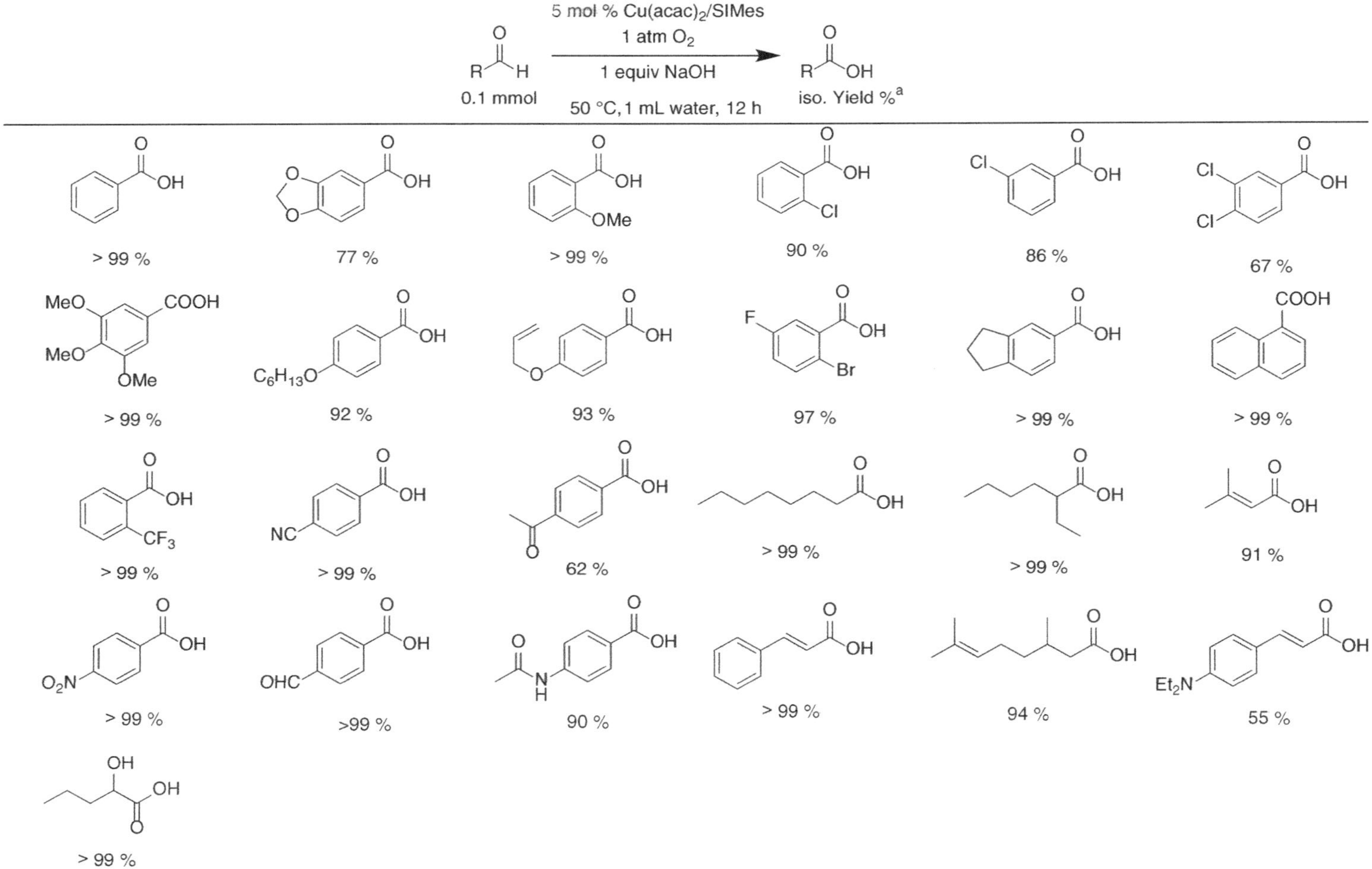

[a] All reactions were conducted with the standard oxidation procedure: Aldehyde (0.1 mmol), Cu catalyst (0.005 mmol, 5 mol %), NaOH (0.1 mmol, 1 equiv) was mixed in 1 mL water and non-pressurized oxygen with stirring at 50 °C for 12 h

Scheme 5.9 Li's copper-catalyzed aerobic aldehyde oxidation.

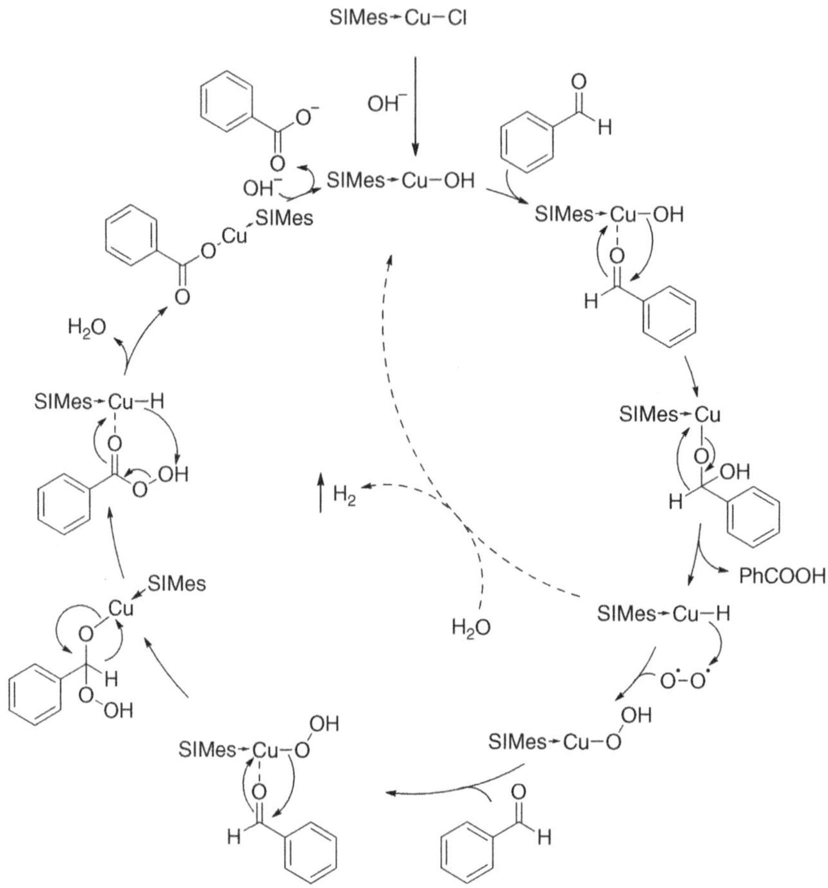

Figure 5.11 Mechanism of Li's copper-catalyzed aerobic aldehyde oxidation.

was used as catalyst, with TEMPO as additive. The catalyst system achieved selective oxidation of alcohol or aldehyde with different conditions. Many natural products and their derivatives were also successfully demonstrated. It was also proposed that β-hydride elimination of aldehyde intermediate was also the crucial step in the mechanism (Figure 5.12).

5.2 Oxidation of C—H

Due to the drastic oxidation state change, oxidation of C—H to directly produce carboxylic acid is rarely adopted for lab or pharmaceutical scale synthesis, because a strong oxidant is often needed and the reaction often lacks selectivity.

It has long been known that toluene and its related compounds such as xylene, mesitylene, and so on, can bleach acidic $KMnO_4$ solution, indicating the presence of an oxidation reaction. It was later realized that carboxylic acid (benzoic acid, terephthalic acid, etc.) is the main product in this reaction. Nowadays, a similar method has been modified

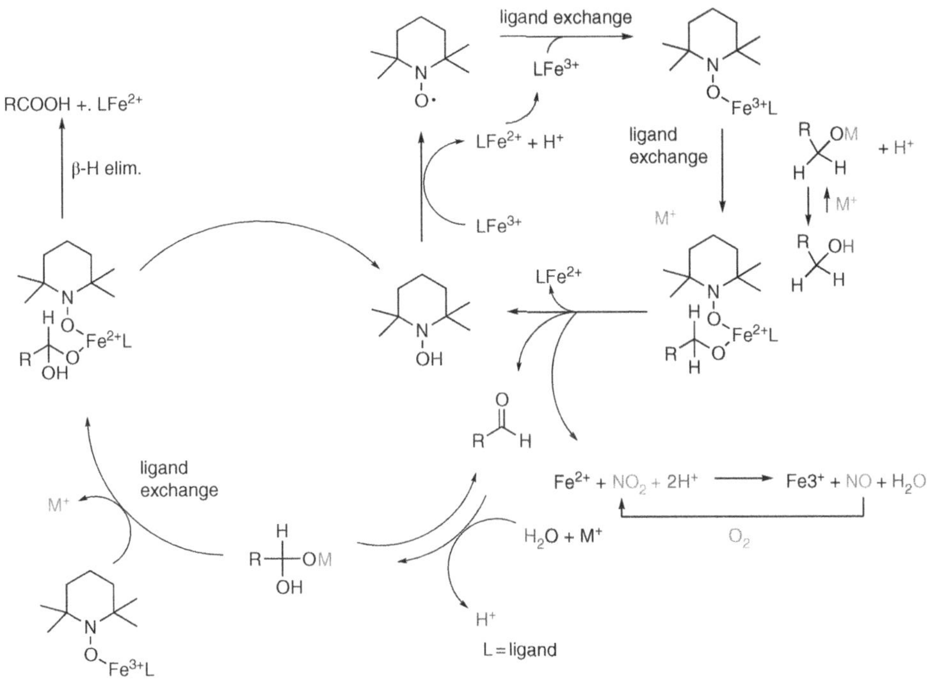

Figure 5.12 The mechanism of Ma's aldehyde and alcohol oxidation.

to accommodate the major industrial production of benzoic acid (Figure 5.13). By using cobalt or manganese naphthenate as the catalyst and oxygen as the oxidant, oxidation of toluene directly into benzoic acid can give a very high yield on a large scale [34].

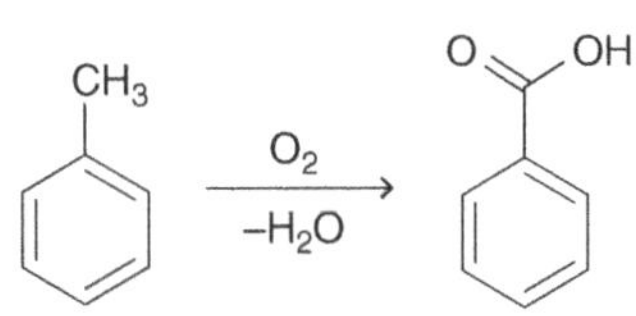

Figure 5.13 Catalytic aerobic oxidation of toluene to produce benzoic acid.

Recently, direct C—H activation has become a hotspot in chemistry. In 2014, Dong and co-workers reported an oxidative C—H activation to achieve carboxylic acid derivatives in a cross-coupling fashion [35]. By using nickel catalyst oxidative addition to the C—H bond of an aldehyde, followed by transmetallation of an alkoxide or alkylamino anion, carboxylic derivatives such as esters and amide were readily achieved. The hydride was extracted using 2,2,2-trifluoroacetophenone, which formally serves as the oxidant in the reaction (Figure 5.14). Although the overall reaction was still the oxidation of aldehyde, it was classified here as a C—H oxidation method due to its unique –H transfer mechanistic pathway.

5.3 Oxidative Cleavage of C—C or C=C

5.3.1 Oxidative Cleavage of C=C

Similar to toluene, it has long been known that olefin/alkene can also bleach acidic $KMnO_4$ solution to produce carboxylic acid. Although poor selectivity has been

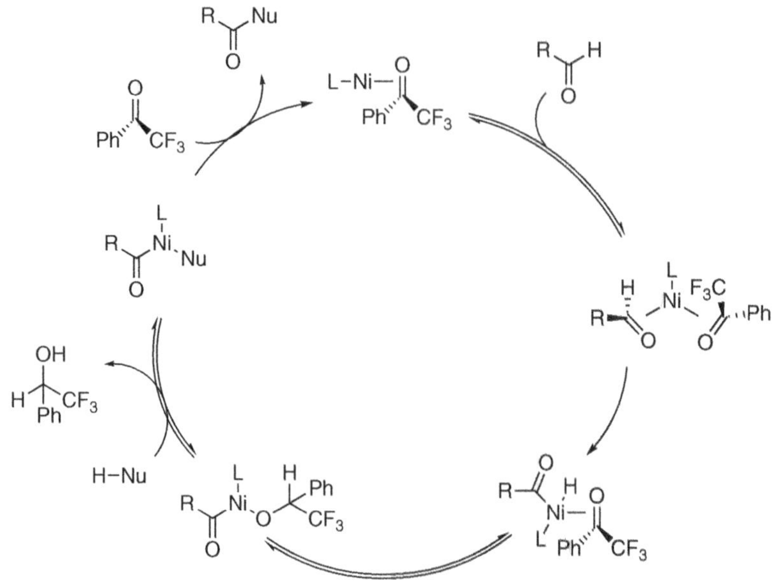

Figure 5.14 Dong's C—H activating aldehyde oxidation.

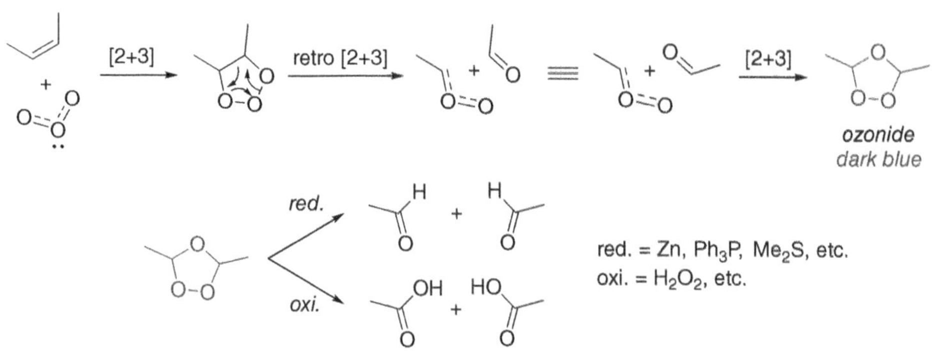

Figure 5.15 Ozonolysis and its reductive and oxidative work-ups.

achieved for $KMnO_4$ due to its strong oxidizing potential, the concept of oxidative C=C oxidative cleavage is still promising. Ozonolysis is among the most promising and clean classical methods to achieve C=C oxidative cleavage [36]. The reaction went through a crucial five-membered ring, named ozonide, after C=C reacted with ozone. Then, depending on the type of work-ups used to cleave the ozonide ring, the reaction selectively generated either aldehyde/ketone (reductive) or carboxylic acid (oxidative). Although reductive work-ups, such as triphenylphosphine, zinc, thiolether/thiourea, are much more common in ozonolysis, oxidative work-up using peroxides such as H_2O_2 produced carboxylic acid efficiently and selectively (Figure 5.15).

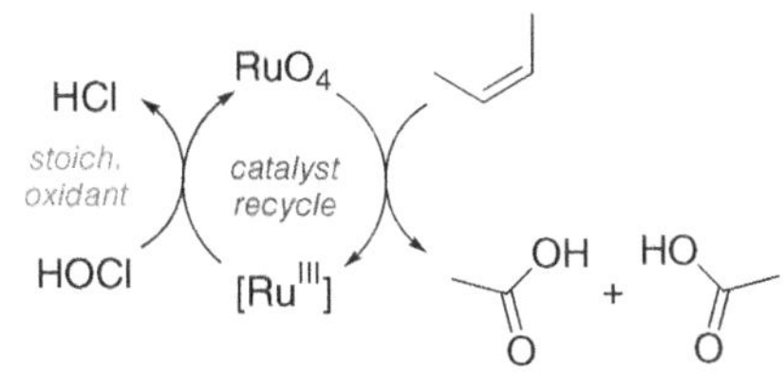

Figure 5.16 Wolfe's Ru-catalyzed oxidative cleavage of olefin.

RuO_4 is known to be a powerful oxidant for the oxidation of organic compounds including C═C [37], despite being expensive and hazardous itself. It was then noticed by chemists that RuO_4 can "be kept unchanged for years if a few drops of chlorine water are added to it" [38], indicating the potential to use RuO_4 catalytically with hypochlorite as an oxidant (Figure 5.16). In 1970, Wolfe and co-workers reported the oxidation of organic compounds using hypochlorite catalyzed by Ru [39]. The reaction gave selective cleavage of C═C into the corresponding carboxylic acids in good efficiency, indicating an early example of green oxidation of C═C.

This strategy was shown to be effective in many other systems, by using a greener oxidant to regenerate the catalytic amount of metal-oxo active species to achieve oxidation. In 1998, a poly-tungstate catalyst was reported by Venturello and co-workers for the oxidative cleavage of C═C using hydrogen peroxide as the oxidant [40]. The poly-peroxo-tungstate was shown before to be a suitable oxidant for C═C oxidative cleavage [41]. The catalytic version was also demonstrated to be efficient and a range of linear or cyclic olefins were successfully cleaved into the corresponding carboxylic acids. Mechanistic studies suggested the reaction to proceeded by initial epoxidation of C═C followed by ring-opening to give 1,2-diol, which was then cleaved to give the product (Figure 5.17).

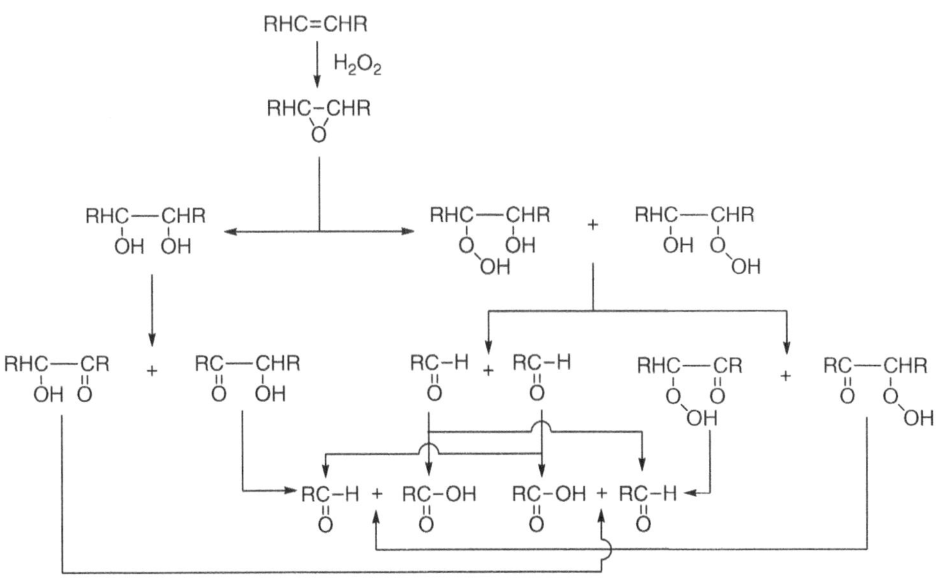

Figure 5.17 Mechanistic pathways for W-catalyzed oxidative olefin cleavage.

5.3.2 Oxidative Cleavage of 1,2-Diol

In many cases, the oxidative cleavage of C=C proceeds via a 1,2-diol intermediate. Therefore, many conditions that afford C=C oxidative cleavage can also cleave 1,2-diols. Pioneering studies in the 1,2-diol oxidative cleavage area were represented by the Malaprade reaction [42] and the Criegee oxidation [43], using stoichiometric amounts of periodate (H_5IO_6) derivatives or lead ($Pb(OAc)_4$), respectively. These pioneering methods are very frequently applied to the oxidative cleavage of diol even today, as one of the primary methods to synthesize carboxylic acids. Later, many methods, especially catalytic ones, were developed to advance the sustainability of 1,2-diol cleavage by avoiding the use of hazardous oxidant. As most of those methods achieve oxidative cleavage of 1,2-diol into the corresponding aldehyde or ketone product, only representative methods affording carboxylic acids will be discussed as follows.

5.3.2.1 Oxygen as Oxidant

The first catalytic aerobic oxidative cleavage of 1,2-diol was reported in 1968 by Schors and co-workers [44]. Co(II) was demonstrated to be a good catalyst in an aprotic, polar solvent. Although only two substrate examples were shown, the reaction could afford good carboxylic acid selectivity without the need for separation of the aldehyde intermediate (Figure 5.18).

As the cleavage reaction requires an efficient catalyst interaction with oxygen, a catalyst with high surface area is obviously beneficial. In 1987, Felthouse reported a class of ternary ruthenium oxide catalyst, represented by $Pb_{2.62}Ru_{1.38}O_{6.5}$, as a good catalyst for aerobic oxidative cleavage of 1,2-diols [45]. It was suggested that the Pb(IV) and oxo-Ru(IV) center on the catalyst surface promotes the cleavage (Figure 5.19).

The adaptability of Ru as aerobic oxidative 1,2-diol cleavage catalyst was greatly enhanced in 1999 by Ishii and co-workers [46]. The Ru catalyst was coordinated by a triphenylphosphine (PPh_3) ligand as good electron donor and stabilized by the ligand.

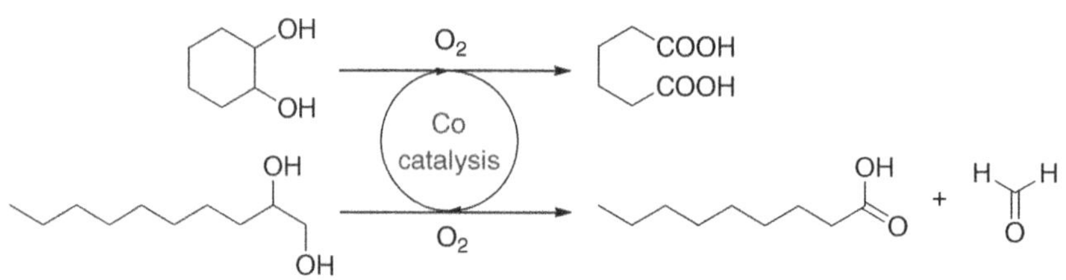

Figure 5.18 Schors' aerobic oxidative cleavage of 1,2-diol in 1968.

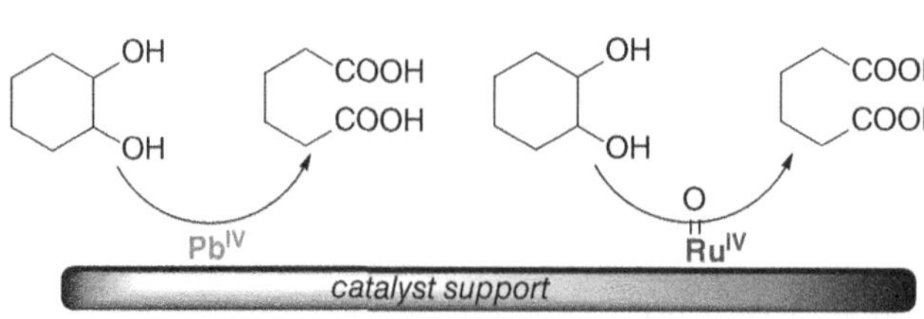

Figure 5.19 Aerobic oxidative 1,2-diol cleavage by Felthouse's heterogeneous catalyst.

Active carbon was also used to enhance the surface area and oxygen adsorption of the catalyst. The catalyst gave both carbonyl and carboxylic products with a range of different 1,2-diols. An RuH_2 intermediate was proposed for the reaction mechanism, which was generated by hydride extraction of alcohol by Ru and oxidized by oxygen (Figure 5.20).

In 2000, Ishii reported a new aerobic oxidative cleavage of 1,2-diols into carboxylic acids [47]. By combining a Co(II) catalyst, which was known to conduct the oxidative 1,2-diol cleavage, and a N-hydroxyphalimide (NHPI) co-catalyst, a substrate scope with aliphatic-, aromatic-, terminal-, internal-, and cyclic-1,2-diols was achieved. It was suggested that the Co(II) and NHPI conducted the cleavage by extracting α-hydrogen atom from alcohol, generating an α-radical and promoting cleavage (Figure 5.21).

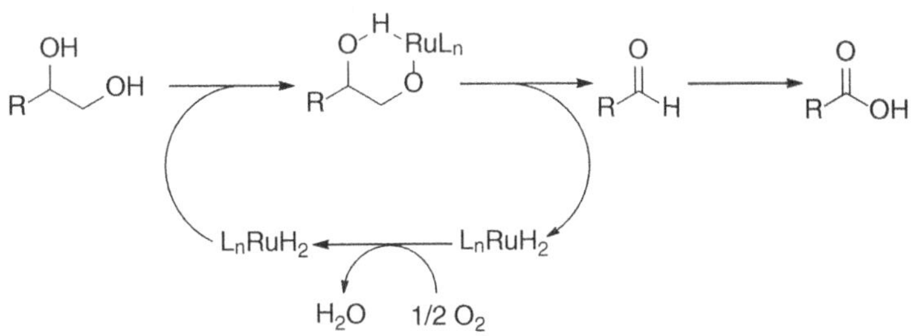

Figure 5.20 Ishii's aerobic 1,2-diol oxidative cleavage catalyzed by Ru in 1999.

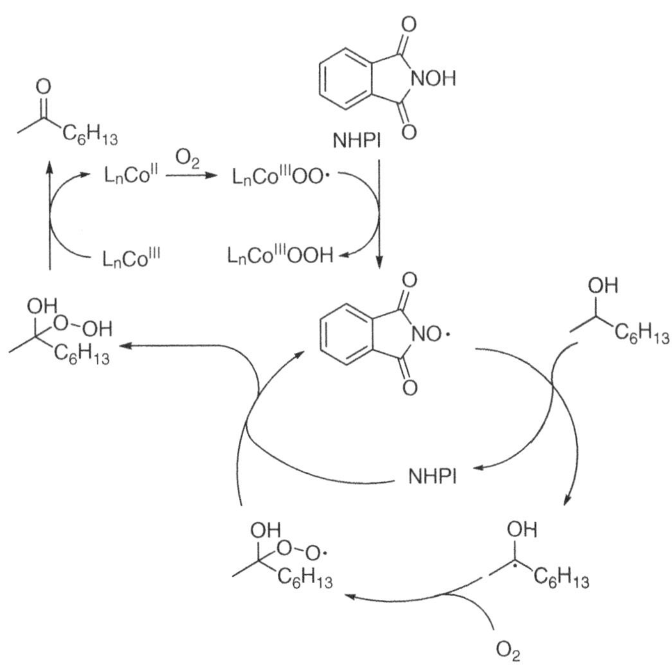

Figure 5.21 The initial step of Ishii's aerobic oxidative cleavage of 1,2-diol catalyzed by Co and NHPI.

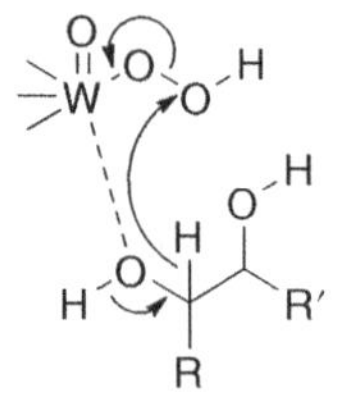

Figure 5.22 Transition state of 1,2-diol oxidative cleavage by H_2O_2 in 1986.

5.3.2.2 Peroxides as Oxidants

Similar to oxygen, peroxides can also be good greener oxidants for oxidative cleavage of 1,2-diol. The simplest peroxide, hydrogen peroxide (H_2O_2), was demonstrated to serve as oxidant for 1,2-diol oxidative cleavage in 1986 [41b]. Over 10 examples of different 1,2-diols have been successfully demonstrated to be cleaved into their corresponding carboxylic acids in excellent selectivity with tungstate (WO_4^{2-}) as the catalyst. Similar to aerobic cleavage, the oxidative cleavage of 1,2-diol by H_2O_2 also proceeds via the initial oxidation of alcohol to carbonyl (Figure 5.22).

Since WO_4^{2-} can easily aggregate to produce polyoxometalate (POM), in 1988 Ishii examined different poly-molybdic acid and poly-tungstic acid as catalyst for oxidative 1,2-diol cleavage with H_2O_2 [48]. Improved substrate adaptability was achieved as well as catalyst recyclability. It was found the metal peroxo-species serves as a key intermediate in the reaction mechanism. Similar to the WO_4^{2-} catalyzed reaction, the reaction catalyzed by POM also proceeds by initial oxidation of alcohol to carbonyl (Figure 5.23).

As the feasibility of poly-molybdic acid to serve as a catalyst for oxidative 1,2-diol cleavage was demonstrated, in the same year, 1988, mono-molybdate, $MoO_2(acac)_2$, was also reported as an effective catalyst for oxidative 1,2-diol cleavage [49]. Here, *tert*-butylhydroperoxide (TBHP) was used as the oxidant and gave a good selectivity toward carboxylic acid.

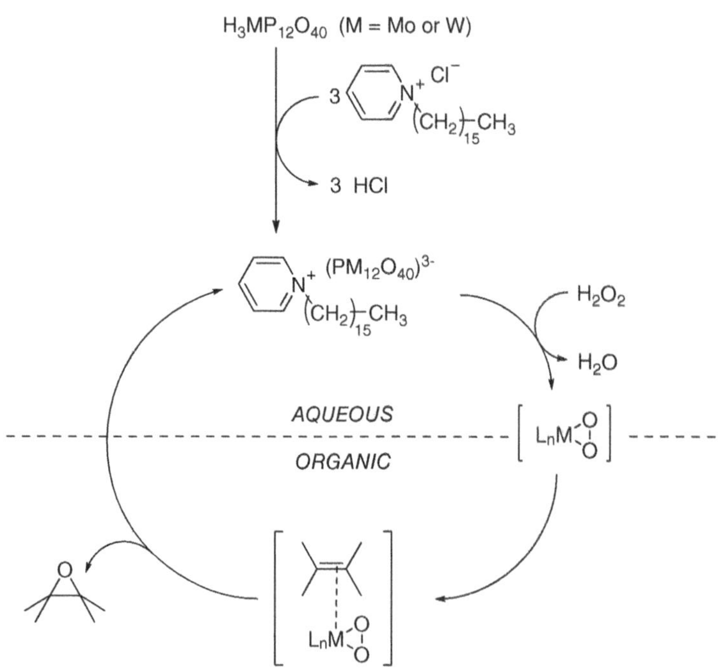

Figure 5.23 Ishii's POM-catalyzed oxidative 1,2-diol cleavage.

5.4 Conclusion

With precise design of the catalyst, especially via utilization of various functionalized ligands for homogeneous catalysis or specific doping/geometry/support for heterogeneous catalysis, the desired selectivity and efficiency of the aldehyde oxidation by using a greener oxidant (O_2) can be achieved via multiple methods. Furthermore, the development of innovative aldehyde oxidation with special regio-, chemo- and stereoselectivity, and adaptability remains highly challenging. The development of those reactions will not only facilitate the transformation process of aldehyde into various everyday products, but will also be likely to initiate the discovery of novel aldehyde reduction/oxidation mechanisms, which can enable further innovative chemical transformations and contribute to fundamental chemistry knowledge.

References

1 Sano, K.-I., Uchida, H., and Wakabayashi, S. (1999). *Catal. Surv. Jpn.* 3: 55–60.
2 Johnston, E.V. and Backvall, J.-E. (2011). *Oxidation of Carbonyl Compounds in Modern Oxidation Methods*, 2nd Ed. (ed. J.-E. Backvall), 353–355. New York: Wiley-VCH.
3 Rahimi, A., Azarpira, A., Kim, H. et al. (2013). *J. Am. Chem. Soc.* 135: 6415–6418.
4 Fehling, H. *Ann. Chem. Pharm.* 1849 (72): 106–113.
5 Oshitna, K. and Tollens, B. (1901). *Ber. Dtsch. Chem. Ges.* 34: 1425.
6 Benet, W.E., Lewis, G.S., Yang, L.Z., and Hugn, D.E.P. (2011). *J. Chem. Res.* 675–677.
7 Heilbron, I.M., Jones, E.R.H., and Sondheimer, F. (1949). *J. Chem. Soc.* 604.
8 Lv, C., Yan, X., Tu, Q. et al. (2016). *Angew. Chem. Int. Ed.* 55: 7539–7543.
9 Lindgren, B.O. and Nilsson, T. (1973). *Acta Chem. Scand.* 27: 888.
10 Bal, B.S., Childers, W.E. Jr., and Pinnick, H.W. (1981). *Tetrahedron* 37: 2091–2096.
11 Maehara, T., Motoyama, K., Toma, T. et al. (2017). *Angew. Chem. Int. Ed.* 56: 1549–1552.
12 van Effen, R.M. and Evans, D.H. (1979). *J. Electroanal. Chem.* 103: 383–397.
13 (a) Liu, R., Liang, X., Dong, C., and Hu, X. (2004). *J. Am. Chem. Soc.* 126: 4112. (b) Liu, R., Dong, C., Liang, X. et al. (2005). *J. Org. Chem.* 70: 729. (c) Xie, Y., Mo, W., Xu, D. et al. (2007). *J. Org. Chem.* 72: 4288. (d) Wang, X., Liu, R., Jin, Y., and Liang, X. (2008). *Chem. Eur. J.* 14: 2679. (e) Tao, J., Lu, Q., Chu, C. et al. (2010). *Synthesis* 3974. (f) He, X., Shen, Z., Mo, W. et al. (2009). *Adv. Synth. Catal.* 351: 89. (g) Miao, C.-X., He, L.-N., Wang, J.-L., and Wu, F. (2010). *J. Org. Chem.* 75: 257. (h) Kuang, Y., Rokubuichi, H., Nabae, Y. et al. (2010). *Adv. Synth. Catal.* 352: 2635. (i) Shibuya, M., Osada, Y., Sasano, Y. et al. (2011). *J. Am. Chem. Soc.* 133: 6497. (j) Lauber, M.B. and Stahl, S.S. (2013). *ACS Catal.* 3: 2612.
14 Anelli, P.L., Biffi, C., Montanari, F., and Quici, S. (1987). *J. Org. Chem.* 52: 2559.
15 Sladek, N.E. (2003). *J. Biochem. Mol. Toxicol.* 17: 6–23.
16 Yoshida, A., Rzhetsky, A., Hsu, L.C., and Chang, C. (1998). *Eur. J. Biochem.* 251: 549–557.
17 Travis, B.R., Sivakumar, M., Hollist, G.O., and Barhan, B. (2003). *Org. Lett.* 5: 1031–1034.
18 Hunsen, M. (2005). *Synthesis* 15: 2487–2490.

19 Malaprade, L. (1934). *Bull. Soc. Chim. Fr.* 3: 833.
20 Harries, C. and Langheld, K. (1905). *Liebigs Ann. Chem.* 343: 311.
21 Voukides, A.C., Konrad, K.M., and Johnson, R.P. (2009). *J. Org. Chem.* 74: 2108–2113.
22 Corona, T. and Company, A. (2016). *Dalton Trans.* 45: 14530–14533.
23 Murray, A.T., Matton, P., Fairhurst, N.W.G. et al. (2012). *Org. Lett.* 14: 3656–3659.
24 Okorochenkov, S., Burglova, K., Popa, I., and Hlavac, J. (2015). *Org. Lett.* 17: 180–183.
25 Burglova, K., Okorochenkov, S., Budesinsky, M., and Hlavac, J. (2017). *Eur. J. Org. Chem.* 2017: 389–396.
26 Marteau, C., Ruyffelaere, F., Aubry, J.-M. et al. (2013). *Tetrahedron* 69: 2268–2275.
27 Tada, M., Muratsugu, S., Kinoshita, M. et al. (2010). *J. Am. Chem. Soc.* 132: 713–724.
28 Conte, M., Miyamura, H., Kobayashi, S., and Chechik, V. (2010). *Chem. Commun.* 46: 145–147.
29 Tian, Q., Shi, D., and Sha, Y. (2008). *Molecules* 13: 948–957.
30 Yoshida, M., Katagiri, Y., Zhu, W.-B., and Shishido, K. (2009). *Org. Biomol. Chem.* 7: 4062–4066.
31 Liu, M., Wang, H., Zeng, H., and Li, C.-J. (2015). *Sci. Adv.* 1: e1500020.
32 Liu, M. and Li, C.-J. (2016). *Angew. Chem. Int. Ed.* 55: 10806–10810.
33 Jiang, X., Zhang, J., and Ma, S. (2016). *J. Am. Chem. Soc.* 138: 8344–8347.
34 Partenheimer, W. (1995). *Catal. Today* 23: 69.
35 Whittaker, A.M. and Dong, V.M. (2015). *Angew. Chem. Int. Ed.* 54: 1312–1315.
36 Bailey, P.S. and Erickson, R.E. (1973). *Org. Synth.* 5: 489.
37 Djerassi, C. and Engle, R.R. (1953). *J. Am. Chem. Soc.* 75: 3838.
38 Sidgwiclc, N.V. (1950). *The Chemical Elements and Their Compounds, vol. II*, 1482. Oxford: Clarendon Press.
39 Wolfe, S., Hasan, H.K., and Campbell, J.R. (1970). *Chem. Commun.* 23: 1420–1421.
40 Antonelli, E., D'Aloisio, R., Gambaro, M. et al. (1998). *J. Org. Chem.* 63: 7190–7206.
41 (a) Venturello, C., Alneri, E., and Ricci, M. (1983). *J. Org. Chem.* 48: 3831. (b) Venturello, C. and Ricci, M. (1986). *J. Org. Chem.* 51: 1599.
42 Malaprade, L. (1934). *Bull. Soc. Chim. Fr.* 3: 833–852.
43 Criegee, R. (1931). *Ber. Dtsch. Chem. Ges.* 64: 260–266.
44 de Vries, G. and Schors, A. (1968). *Tetrahedron Lett.* 54: 5689–5690.
45 Felthouse, T.R.J. (1987). *Am. Chem. Soc.* 109: 7566–7568.
46 Takezawa, E., Sakaguchi, S., and Ishii, Y. (1999). *Org. Lett.* 1 (713): 715.
47 Iwahama, T., Yoshino, Y., Keitoku, T. et al. (2000). *J. Org. Chem.* 65: 6502–6507.
48 Ishii, Y., Yamawaki, K., Ura, T. et al. (1988). *J. Org. Chem.* 53: 3587–3593.
49 Kaneda, K., Morimoto, K., and Imanaka, T. (1988). *Chem. Lett.* 17: 1295–1296.

6

Green Oxidative Synthesis of Amines, Amides, and Imines

Hong Yi[1], *Tianyi Liu*[1], *Zilu Tang*[1], *Atul K. Singh*[1], *and Aiwen Lei*[1,2]

[1] *College of Chemistry and Molecular Sciences, the Institute for Advanced Studies (IAS), Wuhan University, Wuhan, China*
[2] *State Key Laboratory for Oxo Synthesis and Selective Oxidation, Lanzhou Institute of Chemical Physics, Chinese Academy of Sciences, Lanzhou, China*

6.1 Brief Introduction

A major challenge of modern synthesis is to design not only efficient and selective reactions, but also to be eco-compatible, which provides maximum structural diversity and complexity with a minimum number of synthetic steps. The development of highly efficient, mild, and selective oxidation reactions for the synthesis of useful organic substrates is the central topic in synthetic chemistry [1, 2]. Molecular oxygen is considered as an ideal oxidant and offers attractive academic and industrial prospects. Aerobic oxidative reactions using molecular oxygen as the oxidant are of great interest because the use of stoichiometric amounts of explosive oxidants can be avoided and water is the byproduct [3]. In this chapter, we sum up the recent green synthesis of amines, amides, and imines using oxygen or hydrogen peroxide.

6.2 Green Oxidative Synthesis of Amines

The amine functionality is ubiquitously found in pharmaceuticals, specialty chemicals, and biologically important compounds [4]. As a result, the development of efficient and step-economical methods for C—N bonds construction has been recognized as one of the key issues in synthetic chemistry community. In 1903, Ullman reported the pioneering work of amination of aryl halides [5]. Later, Buchwald and Hartwig reported the palladium-catalyzed amination of aryl halides with simple amines, which allowed the straightforward access to various arylamines (Scheme 6.1) [6, 7]. Although these transformations are efficient and reliable, some shortages still exist, such as the need for aryl halides (or pseudohalides) and the generation of stoichiometric amounts of metal halide wastes. Notably, transition-metal-catalyzed C–H/N–H dehydrogenative cross-coupling reaction using O_2 as the sole oxidant is regarded as an ideal approach to facilely assemble carbon–nitrogen bonds due to its high atom-economy, efficiency, and environmental friendliness [8]. Such direct C–H amination is of great importance to supply various

Green Oxidation in Organic Synthesis, First Edition. Edited by Ning Jiao and Shannon S. Stahl.

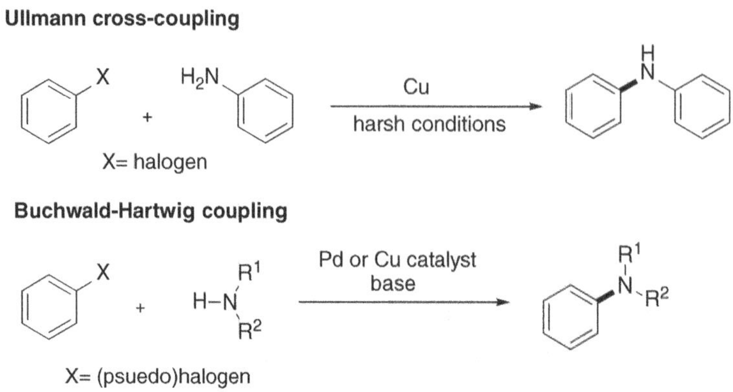

Scheme 6.1 Ullmann cross-coupling and Buchwald–Hartwig amination.

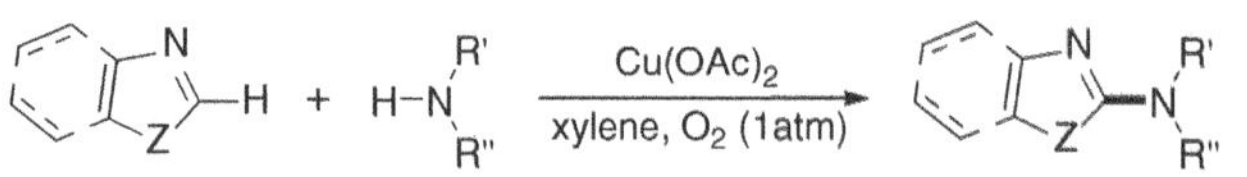

Scheme 6.2 Copper-catalyzed oxidative C—H/N—H coupling of azoles.

amino compounds with high step efficiency and waste-free strategy. In this section, we sum the recent development of oxidative C—H amination reactions using O_2, which is divided into three parts: oxidative amination of C(sp^2)—H bonds, oxidative amination of C(sp^3)—H bonds and oxidative amination of C(sp)—H bonds.

6.2.1 Direct Amination of C(sp^2)—H Bonds

Arylamines are ubiquitous in a range of biology-oriented and materials-oriented aromatics [9, 10]. In particular, N-heteroaryl derivatives have received much attention due to their unique biological and physical properties [11]. Therefore, the direct C—H amination of heteroarenes provided a rapid and straightforward access to the heteroarylamines. Among heteroaromatic compounds, five-membered heteroarenes are known to be reactive, and several C—H amination reactions have therefore been developed. In 2009, Mori and co-workers developed a copper-catalyzed oxidative C–H/N–H coupling of azoles (Scheme 6.2). These aminated products were afforded in the presence of sodium acetate and 20 mol% $Cu(OAc)_2$ in xylene under an oxygen atmosphere [12]. Later, a copper-catalyzed direct C–H amination of benzoxazoles with formamides or secondary amines was achieved by Duan. A variety of substituted benzoxazole-2-amines could be synthesized in moderate to excellent yields [13]. Using iron catalyst/aqueous H_2O_2 system, a similar coupling of benzoxazoles with secondary amines was also developed toward the synthesis of 2-aminobenzoxazoles [14].

Using tertiary amine as the amine source, Huang and co-workers developed a copper-catalyzed oxidative amination of azoles via C—H and C—N bonds activation (Scheme 6.3) [15]. The reaction could be realized in the absence of an external base and only required atmospheric oxygen as an oxidant. The protocol opens a new method for C—N bond construction using tertiary amines as nitrogen group sources.

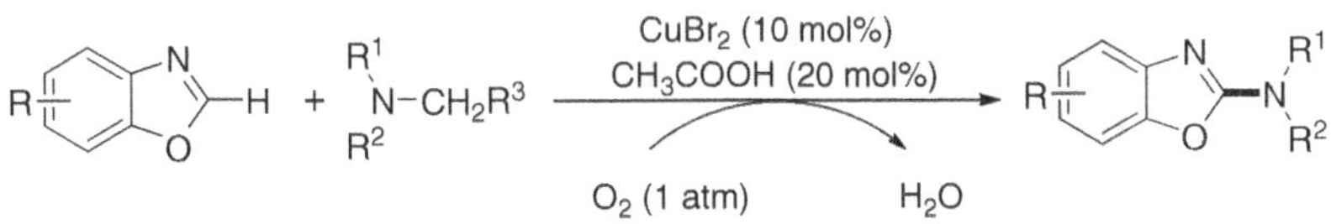

Scheme 6.3 Copper-catalyzed oxidative amination of azoles and tertiary amines.

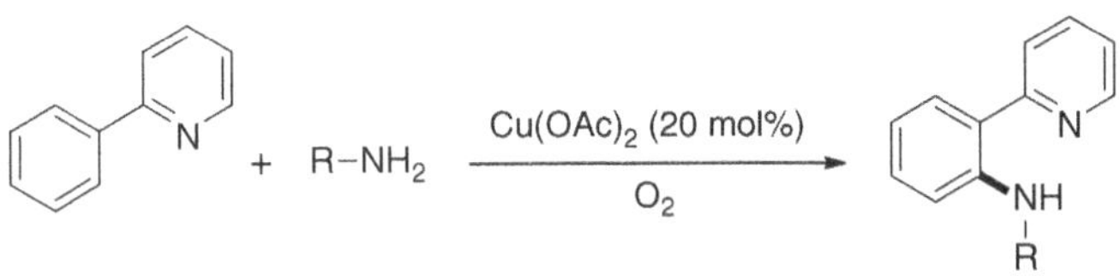

Scheme 6.4 Copper-catalyzed oxidative C—H/N—H coupling of 2-phenylpyridines.

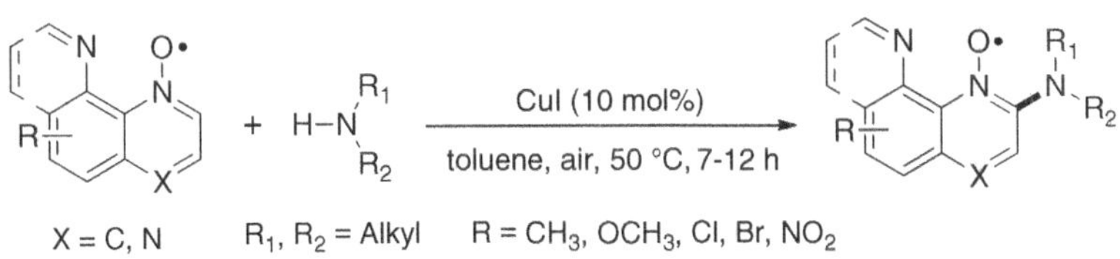

Scheme 6.5 Copper-catalyzed oxidative amination of quinoline N-oxides.

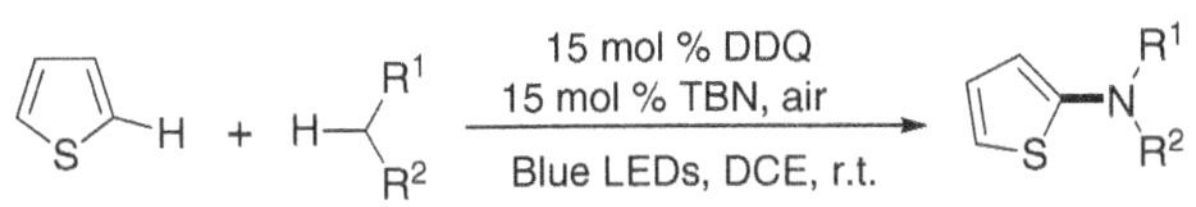

Scheme 6.6 Oxidative amination of thiophenes via photoredox catalysis.

Nicholas and co-workers reported an oxidative C–H/N–H coupling of 2-phenylpyridines using $Cu(OAc)_2$ as the catalyst and O_2 as the green oxidant (Scheme 6.4) [16]. A variety of nitrogen partners including sulfonamides, carboxamides, and anilines were tolerated and transformed into the desired products in moderate to good yields.

The 2-aminoquinoline derivatives widely exist in pharmaceuticals and biological activity antagonists. Cui and co-workers reported a copper-catalyzed direct amination of quinoline N-oxides for C–N formation (Scheme 6.5) [17]. A variety of quinolone N-oxides and secondary aliphatic amines were well-tolerated in this system. This methodology provides a practical system and environmental friendliness pathway to 2-aminoquinolines.

Recently, Lei and co-worked have achieved a visible light-mediated direct C–H amination of thiophenes via oxidation pathway (Scheme 6.6) [18]. Various thiophenes and azoles were well-tolerated and transformed into the desired amination products and water was the sole byproduct. In the reaction system, *tert*-butyl nitrite (TBN) served as the electron transfer mediator and O_2 was the terminal oxidant to regenerate photocatalyst DDQ.

Due to the low reactivity, the C—H bond of simple arene is relatively hard to activate. Recent research interest of C—H amination is focused on the transformation of simple arenes. In 2010, Su and co-workers reported a copper-catalyzed oxidative C—H bond

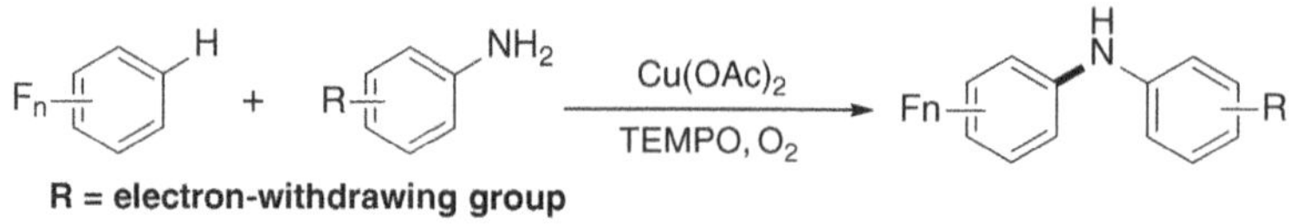

Scheme 6.7 Oxidative amination of polyfluorobenzenes.

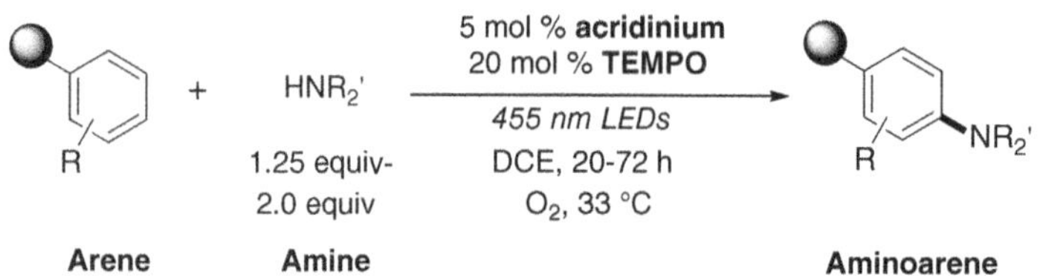

Scheme 6.8 Photo-induced oxidative amination of simple arenes.

amination of simple arenes using O_2 as the green oxidant (Scheme 6.7) [19]. Various polyfluorobenzenes and primary aromatic amines were the suitable substrates in this system. The reactivity of the polyfluorobenzenes depended on the acidity of C—H bonds. The direct amination of azoles could also be realized in this system.

Using the photocatalysis system, Nicewicz and co-workers reported a direct amination of simple arenes via photoredox catalysis (Scheme 6.8) [20]. An organic photoredox-based catalyst system, consisting of an acridinium photooxidant and a nitroxyl radical, promoted this site-selective amination. A variety of simple and complex aromatics with heteroaromatic azoles were transformed into the corresponding amines. This system could also be used to aniline synthesis without the need for prefunctionalization of the aromatic component.

Metal-catalyzed cross-coupling between nitrogen nucleophiles (e.g. amines, amides, carbamates) and alkenes represents a longstanding challenge in organic chemistry and an attractive target for the preparation of nitrogen-containing organic molecules [21]. In 2007, Stahl and co-workers reported an intermolecular aerobic oxidative amination of alkenes mediated by palladium catalyst [22]. The olefin was used as the limiting reagent, and a variety of functional groups were tolerated. Later, Su and co-workers also developed two catalytic systems for the Pd-catalyzed amination of alkenes with indoles (Scheme 6.9) [23]. This process proceeds in generally good yields and is compatible with a broad range of functional groups.

6.2.2 Direct Amination of C(sp^3)—H Bonds

Although a significant progress has been achieved in transition-metal-catalyzed aerobic oxidative functionalization of C(sp^2)—H bonds over the past decades, the analogous transformations of unactivated C(sp^3)—H bonds with air or O_2 as the oxidant are still rare. In 2015, Pandey and co-workers reported a visible light-mediated cross dehydrogenative benzylic C(sp^3)—H amination reaction for amides synthesis [24]. Using a photocatalytic system, Lei and co-workers reported a visible light promoted oxidative amination of tetrahydrofuran using azoles as the amine sources under an air atmosphere (Scheme 6.10) [25]. The method provides a simple and mild pathway for N-alkylation of azoles.

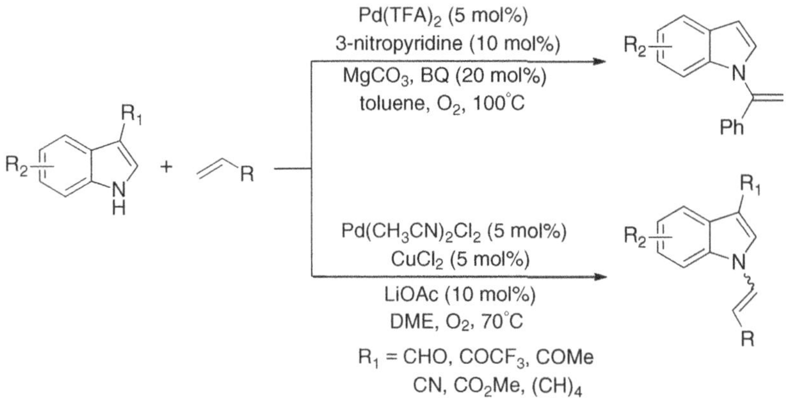

Scheme 6.9 Pd-catalyzed oxidative amination of alkenes.

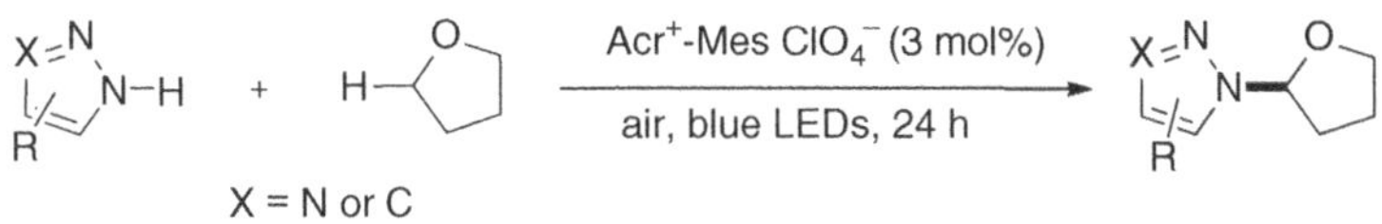

Scheme 6.10 Oxidative amination of tetrahydrofuran via photoredox catalysis.

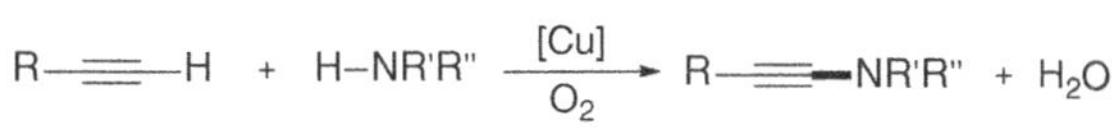

Scheme 6.11 Copper-catalyzed oxidative amination of alkynes.

6.2.3 Direct Amination of C(*sp*)—H Bonds

The ynamides are widely served as the synthons in organic synthesis. Direct coupling of terminal alkynes and amines represents an ideal way for ynamides preparation. Stahl and co-workers developed a new Cu-catalyzed method for the synthesis of ynamides via direct amidation of terminal alkynes using O_2 as the terminal oxidant (Scheme 6.11) [26]. The reaction provides a direct environment-friendly way to ynamides via aerobic oxidative coupling strategy.

6.3 Green Oxidative Synthesis of Amides

Amide moiety can be widely found in a variety of pharmaceuticals, agrochemicals, and polymeric materials [27]. Traditional methods for amide constructions involve the reaction of amines (including ammonia) with activated carboxylic acid derivatives or coupling with carboxylic acids with the use of stoichiometric amounts of coupling agents [28]. Although widely used, these methods have several common drawbacks, such as long reaction times, poor atom-efficiency, or the use of hazardous reagents. Therefore, the development of novel strategies for amide synthesis is highly desirable. In the past few years, a new brand of catalytic oxidative strategies has been developed to provide efficient and environmentally friendly sustainable methods for amide synthesis under mild conditions [29].

6.3.1 Oxidative Amidation of Aldehydes with Amines

The oxidative coupling between aldehydes and amine salts is highly synthetically useful to amide synthesis as both starting materials are readily available and less hazardous than those traditionally used, such as acid halides. Several groups have developed a series of methods for the direct synthesis of amides from aldehydes using stoichiometric oxidants, such as iodine and NBS, leading to waste after the reactions [30–33]. Therefore, the development of oxidative amidation of aldehydes and amines using the green oxidant hydrogen peroxide or molecular oxygen provides a more economical synthetic route to amides. In 2011, Pathak and co-workers reported an efficient method for the amidation and esterification of aldehydes using hydrogen peroxide (Scheme 6.12) [34]. In the transformation, cyclic amines selectively reacted with aromatic aldehydes under mild conditions to yield the corresponding amides. In addition, the selective coupling of primary alcohols and aldehydes could be also achieved to afford esters. Using continuous flow microreactor systems, the direct oxidative amidation of aromatic aldehydes with secondary amines to amides was improved and achieved in a single operation [35].

The amidation of aromatic aldehydes with aminopyridines was also accomplished using simple aqueous hydrogen peroxide as the oxidant (Scheme 6.13) [36]. This process proceeded in water without the need of any catalyst or additive. Green oxidant and mild reaction conditions rendered this method a practical strategy for the synthesis of diverse *N*-(pyridin-2-yl)benzamides.

In 2012, Fujita, Yamaguchi, and co-workers reported a copper-catalyzed oxidative amidation of aldehydes with secondary amines using air the sole oxidant (Scheme 6.14) [37]. Using the CuI/2-pyridonatecatalytic system, a variety of useful arylamides could be synthesized in moderate to good yields, generating only water as a byproduct.

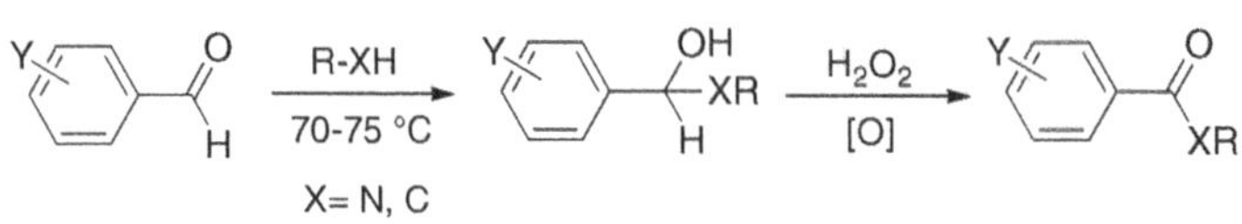

Scheme 6.12 Amidation and esterification of aldehydes using hydrogen peroxide.

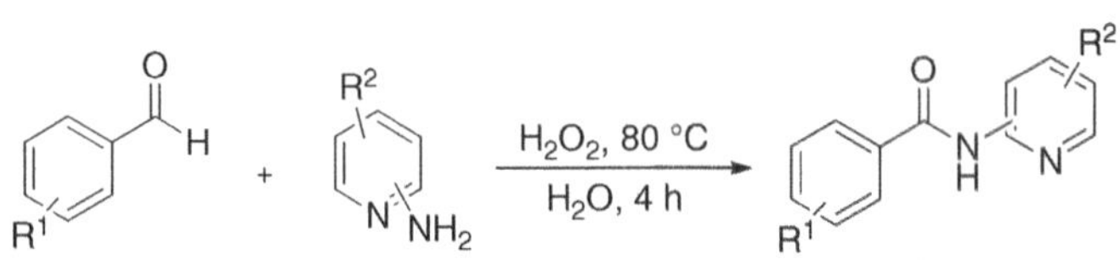

Scheme 6.13 Amidation of aromatic aldehydes with aminopyridines.

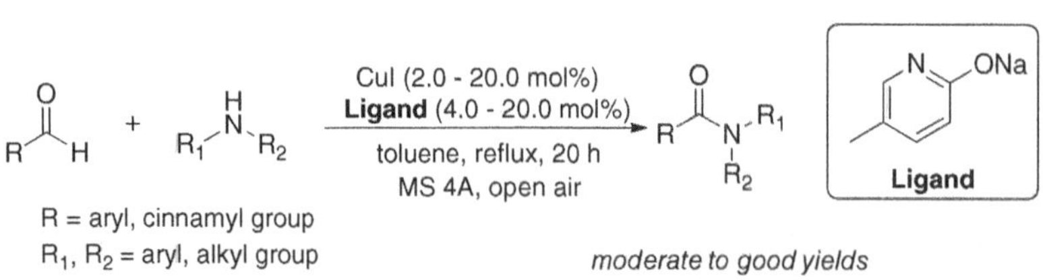

Scheme 6.14 Copper-catalyzed oxidative amidation.

In the same year, an efficient gold-catalyzed oxidative coupling of aldehydes with amines under mild reaction conditions was achieved using O_2 as the oxidant (Scheme 6.15) [38]. This method provides an easy access to amides from the reaction of aromatic, aliphatic, and polyhydroxyl oligosaccharide-based aldehydes with secondary amines. Later, Kobayashi and co-workers also developed polymer-incarcerated gold (Au) nanoparticles (NPs) that were catalyzed directly from aerobic oxidative amide synthesis from aldehydes and amines [39].

6.3.2 Oxidative Amidation of Alcohols with Amines

The aldehydes are normally synthesized through the selective oxidation of alcohols. Therefore, the catalytic amidation of alcohols has emerged as an atom- and step-economical alternative protocol [40]. Kobayashi and co-workers achieved the direct oxidative coupling of alcohols with amines for amide formation in which heterogeneous gold and gold/iron, -nickel, or -cobalt nanoparticles were used as catalysts and molecular oxygen was used as a terminal oxidant (Scheme 6.16a). A variety of alcohols and amines, including aqueous ammonia and amino acids, could be used as the starting materials for the amide synthesis. In addition, this system could be recovered and reused several times without loss of activity [41]. In the same year, Wang and co-workers prepared a water-soluble gold catalyst, which showed good reactivity in the direct amidation of alcohols and amines (Scheme 6.16b) [42]. This reaction was carried out smoothly under mild reaction conditions in the neat water system, and the aromatic amines could also be transformed into the amides with good to excellent yields.

Using the copper catalytic system, Krabbe and co-workers developed a Cu-catalyzed amidation between alcohols and secondary amines using the oxygen in air as the terminal oxidant (Scheme 6.17) [43]. The commercially available and nonprecious metal catalyst, $Cu(phen)Cl_2$, in combination with di-*tert*-butyl hydrazine dicarboxylate ($DBADH_2$) and an inorganic base provides a variety of benzamides in moderate to excellent yields.

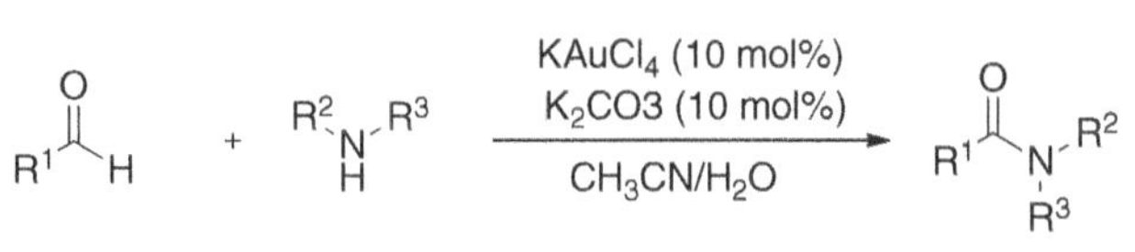

Scheme 6.15 Gold-catalyzed oxidative coupling of aldehydes with amines.

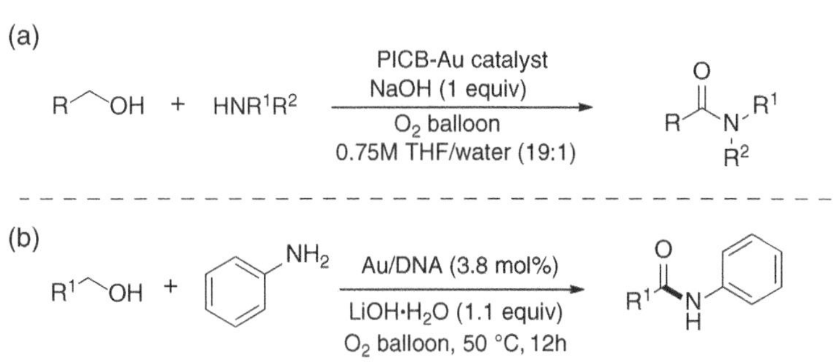

Scheme 6.16 Gold-catalyzed oxidative coupling of alcohols with amines.

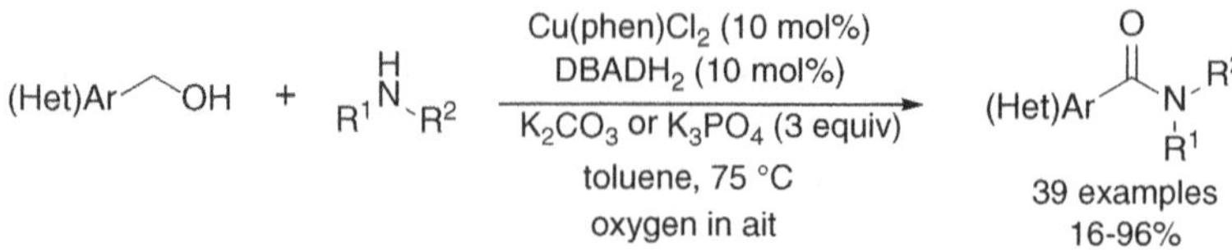

Scheme 6.17 Copper-catalyzed amidation between alcohols and secondary amines.

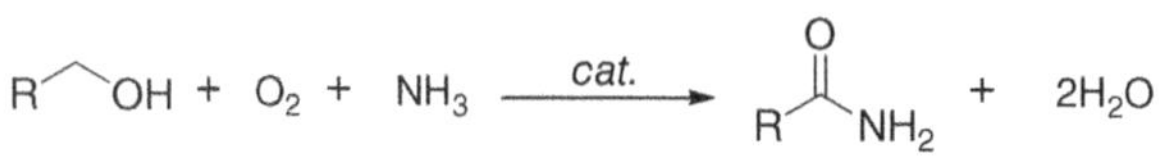

Scheme 6.18 Oxidative coupling of primary alcohols and aqueous ammonia.

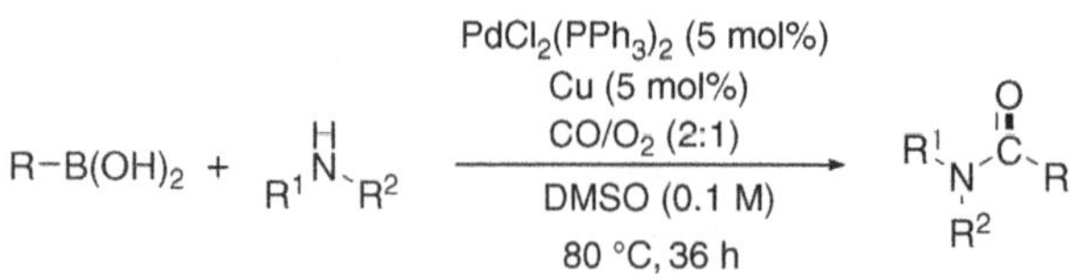

Scheme 6.19 Pd-catalyzed aerobic aminocarbonylation of organoboronic acids and amines.

The direct synthesis of primary amides from primary alcohols and ammonia is very challenging because dehydration of the hemiaminal from ammonia readily occurs and/or catalysts are deactivated in the presence of ammonia [44]. In 2012, Mizuno and co-workers developed an efficient heterogeneous catalyst for a widely applicable synthesis of primary amides directly from primary alcohols using aqueous ammonia as the amine source (Scheme 6.18) [45]. This heterogeneous system provided a new route for sustainable amide synthesis, which can completely avoid the need for conventional stoichiometric reagents and the formation of inorganic waste.

6.3.3 Oxidative Carbonylation for Amide Synthesis

Carbonylation, the incorporation of CO into one organic molecule, is now widely recognized as a very important tool in industrial and organic chemistry [46, 47], which allows the direct synthesis of important carbonyl compounds starting from the simplest C-1 unit. Jiao and co-workers reported a direct approach from organoboronic acids and amines to tertiary amides via Pd-catalyzed aerobic aminocarbonylation (Scheme 6.19) [48]. This method provides an easy route for amide synthesis with broad substrate scope and high functional group tolerance from commercially available organoboronic acids and cheap CO and O_2 (1 atm). The presence of O_2 significantly promoted the efficiency of this transformation.

In 2013, Lei and co-workers developed an aerobic palladium/copper-catalyzed oxidative C—H alkenylation/N-dealkylative carbonylation of tertiary anilines to form 3-methyleneindolin-2-one derivatives (Scheme 6.20) [49]. This transformation provides an effective and straightforward method toward the synthesis of biologically and medicinally useful 3-methyleneindolin-2-ones in moderate to good yields from commercial and simple substrates.

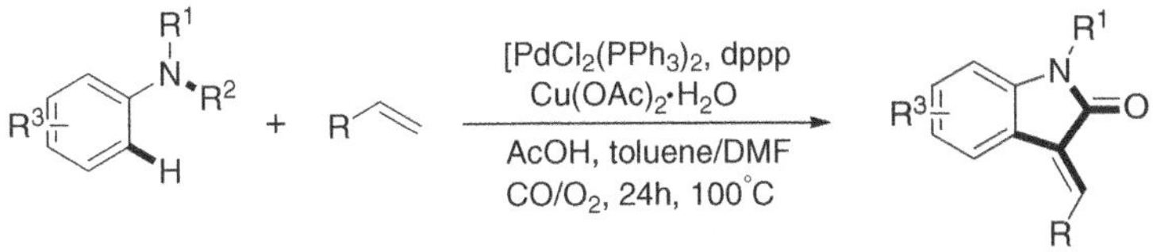

Scheme 6.20 Palladium/copper-catalyzed carbonylation of tertiary anilines.

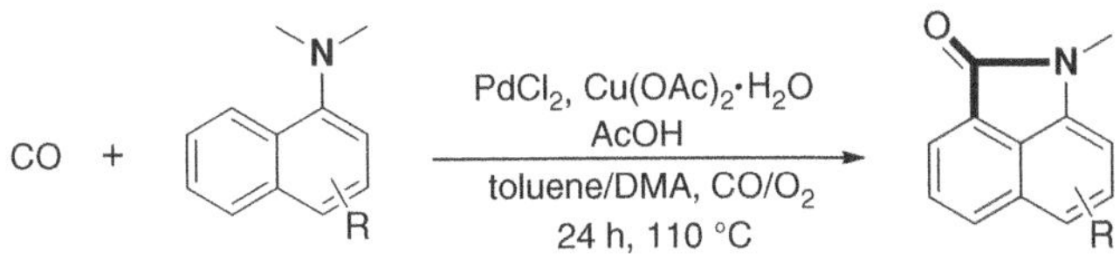

Scheme 6.21 Pd-catalyzed selective carbonylation of naphthyl amines.

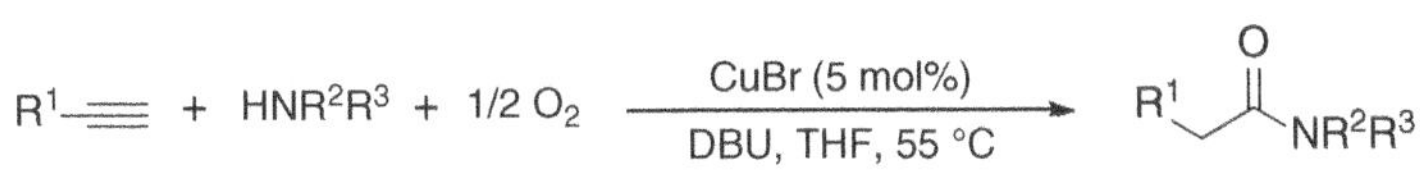

Scheme 6.22 Copper-catalyzed oxidative amidation of alkynes.

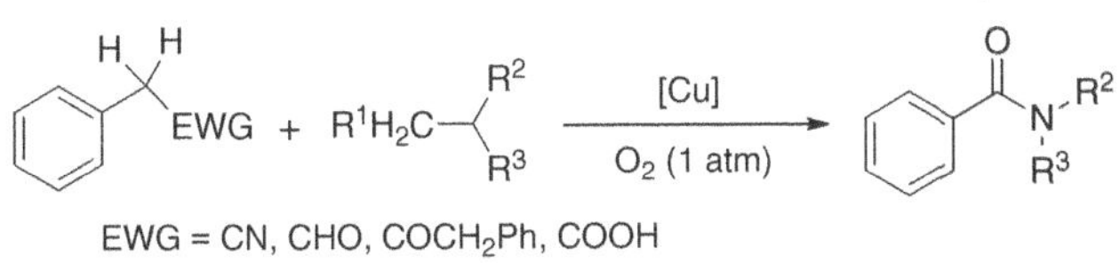

Scheme 6.23 Copper-catalyzed aerobic oxidative amidation through C—C/C—N bond cleavage.

With the use of C—H compound, Lei group also achieved a Pd-catalyzed selective C8—H bond activation/N-dealkylative carbonylation of naphthyl amines (Scheme 6.21) [50]. The amine group from dealkylation was proposed to be the directing group for promoting this process. This method represented a straightforward and easy route to afford various biologically important benzo[*cd*]indol-2(1*H*)-one derivatives.

6.3.4 Others

There were other representative examples for amide synthesis using the available materials by employing O_2 as the oxidant. In 2012, Ji and co-workers developed a copper-catalyzed direct oxidative synthesis of amides from alkynes and amines under mild conditions, in which DBU was used as the key additive (Scheme 6.22) [51]. The present methodology, which utilizes dioxygen as an oxidant and oxygen source, and cheap copper salts as catalysts, opens up an interesting and attractive avenue for the synthesis of amide functionality.

Yin and co-workers developed a copper-catalyzed aerobic oxidative amidation reaction of inert C—C bonds with tertiary amines (Scheme 6.23) [52]. This method combines C—C bond activation, C—N bond cleavage, and C—H bond oxygenation in a

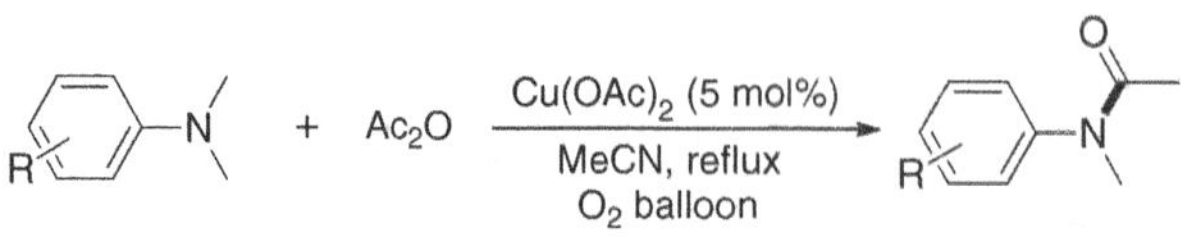

Scheme 6.24 Catalytic oxidative amidation of tertiary amines.

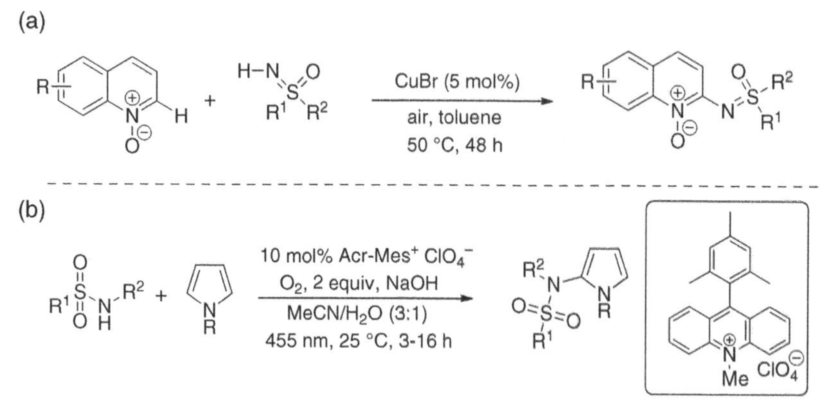

Scheme 6.25 Oxidative C—H/N—H coupling for amide synthesis.

one-pot protocol, using molecular oxygen as the sole oxidant without any additional ligands. This protocol also provides a practical, neutral, and mild synthetic approach to tertiary amides, which are important units in biologically active molecules.

In 2015, Guan and co-workers achieved a copper-catalyzed aerobic oxidative amidation of tertiary amines and acetic anhydride for the synthesis of tertiary amides (Scheme 6.24) [53]. Various functional groups were well-tolerated using the dioxygen as the terminal oxidant. The extensive substrates studies demonstrated its potential as a practical approach for the synthesis of tertiary amides.

The direct oxidative C—H/N—H coupling can also provide a direct route for amide synthesis. Bolm and co-workers reported a similar C—H/N—H dehydrogenative coupling of quinoline-type N-oxides with sulfoximines to *N*-(hetero)arylsulfoximines using CuBr as the catalyst and air as the green oxidant (Scheme 6.25a) [54]. Later, König and co-workers developed a photo-induced coupling of sulfonamides and pyrroles for the preparation of *N*-(2-pyrrole)-sulfonamides [55]. The reaction used an acridinium dye as photocatalyst and oxygen as the terminal oxidant for the oxidative C—N bond formation (Scheme 6.25b).

6.4 Green Oxidative Synthesis of Imines

Imines, also known as Schiff bases, are widely applied in pharmaceutical synthesis and organic transformations [56]. The traditional imine synthesis involves the using of unstable aldehydes, dehydrating agents, and Lewis acid catalysts. The efficient and diverse synthesis of imines is a consistently demanding research topic in the organic synthetic community [57, 58]. During the past decade, considerable efforts have been devoted to the direct synthesis of imines, in particular via one-pot procedures through

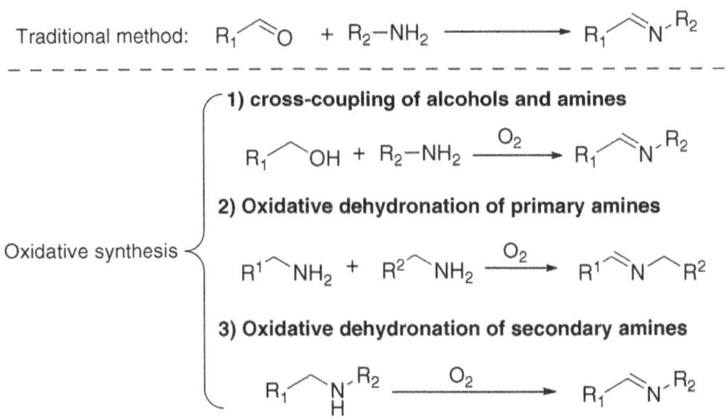

Scheme 6.26 Oxidative synthesis of imines using dioxygen.

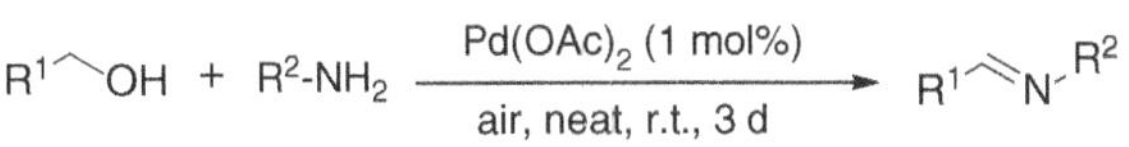

Scheme 6.27 Pd-catalyzed oxidative coupling between alcohols and amines.

an oxidative process to reduce the energy consumption, waste emission, operating, and purification steps. Among the various methodologies for imine formation, the following three approaches have received much more attention and have been greatly studied because the starting materials are readily available and green molecular oxygen or air can serve as the terminal oxidant (Scheme 6.26) [59]. Since great progress has been achieved using heterogeneous and homogenous catalysis, we have just reviewed here the representative green oxidative imine syntheses using molecular oxygen or air in homogenous catalysis.

6.4.1 Oxidative Cross-Coupling of Alcohols with Amines

Since aldehydes and ketones are usually obtained via the oxidation of alcohols, and alcohols are generally greener than the carbonyl compounds, due to being wider in scope, more readily available, more stable, cheaper, less toxic, and easier to handle [60]. A potentially more general and more advantageous protocol to imines would be the one-pot oxidative tandem synthesis from alcohols and amines. In 2011, Yu, Xu, and co-workers developed a mild and green one-pot imine synthesis from alcohols and amines (Scheme 6.27) [61]. This reaction was performed via a low-loading palladium-catalyzed tandem aerobic alcohol oxidation–dehydrative condensation reaction in the open air at room temperature.

Using the copper salt as the catalyst, a highly efficient and green tandem imine synthesis from alcohol and amine was also achieved with molecular oxygen as the terminal oxidant (Scheme 6.28a). A low catalyst-loading, with a commercially available and environmentally benign copper catalyst was used in the reaction system [62]. In the same year, Xu and co-workers also developed a copper-catalyzed aerobic oxidative reaction of alcohols and amines in the open air at room temperature under base- and dehydrating reagent-free conditions (Scheme 6.28b) [63]. This method provides a general, green, and scalable synthesis of the useful imines and α, β-unsaturated imines. A wide range of substrates was tolerated, affording high yields of the target imines on a large scale.

6.4.2 Oxidative Coupling of Primary Amines to Imines

Oxidative coupling of primary amines provides an alternative strategy to the synthesis of imines. In 2011, Fu and co-workers developed a metal-free oxidative coupling of primary amines, providing a green protocol for imine synthesis (Scheme 6.29). This reaction was performed by refluxing suspensions of amines and water using dioxygen as the sole oxidant without any additives [64].

6.4.2.1 Copper Catalysis

Adimurthy and co-workers described a copper (I)-catalyzed direct synthesis of imines from amines under mild aerobic conditions [65]. The amines such as benzylic, aliphatic, cyclic secondary, heteroaromatic species were well transformed into the corresponding imines. In addition, the oxidative condensation of benzylamines with anilines extended the scope using the CuCl catalytic system (Scheme 6.30). It is worth noting that the solvent-free procedure, use of air as a benign oxidant, and the cheap and easy availability of the catalyst were the advantages of the protocol. Later, the aerobic dehydrogenation of benzylic amines and anilines was achieved with a simple $CuBr_2$-TEMPO catalyst system [66]. In the same year, Largeron and co-workers achieved chemoselective biologically inspired aerobic oxidation of primary amines using a copper catalyst [67]. This atom-economical green process tolerated the presence of various reactive functional groups and was highly selective in the case of cross-coupling of two amines.

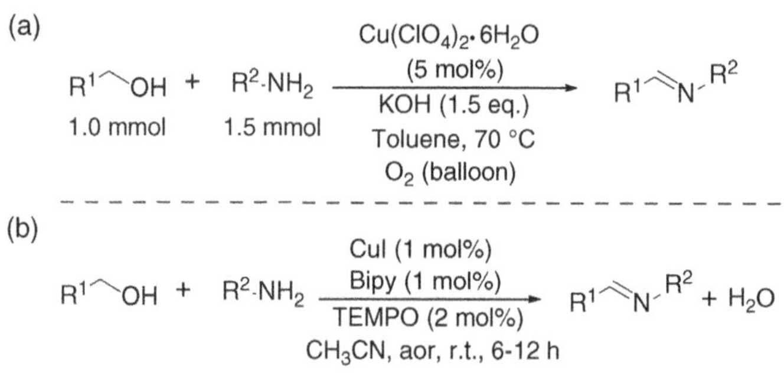

Scheme 6.28 Copper-catalyzed oxidative coupling between alcohols and amines.

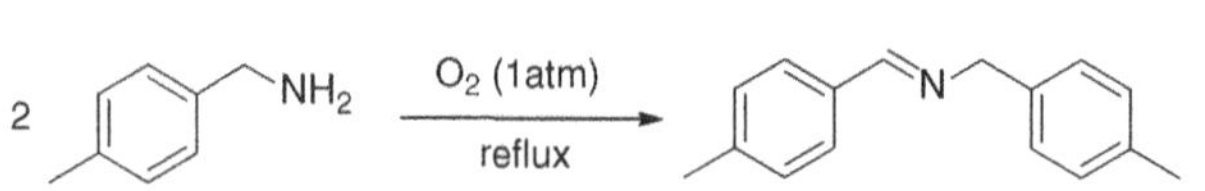

Scheme 6.29 Metal-free oxidative coupling of primary amines.

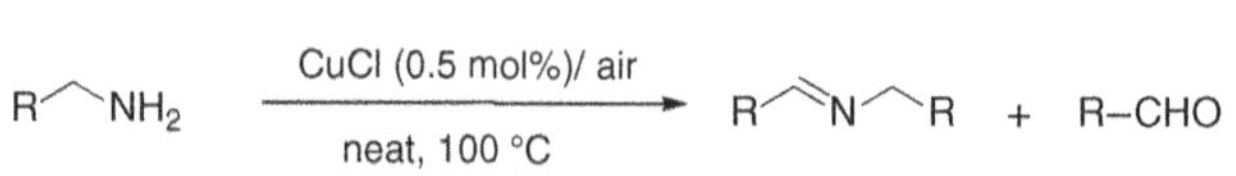

Scheme 6.30 Copper-catalyzed direct synthesis of imines from amines.

6.4.2.2 Iron Catalysis

By using the general $Fe(NO_3)_3$/TEMPO catalyst system, a Fe-catalyzed aerobic oxidation reaction of primary amines was realized by Xu and co-workers using the air as the green oxidant (Scheme 6.31) [68]. The useful imines could be prepared by, secondary amines, benzylamines with anilines, and alcohols with amines. Since abundant, cheap, and relatively more biofriendly iron can be readily used as the catalyst and air as the economic and safe oxidant, these aerobic methods may become green and practical alternatives for the direct synthesis of imines.

6.4.2.3 Cobalt Catalysis

Using Co(II) β-tetrakis(trifluoromethyl)-meso-tetraphenylporphyrin as the catalyst, Chen and co-workers developed an efficient and practical method for oxidative coupling of benzylamines to imines by molecular oxygen (Scheme 6.32) [69]. The reaction showed good selectivity and high turnover number, which may be due to the effect of the four bulky electron-withdrawing β-CF3 groups on the catalyst. In addition, the reaction could be easily scaled up and generated high yields of the useful imines, which made it applicable for large-scale preparation.

6.4.2.4 Quinone-Based Catalysts

Stahl and co-workers also achieved a biomimetic aerobic oxidation of primary benzylic amines by using a quinone catalyst using O_2 as the oxidant, which was favorable with the previously metal-based conditions (Scheme 6.33) [70]. The oxidative coupling of primary, unbranched benzylic amines relative to secondary/tertiary amines, branched

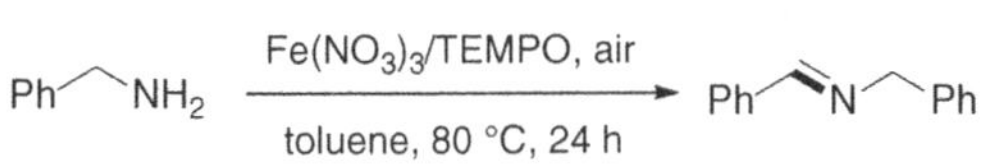

Scheme 6.31 Iron-catalyzed direct synthesis of imines from amines.

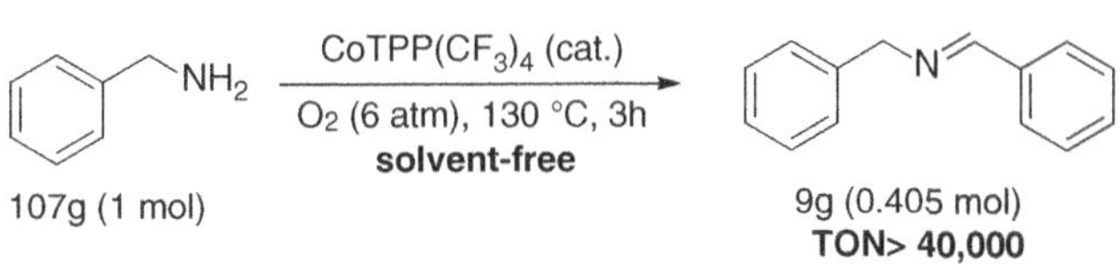

Scheme 6.32 Cobalt-catalyzed direct synthesis of imines from amines.

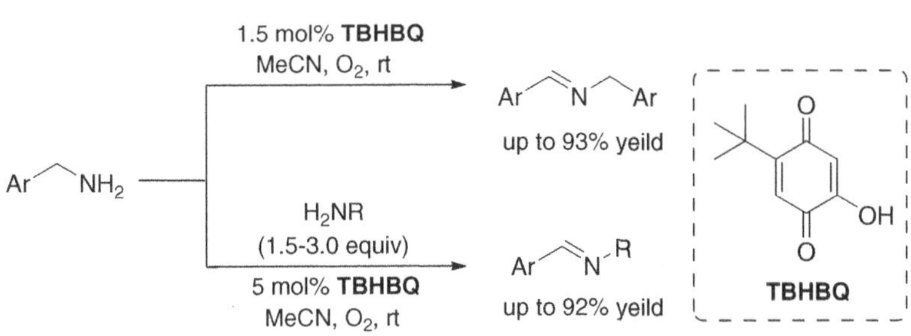

Scheme 6.33 Direct oxidation of imines from amines by quinone-based catalysts.

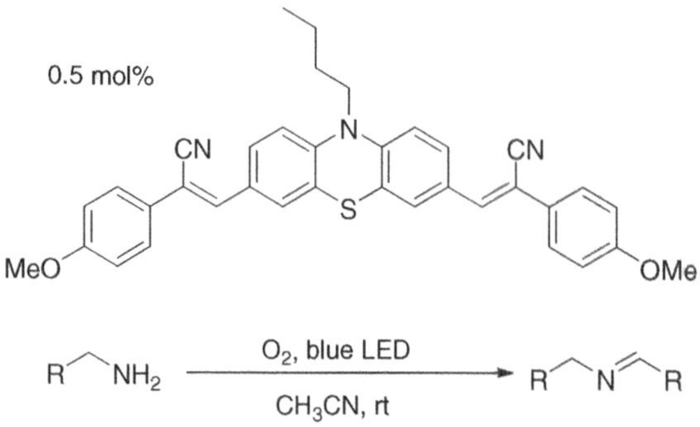

Scheme 6.34 Photo-induced oxidation of imines from amines.

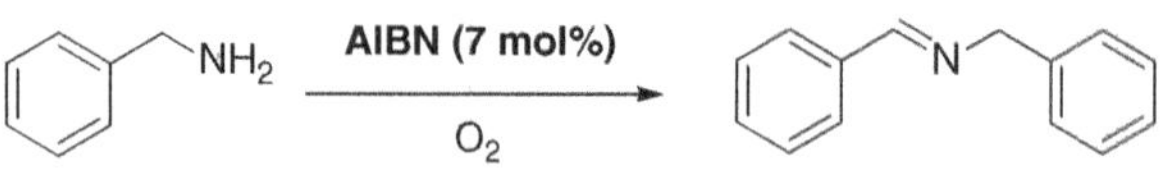

Scheme 6.35 AIBN-initiated oxidation of imines from amines.

benzylic amines, and aliphatic amines were achieved with high efficiency and excellent selectivity. The exquisite selectivity for benzylic amines enabled the oxidative self-sorting within dynamic mixtures of amines and imines to afford cross-coupled imine products with high yields.

6.4.2.5 Photocatalysis

Lee, Son, and co-workers prepared a new phenothiazine based organic dyes and applied the dyes to the visible-light-driven oxidative coupling of amines to imines (Scheme 6.34) [71]. The 3,7-disubstituted phenothiazine derivatives showed visible light absorption and reversible one-electron oxidation behavior. In the presence of catalytic 3,7-disubstituted phenothiazines, a variety of primary benzylamines showed good reactivity under visible light irradiation.

6.4.2.6 AIBN as the Catalyst

In the presence of a catalytic amount of radical initiator azobisisobutyronitrile (AIBN), Fu and co-workers developed a metal-free aerobic oxidative coupling of primary amines to imines (Scheme 6.35) [72]. A catalytic amount of AIBN was used, and dioxygen was the sole oxidant in this metal-free system. The cyanation of tertiary amines to α-aminonitriles could be achieved under these conditions.

6.4.3 Oxidative Dehydrogenation of Secondary Amines

The direct oxidation of secondary amines with the loss of two protons can also afford the imine derivatives. In 2012, Che and co-workers prepared two organogold(III) complexes, which showed a long-lived and highly emissive triplet excited state in solutions under ambient conditions (Scheme 6.36) [73]. The Au complex also showed high reactivity in the oxidation of secondary amines to imines and a-substitution of tertiary amines using O_2 as the green oxidant.

In 2014, Stahl and co-workers reported a novel route for aerobic dehydrogenation of secondary amine substrates that employed 1,10-phenanthroline-5,6-dione (phd) as a bifunctional *o*-quinone catalyst using O_2 as the green oxidant (Scheme 6.37) [74].

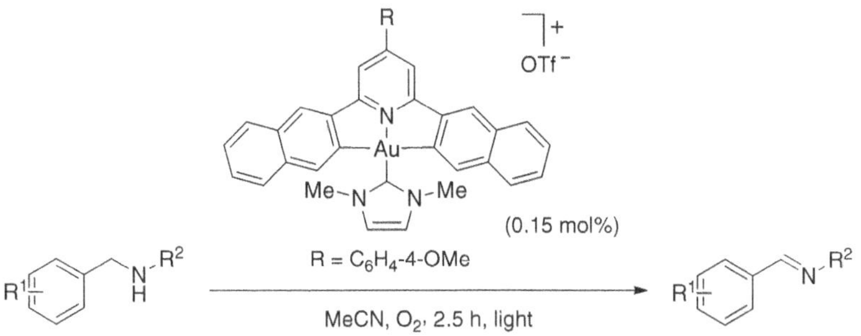

Scheme 6.36 Au-catalyzed oxidative dehydrogenation of secondary amines.

Scheme 6.37 Oxidative dehydrogenation of secondary amines using the quinone catalyst.

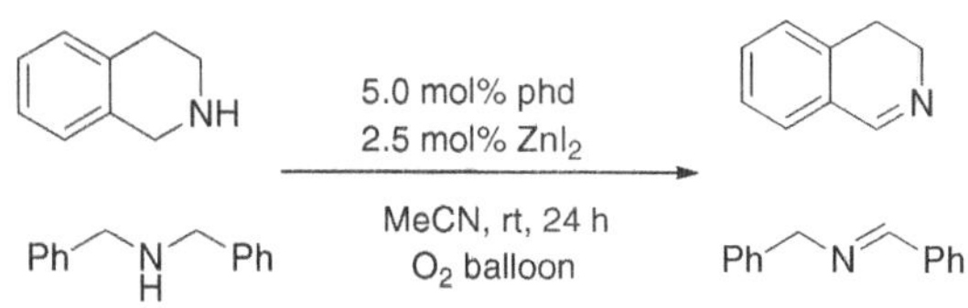

The non-native secondary amine substrates, including pharmaceutically relevant nitrogen heterocycles, were well-tolerated in the reaction system. Mechanistic studies revealed the reaction might be through an abiological pathway involving a hemiaminal intermediate. This protocol provided the foundation for broader exploration of quinones and related redox-active organic catalysts in selective aerobic oxidation reactions.

6.5 Conclusions

The green oxidative synthesis using O_2 or hydrogen peroxide has been widely explored during the past several years, providing a mild and clean route for organic synthesis. This chapter provides an updated summary of a green synthesis of amines, amide, and imines via green oxidative synthesis strategies. Compared with the traditional synthetic methods, these oxidative syntheses provide new ways for the useful compounds synthesis (amines, amides, and imines). There are still many unexplored methods and unknown insightful mechanism left, such as the improvement of efficiency of the oxidation process, the extension of substrate scope, the understanding the dioxygen activation. Despite these challenges, oxidative synthesis is still important in organic synthesis and will achieve better improvements in the future.

References

1 McCann, S.D. and Stahl, S.S. (2015). *Acc. Chem. Res.* 48: 1756–1766.
2 Yi, H., Zhang, G., Wang, H. et al. (2017). *Chem. Rev.* 117: 9016–9085.
3 Cavani, F. and Teles, J.H. (2009). *ChemSusChem.* 2: 508–534.
4 Muller, P. and Fruit, C. (2003). *Chem. Rev.* 103: 2905–2920.
5 Ullmann, F. (1903). *Berichte der deutschen chemischen Gesellschaft* 36: 2382–2384.

6 Guram, A.S. and Buchwald, S.L. (1994). *J. Am. Chem. Soc.* 116: 7901–7902.
7 Paul, F., Patt, J., and Hartwig, J.F. (1994). *J. Am. Chem. Soc.* 116: 5969–5970.
8 Jiao, J., Murakami, K., and Itami, K. (2016). *ACS. Catal.* 6: 610–633.
9 Kim, J. and Movassaghi, M. (2015). *Acc. Chem. Res.* 48: 1159–1171.
10 Walzer, K., Maennig, B., Pfeiffer, M., and Leo, K. (2007). *Chem. Rev.* 107: 1233–1271.
11 Seregin, I.V. and Gevorgyan, V. (2007). *Chem. Soc. Rev.* 36: 1173–1193.
12 Monguchi, D., Fujiwara, T., Furukawa, H., and Mori, A. (2009). *Org. Lett.* 11: 1607–1610.
13 Li, Y., Xie, Y., Zhang, R. et al. (2011). *J. Org. Chem.* 76: 5444–5449.
14 Xu, D., Wang, W., Miao, C. et al. (2013). *Green. Chem.* 15: 2975.
15 Guo, S., Qian, B., Xie, Y. et al. (2011). *Org. Lett.* 13: 522–525.
16 John, A. and Nicholas, K.M. (2011). *J. Org. Chem.* 76: 4158–4162.
17 Zhu, C., Yi, M., Wei, D. et al. (2014). *Org. Lett.* 16: 1840–1843.
18 Song, C., Yi, H., Dou, B. et al. (2017). *Chem. Commun.* 53: 3689–3692.
19 Zhao, H., Wang, M., Su, W., and Hong, M. (2010). *Adv. Synth. Catal.* 352: 1301–1306.
20 Romero, N.A., Margrey, K.A., Tay, N.E., and Nicewicz, D.A. (2015). *Science* 349: 1326–1330.
21 Hong, S. and Marks, T.J. (2004). *Acc. Chem. Res.* 37: 673–686.
22 Rogers, M.M., Kotov, V., Chatwichien, J., and Stahl, S.S. (2007). *Org. Lett.* 9: 4331–4334.
23 Wu, G. and Su, W. (2013). *Org. Lett.* 15: 5278–5281.
24 Pandey, G. and Laha, R. (2015). *Angew. Chem. Int. Ed.* 54: 14875–14879.
25 Zhang, L., Yi, H., Wang, J., and Lei, A. (2017). *J. Org. Chem.* https://doi.org/10.1021/acs.joc.7b01841.
26 Hamada, T., Ye, X., and Stahl, S.S. (2008). *J. Am. Chem. Soc.* 130: 833–835.
27 Humphrey, J.M. and Chamberlin, A.R. (1997). *Chem. Rev.* 97: 2243–2266.
28 Valeur, E. and Bradley, M. (2009). *Chem. Soc. Rev.* 38: 606–631.
29 Ojeda-Porras, A. and Gamba-Sánchez, D. (2016). *J. Org. Chem.* 81: 11548–11555.
30 Ghosh, S.C., Ngiam, J.S., Seayad, A.M. et al. (2012). *J. Org. Chem.* 77: 8007–8015.
31 Shie, J.-J. and Fang, J.-M. (2003). *J. Org. Chem.* 68: 1158–1160.
32 Wang, L., Fu, H., Jiang, Y., and Zhao, Y. (2008). *Chem. Eur. J.* 14: 10722–10726.
33 Wang, J., Liu, C., Yuan, J., and Lei, A. (2014). *Chem. Commun.* 50: 4736–4739.
34 Tank, R., Pathak, U., Vimal, M. et al. (2011). *Green. Chem.* 13: 3350.
35 Liu, X. and Jensen, K.F. (2012). *Green. Chem.* 14: 1471–1474.
36 Sankari Devi, E., Alanthadka, A., Tamilselvi, A. et al. (2016). *Org. Biomol. Chem.* 14: 8228–8231.
37 Zhu, M., Fujita, K., and Yamaguchi, R. (2012). *J. Org. Chem.* 77: 9102–9109.
38 Li, G.L., Kung, K.K., and Wong, M.K. (2012). *Chem. Commun.* 48: 4112–4114.
39 Miyamura, H., Min, H., Soule, J.F., and Kobayashi, S. (2015). *Angew. Chem. Int. Ed.* 54: 7564–7567.
40 Chen, C. and Hong, S.H. (2011). *Org. Biomol. Chem.* 9: 20–26.
41 Soule, J.F., Miyamura, H., and Kobayashi, S. (2011). *J. Am. Chem. Soc.* 133: 18550–18553.
42 Wang, Y., Zhu, D., Tang, L. et al. (2011). *Angew. Chem. Int. Ed.* 50: 8917–8921.
43 Krabbe, S.W., Chan, V.S., Franczyk, T.S. et al. (2016). *J. Org. Chem.* 81: 10688–10697.
44 Zweifel, T., Naubron, J.V., and Grutzmacher, H. (2009). *Angew. Chem. Int. Ed.* 48: 559–563.
45 Yamaguchi, K., Kobayashi, H., Oishi, T., and Mizuno, N. (2012). *Angew. Chem. Int. Ed.* 51: 544–547.
46 Wu, X.F., Neumann, H., and Beller, M. (2011). *Chem. Soc. Rev.* 40: 4986–5009.

47 Liu, Q., Zhang, H., and Lei, A. (2011). *Angew. Chem. Int. Ed.* 50: 10788–10799.
48 Ren, L., Li, X., and Jiao, N. (2016). *Org. Lett.* 18: 5852–5855.
49 Shi, R., Lu, L., Zhang, H. et al. (2013). *Angew. Chem. Int. Ed.* 52: 10582–10585.
50 Shi, R., Lu, L., Xie, H. et al. (2016). *Chem. Commun.* 52: 13307–13310.
51 Wei, W., Hu, X.Y., Yan, X.W. et al. (2012). *Chem. Commun.* 48: 305–307.
52 Chen, X., Chen, T., Li, Q. et al. (2014). *Chem. Eur. J.* 20: 12234–12238.
53 Cheng, H.C., Hou, W.J., Li, Z.W. et al. (2015). *Chem. Commun.* 51: 17596–17599.
54 Yu, H., Dannenberg, C.A., Li, Z., and Bolm, C. (2016). *Chem. Asian. J.* 11: 54–57.
55 Meyer, A.U., Berger, A.L., and Konig, B. (2016). *Chem. Commun.* 52: 10918–10921.
56 Yao, S., Saaby, S., Hazell, R.G., and Jørgensen, K.A. (2000). *Chem. Eur. J.* 6: 2435–2448.
57 Liu, G., Cogan, D.A., Owens, T.D. et al. (1999). *J. Org. Chem.* 64: 1278–1284.
58 Westheimer, F.H. and Taguchi, K. (1971). *J. Org. Chem.* 36: 1570–1572.
59 Chen, B., Wang, L., and Gao, S. (2015). *ACS. Catal.* 5: 5851–5876.
60 Guillena, G., JR, D., and Yus, M. (2010). *Chem. Rev.* 110: 1611–1641.
61 Jiang, L., Jin, L., Tian, H. et al. (2011). *Chem. Commun.* 47: 10833–10835.
62 Kang, Q. and Zhang, Y. (2012). *Green. Chem.* 14: 1016–1019.
63 Tian, H., Yu, X., Li, Q. et al. (2012). *Adv. Synth. Catal.* 354: 2671–2677.
64 Liu, L., Zhang, S., Fu, X., and Yan, C.H. (2011). *Chem. Commun.* 47: 10148–10150.
65 Patil, R.D. and Adimurthy, S. (2011). *Adv. Synth. Catal.* 353: 1695–1700.
66 Hu, Z. and Kerton, F.M. (2012). *Org. Biomol. Chem.* 10: 1618–1624.
67 Largeron, M. and Fleury, M.B. (2012). *Angew. Chem. Int. Ed.* 51: 5409–5412.
68 Zhang, E., Tian, H., Xu, S. et al. (2013). *Org. Lett.* 15: 2704–2707.
69 Zhao, S., Liu, C., Guo, Y. et al. (2014). *J. Org. Chem.* 79: 8926–8931.
70 Wendlandt, A.E. and Stahl, S.S. (2012). *Org. Lett.* 14: 2850–2853.
71 Park, J.H., Ko, K.C., Kim, E. et al. (2012). *Org. Lett.* 14: 5502–5505.
72 Liu, L., Wang, Z., Fu, X., and Yan, C.H. (2012). *Org. Lett.* 14: 5692–5695.
73 To, W.P., Tong, G.S., Lu, W. et al. (2012). *Angew. Chem. Int. Ed.* 51: 2654–2657.
74 Wendlandt, A.E. and Stahl, S.S. (2014). *J. Am. Chem. Soc.* 136: 506–512.

7

Green Oxidative Synthesis of Nitriles

Laura M. Dornan, Qun Cao, and Mark J. Muldoon

Queen's University Belfast, Belfast, United Kingdom

7.1 Introduction

The nitrile group is an important functionality. Nitriles are intermediates in the synthesis of a wide variety of organic compounds, as well as being present in products ranging from pharmaceuticals [1] to materials [2]. Figure 7.1 illustrates a few examples of nitrile containing compounds and details of their applications.

Unfortunately, the synthesis of the nitrile group often poses significant challenges. For example, methods that have typically been used by synthetic chemists for the preparation of aromatic nitriles such as the Rosenmund–von Braun reaction and the Sandmeyer reaction rely on stoichiometric metal cyanides. Due to the problems associated with such methods, there has been a move toward catalytic approaches [3] and the use of alternative (non-metallic) sources of the cyano groups [4]. Nonetheless, a further improvement would be to move away from methods that rely on aryl halides all together and that utilize readily accessible substrates. Catalytic oxidation methods are highly attractive from this point of view and in fact many important nitriles are produced in such a manner on a very large scale [5]. So called "ammoxidation" normally involves the catalytic oxidation of an alkene or aromatic compound in the presence of ammonia. The production of acrylonitrile (Figure 7.1) from propene (using the Sohio Process), is perhaps the best-known example of industrial ammoxidation, producing this commodity chemical in the multi-million-ton scale. Typically, ammoxidation reactions are performed at high temperature between 400 and 550 °C with high pressure and large excess of ammonia; conditions that limit the substrate scope to relatively simple substrates such as toluene, xylenes, and methyl pyridines. Therefore, such methods have limited applicability for laboratory synthesis of nitriles or indeed the production of functionalized molecules commonly found in pharmaceuticals and agrochemicals. This chapter will focus on recent developments in the area of oxidation catalysis for nitrile synthesis, where the conditions are milder and readily accessible for synthetic chemists. In recent years, there has been an incredible growth in the number of such publications, and there is now a plethora of new synthetic methods for the preparation of nitriles. Indeed, there have been a number of review articles published recently that discuss a range of approaches [3, 4, 6–8]. Here, we will examine catalytic oxidative methods that avoid the

Green Oxidation in Organic Synthesis, First Edition. Edited by Ning Jiao and Shannon S. Stahl.

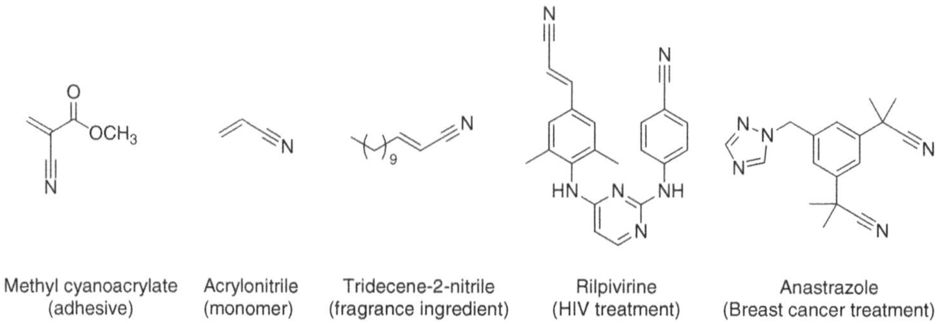

Figure 7.1 Examples of chemicals containing the nitrile functional group.

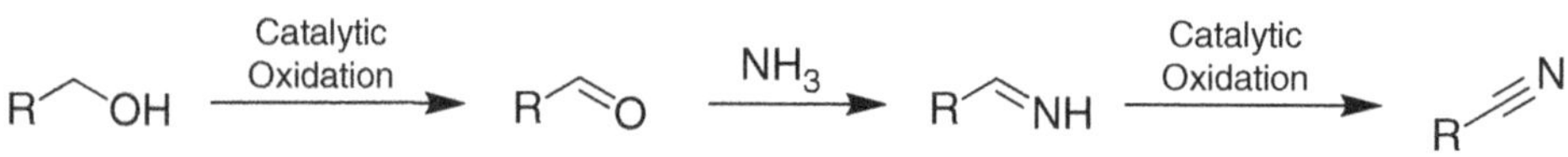

Figure 7.2 Overall pathway for nitrile synthesis from alcohols (or aldehydes) using ammonia.

use of stoichiometric metal cyanide salts and in-line with the ethos of this book, we will focus on methods that use O_2 as the terminal oxidant, as this is the most sustainable oxidant available. Even within these parameters, some methods will arguably be "greener" than others, however, we wish to highlight methods that allow nitriles to be prepared from a variety of substrates. Such substrate flexibility is highly beneficial when it comes to designing synthetic pathways for nitrile containing products.

7.2 Nitriles from Alcohols and Aldehydes

7.2.1 Ammonia as the Nitrogen Source

Alcohols and aldehydes are particularly attractive substrates as these are readily accessible. As shown in Figure 7.2, when starting from an alcohol it is first oxidized to form an aldehyde. This can then react with ammonia to generate an imine intermediate that then undergoes a further catalytic oxidative dehydrogenation to produce the nitrile. Utilizing ammonia as an inexpensive nitrogen source is highly desirable [9], and in many cases, aqueous ammonia, or ammonium salts can be employed, which is advantageous as these forms are easy to handle. In recent years, there has been a significant number of reports detailing catalytic oxidation of alcohols and/or aldehydes to nitriles. Here, we will highlight some examples, using both heterogeneous and homogeneous catalysts.

7.2.1.1 Heterogeneous Catalysis

In 2009, Mizuno and co-workers reported perhaps one of the first examples of a method that operated under relatively mild conditions (Figure 7.3) [10]. A ruthenium hydroxide catalyst was employed and a range of primary alcohols and aldehydes including benzylic, allylic, and those containing heteroatoms could be converted to their corresponding nitriles. However, in the case of aliphatic substrates, this method was only able to convert aliphatic aldehydes to the corresponding nitrile.

Figure 7.3 General conditions for ruthenium catalyzed oxidation by Mizuno and co-workers [10]. Reproduced with permission of John Wiley & Sons.

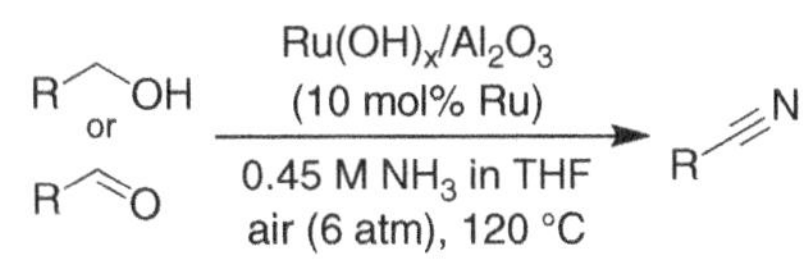

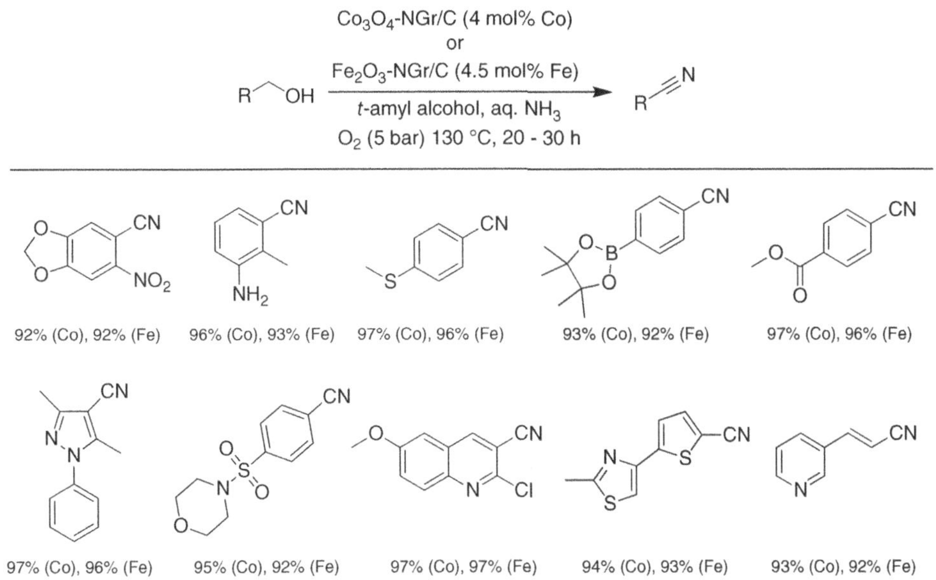

Figure 7.4 Exemplar products using Co and Fe based nanocatalysts [12]. Reproduced with permission of Springer Nature.

Subsequently, Ishida and Haruta discussed the use of metal oxides as catalysts for the oxidation of various alcohols to their corresponding nitriles with gaseous ammonia [11]. They found that MnO_2, Co_3O_4, and NiO were catalytically active in the preparation of nitriles, with MnO_2 providing the best results. A range of benzylic, allylic, and aliphatic alcohols were converted and the system was relatively tolerant of heteroatoms. In this case, aliphatic alcohols could be converted, for example, the nitrile could be produced from 1-octanol in 94% yield, after reacting for 10 hours at 100 °C. They also found that the addition of gold nanoparticles to the metal oxide catalyst led to hydration of the nitrile and produced the corresponding amides.

In 2014, Beller and co-workers described the development of nanocatalysts prepared from either cobalt or iron precursors [12]. Not only does this method use non-precious metals, the resultant catalysts delivered excellent performance for an extensive range of substrates, and they demonstrated that the catalysts could be recycled and reused. A detailed guide for the preparation of these catalysts was subsequently published [13], but in short, they were prepared by taking phenanthroline complexes of either $Co(OAc)_2.4H_2O$ or $Fe(OAc)_2$ and carrying out the pyrolysis of these complexes on carbon powder (Vulcan XVC72R) at 800 °C under argon. The resultant nanocatalysts, denoted as Co_3O_4-NGr/C and Fe_2O_3-NGr/C, have nitrogen-doped graphene layers (NGr) and nanoparticles of metal oxides. The general conditions are shown in Figure 7.4 along with some exemplar products. The catalysts had a broad substrate scope and they demonstrated the usefulness of their catalysts with the preparation of over 80 structurally diverse nitriles. In the majority of cases, products were produced in good to excellent yields. The catalysts demonstrate exceptional functional group tolerance. Aliphatic nitriles could be produced, with only slight modification of the reaction conditions, for example, these generally required an increase in catalyst loading (to 5 mol% Co or 6 mol% Fe), change of the solvent and an increase in temperature to 140 or 150 °C. Most substrates in this study were reacted on a 0.5 mmol scale, however, for a number of benzylic substrates they showed that these reactions could be scaled up (5 g of substrate) and similar yields could be obtained. Additionally, the catalyst has the potential to be recycled, and they demonstrated that the same catalyst could be used for four reactions (using benzyl alcohol as the substrate), delivering similar yields without the need for any reactivation procedures.

The methods reported by Beller and co-workers are the best reported to-date for heterogeneous catalysts, in terms of substrate scope and general utility. Nonetheless it is worth mentioning some recent more reports by Gao and co-workers, which follows on in this area. They initially reported the use of mesoporous cobalt, nitrogen-doped carbon catalyst [14]. This involved the pyrolysis of cobalt complexed with 11,11′-bis(dipyrido[3,2-*a*,2′,3′-c]phenazinyl), but in this case, a mesoporous structure could be obtained by using a silica colloid or mesoporous silica (SBA-15) as a template. The catalytic reaction conditions were similar to that used in the Beller work, albeit with a lower loading of Co (0.5 mol%). It was shown that a range of aromatic nitriles could be produced in excellent yields, although the catalyst was not effective for aliphatic substrates. Subsequently they demonstrated that metal-free, nitrogen-doped carbons could act as an effective catalyst [15]. Good to excellent yields of aromatic nitriles were obtained, however, the catalyst was not effective for aliphatic alcohols.

7.2.1.2 Homogeneous Catalysis

Homogeneous copper/stable radical catalyst systems for the synthesis of nitriles date back to 1960s [16], but more recently this area has matured. Such catalyst systems have been well studied for alcohol oxidation [17], and in 2013 there were reports by Tao [18], Huang [19], Muldoon [20], and Stahl [21], describing the use of copper catalysts and the stable radical TEMPO (2,2,6,6-tetramethylpiperidinyloxyl). These methods are attractive as they demonstrate wide substrate scope, use mild conditions, and the catalyst is comprised of commercially available components. Here we will highlight the work by Huang and co-workers, which employed 5 mol% of copper iodide salt, 5 mol% 2,2′-bipyridine (bpy) ligand in conjunction with 5 mol% TEMPO and aqueous ammonia as the nitrogen source to synthesize nitriles from alcohols under an oxygen atmosphere. Aromatic alcohols were converted to nitriles in ethanol at room temperature, while aliphatic alcohols required temperatures of 50 °C and used acetonitrile as the solvent. They demonstrated that a wide range of substrates could yield products, mostly in good to excellent yields. Some examples are shown in Figure 7.5, where it can be seen that heteroatom-containing molecules were tolerated including pyridines, triazoles, piperazines, tetrahydropyrans, and morpholines.

The group also demonstrated the one-pot synthesis of various heterocycles, whereby the reagents for the cyclization reactions were added once the alcohol to nitrile reaction was complete (Figure 7.6). In preparation of molecules such as imidazolines, oxazolines,

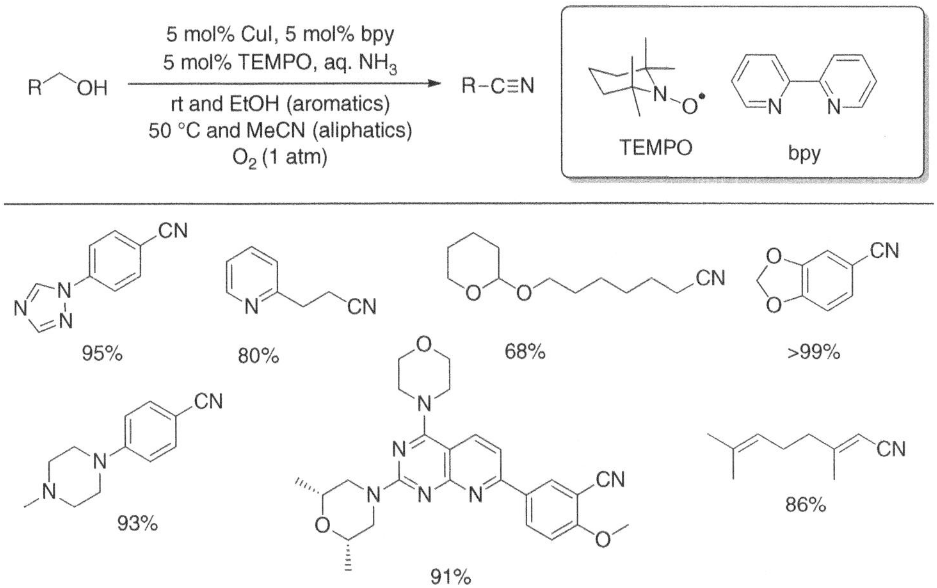

Figure 7.5 Copper/TEMPO oxidation of alcohols by Huang and co-workers [19]. Reproduced with permission of The American Chemical Society.

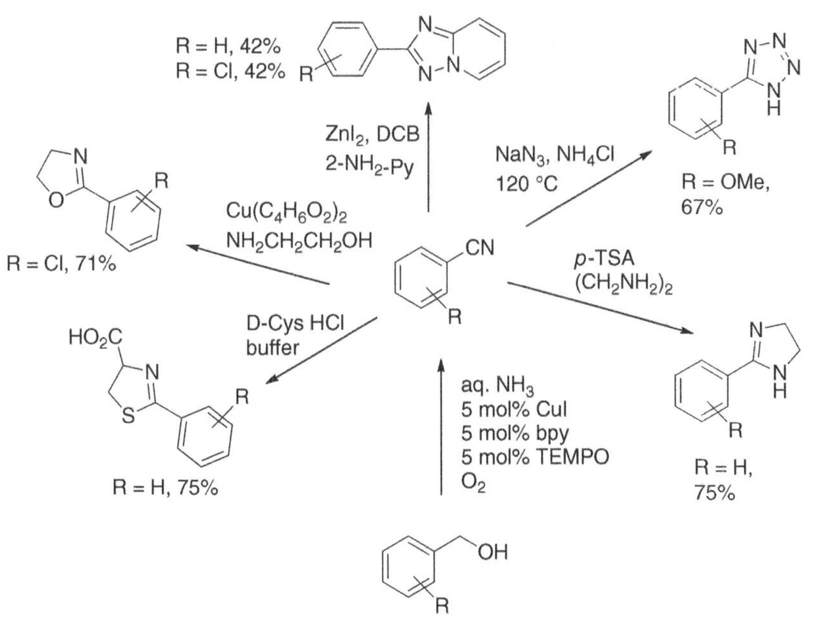

Figure 7.6 One-pot synthesis of heterocycles starting from alcohols [19]. Reproduced with permission of The American Chemical Society.

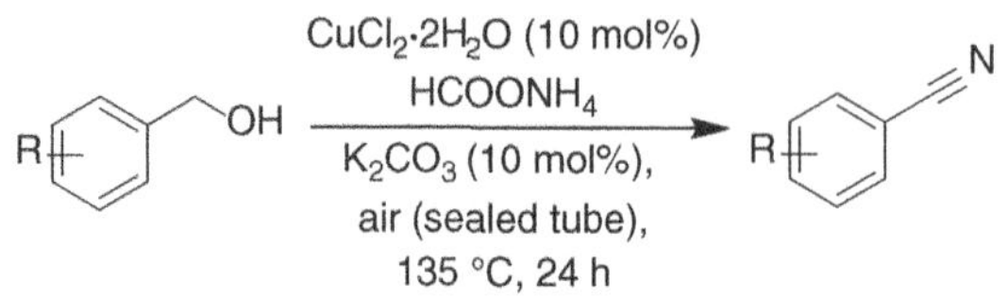

Figure 7.7 TEMPO free catalysis of activated alcohols [24]. Reproduced with permission of John Wiley & Sons.

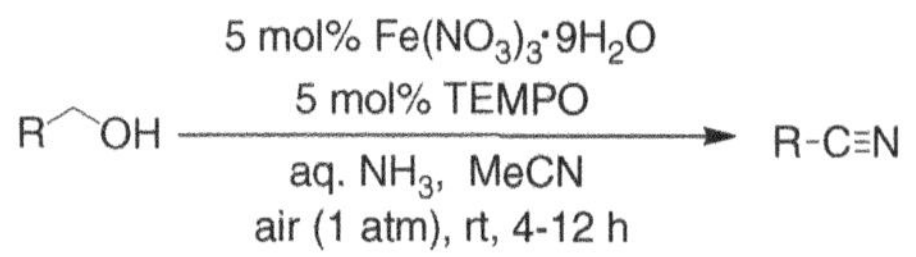

Figure 7.8 General conditions for iron/TEMPO oxidation of alcohols to nitriles [25]. Reproduced with permission of John Wiley & Sons.

and thiazolines, it was found that this protocol did not lead to further oxidation to the corresponding heteroaromatic compounds. Overall, this study demonstrates that Cu/TEMPO has the potential to prepare functionalized molecules under mild conditions using commercially available reagents.

There have been reports of copper systems that do not utilize TEMPO as a co-catalyst [22, 23], however, in these instances the substrate scope is limited to aromatic alcohols. For example, Bhanage and co-workers reported the use of $CuCl_2$ as the catalyst for synthesis of nitriles from alcohols using ammonium formate as the nitrogen source (Figure 7.7) [24]. A range of benzylic alcohols were converted to nitriles in good yields, although it was found that yields were significantly lower when there were substituents in the *ortho* position. The method was also shown to work for the allylic alcohol cinnamyl alcohol, which was converted to cinnamonitrile with 80% yield.

In 2014, Batra and co-workers reported an iron nitrate/TEMPO catalyst system for the oxidation of alcohols to nitriles (Figure 7.8) [25]. The Fe/TEMPO system could produce nitriles under mild conditions using aqueous ammonia as the nitrogen source. It was shown to deliver good to excellent yields for a range of benzylic and allylic alcohols and was tolerant of heteroatoms. It could also oxidize aliphatic alcohols, with octanenitrile produced in 77% yield and adiponitrile produced in 58% yield.

The use of a simple, inexpensive iron salt and no additional ligand is an attractive feature of this system. Such catalysts have previously been used for alcohol oxidation [17a] and as discussed in this paper, it operates via a different mechanism to the copper/TEMPO systems, with the $Fe(NO_3)_3$ likely leading to NOx that can result in the oxoammonium cation. While copper/TEMPO operate cooperatively as one-electron oxidants, the oxoammonium cation can act as a two-electron oxidant. In 2015, Kim and co-workers have reported a metal-free system for the conversion of aldehydes to nitriles using a nitroxyl radical/NOx system [26]. Figure 7.9 shows the conditions and also demonstrates the proposed mechanism that involves NOx and the oxidation of the imine by oxoammonium salt. In this case, the nitroxyl radical used was 4-acetamido-2,2,6,6-tetramethylpiperidine-*N*-oxyl and it was combined with $NaNO_2$, HNO_3 (to generate NOx) and ammonium acetate as the nitrogen source. A range of aromatic aldehydes were converted to aromatic nitriles including those with electron-donating and electron-withdrawing substituents. 4-Hydroxybenzyaldehyde was converted with moderate yield to the nitrile, a substrate that has previously been problematic in Cu/TEMPO systems [27].

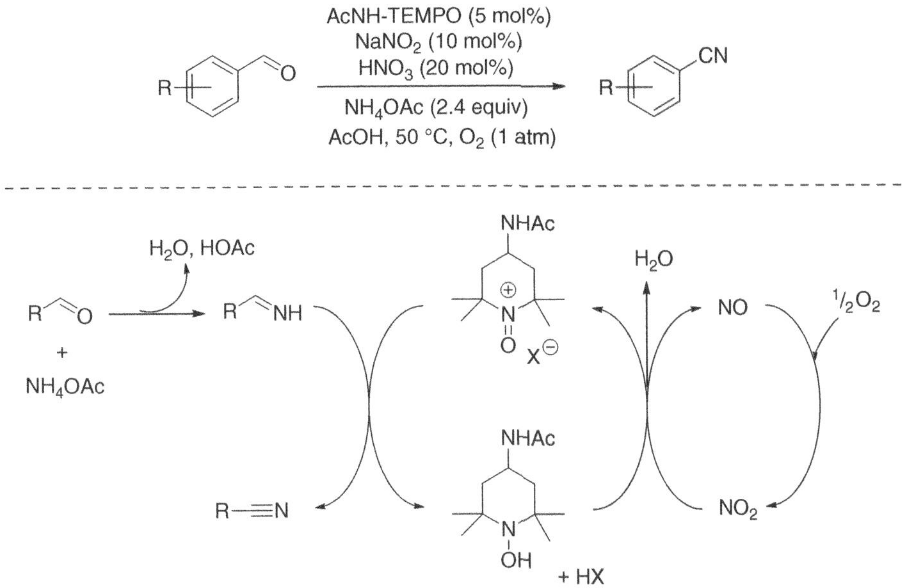

Figure 7.9 AcNH-TEMPO conditions and proposed mechanism by Kim and co-workers [26]. Reproduced with permission of The American Chemical Society.

Shen and co-workers also explored this avenue using hexamethyldisilazane (HMDS) as the nitrogen source and TEMPO as the nitroxyl radical [28]. In this case either $NaNO_2$ or *tert*-butyl nitrite can be used as the NOx source. They also discovered that using an additive of KPF_6 increased the yield of nitrile product, and postulated that the positive effect may be due to alteration of the $TEMPO^+$/TEMPOH redox potential.

There are also reports that use oxoammonium salts and do not use O_2 as the terminal oxidant. For example, the oxoammonium salt can be prepared from the nitroxyl radical using tetrafluoroboric acid and bleach. The oxoammonium tetrafluoroborate salt can be isolated and then as used as a stoichiometric oxidant for the synthesis of nitriles from aldehydes [29, 30]. It is also possible to use electrochemical oxidation to generate the oxoammonium salt *in situ* and Li and co-workers have reported electrochemical methods for producing nitriles from aldehydes and alcohols [31–33].

7.2.2 Oxidative Reactions of Aromatic Alcohols or Aldehydes with Acetonitrile

In 2014, Liu and co-workers reported a method for the oxidative coupling of aromatic alcohols (or aldehydes) with acetonitrile to prepare β-ketonitriles [34]. The method utilizes $CuCl_2$ as an inexpensive catalyst and operates under mild conditions. In the case of alcohols, it is thought these are first oxidized to the corresponding aldehyde. In terms of a mechanism, the authors postulate that the aldehyde is oxidized to a radical intermediate, which then reacts with a $Cu(NCCH_2)_2$ species that is generated under the reaction conditions. Some examples from their substrate studies are shown Figure 7.10 and it can be seen that the method could deliver good yields for a number of substrates.

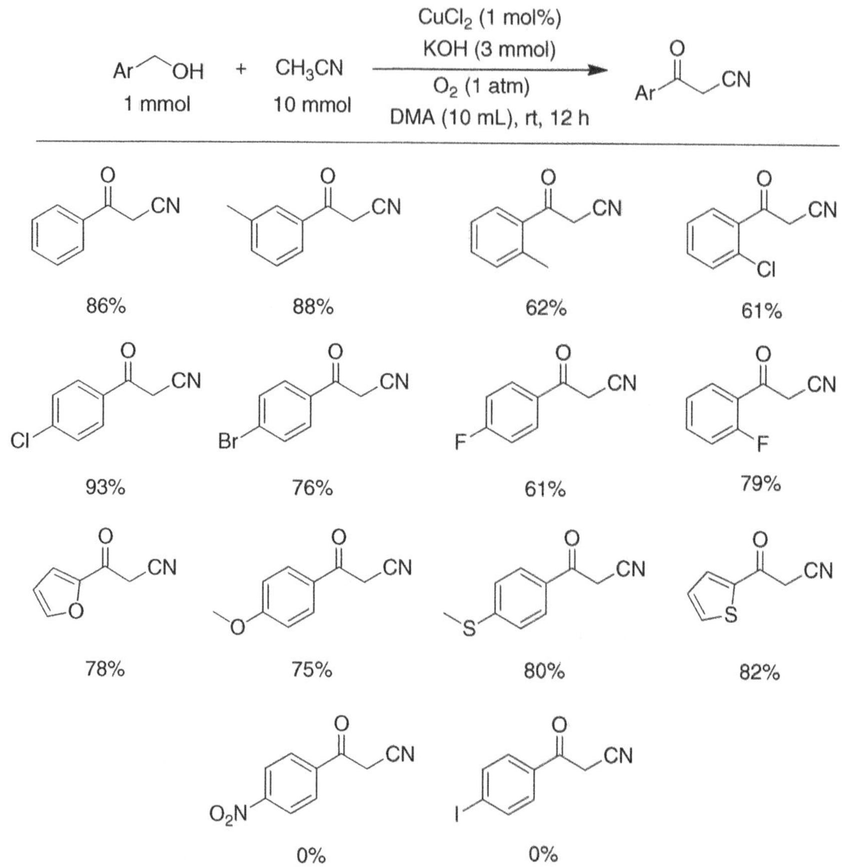

Figure 7.10 Examples of β-ketonitriles from aromatic alcohols and acetonitrile [34]. Reproduced with permission of The American Chemical Society.

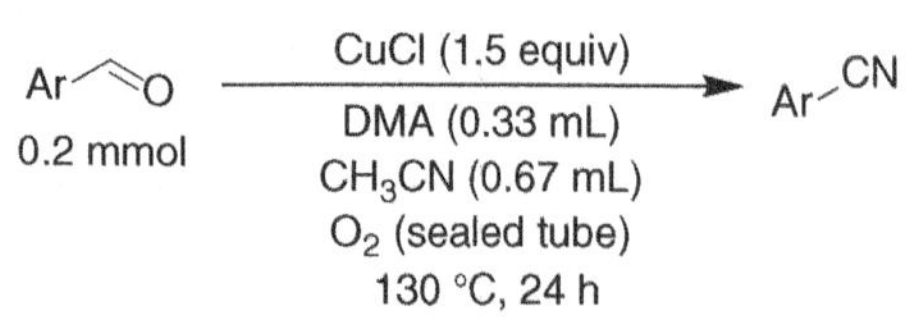

Figure 7.11 Copper mediated synthesis of nitriles from aldehydes using acetonitrile as a reagent [35]. Reproduced with permission of The Royal Society of Chemistry.

Halogens other than iodide were tolerated and it was found that a nitro group also resulted in none of the desired product.

Although not catalytic in copper, given the similar reagents, it is also worth mentioning a report by Yin and co-workers, which described a method for converting aromatic aldehydes to aromatic nitriles [35]. The general conditions are shown in Figure 7.11 and unlike the method discussed before that results in β-ketonitriles, there is no base present and a temperature of 130 °C is employed. The authors hypothesized a free radical based mechanism as both TEMPO and 2,6-di-*tert*-butyl-4-methylphenol (BHT) stopped the reaction. In the case of TEMPO, they managed to detect the TEMPO-acetonitrile radical adduct. They believed that the reaction involves N atom transfer via cleavage of the C—N triple bond and the C—CN carbon bond of acetonitrile. Although not catalytic, this is an interesting report, which could lead to catalytic methods in the future.

Given the cost and availability of acetonitrile, it is an attractive reagent for the preparation of nitriles and we discuss a number of other methods that use acetonitrile in Section 7.4.

7.3 Nitriles from Amines

The catalytic aerobic oxidation of amines is an important area of research, as these reactions can produce highly valuable intermediates and products [36]. The oxidation of primary amines can potentially lead to a number of different products, such as oximes, imines, amides, and nitriles, therefore developing selective catalysts is vitally important when using such substrates. The coupling of the primary imine intermediate is often the major competing reaction with the oxidation of the primary imine to the nitrile (Figure 7.12). In some cases, certain conditions and catalysts can also result in the hydration of the nitrile leading to the corresponding primary amide.

As discussed in a 2012 review by Hermans and co-workers [36a], ruthenium-based catalysts have traditionally been the preferred choice for the selective oxidation of amines to nitriles. At the turn of the millennium, important developments in this area were made by the groups of Kaneda [37] and Mizuno [38], who reported efficient heterogeneous catalysts. The use of ruthenium-based catalysts still gathers attention for the oxidation of amines to nitriles [39–41], however, here we will mostly focus on other metals that have come to the fore in recent years. In the last few years there have been a number of reports of very effective catalysts that use inexpensive, first-row transition metals.

In terms of heterogeneous catalysis, the best performing catalysts are once again the nanoscale metal oxide catalysts developed by Beller and co-workers [42, 43]. As we discussed earlier (in Section 7.2.1.1) for the oxidation of alcohols to nitriles, Beller and co-workers developed highly efficient catalysts based on nano-sized metal oxides and nitrogen-doped carbon. In 2015, they demonstrated that their Fe_2O_3-based nanocatalysts were excellent for the oxidation of amines to nitriles [42]. In 2016, they then demonstrated the use of their nanoscale Co_3O_4-based catalysts for this transformation too [43]. The general conditions for both these catalysts are shown in Figure 7.13. In the case of their iron catalyst, they demonstrated that this catalyst could be used for a wide range of substrates (around 60 were shown). Many substrates were shown to deliver yields of over 80%, with many products produced in quantitative yield. They also

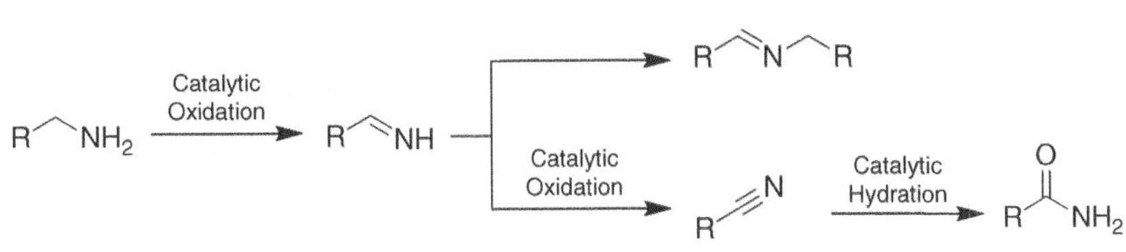

Figure 7.12 Simplified overview of the reaction pathways during oxidation of amines.

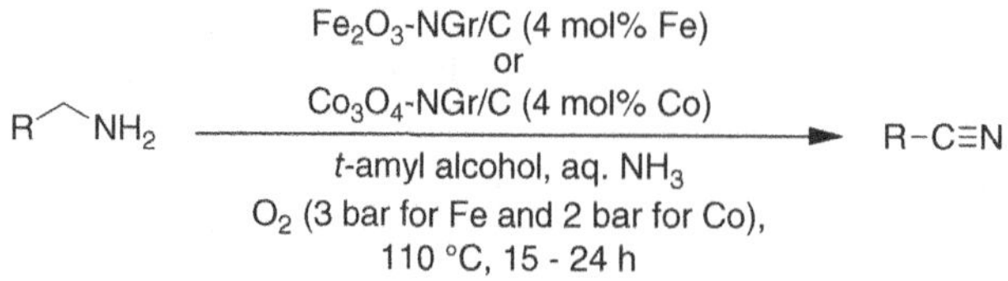

Figure 7.13 Amine oxidation using iron [42] and cobalt [43] based catalysts by Beller and co-workers.

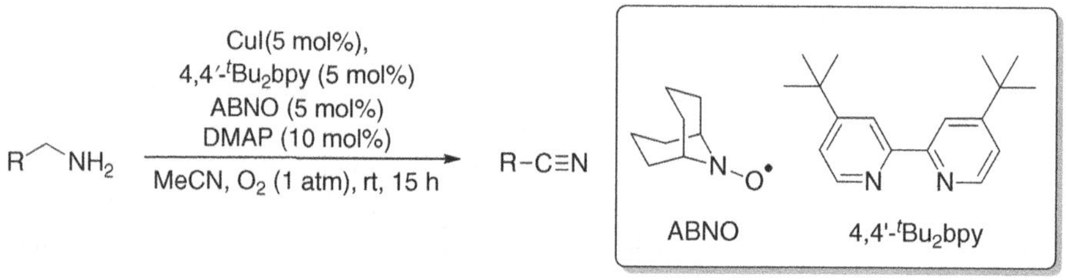

Figure 7.14 General conditions for the oxidation of amines to nitriles using Cu/ABNO [21]. Reproduced with permission of The American Chemical Society.

demonstrated that this catalyst was effective when used on a larger scale (2 g of substrate). Similar excellent performance was exhibited for the cobalt based catalyst and they prepared nitriles from around 40 diverse substrates. In both these studies, they showed that these heterogeneous catalysts exhibit impressive stability and they could be readily recycled and reused.

Next, we will highlight homogeneous based systems, which have advantages in terms of using commercially available catalysts and operating under very mild conditions. We previously discussed (Section 7.2.1.2) how nitroxyl radicals (such as TEMPO) can be employed in a number of different ways to convert alcohols or aldehydes into nitriles and such radicals can also be applied to amine oxidation. For example, Semmelhack, and Schmid reported the electrochemical oxidation of amines to nitriles using TEMPO (via the oxoammonium salt) back in 1983 [44]. Bailey and co-workers demonstrated that isolated oxoammonium salts can be used as a stoichiometric oxidant to convert amines to nitriles [45]. In terms of aerobic methods, the use of a nitroxyl radical in tandem with a copper co-catalyst, (acting together as 1-electron oxidants) is an attractive option. As discussed earlier, this system works well for oxidation of alcohols in the presence of ammonia. However, in the case of amine oxidation, most copper/nitroxyl systems result in the homocoupled imine as the major product [46]. In 2013, Kim and Stahl not only reported the use of Cu/TEMPO for oxidation of alcohols to nitriles, they also disclosed that if sterically less hindered radicals were employed, then the nitrile could be produced as the major product in amine oxidation [21]. The optimized conditions for amine oxidation (Figure 7.14) utilized ABNO (9-azabicyclo[3.3.1]nonane-N-oxyl) in conjunction with CuI and 4,4'-^{t}Bu$_2$bpy (4,4''-di-*tert*-butyl-2,2'-bipyridyl) with DMAP (4-dimethylaminopyridine) as a base. These conditions facilitated the oxidative dehydrogenation of a large scope of amines to their nitrile counterparts. Primary benzylic amines underwent the transformation readily at room temperature with an O_2 balloon, producing products in good to excellent yields, for substrates with a wide range of functional groups. Aliphatic amines were slower to react but good to excellent

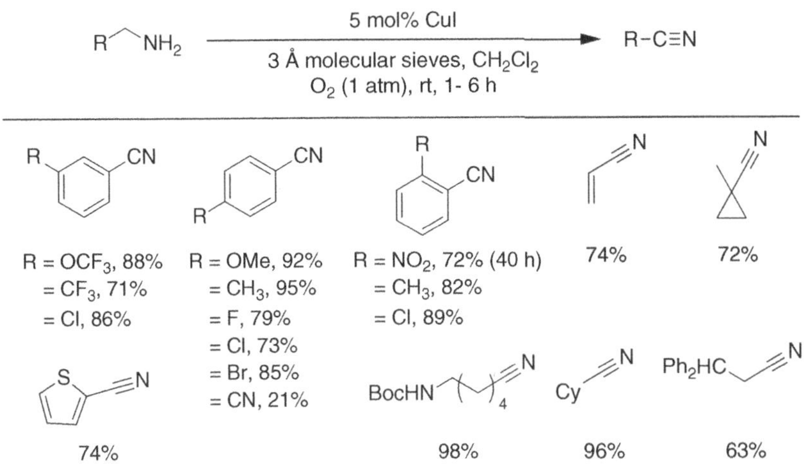

Figure 7.15 CuI catalyzed oxidation of amines to nitriles [48]. Reproduced with permission of John Wiley & Sons.

yields were observed when the catalyst loading doubled from 5 to 10 mol% (and 20 mol% DMAP). Usefully, they also highlighted substrates that were problematic. It was found that tetrahydrofurfurylamine, gave a lower yield (47%) even with a higher loading of catalyst and this may be due to the fact that the substrate could chelate the copper complex. 4-bromophenethylamine gave a mixture of products, while 4-hydroxybenzylamine produced no product. In the former case, this is likely due to the catalyst oxidizing the benzylic position while phenols inhibit these catalysts, both of these types of substrate limitations were also observed previously for Cu/nitroxyl oxidation of alcohols [47].

In 2016, Lumb and Arndtsen and co-workers developed a system that avoids the need for a nitroxyl radical and uses simple Cu(I) salts [48]. They postulate that the amine substrate could have a secondary role in the reaction and also act as a ligand. Water appears to be detrimental to the reaction, and of course it will be produced in such an aerobic oxidation, therefore molecular sieves are used to improve catalyst performance. They also found that the selectivity of the reaction was greatly influenced by the counterion on the copper catalyst. The use of copper(I) iodide formed predominantly the nitrile product whereas the use of copper(I) hexafluorophosphate formed the imine product preferentially. A variety of aromatic amines were converted to nitriles, as well as aliphatic amines and some containing heteroatoms, with good to excellent yields in most cases (some examples are shown in Figure 7.15). This method is highly attractive, as it employs an inexpensive copper salt without the need for any additional ligands or a nitroxyl catalyst.

7.4 Nitriles via C—H and C—C Bond Cleavage Reactions

The direct functionalization of hydrocarbons is a desirable but challenging route for the synthesis of higher value products. In recent years, there has been good progress in this area and there have been numerous reports of methods for the synthesis of nitriles that proceed via the cleavage of C—H or C—C bonds. Many of the methods fall outside the scope and limits of this chapter and for further reading in this area, we would direct you to some excellent review articles that have been published recently[7, 8].

7.4.1 Nitriles from Alkynes

Alkynes are readily available and established as valuable building blocks in organic chemistry. A number of research groups have reported methods for the conversion of alkynes to nitriles (via cleavage of the C≡C triple bond). The work to-date for these transformations does not really fall into the category of aerobic oxidation catalysis. For example, Jiao and co-workers demonstrated a novel route to preparing nitriles from terminal alkynes using Ag_2CO_3 as the catalyst along with azidotrimethylsilane ($TMSN_3$) [49]. Although this reaction is conducted in air, there is not a clear role for O_2 in the proposed reaction mechanism. Yanada and co-workers disclosed that internal alkynes could be cleaved to prepare nitriles using *N*-iodosuccinimide (oxidant) and $TMSN_3$ (as the nitrogen source) [50]. Maiti and co-workers used *tert*-butyl nitrite as the nitrogen source and 2-picoline-*N*-oxide as the oxidant, for the synthesis of aryl nitriles from the corresponding terminal alkynes [51]. Subsequently, Lin and Song reported a method that also used *tert*-butyl nitrite, finding that organic bases improved the yield, with the optimal being 40 mol% of *N*-benzylmethylamine [52] In this case, they found that the method required O_2. They proposed that O_2 produced NO_2, which then reacted with the terminal alkyne. NOx-based systems often have quite complex mechanisms and O_2 dependent equilibria and this may explain why they found that using air was superior to a pure O_2 atmosphere.

In terms of an aerobic catalytic system for alkynes, Jiao and co-workers reported a method for transforming aliphatic alkynes into alkenyl nitriles [53]. Such unsaturated nitriles are useful building blocks in organic synthesis [54], as well as being present in some drug molecules [1]. This protocol offers a simple method for preparing these valuable compounds, and it was found to be tolerant of a range of functional groups and they demonstrated that it had the potential for late-stage modification. Figure 7.16 gives details of the reaction conditions and some examples of the products they produced. They carried some experimental and theoretical (DFT) studies to try and gain some insights into the mechanism. Further studies are needed, but it was suggested that a 1,2-hydride shift could be a key step in the mechanism. DFT calculations found little energy difference between the *Z* and E isomers of the products, and predicted a ratio that was in good agreement with the ratios found experimentally (Figure 7.16).

7.4.2 Nitriles from Alkenes

There are a number of reports for the conversion of aromatic alkenes to nitriles, which involves cleavage of the C=C double bond, however, most examples are outside the remit of this chapter. For example, Jiao utilized *N*-bromosuccinimide, $TMSN_3$ and $Cu(OAc)_2$ [55], Guo employed (diacetoxyiodo)benzene as the oxidant and ammonium bicarbonate as the nitrogen source [56], and then later reported the use of $NaNO_2$ as the oxidant and nitrogen source [57]. In 2013, Wang and Jiao reported a novel aerobic

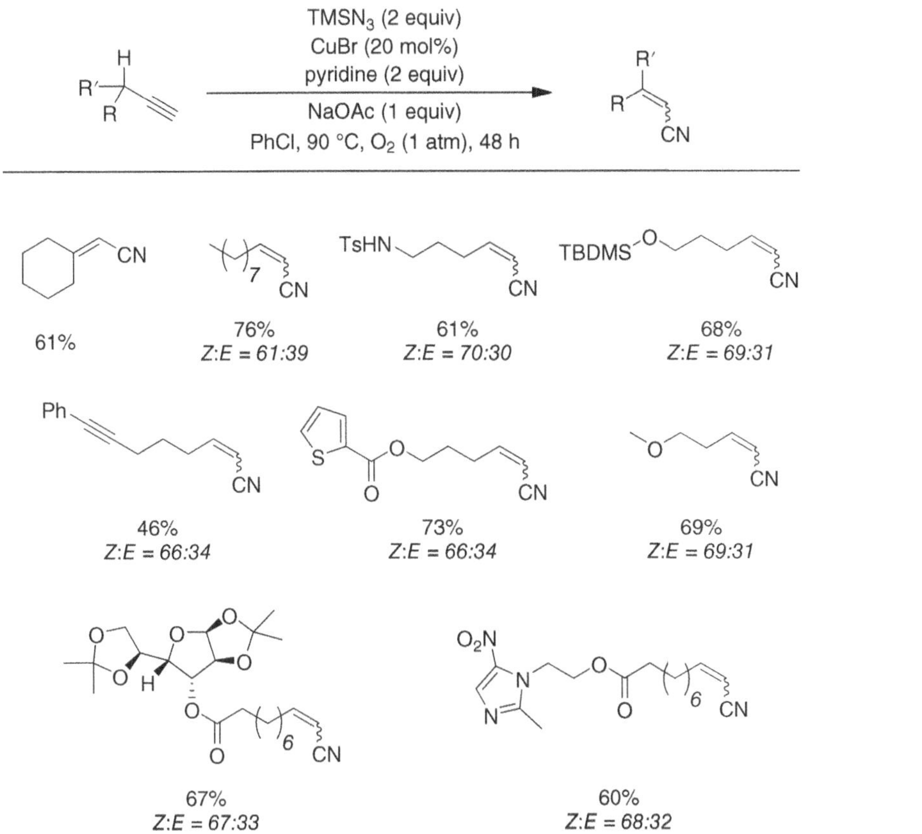

Figure 7.16 Synthesis of alkenyl nitriles from terminal alkynes [53]. Reproduced with permission of The Royal Society of Chemistry.

method that cleaved C=C double bonds to produce oxo-nitriles [58]. The method employs TEMPO as the catalyst and $TMSN_3$ as the nitrogen source and in the case of aliphatic alkenes it is also necessary to add (diacetoxyiodo)benzene (also known as phenyliodine(III) diacetate (PIDA)) (Figure 7.17). This method could enable the more efficient synthesis of high value products, as oxo-nitriles are valuable building blocks for the synthesis of *N*-heterocycles, alkenenitriles, and α-hydroxyketones. This method enables these important intermediates to be prepared in a more direct manner compared to conventional routes. For example, they demonstrated the synthesis of an oxo-nitrile in high yield (86%) that can be used for the efficient synthesis of (*S*)-ketamine, which is used widely as an anesthetic. Mechanistic evidence suggested that the reaction proceeds via a free radical mechanism, with the $TMSN_3$ initially generating an azido free radical that then attacks the alkene. They also carried out ^{18}O-labeling experiments to show that O_2 is not only the oxidant for the catalyst but also the source of oxygen in the oxo-nitrile product.

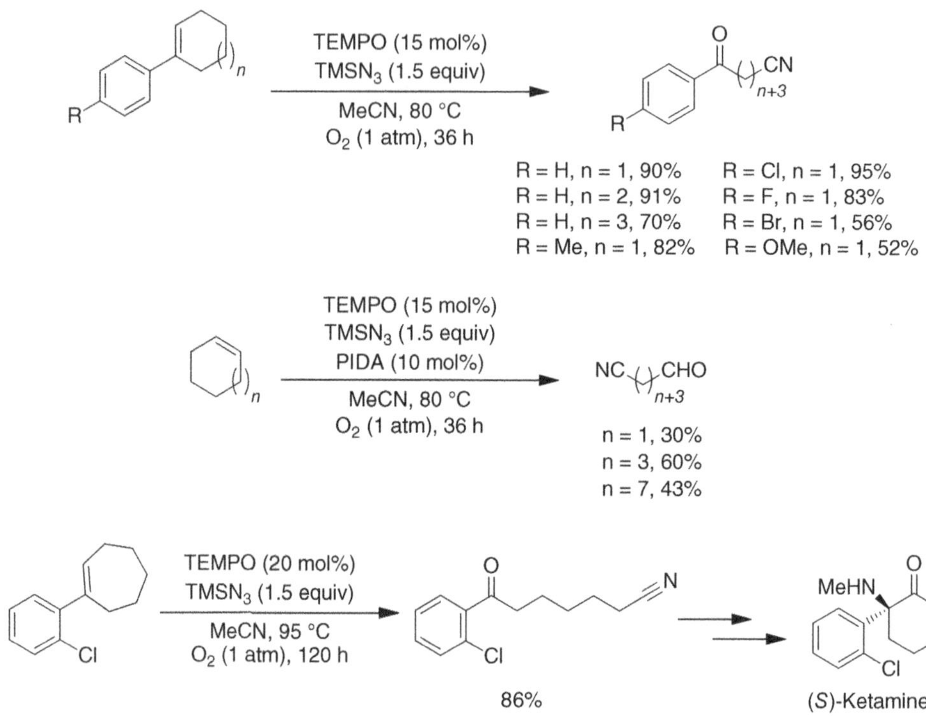

Figure 7.17 TEMPO-catalyzed synthesis of oxo-nitriles from alkenes [58]. Reproduced with permission of The American Chemical Society.

7.4.3 Cleavage of C—C Bonds in Oxygenated Substrates

Carboxylic acids are often readily accessible and/or naturally occurring, therefore being able to prepare nitriles via decarboxylation is a valuable synthetic route. In 2014, Song and co-workers developed a method of deriving benzonitriles from decarboxylation and subsequent oxidation and ammoxidation of phenylacetic acids [59]. Using 20 mol% $Cu(TFA)_2$ as a catalyst and urea as a nitrogen source in DMSO, various phenylacetic acids could be oxidized to the corresponding nitrile (Figure 7.18). The method tolerated electron-withdrawing and -donating groups, and in most cases, yields of above 70% were obtained and in numerous examples, yields of above 90% were achieved. Halogens (Br, Cl, and F) were tolerated that could enable further functionalization of the resultant nitrile molecule. 4-Nitrophenylacetic acid was the most challenging substituted phenylacetic acid and the reaction required heating to 150 °C (in most cases temperatures of 110 or 120 °C were employed) in order to obtain a yield of 56%. They also demonstrated that heteroaromatic acetic acids could be converted, albeit with slightly lower yields on average such as nicotinonitrile (55% yield), thiophene-2-carbonitrile (45% yield), and thiophene-3-carbonitrile (53% yield). 6-chloronicotinonitrile and benzothiophene-3-carbonitrile could be obtained in better yields (75 and 77%, respectively). The mechanism is believed to occur through catalytic oxidative decarboxylation and subsequent C—H oxidation to form the benzaldehyde that then reacts with ammonia (generated from the urea) to form the imine, which is then oxidized by the copper catalyst to the desired nitrile.

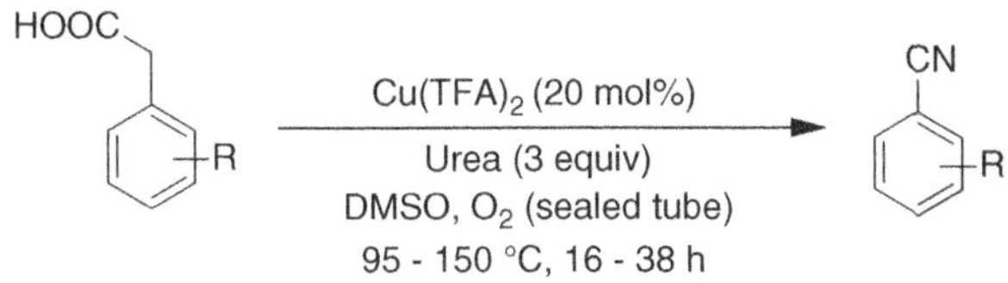

Figure 7.18 Synthesis of aromatic nitriles from phenylacetic acids [59]. Reproduced with permission of John Wiley & Sons.

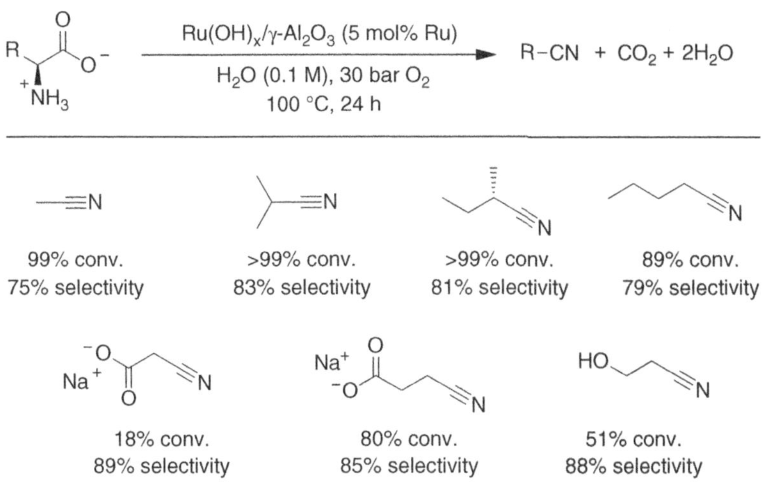

Figure 7.19 Ruthenium catalyzed synthesis of nitriles from amino acids [60]. Reproduced with permission of The Royal Society of Chemistry.

The following year, De Vos and co-workers reported a method that used a ruthenium catalyst to synthesize nitriles from amino acids [60]. As mentioned in Section 7.3, ruthenium catalysts have long been used for the oxidation of amines to nitriles, but in this case the reaction also involves the decarboxylation of the substrate. As the authors discuss, such amino acids can be obtained from proteins that can be recovered from (non-edible) biomass, which is currently in the waste streams from a number of industries (e.g. agriculture). Obtaining chemicals from renewable sources such as this is important if we are to improve the sustainability of chemical production and move toward a circular economy. Figure 7.19 gives the catalytic conditions along with some examples of the products that they produced. As can be seen, the substrates reacted with high conversion and with good selectivity in most cases.

As discussed in Section 7.2, aldehydes readily react with ammonia to form the primary imine that can then be oxidized to the nitrile. In 2015, Guo and co-workers reported the copper catalyzed oxidation of ketones to nitriles (Figure 7.20) [61]. Mechanistic studies suggested that cleavage of the C—C bond produces the aldehyde that can then produce the nitrile via the imine, as ammonium carbonate is added as a source of ammonia. Aryl methyl ketones delivered products in generally good to excellent yields and the reaction tolerated electron-donating and withdrawing (including halogens) groups. In the case of aryl substrates with longer alkyl chains, these were

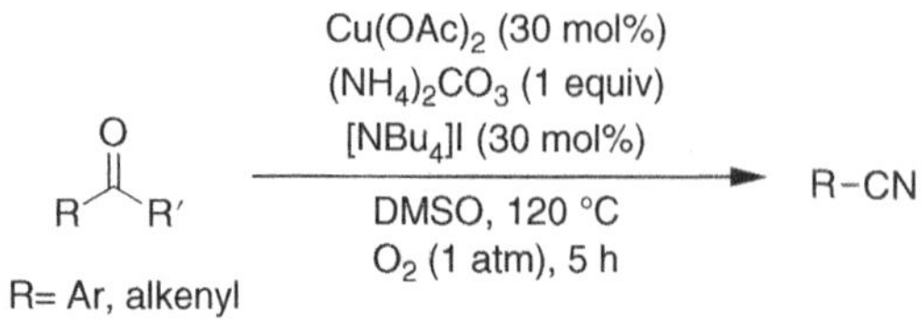

Figure 7.20 Copper catalyzed synthesis of nitriles from ketones [61]. Reproduced with permission of The Royal Society of Chemistry.

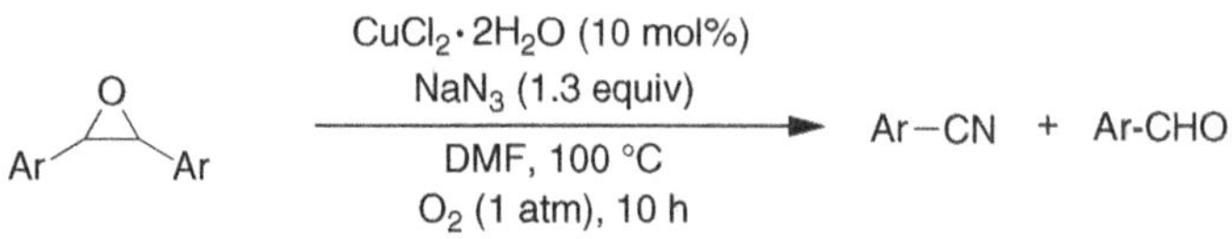

Figure 7.21 Copper catalyzed synthesis of nitriles (and aldehydes) from epoxides [63]. Reproduced with permission of The Royal Society of Chemistry.

more challenging and resulted in nitriles in moderate yields. Related to this work, it is worth mentioning a 2016 report by Kang and co-workers, which described the synthesis of nitriles from both aromatic and aliphatic aldehydes and ketones [62]. This was an effective and accessible method; however, it is outside the main remit of this chapter as it uses $AlCl_3$ (2 equivalents) and $NaNO_2$ as the oxidant and source of nitrogen.

In 2015, Gu and Jin reported a novel method that cleaves the C—C bond of epoxides and produces aromatic nitriles and aldehydes (Figure 7.21) [63]. In the case of symmetrical aromatic epoxides, nitriles could be obtained in yields ranging from 53 to 72%, but lower yields were obtained for asymmetric substrates. Epoxides with one or two aliphatic groups were not suitable substrates and did not produce any nitrile or aldehyde products.

7.4.4 Cyanation of Arenes Using Acetonitrile or *N,N*-Dimethylformamide as a Reagent

Utilizing common solvents such as acetonitrile or *N,N*-dimethylformamide (DMF) as a reagent for preparing nitriles is an attractive option. In recent years, a number of oxidative methods have been reported that use these solvents as a reagent for the synthesis of nitriles and we have already highlighted one example that used acetonitrile in Section 7.2.2. In this section, many of the reports that we highlight propose that these solvents result in the formation of metal-CN species, but generating these *in situ* is arguably very advantageous in terms of safety when compared to using highly toxic metal cyanide reagents (e.g. KCN, CuCN, and TMSCN) as stoichiometric reagents.

In 2012, Li and co-workers reported a method for the transformation of aryl iodides to aryl nitriles using acetonitrile [64]. This method utilizes a catalytic amount of a copper complex as well as Ag_2O (stoichiometric) and air as the oxidants. Following this study, there were two reports of copper-catalyzed cyanation of aromatic C—H bonds that utilized acetonitrile as a source of CN [65, 66]. Zhu and co-workers reported a method (Figure 7.22) that could be employed for the preparation of 2-cyanoindoles [65]; compounds that are key intermediates in the synthesis of pharmaceuticals. Selectivity is obtained by using a pyrimidyl (or pyridine) directing group. Substrates with both electron-donating and withdrawing groups (including nitrile) were tolerated

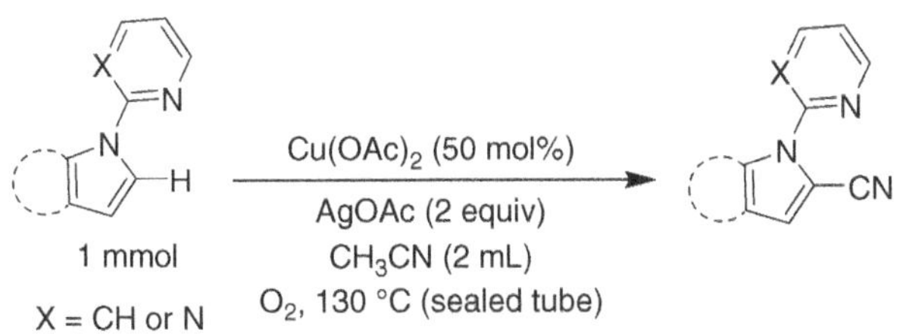

Figure 7.22 C2 cyanation of indoles using Cu and Ag by Zhu and co-workers [65]. Reproduced with permission of The American Chemical Society.

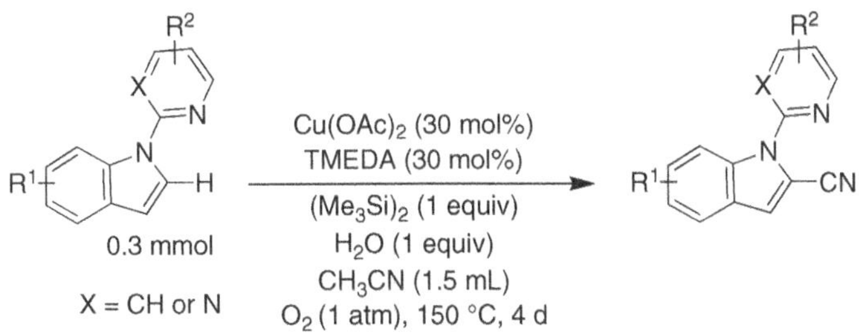

Figure 7.23 C2 cyanation of indoles using Cu and $(Me_3Si)_2$ by Shen and co-workers [66]. Reproduced with permission of John Wiley & Sons.

and yields of between 50 and 73% were obtained. In one example, they demonstrated that the pyrimidyl group can be subsequently removed using NaOEt in DMSO at 110 °C to leave the desired 2-cyanoindole in high yield.

Shortly after this report, a study by Shen and co-workers reported an alternative copper catalyst system that could be used for the cyanation of indoles [66]. The method utilizes $Cu(OAc)_2$, *N,N,N′,N′*-tetramethylethylenediamine (TMEDA) and hexamethyldisilane (Figure 7.23). Overall, the yields were moderate to good although for some substrates the yields were lower when compared directly with the Cu/Ag method.

They also demonstrated that the pyrimidyl directing group could be effectively removed using the same conditions as Zhu and co-workers. In this case, they demonstrated how the resultant 2-cyanoindole could be utilized, for example, transforming the nitrile functionality into an oxazole, thiazole, imidazole, or tetrazole in one additional step. They also demonstrated that a biologically active compound could be produced in two steps from the cyanoindole.

In addition to the preparation of 2-cyanoindoles, they showed that the same method could be employed for cyanation of a range of 2-phenyl-pyridines, as shown in Figure 7.24. The reaction proceeded well when R^1 was either electron-withdrawing or -donating in nature and impressively, both aldehyde and ester groups were also tolerated. These reactions have the potential to result in the incorporation of two nitrile groups, however, it was found that if the reaction time was kept to 10 hours, good selectivity to the mono-cyanated product could be achieved, and yields of between 62 and 76% were obtained for these products. If R^2 is a methyl group in the 3-position, this results in only the mono-cyanated product being produced, although extended reaction times of three to four days were required. For this type of reaction, they also

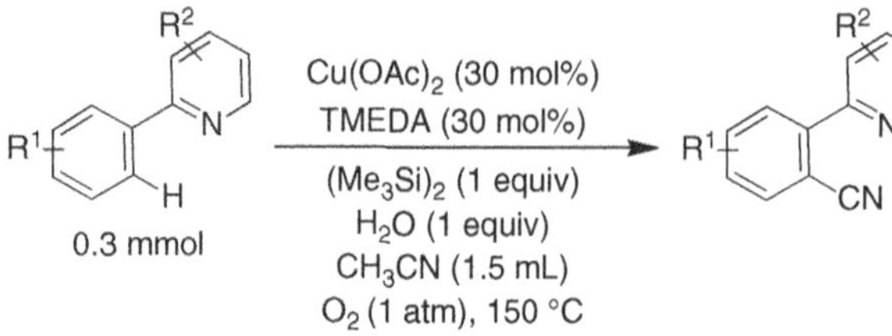

Figure 7.24 Copper catalyzed cyanation of aromatic C—H bonds by Shen and co-workers [66]. Reproduced with permission of John Wiley & Sons.

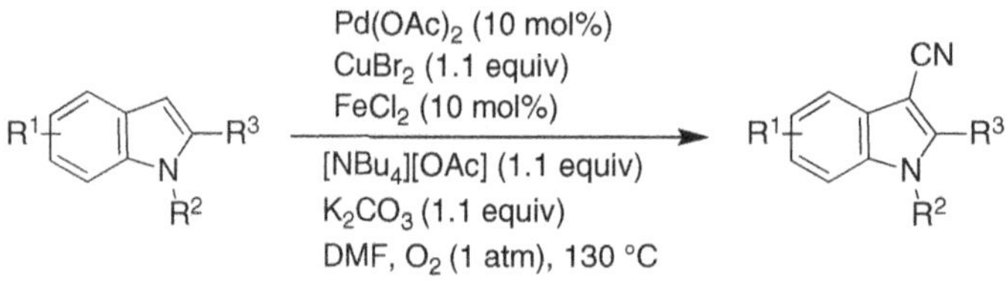

Figure 7.25 Cyanation of indoles using DMF as a source of CN [69]. Reproduced with permission of The American Chemical Society.

demonstrated that acetonitrile could be substituted with pentanenitrile or isobutyronitrile as the cyanation reagent.

For both these studies, the groups postulated possible mechanisms, but further work is needed to provide more concrete evidence for a detailed mechanism. In both cases, it was suggested that the C—CN bond of acetonitrile is cleaved and that copper cyanide species are produced *in situ*.

Related to this work, Jiang and co-workers reported the use of ethyl (ethoxymethylene)cyanoacetate as a reagent for the cyanation of aromatic C—H bonds [67]. Although this reaction utilizes O_2, it was not catalytic in copper and used one equivalent of $Cu(OAc)_2$. While Yu and co-workers have also demonstrated that such a reaction could be performed with nitromethane ($MeNO_2$) as the source of CN this also relied on one equivalent of $Cu(OAc)_2$ [68].

In 2011, Ding and Jiao disclosed a method for the cyanation of indoles and benzofurans that used DMF as the solvent and source of CN [69], and the general conditions are shown in Figure 7.25. The yields obtained were very substrate dependent, for example, good yields could be obtained for a number of 2-aryl indoles, but the performance was not as good for simple *N*-protected indoles. Importantly, this is perhaps the first example of utilizing DMF as the sole source of CN, with isotopic labeling showing that both the carbon and nitrogen of the cyano group are from the solvent. More mechanistic studies are required but these initial studies suggested that Cu–CN species were not involved in these reactions.

In 2015, Shen and co-workers reported a series of papers, where acetonitrile was used as the solvent and cyano source. Similar catalytic methods were employed for the cyanation of indoles [70], arenes [71] and arylboronic acids [72], all of which are outlined in Figure 7.26. For all of these transformations, a wide range of substrates were tested and good to excellent yields could be obtained for many of these. Reaction times were quite long, typically two to three days, and as can be seen from Figure 7.26, a number of

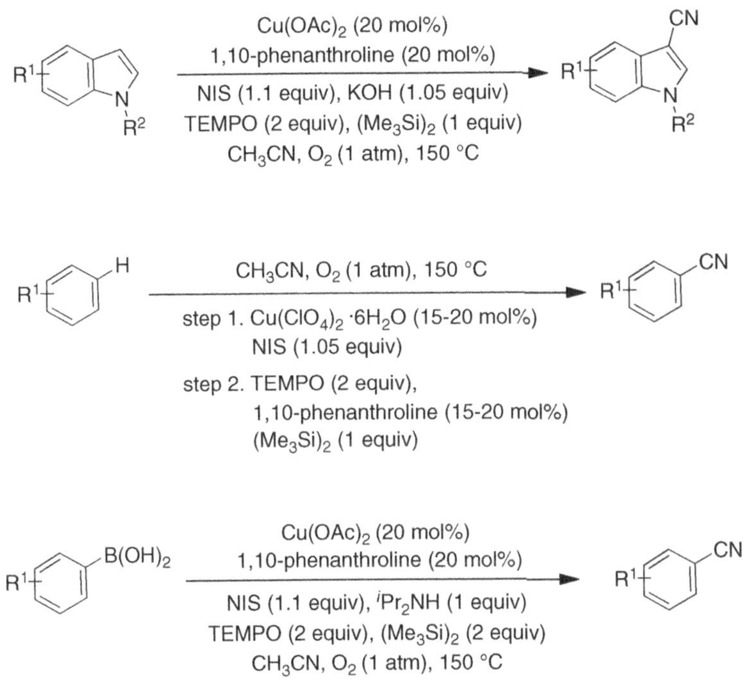

Figure 7.26 Cyanation of indoles [70], arenes [71], and arylboronic acids [72] using acetonitrile. Reproduced with permission of The American Chemical Society.

the key reagents were used in stoichiometric quantities, which arguably reduces the method's green credentials or the application on larger scales. Nonetheless, these reactions are undoubtedly very challenging and more efficient catalysts will hopefully be developed in the future. The authors proposed that the reactions first involve iodination of the substrate by *N*-iodosuccinimide (NIS). The role of oxygen appeared to be in producing TEMPO-CH_2CN species, and the generation of this may involve $Cu(NCCH_2)_2$ species (a species also suggested in the method that oxidized alcohols [34], discussed in Section 7.2.2). The formation of the cyanide ion via cleavage of the C—CN bond in TEMPO-CH_2CN then occurs due to copper species and/or $(Me_3Si)_2$. The copper mediated cyanation of the aryl iodide can then occur, with aryl–Cu–CN species as a likely intermediate.

7.5 Concluding Remarks

In recent years there has been real progress in this area of oxidation catalysis. When using amines or alcohols as substrates, there now exist very efficient methods, which can use low loadings of heterogeneous catalysts based on abundant metals such as iron. Given the performance of these catalysts, we envisage that such catalysts could well be used for industrial applications. There are also homogeneous methods for these substrates that employ commercially available catalyst components and operate under mild conditions (e.g. room temperature and 1 atm pressure). These homogeneous methods are therefore readily accessible and attractive for synthetic chemists. There have also been a wide range of new methods reported for more challenging routes to nitriles, for example those involving C—H or C—C bond cleavage. At the present time, these methods are less mature and understandably less "green," when compared to some of the methods that use amines and alcohols as substrates. The synthesis of nitriles via such direct routes is highly desirable, however, and these recent reports have demonstrated the potential of these new synthetic routes. We are confident that this is an area that will continue to develop and more efficient, greener catalytic methods will be reported in the future.

References

1 Review of nitrile containing pharmaceuticals: Fleming, F.F., Yao, L., Ravikumar, P.C. et al. (2010). *J. Med. Chem.* 53: 7902.
2 Pollak, P., Romeder, G., Hagedorn, F., and Gelbke, H.-P. (2000). *Nitriles, in Ullmann's Encyclopedia of Industrial Chemistry*. Wiley-VCH.
3 Review: Anbarasan, P., Schareina, T., and Beller, M. (2011). *Chem. Soc. Rev.* 40: 5049.
4 Review: Kim, J., Kim, H.J., and Chang, S. (2012). *Angew. Chem. Int. Ed.* 51: 11948.
5 Review: Martin, A. and Kalevaru, V.N. (2010). *ChemCatChem* 2: 1504.
6 Review on synthesis of aryl nitriles: Yan, G., Zhang, Y., and Wang, J. (2017). *Adv. Synth. Catal.* 359: 4068.
7 Review on C–CN bond formation via C–H bond activation: Ping, Y., Ding, Q., and Peng, Y. (2016). *ACS Catal.* 6: 5989.
8 Review on nitriles via C–H or C–C bond cleavage: Wang, T. and Jiao, N. (2014). *Acc. Chem. Res.* 47: 1137.
9 Reviews on catalytic reactions which use ammonia: (a) Klinkenberg, J.L. and Hartwig, J.F. (2011). *Angew. Chem. Int. Ed.* 50: 86. (b) Kim, J., Kim, H.J., and Chang, S. (2013). *Eur. J. Org. Chem.* 2013: 3201.
10 (a) Oishi, T., Yamaguchi, K., and Mizuno, N. (2009). *Angew. Chem. Int. Ed.* 48: 6286. (b) Oishi, T., Yamaguchi, K., and Mizuno, N. (2010). *Top. Catal.* 53: 479.
11 Ishida, T., Watanabe, H., Takei, T. et al. (2012). *Appl. Catal. A* 425–426: 85.
12 Jagadeesh, R.V., Junge, H., and Beller, M. (2014). *Nat. Commun.* 5: 4123.
13 Jagadeesh, R.V., Stemmler, T., Surkus, A.-E. et al. (2015). *Nat. Protoc.* 10: 916.
14 Shang, S., Wang, L., Dai, W. et al. (2016). *Catal. Sci. Technol.* 6: 5746.
15 Shang, S., Dai, W., Wang, L. et al. (2017). *Chem. Commun.* 53: 1048.
16 (a) Brackman, W. and Gaasbeek, C.J. (1966). *Rec. Trav. Chim.* 85: 221. (b) Brackman, W. and Gaasbeek, C.J. (1966). *Rec. Trav. Chim.* 85: 257.
17 Reviews: (a)Cao, Q., Dornan, L.M., Rogan, L. et al. (2014). *Chem. Commun.* 50: 4524. (b) Seki, Y., Oisaki, K., and Kanai, M. (2014). *Tetrahedron Lett.* 55: 3738. (c) Ryland, B.L. and Stahl, S.S. (2014). *Angew. Chem. Int. Ed.* 53: 8824.
18 Tao, C., Liu, F., Zhu, Y. et al. (2013). *Org. Biomol. Chem.* 11: 3349.
19 Yin, W., Wang, C., and Huang, Y. (2013). *Org. Lett.* 15 (8): 1850.
20 Dornan, L.M., Cao, Q., Flanagan, J.C.A. et al. (2013). *Chem. Commun.* 49: 6030.
21 Kim, J. and Stahl, S.S. (2013). *ACS Catal.* 3: 1652.
22 Wang, J., Lu, S., Cao, X., and Gu, H. (2014). *Chem. Commun.* 50: 5637.

23 Molla, R.A., Ghosh, K., Tuhina, K., and Islam, S.M. (2015). *New J. Chem.* 39: 921.
24 Yadav, D.K.T. and Bhanage, B.M. (2013). *Eur. J. Chem.* 2013: 5106.
25 Dighe, S.U., Chowdhury, D., and Batra, S. (2014). *Adv. Synth. Catal.* 356: 3892.
26 Noh, J.-H. and Kim, J. (2015). *J. Org. Chem.* 8: 11624.
27 Hoover, J.M. and Stahl, S.S. (2011). *J. Am. Chem. Soc.* 133: 16901.
28 Fang, C., Li, M., Hu, X. et al. (2016). *Adv. Synth. Catal.* 358: 1157.
29 Kelly, C.B., Lambert, K.M., Mercadante, M.A. et al. (2015). *Angew. Chem. Int. Ed.* 54: 4241.
30 Kim, M.J., Mun, J., and Kim, J. (2017). *Tetrahedron Lett.* 58: 4695.
31 Chen, Q., Fang, C., Shen, Z., and Li, M. (2016). *Electrochem. Commun.* 64: 51.
32 Yang, X., Fan, Z., Shen, Z., and Li, M. (2017). *Electrochim. Acta* 226: 53.
33 Fan, Z., Yang, X., Chen, C. et al. (2017). *J. Electrochem. Soc.* 164: G54.
34 Shen, J., Yang, D., Liu, Y. et al. (2014). *Org. Lett.* 16: 350.
35 Liu, L., Dong, J., Zhang, Y. et al. (2015). *Org. Biomol. Chem.* 13: 9948.
36 Reviews: (a)Schümperli, M.T., Hammond, C., and Hermans, I. (2012). *ACS Catal.* 2: 1108. (b) Largeron, M. (2013). *Eur. J. Org. Chem.* 2013: 5225. (c) Chen, B., Wang, L., and Gao, S. (2015). *ACS Catal.* 5: 5851.
37 Mori, K., Yamaguchi, K., Mizugaki, T. et al. (2001). *Chem. Commun.* 461.
38 Yamaguchi, K. and Mizuno, N. (2003). *Angew. Chem. Int. Ed.* 42: 1480.
39 Some recent examples of heterogeneous ruthenium catalysts: (a) Corker, E.C., Mentzel, U.V., Mielby, J. et al. (2013). *Green Chem.* 15: 928. (b) Niu, B., Lu, F., Zhang, H.-Y. et al. (2017). *Chem. Lett.* 46: 330. (c) Ovoshchnikov, D.S., Donoeva, B.G., and Golovko, V.B. (2015). *ACS Catal.* 5: 34.
40 Some recent examples of homogeneous ruthenium catalysts: (a) Tseng, K.-N.T., Rizzi, A.M., and Szymczak, N.K. (2013). *J. Am. Chem. Soc.* 135: 16352. (b) Cristian, L., Nica, S., Pavel, O.D. et al. (2013). *Catal. Sci. Technol.* 3: 2646. (c) Tao, C., Wang, B., Sun, L. et al. (2017). *Org. Biomol. Chem.* 15: 328.
41 Review on ruthenium catalyzed oxidation of amines: Maiti, D., Lahiri, G.K., Ray, R., and Hazari, A.S. (2018). *Chem. Asian J.* https://doi.org/10.1002/asia.201701748.
42 Jagadeesh, R.V., Junge, H., and Beller, M. (2015). *ChemSusChem* 8: 92.
43 Natte, K., Jagadeesh, R.V., Sharif, M. et al. (2016). *Org. Biomol. Chem.* 14: 3356.
44 Semmelhack, M.F. and Schmid, C.R. (1983). *J. Am. Chem. Soc.* 105: 6732.
45 Lambert, K.M., Bobbitt, J.M., Eldirany, S.A. et al. (2014). *Org. Lett.* 16: 6484.
46 Review: Bartelson, A.L., Lambert, K.M., Bobbitt, J.M., and Bailey, W.F. (2016). *ChemCatChem* 8: 3421.
47 (a) Hoover, J.M. and Stahl, S.S. (2011). *J. Am. Chem. Soc.* 133: 16901. (b) Steves, J.E. and Stahl, S.S. (2013). *J. Am. Chem. Soc.* 135: 15742.
48 Xu, B., Hartigan, E.M., Feula, C. et al. (2016). *Angew. Chem. Int. Ed.* 5: 15802.
49 Shen, T., Wang, T., Qin, C., and Jiao, N. (2013). *Angew. Chem. Int. Ed.* 52: 6677.
50 Okamoto, N., Ishikura, M., and Yanada, R. (2013). *Org. Lett.* 15: 2571.
51 Dutta, U., Lupton, D.W., and Maiti, D. (2016). *Org. Lett.* 18: 860.
52 Lin, Y. and Song, Q. (2016). *Eur. J. Org. Chem.* 2016: 3056.
53 Huang, X., Li, X., and Jiao, N. (2015). *Chem. Sci.* 6: 6355.
54 Fleming, F.F. and Wang, Q. (2003). *Chem. Rev.* 103: 2035.
55 Zong, X., Zheng, Q.-Z., and Jiao, N. (2014). *Org. Biomol. Chem.* 12: 1198.
56 Xu, J.-H., Jiang, Q., and Guo, C.-C. (2013). *J. Org. Chem.* 78: 11881.
57 Liu, Q., Fang, B., Bai, X. et al. (2016). *Tetrahedron Lett.* 57: 2620.

58 Wang, T. and Jiao, N. (2013). *J. Am. Chem. Soc.* 135: 11692.
59 Feng, Q. and Song, Q. (2014). *Adv. Synth. Catal.* 356: 1697.
60 Claes, L., Verduyckt, J., Stassen, I. et al. (2015). *Chem. Commun.* 51: 6528.
61 Xin, B., Jiang, Q., Zhao, A. et al. (2015). *Chem. Commun.* 52: 11264.
62 Ge, J.-J., Yao, C.-Z., Wang, M.-M. et al. (2016). *Org. Lett.* 18: 228.
63 Gu, L. and Jin, C. (2015). *Chem. Commun.* 51: 6572.
64 Song, R.-J., Wu, J.-C., Liu, Y. et al. (2012). *Synlett* 23: 2491.
65 Pan, C., Jin, H., Xu, P. et al. (2013). *J. Org. Chem.* 78: 9494.
66 Kou, X., Zhao, M., Qiao, X. et al. (2013). *Chem. Eur. J.* 19: 16880.
67 Qi, C., Hu, X., and Jiang, H. (2017). *Chem. Commun.* 53: 7994.
68 Chen, X., Hao, X.-S., Goodhue, C.E., and Yu, J.-Q. (2006). *J. Am. Chem. Soc.* 128: 6790.
69 Ding, S. and Jiao, N. (2011). *J. Am. Chem. Soc.* 133: 12374.
70 Zhao, M., Zhang, W., and Shen, Z. (2015). *J. Org. Chem.* 80: 8868.
71 Zhu, Y., Zhao, M., Lu, W. et al. (2015). *Org. Lett.* 17: 2602.
72 Zhu, Y., Li, L., and Shen, Z. (2015). *Chem. Eur. J.* 21: 13246.

8

Green Oxidative Synthesis of Azo, Diazo, and Azido Compounds

Yujie Liang, Xiaoyu Geng, and Ning Jiao

State Key Laboratory of Natural and Biomimetic Drugs, School of Pharmaceutical Sciences, Peking University, Beijing, China

8.1 Introduction

Azo, diazo, and azido compounds are useful synthetic intermediates and building blocks. Their wide applications in organic synthesis, chemical biology, and materials science have attracted considerable attention [1–6]. Some representative bioactive organic azo, diazo, and azido compounds are displayed in Figure 8.1 [1–6]. Aromatic azo compounds are an important class of molecules used as dyes, pigments, indicators, and drug-coating material [7–10] In addition, azo compounds have potential for use in electronic [11] and drug delivery applications [12]. Diazo compounds are widely used synthetic blocks in various chemical transformations, particularly in the transition-metal-catalyzed carbene transformations, which represents one of the major reaction types in organometallic chemistry and organic synthesis [13]. The generated metal carbene species can undergo diverse reactions, such as cross coupling [13], cyclopropanation of alkenes [14–16], alkynes [17–20], and aromatic rings [16, 21, 22], X-H (X = C, O, S, N, etc.) bond insertion reactions [23]. Azido compounds are versatile synthetic intermediates and have been widely used in the synthesis of N-containing compounds by a series of chemical transformations [24]. Besides, azide moiety has been investigated in lead compounds design for drug discovery and applied as prodrugs of primary amine [25–27]. The azides could also serve as a point of attachment for various probes in bioconjugation reaction, such as click chemistry [28, 29]. In light of their importance, the development of highly efficient methodologies for the synthesis of these azo, diazo, and azido compounds is of great significance. Indeed, in recent years, these areas have witnessed dramatic growth and many novel and useful methods have been developed. In this chapter, we will focus on recent advances in the synthesis of organic azo, diazo, and azido compounds with green oxidants as well as some oxidation reactions with good selectivity and high atom economy.

Green Oxidation in Organic Synthesis, First Edition. Edited by Ning Jiao and Shannon S. Stahl.

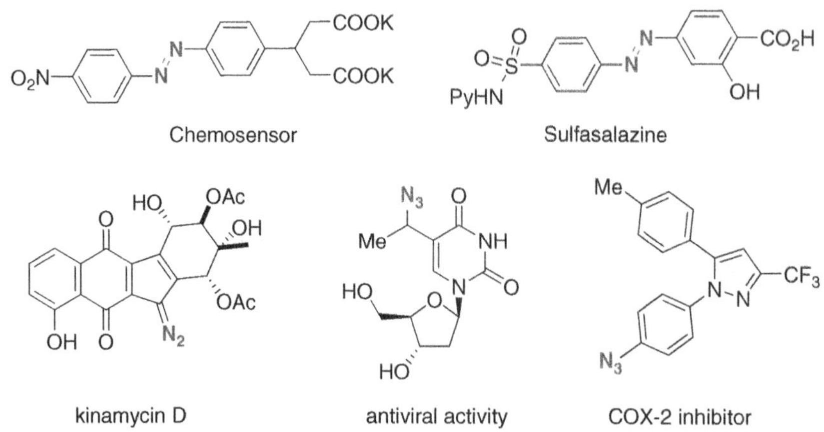

Figure 8.1 Representative bioactive organic azo, diazo, and azido compounds.

8.2 Green Oxidative Synthesis of Azo Compounds

Aromatic azo compounds have wide applications in chemical industry, medicinal chemistry, and material science [30]. Moreover, due to their ability to undergo photochemical or thermal switching between the *cis*- and *trans*-isomeric forms, they also play an important role in nonlinear optics [31]. Their unique properties make the development of new methods for their preparation important in chemical synthesis [32]. Conventional synthetic routes toward azo compounds include: (i) catalytic reduction of nitrobenzenes [33, 34]; (ii) coupling of primary aryl amines with nitroso compounds [35]; (iii) electrophilic reactions of diazonium salts with activated aromatics [36]; and (iv) deoxygenation of azoxybenzene [37]. However, these approaches typically require harsh or special reaction conditions, which often produce unwanted byproducts and lead to low overall yields. Compared to the classic prevailing methods, the direct dehydrogenation of hydrazo derivatives and oxidative coupling of aromatic anilines are without doubt the most straightforward and green methods. Previous works employed $KMnO_4$ [38], MnO_2 [39], $Pb(OAc)_4$ [40], or HgO [41] as stoichiometric metal oxidants in the oxidation of aniline derivatives, but the selectivity issues and the inorganic side-product formation make these methods less practical. The application of mild oxidants, such as *tert*-butyl hypoiodite (*t*BuOI) [42, 43] and phenyliodine(III) diacetate (PIDA) [44], can afford improved selectivities, however, these processes typically require precise temperature control, and the byproducts formed via the reduction of the oxidants would cause purification issues. In light of the current interest in developing mild and environmentally friendly methods for chemical transformations, the utilization of O_2 or H_2O_2 as the oxidizing agents have clearly the environmental advantages due to their inexpensiveness, abundance, and nontoxicity [45, 46]. Therefore, this part will focus on recent advances in the preparation of azo compounds using O_2 or H_2O_2 as oxidants.

8.2.1 Synthesis of Azo Compounds via Aerobic Oxidation of Anilines

The catalytic aerobic oxidation of aniline derivatives has recently come into the focus of interest in the preparation of aromatic azo compounds. In 2008, García's group reported that gold nanoparticles supported on titanium dioxide (TiO_2) can catalyze the aerobic

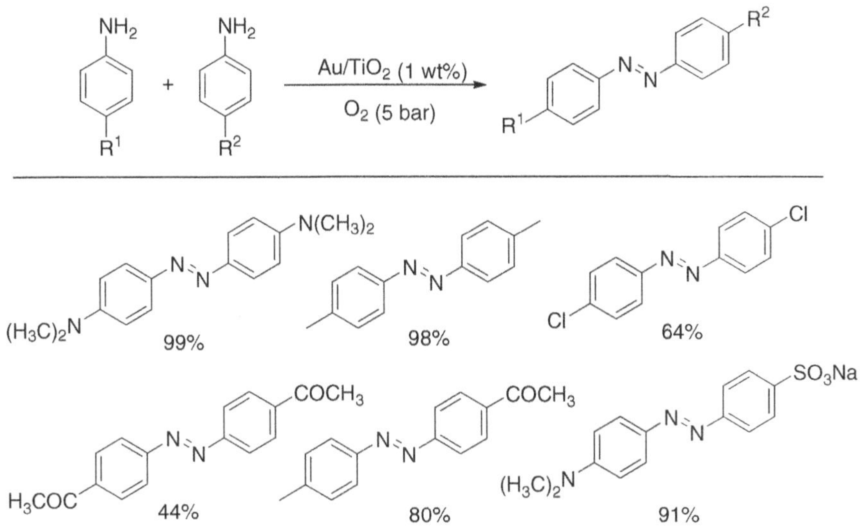

Scheme 8.1 Au-nanoparticles-catalyzed synthesis of azobenzenes.

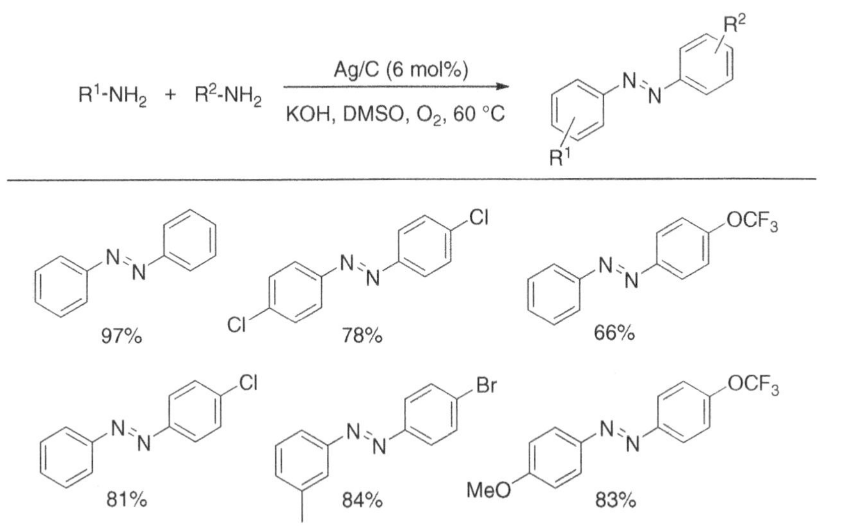

Scheme 8.2 Ag-nanoparticles-catalyzed synthesis of azobenzenes.

oxidation of anilines to the corresponding aromatic azo compounds (Scheme 8.1) [47]. Control experiments indicated that the nature of the support has a strong influence on the activity of the gold nanoparticles. Whereas gold nanoparticles supported on titanium dioxide (TiO_2) proved to be active and selective catalysts for formation of azobenzenes, gold with similar crystallite size on carbon or ferric oxide was not active. This catalytic system shows promise for efficient synthesis of symmetric and unsymmetric azo compounds.

Later, Li and co-workers developed an efficient monodispersed metal nanoparticle catalyzed aerobic oxidation of aniline to form both symmetric and asymmetric aromatic azo compounds under mild conditions (Scheme 8.2) [48]. It was found that AgNPs is critical to the generation of superoxide anion radical using electron spin

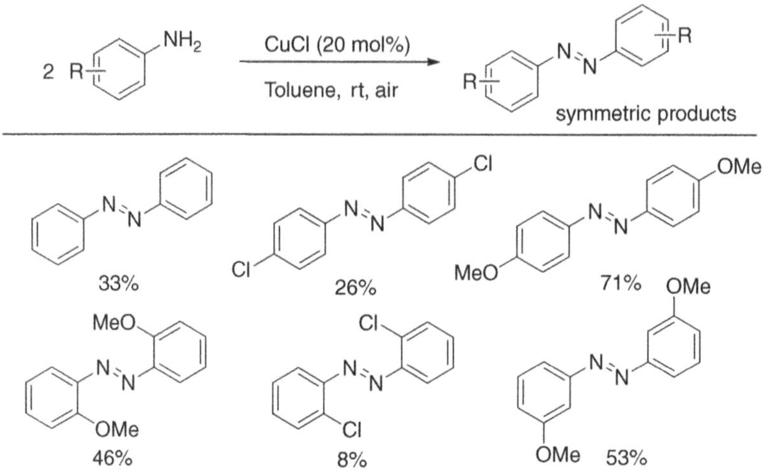

Scheme 8.3 Cu-catalyzed aerobic oxidative synthesis of azobenzenes.

resonance (ESR) because no superoxide radical anion was detected in the absence of AgNPs. This is a nice example showing that exploiting the catalytic power of nanoparticles could enable sustainable chemistry suitable for oxidation reactions.

Despite the elegance of these works, the high cost of gold and silver, to a large extent, limit the application of these processes. The recognition that copper can efficiently catalyze the oxidative coupling of arylamines has opened up new avenues for the synthesis of azobenzenes. In 2008, Xi and co-workers reported a copper-catalyzed aerobic oxidation of anilines to prepare azobenzenes at room temperature (Scheme 8.3) [49]. Control experiments showed that no oxidation of azo compounds occurred in the absence of copper catalyst. In most cases, the yields were inferior and the substrate scope was narrow.

To address this limitation, Jiao and co-workers reported an improved catalytic system. The substrates scope was significantly enlarged and the corresponding azobenzenes can be produced in good to excellent yields (Scheme 8.4a) [50]. Moreover, in addition to the symmetric substrates, the oxidative coupling of anilines for asymmetric azo compounds can be efficiently realized by this method. The mechanistic studies suggested that a single-electron oxidation was involved in the reaction (Scheme 8.4b). First, the copper(I) was coordinated by a pyridine ligand and oxidized by dioxygen to form a copper-ligand-oxygen complex **A**. Then, a single-electron oxidation of aniline, mediated by the copper(II) complex, into corresponding radical cation **B**, followed by subsequent coupling of **B** with azobenzene formed a three-electron sigma bond **C**, which consecutively donated two protons and one electron leading to hydrazine **D**. Hydrazine was further oxidized by the complex **A** or dioxygen to generate the corresponding aromatic azo products. The whole process renders water as the sole byproduct, instead of halogen compounds, making this method cleaner and more practical.

Since the work by Xi and Jiao's groups, some studies toward aerobic oxidation of anilines were also reported. For example, Suib reported mesoporous manganese oxide catalyzed aerobic oxidative coupling of anilines to azobenzenes (Scheme 8.5, a) [51]. In the same year, Mukhopadhyay's group developed a silver catalyzed aerobic synthesis of

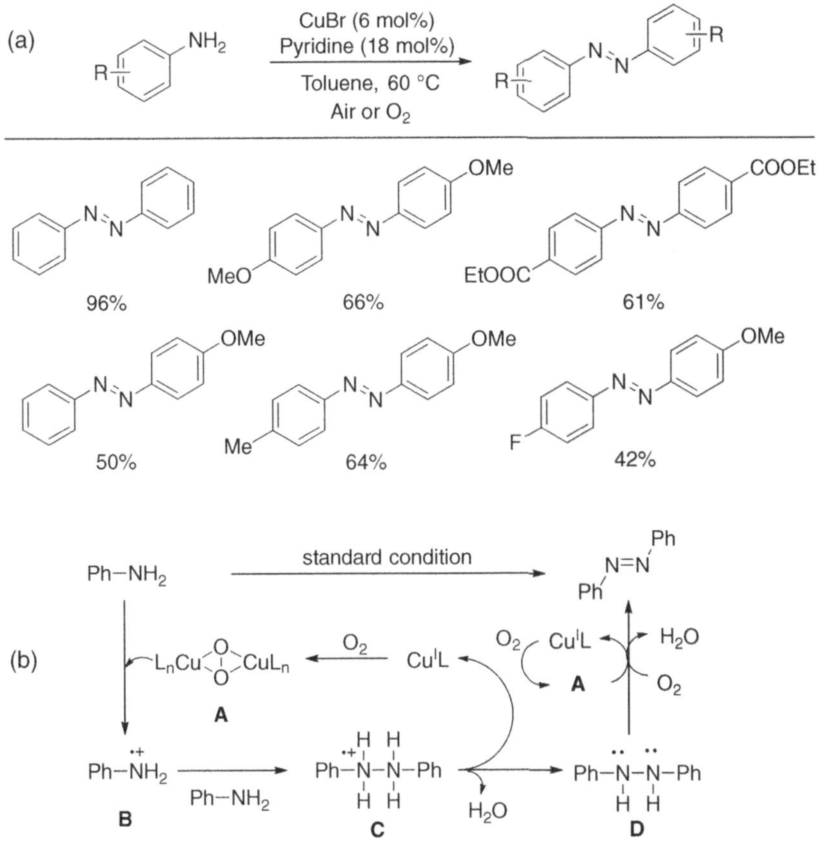

Scheme 8.4 Cu-catalyzed aerobic oxidative synthesis of azobenzenes from anilines.

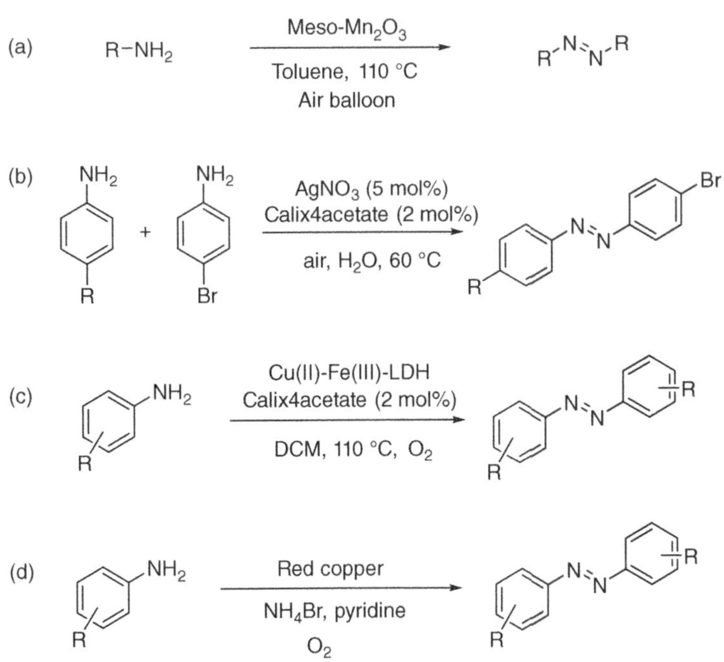

Scheme 8.5 Metal-catalyzed aerobic oxidative synthesis of azobenzenes.

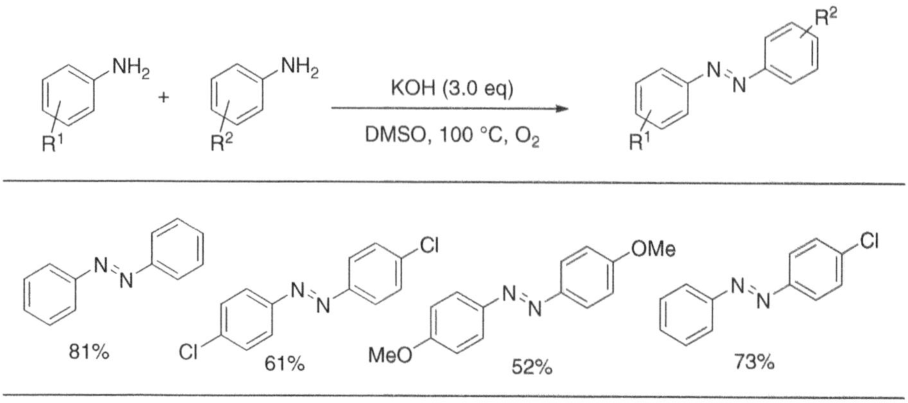

Scheme 8.6 KOH-mediated aerobic oxidation of aniline.

azobenzene with nanoreactor (Scheme 8.5, b) [52]. Recently, Fülöp disclosed bimetallic (Cu—Fe) catalyzed homocoupling of anilines using continuous flow chemistry (Scheme 8.5, c) [53]. Moreover, the common metal copper(0) can also be directly used in the oxidative coupling reactions (Scheme 8.5, d) [54]. In most cases of the well-established dehydrogenative oxidation of aniline to azobenzene derivatives, a metal-peroxo species that could efficiently oxidize aniline is probably involved.

In addition to metal catalysis, base-mediated oxidative coupling of anilines has recently emerged as a powerful strategy in the preparation of azobenzenes. Fu and co-workers reported a simple and efficient protocol for aerobic oxidative coupling of amines to form imines and azo compounds using KOH as the base and DMSO as the solvent (Scheme 8.6) [55]. The azo compounds could be easily prepared simply by heating the mixture of amines and DMSO under aerobic atmosphere. It was noted that excess KOH was applied to take the deprotonation and aerobic hydrogenation could be initiated to deliver azo compounds.

8.2.2 Synthesis of Azo Compounds via Aerobic Oxidation of Hydrazobenzene

Oxidative dehydrogenation of hydrazo derivatives has been utilized to synthesize azo compounds. Pioneering work was reported by Sawyer by using a Fe^{II}(DPAH)-catalyzed activation of O_2 strategy for the preparation of azobenzene [56]. Additionally, Simándi and co-workers reported a mild and efficient aerobic oxidative dehydrogenation of hydrazobenzene using $Co^{II}(Hdmg)_2(PPh_3)_2$ as the catalyst under an O_2 atmosphere (Scheme 8.7) [57]. The catalytic activity of cobaloxime(II) complexes depends on the nature of the axial group. It should be noted that the reaction could be restarted by adding a new portion of the hydrazobenzene after the original substrate had been totally consumed in a run. The author conceived that mononuclear superoxo or dinuclear peroxo species derived from interaction of Co(II) complexed with O_2 were involved in catalytic cycle.

Recently, Cannon and co-workers reported a Co^{II}(3-OMe-salen) catalyzed dehydrogenation of hydrazobenzenes with similar O_2-based strategy (Scheme 8.8) [58]. The kinetic studies showed that a ternary complex involving Co(3-MeO-salen), PPh_3 and O_2

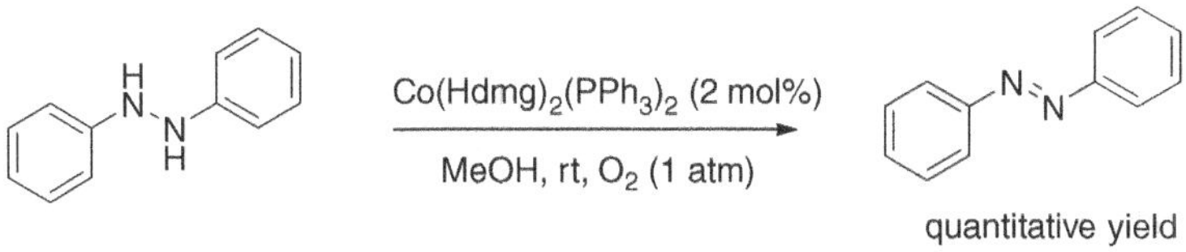

Scheme 8.7 Cobalt-catalyzed aerobic oxidation of hydrazobenzene.

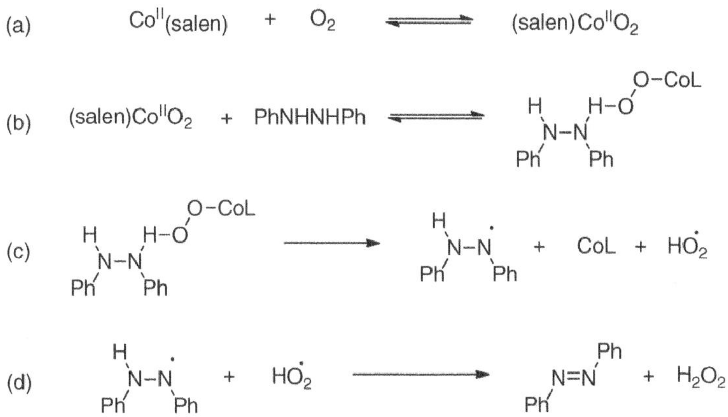

Scheme 8.8 Co^{II}(salen)-catalyzed oxidation of hydrazobenzene.

was the catalytically active species. First, the coordination of the cobalt(II) complex with methanol affects the electron density of the $d_z{}^2$ orbital of cobalt, increasing the charge on the dioxygen moiety (Scheme 8.8, a). The N—H—O bridging structure lead to the abstraction of the hydrogen atom from hydrazobenzene (Scheme 8.8b). After abstraction of the hydrogen atom, a nitrogen radical and HO_2 radical may be formed (Scheme 8.8c). If a *cis*-isomer is formed, it could rearrange rapidly to the *trans*-form by rotation about the N—N bond before reacting with HO_2 to produce a *trans*-isomer and hydrogen peroxide (Scheme 8.8d). The hydrogen peroxide may undergo further oxidation of substrate to produce azobenzene and water.

Huang and co-workers described a palladium-catalyzed oxidative procedure to prepare azobenzenes using air as the oxidant (Scheme 8.9) [59]. This reaction employed commercially available palladium acetate and cheap inorganic base K_2CO_3, obviating the use of a complex ligand. In the presence of a catalytic amount of K_2CO_3, hydrazobenzenes can be converted to azobenzenes efficiently in an open-flask fashion under mild conditions.

Nanostructured carbon allotropes, in particular carbon nanotubes (CNTs), have emerged as highly promising platforms that provide unique advantages such as chemical, thermal, and mechanical stability in liquid media. Doris reported a carbon nanotube-rhodium nanohybrid as an efficient catalytic system to perform mild and selective oxidation of various substrates including hydrazines, hydroxylamines, hydroquinones, silanes, and thiols topography [60]. The reaction of hydrazobenzenes clearly afforded azobenzenes in excellent yields (Scheme 8.10). Besides, the nanohybrid can be readily recycled.

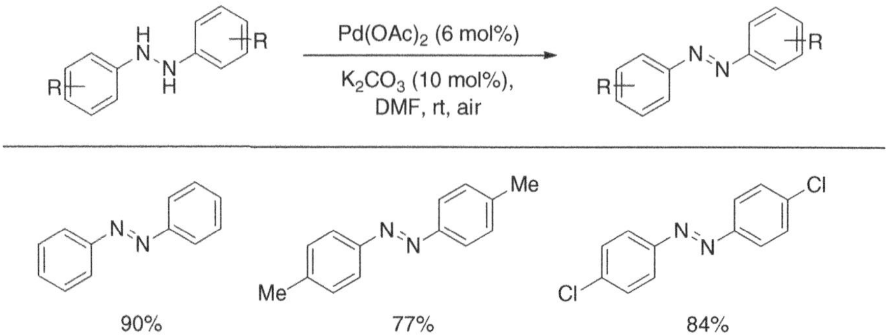

Scheme 8.9 Pd-catalyzed oxidative synthesis of azobenzenes.

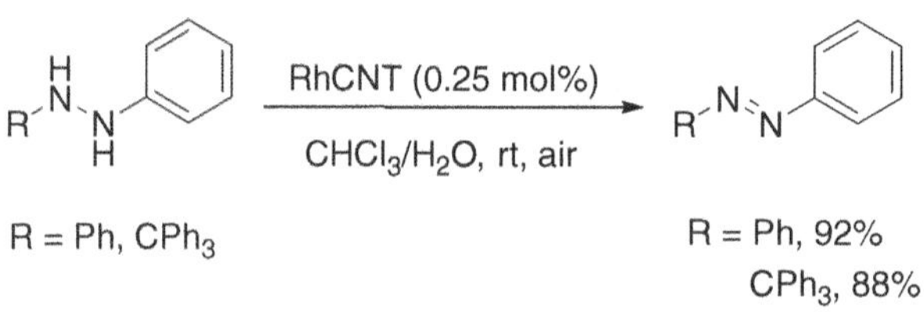

Scheme 8.10 RhCNT-catalyzed synthesis of azoxys or azobenzenes.

In addition to the oxidation of diaryl azobenzenes, the dehydrogenation of arylhydrazinecarboxylate has also been reported. Pifferi and co-workers reported a direct method for the synthesis of several 2-arylazocarboxylates from the corresponding 2-arylhydrazinecarboxylates with atmospheric oxygen as oxidant and 10% palladium on charcoal as the catalyst (Scheme 8.11, a) [61]. Recently, Taniguchi and co-workers reported that iron phthalocyanine can catalyze aerobic oxidation of arylhydrazinecarboxylates to give arylazocarboxylates (Scheme 8.11, b) [62]. The substrate scope was broad and phenyl rings bearing both electron-donating and electron-withdrawing groups were tolerated in this protocol. Kinetic studies indicated a single-electron transfer process between the iron complex and oxygen may be involved.

Transition metal-free aerobic oxidation toward the synthesis of azo compounds has also been reported. Substrate 5,6-dihydrodibenzo[*b*,*f*][1,4,5]thiadiazepin was reported to be readily oxidized to the dehydrogenative product by blubbing air into the ethanolic solution of this compound with catalytic amount of NaOH. (Scheme 8.12) [63]. In contrast to the simple and stable azobenzene, the easy dehydrogenative nature of dihydrodibenzo[*b*,*f*][1,4,5]thiadiazepin may be attributed to the aromatization tendency ability.

Recently, Hashimoto's group realized the dehydrogenation of the NH—NH bond using *t*BuOK in liquid ammonia at room temperature (Scheme 8.13) [64]. This reaction is fast and numerous aromatic azo compounds could be synthesized in excellent yields. This reaction could also be scaled up to gram scale. Moreover, the strategy could be applied to the synthesis of aliphatic diazirines, which are important photoaffinity labeling agents in chemical biology. It was considered that the employment of *t*BuOK triggered the dehydrogenation process, and thus facilitated oxidation to form the azo products.

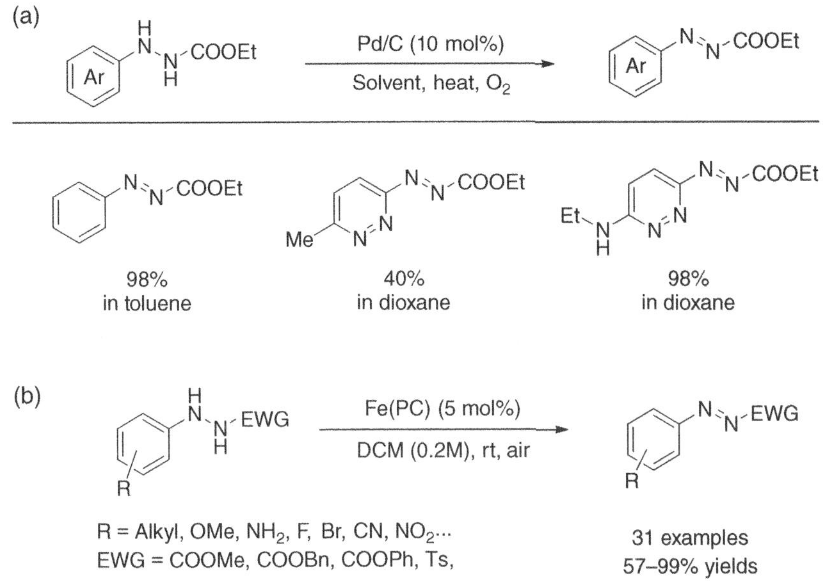

Scheme 8.11 Green oxidation synthesis of arylazocarboxylates.

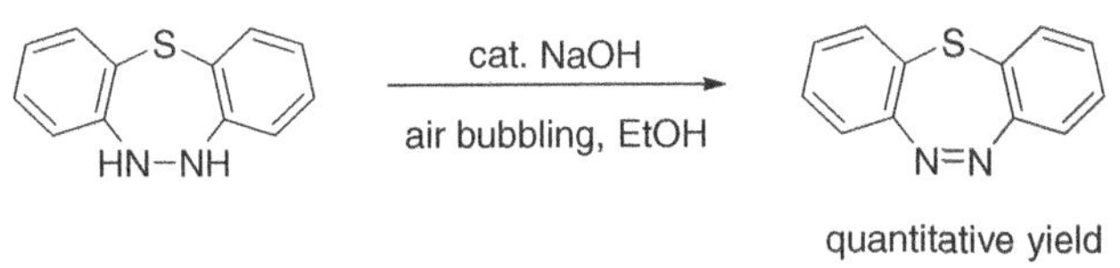

Scheme 8.12 NaOH-Catalyzed aerobic oxidation of dihydrothiadiazepin.

It should be noted that, H_2O_2 can also be used as the oxidant in the dehydrogenation of hydrazobenzene. Trischle reported the first $NaVO_3$ catalyzed dehydrogenation of hydrazobenzene using H_2O_2 as the sole oxidant (Scheme 8.14, a) [65]. They considered that hydrazobenzene could not be auto-oxidized to azobenzene by hydrogen peroxide in neutral ethanol, but the reaction was found to be rapid in the presence of a suitable catalyst. According to previous work, the author conceived that the oxidizing agent is not hydrogen peroxide itself but the ions and radicals (OH^+, $O^\bullet$, $OH^\bullet$, $HO_2^\bullet$) produced from interaction of the catalyst with H_2O_2. In 2007, Gozin and co-workers reported that the successful application of a novel catalytic system merging $TiCl_3$ and HBr catalysis is capable of oxidating hydrazo compounds into the corresponding azo derivatives with high efficiency and selectivity (Scheme 8.14b) [66]. This new protocol tolerated a range of substrates, including aryl, methoxyaryl, nitroaryl, and acetyl on the hydrazo functional group. The author proposed a hypobromous acid specie derived from the oxidation of HBr by hydroxyl radicals as the key intermediate. Then HOBr was reduced to the HBr and water by the hydrazo substrate, concurrently producing the desired azo derivative (Scheme 8.14c).

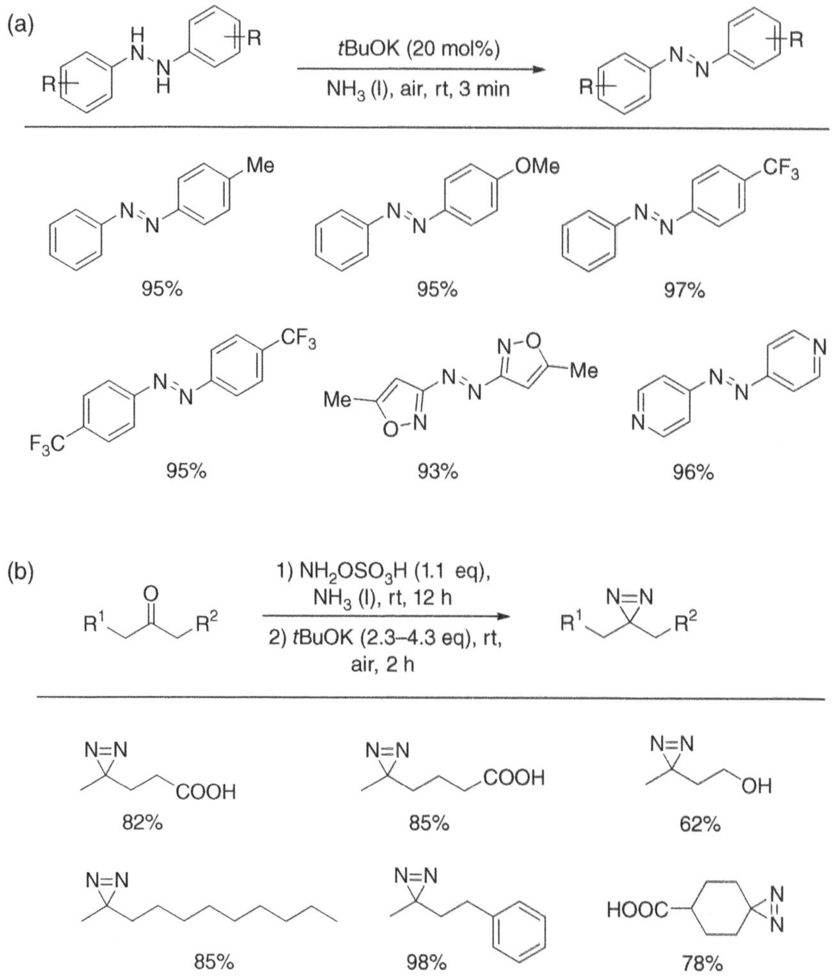

Scheme 8.13 *t*BuOK-catalyzed aerobic oxidation of hydrazobenzenes.

8.3 Green Oxidative Synthesis of Diazo Compounds

Diazo compounds (R^1R 2=N_2) are versatile and useful building blocks in organic synthesis. The development of efficient methods for their access remains the long-standing interest of organic chemists. The first diazo compounds diazomethane (CH_2=N_2) [67] and ethyl diazoacetate (EDA) [68] were prepared more than 100 years ago and have been used as common reagents in organic synthesis. Since then, considerable effective synthetic methods have been reported for the preparation of various diazo compounds. To date, the major routes toward diazo compounds are as follows: (i) diazo-group transfer; (ii) diazotization (iii) base-mediated deprotonation of sulfonyl hydrazones; (iv) alkaline cleavage of N-alkyl-N-nitroso compounds, (v) triazene fragmentation, and (vi) elaboration of diazo compounds. It should be noted that most of these routes have been reviewed extensively for their merits in organic synthesis [69, 70]. Among them, oxidative dehydrogenation reactions of N—H or C—H bonds (such as diazo-group transfer)

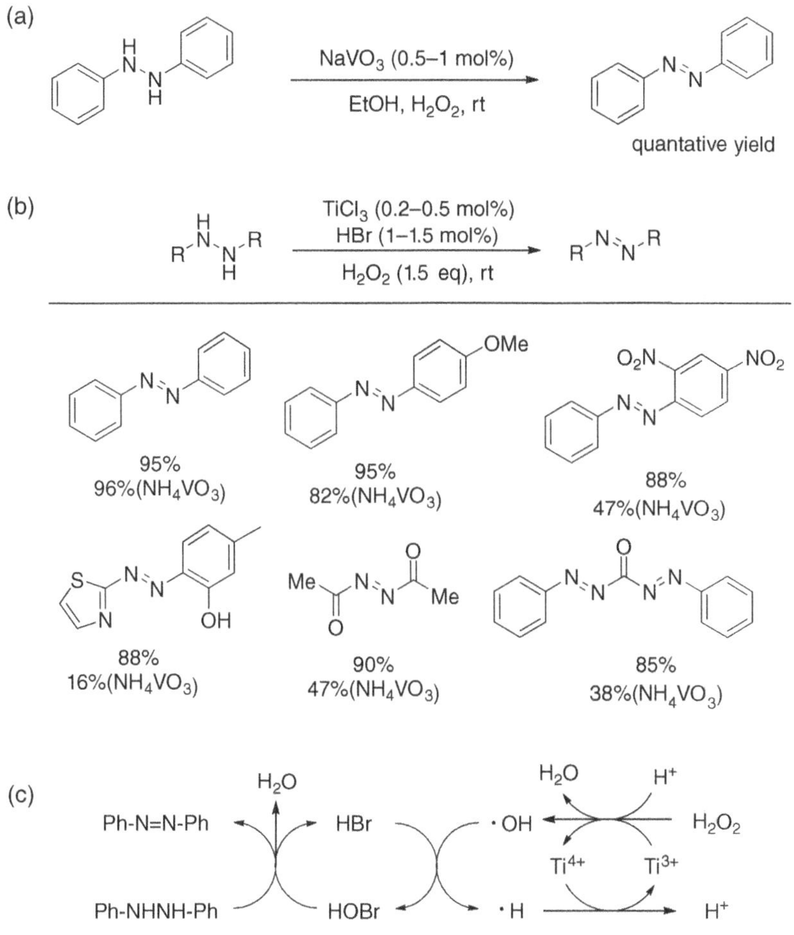

Scheme 8.14 H_2O_2-mediated oxidative synthesis of azobenzenes.

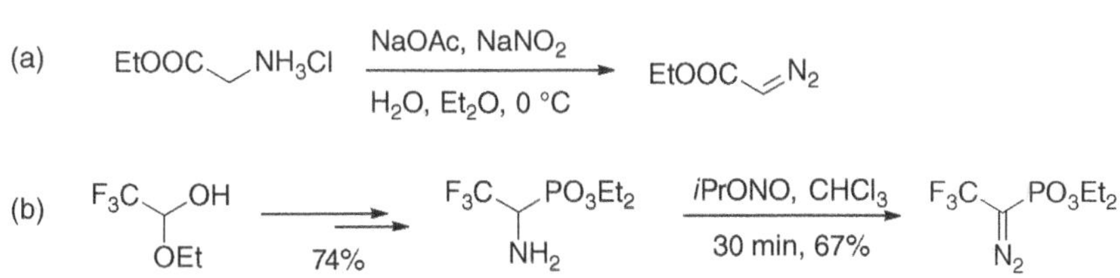

Scheme 8.15 $NaNO_2$ and IBN-mediated synthesis of diazo compounds.

represent simple and straightforward methods for the preparation of diazo compounds. For example, ethyl diazoacetate (EDA) could be prepared from dehydrogenation reaction of amino acetate with sodium nitrite (Scheme 8.15a) [71]. In this case, the sodium nitrite serves as both oxidant and diazo source. To address the solubility problem of $NaNO_2$, alkyl nitrites (RONO) have been used as an alternative for diazotization of amine in neutral condition. Beletskaya and co-workers reported a diazotization of amino phosphonic acid derivatives using *iso*-propyl nitrite as oxidant (Scheme 8.15b) [72]. Compared to the well-established $NaNO_2$ oxidative strategy, this reaction featured short reaction time, neutral conditions, simple operation, and no additional acid was required to stabilize the unreacted amine.

Diazo transfer is a simple and effective way to introduce the diazo group onto activated methylene compounds when the C—H bond on the acceptor carbon is acidic enough to be extracted with a mild base. Several sulfonyl azides and some electron-deficient organoazide derivatives have proven to be efficient in diazo transfer onto activated methylene compounds [73]. Among the various azides used in diazo transfer reactions, *p*-tosyl azide has been the prevailing reagent for many years. However, compared to the frequently used *p*-tosyl azide, the identification of an azide that is less likely to explode and be transformed into a sulfonamide derivative that can be easily separated from the diazo products is still important. After a thorough examination of the azide reagents, it was found that methanesulfonyl azide [74] and *p*-acetamidobenzenesulfonyl azide [75] displayed some advantages. The diazo-group transfer reaction succeeds for a range of substrates bear two strong acceptor substituents (cyano, nitro, acyl, sulfonyl, phosphoryl), and (hetero)aryl substituents [76]. The combination of *p*-acetamidobenzenesulfonyl azide and DBU has been widely used for these applications.

In general, (diazomethyl)ketone and α-diazocarboxylate derivatives are not accessible by direct diazo-group transfer onto simple enolizable ketones and carboxylic esters. These substrates need to be activated beforehand by introducing an additional electron-withdrawing group, which is removed in the course of the diazo-transfer reaction. To this end, Regitz and co-workers developed a so-called "deformylating diazo-group-transfer" strategy [77]. Later, Danheiser reported a more efficient activation strategy by using the trifluoroacetyl group as the activated group (Scheme 8.16, a) [78] Taber and co-workers reported a simple method for preparation of various a-diazocarboxylate compounds by a $TiCl_4$-mediated benzoylation and subsequent diazo transfer reaction (Scheme 8.16, b) [79]. Moreover, α-aryl-α-diazoketones can also be prepared using sulfonyl azides and a mild base. By simply stirring the mixture of 1,3-diketone,

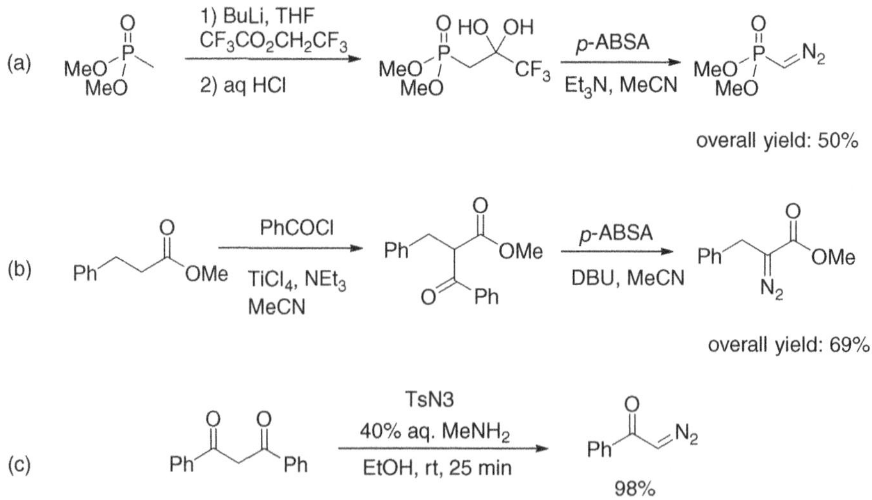

Scheme 8.16 Diazo transfer via dehydrogenation of methylene or methane.

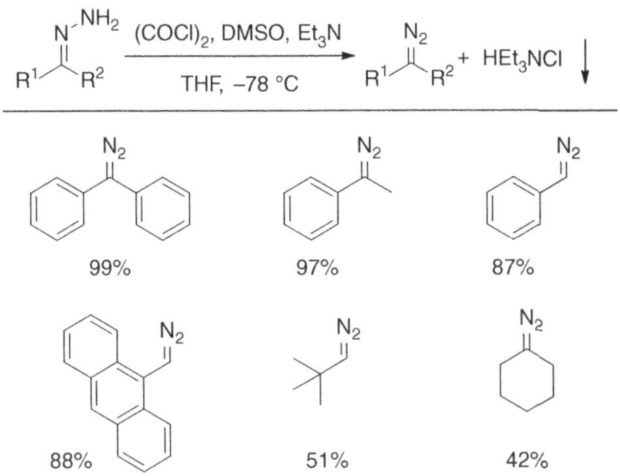

Scheme 8.17 Diazo synthesis via oxidation of hydrazones with Swern reagent.

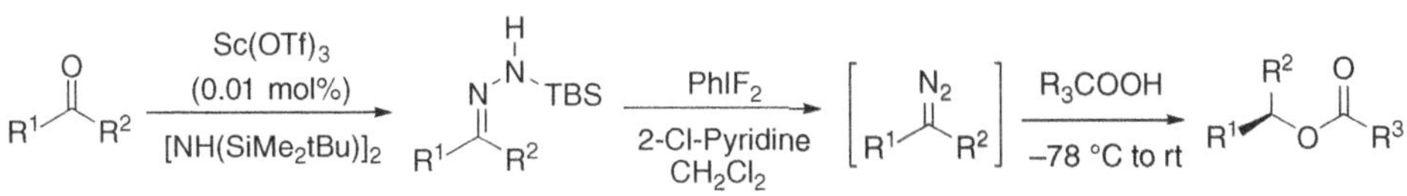

Scheme 8.18 Diazo synthesis via dehydrogenation of hydrazones with $PhIF_2$.

TsN_3, and aqueous $MeNH_2$ in EtOH, α-aryl-α-diazoketones can be produced rapidly with high efficiency (Scheme 8.16, c) [80]

Green oxidative synthesis of diazo compounds by oxidation of hydrazones is another useful strategy. Brewer et al. reported the preparation of diazo compounds via a dehydrogenation reaction with DMSO as oxidant (Scheme 8.17) [81]. This method allows the preparation of various aryl- or alkylsubstituted derivatives of diazomethane in moderate to good yields and even the rather unstable diazocyclohexane can also be obtained.

Recently, a novel transition-metal-free synthesis of diazo compounds was reported by Myers' group using (difluoroiodo)benzene as oxidant (Scheme 8.18) [82]. When a carboxylic acid was present during the oxidation step, the generated diazoalkane was immediately consumed to form the carboxylic ester.

Few catalytic methods have been developed for the dehydrogenation of hydrazones with molecular oxygen as oxidant. Nishinaga and co-workers reported a cobalt(II) Schiff base complex catalyzed aerobic oxidation of hydrazones under neutral conditions to give diazo compounds (Scheme 8.19) [83]. Although the substrate scope was not significantly examined, this represents a green and environmentally friendly method for the preparation of diazo compounds, since water was produced as the sole byproduct.

8.4 Green Oxidative Synthesis of Azido Compounds

Azido compounds are versatile intermediates in organic synthesis and the development of efficient synthetic methods for their access has attracted extensive attention [29, 84–93]. Conventional routes for the preparation of organic azides mainly rely on a

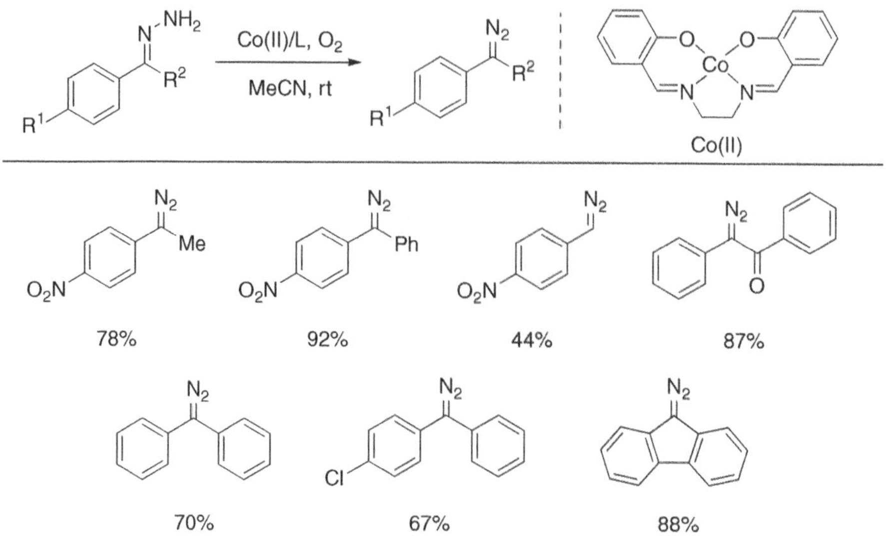

Scheme 8.19 Aerobic dehydrogenative of hydrazones via Co(II) catalysis.

nucleophilic addition or substitution reaction [85]. However, these methods require tedious substrate prefunctionalizations, thus restricting the scope of application. Therefore, the development of green oxidative strategies for the incorporation of a N_3 group into readily available organic molecules is highly desirable. Recently, transition-metal-catalyzed difunctionalization of alkenes and alkynes has emerged as a hot researching area [94–105], which has been elegantly utilized for the preparation of alkyl azides and vinyl azides through the hydroazidation of alkenes and alkynes [106–112]. Significantly, the direct azidation via C—H or C—C bonds cleavage provides a powerful strategy for C—N_3 bond formation, and has witnessed surprising advances. C—H and C—C bonds are ubiquitous in organic compounds, since no prefunctionalization of substrates require, the direct azidation of C—H and C—C bond represents without doubt the most straightforward, atom-, and step-economic methods [113–115]. Therefore, this part will focus on the green synthesis of azido compounds via C—H or C—C bond cleavage strategies, including some recent important pioneering studies.

8.4.1 Green Oxidative Synthesis of Organic Azides via C—H Bond Cleavage

8.4.1.1 Synthesis of Aryl and Acyl Azides via *sp*^{2} C—H Bond Azidation

Since the first preparation of aryl azides by Peter Grieβ in 1864 [116], considerable interest has been aroused in the synthetic community. However, the considerable strength of the arene C—H bond (472.4 kJ mol^{-1} for PhH) makes their direct azidation difficult [117]. A pioneering study was conducted by Kita, who first reported the direct azidation of electron-rich aromatic rings by using stoichiometric amounts of hypervalent iodine (III) reagents [bis(trifluoroacetoxy)iodo] benzene (PIFA) as the oxidant and $TMSN_3$ as the N_3 source in polar protic solvents (Scheme 8.20a) [118]. A possible mechanism was proposed. It includes the interaction of arene with PIFA to form the π-complex **A**, then the substitution of $TMSN_3$ of A generating the π-complex **B**, which

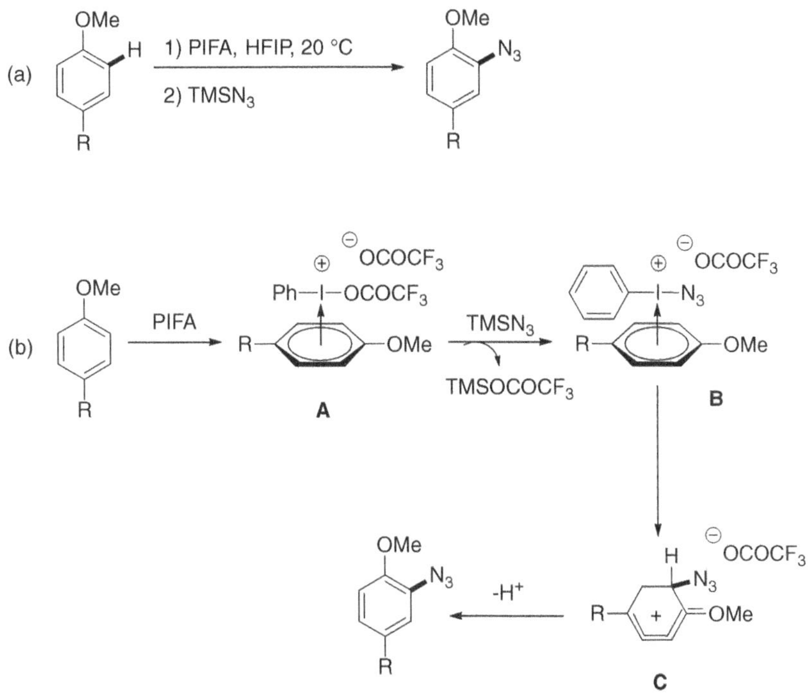

Scheme 8.20 Direct azidation of electron-rich arenes with PIFA and $TMSN_3$.

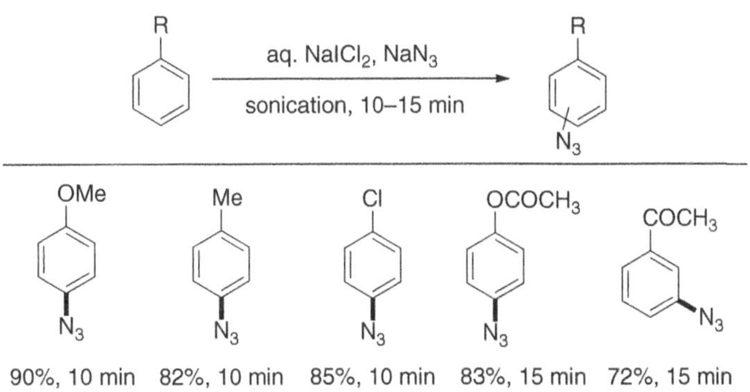

Scheme 8.21 Sonication-mediated direct azidation of arenes.

subsequently undergoes reductive elimination and deprotonated step to form the aryl azides (Scheme 8.20b) [119].

To broaden the substrates scope, in 2012, Telvekar and Sasane reported an efficient procedure for aryl C—H azidation. By using aqueous sodium dichloroiodate and NaN_3 in combination with sonication, both electron-rich and electron-deficient aryl compounds can be smoothly azidated (Scheme 8.21) [120]. This reaction proceeds rapidly with good yields and excellent regioselectivity. Moreover, a Friedel–Crafts type reaction pathway has been suggested.

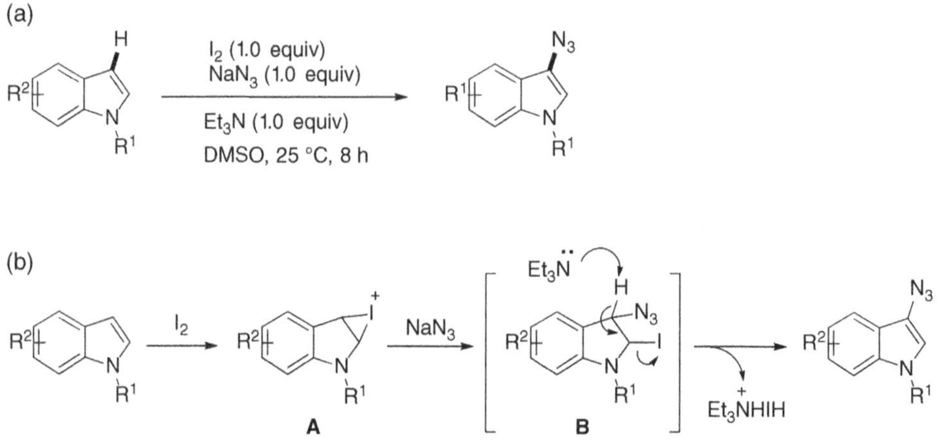

Scheme 8.22 I_2-mediated regioselective C-3 azidation of indoles.

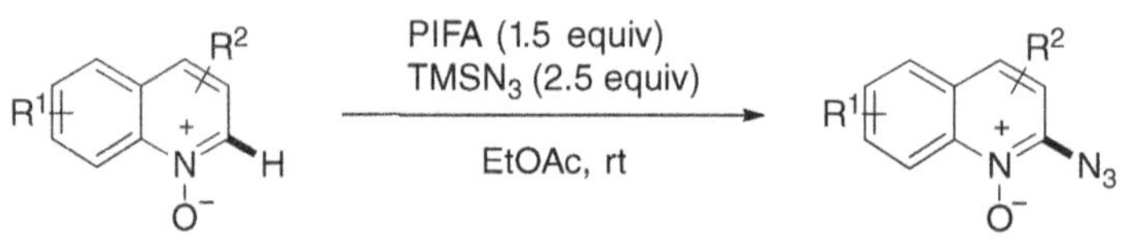

Scheme 8.23 The C2-azidation of quinoline *N*-oxides with PIFA and $TMSN_3$.

Moreover, azidation of heteroaromatic substrates has also been reported. Sudalai and co-workers disclosed an I_2-mediated regioselective C-3 azidation of indoles with NaN_3 under mild reaction conditions (Scheme 8.22) [121]. The functional group tolerance of this method was excellent and a putative pathway was proposed. The indole reacts initially with I_2 to form the iodonium ion **A**, which undergoes ring opening with a N_3 anion to generate intermediate **B**. Elimination of HI of intermediate **B** with the aid of Et_3N leads to the final product.

In addition to electron-rich substrates, azidation at an electron-deficient site of heterocycle has also been achieved. By using a novel carbene-stabilized N—O radical coupling strategy, Li and co-workers realized the regioselective C2-azidation of quinoline N-oxides (Scheme 8.23) [122].

Transition-metal-catalyzed directed C(sp^2)-H azidation of arenes has proven to be a powerful tool in the preparation of aryl azides. In 2012, Suna and co-workers reported a copper-catalyzed one-pot multistep protocol for heteroaryl azide synthesis (Scheme 8.24) [123]. The addition of copper catalyst considerably decreased the reaction time, and ensured complete regiocontrol in the fragmentation step. In this reaction, heteroaryl(phenyl)iodonium azides **A** are generated in situ, which undergo a copper(I)-catalyzed regioselective fragmentation to form the heteroaryl azides **B**.

Significantly, Jiao and co-workers developed a copper(I)-catalyzed C—H azidation of aniline derivatives using the amino group as the directing group (Scheme 8.25) [124]. In this reaction, *tert*-butyl hydroperoxide (TBHP) is used as a stoichiometric oxidant and $TMSN_3$ as the N_3 source. This is a very useful reaction and the products can be readily

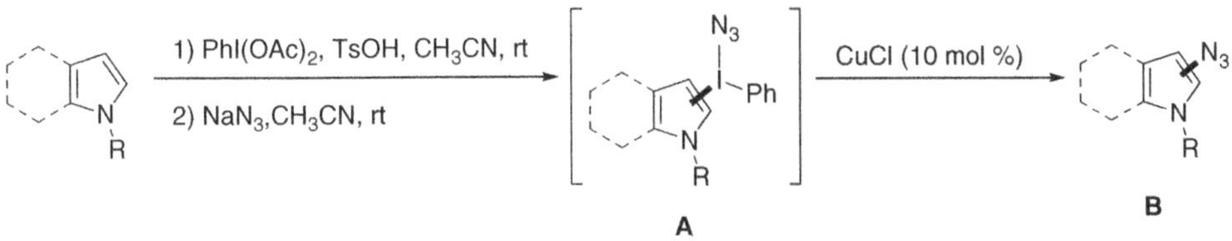

Scheme 8.24 Copper(I)-catalyzed synthesis of heteroaryl azides.

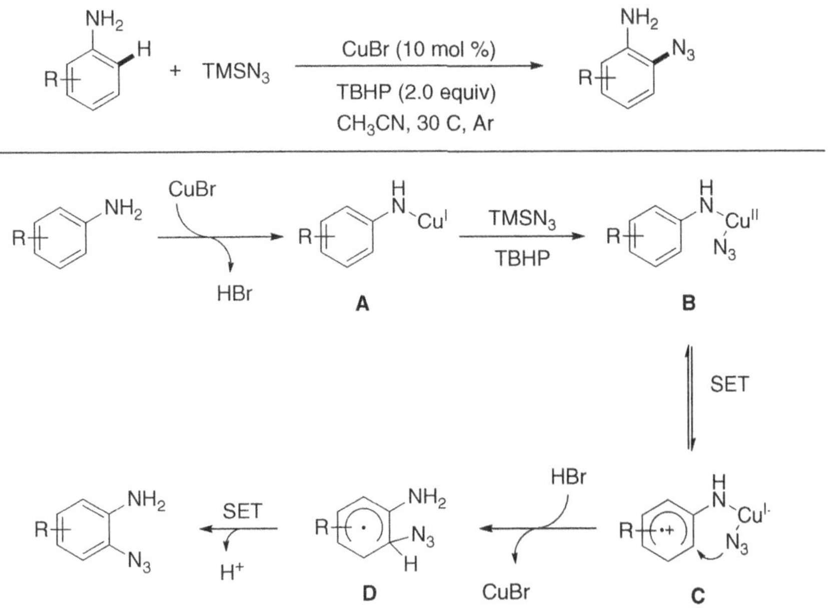

Scheme 8.25 Copper(II)-catalyzed regioselective C—H azidation of anilines.

diversified. Control experiments suggest that a radical process may be involved. Initially, the coordination of copper(I) ion to the aniline forms the intermediate **A**, which combines with an azide radical generated in situ from TBHP and $TMSN_3$ to deliver the intermediate **B**. The intramolecular single-electron transfer (SET) from the aryl ring to copper(II) produces the radical cation **C**. Then, the intramolecular azido group transfers into the aryl ring with release of CuBr providing the radical intermediate **D**. Finally, the radical intermediate **D** undergoes deprotonation via a SET process to give the product (Scheme 8.25). Despite the elegance of this work, one limitation is that, for mono-substituted substrates, typically two products were obtained (mono-azidation and di-azidation product). After this work, Hao and co-workers reported an Cu(II)-catalyzed regioselective *ortho*-C—H azidation of anilines. By combining the stable azidoiodinane reagent and the copper catalysis, the primary amines display an excellent *ortho*-directing effect and various *ortho*-azidated anilines can be produced in high selectivity [125]. Compared to these frequently used azidoiodinane reagen0st and other organic oxidants, the use of molecular oxygen as an oxidant in direct C—H azidation is very rare. Remarkably, Wang and co-workers recently realized an Cu(II)-catalyzed oxidative

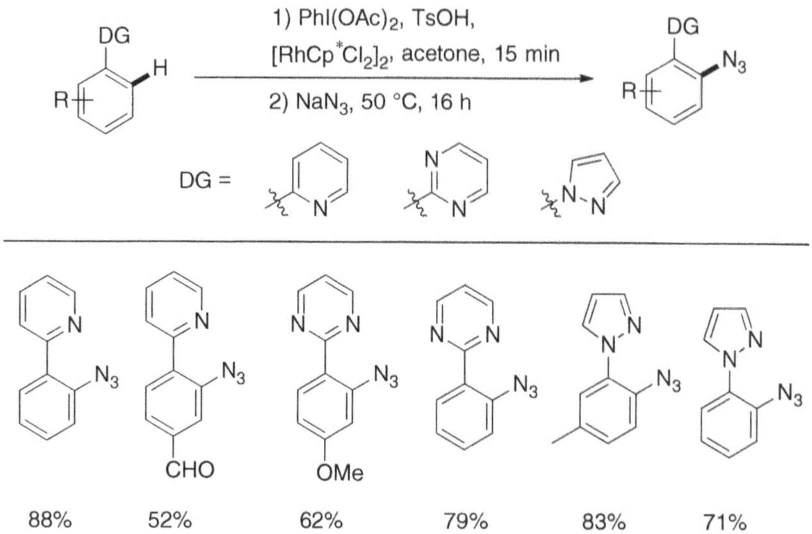

Scheme 8.26 Rhodium-catalyzed aryl C—H azidation.

azidation of arene C—H bonds using molecular oxygen as the oxidant and the reaction was proposed to involve a Cu(II)-ArCu(II)-ArCu(III)-Cu(I) catalytic cycle [126].

In addition to copper catalysis, rhodium(III)-catalyzed C—H azidation of arenes through C—H activation has also been accomplished. By using NaN_3 as the azide source and hypervalent iodine reagent (PIDA) as an oxidant, azidation can be efficiently performed with substrates bearing a pyridine, pyrimidine, or pyrazole heterocycle as a directing group (Scheme 8.26) [127].

In addition to arene C—H bonds, direct aldehyde C—H bond azidation has also been achieved for the preparation of acyl azides, a class of useful and versatile synthetic intermediates that can convert to isocyanates, amides, heterocycles, and other important compounds. Traditional methods require acyl chlorides or mixed anhydrides and NaN_3 for the preparation of acyl azides, without doubt, the direct aldehyde C—H bond azidation is a more efficient and atom-economical approach for their access (Scheme 8.27). The strategy mainly relies on the combinations of various oxidants with $TMSN_3$ or NaN_3, such as $CrO_3/TMSN_3$ [128], $SiCl_4/MnO_2/TMSN_3$ [129], PIDA/NaN_3 [130], *t*BuOCl/NaN_3 [131], and DMP/NaN_3 [132] combined reagents, and even the IN_3 can easily transform the aldehydes into the corresponding acyl azides too [133]. Other catalytic systems were also reported, for instance, the employing of 3,3′5,5′-tetra-*tert*-butyldiphenoquinone as the oxidant under the *N*-heterocyclic carbene (NHC) catalysis [134], and the use of azidoiodinane reagent with a *tetra-n*-butylammonium iodide (TBAI) catalyst [135].

8.4.1.2 Synthesis of Alkyl Azides via sp^3 C—H Bond Azidation

The synthesis of alkyl azides via direct azidation of aliphatic C—H bonds is the most straightforward method, yet is very challenging. The low acidity (p*K*a = 43–59 in DMSO), considerable strength (438.9 kJ mol^{-1} for CH_4), and weak coordination ability of aliphatic C—H bonds make it very difficult to functionalize [117, 136, 137]. Despite

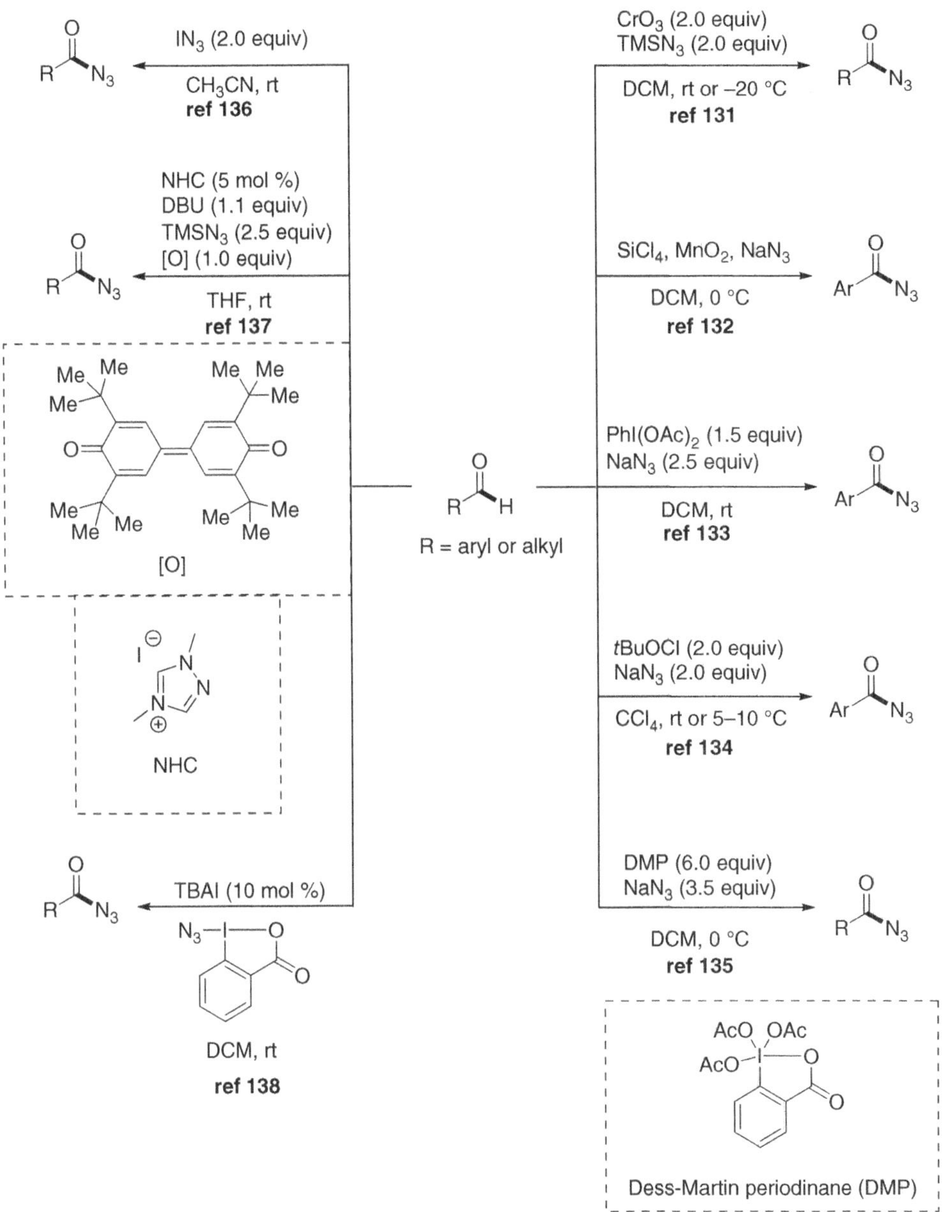

Scheme 8.27 Aldehyde C—H azidation under oxidative conditions.

these difficulties, by using radical or non-radical mediated sp^3 C—H bond activation through hydrogen atom abstraction strategy, the direct sp^3 C—H azidations can be achieved. Pioneering works were reported by Magnus' and Kita's groups. In 1992, Magnus and co-workers revealed that the PhIO/$TMSN_3$ system can efficiently convert a range of triisopropylsilyl (TIPS) enol ethers to β-azido triisopropylsilyl enol ethers (Scheme 8.28a) [138]. Under the similar reaction conditions, the direct *N*-alkyl

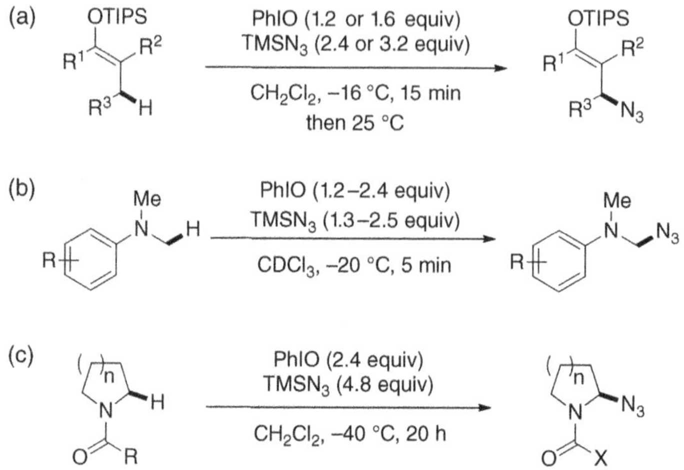

Scheme 8.28 Aliphatic C—H azidation with PhIO and $TMSN_3$.

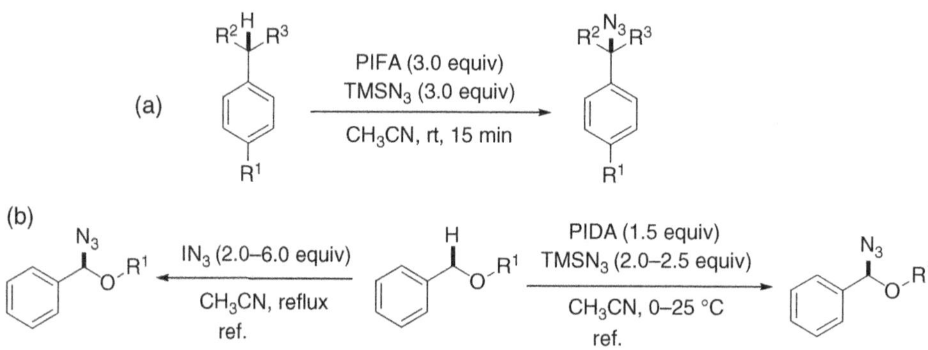

Scheme 8.29 Metal- free benzylic C—H azidation.

azidation of *N,N*-dialkylarylamines (Scheme 8.28b) [139] and α-azidation of amides, carbamates, and ureas (Scheme 8.28c) [140] can also be accomplished.

Almost simultaneously, Kita and co-workers disclosed that the PIFA/$TMSN_3$ system is capable of performing direct azidation of benzylic C—H bonds at room temperature (Scheme 8.29a) [141]. Inspired by these works, Bols and co-workers reported that iodonium azide (IN_3) could be employed to azidate ethereal C—H bonds (Scheme 8.29b) [142], and it was found that the IN_3 can be replaced by using a combination of PIDA/$TMSN_3$ [143]. Moreover, direct azidation of 1,3-dicarbonyl compounds has also been achieved [144–146] and effort has also been made to obtain excellent enantioselectivity [147].

Despite their significance, these reaction systems are limited to the reactive substrates, probably due to the low thermal stability of the *in situ* generating reactive azidoiodinanes and the relatively weak hydrogen-abstracting ability of the azidyl radical. To overcome these limitations, Zhdankin and co-workers reported that, by using the azidoiodinane reagent with a catalytic amount of benzoyl peroxide at elevated

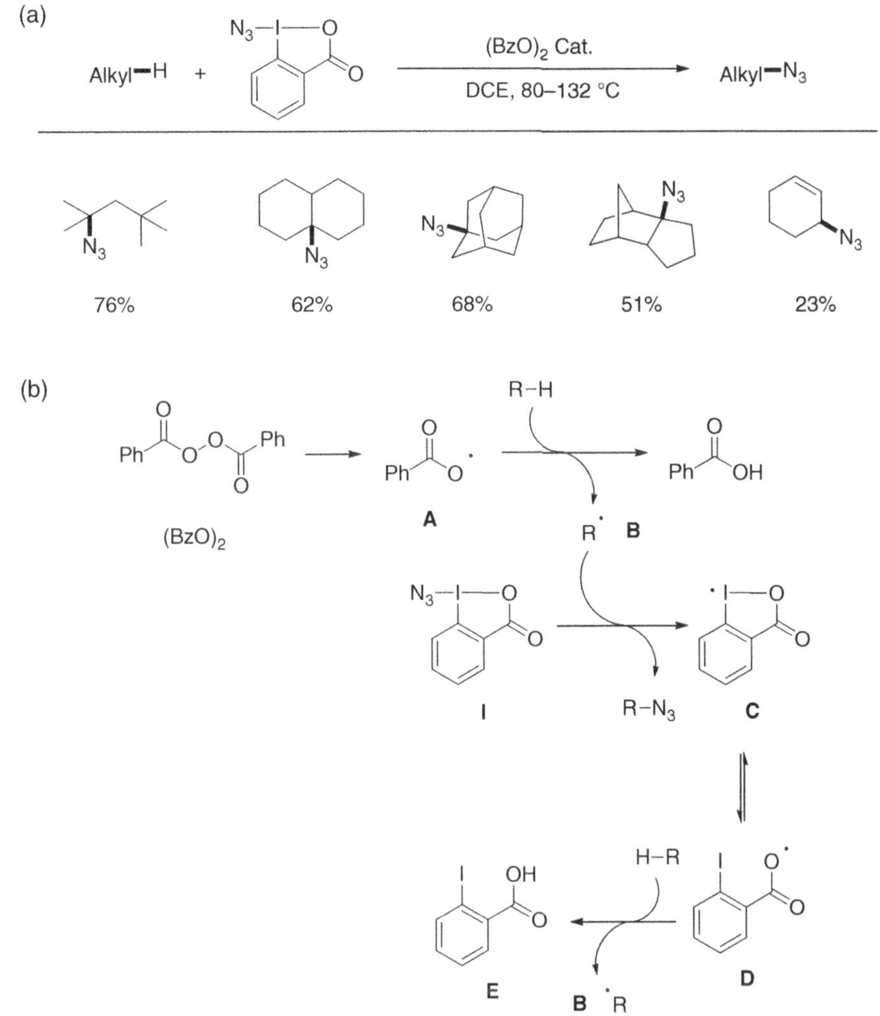

Scheme 8.30 Aliphatic C—H azidation with stable azidoiodinane reagent and radical initiators $(BzO)_2$.

temperatures, various simple hydrocarbons could be smoothly azidated (Scheme 8.30a) [148]. It should be noted that the azidation preferably occurred at the tertiary carbon. A plausible mechanism was proposed. Initially, the catalytic amount benzoyl radical **A** was generated by the thermal decomposition of benzoyl peroxide, which abstracted a hydrogen atom from the alkane to form the alkyl radical **B**. Subsequently, the alkyl radical **B** reacted with the azidoiodinane **I** to give the alkyl azide product along with the 2-iodobenzoate radical **D**, which could further abstract a hydrogen atom from the alkane to form the alkyl radical **B** to sustain the radical chain (Scheme 8.30b).

Despite the elegance of Zhdankin's method, however, the requirement for a high temperature restricts its application, particularly in the late-stage azidation of complex molecules. In this regard, Hartwig and co-worker developed an efficient iron(II)-catalyzed regioselective aliphatic C—H azidation reaction under mild conditions

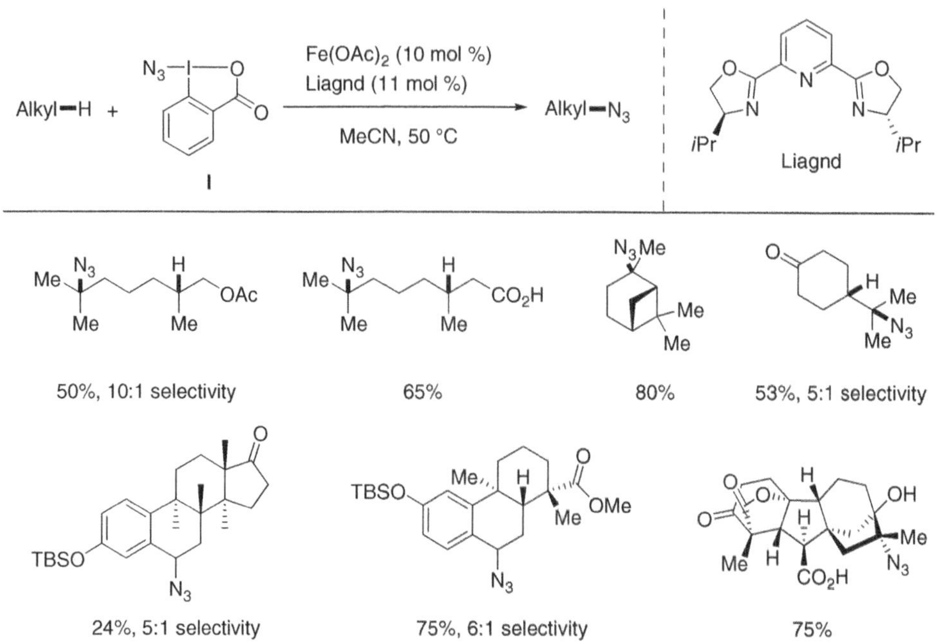

Scheme 8.31 Iron(II)-catalyzed aliphatic C—H azidation with stable azidoiodinane reagent and PyBOX ligand.

(Scheme 8.31) [149]. By using commercially available PyBOX as the ligand, and the azidoiodinane reagent **I** as the azidating reagent and oxidant under iron catalysis, a wide range of aliphatic C—H bonds of simple hydrocarbons and complex molecules could be successfully azidated. Preliminary mechanistic studies suggest a radical process was involved.

Almost simultaneously, Groves and co-workers reported an efficient manganese(III)-catalyzed aliphatic C—H azidation reaction by using NaN_3 as the N_3 source (Scheme 8.32a) [150]. This method could efficiently azidate secondary, tertiary, and benzylic C—H bonds and could be applied in the late-stage azidation of various bioactive molecules. A catalytic enantioselective process with chiral manganese salen Jacobsen catalyst was also presented. Based on mechanistic studies and DFT calculations, a plausible mechanism was proposed. The resting Mn(III) catalyst was initially oxidized by PhIO to produce the hydrogen-abstracting oxoMn(V) intermediate **B**. The substrate radical **C** was formed after hydrogen abstraction by intermediate **B**, which was then trapped by Mn(IV)-N_3 intermediate **E** to give the alkyl azide along with the regeneration of Mn(III) catalyst (Scheme 8.32b).

In addition to NaN_3, $TMSN_3$, and azidoiodinane reagents, sulfonyl azide has also been used as a radical azidating reagent. Recently, Tang and co-workers developed an efficient method for selective azidaiton of unactivated aliphatic C—H bonds with readily available sulfonyl azide as the azidating reagent and $K_2S_2O_8$ as the oxidant (Scheme 8.33a) [151]. This reaction required no assistance from a transition metal and can be applied in the late-stage azidation of complex molecules. The mechanism of this simple yet powerful reaction was also proposed. To begin with, peroxydisulfate

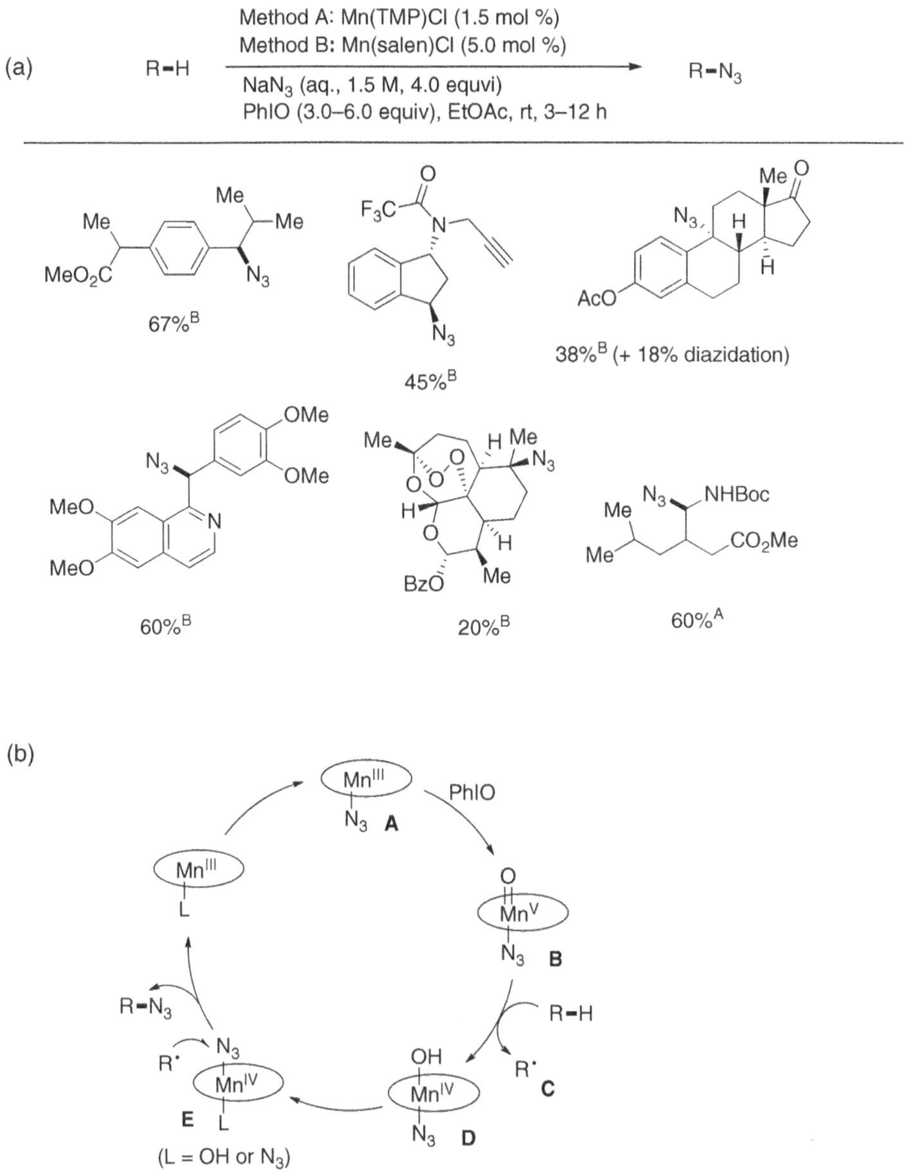

Scheme 8.32 Manganese(III)-catalyzed aliphatic C—H azidation.

anion decomposed into sulfate radical anion upon heating. Then, the sulfate radical anion abstracted a hydrogen atom from the substrate to produce the alkyl radical **A**. Finally, the alkyl radical **A** is trapped by sulfonyl azide to afford the alkyl azides (Scheme 8.33b).

Recently, visible-light photoredox catalysis has also been utilized in aliphatic C—H azidation reactions. In this context, Chen's group (Scheme 8.34, a) [152] and Greaney's group (Scheme 8.34, b) [153] independently reported the visible-light-mediated

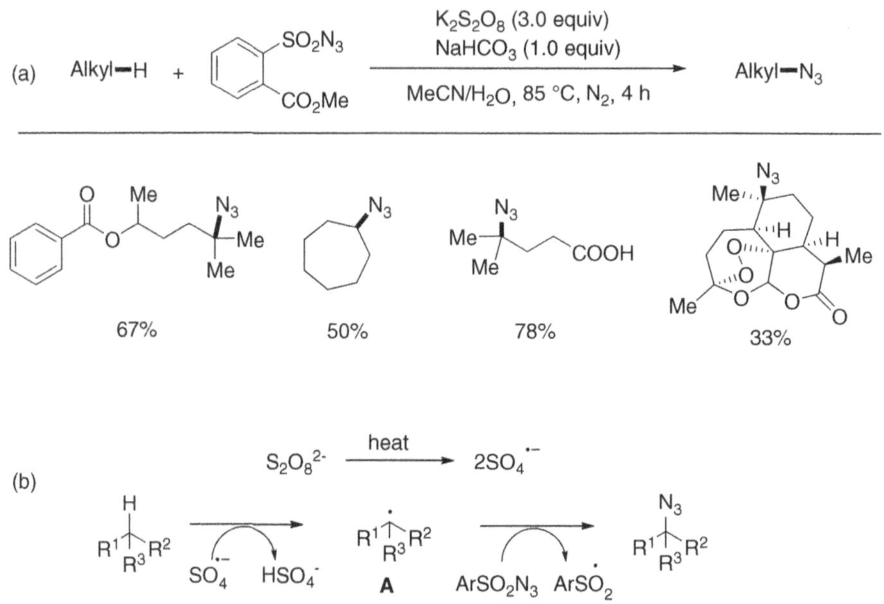

Scheme 8.33 Transition-metal-free oxidative aliphatic C—H azidation.

aliphatic C—H azidation by using Ru complexes and copper complexes as the photocatalysts, respectively. In these reactions, azidoiodinane reagent **I** was used as the N_3 source. The mechanism was similar in both transformations. First, in the presence of a photocatalyst, the 2-iodobenzoate radical **A** was generated from the azidoiodinane reagent **I** through an SET process upon visible-light irradiation. The 2-iodobenzoate radical **A** then abstracted a hydrogen atom from the substrate to form the alkyl radical **C**, which attacked the azidoiodinane reagent **I** to furnish the azides and regenerate the 2-iodobenzoate radical **A** to sustain the radical chain reaction (Scheme 8.34c).

8.4.2 Green Oxidative Synthesis of Organic Azides via C—C Bond Cleavage

C—C bonds are ubiquitous in organic compounds. The selective azidation of C—C bonds is very intriguing, yet is very challenging. This area mainly concerns the azidation of highly strained rings, such as cyclopropanes and cycloalkanols, as well as the decarboxylative azidation of carboxylic acids. Particularly, the ring opening of cyclopropanes or cycloalkanols and the decarboxylative process of carboxylic acids provide an efficient route for the site-selective introduction of a N_3 group. In 2012, Kerr and co-workers disclosed a tandem ring-opening/decarboxylation of cyclopropane hemimalonates with NaN_3 as an azide source to prepare γ-azidocarboxylic acid esters (Scheme 8.35a) [154]. A possible pathway of this reaction is depicted in the scheme. Cyclopropane hemimalonate reacted initially with HN_3 to generate the acyl azide **A**, which then underwent a [3]-sigmatropic rearrangement to afford the ketene **B**, followed by the nucleophilic attraction of H_2O to form the acid **C**. Decarboxylation of the acid **C** gave the final product (Scheme 8.35b). Shortly afterward, the electrophilic ring-opening of donor-acceptor cyclopropanes was also reported [155, 156].

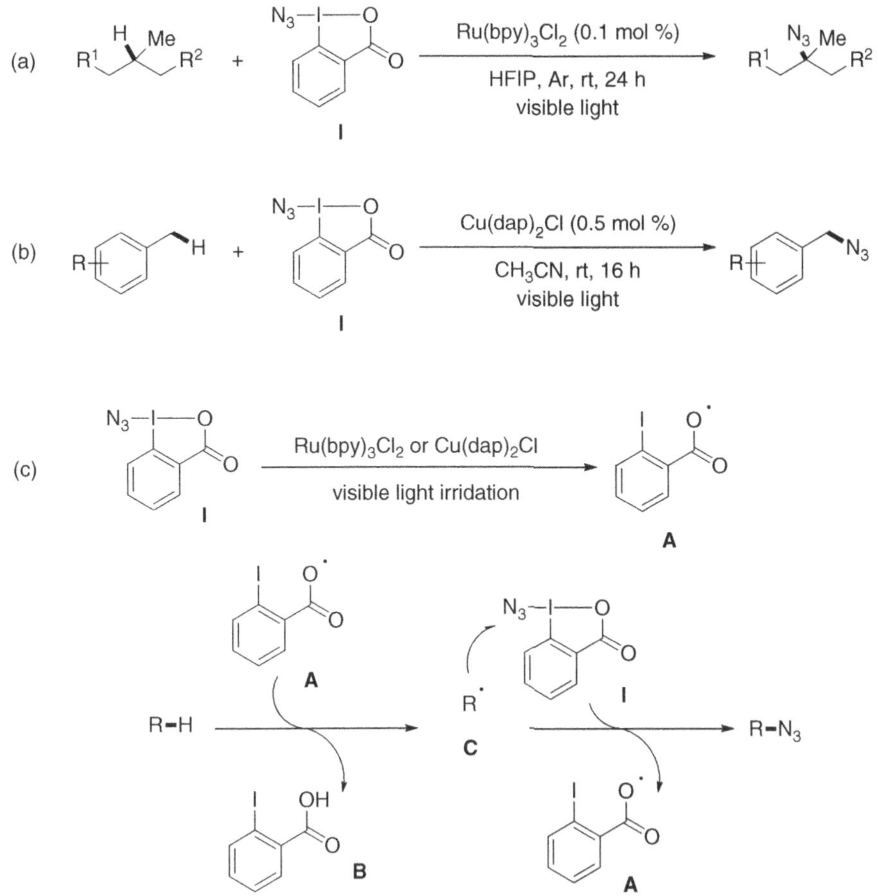

Scheme 8.34 Aliphatic C—H azidation through visible-light photoredox catalysis.

Cycloalkanols are frequently used compounds in organic synthesis, the ring-opening of cycloalkanols through C—C bond cleavage would provide a valuable method for the preparation of distally functionalized ketones that were not easily obtained by conventional methods. In 2007, Flowers II and co-workers reported a novel synthetic method for β-azido ketones via cerium(IV) ammonium nitrate (CAN) mediated oxidative azidation of cyclopropanols (Scheme 8.36a) [157]. First, the CAN oxidated N_3 anion to a N_3 radical, which was trapped by the cyclopropanol to generate the radical intermediate **A** and was further oxidized by CAN to give the final product (Scheme 8.36b).

In addition to cyclopropanols, cyclobutanols have also been successfully applied in the ring-opening reaction. In 2015, Zhu and co-workers developed an efficient manganese-catalyzed oxidative azidation of cyclobutanols using 2,2′-bipyridine as a ligand and BI-OH as an oxidant (Scheme 8.37a) [158]. Various primary, secondary, and tertiary as well as medium-sized cyclic azides with various functional groups were smoothly obtained with excellent regioselectivity. Based on the mechanism studies, a radical process was proposed. To begin with, the combination of $Mn(OAc)_3$, BI-OH, and $TMSN_3$ gave the high-valent Mn(V)-N_3 specie **A**. The specie **A** reacted with the

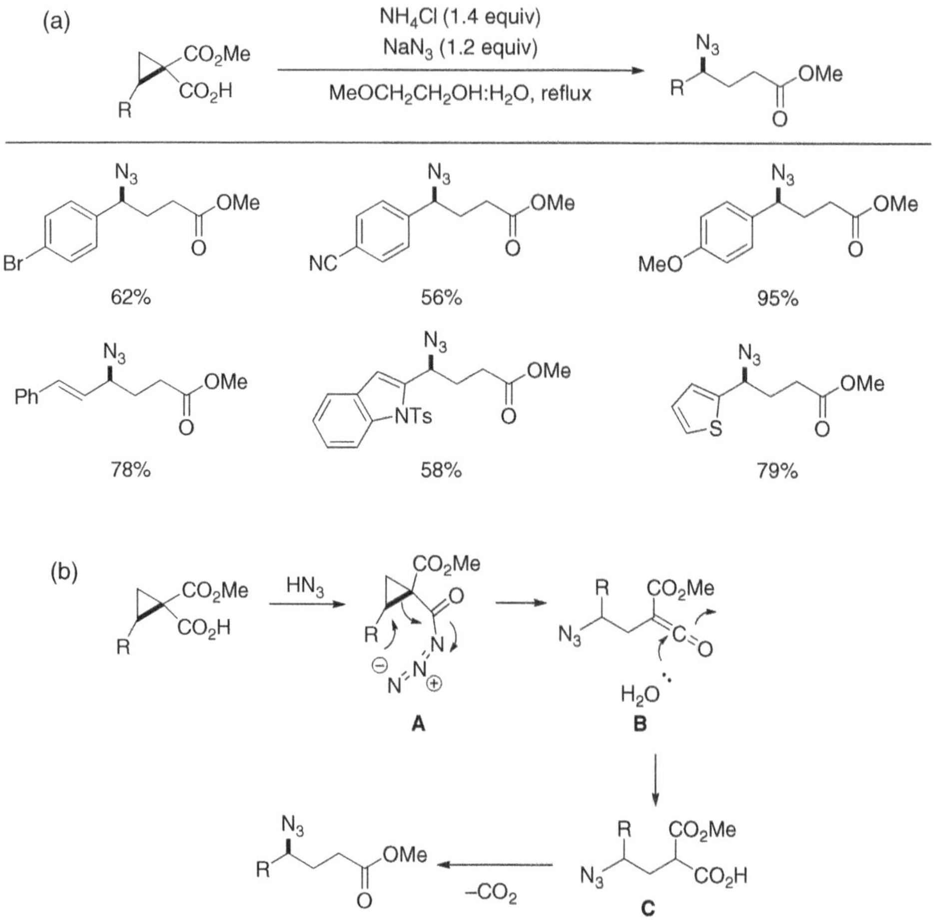

Scheme 8.35 Tandem ring-opening/decarboxylation of cyclopropane hemimalonates with NaN_3.

cyclobutanol and led to the complex **B**, which underwent an SET process to generate the Mn(IV)-N_3 specie **C** and the cyclobutyloxy radical **D**. The cyclobutyloxy radical **D** underwent rearrangement to give the alkyl radical **E**, which was intercepted by the Mn(IV)-N_3 specie **C** to give the alkyl azide, along with the regeneration of Mn catalyst (Scheme 8.37b).

Carboxylic acids are readily available and are frequently used in organic synthesis. The transformations of these compounds are of great importance, particularly the decarboxylative azidation involving C(sp^3)-COOH bond cleavage, therefore the introduction of a N_3 group is site-specific. Pioneering work in this area included the decarboxylative azidation of β-silyl carboxylic acids to deliver β-silyl azides, as reported by Porter's group in 2002. This method operated by transforming β-silylcarboxylic acids into the corresponding PTOC (pyridine-2-thione-*N*-oxy-carbonyl) esters, followed by the reaction with ethanesulfonyl azide upon ultraviolet irradiation [159]. Renaud and co-workers reported a decarboxylative azidation of aliphatic carboxylic acids via

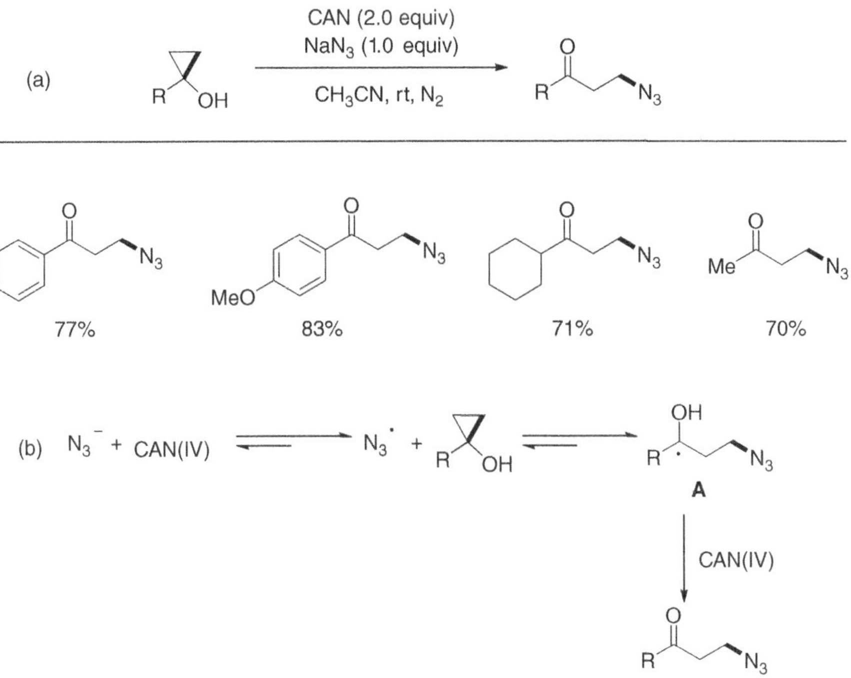

Scheme 8.36 CAN-mediated oxidative azidation of cyclopropanols.

MPDOC (*S*-methyl-*N*-phenyl-1,3-dithiocarbamoyloxycarbonyl) esters and MMODC (*N*,*S*-dimethyldithio-carbamoyl-*N*-oxy-carbonyl) esters using sulfonyl azide as a N_3 radical source to prepare the alkyl azides in moderate yields [160]. Despite their significance, these reactions generally required multiple steps, and produced unwanted waste that would hinder their applications. In 2014, Kirsch and co-workers reported a direct decarboxylative azidation of malonic acid monoesters to synthesize the corresponding α-azidoesters with NaN_3 and substoichiometric amounts of NaI in aqueous DMSO [161]. The operation of this process was very simple and yields were high. Despite their elegance, the substrates are still confined to some certain compounds. Therefore, a general and efficient method for the decarboxylative azidation of all kinds of carboxylic acid is still required. In 2015, Li and co-workers reported a direct decarboxylative azidation of simple aliphatic carboxylic acids (Scheme 8.38, a) [162]. By using $AgNO_3$ as the catalyst and $K_2S_2O_8$ as the oxidant, the reactions of various aliphatic carboxylic acids with tosyl azide or pyridine-3-sulfonyl azide in aqueous CH_3CN solution afforded the corresponding alkyl azides in good yields. Moreover, a radical mechanism was proposed for this site-specific azidation reaction. First, persulfate oxidated the Ag(I) to the Ag(II) intermediate, which underwent SET with the aliphatic carboxylic acid to generate a carboxyl radical **A**. Carboxyl radical **A** underwent fast decarboxylation giving the alkyl radical, which reacted with a sulfonyl azide to afford the corresponding alkyl azide (Scheme 8.38b). Almost simultaneously, Jiao and co-workers reported a similar strategy by using AgF as the catalyst [163].

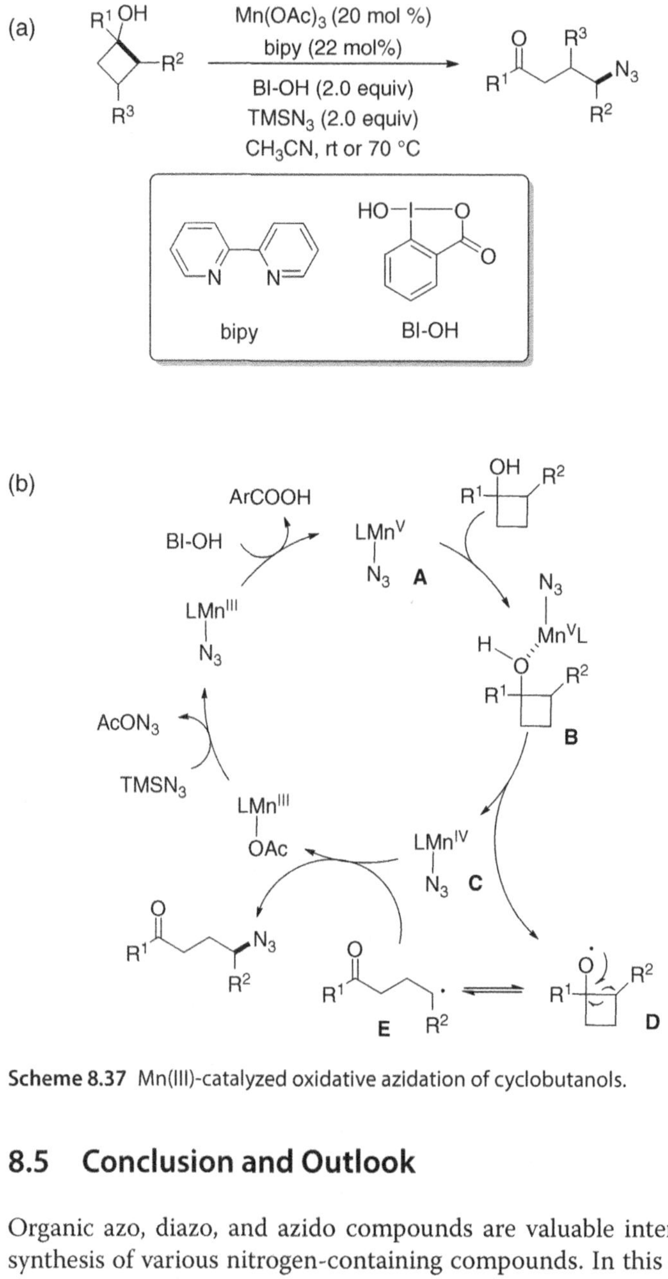

Scheme 8.37 Mn(III)-catalyzed oxidative azidation of cyclobutanols.

8.5 Conclusion and Outlook

Organic azo, diazo, and azido compounds are valuable intermediates applied in the synthesis of various nitrogen-containing compounds. In this chapter, recent advances in the preparation of these compounds via green oxidative strategies have been displayed. These contributions presented in this chapter exhibit great potential in organic synthesis. Despite their significance, there are still some challenges that need to be addressed. The employment of strong oxidants in the metal-free synthesis of azobenzenes often meets with problems, mainly because of the ease of overreaction (such as overreaction to azoxy products) leading to nonoptimal yields and product selectivity,

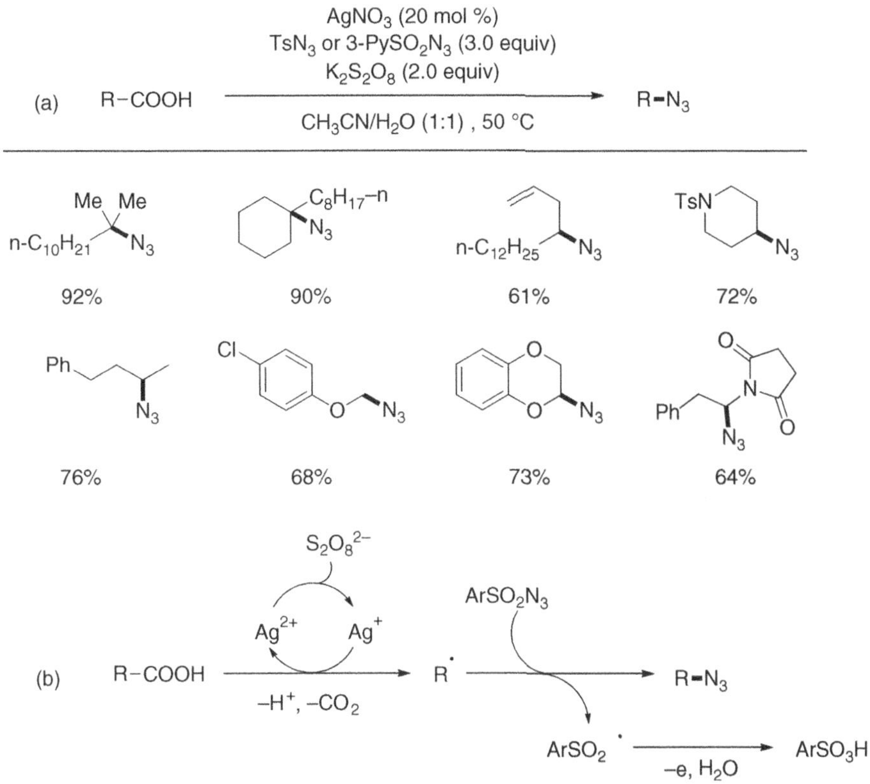

Scheme 8.38 Silver(I)-catalyzed decarboxylative radical azidation of aliphatic carboxylic acids.

and also because of reactivity issues, long reaction times and special reaction conditions are required. Besides, most synthetic methods of azo compounds focus on the symmetric reactions, and reports on the asymmetric version with high yield and selectivity are still rare. Thus, it is necessary to continue developing new and efficient methods for the synthesis of asymmetric azo compounds from simple anilines. In diazo synthesis, the unsatisfying atom economy and/or the low cost-effectiveness of several methods is still hard to eliminate. The use of simple azides such as $TMSN_3$ and NaN_3 in combination with a benign oxidant (particularly with molecular oxygen) in the direct diazotization of methylene compounds remains unresolved. In the synthesis of azido compounds, the azidation of aryl C—H bond is not general. The substrate scope is still limited; only some activated compounds, such as anilines, indoles, or substrates with a directing group are suitable. Moreover, azidation of non-activated aliphatic C—H bonds mainly rely on the use of hypervalent iodine reagents, the utilization of benign oxidants, particularly molecular oxygen, is still challenging. On the whole, the discovery of more powerful catalytic systems for the green synthesis of these compounds is still needed. We hope this chapter will stimulate more inspiration for the development of greener and more environmentally friendly protocols for the preparation of azo, diazo, and azido compounds.

References

1 Natansohn, A. (2002). Photoinduced motions in azo-containing polymers. *Chem. Rev.* 102 (11): 4139–4176.
2 Broichhagen, J., Frank, J.A., and Trauner, D. (2015). A roadmap to success in photopharmacology. *Acc. Chem. Res.* 48 (7): 1947–1960.
3 Dong, M., Babalhavaeji, A., Samanta, S. et al. (2015). Red-shifting azobenzene photoswitches for in vivo use. *Acc. Chem. Res.* 48 (10): 2662–2670.
4 Mix, K.A., Aronoff, M.R., and Raines, R.T. (2016). Diazo compounds: versatile tools for chemical biology. *ACS Chem. Biol.* 11 (12): 3233–3244.
5 Kumar, R., Wiebe, L.I., and Knaus, E.E. (1993). Synthesis and antiviral activity of novel 5-(1-azido-2-haloethyl) and 5-(1-azido-, amino-, or methoxyethyl) analogs of 2′-deoxyuridine. *J. Med. Chem.* 36 (17): 2470–2474.
6 Habeeb, A.G., Rao, P.N.P., and Knaus, E.E. (2001). Design and synthesis of celecoxib and rofecoxib analogues as selective cyclooxygenase-2 (COX-2) inhibitors: replacement of sulfonamide and methylsulfonyl pharmacophores by an azido bioisostere. *J. Med. Chem.* 44 (18): 3039–3042.
7 Merino, E. (2011). Synthesis of azobenzenes: the coloured pieces of molecular materials. *Chem. Soc. Rev.* 40 (7): 3835–3853 and references cited therein.
8 Hunger, K. (2003). *Industrial Dyes: Chemistry, Properties, Applications*, 20–35. Wiley-VCH.
9 Brenzovich, W.E., Houk, R.J.T., Malubay, S.M.A. et al. (2002). The synthesis and properties of some carboxy-substituted analogs of butter yellow. *Dyes Pigm.* 52 (2): 101–114.
10 Yamaoka, T., Makita, Y., Sasatani, H. et al. (2000). Linear type azo-containing polyurethane as drug-coating material for colon-specific delivery: its properties, degradation behavior, and utilization for drug formulation. *J. Control. Release* 66 (2–3): 187–197.
11 Cisnetti, F., Ballardini, R., Credi, A. et al. (2004). Photochemical and electronic properties of conjugated bis(azo) compounds: an experimental and computational study. *Chem. Eur. J.* 10 (8): 2011–2021.
12 Sandborn, W.J. (2002). Rational selection of oral 5-aminosalicylate formulations and prodrugs for the treatment of ulcerative colitis. *Am. J. Gastroenterol.* 97 (12): 2939–2941.
13 Xiao, Q., Zhang, Y., and Wang, J. (2013). Diazo compounds and n-tosylhydrazones: novel cross-coupling partners in transition-metal-catalyzed reactions. *Acc. Chem. Res.* 46 (2): 236–247.
14 Maas, G. (2004). Ruthenium-catalysed carbenoid cyclopropanation reactions with diazo compounds. *Chem. Soc. Rev.* 33 (3): 183–190.
15 Nishiyama, H., Itoh, Y., Matsumoto, H. et al. (1994). New chiral ruthenium bis(oxazolinyl)pyridine catalyst. efficient asymmetric cyclopropanation of olefins with diazoacetates. *J. Am. Chem. Soc.* 116 (5): 2223–2224.
16 Nani, R.R. and Reisman, S.E. (2013). α-Diazo-β-ketonitriles: uniquely reactive substrates for arene and alkene cyclopropanation. *J. Am. Chem. Soc.* 135 (19): 7304–7311.
17 Nowlan, D.T. and Singleton, D.A. (2005). Mechanism and origin of enantioselectivity in the $Rh_2(OAc)(DPTI)_3$-catalyzed cyclopropenation of alkynes. *J. Am. Chem. Soc.* 127 (17): 6190–6191.

18 Panne, P. and Fox, J.M. (2007). Rh-catalyzed intermolecular reactions of alkynes with α-diazoesters that possess β-hydrogens: ligand-based control over divergent pathways. *J. Am. Chem. Soc.* 129 (1): 22–23.

19 Cui, X., Xu, X., Lu, H. et al. (2011). Enantioselective cyclopropenation of alkynes with acceptor/acceptor-substituted diazo reagents via Co(II)-based metalloradical catalysis. *J. Am. Chem. Soc.* 133 (10): 3304–3307.

20 Goto, T., Takeda, K., Shimada, N. et al. (2011). Highly enantioselective cyclopropenation reaction of 1-alkynes with α-alkyl-α-diazoesters catalyzed by dirhodium(II) carboxylates. *Angew. Chem. Int. Ed.* 50 (30): 6803–6808.

21 Liang, Y., Yu, K., Li, B. et al. (2014). Rh(III)-catalyzed synthesis of 1-aminoindole derivatives from 2-acetyl-1-arylhydrazines and diazo compounds in water. *Chem. Commun.* 50 (46): 6130–6133.

22 Reisman, S.E., Nani, R.R., and Levin, S. (2011). Buchner and beyond: arene cyclopropanation as applied to natural product total synthesis. *Synlett* 2011: 2437–2442.

23 Jia, S. and Ma, S. (2016). New approaches to the synthesis of metal carbenes. *Angew. Chem. Int. Ed.* 55 (32): 9134–9166.

24 Liang, Y., Liang, Y.F., and Jiao, N. (2015). Cu- or Fe-catalyzed C-H/C-C bond nitrogenation reactions for the direct synthesis of N-containing compounds. *Org. Chem. Front.* 2: 403–415.

25 Kotra, L.P., Manouilov, K.K., Cretton-Scott, E. et al. (1996). Synthesis, biotransformation, and pharmacokinetic studies of 9-(β-d-arabinofuranosyl)-6-azidopurine: a prodrug for ara-a designed to utilize the azide reduction pathway. *J. Med. Chem.* 39 (26): 5202–5207.

26 Koudriakova, T., Manouilov, K.K., Shanmuganathan, K. et al. (1996). In vitro and in vivo evaluation of 6-azido-2′,3′-dideoxy-2′-fluoro-β-d-arabinofuranosylpurine and N6-methyl-2′,3′-dideoxy-2′-fluoro-β-d-arabinofuranosyladenine as prodrugs of the anti-HIV nucleosides 2′-F-ara-ddA and 2′-F-ara-ddl. *J. Med. Chem.* 39 (23): 4676–4681.

27 Damen, E.W.P., Nevalainen, T.J., van den Bergh, T.J.M. et al. (2002). Synthesis of novel paclitaxel prodrugs designed for bioreductive activation in hypoxic tumour tissue. *Bioorg. Med. Chem.* 10 (1): 71–77.

28 Wu, P., Feldman, A.K., Nugent, A.K. et al. (2004). Efficiency and fidelity in a click-chemistry route to triazole dendrimers by the copper(I)-catalyzed ligation of azides and alkynes. *Angew. Chem. Int. Ed.* 43 (30): 3928–3932.

29 Thirumurugan, P., Matosiuk, D., and Jozwiak, K. (2013). Click chemistry for drug development and diverse chemical-biology applications. *Chem. Rev.* 113 (7): 4905–4979.

30 Georgiádes, Á., ÖtvÖs, B.S., and FülÖp, F. (2015). Exploring new parameter spaces for the oxidative homocoupling of aniline derivatives: sustainable synthesis of azobenzenes in a flow system. *ACS Sustain. Chem. Eng.* 3 (12): 3388–3397.

31 Yesodha, S.K., Sadashiva Pillai, C.K., and Tsutsumi, N. (2004). Stable polymeric materials for nonlinear optics: a review based on azobenzene systems. *Prog. Polym. Sci.* 29 (1): 45–74.

32 Hamo, F., Djedaini-Pilard, F., Barbot, F., and Len, C. (2009). Azobenzenes-synthesis and carbohydrate applications. *Tetrahedron* 65 (49): 10105–10123.

33 Hu, L., Cao, X.Q., Shi, L.Y. et al. (2011). A highly active nano-palladium catalyst for the preparation of aromatic azos under mild conditions. *Org. Lett.* 13 (20): 5640–5643.

34 Liu, X., Li, H.Q., Ye, S. et al. (2014). Gold-catalyzed direct hydrogenative coupling of nitroarenes to synthesize aromatic azo compounds. *Angew. Chem. Int. Ed.* 53 (29): 7624–7628.

35 Davey, M.H., Lee, V.Y., Miller, R.D., and Marks, T.J. (1999). Synthesis of aryl nitroso derivatives by *tert*-butyl hypochlorite oxidation in homogeneous media. Intermediates for the preparation of high-hyperpolarizability chromophore skeletons. *J. Org. Chem.* 64 (13): 4976–4979.

36 Haghbeen, K. and Tan, E.W. (1998). Facile synthesis of catechol azo dyes. *J. Org. Chem.* 63 (13): 4503–4505.

37 Singh, S.K., Reddy, M.S., Mangle, M., and Ganesh, K.R. (2007). Cu(I)-mediated deoxygenation of N-oxides to amines. *Tetrahedron* 63 (1): 126–130.

38 Thorwirth, R., Bernhardt, F., Stolle, A. et al. (2010). Switchable selectivity during oxidation of anilines in a ball mill. *Chem. Eur. J.* 16 (44): 13236–13242.

39 Wheeler, O.D. and Gonzales, D. (1964). Oxidation of primary aromatic amines with manganese dioxide. *Tetrahedron* 20 (2): 189–193.

40 Baumgarten, H.E., Staklis, A., and Miller, E.M. (1965). Reactions of amines. XIII. the oxidation of N-acyl-N-arylhydroxylamines with lead tetraacetate. *J. Org. Chem.* 30 (4): 1203–1206.

41 Farhadi, S., Zaringhadam, P., and Sahamieh, R.Z. (2007). Photo-assisted oxidation of anilines and other primary aromatic amines to azo compounds using mercury(II) oxide as a photo-oxidant. *Acta Chim. Slov.* 54 (3): 647–653.

42 Okumura, S., Lin, C.H., Takeda, Y., and Minakata, S. (2013). Oxidative dimerization of (hetero)aromatic amines utilizing *t*-BuOI leading to (hetero)aromatic azo compounds: scope and mechanistic studies. *J. Org. Chem.* 78 (23): 12090–12105.

43 Takeda, Y., Okumura, S., and Minakata, S. (2012). Oxidative dimerization of aromatic amines using *t*BuOI: entry to unsymmetric aromatic azo compounds. *Angew. Chem. Int. Ed.* 51 (31): 7804–7808.

44 Monir, K., Ghosh, M., Mishra, S. et al. (2014). Phenyliodine(III) diacetate (PIDA) mediated synthesis of aromatic azo compounds through oxidative dehydrogenative coupling of anilines: scope and mechanism. *Eur. J. Org. Chem.* 1096–1102.

45 Shi, Z., Zhang, C., Tang, C., and Jiao, N. (2012). Recent advances in transition-metal-catalyzed reactions using molecular oxygen as the oxidant. *Chem. Soc. Rev.* 41 (8): 3381–3430.

46 Liang, Y.F. and Jiao, N. (2017). Oxygenation via C-H/C-C bond activation with molecular oxygen. *Acc. Chem. Res.* 50 (7): 1640–1653.

47 Grirrane, A., Corma, A., and García, H. (2008). Gold-catalyzed synthesis of aromatic azo compounds from anilines and nitroaromatics. *Science* 322 (5908): 1661–1664.

48 Cai, S., Rong, H., Yu, X. et al. (2013). Room temperature activation of oxygen by monodispersed metal nanoparticles: oxidative dehydrogenative coupling of anilines for azobenzene syntheses. *ACS Catal.* 3 (4): 478–486.

49 Lu, W. and Xi, C. (2008). CuCl-catalyzed aerobic oxidative reaction of primary aromatic amines. *Tetrahedron Lett.* 49 (25): 4011–4015.

50 Zhang, C. and Jiao, N. (2011). Copper-catalyzed aerobic oxidative dehydrogenative coupling of anilines leading to aromatic azo compounds using dioxygen as an oxidant. *Angew. Chem. Int. Ed.* 49 (35): 6174–6177.

51 Dutta, B., Biswas, S., Sharma, V. et al. (2016). Mesoporous manganese oxide catalyzed aerobic oxidative coupling of anilines to aromatic azo compounds. *Angew. Chem. Int. Ed.* 55 (6): 2171–2175.

52 Sarkar, P. and Mukhopadhyay, C. (2016). First use of p-*tert*-butylcalix[4]arene-tetra-O-acetate as a nanoreactor having tunable selectivity towards cross azo-compounds by trapping silver ions. *Green Chem.* 18 (2): 442–451.
53 Ötvös, S., Georgiádes, Á., Mészáros, R. et al. (2017). Continuous-flow oxidative homocouplings without auxiliary substances: exploiting a solid base catalyst. *J. Catal.* 348: 90–99.
54 Wang, J., He, J., Zhi, C. et al. (2014). Highly efficient synthesis of azos catalyzed by the common metal copper (0) through oxidative coupling reactions. *RSC Adv.* 4 (32): 16607–16611.
55 Lin, M., Wang, Z., Fang, H. et al. (2016). Metal-free aerobic oxidative coupling of amines in dimethyl sulfoxide via a radical pathway. *RSC Adv.* 6 (13): 10861–10864.
56 Sheu, C., Sobkowiak, A., Jeon, S., and Sawyer, D.T. (1990). Activation of dioxygen by bis(2,6-dicarboxylatopyridine)iron(II) for the ketonization of methylenic carbons and the dioxygenation of acetylenes, aryl olefins, and catechols: reaction mimics for dioxygenase proteins. *J. Am. Chem. Soc.* 112 (2): 879–881.
57 Németh, S., Szeverényi, Z., and Simándi, L.I. (1980). Catalytic oxidations with molecular oxygen in the presence of cobaloxime(II) derivatives. *Inorg. Chim. Acta* 44: L107–L109.
58 Kim, S.S.B., Hommer, R.B., and Cannon, R.D. (2006). The oxidation of hydrazobenzene catalyzed by cobalt complexes in nonaqueous solvents. *Bull. Kor. Chem. Soc.* 27 (2): 255–265.
59 Gao, W., He, Z., Qian, Y. et al. (2011). General palladium-catalyzed aerobic dehydrogenation to generate double bonds. *Chem. Sci.* 3 (3): 883–886.
60 Donck, S., Gravel, E., Li, A. et al. (2015). Mild and selective catalytic oxidation of organic substrates by a carbon nanotube-rhodium nanohybrid. *Catal Sci. Technol.* 5 (9): 4542–4546.
61 Gaviraghi, G., Pinza, M., and Pifferi, G. (1981). A mild and convenient synthesis of ethyl 2-phenyl-and 2-(3-pyridazine)-diazenecarboxylates (azocarboxylates). *Synthesis* 1981 (8): 608–610.
62 Hashimoto, T., Hirose, D., and Taniguchi, T. (2015). Catalytic aerobic oxidation of arylhydrazides with iron phthalocyanine. *Adv. Synth. Catal.* 357 (14–15): 3346–3352.
63 Corral, C., Lissavetzky, J., and Quintanilla, G. (1982). New method for the synthesis of chloro-substituted dibenzo[b,f][1,4,5]thiadiazepines and their 5,6-dihydro derivatives. *J. Org. Chem.* 47 (11): 2214–2215.
64 Wang, L., Ishida, A., Hashidoko, Y., and Hashimoto, M. (2017). Dehydrogenation of the NH-NH bond triggered by potassium *tert*-butoxide in liquid ammonia. *Angew. Chem. Int. Ed.* 56 (3): 870–873.
65 Trischler, F. (1979). The determination of hydrzobenzene by the die method with hydrogen peroxide catalyzed with vanadium. *J. Ther. Anal.* 16 (1): 119–122.
66 Drug, E. and Gozin, M. (2007). Catalytic oxidation of hydrazo derivatives promoted by a $TiCl_3$/HBr System. *J. Am. Chem. Soc.* 129 (49): 13784–13785.
67 Pechmann, H.V. (1894). Ueber Diazomethan. *Ber Dtsch. Chem. Ges.* 27 (2): 1888–1891.
68 Curtius, T. (1883). Ueber die Einwirkung von salpetriger Säure auf salzsauren Glycocolläther. *Ber Dtsch. Chem. Ges.* 16 (2): 2230–2231.
69 Ford, A., Miel, H., Ring, A. et al. (2015). Modern organic synthesis with α-diazocarbonyl compounds. *Chem. Rev.* 115 (18): 9981–10080 and references cited therein.

70 Maas, G. (2009). New syntheses of diazo compounds. *Angew. Chem. Int. Ed.* 48 (44): 8186–8195 and references cited therein.
71 Womack, E.B. and Nelson, A.B. (1955). Ethyl diazoacetate. *Org. Synth.* 24: 56–57.
72 Titanyuk, I.D., Vorob'eva, D.V., Osipov, S.N., and Beletskaya, I.P. (2006). 1-Trifluoromethyl-1-diethoxyphosphoryl carbene: a new synthon for the preparation of CF3-containing α-hydroxy and α-amino phosphonic acid derivatives. *Synlett* 2006: 1355–1358 and references cited therein.
73 Heydt, H. (2004). Science of Synthesis, vol. 27, 843–915. Stuttgart: Thieme.
74 Taber, D.F., Ruckle, R.E., and Hennessy, M.J. (1986). Mesyl azide: a superior reagent for diazo transfer. *J. Org. Chem.* 51 (21): 4077–4078.
75 Davies, H.M.L., Cantrell, W.R., Romines, K.R., and JS, B. (1992). Synthesis of furans via rhodium(II) acetate-catalyzed reaction of acetylenes with α-diazocarbonyls: ethyl 2-methyl-5-phenyl-3-furancarboxylate. *Org. Synth.* 70: 93–100.
76 Davies, H.M.L. and Townsend, R.J. (2001). Catalytic asymmetric cyclopropanation of heteroaryldiazoacetates. *J. Org. Chem.* 66 (20): 6595–6603.
77 Regitz, M. and Rüter, J. (1968). Reaktionen CH-aktiver verbindungen mit aziden, XVIII. Synthese von 2-oxo-1-diazo-cycloalkanen durch entformylierende Diazogruppen-Übertragung. *Chem. Ber.* 101 (4): 1263–1270.
78 Danheiser, R.L., Miller, R.F., Brisbois, R.G., and Park, S.Z. (1990). An improved method for the synthesis of alpha-diazo ketones. *J. Org. Chem.* 55 (6): 1959–1964.
79 Taber, D.F., Sheth, R.B., and Joshi, P.V. (2005). Simple preparation of α-diazo esters. *J. Org. Chem.* 70 (7): 2851–2854.
80 Zhang, J., Chen, W., Huang, D. et al. (2017). Tandem synthesis of α-diazoketones from 1,3-diketones. *J. Org. Chem.* 82 (17): 9171–9174.
81 Javed, M.I. and Brewer, M. (2007). Diazo preparation via dehydrogenation of hydrazones with "activated" DMSO. *Org. Lett.* 9 (9): 1789–1792.
82 Furrow, M.E. and Myers, A.G. (2004). A general procedure for the esterification of carboxylic acids with diazoalkanes generated in situ by the oxidation of N-*tert*-butyldimethylsilylhydrazones with (difluoroiodo)benzene. *J. Am. Chem. Soc.* 126 (39): 12222–12223.
83 Yamazaki, S., Matsuura, T., and Nishinaga, A. (1986). A convenient synthesis of diazo compounds by catalytic dehydrogenation of hydrazones with cobalt Schiff base complex-oxygen system. *Chem. Lett.* 15 (4): 505–506.
84 Pinho e Melo, T.M.V.D. (2010). Synthesis of azides. In: *Organic Azides: Syntheses and Applications* (ed. S. Bräse and K. Banert), 53–94. Wiley-VCH.
85 Bräse, S., Gil, C., Knepper, K., and Zimmermann, V. (2005). Organic azides: an exploding diversity of a unique class of compounds. *Angew. Chem. Int. Ed.* 44 (33): 5188–5240.
86 Bräse, S. and Banert, K. (2010). *Organic Azides*. Weinheim: Wiley-VCH.
87 Jung, N. and Bräse, S. (2012). Vinyl and alkynyl azides: well-known intermediates in the focus of modern synthetic methods. *Angew. Chem. Int. Ed.* 51 (49): 12169–12171.
88 Huryn, D.M. and Okabe, M. (1992). AIDS-driven nucleoside chemistry. *Chem. Rev.* 92 (8): 1745–1768.
89 Minozzi, M., Nanni, D., and Spagnolo, P. (2009). From azides to nitrogen-centered radicals: applications of azide radical chemistry to organic synthesis. *Chem. Eur. J.* 15 (32): 7830–7840.

90 Lapointe, G., Kapat, A., Weidner, K., and Renaud, P. (2012). Radical azidation reactions and their application in the synthesis of alkaloids. *Pure Appl. Chem.* 84 (7): 1633–1641.

91 Chiba, S. (2012). Application of organic azides for the synthesis of nitrogen-containing molecules. *Synlett* 23: 21–44.

92 Scriven, E.F.V. and Turnbull, K. (1988). Azides: their preparation and synthetic uses. *Chem. Rev.* 88 (2): 297–368.

93 Driver, T.G. (2010). Recent advances in transition metal-catalyzed *N*-atom transfer reactions of azides. *Org. Biomol. Chem.* 8 (17): 3831–3846.

94 Chemler, S.R. and Bovino, M.T. (2013). Catalytic aminohalogenation of alkenes and alkynes. *ACS Catal.* 3 (6): 1076–1091.

95 McDonald, R.I., Liu, G., and Stahl, S.S. (2011). Palladium(II)-catalyzed alkene functionalization via nucleopalladation: stereochemical pathways and enantioselective catalytic applications. *Chem. Rev.* 111 (4): 2981–3019.

96 Jensen, K.H. and Sigman, M.S. (2008). Mechanistic approaches to palladium-catalyzed alkene difunctionalization reactions. *Org. Biomol. Chem.* 6 (22): 4083–4088.

97 Besset, T., Poisson, T., and Pannecoucke, X. (2015). Direct vicinal difunctionalization of alkynes: an efficient approach towards the synthesis of highly functionalized fluorinated alkenes. *Eur. J. Org. Chem.* 13: 2765–2789.

98 Romero, R.M., Wöste, T.H., and Muñiz, K. (2014). Vicinal difunctionalization of alkenes with iodine(III) reagents and catalysts. *Chem. Asian J.* 9 (4): 972–983.

99 Valiulin, R.A., Mamidyala, S., and Finn, F.G. (2015). Taming chlorine azide: access to 1,2-azidochlorides from alkenes. *J. Org. Chem.* 80 (5): 2740–2755.

100 Prasad, P.K., Reddi, R.N., and Sudalai, A. (2015). Oxidant controlled regio-and stereodivergent azidohydroxylation of alkenes via I_2 catalysis. *Chem. Commun.* 51 (51): 10276–10279.

101 Sun, X., Li, X., Song, S. et al. (2015). Mn-catalyzed highly efficient aerobic oxidative hydroxyazidation of olefins: a direct approach to β-azido alcohols. *J. Am. Chem. Soc.* 137 (18): 6059–6066.

102 Xu, L., Mou, X.Q., Chen, Z.M., and Wang, S.H. (2014). Copper-catalyzed intermolecular azidocyanation of aryl alkenes. *Chem. Commun.* 50: 10676–10679.

103 Matcha, K., Narayan, R., and Antonchick, A.P. (2013). Metal-free radical azidoarylation of alkenes: rapid access to oxindoles by cascade C-N and C-C bond-forming reactions. *Angew. Chem. Int. Ed.* 52 (31): 7985–7989.

104 Nocquet-Thibault, S., Rayar, A., Retailleau, P. et al. (2015). Iodine(III)-mediated diazidation and azido-oxyamination of enamides. *Chem. Eur. J.* 21 (40): 14205–14210.

105 Lu, M.Z., Wang, C.Q., and Loh, T.P. (2015). Copper-catalyzed vicinal oxyazidation and diazidation of styrenes under mild conditions: access to alkyl azides. *Org. Lett.* 17 (24): 6110–6113.

106 Guerin, D.J., Horstmann, T.E., and Miller, S.J. (1999). Amine-catalyzed addition of azide ion to α, β-unsaturated carbonyl compounds. *Org. Lett.* 1 (7): 1107–1109.

107 Yuan, Y.A., Lu, D.F., Chen, Y.R., and Xu, H. (2016). Iron-catalyzed direct diazidation for a broad range of olefins. *Angew. Chem. Int. Ed.* 55 (2): 534–538.

108 Waser, J., Nambu, H., and Carreira, E.M. (2005). Cobalt-catalyzed hydroazidation of olefins: convenient access to alkyl azides. *J. Am. Chem. Soc.* 127 (23): 8294–8295.

109 Kapat, A., König, A., Montermini, F., and Renaud, P. (2011). A radical procedure for the anti-markovnikov hydroazidation of alkenes. *J. Am. Chem. Soc.* 133 (35): 13890–13893.

110 Leggans, E.K., Barker, T.J., Duncan, K.K., and Boger, D.L. (2012). Iron(III)/$NaBH_4$-mediated additions to unactivated alkenes: synthesis of novel 20′-vinblastine analogues. *Org. Lett.* 14 (6): 1428–1431.
111 Liu, Z., Liao, P., and Bi, X. (2014). General silver-catalyzed hydroazidation of terminal alkynes by combining TMS-N_3 and H_2O: synthesis of vinyl azides. *Org. Lett.* 16 (14): 3668–3671.
112 Liu, Z., Liu, J., Zhang, L. et al. (2014). Silver(I)-catalyzed hydroazidation of ethynyl carbinols: synthesis of 2-azidoallyl alcohols. *Angew. Chem. Int. Ed.* 53 (21): 5305–5309.
113 Song, W., Kozhushkov, S.I., and Ackermann, L. (2013). Site-selective catalytic c(sp2)-H bond azidations. *Angew. Chem. Int. Ed.* 52 (26): 6576–6578.
114 Vita, M.V. and Waser, J. (2015). Cyclic hypervalent iodine reagents and iron catalysts: the winning team for late-stage C-H azidation. *Angew. Chem. Int. Ed.* 54 (18): 5290–5292.
115 Huang, X. and Groves, J.T. (2016). Taming azide radicals for catalytic C-H azidation. *ACS Catal.* 6 (2): 751–759.
116 Grieβ, P. (1864). *Philos. Trans. R. Soc. Lond.* 13: 377.
117 Blanksby, S.J. and Ellison, G.B. (2003). Bond dissociation energies of organic molecules. *Acc. Chem. Res.* 36 (4): 255–263.
118 Kita, Y., Tohma, H., Inagaki, M. et al. (1991). A novel oxidative azidation of aromatic compounds with hypervalent iodine reagent, phenyliodine(III) bis(trifluoroacetate) (PIFA) and trimethylsilyl azide. *Tetrahedron Lett.* 32 (34): 4321–4324.
119 Kita, Y., Tohma, H., Hatanaka, K. et al. (1994). Hypervalent iodine-induced nucleophilic substitution of para-substituted phenol ethers. generation of cation radicals as reactive intermediates. *J. Am. Chem. Soc.* 116 (9): 3684–3691.
120 Telvekar, V.N. and Sasane, K.A. (2012). Simple and efficient method for the preparation of aryl azides using sonication. *Synth. Commun.* 42 (7): 1085–1089.
121 Pragati, P.K., Kalshetti, R.G., Reddi, R.N. et al. (2016). I_2-mediated regioselective C-3 azidation of indoles. *Org. Biomol. Chem.* 14: 3027–3030.
122 Li, P., Zhao, J., Xia, C., and Li, F. (2015). The development of carbene-stabilized N-O radical coupling strategy in metal-free regioselective C-H azidation of quinoline N-oxides. *Org. Chem. Front.* 2: 1313–1317.
123 Lubriks, D., Sokolovs, I., and Suna, E. (2012). Indirect C–H azidation of heterocycles via copper-catalyzed regioselective fragmentation of unsymmetrical λ3-iodanes. *J. Am. Chem. Soc.* 134 (37): 15436–15442.
124 Tang, C. and Jiao, N. (2012). Copper-catalyzed C–H azidation of anilines under mild conditions. *J. Am. Chem. Soc.* 134 (46): 18924–18927.
125 Fan, Y., Wan, W., Ma, G. et al. (2014). Room-temperature Cu(II)-catalyzed aromatic C-H azidation for the synthesis of ortho-azido anilines with excellent regioselectivity. *Chem. Commun.* 50: 5733–5736.
126 Yao, B., Liu, Y., Zhao, L. et al. (2014). Designing a Cu(II)-ArCu(II)-ArCu(III)-Cu(I) catalytic cycle: Cu(II)-catalyzed oxidative arene C-H bond azidation with air as an oxidant under ambient conditions. *J. Org. Chem.* 79 (22): 11139–11145.
127 Xie, F., Qi, Z., and Li, X. (2013). Rhodium(III)-catalyzed azidation and nitration of arenes by C-H activation. *Angew. Chem. Int. Ed.* 52 (45): 11862–11866.
128 Lee, J.G. and Kwak, K.H. (1992). Oxidation of aldehydes to acyl azides by chromic anhydride-azidotrimethylsilane. *Tetrahedron Lett.* 33 (22): 3165–3166.

129 Elmorsy, S.S. (1995). Oxidation of aldehydes to acyl azides using triazidochlorosilane (TACS)-active manganese dioxide reagent. *Tetrahedron Lett.* 36 (8): 1341–1342.

130 Chen, D.J. and Chen, Z.C. (2000). Hypervalent iodine in synthesis. Part 54: one-step conversion of aryl aldehydes to aroyl azides using a combined reagent of (diacetoxyiodo)benzene with sodium azide. *Tetrahedron Lett.* 41 (38): 7361–7363.

131 Arote, N.D. and Akamanchi, K.G. (2007). Direct conversion of aldehydes to acyl azides using *tert*-butyl hypochlorite. *Tetrahedron Lett.* 48 (32): 5661–5664.

132 Bose, D.S. and Reddy, A.V.N. (2003). Iodine(V) reagents in organic synthesis. Dess–Martin periodinane mediated efficient one-pot oxidation of aldehydes to acyl azides. *Tetrahedron Lett.* 44 (17): 3543–3545.

133 Marinescu, L., Thinggaard, J., Thomsen, I.B., and Bols, M. (2003). Radical azidonation of aldehydes. *J. Org. Chem.* 68 (24): 9453–9455.

134 Sarkar, S.D. and Studer, A. (2010). Oxidative amidation and azidation of aldehydes by NHC catalysis. *Org. Lett.* 12 (9): 1992–1995.

135 Shinomoto, Y., Yoshimura, A., Shimizu, H. et al. (2015). Tetra-*n*-butylammonium iodide catalyzed C-H azidation of aldehydes with thermally stable azidobenziodoxolon. *Org. Lett.* 17 (21): 5212–5215.

136 McMillen, D.F. and Golden, D.M. (1982). Hydrocarbon bond dissociation energies. *Annu. Rev. Phys. Chem.* 33: 493–532.

137 Bordwell, F.G. (1988). Equilibrium acidities in dimethyl sulfoxide solution. *Acc. Chem. Res.* 21 (12): 456–463.

138 Magnus, P. and Lacour, J. (1992). New trialkylsilyl enol ether chemistry. Direct beta-azido functionalization of triisopropylsilyl enol ethers. *J. Am. Chem. Soc.* 114 (2): 767–769.

139 Magnus, P., Lacour, J., and Weber, W. (1993). Direct N-alkyl azidonation of N, N-dialkylarylamines with the iodosylbenzene/trimethylsilyl azide reagent combination. *J. Am. Chem. Soc.* 115 (20): 9347–9348.

140 Magnus, P., Hulme, C., and Weber, W. (1994). Alpha-azidonation of amides, carbamates, and ureas with the iodosylbenzene/trimethylsilyl azide reagent combination: N-acyliminium ion precursors. *J. Am. Chem. Soc.* 116 (10): 4501–4502.

141 Kita, Y., Tohma, H., Takada, T. et al. (1994). A novel and direct alkyl azidation of p-alkylanisoles using phenyl iodine(III) bis(trifluoroacetate) (PIFA) and trimethylsilyl azide. *Synlett* 1994 (6): 427–428.

142 Viuf, C. and Bols, M. (2001). Radical azidonation of benzylic positions with iodonium azide. *Angew. Chem. Int. Ed.* 40 (3): 623–625.

143 Pedersen, C.M., Marinescu, L.G., and Bols, M. (2005). Radical substitution with azide: TMSN3-PhI(OAc)2 as a substitute of IN3. *Org. Biomol. Chem.* 3: 816–822.

144 Harschneck, T., Hummel, S., Kirsch, S.F., and Klahn, P. (2012). Practical azidation of 1,3-dicarbonyls. *Chem. Eur. J.* 18 (4): 1187–1193.

145 Galligan, M.J., Akula, R., and Ibrahim, H. (2014). Unified strategy for iodine(III)-mediated halogenation and azidation of 1,3-dicarbonyl compounds. *Org. Lett.* 16 (2): 600–603.

146 Vita, M.V. and Waser, J. (2013). Azidation of β-keto esters and silyl enol ethers with a benziodoxole reagent. *Org. Lett.* 15 (13): 3246–3249.

147 Deng, Q.H., Bleith, T., Wadepohl, H., and Gade, L.H. (2013). Enantioselective iron-catalyzed azidation of β-keto esters and oxindoles. *J. Am. Chem. Soc.* 135 (14): 5356–5359.

148 Zhdankin, V.V., Krasutsky, A.P., Kuehl, C.J. et al. (1996). Preparation, X-ray crystal structure, and chemistry of stable azidoiodinanes derivatives of benziodoxole. *J. Am. Chem. Soc.* 118 (22): 5192–5197.

149 Sharma, A. and Hartwig, J.F. (2015). Metal-catalysed azidation of tertiary C-H bonds suitable for late-stage functionalization. *Nature* 517: 600–604.

150 Huang, X., Bergsten, T.M., and Groves, J.T. (2015). Manganese-catalyzed late-stage aliphatic C-H azidation. *J. Am. Chem. Soc.* 137 (16): 5300–5303.

151 Zhang, X., Yang, H., and Tang, P. (2015). Transition-metal-free oxidative aliphatic C-H azidation. *Org. Lett.* 17 (23): 5828–5831.

152 Wang, Y., Li, G.X., Yang, G. et al. (2016). A visible-light-promoted radical reaction system for azidation and halogenation of tertiary aliphatic C-H bonds. *Chem. Sci.* 7: 2679–2683.

153 Rabet, P.T., Fumagalli, G., Boyd, S., and Greaney, M.F. (2016). Benzylic C-H azidation using the ahdankin reagent and a copper photoredox catalyst. *Org. Lett.* 18 (7): 1646–1649.

154 Emmett, M.R., Grover, H.K., and Kerr, M.A. (2012). Tandem ring-opening decarboxylation of cyclopropane hemimalonates with sodium azide: a short route to γ-aminobutyric acid esters. *J. Org. Chem.* 77 (15): 6634–6637.

155 Haveli, S.D., Roy, S., Vibha, G. et al. (2013). Ring opening of activated cyclopropanes with NIS/NaN3: synthesis of C-1 linked pseudodisaccharides. *Tetrahedron* 69 (52): 11138–11143.

156 Ivanov, K.L., Villemson, E.V., Budynina, E.M. et al. (2015). Ring opening of donor-acceptor cyclopropanes with the azide ion: a tool for construction of N-heterocycles. *Chem. Eur. J.* 21 (13): 4975–4987.

157 Jiao, J., Nguyen, L.X., Patterson, D.R., and Flowers, R.A. II (2007). An efficient and general approach to β-functionalized ketones. *Org. Lett.* 9 (7): 1323–1326.

158 Ren, R., Zhao, H., Huan, L., and Zhu, C. (2015). Manganese-catalyzed oxidative azidation of cyclobutanols: regiospecific synthesis of alkyl azides by C-C bond cleavage. *Angew. Chem. Int. Ed.* 54 (43): 12692–12696.

159 Masterson, D.S. and Porter, N.A. (2002). Diastereoselective free radical halogenation, azidation, and rearrangement of β-silyl Barton esters. *Org. Lett.* 4 (24): 4253–4256.

160 Nyfeler, E. and Renaud, P. (2008). Decarboxylative radical azidation using MPDOC and MMDOC esters. *Org. Lett.* 10 (5): 985–988.

161 Klahn, P., Erhardt, H., Kotthaus, A., and Kirsch, S.F. (2014). The synthesis of α-azidoesters and geminal triazides. *Angew. Chem. Int. Ed.* 53 (30): 7913–7917.

162 Liu, C., Wang, X., Li, Z. et al. (2015). Silver-catalyzed decarboxylative radical azidation of aliphatic carboxylic acids in aqueous solution. *J. Am. Chem. Soc.* 137 (31): 9820–9823.

163 Zhu, Y., Li, X., Wang, X. et al. (2015). Silver-catalyzed decarboxylative azidation of aliphatic carboxylic acids. *Org. Lett.* 17 (19): 4702–4705.

9

Green Oxidative Synthesis of Substituted Olefins and Alkynes

Yasushi Obora

Department of Chemistry and Materials Engineering, Faculty of Chemistry, Materials and Bioengineering, Kansai University, Suita, Japan

9.1 Introduction

Functionalization of alkenes and alkynes leading to versatile compounds is an important transformation for organic synthesis. Furthermore, considerable attention has been paid to the development of transition-metal-catalyzed cross-coupling reaction of vinyl halides and alkynyl halides to create a variety of functionalized alkenes and alkynes. This reaction is now well-known as the Mizoroki–Heck reaction [1].

In contrast, development of transition-metal-catalyzed carbon–carbon bond formation by means of carbon–hydrogen bond activation in lieu of the conventional cross-coupling reaction of organic halides and organometallic reagents has been developed as a green alternative for providing a variety of functionalized unsaturated compounds [2].

Thus, this alternative approach to the conventional Mizoroki–Heck type cross-coupling methodology, alkene/alkyne functionalization through an oxidative-coupling reaction, presents a green protocol for providing a wide variety of alkenes/alkynes (Scheme 9.1). For this reaction, various oxidants have been employed, and the greener method is the use of molecular oxygen as the sole oxidant, which produces water as the sole byproduct.

The achievement of a practical methodology of transition-metal-catalyzed C—H bond functionalization through direct C—H bond activation by using a mild oxidant would provide a practical and efficient operation for preparing functionalized alkenes and alkynes from easily accessible feedstock, which would be the practical starting materials. Therefore, for an architecture of organic molecules having unsaturated moiety, functionalization of easily accessible alkenes and alkynes would be useful.

The palladium(II)-catalyzed aerobic oxidative-coupling reaction of ethylene to produce acetone by using copper as co-catalyst is referred to as the Wacker process [3]. This process was successfully extended to the oxidative amination of alkenes, namely Aza–Wacker reactions [4]. As for the oxidative amination reaction of alkenes by means of the Aza–Wacker reaction, after initial reports [5], it has become the focus of increasing research attention (Scheme 9.2) [4, 6].

Green Oxidation in Organic Synthesis, First Edition. Edited by Ning Jiao and Shannon S. Stahl.

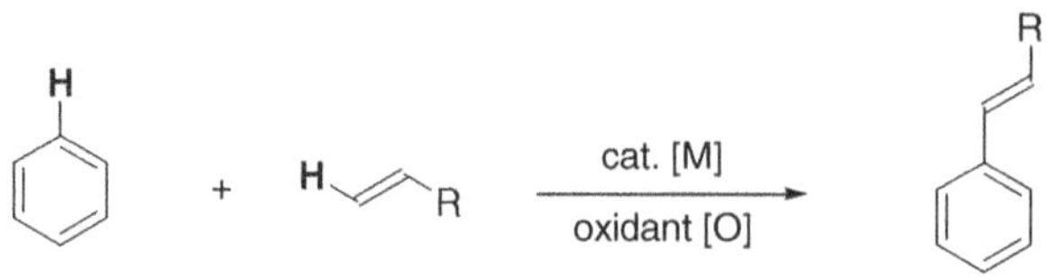

Scheme 9.1 Oxidative/dehydrogenative Heck (Fujiwara–Moritani) reaction.

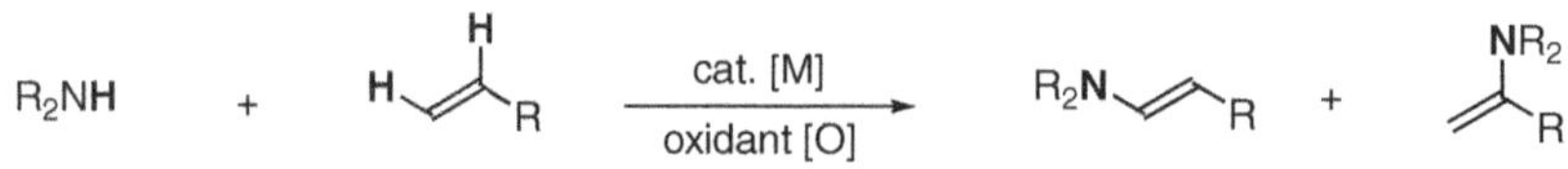

Scheme 9.2 Oxidative amination (Aza–Wacker) reaction.

To achieve these oxidative-Heck/Aza–Wacker reactions, a variety of oxidants has been used. The choice of oxidant is crucial for the evaluation of the green process. Therefore, the use of molecular oxygen (or air) as an oxidant is considered to be ideal and the most preferable alternative from the point of the green and environmentally benign oxidative-coupling protocol for the formation of functionalization of alkenes and alkynes owing to minimized formation of by byproducts in the oxidation process (ideally, water).

In this chapter, we focus on the recent progress of aerobic oxidative alkene and alkyne functionalization by using atmospheric (1 atm) molecular oxygen (or air) as a sole oxidant. This protocol led to the green oxidation methodology for C—C bond formation reaction with water as the sole byproduct. This process mainly involves direct/dehydrogenative Heck and Wacker type processes.

9.2 Oxidative Aryl Functionalization of Alkenes Using Molecular Oxygen as an oxidant

Oxidative aryl functionalizations of alkenes are known as the oxidative Fujihara–Moritani reaction and the direct/dehydrogenative Heck reaction [7]. After the pioneering results of Fujiwara and Moritani's work [8], considerable attention has been paid to oxidative aryl C—H alkenylation. As for oxidative-coupling by using molecular oxygen as the terminal oxidant, the additives include peroxides, benzoquinone, DDQ, $PhI(OAc)_2$, and inorganic salts [9].

The following are selective examples of oxidative alkenylation by using atmospheric dioxygen (O_2) or air as the oxidant.

9.3 Oxidative Aryl Functionalization of Alkenes in the Presence of a Catalytic Amount of Polyoxometalates

Ishii/Obora and Huang/Lin's group reported aerobic arene alkenylations in the presence of small amount of polyoxometalates as a catalyst [10].

As a typical example of Ishii's works, the reaction of benzene (20 equiv.) with ethyl acrylate (1 equiv.) in the presence of a catalytic amount of $Pd(OAc)_2$ (6.7 mol%) combined with $H_4PMo_{11}VO_{40}{\cdot}30H_2O$ (1.3 mol%), actylacetone (6.7 mol%), and NaOAc (5.3 mol%) under 1 atm of O_2 in propionic acid gives the corresponding mono- and diarylated products in 74 and 13% yields, respectively (Scheme 9.3) [11].

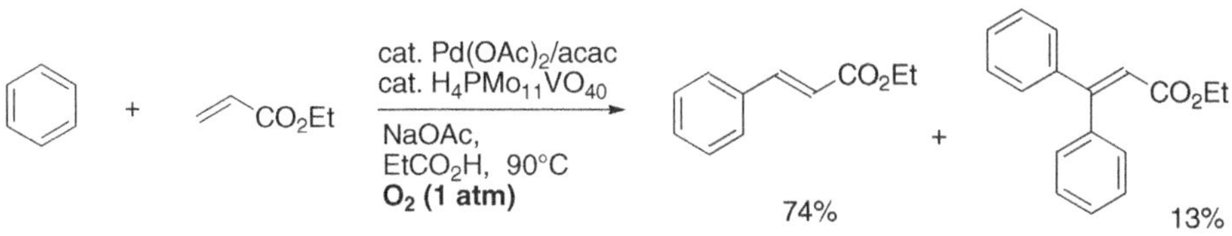

Scheme 9.3 Pd-catalyzed oxidative-Heck reaction of arenes with ethyl acrylate.

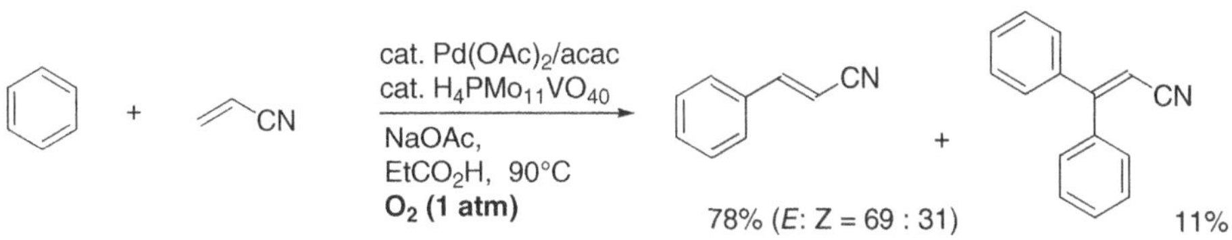

Scheme 9.4 Pd-catalyzed oxidative-Heck reaction of arenes with acrylonitrile.

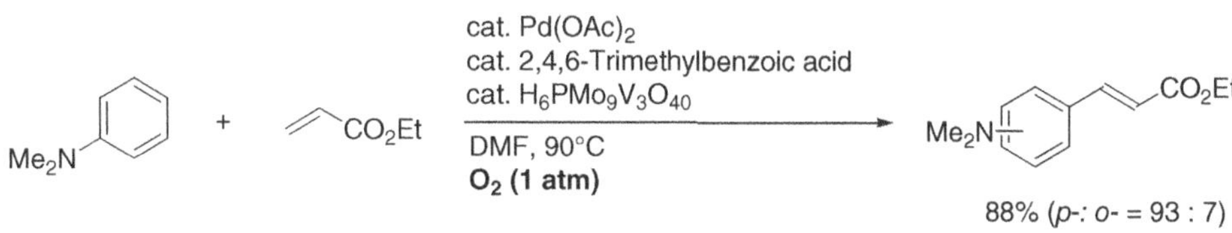

Scheme 9.5 Pd-catalyzed oxidative-Heck reaction of *N,N*-dimethylaniline.

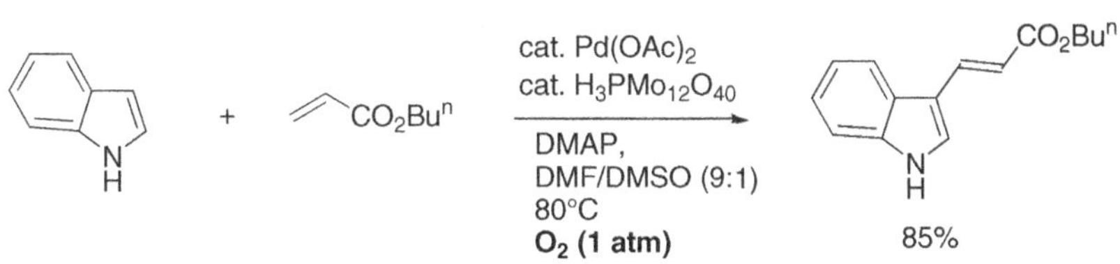

Scheme 9.6 Pd-catalyzed oxidative-Heck reaction of indole.

The Pd(II)/polyoxometalates/O_2 with acid solvent catalyst system can be extended to the oxidative aryl functionalization of acrylonitrile (Scheme 9.4) [12].

Obora and co-workers reported that the use of 2,4,6-trimethylbenzoic acid as an additive effectively enhanced the catalytic activity of Pd(II)/polyoxometalates/O_2 catalyst system even using non-acidic solvents. This reaction system can successfully cause the oxidative-coupling of basic aryl substrate with alkenes under 1 atm of O_2 as the oxidant. Typically, the reaction of *N,N*-dimethylaniline (7.5 equiv.) with ethyl acrylate (1 equiv.) in DMF under 1 atm of O_2 gave the alkenylated product at an 88% yield (Scheme 9.5) [13].

Huang, Lin, and co-workers reported that the oxidative-coupling of indole (1 equiv.) with *n*-butyl acrylate (2 equiv.) in the presence of $Pd(OAc)_2$ (10 mol%) combined with $H_3PMo_{12}O_{40}$ (5 mol%, as polyoxometalate) and 4-dimethylaminopyridine (DMAP) (20 mol%) as the additive under the influence of atmospheric O_2 led to 3-vinylindoles in 85% yield with high regioselectivity (Scheme 9.6) [14].

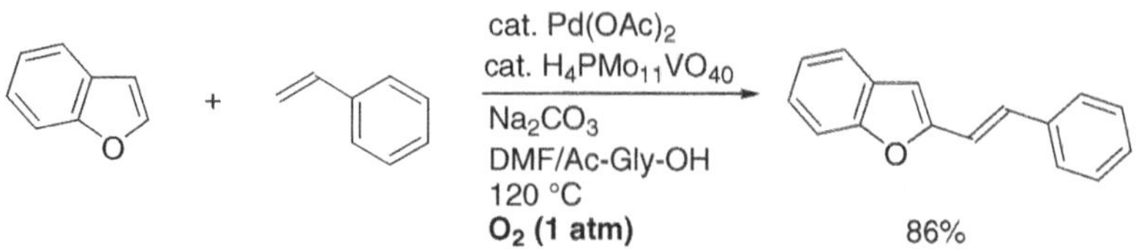

Scheme 9.7 Pd-catalyzed oxidative-Heck reaction of benzofuran.

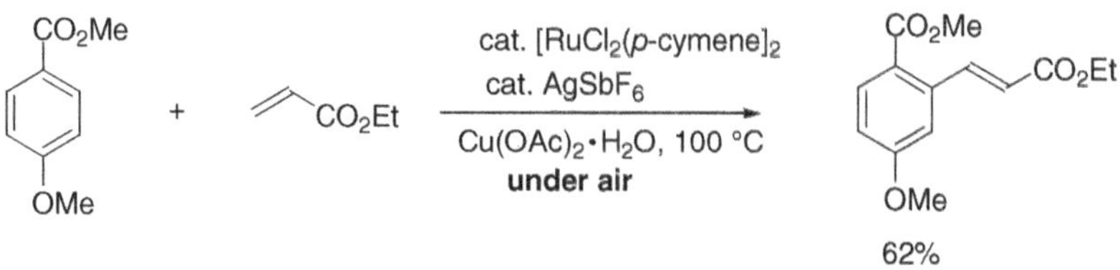

Scheme 9.8 Ru-catalyzed oxidative-Heck reaction of aryl ester.

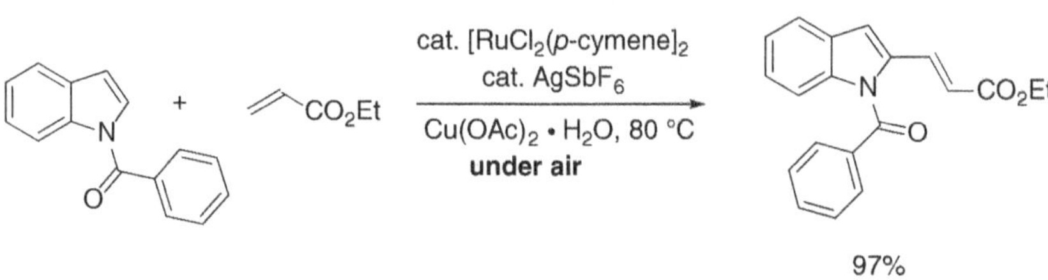

Scheme 9.9 Ru-catalyzed C2-selective oxidative-Heck reaction of indoles.

Oxidative introduction of benzofuranes into alkenes by using 1 atm of O_2 as the oxidant was achieved by the use of a Pd(II)/polyoxometalate catalyst. Here, addition of monoprotected amino acid as the additive effectively catalyzed the reaction (*vide infra)* to afford the corresponding alkenylation product at a high yield (Scheme 9.7) [15].

9.4 Oxidative Aryl Functionalization of Alkenes in the Presence of Metal Additives Using Air as the Oxidant

Aerobic aryl alkenylation by using atmospheric air as the oxidant in the presence of a metal additive has been reported. As for selected examples of this type of reaction system, Ackermann and co-workers reported that the oxidative alkenylation of aryl esters was achieved. Typically, reaction of aryl ester (1 equiv.) with ethyl acrylate (2 equiv.) in the presence of $[RuCl_2(p\text{-cymene})]_2$ (5 mol%), $AgSbF_6$ (40 mol%) and $Cu(OAc)_2{\cdot}H_2O$ (1.0 mmol) in 1,2-dichloroethane gave the alkenylated product in 62% yield (Scheme 9.8) [16].

By using the Ru/$AgSbF_6$/$Cu(OAc)_2$ catalyst system, Prabhu and co-workers reported the regioselective oxidative-coupling of indoles with acrylates to produce a C2-alkenylated product in high yield and regioselectivity (Scheme 9.9) [17].

9.5 Oxidative Aryl Functionalization of Alkenes by Using 1 atm of Dioxygen as the Sole Oxidant with Mono-Protonated Amino Acids (MPAAs) as the Promoter

As an alternative protocol for achieving arylation of alkene by using 1 atm of O_2 as the sole oxidant, the addition of MPAAs is known to be effective for a transition-metal-catalyzed C—H activation reaction [18].

As an example of this reaction, Huang reported the $Pd(OAc)_2$-catalyzed dehydrogenative coupling of polufluoroarenes with alkenes by using molecular oxygen as the sole oxidant. Typically, the reaction of pentafluorobenzene (2.5 equiv.) with ethyl acrylate (1 equiv.) in the presence of $Pd(OAc)_2$ (10 mol%) combined with DL-pGlu-OH as MPAA (10 mol%) and PivOH (PivOH = pivalic acid) (1.5 equiv.) in DMA (DMA = *N*, *N*-dimethylacetamide) at 120 °C gave the arylated product in 91% yield (Scheme 9.10) [19].

The same group also reported that the Pd-catalyzed aerobic C—H alkenylations of uracils and caffeine by 1 atom of O_2 as an oxidant were achieved by using the catalytic amount of MPAAs as an additive (Schemes 9.11 and 9.12) [20].

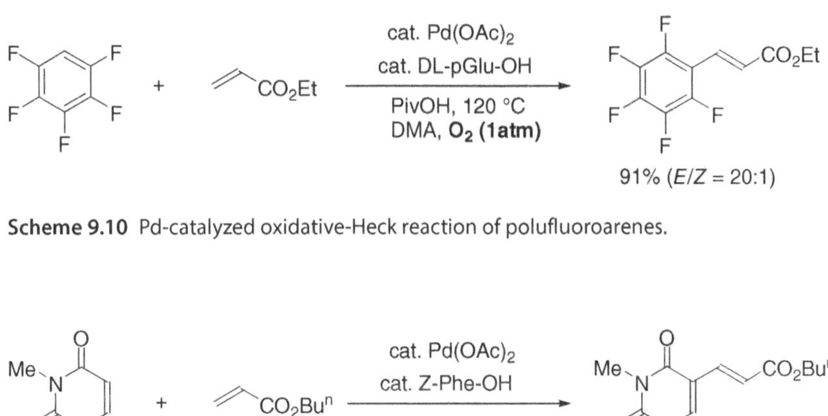

Scheme 9.10 Pd-catalyzed oxidative-Heck reaction of polufluoroarenes.

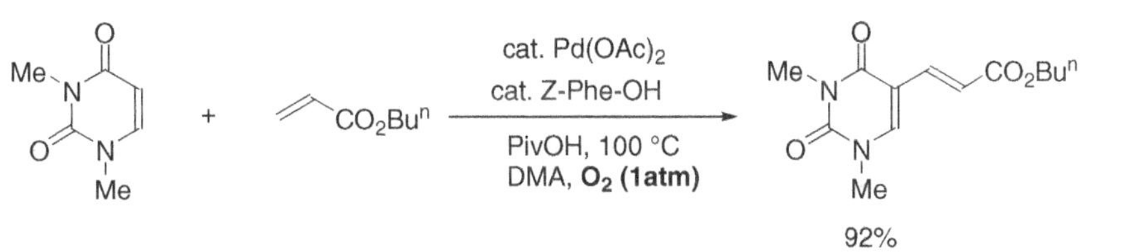

Scheme 9.11 Pd-catalyzed oxidative-Heck reaction of uracils.

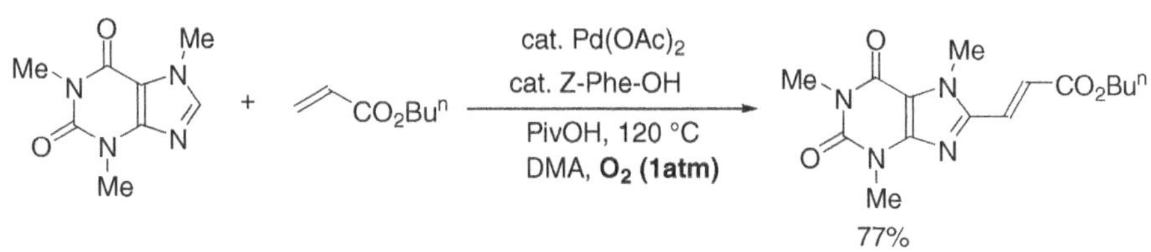

Scheme 9.12 Pd-catalyzed oxidative-Heck reaction of caffeine.

9.6 Oxidative Aryl Functionalization of Alkenes by Using 1 atm of Dioxygen as the Sole Oxidant

Oxidative aryl functionalization of alkenes by using 1 atm O_2 as an oxidant was achieved even in the absence of the cooxidant additives.

As for selective examples of these reactions, Yu and co-workers reported that the reaction of 1,3-bis(trifluoromethyl)benzene (as electron-deficient arene) (20 equiv.) with alkene (1 equiv.) in the presence of $Pd(OAc)_2$ (10 mol%), 2,6-sisubstituted pyridine ligand **L** (20 mol%), and 1 equiv. of Ac_2O under the 1 atm of O_2 at 90 °C gave the oxidative alkenylation product (Scheme 9.13) [21].

Hong and co-workers reported the Pd(II)-catalyzed alkenylation of coumarins by aerobic oxidative-Heck reactions. For example, the reaction of 7-methoxycoumarin (1 equiv.) with *tert*-butyl acrylate (2 equiv.) in the presence of $Pd(OPiv)_2$ (Piv = Pivaloyl) (20 mol%) combined with K_2CO_3 (3 equiv.) as a base under 1 atm of O_2, gave the alkenylated products in 81% yield with regioselectivity at the C3 position of the coumarin (Scheme 9.14) [22].

Shi and co-workers reported oxidative alkenylation of phenol derivatives by using molecular oxygen as an oxidant. For example, the reaction of 2-phenoxypyridine (1 equiv.) with ethyl acrylate (2 equiv.) in the presence of $Pd(OAc)_2$ (10 mol%) as catalyst under 1 atm of O_2 in $MeNO_2$ and DMPU (DMPU = *N,N*-dimethylpropyleneurea) gave alkenylated product in 73% yield (Scheme 9.15). The 2-pyridinyl group can be easily deprotected to convert the corresponding *o*-alkenyl substituted phenol [23].

Wang and co-workers reported oxidative alkenylation of indoles by using molecular oxygen as an oxidant. Typically, the reaction of *N*-methyl indole (1 equiv.) and ethyl acrylate (1.5 equiv.) was carried out under the 1 atm of O_2 as an oxidant in the presence

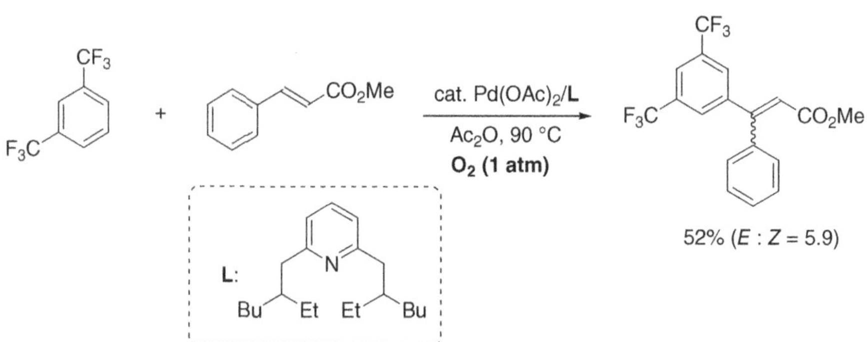

Scheme 9.13 Pd-catalyzed oxidative-Heck reaction of electron-deficient arenes.

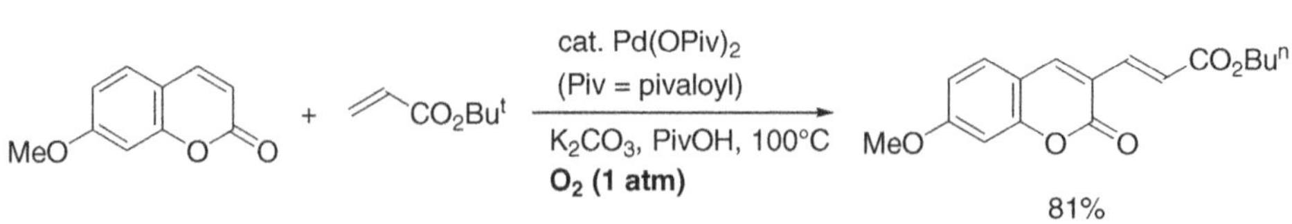

Scheme 9.14 Pd-catalyzed oxidative-Heck reaction of coumarins.

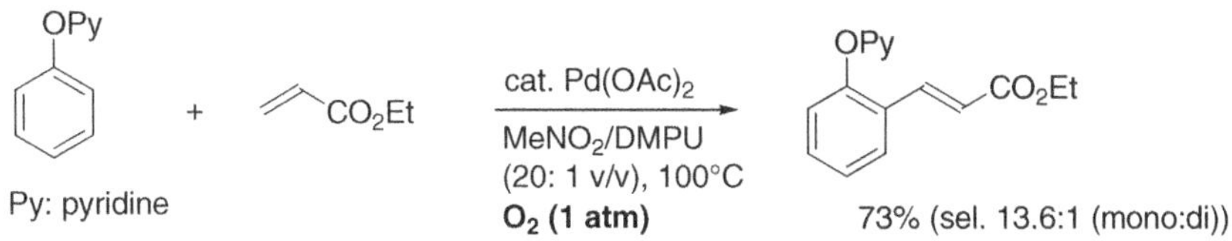

Scheme 9.15 Pd-catalyzed oxidative-Heck reaction of 2-phenoxypyridine.

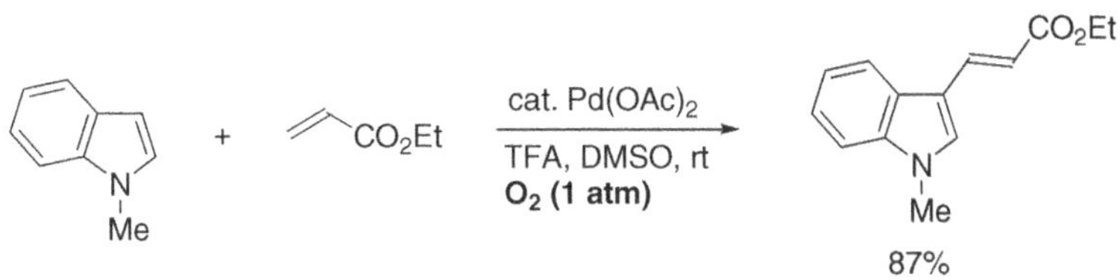

Scheme 9.16 Pd-catalyzed C3-selective oxidative-Heck reaction of indoles.

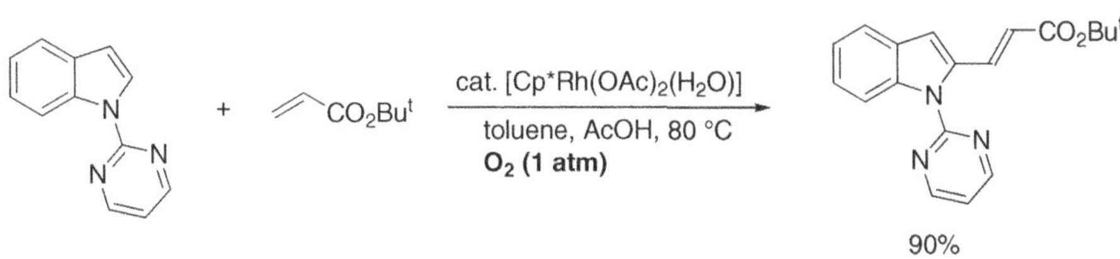

Scheme 9.17 Rh-catalyzed C2-selective oxidative-Heck reaction of indoles.

of $Pd(OAc)_2$ (10 mol%) as a catalyst combined with 8 equiv. of trifluoroacetic acid (TFA) in DMSO at room temperature provided the alkenylation product at a high C3 regioselectivity (Scheme 9.16) [24].

Huang and co-workers reported oxidative C-2 alkenylation of indoles and alkenes using 1 atm of O_2 as an oxidant catalyzed Rh(III) complex. Typically, the reaction of *N*-pyrimidylindole (1 equiv.) with *tert*-butyl acrylate (2 equiv.) was carried out in the presence of $[Cp^*Rh(OAc)_2(H_2O)]$ (1 mol%) combined with AcOH (1 equiv.) in toluene at 80 °C under 1 atm of O_2 and gave the C-2 alkenylated product in 90% yield with high regioselectivity (Scheme 9.17) [25].

Obora and co-workers reported the ligand-controlled chemoselective oxidative aryl alkenylation from benzene and acrylamide by using molecular oxygen as the sole oxidant. Namely, the reaction of benzene (40 equiv.) with *N,N*-dimethylacrylamide (1 equiv.) in the presence of $Pd(DBA)_2$ (6.7 mol%) combined with acetylacetone (acacH) in AcOH under 1 atm of O_2 to produce the alkynylated product. In this reaction the Pd/acacH molar ratio is crucial to the chemoselectivity of the products. Thus, when the reaction was carried out by using a 1:1 ratio of Pd/acacH, diarylsubstituted alkene formed as the main product (82% yield), whereas in the 1:4 ratio of Pd/acacH, a monosubstituted alkene was formed as the main product (71% yield) (Scheme 9.18) [26].

Lu, Sun, Yu, and co-workers reported the substrate-controlled chemoselectivity of the Rh-catalyzed oxidative alkenylation by using air as an oxidant. Namely, the reaction of the benzamide (1 equiv.) with styrene (2.5 equiv.) in the presence of $[Cp^*RhCl_2]_2$ (5 mol%) combined with NaOPiv (1.0 equiv.) in acetonitrile at 80 °C under air as an

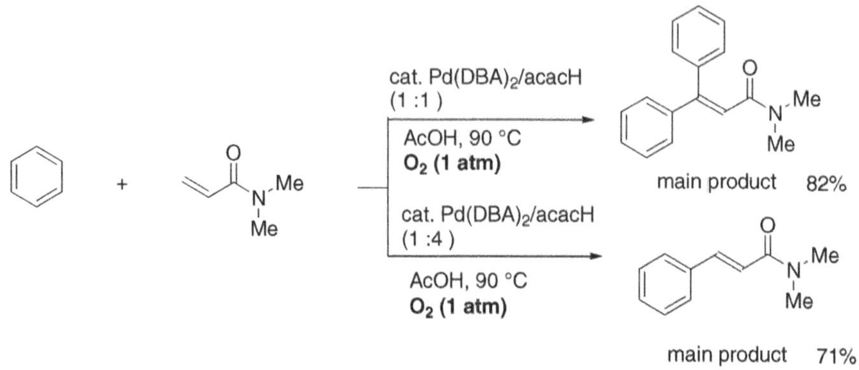

Scheme 9.18 Pd-catalyzed oxidative-Heck reaction of benzene with acrylamide.

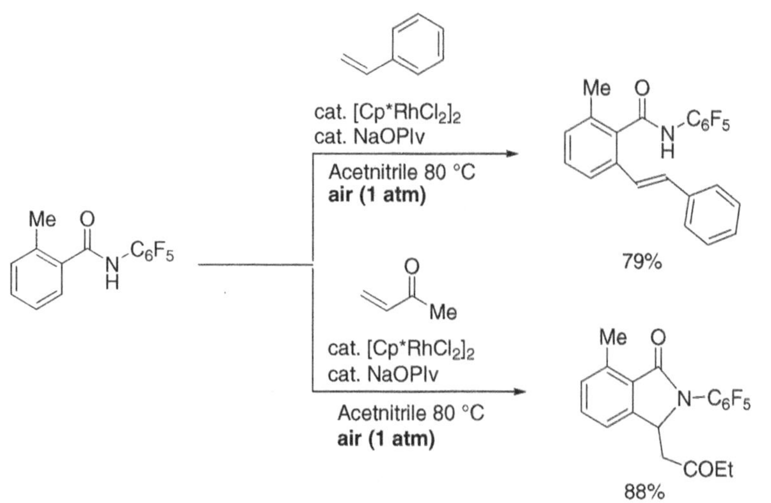

Scheme 9.19 Rh-catalyzed oxidative-Heck reaction of amides.

oxidant gave alkenylated product in 79% yield, whereas the use of ethyl vinyl ketone under these conditions gave the cyclized product in 88% yield (Scheme 9.19) [27].

9.7 Oxidative Aryl Functionalization of Alkenes by Using 1 atm of Dioxygen as the Sole Oxidant with a Palladium-Complex Having a 4,5-Diazafluorenone Ligand

In 2010, Stahl and co-workers reported the Pd-catalyzed allylic C—H acetoxylation reaction by using 1 atm of O_2 as an oxidant [28]. To effectively catalyze this aerobic coupling, 4,5-diazafluorenone (DAF) was found to be an effective ligand (Scheme 9.20) [29].

The $Pd(OAc)_2$-DAF catalyst system is efficient for oxidative-Heck-type reactions.

For example, Zakzewski and co-workers reported oxidative alkenylation of ferrocene derivatives. Following typical experimental procedure, the reaction of ferrocene with

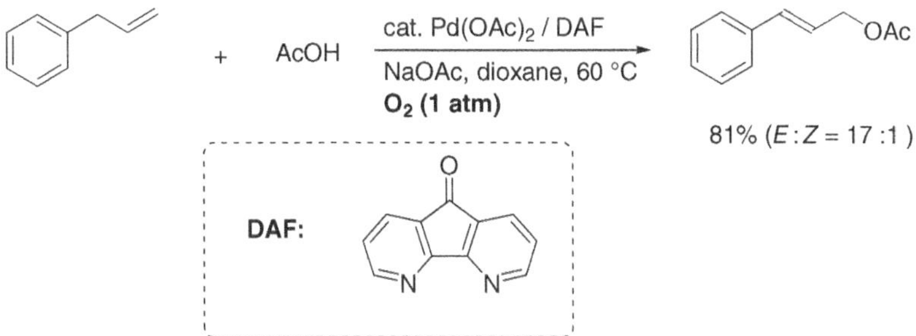

Scheme 9.20 Pd-catalyzed allylic C—H acetoxylatyion.

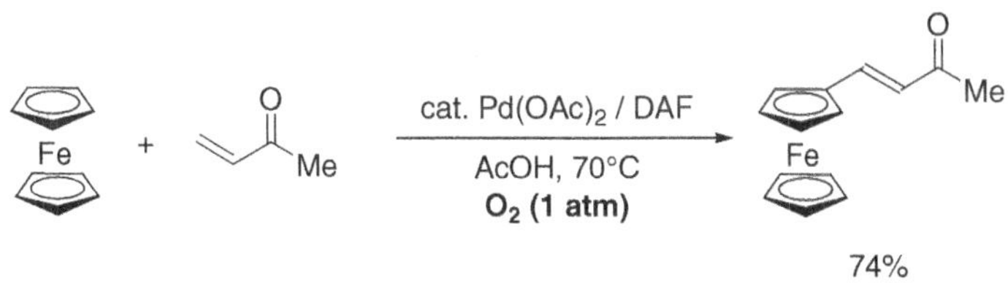

Scheme 9.21 Pd-catalyzed oxidative-Heck reaction of ferrocene.

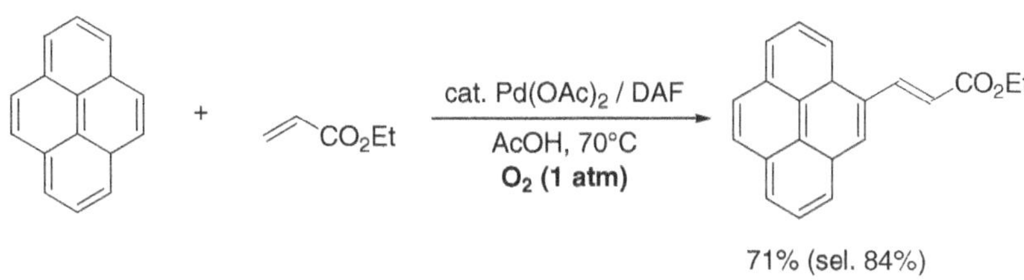

Scheme 9.22 Pd-catalyzed oxidative-Heck reaction of pyrenyl fluorophores.

methyl vinyl ketone in the presence of $Pd(OAc)_2$ (5 mol%) combined with DAF (5 mol%) as ligand under 1 atm of O_2 gave the oxidative alkenylated product in 74% yield (Scheme 9.21) [30].

The same group reported the $Pd(OAc)_2$-DAF catalyzed oxidative alkenylation of pyrenyl fluorophores. Typically, a mixture of pyrene (3 equiv.) and ethyl acrylate (1 equiv.) was allowed to react under 1 atm of O_2 in the presence of $Pd(OAc)_2$ (5 mol%) combined with DAF (5 mol%) and the ligand produced alkenylated pyrene in a good yield (Scheme 9.22) [31]. The resulting product exhibited fluorescence both in solution and solid state.

Le Bras and co-workers reported the $Pd(OAc)_2$-DAF catalyst system as effective for oxidative alkenylation of furans and thiophenes. Following a typical experimental procedure, the reaction of 2-methylthiophene (3 equiv.) with methyl cinnamate (1 equiv.) in the presence of $Pd(OAc)_2$ (10 mol%) combined with a DAF ligand (10 mol%) under 1 atm of O_2 led to the formation of the thiophene-alkenylation product in 90% yield (Scheme 9.23) [32].

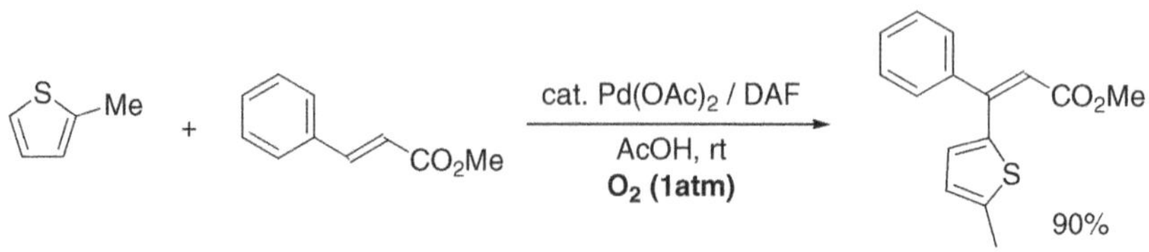

Scheme 9.23 Pd-catalyzed oxidative-Heck reaction of thiophenes.

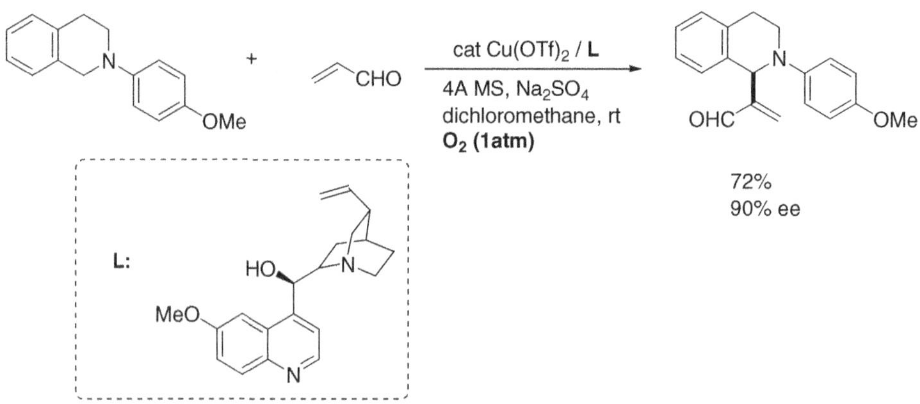

Scheme 9.24 Cu/organo-catalyzed enantioselective oxidative sp^3 C—H alkenylation.

9.8 Oxidative sp^3 C—H Alkenylation by Using 1 atm of O_2 as the Sole Oxidant

Wang and co-workers reported that Cu/organo-catalyzed enantioselective oxidative sp^3 C—H alkenylation of tertiary amines by using 1 atm O_2 as the sole oxidant. A typical reaction of aryl-substituted tetrahydroisoquinolines (1 equiv.) with acrylaldehyde (2 equiv.) in the presence of $Cu(OTf)_2$ (10 mol%) combined with quinine derivatives as organocatalysts (20 mol%), Na_2SO_4 (3 equiv.), and 4 Å MS under 1 atm of O_2 at room temperature gave the coupling products in 72% yield with high enantioselectivity (90% *ee*) (Scheme 9.24) [33].

9.9 Oxidative Cyclization of Alkene by Using 1 atm of O_2 as the Oxidant

Liu and co-workers reported oxidative alkene C—H functionalization-annulation of glycine derivatives by using molecular oxygen as an oxidant. Typically, the reaction of glycine ester (1 equiv.) with styrene (2 equiv.) in the presence of *N*-hydroxyphtalimide (NHPI) (20 mol%) as catalyst combined with $Cu(OTf)_2$ (20 mol%) under 1 atm of O_2 led to the quinoline derivatives as the product in 80% yield (Scheme 9.25) [34].

With regard to the reaction mechanism, Cu-catalyzed oxidation (dehydrogenation) of a glycidine ester was initiated in the reaction to provide the intermediate **A**.

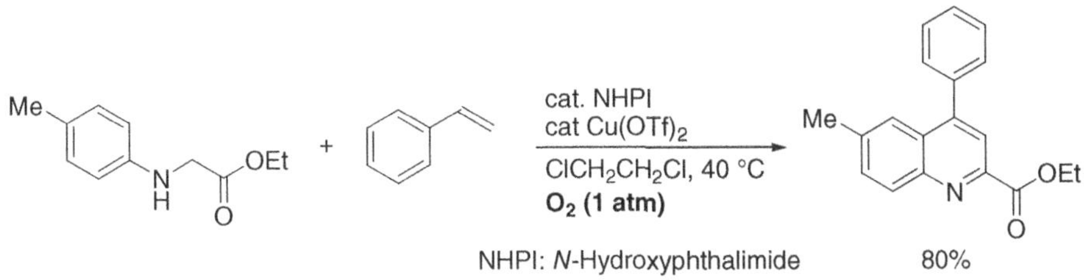

Scheme 9.25 Cu-catalyzed oxidative cyclization.

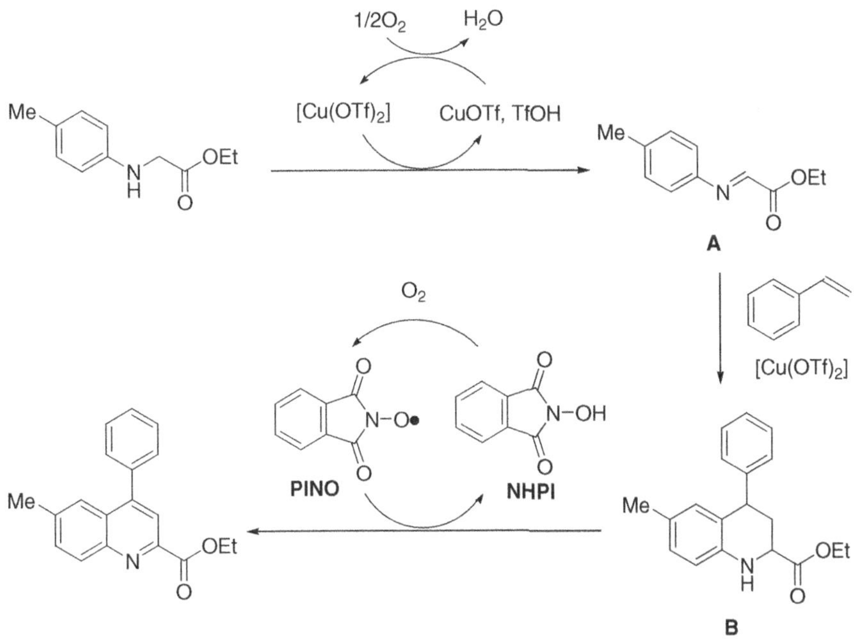

Scheme 9.26 The proposed mechanism.

Subsequent Cu-catalyzed reaction with styrene led to the intermediate **B**, followed by the aromatization by dehydrogenation with a NHPI/O_2 catalyst system, which led to the quinoline derivatives as the product (Scheme 9.26).

9.10 Alkene Functionalization by Using Oxidative Amination of Alkenes

Oxidative functionalization of the Aza–Wacker process provides a useful and green methodology for accessing the amino-group functionalized alkenes.

In 1992, Murahashi and co-workers reported a Pd(II)/Cu(II)-catalyzed enamide synthesis by oxidative amination of activated alkenes using molecular oxygen as an oxidant [35]. Thereafter, Chang and co-workers reported *Z*-selective oxidative amidation of alkenes in the presence of Pd(II)-catalyst combined with a phosphonate ligand by using molecular oxygen as the oxidant [36].

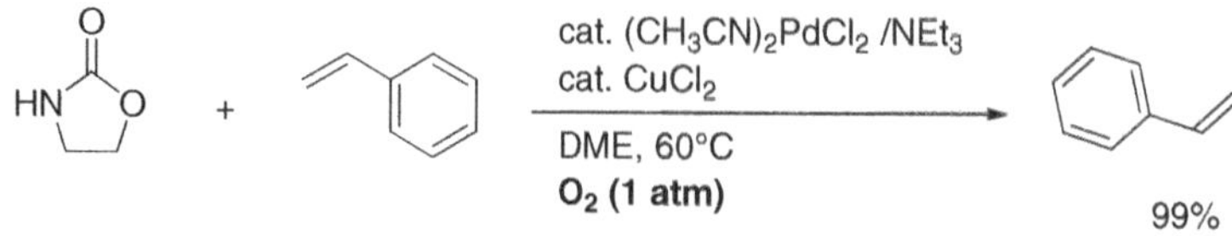

Scheme 9.27 Pd-catalyzed intermolecular oxidative amination of styrene.

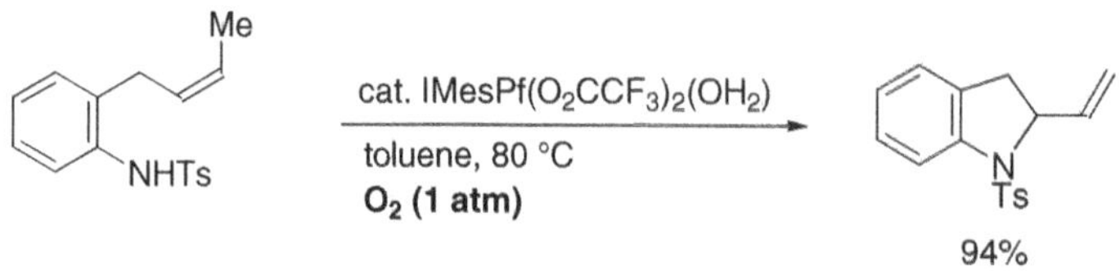

Scheme 9.28 Pd-catalyzed intramolecular oxidative amination of alkenes.

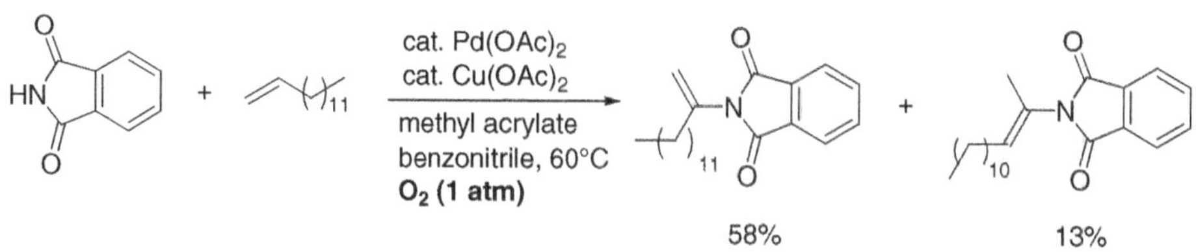

Scheme 9.29 Pd/Cu co-catalyzed oxidative olefination of phthalimide.

In addition, in 2003, Stahl and co-workers reported Pd-catalyzed oxidative amination of styrene by using 1 atm of O_2 as an oxidant. In a typical example, the reaction of oxazolidinone (1 equiv.) with styrene (6 equiv.) was carried out in the presence of $(CH_3CN)_2PdCl_2$ (5 mol%) combined with triethylamine (5 mol%) and $CuCl_2$ (2.5 mol%) as catalysts under 1 atm of O_2 as the oxidant and this produced the alkenylated product in 99% yield (Scheme 9.27) [37].

The same group reported intramolecular oxidative amination of alkenes by using 1 atm of O_2 as the oxidant and $(NHC)Pd(O_2CCF_3)_2(OH_2)$ as a Pd complex catalyst bearing a *N*-heterocyclic carbene ligand (Scheme 9.28) [38].

As for the oxidative amination of unactivated, simple alkenes, Stahl and co-workers reported that the reaction of phthalimide with tetradecene in the presence of $Pd(OAc)_2$ (5 mol%), $Cu(OAc)_2$ (5 mol%), and methylacrylate (1 equiv.) as an additive under 1 atm of O_2 led to oxidative alkenylated products (Scheme 9.29) [39].

Obora, Ishii, and co-workers reported oxidative amination of electron-deficient alkenes by using a secondary amine with 1 atm of molecular oxygen as the oxidant. As usual, the reaction of diphenylamine (1 equiv.) with ethyl acrylate (3 equiv.) was carried out in the presence of $PdCl_2(PhCN)_2$ complex (5 mol%) combined with molybdovanadophosphate (1 mol%), $(NH_4)_5H_4PMo_6V_6O_{40}$ as molybdovanadophosphate (1 mol%), and hydroquinone (20 mol%) in DMF at 60 °C, which gave the corresponding alkenylation product in 90% yield (Scheme 9.30) [40].

The reaction proceeded in an Aza–Wacker type manner, as shown in Scheme 9.31.

Initial ethyl acrylate coordination to $PdCl_2$ generated the palladium(II)-olefin complex. Subsequent reaction of palladium(II)-olefin complex by nucleophilic attack

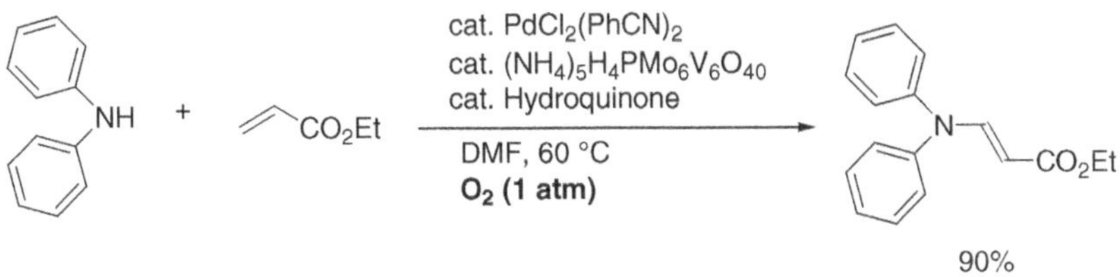

Scheme 9.30 Pd-catalyzed oxidative olefination of secondary amines.

Scheme 9.31 The proposed mechanism.

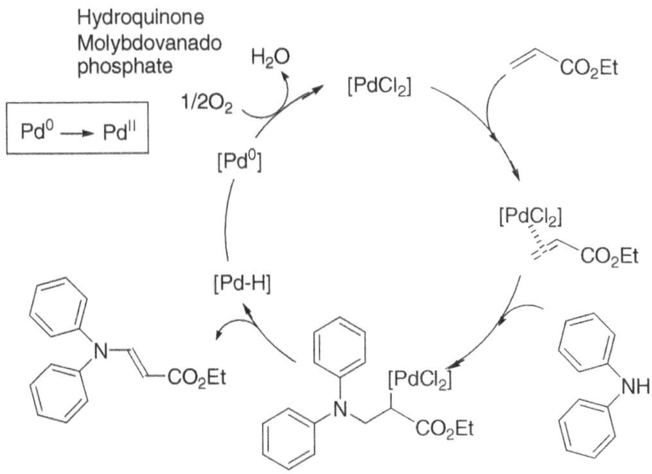

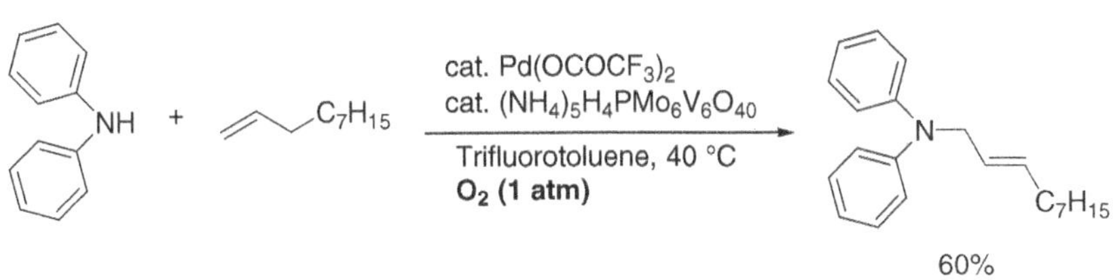

Scheme 9.32 Pd-catalyzed oxidative amination reaction of unactivated alkenes.

of the amine formed an aminopalladation product, which, followed by β-hydride elimination, afforded an amine alkylated product. Hydroquinone and molybdovanadophosphate efficiently served as a promoter of the oxidation step of Pd(0) to Pd(II) by 1 atm of O_2 as an oxidant (Scheme 9.31).

The same group reported the oxidative amination reaction of an unactivated alkene to afford an enamine product. As an example, the reaction of diphenylamine (1 equiv.) with 1-decene (7 equiv.) in the presence of $Pd(OCOCF_3)_2$ (10 mol%) catalyst combined with $(NH_4)_5H_4PMo_6V_6O_{40}$ (2 mol%) in trifluorotoluene at 40 °C gave the corresponding enamine in 60% yield (Scheme 9.32) [41].

As for the oxidative alkene amination by using primary amines with molecular oxygen as a sole oxidant, Obora and Jiang independently reported Pd-catalyzed reaction of aniline derivatives with electron-deficient alkenes. In the example provided by Obora's work, the reaction of *o*-toluidine (3 equiv.) with butyl acrylate (1 equiv.) in the presence

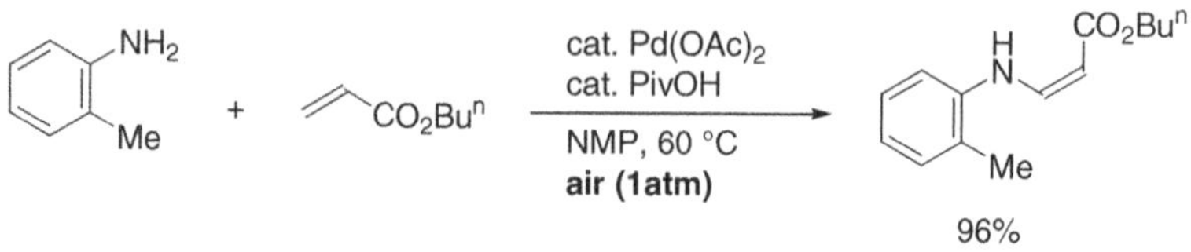

Scheme 9.33 Obora's system on oxidative alkene amination.

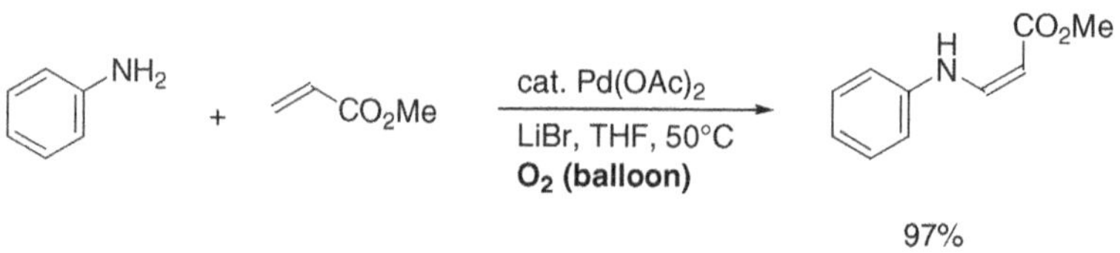

Scheme 9.34 Jiang's system on oxidative alkene amination.

of $Pd(OAc)_2$ (3 mol%) and pivalic acid (PivOH, 2.5 mol%) in NMP as the solvent at 60 °C under open (ambient) air was carried out to provide the alkenylation product in a 96% yield with *Z*-selectively (Scheme 9.33) [42]. In contrast, in Jiang's work, the reaction of aniline (1 equiv.) with methyl acrylate (1.2 equiv.) was carried out in the presence of $Pd(OAc)_2$ (5 mol%), and LiBr (4 equiv.) as the additive in THF under O_2 balloon at 50 °C to provide the alkenylated product in 97% as a *Z*-adduct (Scheme 9.34) [43].

9.11 Oxidative Nitrogen-Sulfur Bond Functionalization of Sulfonyl Hydrazides

As for the reaction of oxidative C—S bond formation methodology, Jiang and co-workers reported the synthesis of sulfones and thioethers by using oxidative C—S bond formation of styrenes. Typically, the reaction of *p*-toluene sulfonohydrazide (3 equiv.) with styrene (1 equiv.) was carried out in the presence of CuCl (10 mol%) and LiBr (30 mol%) in DMSO under O_2 (1 atm) and gave the alkenylated product in a 92% yield (Scheme 9.35) [44].

As for the proposed reaction mechanism, initial oxidation of sulfonohydrazide by Cu catalyst **A** led to decomposition and formed sulfenyl cuprate **B**. Subsequently, the reaction with styrene led to the formation of the intermediate **C** and furnished the sulfones as the products (Scheme 9.36) [45].

9.12 Oxidative Silylation of Alkenes by Means of 1 atm of O_2 as the Sole Oxidant

Obora and co-workers reported the oxidative silylation by using 1 atm of O_2 as a sole oxidant. As a typical example, the reaction of 1-decene (40 equiv.) with hexamethyldisilane (1 equiv.) was carried out in the presence of Pd $(OCOCF_3)_2$ (5 mol%) combined with and acetylacetone (acacH) (20 mol%) under 1 atm of O_2 at 40 °C to give the product in 86% yield (Scheme 9.37) [46].

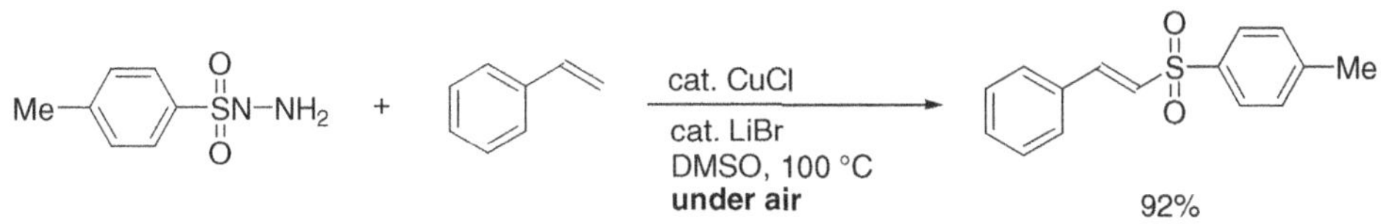

Scheme 9.35 Pd-catalyzed cross-coupling of *p*-toluene sulfonohydrazide with styrene.

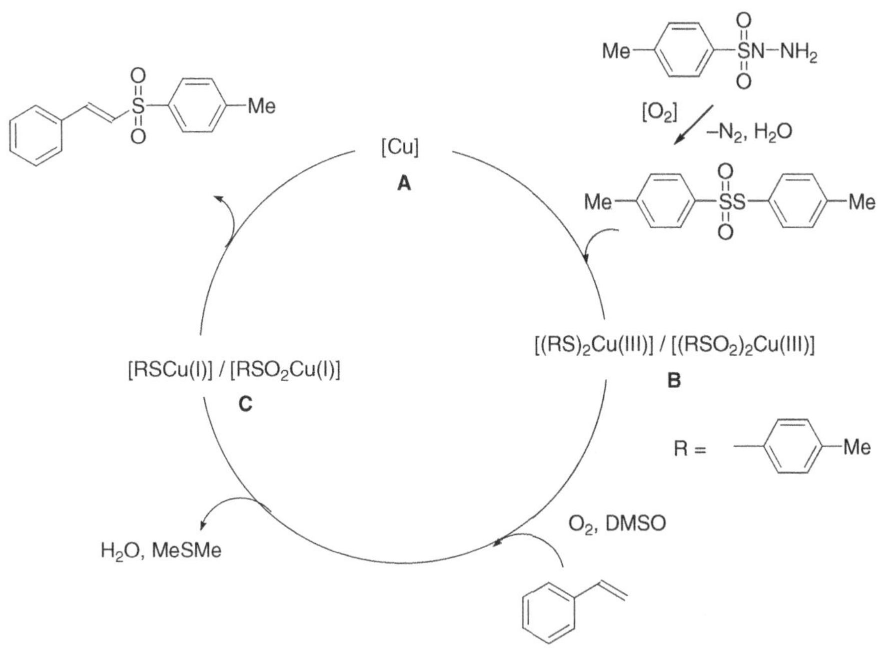

Scheme 9.36 Proposed mechanism.

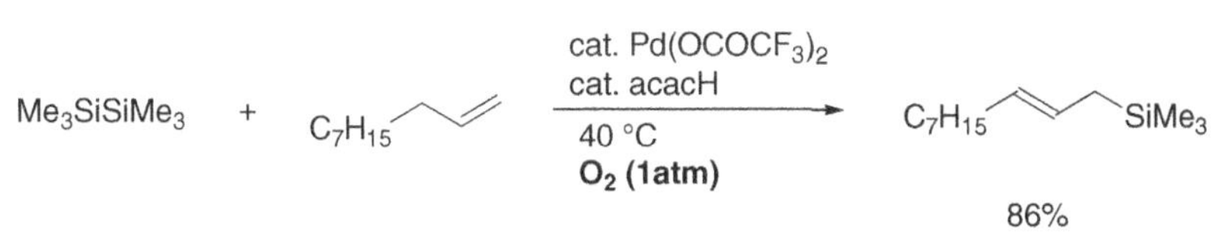

Scheme 9.37 Pd-catalyzed oxidative silylation.

9.13 Oxidative Alkynylation by Using 1 atm of O_2 as an Oxidant

As for oxidative alkyne functionalization by using 1 atm of O_2 as an oxidant, the Pd-catalyzed direct oxidative-Heck–Cassar–Sonogashira type alkynylation was chosen. In representative examples of this type of reaction, Li and Miura reported the oxidative alkynylation reaction of indoles and azoles/perfluoroarenes.

In Li and co-workers' study, the reaction of 1,3-dimethylindole (1 equiv.) with phenylacetylene (2 equiv.) in the presence of K_2PdCl_4 (10 mol%) combined with Cs_2CO_3 (20 mol%)

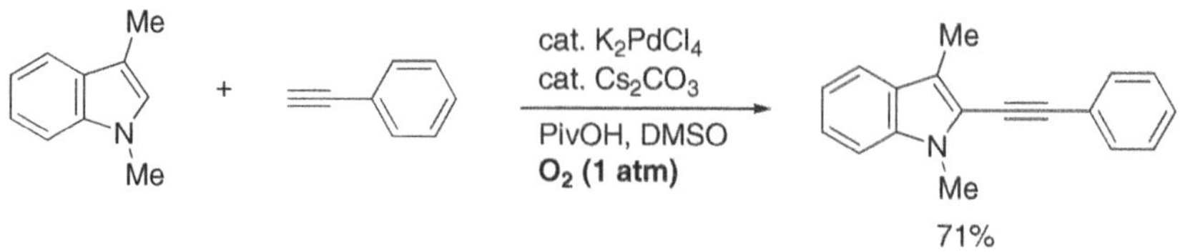

Scheme 9.38 Pd-catalyzed indole C2-alkynylation.

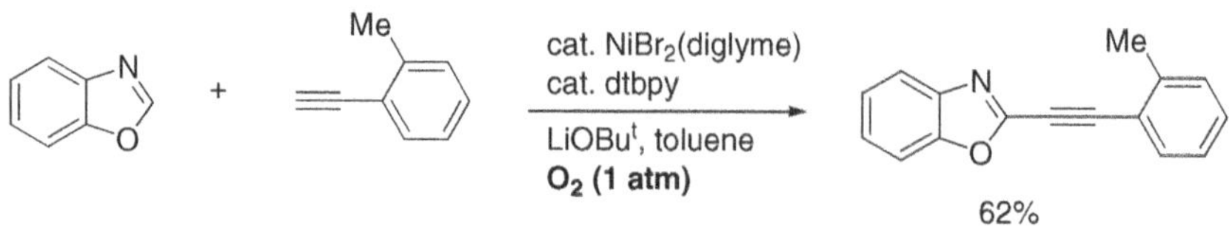

Scheme 9.39 Pd-catalyzed alkynylation of benzoxazole.

and pivalic acid (2 equiv.) in DMSO under 1 atm of O_2 led to the corresponding alkynylation products in 71% yield (Scheme 9.38) [47].

In a study reported by Miura and co-workers, the reaction of benzoxazole (1 equiv.) with (2-methylphenyl)acetylene (2 equiv.) was carried out in the presence of $NiBr_2$·diglyme (5 mol%) combined with 4,4-di(*tert*-butyl)-2,2-bipyridine (dtbpy) (5 mol%) and LiOBu-*t* (3 equiv.) at 100 °C under 1 atm of O_2 and gave the alkynylated product in a 62% yield (Scheme 9.39) [48].

9.14 Oxidative Carbocyclization Alkynylation by Using 1 atm of O_2 as an Oxidant

Bäckvall and co-workers reported Pd-catalyzed oxidative carbocyclization/alkynylation of alleynes by using 1 atm of O_2 as an oxidant. Typically, the reaction of allenyne and phenylacetylene in the presence of $Pd(OCOCF_3)_2$ (5 mol%) combined with benzoquinone (20 mol%) and Co(salophen) (5 mol%) as a co-catalyst under 1 atm of O_2 was carried out to afford the alkynylated product in 74% yield (Scheme 9.40) [45].

9.15 Oxidative Cyclization-Functionalization of Alkene/Alkyne by Using Ambient Oxygen as an Oxidant

Loh and co-workers reported Pd-catalyzed oxidative bisolefination of alkynes to afford naphthalene derivatives. In a typical example, a reaction of 1-(phenylethynyl)-2-vinylbenzene (1 equiv.) with butyl acrylate (3 equiv.) in the presence of $PdCl_2$ (5 mol%) in DMSO under 1 atm of O_2 was performed to provide the naphthalene derivatives in a 75% yield (Scheme 9.41) [49].

With regard to the reaction mechanism, coordination of alkyne to Pd(0) catalyst initiated the reaction. The formed π-coordinated Pd complex **A** was subjected to

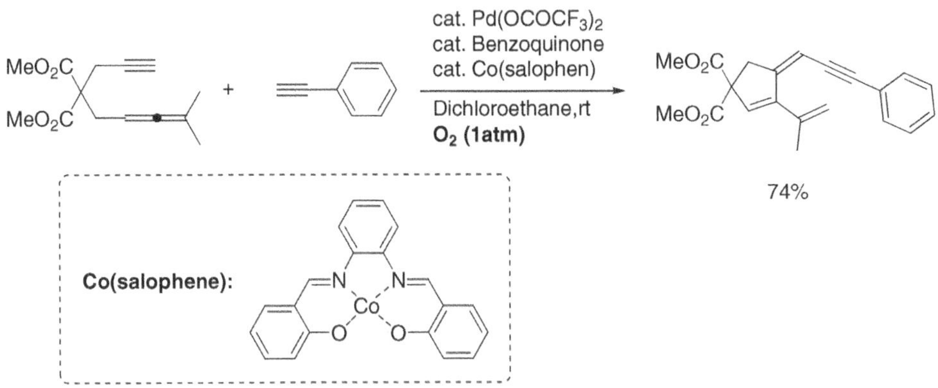

Scheme 9.40 Pd-catalyzed oxidative carbocyclization/alkynylation of alleynes.

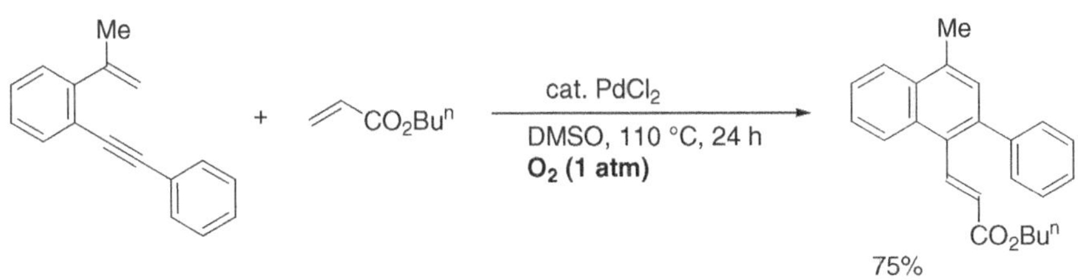

Scheme 9.41 Pd-catalyzed oxidative bisolefination of alkynes.

carbopalladation to form intermediate **B**. Subsequently, deprotonation took place to form intermediate **C**. Then, alkene insertion to **C** to form intermediate **D** was followed by β-H elimination, which led to the naphthalene derivatives (Scheme 9.42).

Ackermann and co-workers reported isocoumarin formation by the oxidation cyclization of benzoic acid derivatives with an alkyne under 1 atm of O_2 as an oxidant. Typically, the reaction of 2-methylbenzoic acid (2 equiv.) with diphenylacetylene (1 equiv.) in the presence of $[Ru(OAc)_2(p\text{-cymene})]_2$ (5 mol%) combined with NaOAc (1 equiv.) in methanol under 1 atm O_2 gave the corresponding isocoumarin derivatives in 83% yield (Scheme 9.43) [50].

Huang and co-workers reported Rh-catalyzed oxidative alkyne functionalization that created isoquinolium salt products. In a typical example of this reaction, reaction of 2-phenylpyridine (1 equiv.) with diphenylacetylene (1 equiv.) in the presence of $[Cp^*Rh(H_2O)_3(OTf)_3]$ (0.2 mol%) with TfOH (1 equiv.) as an additive under 1 atm of O_2 in methanol at 120 °C gave the corresponding isoquinolium salts as a product in 99% yield (Scheme 9.44) [51].

Jiao and co-workers reported Cu-catalyzed oxidative cyclization of pyridinium ylide with alkynes by using 1 atm of O_2 as an oxidant. Typically, the reaction of pyridinium ylide (1 equiv.) with phenylacetylene (3 equiv.) was carried out in the presence of CuI (10 mol%) combined with Ag_2CO_3 (10 mol%) and DABCO (2 equiv.) in chlorobenzene under 1 atm of O_2 as an oxidant to produce the corresponding pyrazolo[1,5-*a*]pyridine derivatives in 74% yield (Scheme 9.45) [52].

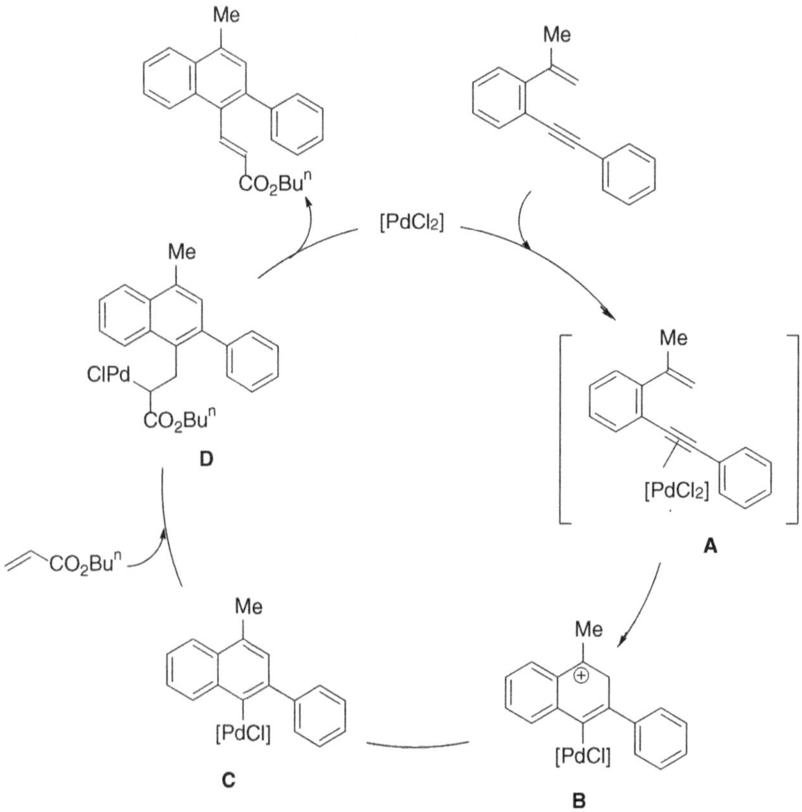

Scheme 9.42 Proposed mechanism of Pd-catalyzed oxidative bisolefination of alkynes.

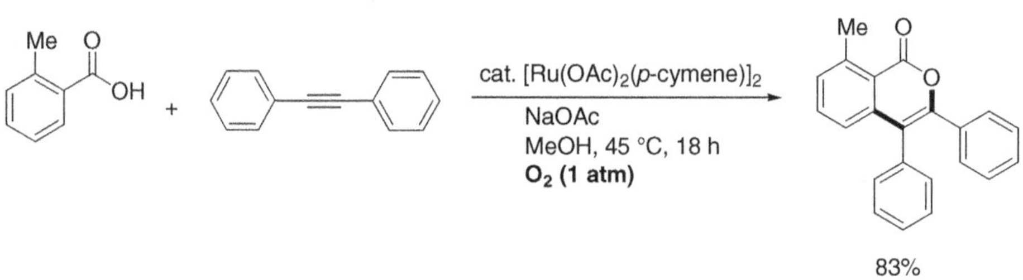

Scheme 9.43 Ru-catalyzed cyclization of benzoic acids with alkynes.

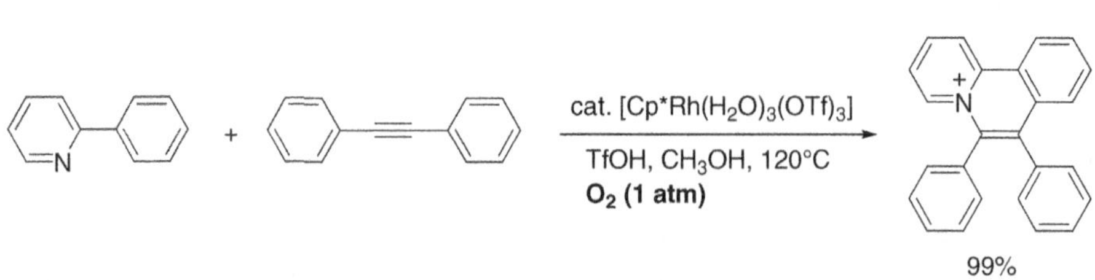

Scheme 9.44 Rh-catalyzed cyclization of 2-phenylpyridines with alkynes.

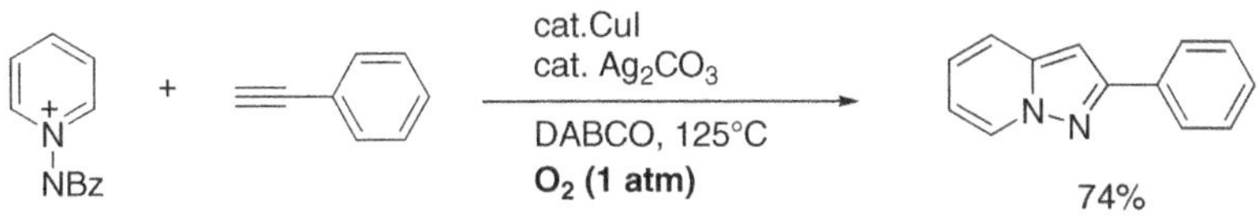

Scheme 9.45 Cu-catalyzed cyclization of pyridinium ylide with alkynes.

9.16 Conclusion and Outlook

In this chapter, selective recent developments of oxidative alkenylation and alkynylation reactions were documented. Here, we focused on the reactions that included use of 1 atm of molecular oxygen as an oxidant. Among the oxidants used in the oxidative-coupling, the use of molecular oxygen at ambient conditions is the most environmentally friendly process, which produces water as the sole byproduct. Furthermore, the use of molecular oxygen as an oxidant in oxidative-coupling is a practical protocol from a safety and non-toxicity point of view. All the listed reactions in this chapter demonstrate transformation by using molecular oxygen as an oxidant.

Inspired by the challenging organic transformations reported in this chapter and from an environmental and economical point of view, green oxidative alkene and alkyne functionalizations will hopefully be further developed.

References

1 Oestreich, M. (2009). *The Mizoroki–Heck Reaction*. Chichester: Wiley.

2 Representative Reviews: (a)Wu, Y., Wang, J., Mao, F., and Kwong, F.Y. (2014). *Chem. Asian J.* 9: 26–47. (b) Girard, S.A., Knauber, T., and Li, C.–.J. (2014). *Angew. Chem. Int. Ed.* 53: 74–100. (c) Engle, K.M., Mei, T.-S., Wasa, M., and Yu, J.-Q. (2012). *Acc. Chem. Res.* 45: 788–802. (d) Yeung, C.S. and Dong, V.M. (2011). *Chem. Rev.* 40: 4740–4761. (e) Wencel-Delord, J., Dröge, T., Liu, F., and Glorius, F. (2011). *Chem. Soc. Rev.* 40: 4740–4761. (f) Cho, S.H., Kim, J.Y., Kwak, J., and Chang, S. (2011). *Chem. Soc. Rev.* 40: 5068–5083. (g) Ackermann, L. and Potukuchi, H.K. (2010). *Org. Biomol. Chem.* 8: 4503–4513. (h) Gaugulis, O. (2010). *Top. Curr. Chem.* 292: 57–84. (i) Lapointe, D. and Fagnou, K. (2010). *Chem. Lett.* 39: 1119–1126. (j) Colby, D.A., Bergman, R.G., and Ellman, J.A. (2010). *J. Am. Chem. Soc.* 110: 624–655. (k) Kuhl, N., Hopkinson, M.N., Wenzel-Delord, J., and Glorius, F. (2012). *Angew. Chem. Int. Ed.* 51: 10236–10254. (l) Liu, C., Zhang, H., Shi, W., and Lei, A. (2011). *Chem. Rev.* 111: 1780–1824. (m) Lyons, T.L. and Sanford, M. (2010). *Chem. Rev.* 110: 1147–1169. (n) Chen, X., Engle, K.M., Wang, D.H., and Yu, J.-Q. (2009). *Angew. Chem. Int. Ed.* 48: 5094–5115. (o) Ackermann, L., Vincent, R., and Kapdi, A.R. (2009). *Angew. Chem. Int. Ed.* 48: 9792–9826. (p) Li, C.-J. (2009). *Acc. Chem. Res.* 42: 335–344.

3 (a) Jira, R. (2002). Oxidation of olefins to carbonyl compounds (Wacker Process). In: *Applied Homogeneous Catalysts with Organometallic Compounds*, Vol. 1 (ed. B. Cornils and W.A. Herrmann), 386–405. New York: Wiley-VCH. (b) Stahl, S.S. (2004). *Angew. Chem. Int. Ed.* 43: 3400–3420. (c) Baiju, T.V., Gravel, E., Doris, E., and Namboothiri, I.N.N. (2016). *Tetrahedron Lett.* 57: 3993–4000.

4 For representative examples, (a) Fix, S.R., Brice, J.L., and Stahl, S.S. (2002). *Angew. Chem. Int. Ed.* 41: 164–166. (b) Trend, R.M., Ramtohul, Y.K., Ferreira, E.M., and Stoltz, B.M. (2003). *Angew. Chem. Int. Ed.* 42: 2892–2895. (c) Kotov, V., Scarborough, C.C., and Stahl, S.S. (2007). *Inorg. Chem.* 46: 1910–1923. (d) Liu, G. and Stahl, S.S. (2007). *J. Am. Chem. Soc.* 129: 6328–6335. (e) Minatti, A. and MuCiz, K. (2007). *Chem. Soc. Rev.* 36: 142–1152. (f) McDonald, R.I., Liu, G., and Stahl, S.S. (2011). *Chem. Rev.* 111: 2981–3019. (g) McDonald, R.I. and Stahl, S.S. (2010). *Angew. Chem. Int. Ed.* 49: 5529–5532. (h) Weinstein, A.B., Schuman, D.P., Tan, Z.X., and Stahl, S.S. (2013). *Angew. Chem. Int. Ed.* 52: 11867–11870.(i) White, P.B., Jaworski, J.N., Zhu, G.H., and Stahl, S.S. (2016). *ACS Catal* 6: 3340–3348.

5 (a) Hegedus, L.S. (1984). *Tetrahedron* 40: 2415–2434. (b) Hegedus, L.S. and McKearin, J.M. (1982). *J. Am. Chem. Soc.* 104: 2444–2451.

6 (a) Hegedus, L.S., Allen, G.F., Bozell, J.J., and Waterman, E.L. (1988). *J. Am. Chem. Soc.* 100: 5800–5807. (b) Hosokawa, T., Takano, M., Kuroki, Y., and Murahashi, S.-I. (1992). *Tetrahedron Lett.* 33: 6643–6646. (c) Hegedus, L.S. and Holden, M.S. (1986). *J. Org. Chem.* 51: 1171–1174. (d) Tamaru, Y., Hojo, M., Higashimura, H., and Yoshida, Z.-I. (1988). *J. Am. Chem. Soc.* 110: 3994–4002. (e) Lee, J.M., Ahn, D.-S., Jung, D.Y. et al. (2006). *J. Am. Chem. Soc.* 126: 12954–12962. (f) Larock, R.C. and Hightower, T.R. (1993). *J. Org. Chem.* 58: 5298–5300. (g) Larock, R.C., Hightower, T.R., Hasvold, L.A., and Peterson, K.P. (1996). *J. Org. Chem.* 61: 3584–3585.

7 Review: Evano, G., Gaumont, A.-C., Alayrac, C. et al. (2014). *Tetrahedron* 70: 1529–1616.

8 (a) Fujiwara, Y., Moritani, I., Matsuda, M., and Teranishi, S. (1968). *Tetrahedron Lett.* 9: 633. (b) Moritani, I. and Fujiwara, Y. (1967). *Tetrahedron Lett.* 8: 1119. (c) Jia, C., Kitamura, T., and Fujiwara, Y. (2001). *Acc. Chem. Res.* 34: 633–639.

9 For selected reviews, see: (a)Zhang, M., Zhang, Y., Jie, X. et al. (2014). *Org. Chem. Front.* 1: 843–895. (b) Liu, C., Yuan, J., Gao, M. et al. (2015). *Chem. Rev.* 115: 12138–12204. (c) Chen, Z., Wang, B., Zhang, J. et al. (2015). *Org. Chem. Front.* 2: 1107–1295. (d) Neufeldt, S.R. and Sanford, M.S. (2012). *Acc. Chem. Res.* 45: 936–946. (e) Le Bras, J. and Muzart, J. (2011). *Chem. Rev.* 111: 1170–1214. (f) Yeung, C.S. and Dong, V.M. (2011). *Chem. Rev.* 111: 1215–1292.

10 (a) Obora, Y. and Ishii, Y. (2010). *Molecules* 15: 1487–1500. (b) Huang, Q., Ke, S., Qiu, L. et al. (2014). *ChemCatChem* 6: 1531–1534.

11 (a) Yamada, T., Sakaguchi, S., and Ishii, Y. (2005). *J. Org. Chem.* 70: 5471–5474. (b) Tani, M., Sakaguchi, S., and Ishii, Y. (2004). *J. Org. Chem.* 69: 1221–1226. (c) Yokota, T., Tani, M., Sakaguchi, S., and Ishii, Y. (2003). *J. Am. Chem. Soc.* 125: 1476–1477.

12 Obora, Y., Okabe, Y., and Ishii, Y. (2010). *Org. Biomol. Chem.* 8: 4071.

13 Mizuta, Y., Obora, Y., Shimizu, Y., and Ishii, Y. (2012). *ChemCatChem* 4: 187–191.

14 Huang, Q., Song, Q., Cai, J. et al. (2013). *Adv. Synth. Catal.* 355: 1512–1516.

15 Huang, Q., Ke, S., Qui, L. et al. (2014). *ChemCatChem* 6: 1531–1534.

16 Graczyk, K., Ma, W., and Ackermann, L. (2012). *Org. Lett.* 14: 4110–4113.

17 Lanke, V. and Prabhu, R. (2013). *Org. Lett.* 15: 2818–2821.

18 (a) Musaev, D.G., Figg, T.M., and Kaledin, A.L. (2014). *Chem. Soc. Rev.* 43: 5009–5031. (b) Haines, B.E. and Musaev, D.G. (2015). *ACS Catal.* 5: 830–840. (c) Li, G., Leow, D., Wan, L., and Yu, J.-Q. (2013). *Angew. Chem. Int. Ed.* 52: 1245–1247. (d) Musaev, D.G., Kaledinm, A.L., Shi, B.-F., and Yu, J.-Q. (2012). *J. Am. Chem. Soc.* 134: 1690–1698. (e) Wang, H.-L., Hu, R.-B., Zhang, H. et al. (2013). *Org. Lett.* 15: 5302–5305.

(f) Cong, X., Tang, H., Wu, C., and Zhang, X. (2013). *Organometallics* 32: 6565–6575. (g) Cheng, G.-J., Yang, Y.-F., Liu, P. et al. (2014). *J. Am. Chem. Soc.* 136: 894–897.
19 Huang, Q., Zhang, X., Qiu, L. et al. (2015). *Adv. Synth. Catal.* 357: 3753–3757.
20 Zhang, X., Su, L., Qui, L. et al. (2017). *Org. Biomol. Chem.* 15: 3499–3506.
21 Zhang, Y.-H., Shi, B.-F., and Yu, J.-Q. (2009). *J. Am. Chem. Soc.* 131: 5072–5074.
22 Min, M., Kim, Y., and Hong, S. (2013). *Chem. Commun.* 49: 196–198.
23 Liu, B., Jiang, H.-Z., and Shi, B.-F. (2014). *J. Org. Chem.* 79: 1521–1526.
24 Chen, W.-L., Gao, Y.-R., Mao, S. et al. (2012). *Org. Lett.* 14: 5920.
25 Yang, L., Zhang, G., and Huang, H. (2014). *Adv. Synth. Catal.* 356: 1509–1515.
26 Harada, S., Yano, H., and Obora, Y. (2013). *ChemCatChem* 5: 121–125.
27 Lu, Y., Wang, H.-W., Spangler, J.E. et al. (2015). *Chem. Sci.* 6: 1925–1927.
28 Campbell, A.N., White, P.B., Guzei, I.A., and Stahl, S.S. (2010). *J. Am. Chem. Soc.* 132: 15116–15119 others in the references.
29 (a) Campbell, A.N. and Stahl, S.S. (2012). *Acc. Chem. Res.* 45: 851–863. (b) Campbell, A.N., White, P.B., Guzel, I.A., and Stahl, S.S. (2010). *J. Am. Chem. Soc.* 132: 15116–15119. (c) Diao, T., Wadzinski, T.J., and Stahl, S.S. (2012). *Chem. Sci.* 3: 887–891. (d) Gao, W., He, Z., Qian, Y., and Huang, Y. (2012). *Chem. Sci.* 3: 883–886.
30 (a) Piotrowicz, M. and Zakrzewski, J. (2013). *Organometallics* 32: 5709–5712. (b) Piotrowicz, M., Zakrzewski, J., Makal, A. et al. (2011). *J. Organomet. Chem.* 696: 3499–3506.
31 Piotrowicz, M., Zakrzewski, J., Métivier, R. et al. (2015). *Organometallics* 80: 2573–2581.
32 Vasseur, A., Laugel, C., Harakat, D. et al. (2015). *Eur. J. Org. Chem.* 2015: 944–948.
33 Zhang, G., Xia, Y., Wang, S. et al. (2012). *J. Am. Chem. Soc.* 134: 12334–12337.
34 Xie, Z., Jia, J., Liu, X., and Liu, L. (2016). *Adv. Synth. Catal.* 358: 919–925.
35 Hosokawa, T., Takano, M., Kuroki, Y., and Murahashi, S.-I. (1992). *Tetrahedron Lett.* 33: 6643–6646.
36 Lee, J.M., Ahn, D.-S., Jung, D.V. et al. (2006). *J. Am. Chem. Soc.* 128: 12954–12962.
37 Timikhin, V.I., Anastasi, N.R., and Stahl, S.S. (2003). *J. Am. Chem. Soc.* 125: 12996–12997.
38 Rogers, M.M., Wendlandt, J.E., Guzei, I.A., and Stahl, S.S. (2006). *Org. Lett.* 8: 2257–2260.
39 Rogers, M.M., Kotov, V., Chatwichien, J., and Stahl, S.S. (2007). *Org. Lett.* 9: 4331–4334.
40 Obora, Y., Shimizu, Y., and Ishii, Y. (2009). *Org. Lett.* 11: 5058–5061.
41 Shimizu, Y., Obora, Y., and Ishii, Y. (2010). *Org. Lett.* 12: 1372–1374.
42 Mizuta, Y., Yasuda, K., and Obora, Y. (2013). *J. Org. Chem* 78: 6332–6337.
43 Ji, X., Huang, H., Wu, W. et al. (2013). *J. Org. Chem* 78: 11155–11162.
44 C-S Li, X., Xu, Y., Wu, W. et al. (2014). *Chem. Eur. J.* 20: 7911–7915.
45 Volla, C.M.R. and Bäckvall, J.-E. (2013). *Angew. Chem. Int. Ed.* 52: 14209–14213.
46 Nakai, S., Matsui, M., Shimizu, Y. et al. (2015). *J. Org. Chem.* 80: 7317–7320.
47 Yang, L., Zhao, L., and Li, C.-J. (2010). *Chem. Commun.* 46: 4184–4186.
48 Matsuyama, N., Kitahara, M., Hirano, K. et al. (2010). *Org. Lett.* 12: 2358–2361.
49 Feng, C. and Loh, T.-P. (2010). *J. Am. Chem. Soc.* 132: 17710–17712.
50 Warratz, S., Kornhaaß, C., Cajaraville, A. et al. (2015). *Angew. Chem. Int. Ed.* 54: 5513–5517.
51 Zhang, G., Yang, L., Wang, Y. et al. (2013). *J. Am. Chem. Soc.* 135: 8850–8853.
52 Ding, S. and Jiao, N. (2013). *Chem. Commun.* 49: 4250–4252.

10

Green Oxidative Synthesis of Substituted Arenes

Arun Jyoti Borah and Zhuangzhi Shi

State Key Laboratory of Coordination Chemistry, School of Chemistry and Chemical Engineering, Nanjing University, Nanjing, China

Aromatic compounds are the universal target of industry and laboratory chemical synthesis with applications ranging from pharmaceuticals and agrochemicals, to electronic materials, as well as commodity chemicals [1]. Development of new methods for the synthesis of substituted aromatic molecules that exhibits high selectivity patterns compared to the traditional methods has a major impact on the chemical industry. This principle is prominently illustrated by the development of metal-catalyzed cross-coupling reactions of aryl halides and related substrates, which have dramatically expanded the scope of substituted aromatic compounds and bypassed many of the regioselectivity constraints and substrate limitations of classical nucleophilic and electrophilic aromatic substitution reactions. The requirement of cross-coupling reactions to use an aromatic substrate that is prefunctionalized with halide, triflate, or metal groups (Li, Mg, B, etc.), however, has paved the way to develop methods for direct functionalization of arene C—H bonds. Many successful examples of arene C—H functionalization reactions have been reported, especially with substrates bearing a chelating directing group to control the regioselectivity. As mentioned in earlier chapters, traditional oxidation reaction conditions involve harmful metal-salts, strong acids, periodides, or quinines that create environmental threat. So, the application of benign oxidants oxygen (or air) and hydrogen peroxide is in high demand in these oxidative reactions.

The common feature among all methods of functionalization of substituted arenes is the use of an aromatic molecule as the starting material for selective incorporation of substituents onto the periphery of the ring. Another efficient and powerful strategy to prepare substituted aromatic compounds that has been the focus of recent investigation in many groups is the construction of substituted arenes from non-arene precursors under aerobic conditions. These materials are readily available and easily derivatized by classical synthetic methods. Therefore, they represent appealing precursors to diversely substituted aromatic compounds. In this chapter, the oxidative synthesis of substituted arenes is portrayed and is discussed in two sections: (i) Functionalization of substituted arenes; and (ii) construction of substituted arenes.

Green Oxidation in Organic Synthesis, First Edition. Edited by Ning Jiao and Shannon S. Stahl.

10.1 Functionalization of Substituted Arenes

10.1.1 Coupling Between Two Carbon–Metal Bonds

The substituted biaryl motif is an important substructure of many bioactive and functional molecules and has been the focus of synthetic chemists for over a hundred years [2]. Homo-coupling reactions of organometallic compounds (Li, Mg, B, etc.) are easy and efficient for symmetrical di- or polyaromatic, olefinic, or acetylenic conjugated compounds. The potential applications of such compounds in optical materials, molecular devices, and organic conductors are well-recognized. However, for industrial applications, such methods are limited by the requirement of harsh organic oxidants. Aryl lithiums readily undergo homo- and hetero-coupling in the presence of stoichiometric copper via formation of a copper(I)-ate complex and oxidation with O_2 [3]. Under catalytic conditions, aryl/hetroaryl zinc reagents generated from aryl halides and zinc(0) undergo highly efficient homo-coupling in the presence of catalytic CuBr, air, or O_2 and catalytic dinitroarene as the cooxidant (Scheme 10.1) [4]. Tethered aryl halides can be effectively cyclized under these reaction conditions.

Cahiez et al. have demonstrated the efficiency of Fe-catalyzed homo-coupling of aryl Grignard reagents using 1,2-dihalogenoethanes as an oxidant [5]. Following that, Knoche and co-workers described an elegant transition-metal-free homo-coupling reaction of organomagnesium compounds via direct oxidation with a quinone oxidant [6]. Later on, two very efficient iron- and manganese-catalyzed procedures were developed to couple Grignard reagents under mild conditions by using atmospheric oxygen as the oxidant (Scheme 10.2) [7]. The key step of this catalytic cycle (Scheme 10.3) is the conversion of the stable diorganomanganese(II) species to a manganese(IV) peroxo complex, which undergoes rapid reductive elimination to afford the homo-coupling product. The formed Mn(II) peroxo complex reacts with the Grignard reagent to recover the organomanganese. It is necessary to achieve a very rapid catalytic process to avoid the direct oxidation of the Grignard reagent by an oxygen molecule.

Several examples of homo-coupling reactions between two organometallic compounds have been reported, but oxidative cross-couplings are very challenging. Atmospheric oxygen as cooxidant in combination with benzoquinone enables the cross-coupling of aryl silanes with aryl boronic acids in the presence of palladium catalysis (Scheme 10.4) [8].

Mechanistically, this reaction starts with transmetallation of Pd^{II} with either aryl silane or aryl boronic acid. The formed $Pd^{II}Ar$ species undergoes a second transmetallation to form $Pd^{II}Ar^1Ar^2$ that, on reductive elimination, generates the biaryl product and a Pd^0 complex. Subsequently, the Pd^0 species is oxidized to Pd^{II} by BQ or O_2 to close the catalytic cycle (Scheme 10.5).

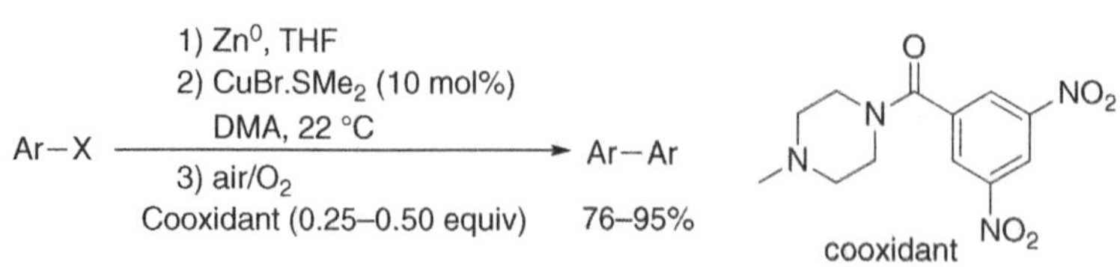

Scheme 10.1 Intermolecular catalytic zinc organocuprate oxidation.

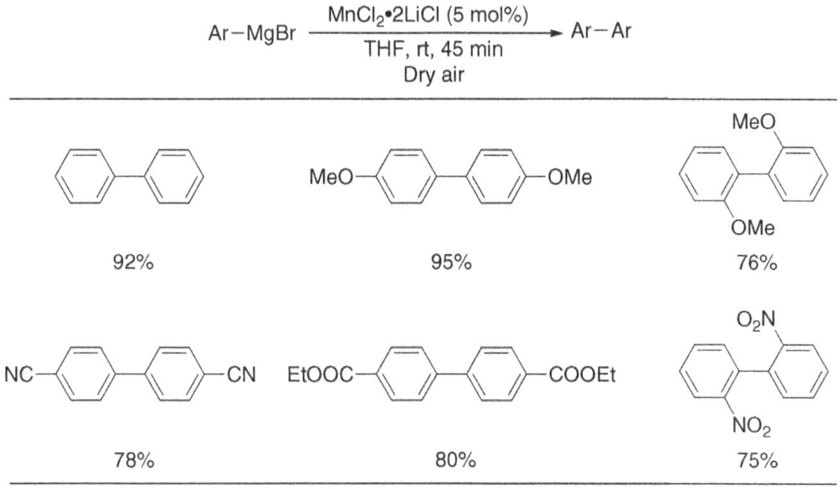

Scheme 10.2 Mn-catalyzed homo-coupling of Grignard reagents using O_2 as the oxidant.

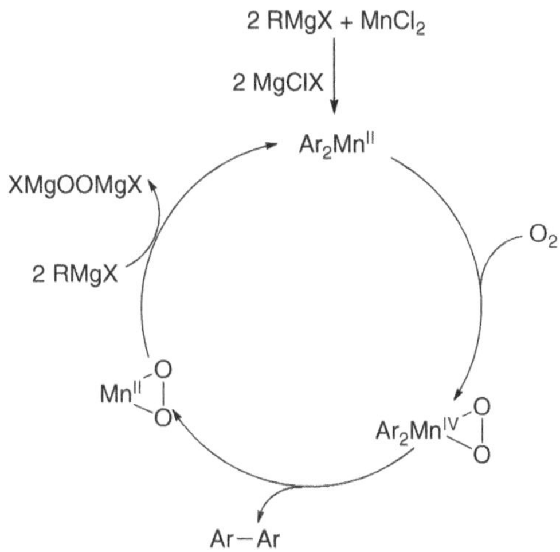

Scheme 10.3 Proposed mechanism for the Mn-catalyzed reaction.

10.1.2 Coupling Between Carbon–Metal and Carbon–Hydrogen Bonds

While boronic acids are often prepared from aryl halides, the absence of aryl halides in end-game reactions is advantageous from an environmental point of view. Oxidative coupling of electron-rich arenes and heteroarenes with aromatic boronic acids can be accomplished under acidic condition using O_2 as the terminal oxidant. Shi and co-workers have achieved the heteroarenes with phenyl boronic acid in the presence of $Pd(OAc)_2$ in AcOH at room temperature under aerobic conditions. Oxidative coupling of electron-rich arenes with aromatic boronic acids can also be accomplished at room temperature under acidic conditions and application of O_2 as the cooxidant with $Cu(OAc)_2$ (Scheme 10.6) [9]. These reactions proceed via electrophilic metallation and a transmetallation pathway. Under acidic conditions, the homo-coupling of boronic acids get retarded and facilitate electrophilic attack.

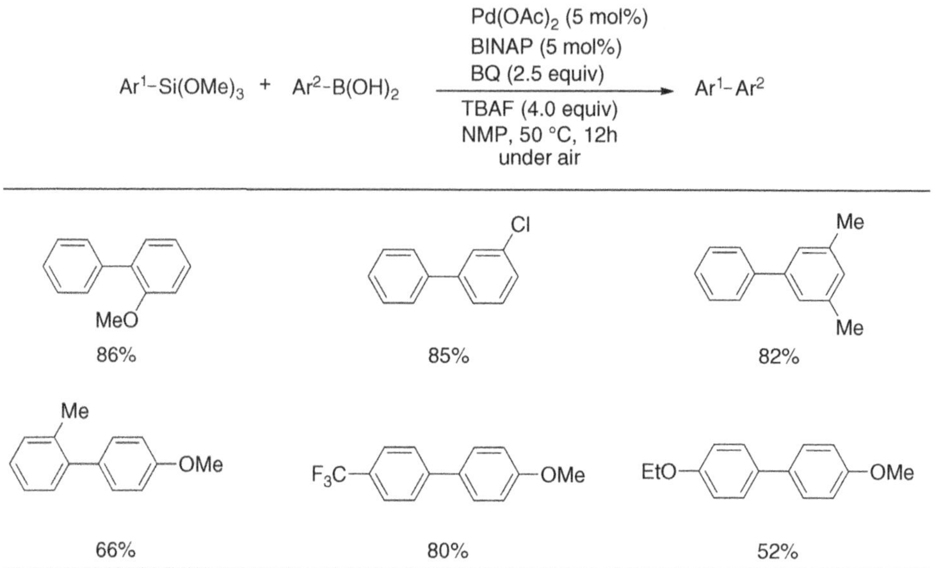

Scheme 10.4 Pd-catalyzed cross-coupling of aryl boronic acids and aryl silanes.

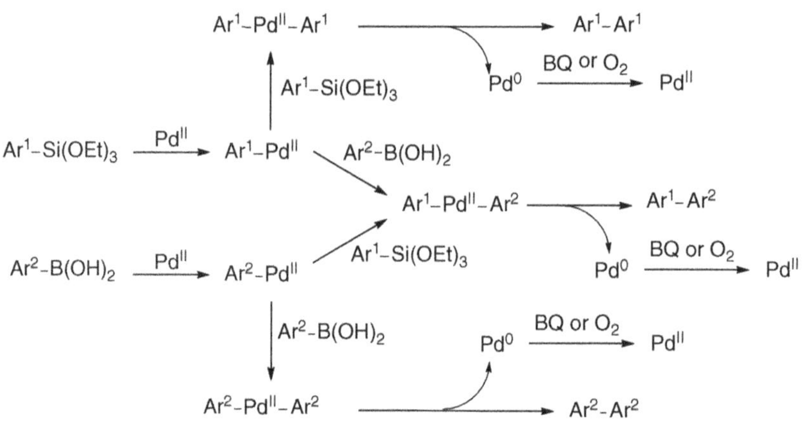

Scheme 10.5 Proposed mechanisms for the cross-coupling and homo-coupling products.

Aryl-substituted benzoic acids can also be synthesized using Pd(II)/Pd(0) catalysis. Using aryltrifluoroborates as the nucleophilic arylating partners, benzoic acids (*in situ* formation of carboxylates as the active substrate in presence of K_2HPO_4 base) can be subjected to *ortho*-arylation in presence of O_2/air as the terminal oxidant. Use of 20 atm air or O_2 is beneficial for a short reaction time (Scheme 10.7) [10].

10.1.3 Coupling Between Two Carbon–Hydrogen Bonds

Step-economic direct functionalization of C—H bond provides significant advantages over traditional cross-coupling methods. It is more beneficial to carry out coupling between two C—H bonds. Daugulis et al. have developed a general method for

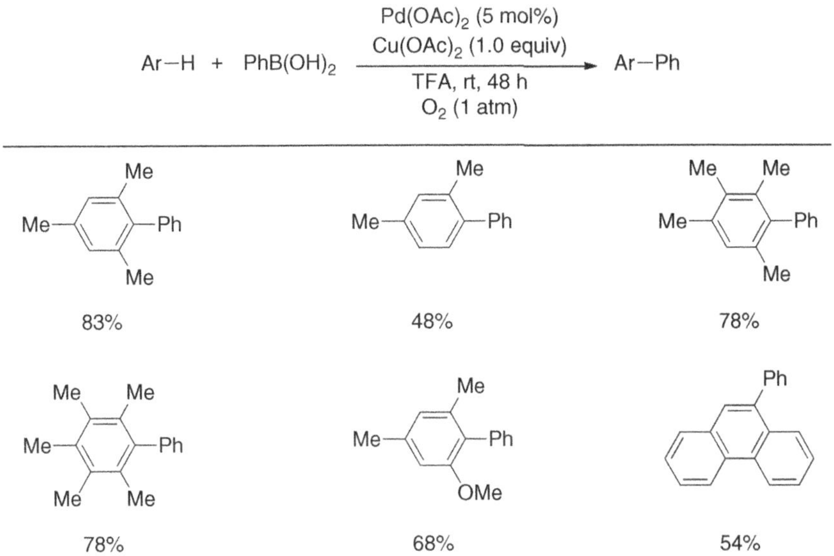

Scheme 10.6 Pd-catalyzed direct arylation of arenes with aryl boronic acids.

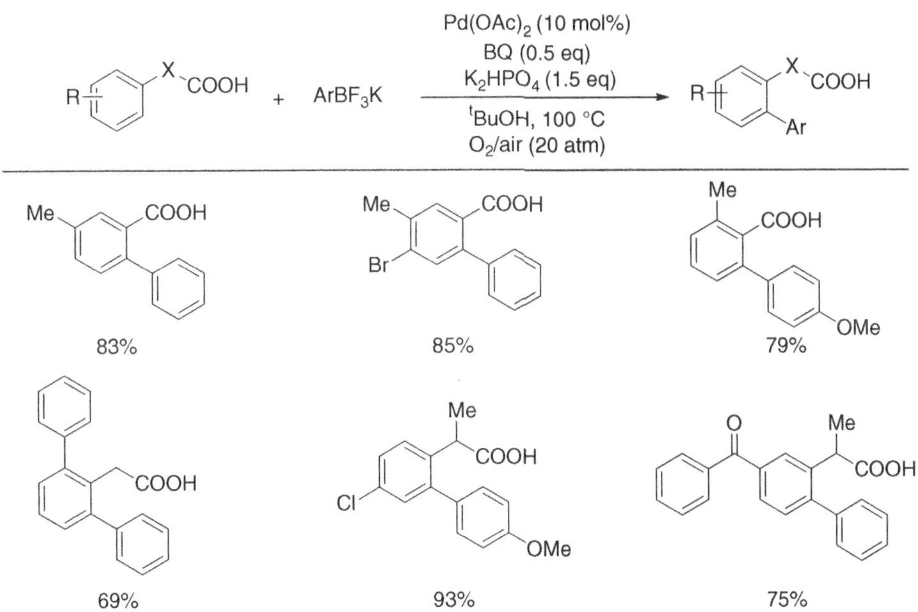

Scheme 10.7 Coupling of benzoic acids with potassium aryltrifluoroborates.

copper-catalyzed, deprotonative dimerization of arenes by employing oxygen as the terminal oxidant (Scheme 10.8) [11]. This process involves generation of an organocopper intermediate that can react with an aryl halide to afford a biaryl product. Under an oxygen atmosphere, the intermediate arylcopper species is expected to form

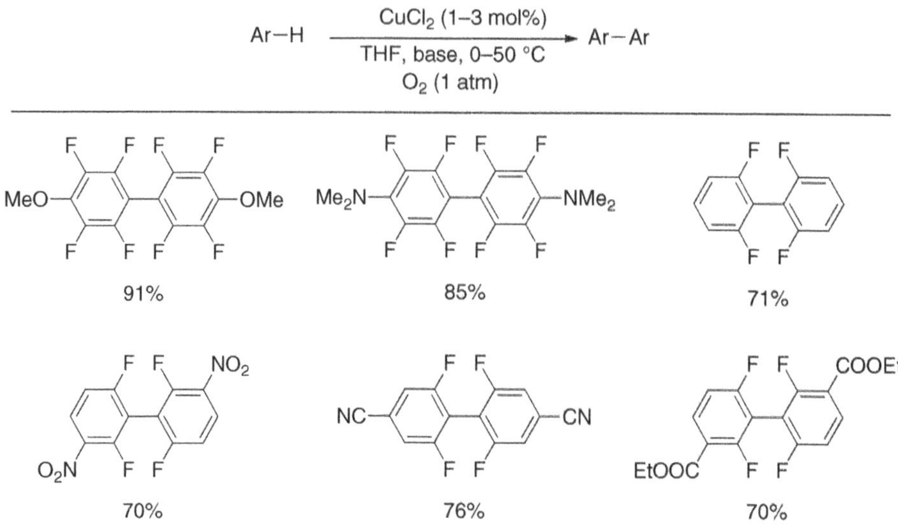

Scheme 10.8 Cu-catalyzed aromatic Glaser–Hay reaction.

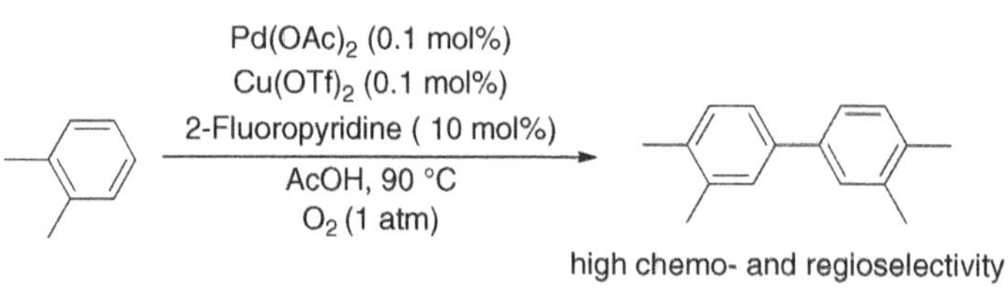

Scheme 10.9 Pd-catalyzed aerobic oxidative coupling of *o*-xylene.

biaryl and a low-valent copper complex. Regeneration of arylcopper by reaction of arylmetal with the low-valent Cu species closes the catalytic cycle.

An improved method for the direct oxidative coupling of *o*-xylene could provide streamlined access to an important monomer used in polyimide resins. The use of 2-fluoropyridine as a ligand enables unprecedented levels of chemo- and regioselectivity in palladium-catalyzed aerobic oxidative coupling reaction of *o*-xylenes (Scheme 10.9) [12]. Preliminary insights have been obtained into the origin of the effectiveness of 2-fluoropyridine as a ligand. Evidence of a mechanism via transmetallation between two separate palladation steps [Pd(II)-aryl intermediates] during biaryl coupling of such arenes under aerobic condition was established by Stahl [12d]. A significant rate acceleration in the coupling of *o*-xylene could be achieved by using catalytic metal-triflate salts [$Cu(OTf)_2$, $Fe(OTf)_3$], which activates the Pd^{II} catalyst for C—H activation via a non-redox pathway [12c].

1,1′-Bi-2-naphthol (BINOL) derivatives including chirals can be synthesized under catalytic aerobic oxidative conditions via a one-electron oxidation process. Various metals like V [13], Fe [14], and Cu [15], have been utilized for their synthesis. Conditions for the aerobic, copper-catalyzed oxidative dimerization of naphthols are quite mild and chiral amines are used under an O_2 or air atmosphere at ambient temperatures to

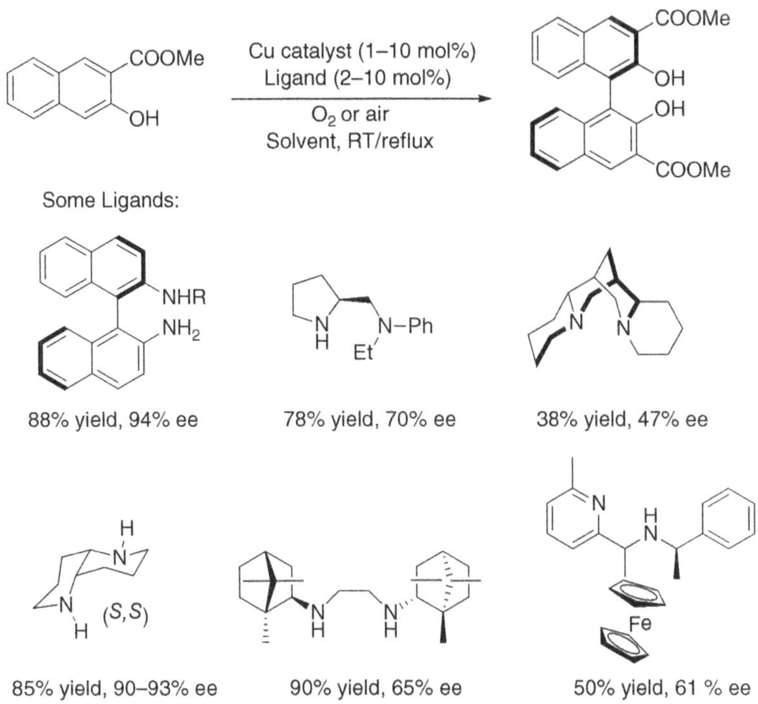

Scheme 10.10 Asymmetric synthesis of BINOL.

achieve high enantio-rich products (Scheme 10.10) [15a–f, k–n]. Naphthol substrates containing a methyl ester substituent at the C3 position can provide high enantioselectivity. The mechanistic path of these reactions features formation of a Cu^{II}-naphthoxide species that undergoes intramolecular electron transfer from the coordinated naphthoxide to Cu^{II} to form a Cu^{I}-naphthoxyl radical, which subsequently undergoes either attack from a second naphthol substrate on the naphthoxyl radical or bimolecular coupling of two Cu^{I}-naphthoxyl radicals [15f]. In the gas phase, evidence for binuclear Cu intermediates and their potential involvement in the C—C coupling step have been found [16].

Cross-dehydrogenative coupling (CDC) of different arenes is a big challenge for organic chemists, especially to control chemo-, regio-, and even stereoselectivity. Shi and co-workers have made remarkable progress toward environmentally friendly routes for biaryls. Biologically active molecules were constructed completely free of organohalogen and organometallic components (Scheme 10.11) [17]. With a acetamido directing group, Pd(II) can catalyze coupling of arenes with other arenes in the presence of $Cu(OTf)_2$ as the catalytic-cooxidant and O_2 as the terminal oxidant. The proposed mechanism includes acetamido-directed C—H activation followed by sterically controlled regioselectivity in a second activation step (Scheme 10.12).

Such twofold oxidative arene coupling can be achieved in the absence of a cooxidant using different amido directing groups, namely NHPiv. DMSO as an additive is needed to prevent palladium black formation in such reactions (Scheme 10.13) [18].

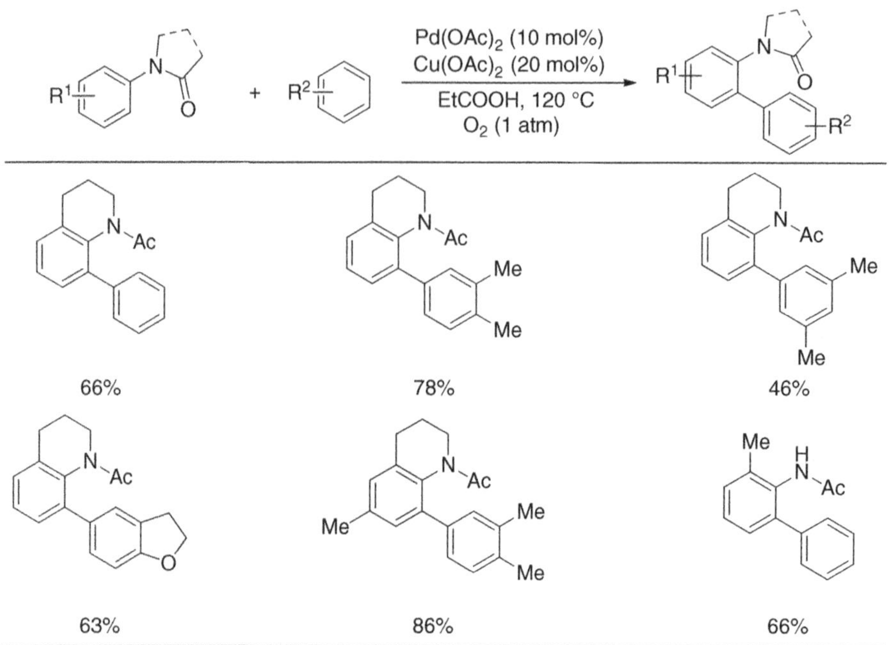

Scheme 10.11 Shi's cross dehydrogenative coupling.

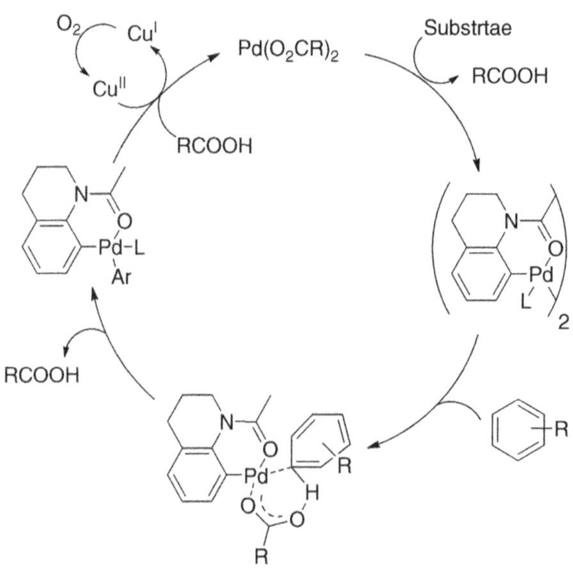

Scheme 10.12 Twofold C—H activation process.

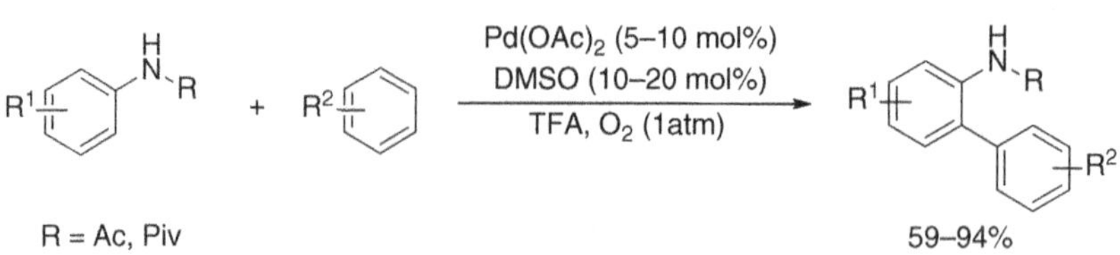

Scheme 10.13 Buchwald's cross dehydrogenative coupling.

10.1.4 Oxidative Heck Reaction

The Mizoroki–Heck reaction, discovered by Mizoroki and Heck in 1972, has been a powerful method to generate various alkenyl-substituted arenes [19]. Compared to traditional the Mizoroki–Heck reaction, the oxidative variation of organometallic compounds or arenes with alkenes can afford the corresponding products [20]. Thus, the oxidative Heck reaction implies catalysis with Pd(II), requires a reoxidant to regenerate the active catalytic species (Scheme 10.14). It is conceivable that use of oxygen as the terminal oxidant is very promising.

The Pd(II)/molybdovanadophosphoric acid (HPMoV) catalytic system is very effective for such transformations in the presence of O_2 as the terminal oxidant. However, positional selectivity of alkenylation in substituted arenes is low and all *ortho*, *meta*, and *para*-isomers can be obtained under the reaction conditions (Scheme 10.15) [21]. Comparatively, a high catalytic activity (higher in both turnover number (TON) and turnover frequency (TOF)) in the reaction between benzene derivatives and electron-deficient alkenes can be achieved using a $Pd(OAc)_2$/PhCOOH catalytic system [22].

Most of the arenes employed in the oxidative C—H olefinations are electron-rich, as the electronic palladation process is the main pathway for those transformations. Electron-deficient arenes can be subjected to corresponding oxidative coupling by using suitable ligands. 2,6-dialkylpyridine ligands can mediate Pd(II)-catalyzed meta-olefination of electron poor arenes under O_2 atmosphere (Scheme 10.16) [23].

Scheme 10.14 Proposed catalytic cycle of oxidative Heck reaction.

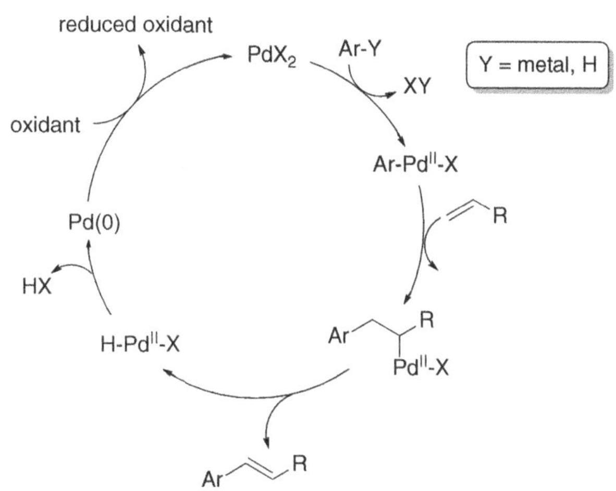

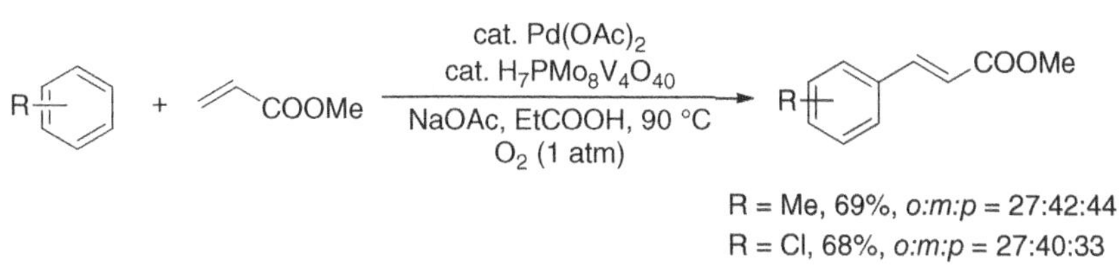

Scheme 10.15 Oxidative Heck type alkenylation of simple arenes.

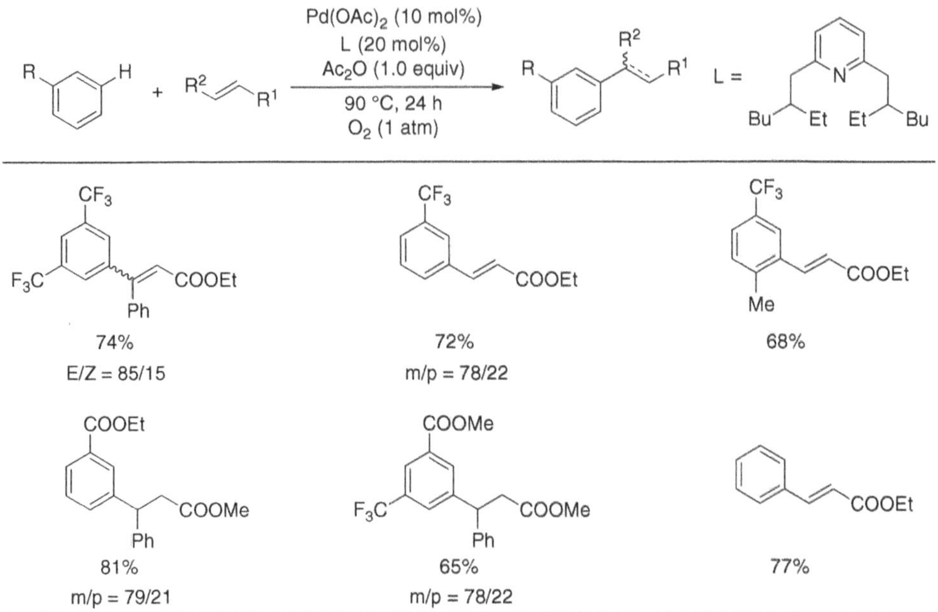

Scheme 10.16 *meta*-Olefination of electron-deficient arenes.

Such ligands with steric bulk at the two and six positions have a very important role in reactivity as well as in the reoxidation of Pd(0) by molecular O_2. In the presence of a pyridine-type ligand, arenes bearing electron-withdrawing groups (e.g. trifluoromethyl, nitro, ester, ketone) are able to coordinate to the Pd catalyst (a strong Pd—N interaction is maintained while imparting the necessary destabilization in dimers required to promote substrate coordination) and undergo sp^2 C—H bond activation. Oxidative alkenylation occurs regioselectively due to enhanced acidity of the C—H bonds *meta*- to the electron-withdrawing groups via an electrophilic substitution mechanism. Pd(II)/mono-protected amino acid (MPPA) catalytic system is suitable for olefination of polyfluoroarenes [24].

The Pd(II)–/benzoquinone catalytic system is very effective for *ortho*-olefination of phenylacetic acids in presence of O_2 as the oxidant. The benzoquinone ligand can control the low level of the diolefination product formation (Scheme 10.17) [25]. The positional selectivity of the olefination of multiply substituted phenylacetic acids can be controlled by utilizing MPPA. The MPPA ligand has also significant control over reactivity for oxidative olefination of 3-phenylpropionic acid and electron-deficient phenylacetic acids. In the presence of a Ac-Val-OH ligand, the reactivity of the phenyl acetic acid can be tuned to a high diolefination product.

The reactivity/selectivity balance in this Pd-catalyzed *ortho*-C—H olefination can be modulated to enable sequential C—H functionalization for the rapid preparation of 1,2,3-trisubstituted arenes. Additionally, there is a rare example of iterative C—H activation, in which a newly installed functional group directs subsequent C—H activation to form 1,2,3,4-tetrasubstituted arenes (Scheme 10.18) [26].

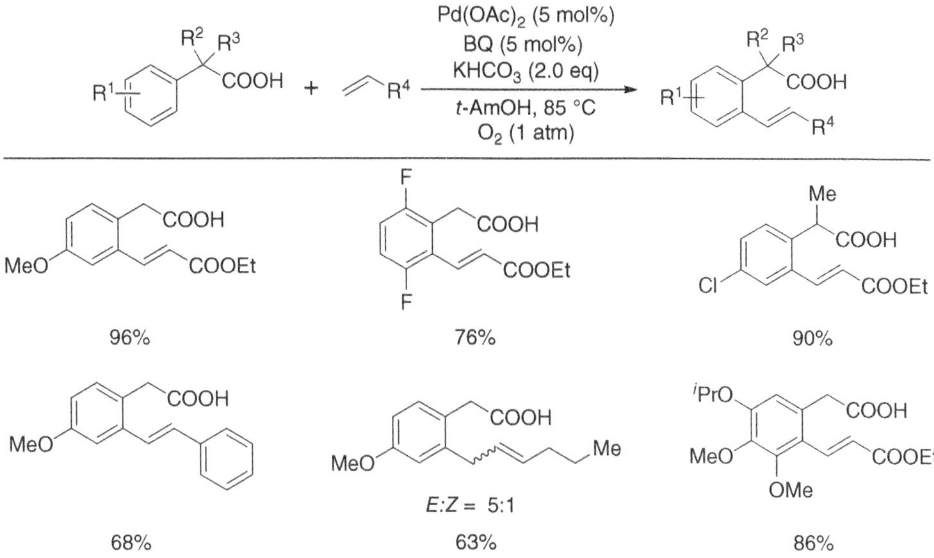

Scheme 10.17 C—H olefination of phenylacetic acid substrates with olefins.

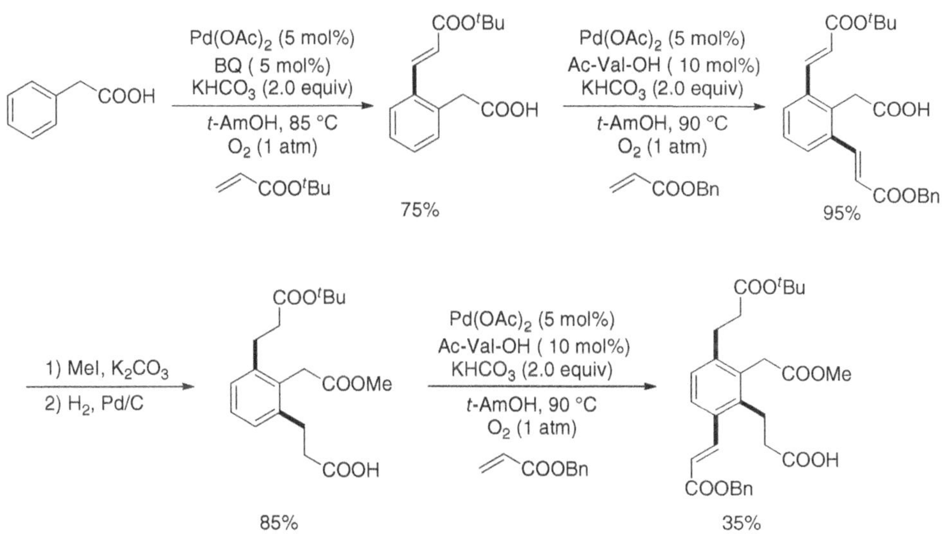

Scheme 10.18 Constructing multiply substituted arenes using sequential C—H olefination.

10.1.5 Diversity of Reaction Partners

Arenes can be substituted with acetoxy functional groups using a pyridyl directing group in the presence of catalytic $Cu(OAc)_2$. A mixture of mono- and bis-acetoxylated products can be obtained by carrying out the reaction at 130 °C in a $Ac_2O/AcOH$ solvent system. The pyridyl directing group is also effective for other nucleophilic

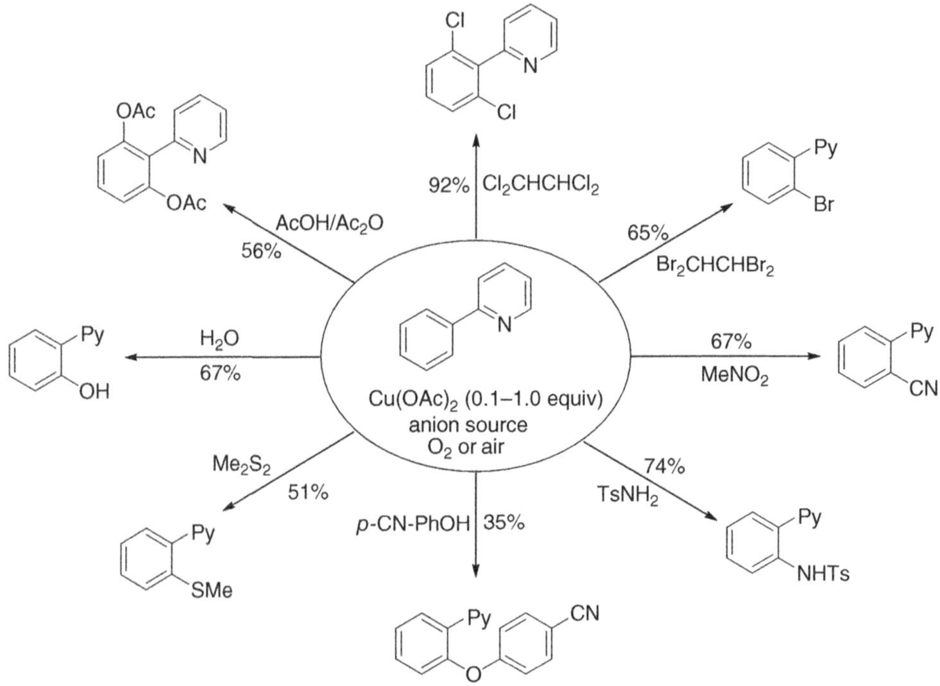

Scheme 10.19 Cu(II)-catalyzed functionalizations of aryl C—H bonds using O_2 as an oxidant.

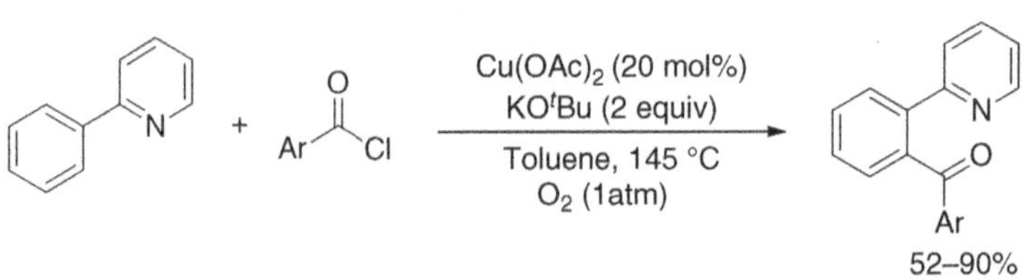

Scheme 10.20 *Ortho*-acyloxy and benzoxylation of 2-arylpyridines.

process such as bromination, iodination, cyanation, alkoxylation, hydroxylation, thiolation, and amination, but stoichiometric amounts of Cu(OAc)$_2$ are required (Scheme 10.19) [27]. However, catalytic turn over for the amination is operative at a higher temperature [28]. Mechanistically, these reactions follow a different mechanism from similar commonly observed Pd-catalyzed C—H activation processes and can be explained by a radical-cation pathway (single-electron transfer process).

The acetoxylation process can be extended with utilization of a wide range of alkyl and aryl anhydrides as well as acyl chlorides by electrophilic attack of Cu(II)- on a phenyl ring of the 2-arylpyridines (Scheme 10.20) [29]. Acyl chlorides are good precursors of anhydrides in the presence of a base and moisture. However, in such reactions with acyl chlorides, the presence of Li_2CO_3 as the base can provide *ortho*-chlorination products [29b].

The Pd(II)/Pd(0) catalytic cycle is also operative for regioselective acylations with aldehydes in the presence of air as the terminal oxidant to afford aromatic ketones in moderate to good yields. However, alkyl aldehydes cannot work under such conditions (Scheme 10.21) [30]. This reaction is a very convenient and atom-economic method for the synthesis of aromatic ketones.

Utilizing substrates with weak coordinating directing groups (e.g. ketones, carboxylic acids, and ethers) for Pd(II)-catalyzed C—H functionalization is a challenging approach for developing synthetically versatile reactions [31]. Because of the higher reactivity, a range of different nucleophiles and electrophiles can be successfully used as reaction partners. Yu and co-workers have developed a versatile Pd-catalyzed *ortho* hydroxylation of benzoic acids by utilizing 1 atm of O_2 or air under nonacidic conditions (Scheme 10.22). Labeling studies with both $^{18}O_2$ and $H_2{}^{18}O$ support a direct oxygenation of the arylpalladium intermediates instead of an acetoxylation/hydrolysis sequence [32]. Oxygenation of the aryl-Pd species by molecular oxygen may either form Pd^{IV} intermediates or oxygen atoms can directly insert to Ar—Pd bonds to generate $ArO\text{-}Pd^{II}$ or $ArO_2\text{-}Pd^{II}$ complexes.

Stahl and co-workers observed different products from Cu^{II}-mediated oxidation of *N*-(8-quinolinyl)benzamide, depending on the reaction conditions. Under basic conditions, the benzamide group undergoes directed C—H methoxylation or chlorination (Scheme 10.23). Under acidic conditions, the quinoline group undergoes

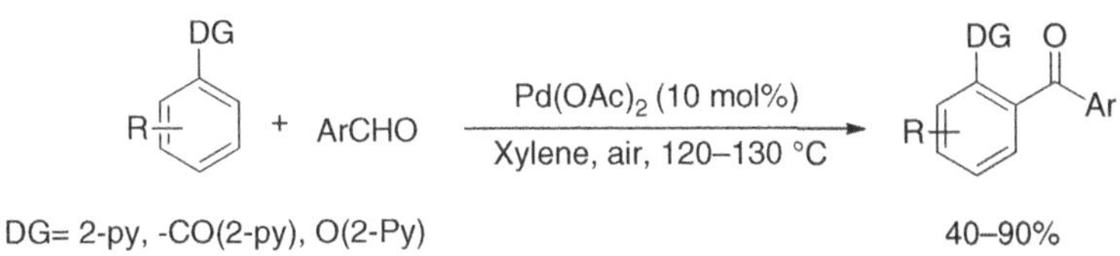

Scheme 10.21 Acylation reaction with benzaldehydes.

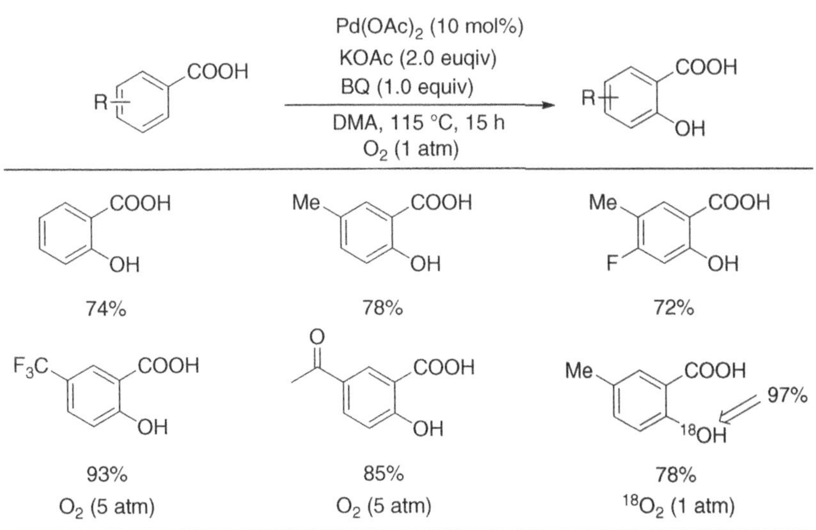

Scheme 10.22 Pd(II)-catalyzed hydroxylation of arenes with O_2.

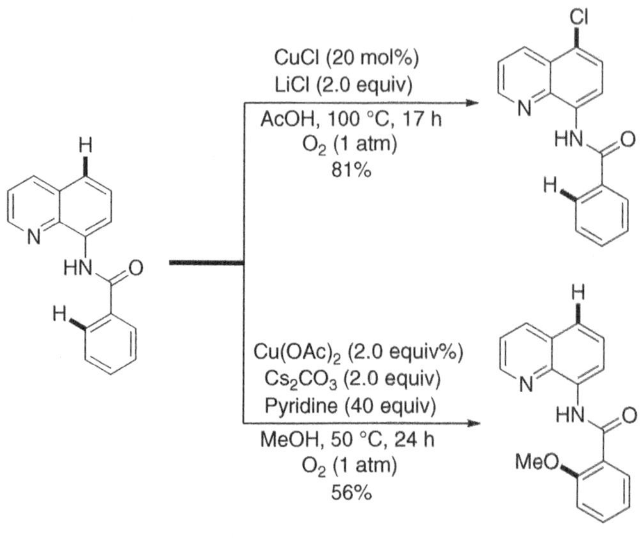

Scheme 10.23 Cu-mediated C—H oxidation of the amidoquinoline substrate.

nondirected chlorination. Experimental and computational mechanistic studies implicate an organometallic C—H activation/functionalization mechanism under the former conditions and a single-electron-transfer mechanism under the latter conditions [33].

10.2 Construction of Substituted Arenes

10.2.1 Cycloaddition

C—H functionalization using the directing group also presents an attractive and powerful strategy for the generation of heteroaromatic compounds, such as indoles, isoquinolines, carbazoles, benzothiazoles, pyridines, and so on. However, cycloaromatization of alkynes with arenes through C—H bond activation forming benzene rings still poses a challenge. A major advance in this area is the synthesis of polycyclic aromatic described by Miura et al. through Rh-catalyzed annulation of phenylazoles with internal alkynes via dual C—H cleavage directed by an *ortho* azole group using $Cu(OAc)_2$ as an oxidant [34]. In 2009, Jiao and co-workers developed a Pd-catalyzed intermolecular cycloaromatization of biaryls with alkynes through dual C—H activation to build benzene cores in the presence of O_2 as the sole oxidant (Scheme 10.24). This observation not only provides a new strategy for constructing aromatics from biaryls and internal alkynes, but also offers an efficient approach for synthetically and medicinally important polycyclic carbazoles [35].

The use of an alkene as the initiator for nucleopalladation of an alkyne followed by coupling with another alkene also allows the concise construction of benzene rings (Scheme 10.25). Loh and co-workers found that both acrylate and styrene derivatives reacted efficiently to afford the desired products in good to excellent yields catalyzed by 5 mol% $PdCl_2$ using molecular oxygen as the oxidant in DMSO at 110 °C [36]. This transformation is believed to be an efficient and sustainable process in which molecular oxygen is the exclusive oxidant with water as the only byproduct generated.

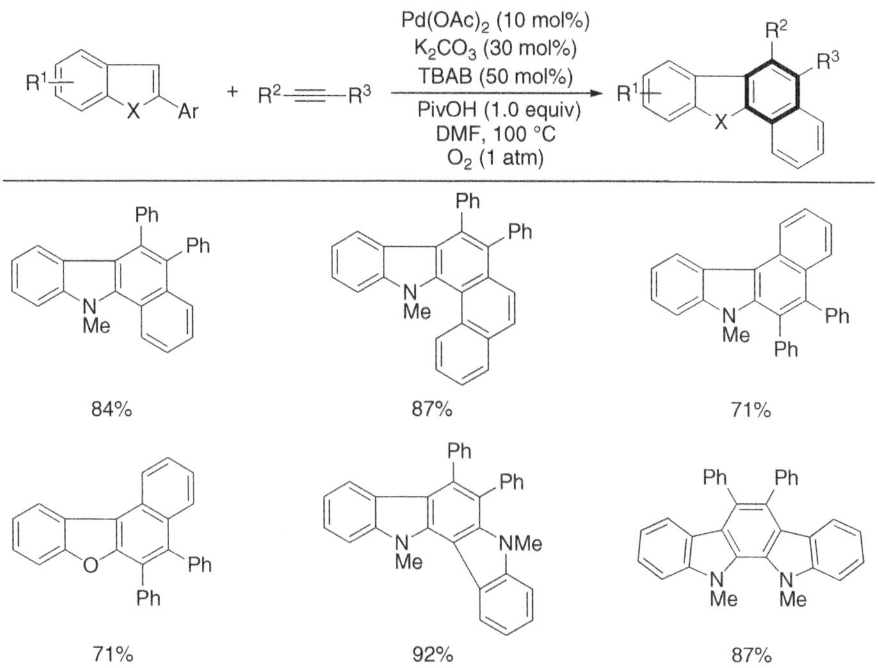

Scheme 10.24 Palladium-catalyzed intermolecular cycloaromatization of biaryls with alkynes.

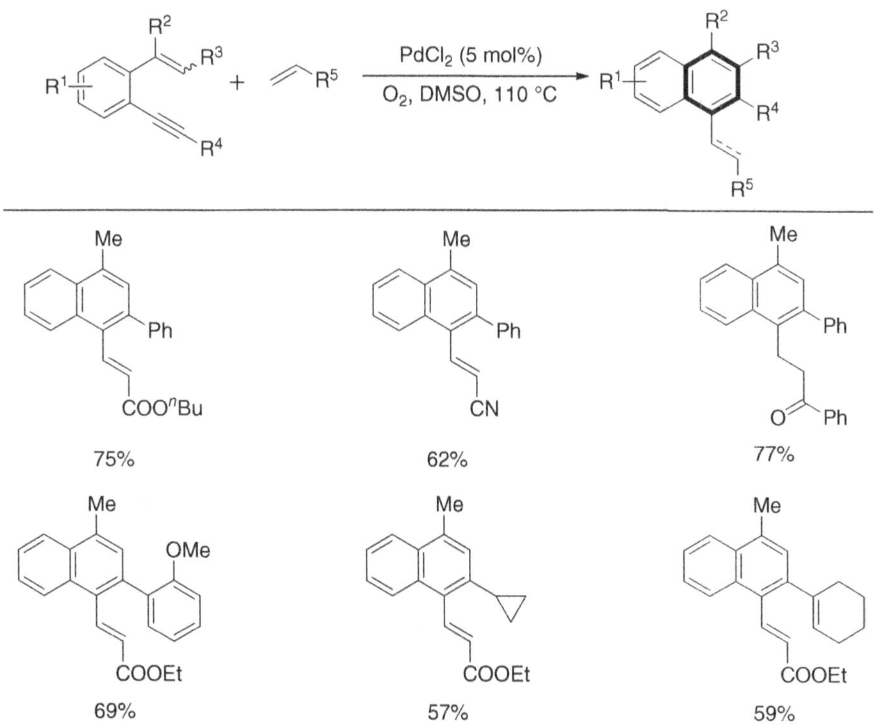

Scheme 10.25 Palladium-catalyzed synthesis of naphthalene derivatives.

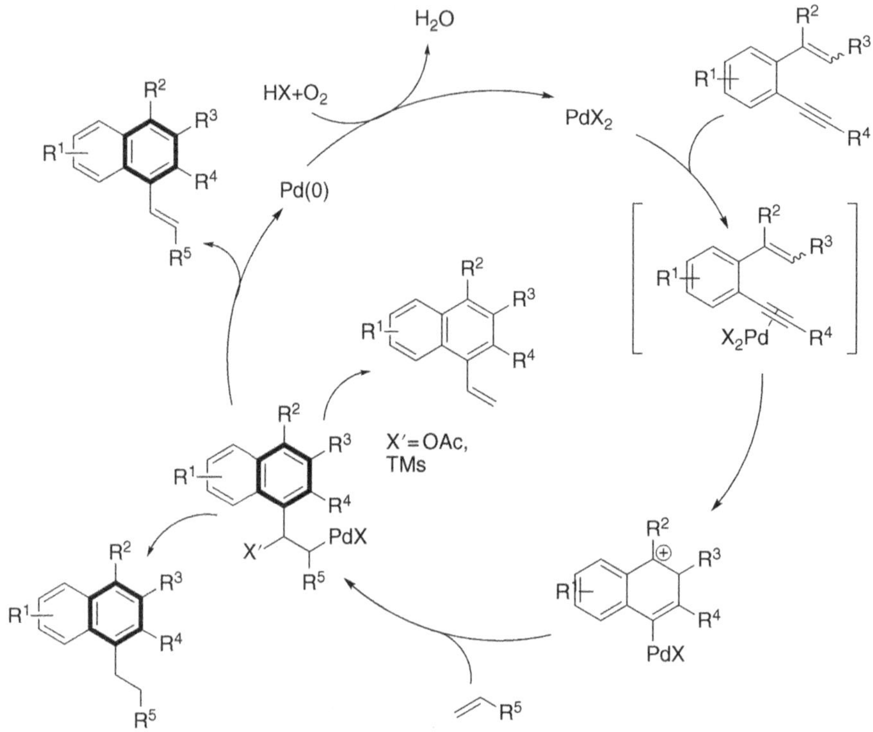

Scheme 10.26 Proposed mechanism for benzene construction.

Mechanistically, the reaction proceeds via intramolecular carbopalladation followed by insertion with an external olefin. It is believed that the success of this transformation relies on the dual roles of the palladium catalyst; namely, its Lewis acidity and late-transition-metal character (Scheme 10.26).

A tandem coupling-ketooxygenation reaction of readily accessible enediyne-carboxylic compounds in the presence of $PdCl_2/CuBr_2$ catalytic system can provide synthesis of substituted aromatics where O_2 both serves as the terminal oxidant and a source of oxygen atoms into the product (Scheme 10.27) [37]. These reactions proceed via an intramolecular anti-aminopalladation process that can undergo *syn*-palladation with an external alkyne [37a] or an enedyine-carboxylic compound [37b] leading to the formation of O-acyl benzoic acids and 2,6-diacylnapthalenes. Incorporation of two oxygen atoms in these products occurs from atmospheric molecular oxygen and H_2O. The aromatization in these systems takes place via a Cu-catalyzed one-electron oxidation process (Scheme 10.28).

Base mediated annulation reaction followed by aerobic oxidative aromatization leading to benzannulation product in one pot generates highly functionalized substituted arenes (Scheme 10.29) [38]. The aromatization is operative in open air reaction conditions. Polysubstituted phenols are efficiently assembled from an aldehyde and two different methyl ketones through a regiospecific [4 + 1 + 1] annulations process (Scheme 10.29a) [38a]. [3 + 3] annulations of α,β-unsaturated carbonyl compounds with dimethyl glutaconate [38b] or 4-sulfonylcrotonates or 1,3-bisphenylsulfonylpropene [38c] can

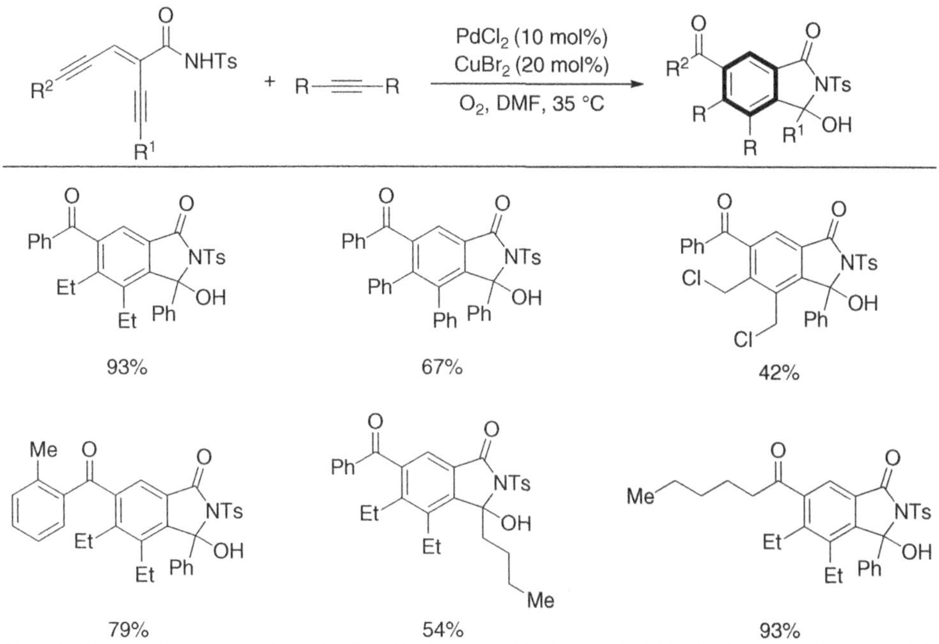

Scheme 10.27 Pd/Cu-catalyzed aerobic cyclization of enediyne compounds and alkynes.

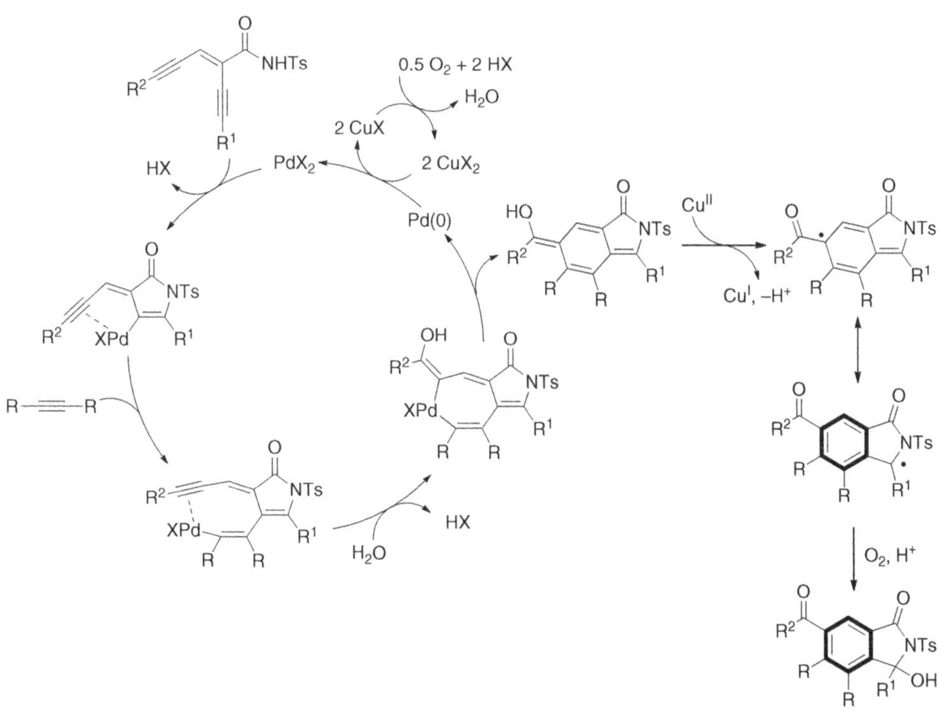

Scheme 10.28 Proposed mechanism for aerobic cyclization of enediyne compounds and alkynes.

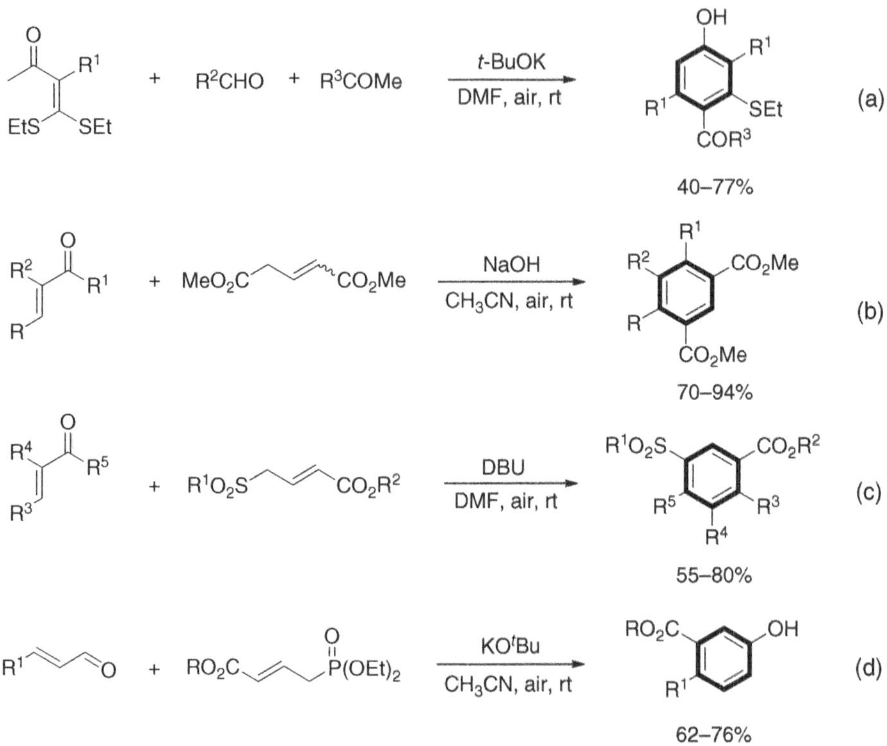

Scheme 10.29 Synthesis of substituted arenes via aerobic oxidative benzannulation process.

provide highly substituted benzene derivatives (Scheme 10.29b–c). In a similar [3 + 3] benzannulation reaction of *α*,*β*-unsaturated aldehydes and *γ*-phosphonyl crotonates (Scheme 10.29d) [39], oxygen atom from air can also be incorporated into the product.

10.2.2 Oxidative Dehydrogenation of Non-Arene Precursors

The dehydrogenation of cyclohexenones to phenols is usually achieved with a stoichiometric application of 2,3-dichloro-5,6-dicyano-*p*-benzoquinone (DDQ) or via bromination/dehydrobromination procedures [40]. Catalytic aerobic dehydrogenation methods represent an appealing alternative; however, earlier methods were limited to high temperature, gas-phase reactions of cyclohexanone itself, or methods that employ stoichiometric palladium salts and proceed in a relatively low (e.g. <30%) yield [41, 42]. In 2011, Stahl and co-workers reported the discovery of a palladium(II) catalytic system, incorporating a pyridine-type ligand underneath for the conversion of substituted cyclohexanones to the corresponding phenols using molecular oxygen as the hydrogen acceptor. This transformation proved to be effective in the preparation of a number of substituted phenol derivatives (Scheme 10.30) [43]. Successful catalysts for this class of reactions find broad utility owing to the numerous straightforward chemical reactions that provide access to substituted cyclohexanones, including enolate arylation and

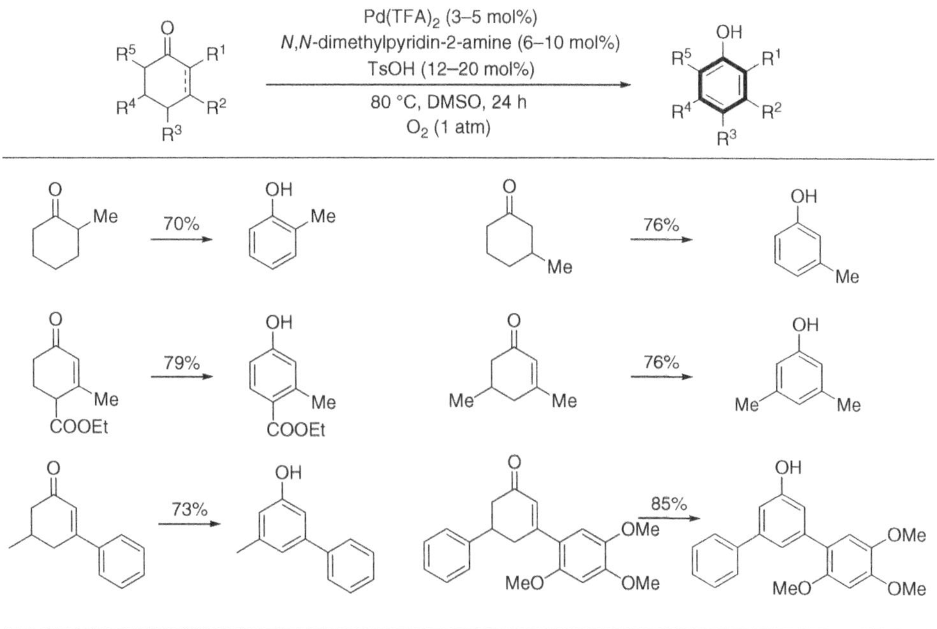

Scheme 10.30 Aerobic dehydrogenation of substituted cyclohexanones.

alkylation methods, conjugate addition to cyclohexenones, Robinson annulation, Diels–Alder reactions, and so on. Readily accessible 3-substituted cyclohexanones are direct precursors to *meta*-substituted phenols that are challenging to prepare via traditional aromatic substitution methods.

Numerous experimental methods, including kinetic studies, filtration tests, Hg poisoning experiments, transmission electron microscopy, and dynamic light scattering, provide compelling evidence that the initial Pd^{II} catalyst mediates the first dehydrogenation of cyclohexanone to cyclohexenone, after which it evolves into soluble Pd nanoparticles that retain catalytic activity. This nanoparticle formation and stabilization is facilitated by each of the components in the catalytic reaction, including the ligand, TsOH, DMSO, substrate, and cyclohexanone intermediate (Scheme 10.31) [44]. By introducing an appropriate nucleophile into the catalytic system, through the selective control of nucleophilic addition or conjugate addition cyclohexanone and cyclohexenone, respectively, followed by catalytic aerobic dehydrogenation, various substituted arene-like aryl ethers [45] and anilines [46] can be synthesized.

Initially, Li and co-workers described a copper-catalyzed aerobic oxidative dehydrogenation strategy for the preparation of aryl ethers from cyclohexenones. The reaction can be accomplished with stoichiometric $CuCl_2{\cdot}H_2O$ and also under aerobic conditions with catalytic $Cu(OTf)_2$ in the presence of cocatalytic *N*-hydroxyphthalimide (NHPI) and stoichiometric KI (Scheme 10.32) [47]. The reaction has a relatively broad substrate scope with respect to the alcohol coupling partner, while the cyclohexenone scope was explored only with stoichiometric $CuCl_2$, which resulted in low to moderate yields of aromatic ethers. With a Pd/C catalytic system, Lemaire and co-workers reported a similar

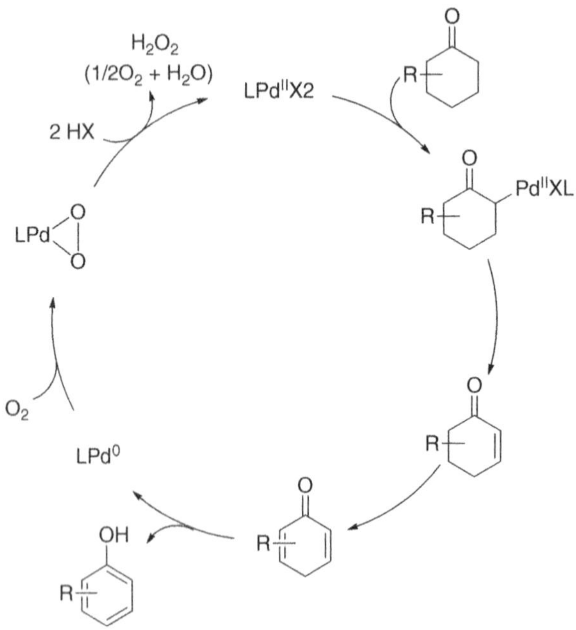

Scheme 10.31 Proposed catalytic mechanism.

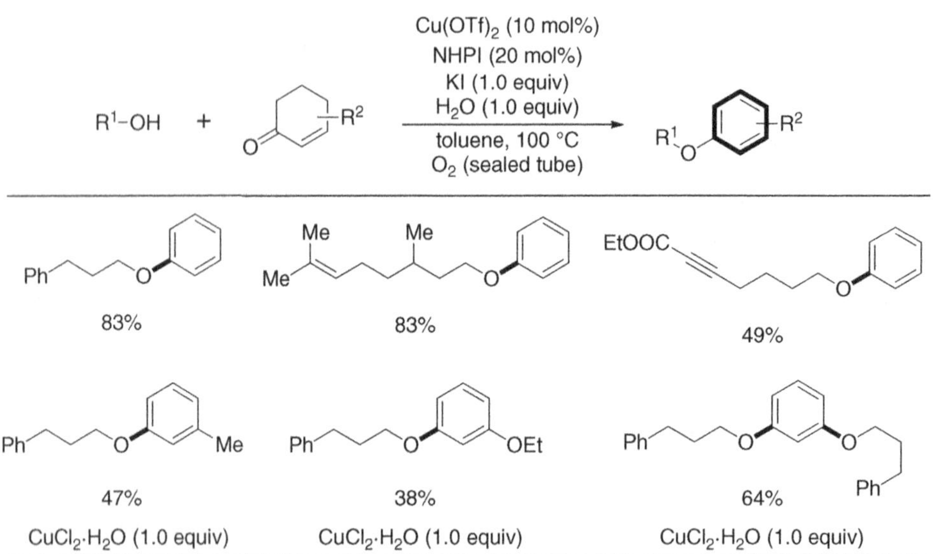

Scheme 10.32 Aerobic synthesis of aromatic ethers.

reaction typically with neat substrate and excess alcohol (5 equiv.) and open to air [45]. Both cyclohexanones and cyclohexenones were effective substrates as well as tetralones.

The groups of Deng/Li [46a–c] and Yoshikai [46d] developed aerobic Pd-catalyzed methods for the synthesis of secondary and tertiary anilines, as well as aromatic sulfonamides via this approach. The catalyst developed by Yoshikai and co-workers was primarily active when primary amines or electron-rich primary anilines were used as substrates in combination with cyclohexanones (Scheme 10.33, condition B). The Deng/Li catalyst system was employed with primary anilines as nucleophilic partners (Scheme 10.33, condition A).

Hong and co-workers reported the Pd-catalyzed method for the synthesis of coumarins from cyclohexanones that proceeds via dehydrogenation of cyclohexanones to phenols, followed by a *ortho*-selective oxidative Heck reaction with acrylates and lactonization [48]. This reaction is effective with diverse substituted cyclohexanones (Scheme 10.34).

Deng and co-workers demonstrated a compelling cyclohexanone condensation route to 3-aryl indoles that proceeds via aerobic dehydrogenation of cyclohexene intermediate (Scheme 10.35) [49]. The cyclohexene intermediate was independently synthesized and shown to convert into the aryl-substituted indole product. The cyclohexanones represent arene surrogates, similar to the aryl ether and aniline synthetic routes described before. The catalyst system, consisting of $Pd(TFA)_2$ and *bis*[2-(diphenylphophino)phenyl]ether (DPEPhos), was effective for the formation of a number of different 3-aryl indoles utilizing substituted indoles and 3- and 4-substituted cyclohexanones.

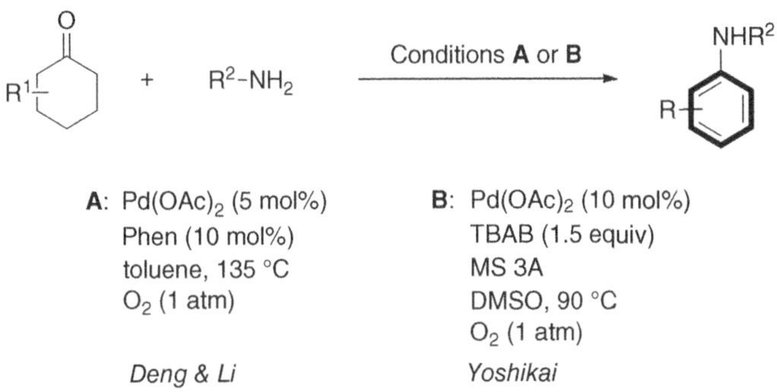

Scheme 10.33 Aerobic synthesis of anilines.

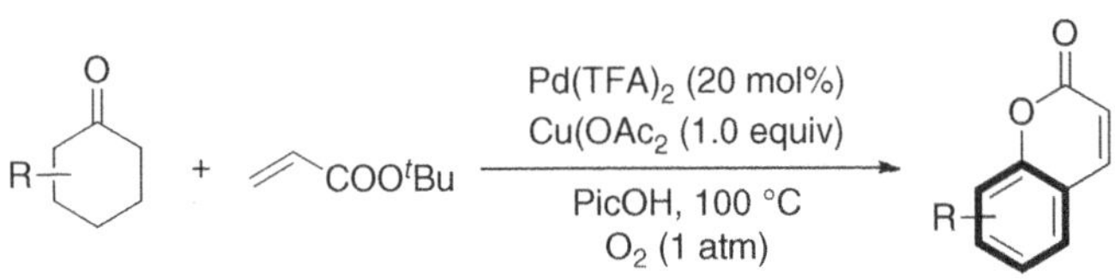

Scheme 10.34 Aerobic synthesis of coumarins.

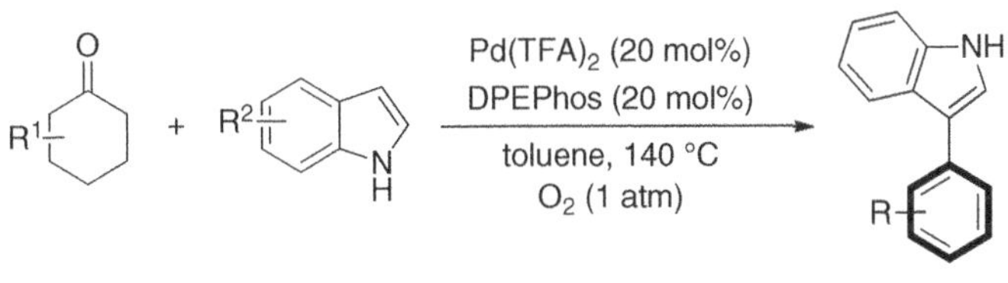

Scheme 10.35 Aerobic construction of indole-substituted arenes.

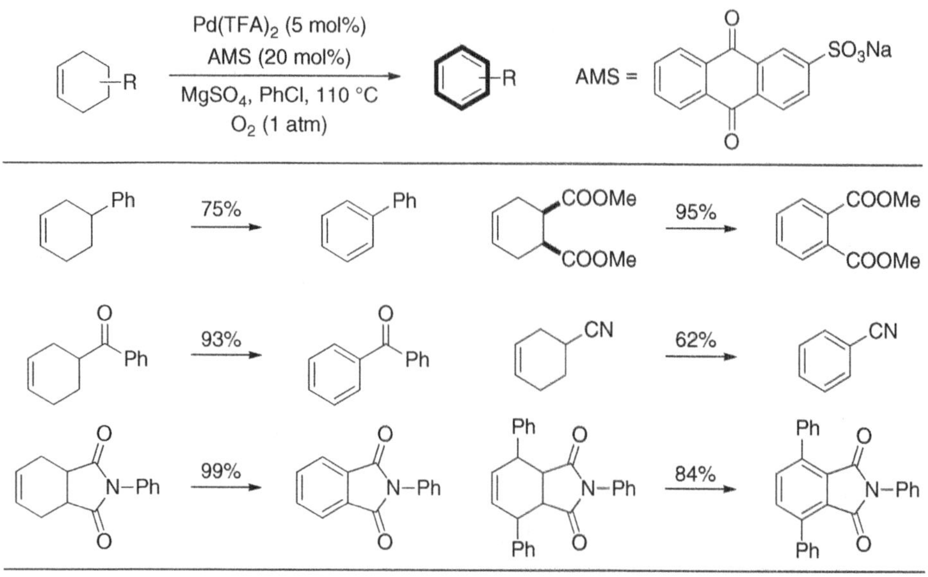

Scheme 10.36 Aerobic dehydrogenation of substituted cyclohexenes.

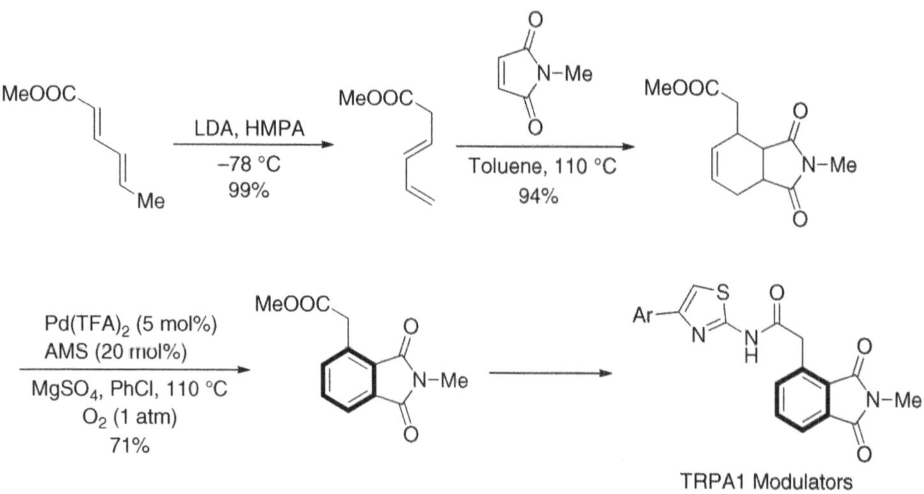

Scheme 10.37 Synthesis of a precursor to a TRPA1 modulator via oxidative dehydrogenation.

Aerobic oxidative dehydrogenation of cyclohexene under palladium catalysis can also provide synthesis of substituted arenes. The dehydrogenation of cyclohexenes could proceed via a Pd^{II}-mediated allylic C—H activation, followed by β-H elimination and aerobic regeneration of Pd^0 (Scheme 10.36) [50]. Successive oxidative aromatization results in synthesis of substituted arenes. The β-H elimination step in this process differs from the more established reactivity of π-allyl-Pd^{II} species with nucleophiles. Using sodium anthraquinone-2-sulfonate (AMS) as a cocatalyst, aerobic dehydrogenation of substituted cyclohexene leads to the synthesis of arene derivatives. AMS is efficient for aerobic turnover and avoids disproportionation reactivity of the cyclohexene. An alternative catalyst, generated *in situ* from $Pd(OAc)_2$ and *p*-TsOH, allows the synthesis of tri- and tetrasubstituted aromatic rings.

The utility of this aromatization strategy offers a three-step synthetic route toward phthalimide-based modulators of TRPA1, an ion channel that has been implicated in the pain response (Scheme 10.37). The methyl sorbate and *N*-methylmaleimide starting materials are very inexpensive, and the carbocyclic core with the desired functionality may be accessed in a single step via Diels–Alder cycloaddition. Pd-catalyzed aerobic dehydrogenation of the resulting cyclohexene provides efficient access to the desired phthalimide product. The previously reported route starts from 3-nitrophthalimide, featured a series of functional group interconversions, including a Suzuki cross-coupling step and Ru-catalyzed oxidative cleavage of an alkene, to access the desired precursor in five steps [51].

The efficient and selective synthesis of aromatic molecules with nonconventional substitution patterns remains a challenging and active area of research. Two strategies herein provide a compelling approach for the conversion of readily accessible substrates into various substituted arenes. Particularly, transition-metal-catalyzed cross-coupling reactions dramatically expand the scope of accessible substituted aromatic compounds, and bypass many of the regioselectivity constraints and substrate limitations of classical nucleophilic and electrophilic aromatic substitution reactions. The (partially) saturated substrates for aerobic oxidative dehydrogenation are often readily assembled or derivatized by straightforward synthetic methods and thereby provide a means to access aromatic compounds with substitution patterns that are not readily available by existing cross-coupling or aromatic C—H functionalization methods. Furthermore, applications of these methods in the preparation of various pharmaceutical and natural product intermediates and targets show that these new strategies can provide compelling solutions to synthetic challenges. The development of new catalytic systems under the benign oxidants providing highly functionalize substituted arenes will undoubtedly play an important role in continued expansion of this field.

References

1 (a) Smith, M.B. and March, J. (2007). *Advanced Organic Chemistry Reactions, Mechanisms and Structure*, 6e, 657–751. NJ: Wiley. (b) Astruc, D. (ed.) (2002). *Modern Arene Chemistry*. Weinheim, Germany: Wiley-VCH.

2 Ullmann, F. and Bielecki, J. (1901). *Chem. Ber* 34: 2174.

3 (a) Whitesides, G.M., San Filippo, J., Casey, C.P., and Panek, E.J. (1967). *J. Am. Chem. Soc.* 89: 5302. (b) Whitesides, G.M., Fischer, W.F., San Filippo, J. et al. (1969). *J. Am. Chem. Soc.* 91: 4871. (c) Lipshutz, B.H., Siegmann, K., and Garcia, E. (1991). *J. Am. Chem. Soc.* 113: 8161. (d) Lipshutz, B.H., Siegmann, K., Garcia, E., and Kayser, F. (1993). *J. Am. Chem. Soc.* 115: 9276. (e) Lipshutz, B.H., Kayser, F., and Maullin, N. (1994). *Tetrahedron Lett.* 35: 815.

4 Su, X., Fox, D.J., Blackwell, D.T. et al. (2006). *Chem. Commun.* 42: 3883.

5 Cahiez, G., Chaboche, C., Mahuteau-Betzer, F., and Ahr, M. (2005). *Org. Lett.* 7: 1943.

6 Krasovskiy, A., Tishkov, A., del Amo, V. et al. (2006). *Angew. Chem. Int. Ed.* 45: 5010.

7 (a) Cahiez, G., Moyeux, A., Buendia, J., and Duplais, C. (2007). *J. Am. Chem. Soc.* 129: 13788. (b) Maji, M.S., Pfeifer, T., and Studer, A. (2008). *Angew. Chem. Int. Ed.* 47: 9547.

8 Yu, J., Liu, J., Shi, G. et al. (2015). *Angew. Chem. Int. Ed.* 54: 4079.

9 Yang, S.-D., Sun, C.-L., Fang, Z. et al. (2008). *Angew. Chem. Int. Ed.* 47: 1473.

10 Wang, D.-H., Mei, T.-S., and Yu, J.-Q. (2008). *J. Am. Chem. Soc.* 130: 17676.

11 (a) Do, H.-Q. and Daugulis, O. (2009). *J. Am. Chem. Soc.* 131: 17052. (b) Monguchi, D., Yamamura, A., Fujiwara, T. et al. (2010). *Tetrahedron Lett.* 51: 850. (c) Li, Y., Jin, J., Qian, W., and Bao, W. (2010). *Org. Biomol. Chem.* 8: 326. (d) Zhu, M., Fujita, K., and Yamaguchi, R. (2011). *Chem. Commun.* 47: 12876.

12 (a) Izawa, Y. and Stahl, S.S. (2010). *Adv. Synth. Catal.* 352: 3223. (b) Erdmann, N., Su, Y., Bosmans, B. et al. (2016). *Org. Process Res. Dev.* 20: 831. (c) Wang, D. and Stahl, S.S. (2017). *J. Am. Chem. Soc.* 139: 5704. (d) Wang, D., Izawa, Y., and Stahl, S.S. (2014). *J. Am. Chem. Soc.* 136: 9914.

13 (a) Hon, S.-W., Li, C.-H., Kuo, J.-H. et al. (2001). *Org. Lett.* 3: 869. (b) Barhate, N.B. and Chen, C.-T. (2002). *Org. Lett.* 4: 2529. (c) Guo, Q.X., Wu, Z.-J., Luo, Z.-B. et al. (2007). *J. Am. Chem. Soc.* 129: 13927. (d) Chu, C.-Y., Hwang, D.-R., Wang, S.-K., and Uang, B.-J. (2001). *Chem. Commun.* 37: 980. (e) Takizawa, S., Katayama, T., and Sasai, H. (2008). *Chem. Commun.* 44: 4113.

14 (a) Egami, H. and Katsuki, T. (2009). *J. Am. Chem. Soc.* 131: 6082. (b) Egami, H., Matsumoto, K., Oguma, T. et al. (2010). *J. Am. Chem. Soc.* 132: 13633.

15 (a) Nakajima, M., Kanayama, K., Miyoshi, I., and Hashimoto, S.-I. (1995). *Tetrahedron Lett.* 36: 9519. (b) Nakajima, M., Miyoshi, I., Kanayama, K. et al. (1999). *J. Org. Chem.* 64: 2264. (c) Noji, M., Nakajima, M., and Koga, K. (1994). *Tetrahedron Lett.* 35: 7983. (d) Kim, K.H., Lee, D.-W., Lee, Y.-S. et al. (2004). *Tetrahedron* 60: 9037. (e) Li, X., Yang, J., and Kozlowski, M.C. (2001). *Org. Lett.* 3: 1137. (f) Li, X., Hewgley, J.B., Mulrooney, C. et al. (2003). *J. Org. Chem.* 68: 5500. (g) Sakamoto, T., Yonehara, H., and Pac, C. (1997). *J. Org. Chem.* 62: 3194. (h) Sharma, V.B., Jain, S.L., and Sain, B. (2004). *J. Mol. Catal. A: Chem.* 219: 61. (i) Yekta, S., Krasnova, L.B., Mariampillai, B. et al. (2004). *J. Fluor. Chem.* 125: 517. (j) Karikomi, M., Yamada, M., Ogawa, Y. et al. (2005). *Tetrahedron Lett.* 46: 5867. (k) Caselli, A., Giovenzana, G.B., Palmisano, G. et al. (2003). *Tetrahedron: Asymmetry* 14: 1451. (l) Grach, G., Pieters, G., Dinut, A. et al. (2011). *Organometallics* 30: 4074. (m) Majumder, P.L., Bandyopadhyay, S., and Pal, S. (2008). *J. Indian Chem. Soc.* 85: 1116. (n) Gao, J., Reibenspies, J.H., and Martell, A.E. (2003). *Angew. Chem. Int. Ed.* 42: 6008.

16 Roithov, J. and Milko, P. (2010). *J. Am. Chem. Soc.* 132: 281.

17 Li, B.-J., Tian, S.-L., Fang, Z., and Shi, Z.-J. (2008). *Angew. Chem. Int. Ed.* 47: 1115.

18 Brasche, G., Garcia-Fortanet, J., and Buchwald, S.L. (2008). *Org. Lett.* 10: 2207.

19 (a) Heck, R.F. and Nolley, J.P. (1972). *J. Org. Chem.* 37: 2320. (b) Mizoroki, T., Mori, K., and Ozaki, A. (1971). *Bull. Chem. Soc. Jpn.* 44: 581.

20 Bras, J.L. and Muzart, J. (2008). *Chem. Rev.* 111: 1170.

21 (a) Yokota, T., Tani, M., Sakaguchi, S., and Ishii, Y. (2003). *J. Am. Chem. Soc.* 125: 1476. (b) Tani, M., Sakaguchi, S., and Ishii, Y. (2004). *J. Org. Chem.* 69: 1221. (c) Yamada, T., Sakaguchi, S., and Ishii, Y. (2005). *J. Org. Chem.* 70: 5471. (d) Obora, Y., Okabe, Y., and Ishii, Y. (2010). *Org. Biomol. Chem.* 8: 4071.

22 Dams, M., De Vos, D.E., Celen, S., and Jacobs, P.A. (2003). *Angew. Chem. Int. Ed.* 42: 3512.

23 Zhang, Y.-H., Shi, B.-F., and Yu, J.-Q. (2009). *J. Am. Chem. Soc.* 131: 5072.

24 Huang, Q., Zhang, X., Qiu, L. et al. (2015). *Adv. Synth. Catal.* 357: 3753.
25 Wang, D.-H., Engle, K.M., Shi, B.-F., and Yu, J.-Q. (2010). *Science* 327: 315.
26 Engle, K.M., Wang, D.-H., and Yu, J.-Q. (2010). *Angew. Chem. Int. Ed.* 49: 6169.
27 Chen, X., Hao, X.-S., Goodhue, C.E., and Yu, J.-Q. (2006). *J. Am. Chem. Soc.* 128: 6790.
28 John, A. and Nicholas, K.M. (2011). *J. Org. Chem.* 76: 4158.
29 (a) Wang, W., Luo, F., Zhang, S., and Cheng, J. (2010). *J. Org. Chem* 75: 2415. (b) Wang, W., Pan, C., Chen, F., and Cheng, J. (2011). *Chem. Commun.* 47: 3978.
30 Jia, X., Zhang, S., Wang, W. et al. (2009). *Org. Lett.* 11: 3120.
31 Engle, K.M., Mei, T.-S., Wasa, M., and Yu, J.-Q. (2012). *Acc. Chem. Res.* 45: 788.
32 Zhang, Y.-H. and Yu, J.-Q. (2009). *J. Am. Chem. Soc.* 131: 14654.
33 Suess, A.M., Ertem, M.Z., Cramer, C.J., and Stahl, S.S. (2013). *J. Am. Chem. Soc.* 135: 9797.
34 (a) Umeda, U., Tsurugi, H., Satoh, T., and Miura, M. (2008). *Angew. Chem. Int. Ed.* 47: 4019. (b) Morimoto, K., Hirano, K., Satoh, T., and Miura, M. (2010). *Org. Lett.* 12: 2068.
35 Shi, Z., Ding, S., Cui, Y., and Jiao, N. (2009). *Angew. Chem. Int. Ed.* 48: 7895.
36 Feng, C. and Loh, T.-P. (2010). *J. Am. Chem. Soc.* 132: 17710.
37 (a) Ling, F., Li, Z., Zheng, C. et al. (2014). *J. Am. Chem. Soc.* 136: 10914. (b) Ling, F., Wan, Y., Wang, D., and Ma, C. (2016). *J. Org. Chem.* 81: 2770.
38 (a) Wang, M., Fu, Z., Feng, H. et al. (2010). *Chem. Commun.* 46: 9061. (b) Diallo, A., Zhao, Y.-L., Wang, H. et al. (2012). *Org. Lett.* 14: 5776. (c) Joshi, P.R., Undeela, S., Reddy, D.D. et al. (2015). *Org. Lett.* 17: 1449.
39 Joshi, P.R., Nanubolu, J.B., and Menon, R.S. (2016). *Org. Lett.* 18: 752.
40 (a) Walker, D. and Hiebert, J.D. (1967). *Chem. Rev.* 67: 153. (b) Buckle, D.R. (2010). *Encyclopedia of Reagents for Organic Synthesis*. Hoboken, NJ: Wiley.
41 Muzart, J. (2010). *Eur. J. Org. Chem.* 2010: 3779.
42 (a) Yi, C.S. and Lee, D.W. (2009). *Organometallics* 28: 947. (b) Muzart, J. and Pete, J.P. (1982). *J. Mol. Catal.* 15: 373. (c) Wenzel, T.T.J. (1989). *Chem. Commun.* 932.
43 Izawa, Y., Pun, D., and Stahl, S.S. (2011). *Science* 333: 209.
44 (a) Pun, D., Diao, T., and Stahl, S.S. (2013). *J. Am. Chem. Soc.* 135: 8213. (b) Diao, T., Pun, D., and Stahl, S.S. (2013). *J. Am. Chem. Soc.* 135: 8205.
45 Sutter, M., Sotto, N., Raoul, Y. et al. (2013). *Green Chem.* 15: 347.
46 (a) Girard, S.A., Hu, X., Knauber, T. et al. (2012). *Org. Lett.* 14: 5606. (b) Xie, Y., Liu, S., Liu, Y. et al. (2012). *Org. Lett.* 14: 1692. (c) Cao, X., Bai, Y., Xie, Y., and Deng, G.-J. (2014). *J. Mol. Catal. A: Chem.* 94: 383. (d) Hajra, A., Wei, Y., and Yoshikai, N. (2012). *Org. Lett.* 14: 5488.
47 Simon, M.-O., Girard, S.A., and Li, C.-J. (2012). *Angew. Chem. Int. Ed.* 51: 7537.
48 Kim, D., Min, M., and Hong, S. (2013). *Chem. Commun.* 49: 4021.
49 Chen, S., Liao, Y., Zhao, F. et al. (2014). *Org. Lett.* 16: 1618.
50 Iosub, A.V. and Stahl, S.S. (2015). *J. Am. Chem. Soc.* 137: 3454.
51 Muthuppal-Niappan, M., Kumar, S., Thomas, A., KhairatkarJoshi, N., and Mukhopadhyay, I. World Patent WO 09/118596, 2009, October 1.

11

Green Oxidative Synthesis of Heterocyclic Compounds

Sreekumar Pankajakshan and Shunsuke Chiba

Nanyang Technological University, Singapore

11.1 Introduction

The importance of heterocycles cannot be overstated owing to their ubiquitous presence in natural products, pharmaceuticals, and functional materials [1]. Due to their widespread applications, developments of new methods to supply these compounds have always been of high priority in both academic and industrial communities in chemistry. Contemporary organic synthesis greatly emphasizes sustainable protocols that are efficient in energy, resources, and atom economy, and a paradigm shift has taken place in the last two decades where principles of green chemistry have had a profound and continuous influence on the reaction design [2]. Vital innovations in catalysis, technology, and cascade/telescoping approaches have enabled chemical synthesis to reach new greener heights and many classical bond formations could now be efficiently achieved with minimal pre-functionalization, cost, and toxicity concerns. The advancements in new catalytic systems have especially contributed toward the development of sustainable oxidations whereby new covalent bonds could be constructed through direct C—H functionalization [3], cross-dehydrogenative coupling [4], and aerobic oxygenation [5].

This chapter covers selected reports on environmentally benign and operationally facile oxidative methods for the synthesis of various heterocyclic motifs reported in the past 10 years. The green chemical processes for construction of heterocyclic scaffolds with high atom economy and selectivity that feature cheaper and abundant catalysts like copper and iron salts, innocuous and easily available oxidants such as molecular oxygen, or peroxides were chosen. The overview is classified into different sections based on the functional moieties that are oxidatively functionalized including oxidative cyclization/annulation onto alkenes, alkynes, C(sp^2)—H, and C(sp^3)—H bonds.

Green Oxidation in Organic Synthesis, First Edition. Edited by Ning Jiao and Shannon S. Stahl.

11.2 Synthesis of Heterocycles via Oxidative Cyclization onto Alkenes

11.2.1 Diamination of Alkenes

Cyclization/annulation via oxidative difunctionalization of alkenes provides an efficient method for the facile installation of two distinct functional groups across a C—C double bond, resulting in the synthesis of diverse heterocyclic scaffolds with varying molecular complexity. Chiba's research group reported a new approach toward the synthesis of bi-/tricyclic dihydroimidazoles **2** through Cu-catalyzed aerobic [3+2]-annulation of N-alkenyl amidines **1** (Scheme 11.1) [6]. A wide range of bi- and tricyclic dihydroimidazoles could be constructed under mild reaction conditions with molecular oxygen as the terminal oxidant. Both *Z* and *E*-alkene substrates **1a** and **1b** cyclized into single diastereomers **2a** and **2b**, respectively, with retention of the configuration, indicating concerted diamination as a possible reaction pathway. The process should begin with single-electron-oxidation of amidines **1** to form aminidyl radicals **3**, which is further oxidized under the present conditions to form nitrene species **4**. The resulting nitrene species **4** are regarded as an equivalent as 1,3-dipole **5**, which could undergo concerted diamination with pendant alkenes to form **2**. The synthetic appeal of this novel reaction was further enhanced by AlH_3-mediated reductive conversion of **2** to synthetically valuable vicinal diamines **6**.

Hajra and co-workers reported synthesis of 3-aroylimidazo[1,2-a]pyridines **9** by Cu-catalyzed aerobic oxidative coupling of 2-aminopyridines **7** and readily accessible chalcones **8** in 1,2-dichlorobenzene (1,2-DCB) (Scheme 11.2) [7]. The process is

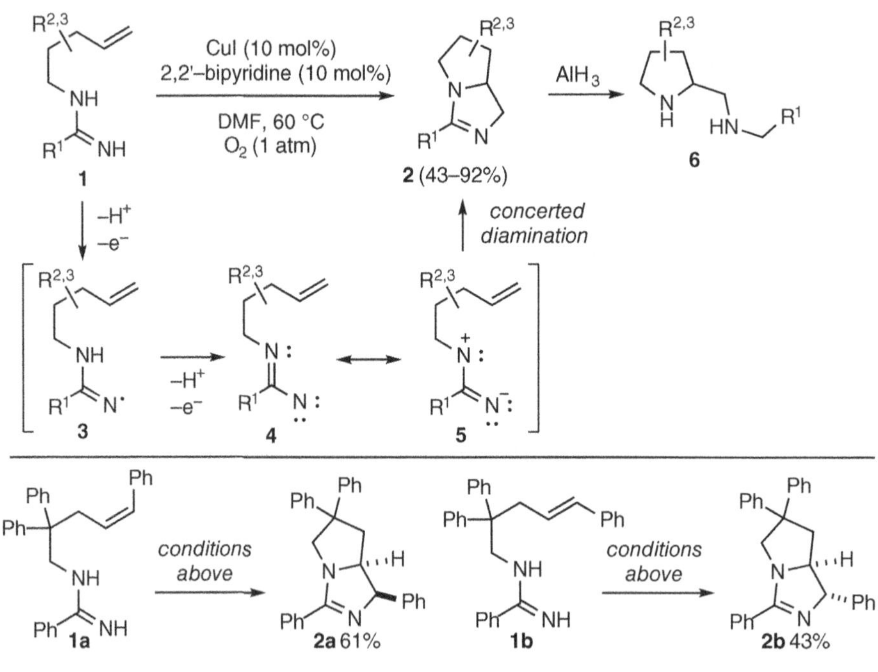

Scheme 11.1 Cu-catalyzed aerobic [3+2]-annulation of N-alkenyl amidines.

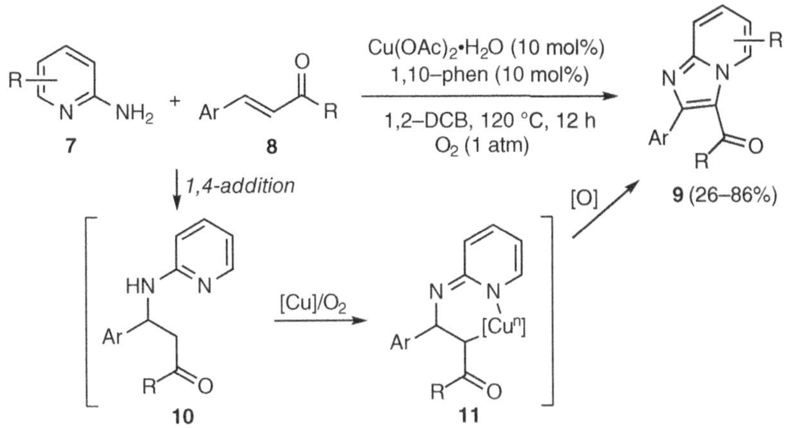

Scheme 11.2 Cu-catalyzed aerobic annulation for synthesis of imidazo[1,2-a]pyridines.

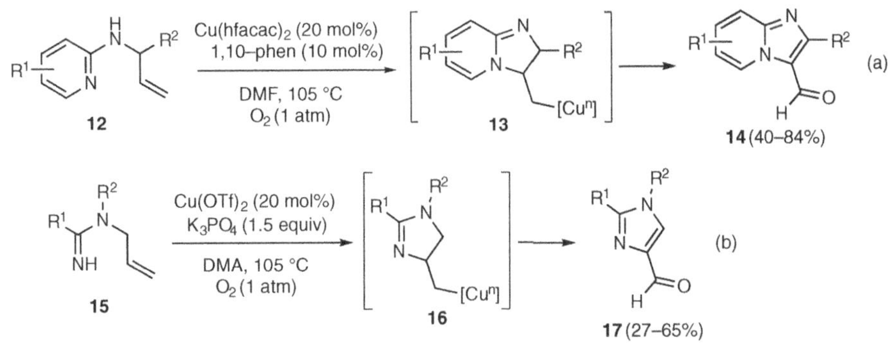

Scheme 11.3 Cu-catalyzed aerobic aminooxygenation for synthesis of imidazo[1,2-a]pyridine-3-carbaldehydes and imidazole-4-carbaldehydes.

initiated by 1,4-addition of 2-aminopyridines 7 onto chalcones **8** to form β-(2-pyridylamino)ketones **10**, which undergo oxidative α-amination through metallacycle intermediates **11** and aromatization to deliver **9**.

11.2.2 Aminooxygenation of Alkenes

Cu-catalyzed aerobic synthesis of imidazo[1,2-a]pyridine-3-carbaldehydes **14** from 2-allylaminopyridines **12** via intramolecular aminooxygenation of alkenes was reported by Zhu (Scheme 11.3a) [8]. The analogous Cu-catalyzed aerobic methodology was also amenable to the synthesis of imidazole-4-carbaldehydes **17** from N-allylamidines **15** (Scheme 11.3b). Aminocupration of the alkene tethers with 5-*exo*-amino-cyclization followed by oxygenation of the resulting organocopper species **13** or **16** were proposed to be the major events in the mechanistic course of these transformations.

Sodeoka and co-workers reported Cu-catalyzed aerobic oxidative cyclization of unsaturated hydrazones **18** (Scheme 11.4) [9]. Interestingly, the product formation was solvent-dependent, namely, with ethanol providing the highest selectivity for the formation

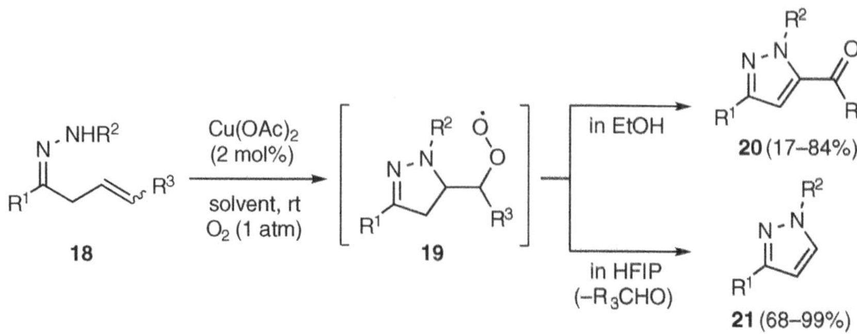

Scheme 11.4 Cu-catalyzed aerobic synthesis of pyrazoles.

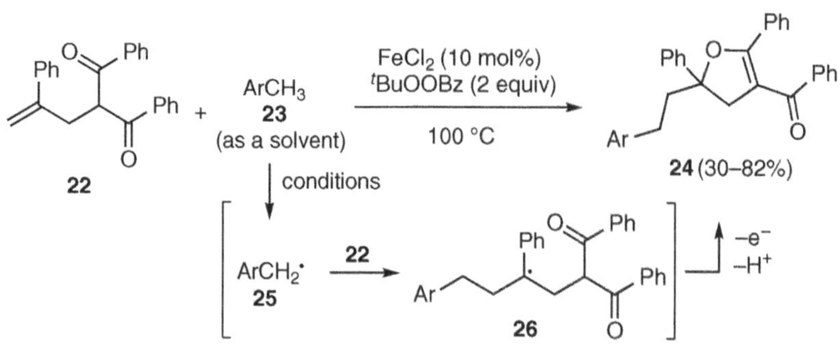

Scheme 11.5 Fe-catalyzed oxidative synthesis of dihydrofurans.

of aminooxygenated pyrazoles **20**, whereas 1,1,1,3,3,3-hexafluoro-2-propanol (HFIP) led to the exclusive formation of the fragmentation pyrazoles **21** through C—C bond cleavage. Oxidative generation of hydrazone radical followed by sequence of C—N bond forming radical cyclization and aerobic oxygenation affords peroxy radical species **19** as a key intermediate. Reduction of the peroxy radical moiety by the Cu(I) species forms the ketone moiety of **20**, whereas oxidative fragmentation of the peroxy species directly delivers **21** and aldehyde.

11.2.3 Carbooxygenation of Alkenes

Guo's group disclosed Fe-catalyzed *tert*-butylperoxy benzoate (*t*-BuOOBz)-mediated carbooxygenation of alkenyl 1,3-dicarbonyl compounds **22** with a solvent amount of hydrocarbons **23**, such as methyl arenes and cycloalkanes, to deliver a range of dihydrofuran structures **24** containing fully substituted carbon centers (Scheme 11.5) [10]. The process was initiated by generation of carbon radicals **25** from the hydrocarbons **23** through decomposition of peroxides and their addition onto the alkenyl moiety of **22**, which was followed by intramolecular C—O bond formation between the resulting carbon radical and the carbonyl group in intermediates **26**. The Fe catalyst could play the role of the electron carrier and acceptor that may facilitate the decomposition of peroxides and the C—O bond formation.

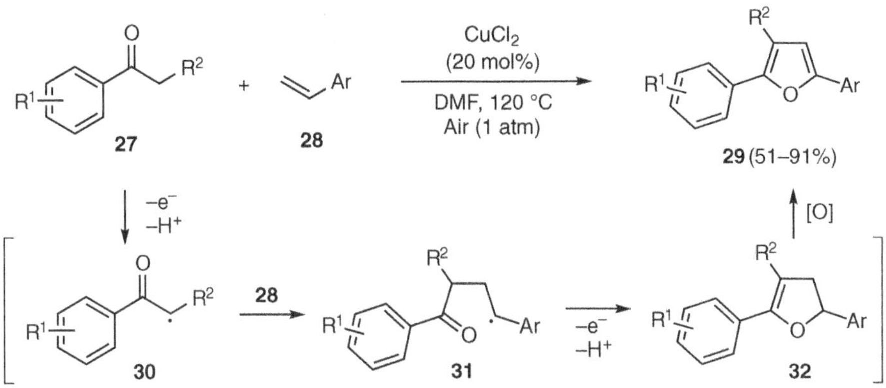

Scheme 11.6 Cu-catalyzed aerobic synthesis of polysubstituted furans.

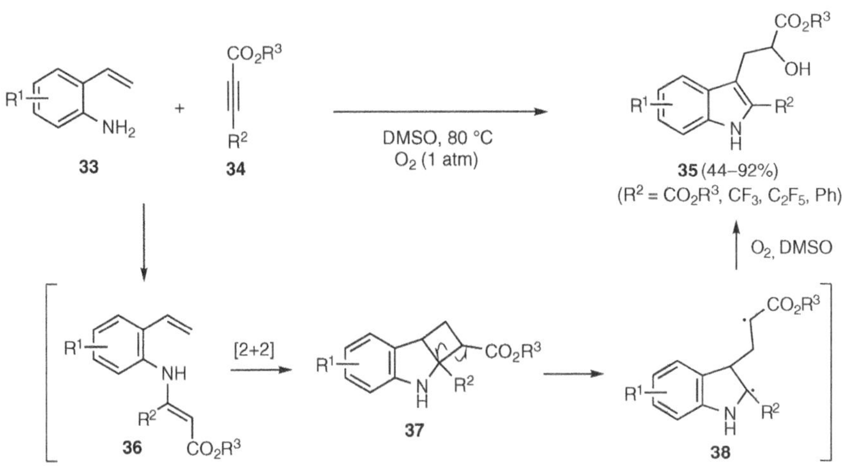

Scheme 11.7 Aerobic oxidative synthesis of tryptophols from 2-vinylanilines and alkynes.

A copper-catalyzed aerobic annulation procedure reported by Hajra's group enabled the regioselective synthesis of polysubstituted furans **29** from aryl ketones **27** and styrenes **28** (Scheme 11.6) [11]. The process could be initiated by generation of α-keto radicals **30** through single-electron-oxidation of ketones **27** that add onto styrenes **28** to form the first C—C bond. Subsequent intramolecular oxidative C—O bond formation of the resulting γ-keto radicals **31** affords dihydrofurans **32**, which is followed by further aromatization to afford furans **29**.

11.2.4 Alkylation/Arylation of Alkenes

Jiao and co-workers developed a metal-free aerobic approach to synthesize highly valuable tryptophol derivatives **35** from readily available *ortho*-vinyl anilines **33** and electron-deficient alkynes **34** including not only symmetrical acetylenedicarboxylates but also unsymmetrical alkynes with perfluoroalkyl and phenyl groups (Scheme 11.7) [12]. The process includes unprecedented C—C bond cleavage of alkyne **34**. It was proposed that the [2+2]-cycloaddition of initially formed enamine **36** produced the key cyclobutane intermediate **37**, which then underwent ring opening through homolysis of the C—C bond to form the biradical species **38**. Further aerobic oxygenation/oxidation proceeded to form tryptophol products **35**.

There have been reported several protocols on metal-catalyzed or metal-free oxidative transformation of readily available N-aryl acrylamides for synthesis of biologically important oxindole skeletons. Li's group reported oxidative tandem coupling of N-arylacrylamides **39** and aldehydes **40** mediated by *tert*-butyl hydroperoxide (^{t}BuOOH) (Scheme 11.8) [13]. The potential aldehyde sources comprised of aryl, heteroaryl, and aliphatic aldehydes as well as formates, allowing facile installation of carbonyl groups to oxindoles **41**. The process was triggered by generation of acyl radicals **42** by H-radical abstraction from the formyl group of **40** with decomposition of peroxide and their subsequent addition onto the alkene. The resulting C-radical intermediates **43** undergo intramolecular radical cyclization and re-aromatization to afford oxindoles **41**. Andrade and co-workers reported conceptually the same strategy for synthesis of oxindoles using secondary formamides as the source of acyl radicals in the presence of an Fe catalyst and hydrogen peroxide in aqueous solvents [14].

Duan and Guo adopted simple alcohols **45** as a C-radical source for synthesis of oxindoles **46** from N-aryl acrylamides **44** using *tert*-butyl hydroperoxide (^{t}BuOOH) as a sole oxidant (Scheme 11.9) [15].

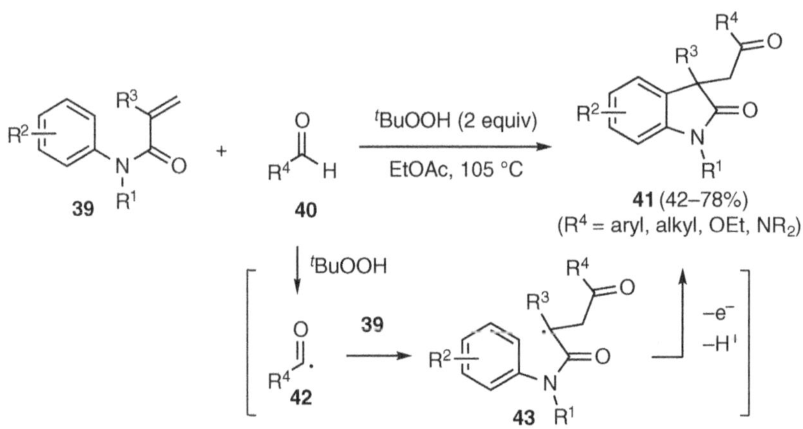

Scheme 11.8 Oxidative coupling of N-aryl acrylamides and aldehydes.

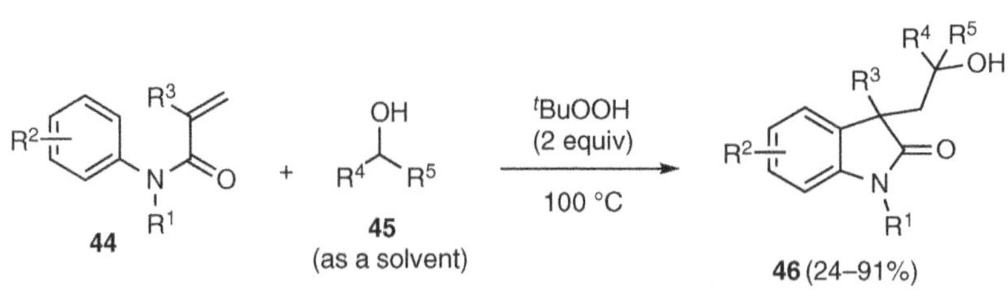

Scheme 11.9 Oxidative coupling of N-aryl acrylamides and alcohols.

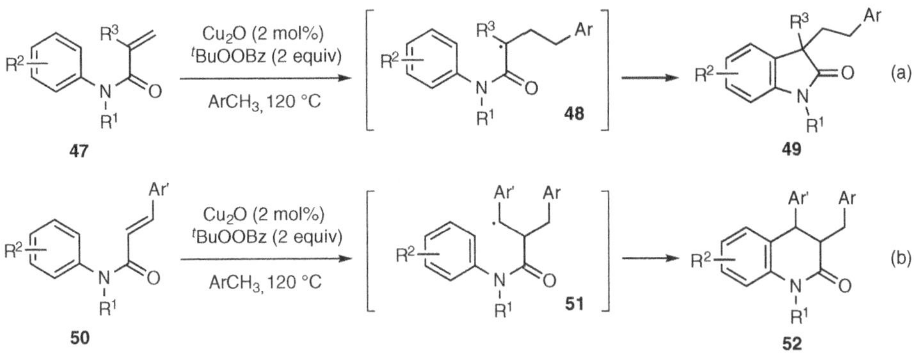

Scheme 11.10 Oxidative coupling of N-aryl acrylamides and arenes.

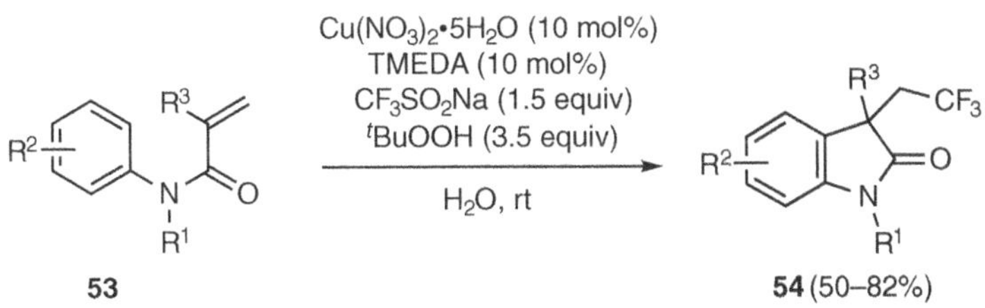

Scheme 11.11 Copper-catalyzed synthesis of 3-trifluoroethyl oxindoles.

Similarly, the same group developed a copper-catalyzed benzylarylation of acrylamides **47** with methyl arenes (employed as a solvent) in the presence of *tert*-butylperoxy benzoate (^{t}BuOOBz) as the stoichiometric oxidant, where the corresponding benzyl radicals add onto **47** to form C-radical intermediates **48**, which cyclize into oxindoles **49** (Scheme 11.10a) [16]. On the other hand, use of cinnamides **50** instead of acrylamides **47** under the same conditions rendered the position of the radical addition α to the amide carbonyl, providing C-radical intermediates **51** that undergo 6-*endo*-trig cyclization to provide dihydroquinolinones **52** (Scheme 11.10b) [17].

Lipshutz utilized the stable and inexpensive Langlois' reagent (CF_3SO_2Na) [18] for the copper-catalyzed oxidative radical trifluoromethylation of acrylamides **53** to 3-trifluoroethyl oxindoles **54** (Scheme 11.11) [19]. This environmentally innocuous process worked under ambient conditions in an aqueous solvent.

11.3 Synthesis of Heterocycles via Oxidative Cyclization onto Alkynes and Nitriles

11.3.1 Diamination of Alkynes

Neuville's group reported a copper-catalyzed aerobic protocol for synthesis of trisubstituted imidazoles **58** from amidines **55** and terminal alkynes **56** (Scheme 11.12) [20]. The reaction involved the sequential formation of two C—N bonds through N-alkynylation of amidines **55** with terminal alkynes **56**; the first intermolecular C—N bond

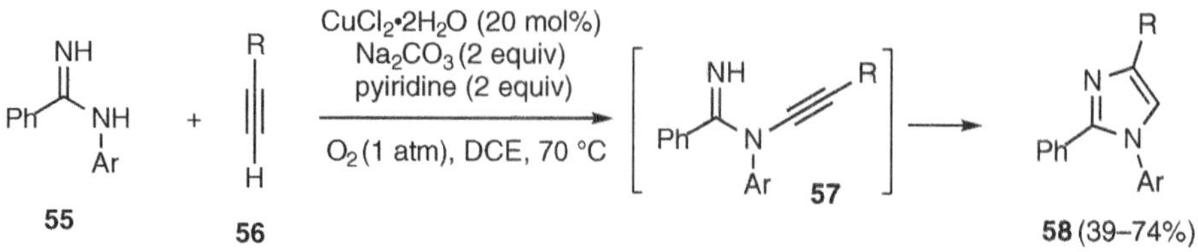

Scheme 11.12 Cu-catalyzed aerobic annulation of amidines and alkynes.

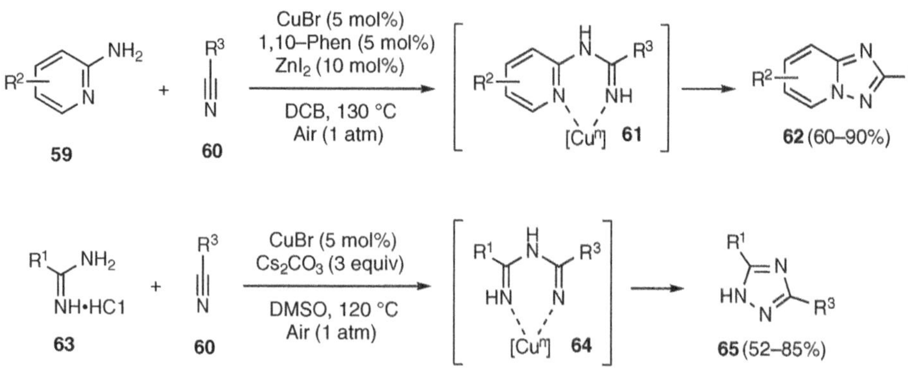

Scheme 11.13 Cu-catalyzed aerobic annulation of 2-aminopyridines/amidines with nitriles.

formation provides a N-alkynylamidine intermediate **57** that is followed by its 5-*endo*-dig cyclization for the second C—N bond formation.

Facile construction of privileged 1,2,4-triazoles **62** or **65** via copper-catalyzed aerobic annulation of 2-aminopyridines **59** or amidines **63** with nitriles **60** was developed by Nagasawa (Scheme 11.13) [21]. Cu-promoted polar addition of the primary amine unit of **59** or **63** to nitriles **60** forms Cu-coordinated amidine intermediates **61** or **64**, which is followed by N—N bond reductive elimination to form the products **62** or **65**.

11.3.2 Carboamination of Alkynes

Glorius reported a regioselective annulation of N-aryl enamine carboxylates **66** and nitriles **67** for the synthesis of tetrasubstituted pyrazoles **69** under Cu-catalyzed aerobic reaction conditions (Scheme 11.14) [22]. The reaction proceeds via addition of enamines **66** to nitriles **67** to form 1,3-diimine intermediates **68** bearing a complexation with the higher valent Cu complex, which undergo N—N bond reductive elimination to give pyrazoles **69**. Use of 2-picolinic acid is a key to render the process catalytic [23].

11.3.3 Carbooxygenation of Alkynes

Liang's [24] and Chiba's [25] groups independently reported copper-catalyzed oxidative cyclization of N-(2-alkynylaryl)enamine carboxylates **70** through intramolecular carbo-cupration of alkynes with the enamine moiety followed by oxygenation of the resulting organocopper intermediates **71**, yielding highly functionalized quinolines **72** (Scheme 11.15a). By taking advantage of the structural analogy of amidines with

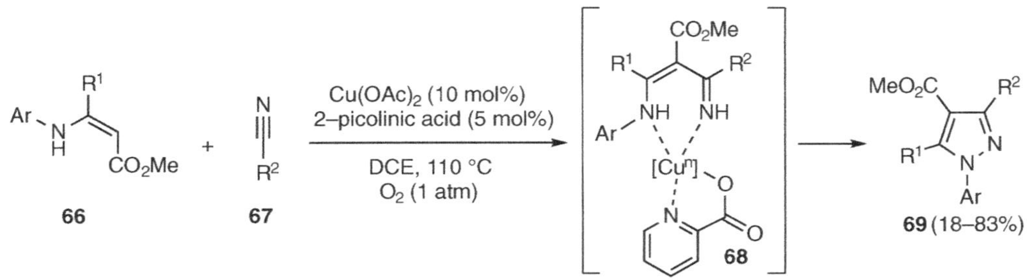

Scheme 11.14 Cu-catalyzed aerobic reactions of enamines and nitriles.

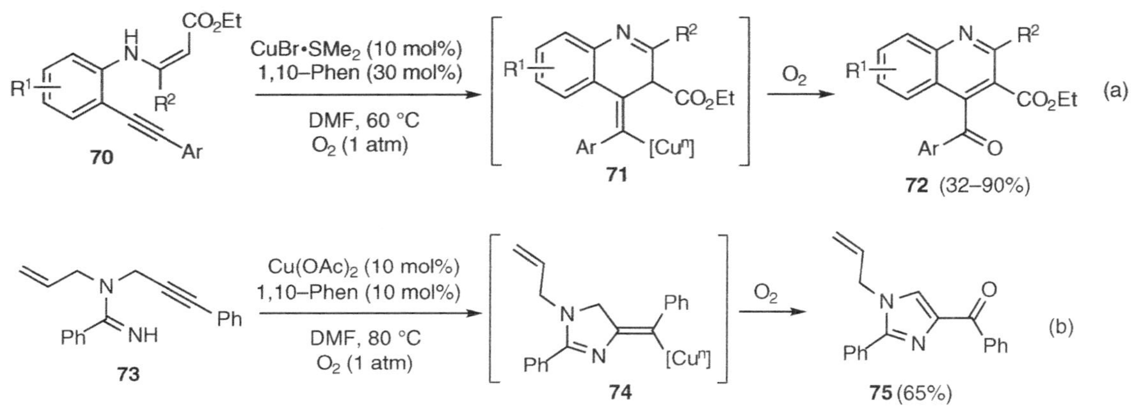

Scheme 11.15 Cu-catalyzed aerobic carbooxygenation of alkynes with enamines/amidines.

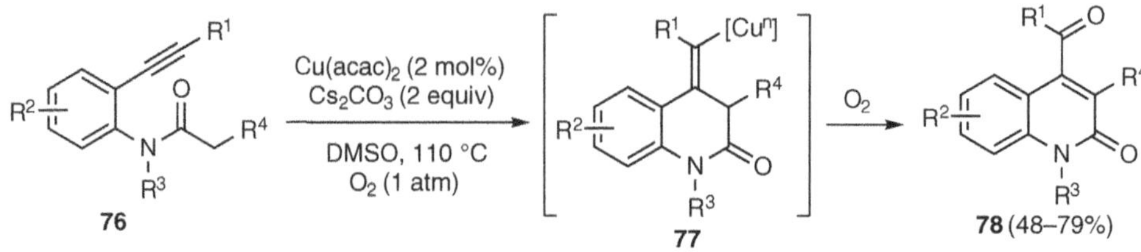

Scheme 11.16 Cu-catalyzed aerobic carbooxygenation of alkynes with amides.

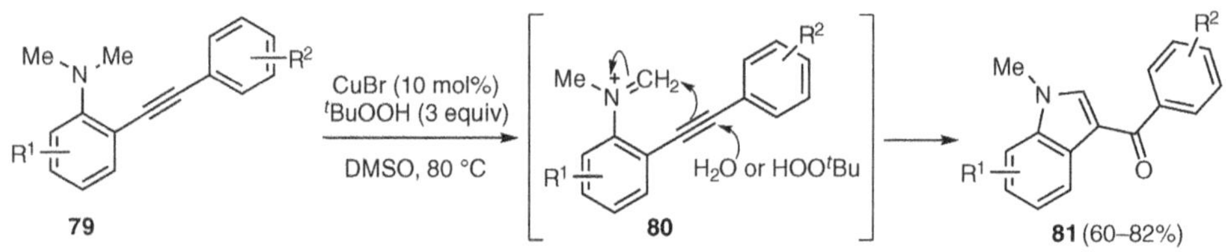

Scheme 11.17 Cu-catalyzed oxidative synthesis of 3-aroylindoles.

enamines, Chiba's group extended the Cu-catalyzed aerobic methodology toward the synthesis of imidazoles **75** from N-propargylamidines **73** through aminooxygenation of alkyne via **74** (Scheme 11.15b).

Li and co-workers reported Cu-catalyzed aerobic synthesis of quinolinones **78** from N-acyl(*o*-alkynyl)anilines **76** through the analogous carbooxygenation of alkynes via organocopper intermediate **77** with the putative Cu-enolate derived from amides **76** (Scheme 11.16) [26].

Patel's group disclosed a method for the synthesis of 3-aroylindoles **81** from *o*-alkynylated *N,N*-dimethylamines **79** through a copper-catalyzed oxidative process in the presence of aqueous ^{t}BuOOH as the oxidant (Scheme 11.17) [27]. The reaction mechanism involves carbooxygenation of the alkynyl moiety via C—C bond formation with in situ generated iminium ion **80** and concurrent oxygenation by nucleophilic attack of either water or peroxide.

11.4 Synthesis of Heterocycles via Oxidation of C(sp^2)—H Bonds

11.4.1 C(sp^2)-N Coupling

Buchwald in 2008 reported a copper-catalyzed aerobic transformation of N-arylamidines **82** to benzimidazoles **84** through aromatic C—H amination (Scheme 11.18) [28], where addition of acetic acid is the key to facilitate the process. Both N—H and N-methylated amidines could be employed. As for the alkyl substituent, only a *tert*-butyl group was allowed to be used. One of the possible mechanisms for the C—N formation is an organometallic process involving C—N reductive elimination from metallacycle intermediates **83**.

A Cu-catalyzed aerobic synthesis of medicinally important pyrido[1,2-a]benzimidazoles **87** from readily available N-aryl-2-aminopyridines **85** was disclosed by Zhu and

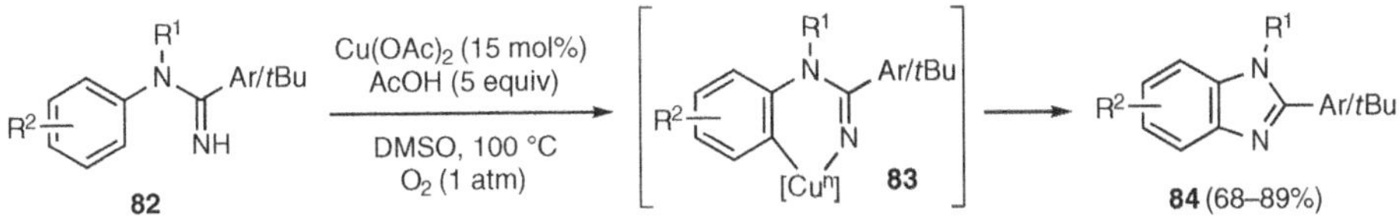

Scheme 11.18 Copper-catalyzed aerobic C–H amination for synthesis of benzimidazoles.

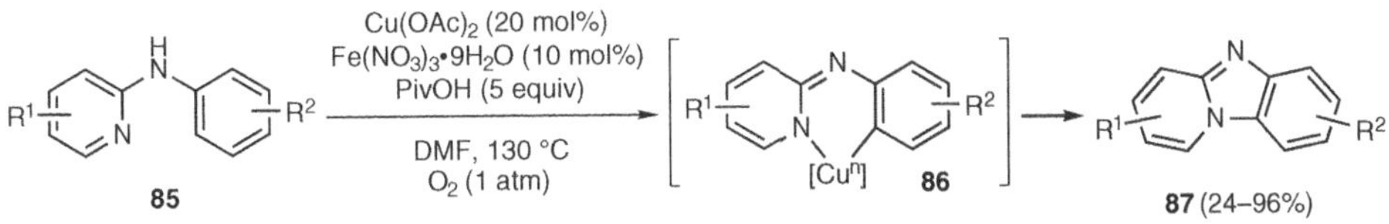

Scheme 11.19 Cu-catalyzed aerobic C–H amination for synthesis of pyrido[1,2-a]benzimidazoles.

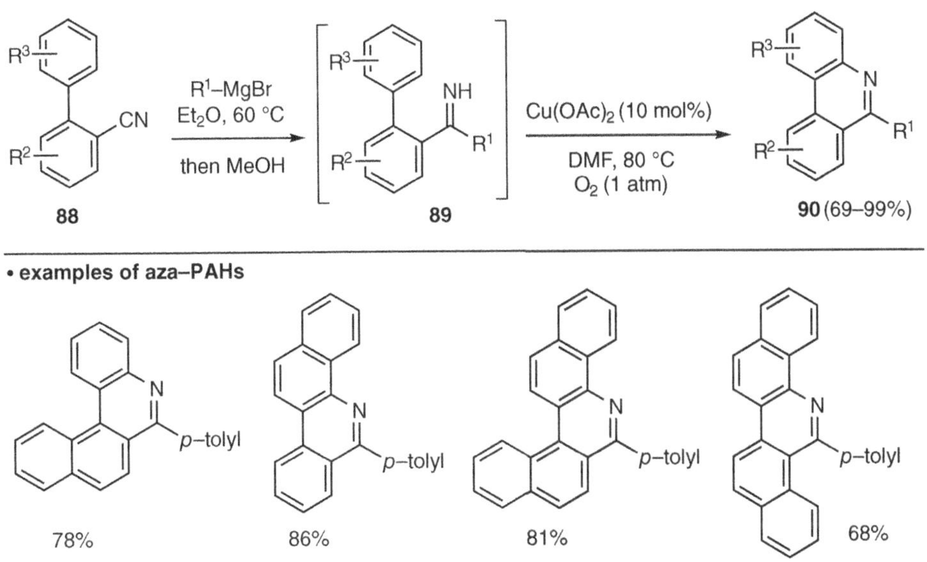

Scheme 11.20 Copper-catalyzed aerobic C–H amination with N–H imines.

co-workers (Scheme 11.19) [29]. The Fe co-catalyst facilitates this intramolecular C–H amination event by generation of more electrophilic copper(III) species for C–H cupration for the formation of metallacycles **86** and the following C–N bond reductive elimination.

Chiba's group developed a one-pot protocol for synthesis of phenanthridines **90** starting from biaryl-2-carbonitriles **88** and Grignard reagents (Scheme 11.20) [30]. The process is initiated with the addition of Grignard reagents to nitriles **88** to form N–H imines **89**, which are subsequently treated under Cu-catalyzed aerobic reaction conditions, resulting in aromatic C–H amination to form phenanthridines **90**. Iminyl radicals, generated via single-electron-oxidation of N–H imines **89**, might be responsible for the C–N bond formation [31]. This method allowed for construction of polynuclear azaaromatic hydrocarbons (aza-PAHs) in a highly efficient manner.

A Cu-catalyzed aerobic rearrangement of *o*-aryl benzaldoxime **91** to 2-arylbenzoxazoles **93** was reported by Punniyamurthy's group (Scheme 11.21) [32]. The skeletal rearrangement involving sequence of N—O bond cleavage and C—O bond formation affords imidates **92**, which undergo aromatic C—H amination by the Cu-catalyst, likely similar to Buchwald's aromatic C—H amination.

An oxidative skeletal rearrangement of readily available 5-aryl-4,5-dihydro-1,2,4-oxadiazoles **94** into quinazolinones **96** was reported by Chiba and co-workers (Scheme 11.22) [33]. The process was mediated just by molecular oxygen in DMSO. Iminyl radical species **95**, formed via oxidative homolysis of the N—O bond of **94**, is supposed to be the key intermediate that subsequently cyclized onto the benzoyl moiety through C—N bond formation that is followed by oxidative rearomatization to complete the process.

11.4.2 C(sp^2)—C Coupling

Cacchi and co-workers reported Cu-catalyzed aerobic synthesis of 3-acylindoles **99** from readily available N-aryl enaminones **97** (Scheme 11.23) [34]. The metallacycle intermediates **98** formed through aromatic C–H cupration undergo C—C reductive elimination to afford indoles **99**.

Li and co-workers reported a copper-catalyzed aerobic synthesis of isatins **101** through intramolecular aromatic C—H acylation of 2-oxo-N-arylacetamides **100** (Scheme 11.24) [35]. The process involves sequence of rate-determining formyl C–H cupration to

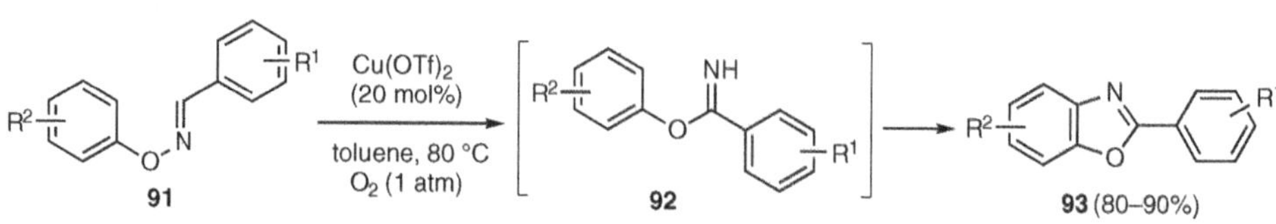

Scheme 11.21 Cu-catalyzed aerobic skeletal rearrangement-C—H amination for synthesis of benzoxazoles.

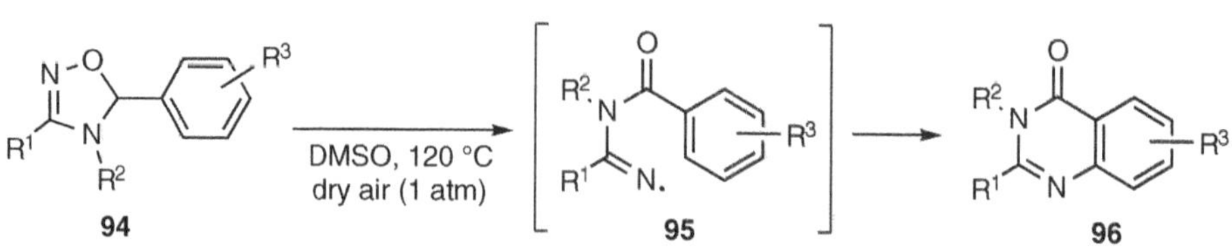

Scheme 11.22 Aerobic skeletal rearrangement of oxadiazoles to quinazolinones.

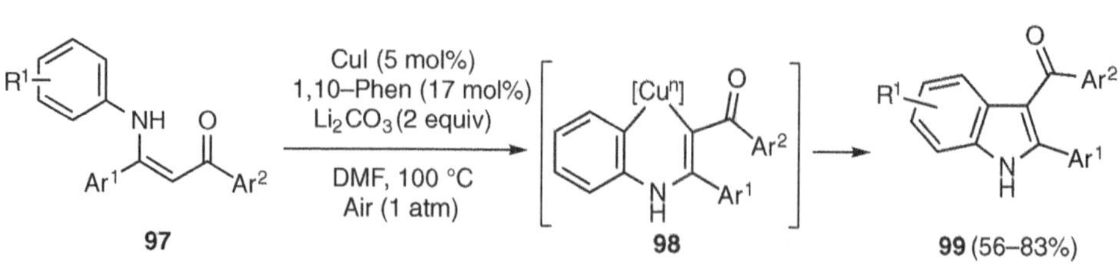

Scheme 11.23 Cu-catalyzed aerobic indole synthesis from N-aryl enaminones.

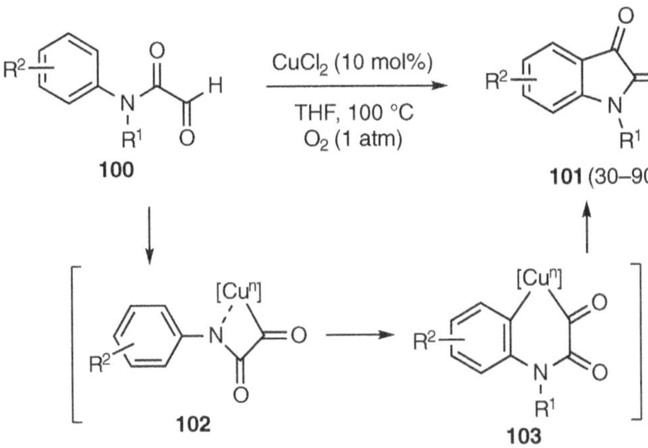

Scheme 11.24 Cu-catalyzed aerobic synthesis of isatins from 2-oxo-N-arylacetamides.

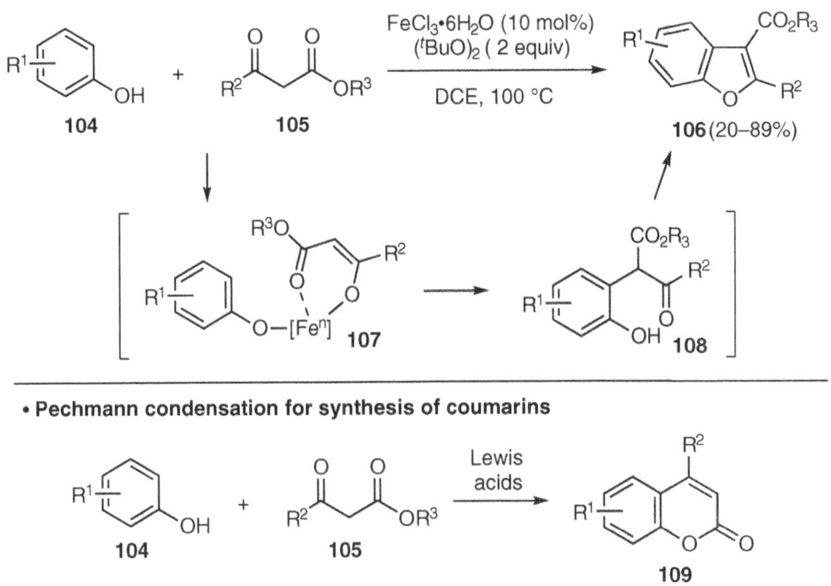

Scheme 11.25 Fe-catalyzed oxidative annulation of phenols and β-keto esters.

generate acylcopper intermediates **102** and their subsequent aromatic C–H cupration to form the metallacycles **103**, which are followed by C—C reductive elimination to give isatins **101**.

Li's group developed Fe-catalyzed oxidative annulation of phenols **104** and β-keto esters **105** using di-*tert*-butyl peroxide as a terminal oxidant for the synthesis of substituted benzofurans **106** (Scheme 11.25) [36]. The detailed mechanistic investigation suggested that the process is initiated by the addition of β-keto esters **105** at the *ortho*-position of phenols **104** through the higher valent iron chelate intermediates **107** to form α-(2-hydroxyaryl)-β-keto esters **108**, which undergo dehydrative condensation to deliver benzofuran **106**. The method is complementary to the Pechmann condensation of phenols **104** and β-keto esters **105** in the presence of Lewis acids to provide coumarins **109** [37].

11.4.3 C(sp^2)-O Coupling

Martin and co-workers reported a new protocol for intramolecular aromatic C—H oxygenation of *ortho*-arylbenzoic acids **110** in the presence of copper catalyst and benzoyl peroxide as a stoichiometric oxidant, resulting in the formation of lactones **112** (Scheme 11.26) [38]. Formation of the organocopper cycle **111** followed by its C—O bond reductive elimination was proposed as one of the possible processes for the C—H oxygenation.

Nagasawa's group developed Cu-catalyzed aerobic aromatic C—H oxygenation of readily available N-aryl benzamides **113**, which provides a facile access to substituted benzoxazoles **114** (Scheme 11.27) [39]. Investigation of the kinetic isotope effect revealed that the aromatic C—H bond cleavage process might not be involved in the rate-determining step.

Patel's group disclosed a Cu-catalyzed aerobic synthesis of 2,5-disubstituted 1,3,4-oxadiazoles **116** from arylaldehyde N-benzoylhydrazones **115** through C—H oxygenation of the hydrazonyl C(sp^2)—H bond (Scheme 11.28) [40]. The detailed mechanism was not disclosed, while the kinetic isotope study implies that the C—H bond cleavage step is not involved in the rate-limiting step of the process.

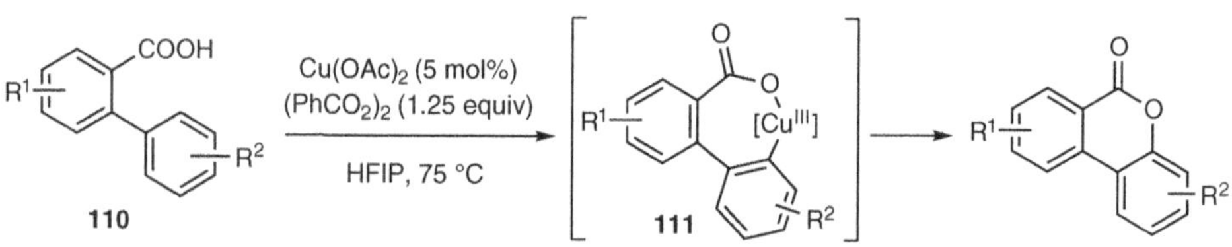

Scheme 11.26 Cu-catalyzed aromatic C—H oxygenation of 2-arylbenzoic acids.

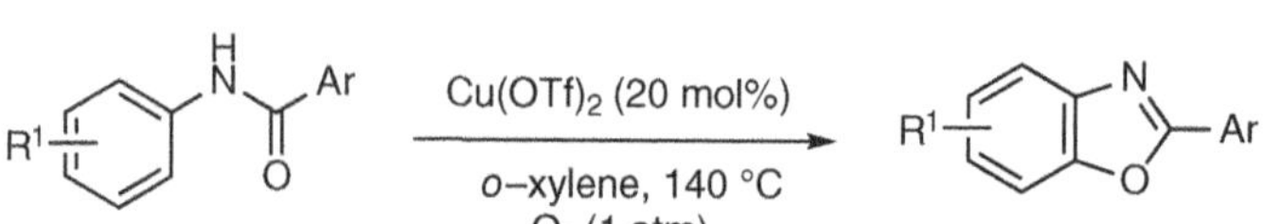

Scheme 11.27 Cu-catalyzed aerobic C—H oxygenation of benzanilides.

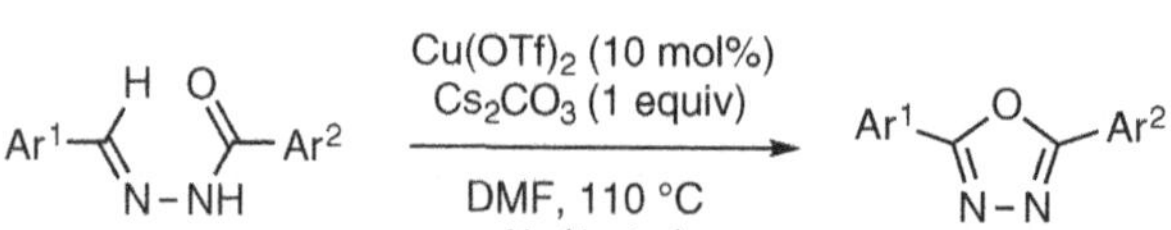

Scheme 11.28 Cu-catalyzed aerobic synthesis of 2,5-disubstituted [1,3,4]-oxadiazoles.

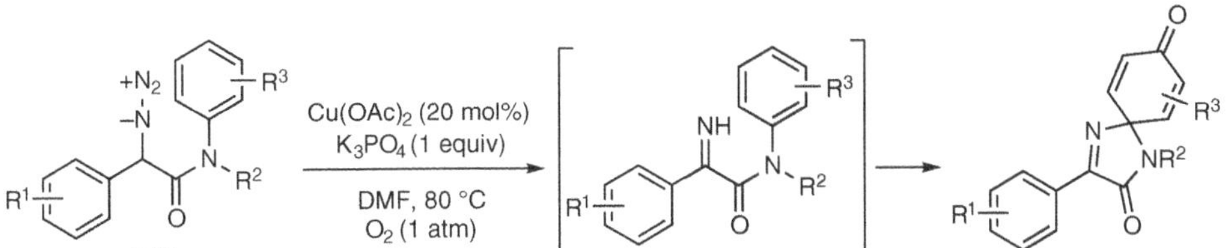

117 118 119(18–83%)

Scheme 11.29 Cu-catalyzed aerobic spirocyclization of α-azido-N-arylamides.

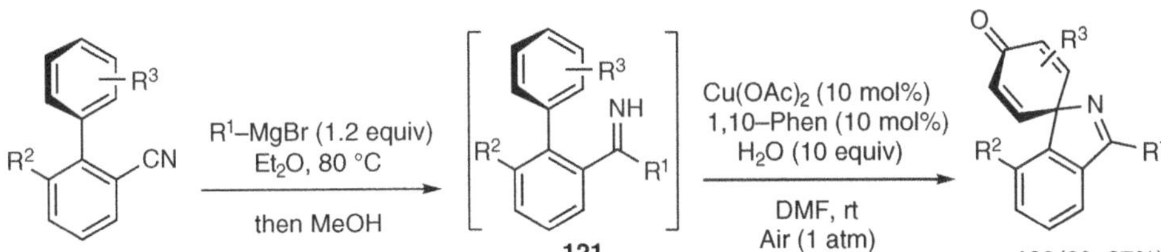

120 121 122(29–87%)

Scheme 11.30 Cu-catalyzed aerobic spirocyclization of biaryl N—H imines.

11.4.4 Miscellaneous Oxidative Cyclization

Chiba reported an unprecedented and mechanistically distinct copper-catalyzed aerobic oxidative spirocyclization for the synthesis of azaspirocyclohexadienones **119** from α-azido-N-arylamides **117** via aminooxygenation of arenes (Scheme 11.29) [41]. Utilization of $^{18}O_2$ showed that the incorporated oxygen atom of the product was derived from molecular oxygen. The transformation is thought to proceed through denitrogenative formation of N–H imines **118** before triggering a radical pathway consisting of single-electron-oxidation to generate iminyl radical and subsequent 1,4-aminooxygenation of arenes leading to the formation of azaspirocyclohexadienones **119**.

The reaction of carbonitriles with Grignard reagents and subsequent protonation are an alternative way to form N—H imines. By taking advantage of the restricted rotation of the biaryl axis in substituted biaryl rings, Chiba's group successfully extended the Cu-catalyzed aerobic spirocyclization to biaryl N—H imines **121** generated from biaryl 2-carbonitriles **120** and Grignard reagents for synthesis of azaspirocyclohexadienones **122** (Scheme 11.30) [42].

11.5 Heterocycles via Oxidative Cyclization onto C(sp^3)—H Bonds

11.5.1 C(sp^3)-N Coupling

Zeng's and the Xu's groups independently reported Cu-catalyzed aerobic construction of imidazo[1,5-a]pyridines **125** from 2-formyl or 2-acylpyridines **123** and alkyl amines **124** (Scheme 11.31) [43, 44]. The process involves sp^3-C-H amination of in situ generated imine intermediates **126** probably through formation of copper metallacycles **127**.

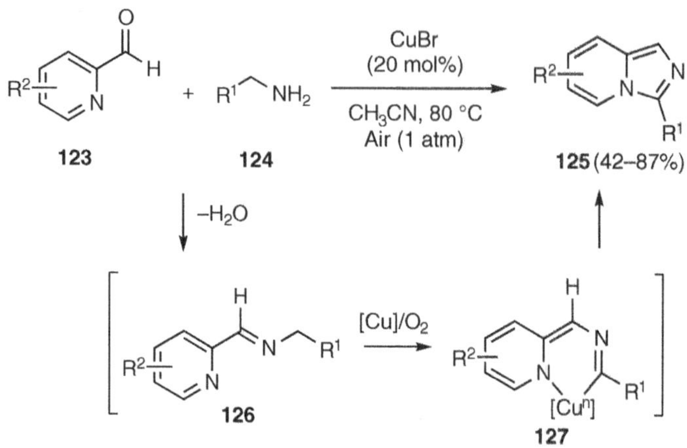

Scheme 11.31 Cu-catalyzed aerobic synthesis of imidazo [1,5-a]pyridines.

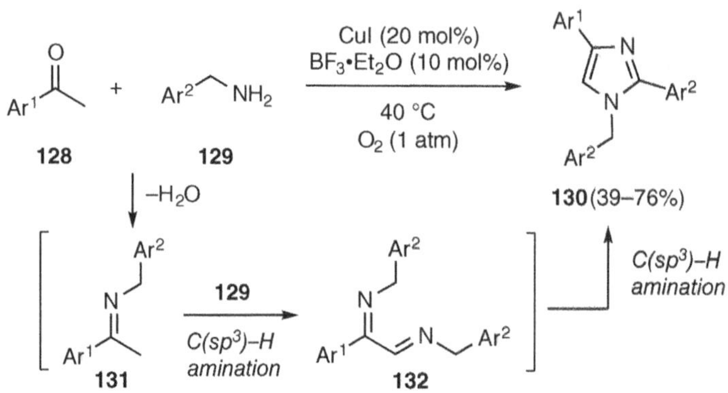

Scheme 11.32 Cu-catalyzed aerobic synthesis of imidazoles.

Ji's group disclosed dehydrogenative annulation of aromatic ketones **128** with benzylamines **129** for the synthesis of highly substituted imidazoles **130** under Cu-catalyzed aerobic reaction conditions (Scheme 11.32) [45]. The process includes two sequential C(sp^3)—H amination events; first at the α-position of initially formed ketimine intermediates **131** with another benzylamine **129** and following with cyclization of the resulting 1,2-dimine intermediate **132**.

11.5.2 C(sp^3)-O Coupling

Copper-catalyzed aerobic C—H oxygenation of aliphatic tertiary C—H bonds of N-alkylamidines **133** for synthesis of dihydrooxazoles **134** was developed by Chiba (Scheme 11.33) [46]. The process proceeds through generation of aminidyl radicals **135** by single-electron-oxidation of amidines **133** followed by 1,5-H transfer to form tertiary C-radicals, which are oxygenated with atmospheric oxygen to afford peroxy radicals **136**. Further reduction of the peroxy moiety gave Cu(II)-alkoxy species **137** that undergo cyclization with the amidine moiety to deliver dihydrooxazoles **134**.

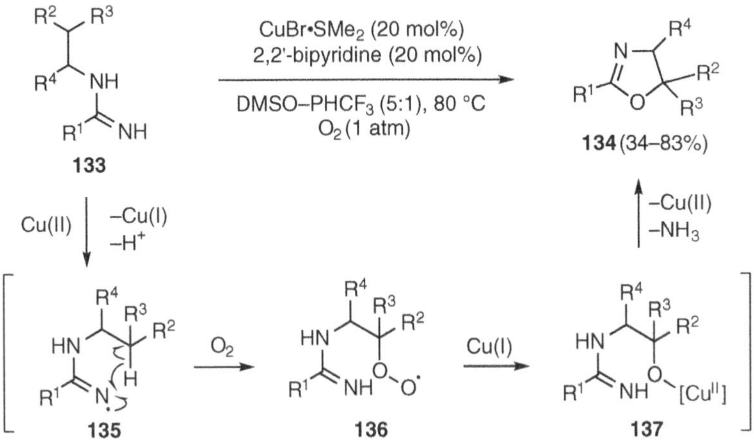

Scheme 11.33 Cu-catalyzed aerobic aliphatic C–H oxygenation of N-alkylamidines.

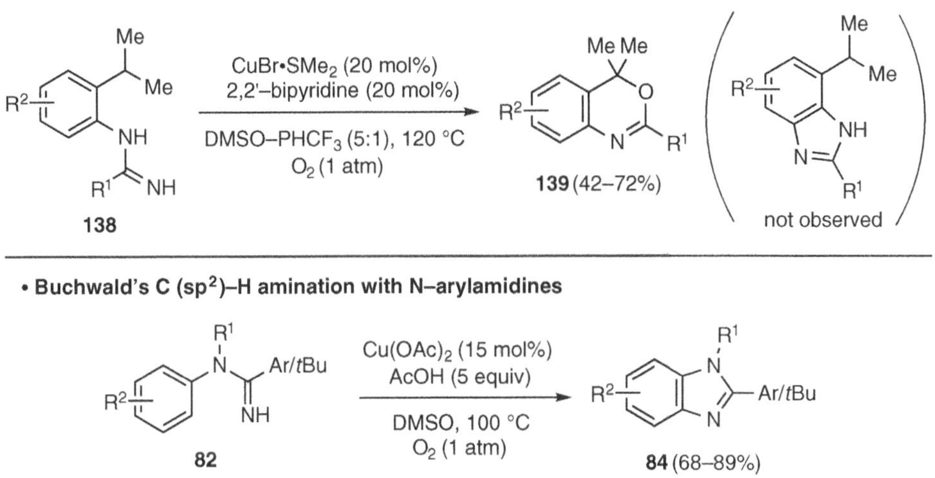

Scheme 11.34 Cu-catalyzed aerobic synthesis of 4H-1,3-benzoxazines.

N-(2-isopropylphenyl)amidines **138** were cyclized to 4H-1,3-benzoxazine **139** via 1,6-H-radical transfer followed by oxygenation and cyclization under the reaction conditions (Scheme 11.34). This reactivity is complementary to the Buchwald's aromatic C–H amination of N-arylamidines for the synthesis of benzimidazoles (see Scheme 11.18).

Maycock's group reported Cu-catalyzed aerobic transformation of 2-hydroxybenzyl amine derivatives **140** into 3,4-dihydro-2H-benzo[*e*][1,3]oxazines **141** via formal intramolecular α-C-H oxygenation (Scheme 11.35a) [47]. The process proceeds via formation of iminium ions **142** via single-electron-oxidation of the amine moiety. As a major pathway toward the final products, ionic fragmentation of **142** delivers *ortho*-quinone methide **143** and imines **144**, which undergo *endo*-[4+2] cycloaddition to form **141** in a *trans*-selective manner. Similarly, Cu-catalyzed aerobic C–H oxygenation of

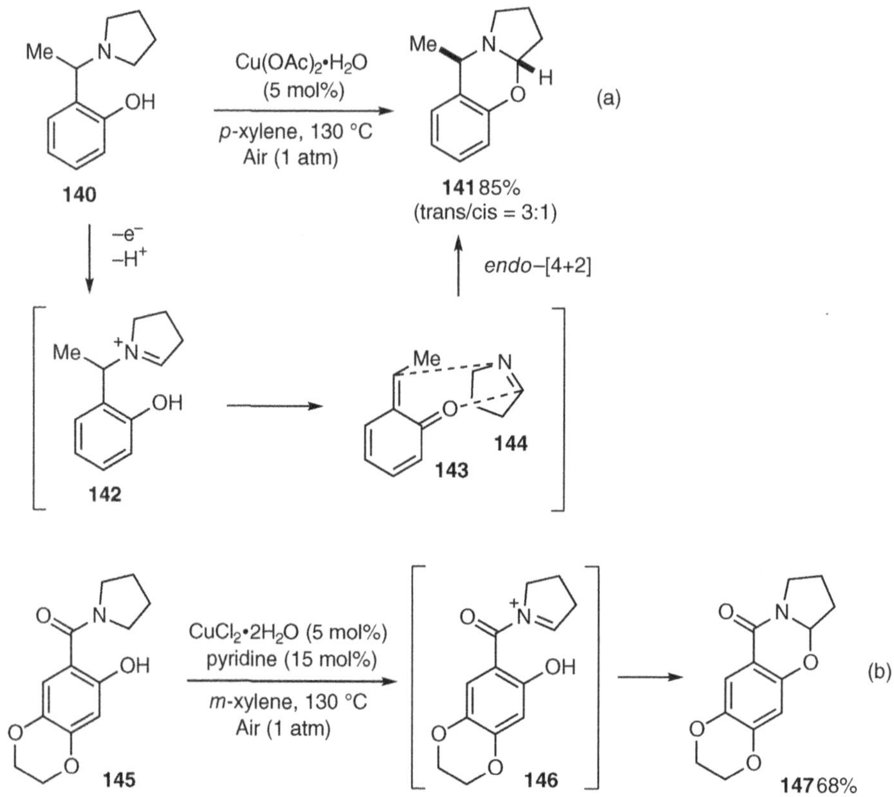

Scheme 11.35 Cu-catalyzed aerobic α-C—H oxygenation of *tert*-amines.

salicylamide **145** into dihydrooxazinone **147** via iminium ion **146** was reported by Maiti's group (Scheme 11.35b) [48].

11.5.3 C(sp^3)-C Coupling

Ge's group developed a method for the synthesis of pyrazoles **149** from N,N-disubstituted hydrazones **148** derived from aryl methyl ketones under Cu-catalyzed aerobic reaction conditions (Scheme 11.36) [49]. The process is mediated by sequence of formation of iminium cation **150**, imine-enamine tautomerization, and C—C bond forming cyclization of thus formed N-iminium enamines **151** to form dihydropyrazole (pyrazoline) intermediate **152** that is finally aromatized under the present reaction conditions to afford pyrazoles **149**.

The fine-tuning of the reaction conditions allowed for development of a complementary way to construct pyrazolines **155** and **158** from not only aromatic ketone hydrazones but also aliphatic (both acyclic and cyclic) ketone hydrazones **153** and **156** by prevention of the oxidative aromatization (Scheme 11.37) [50]. The reactions showed a substituent-dependent diastereoselectivity: aromatic ketone hydrazones such as **153** produce 4,5-*anti*-pyrazolines such as **155** as a major isomer, whereas those derived from aliphatic ketones such as **156** yielded 4,5-*syn*-isomers **158**, through the

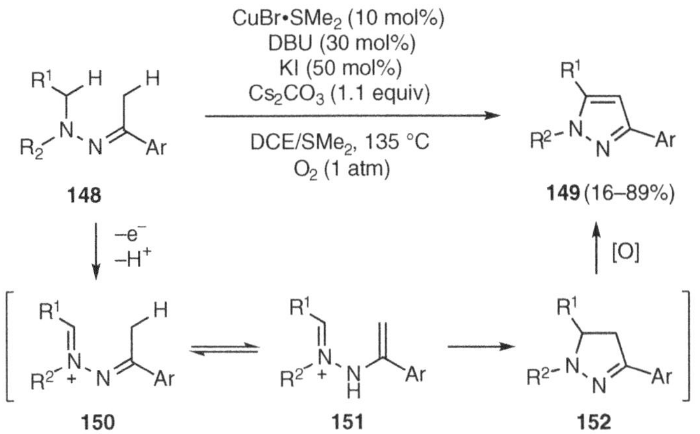

Scheme 11.36 Cu-catalyzed aerobic synthesis of pyrazoles from hydrazones.

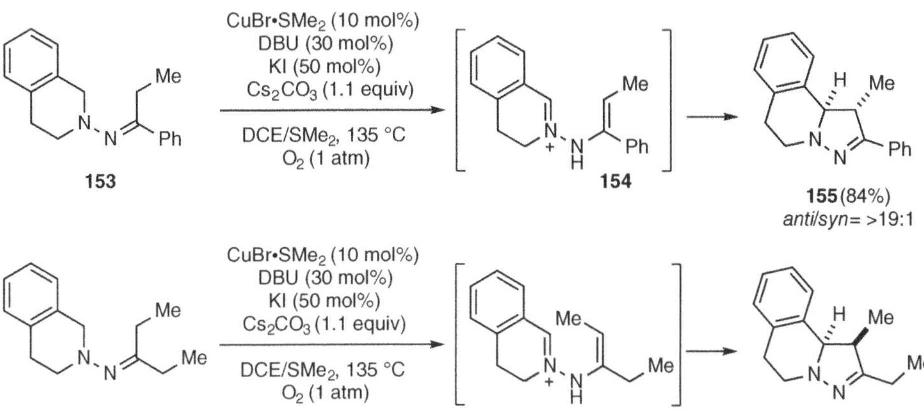

Scheme 11.37 Cu-catalyzed aerobic diastereoselective synthesis of pyrazolines from hydrazones.

stereoselective formation of N-iminium enamine intermediates **154** and **157** and their subsequent 5-center-6-electron disrotatory cyclization.

11.6 Summary and Outlook

A selected summary of oxidative molecular transformations that constituted mild and environmentally benign synthesis of various heterocyclic systems were presented. A substantial number of these methods relied on catalytic direct-oxidation of C—H bonds and oxidative difunctionalization of C—C/N unsaturated bonds using combinations of ubiquitous Cu/Fe catalysts and/or molecular oxygen/simple peroxides. These protocols, especially those using molecular oxygen as the oxidant, have vastly obviated the necessity for strong oxidants and stoichiometric metal salts thereby boosting the operational ease and environmental appeal of new C—C and C—heteroatom bond formations. While the quest for sustainable synthesis is in its ascendency, a few major challenges remain. Efficiency in combination with mildness and selectivity forms the central tenets of green chemistry but the higher temperature and non-neutral conditions usually associated with the cleavage of inert yet ubiquitous C—H bonds often compromise the selectivity and functional group compatibility of these reactions. Reducing the catalytic loading, development of chiral protocols, and improved scalability potential are also desired. New opportunities lie within these challenges and, with the advent of new powerful catalysts and a better understanding of mechanistic pathways and reactivity nuances of organometallic intermediates at various oxidation states, finer processes could be anticipated.

References

1 (a) Kim, J. and Movassaghi, M. (2009). *Chem. Soc. Rev.* 38: 3035–3050. (b) Bagley, M.C., Dale, J.W., Merritt, E.A., and Xiong, X. (2005). *Chem. Rev.* 105: 685–714. (c) Mishra, A., Ma, C.-Q., and Bäuerle, P. (2009). *Chem. Rev.* 109: 1141–1276.

2 (a) Anastas, P. and Eghbali, N. (2010). *Chem. Soc. Rev.* 39: 301–312. (b) Sneddon, H.F. (2015). *Green Chemistry Strategies for Drug Discovery*, vol. 46, 14. Cambridge, UK: The Royal Society of Chemistry.

3 (a) Wencel-Delord, J., Dröge, T., Liu, F., and Glorius, F. (2011). *Chem. Soc. Rev.* 40: 4740–4761. (b) Colby, D.A., Tsai, A.S., Bergman, R.G., and Ellman, J.A. (2012). *Acc. Chem. Res.* 45: 814–825. (c) Lyons, T.W. and Sanford, M.S. (2010). *Chem. Rev.* 110: 1147–1169. (d) Zhao, B., Shi, Z., and Yuan, Y. (2016). *Chem. Rec.* 16: 886–896.

4 (a) Yeung, C.S. and Dong, V.M. (2011). *Chem. Rev.* 111: 1215–1292. (b) Yang, Y., Lan, J., and You, J. (2017). *Chem. Rev.* 117: 8787–8863.

5 (a) Allen, S.E., Walvoord, R.R., P-Salinas, R., and Kozlowski, M.C. (2013). *Chem. Rev.* 113: 6234–6358. (b) Wendlandt, A.E., Suess, A.M., and Stahl, S.S. (2011). *Angew. Chem. Int. Ed.* 50: 11062–11087. (c) Zhang, C., Tang, C., and Jiao, N. (2012). *Chem. Soc. Rev.* 41: 3464–3484. (d) Chiba, S. (2013). *Bull. Chem. Soc. Jpn.* 86: 1400–1411. (e) Zhang, C., Tang, C., and Jiao, N. (2012). *Chem. Soc. Rev.* 41: 3381–3430.

6 Wang, Y.-F., Zhu, X., and Chiba, S. (2012). *J. Am. Chem. Soc.* 134: 3679–3682.

7 Monir, K., Bagdi, A.K., Mishra, S. et al. (2014). *Adv. Synth. Catal.* 356: 1105–1112.

8 Wang, H., Wang, Y., Liang, D. et al. (2011). *Angew. Chem. Int. Ed.* 50: 5678–5681.

9 Pünner, F., Sohtome, Y., and Sodeoka, M. (2016). *Chem. Commun.* 52: 14093–14096.

10 Guo, L.-N., Wang, S., Duan, X.-H., and Zhou, S.-L. (2015). *Chem. Commun.* 51: 4803–4806.

11 Dey, A., Ali, M.A., Jana, S., and Hajra, A. (2017). *J. Org. Chem.* 82: 4812–4818.

12 Shen, T., Zhang, Y., Liang, Y.-F., and Jiao, N. (2016). *J. Am. Chem. Soc.* 138: 13147–13150.

13 Zhou, M.-B., Song, R.-J., Ouyang, X.-H. et al. (2013). *Chem. Sci.* 4: 2690–2694.

14 Correia, V.G., Abreu, J.C., Barata, C.A.E., and Andrade, L.H. (2017). *Org. Lett.* 19: 1060–1063.

15 Meng, Y., Guo, L.-N., Wang, H., and Duan, X.-H. (2013). *Chem. Commun.* 49: 7540–7542.

16 Zhou, S.-L., Guo, L.-N., Wang, H., and Duan, X.-H. (2013). *Chem. Eur. J.* 19: 12970–12973.

17 Zhou, S.-L., Guo, L.-N., Wang, S., and Duan, X.-H. (2014). *Chem. Commun.* 50: 3589–3591.
18 Zhang, C. (2014). *Adv. Synth. Catal.* 356: 2895–2906.
19 Yang, F., Klumphu, P., Liang, Y.-M., and Lipshutz, B.H. (2014). *Chem. Commun.* 50: 936–938.
20 Li, J. and Neuville, L. (2013). *Org. Lett.* 15: 1752–1755.
21 Ueda, S. and Nagasawa, H. (2009). *J. Am. Chem. Soc.* 131: 15080–15081.
22 Suri, M., Jousseaume, T., Neumann, J.J., and Glorius, F. (2012). *Green Chem.* 14: 2193–2196.
23 Neumann, J.J., Suri, M., and Glorius, F. (2010). *Angew. Chem. Int. Ed.* 49: 7790–7794.
24 Xia, X.-F., Zhang, L.-L., Song, X.-R. et al. (2012). *Org. Lett.* 14: 2480–2483.
25 Toh, K.K., Sanjaya, S., Sahnoun, S. et al. (2012). *Org. Lett.* 14: 2290–2292.
26 Xie, P., Wang, Z.-Q., Deng, G.-B. et al. (2013). *Adv. Synth. Catal.* 355: 2257–2262.
27 Gogoi, A., Guin, S., Rout, S.K., and Patel, B.K. (2013). *Org. Lett.* 15: 1802–1805.
28 Brasche, G. and Buchwald, S.L. (2008). *Angew. Chem. Int. Ed.* 47: 1932–1934.
29 Wang, H., Wang, Y., Peng, C. et al. (2010). *J. Am. Chem. Soc.* 132: 13217–13219.
30 Zhang, L., Ang, G.Y., and Chiba, S. (2010). *Org. Lett.* 12: 3682–3685.
31 Zhang, L., Ang, G.Y., and Chiba, S. (2011). *Org. Lett.* 13: 1622–1625.
32 Guru, M.M., Ali, M.A., and Punniyamurthy, T. (2011). *Org. Lett.* 13: 1194–1197.
33 Wang, Y.-F., Zhang, F.-L., and Chiba, S. (2013). *Org. Lett.* 15: 2842–2845.
34 Bernini, R., Fabrizi, G., Sferrazza, A., and Cacchi, S. (2009). *Angew. Chem. Int. Ed.* 48: 8078–8081.
35 Tang, B.-X., Song, R.-J., Wu, C.-Y. et al. (2010). *J. Am. Chem. Soc.* 132: 8900–8902.
36 Guo, X., Yu, R., Li, H., and Li, Z. (2009). *J. Am. Chem. Soc.* 131: 17387–17393.
37 (a) Sethna, S. and Shah, N. (1945). *Chem. Rev.* 36: 1–62. (b) De, S.K. and Gibbs, R.A. (2005). *Synthesis* 8, 1231–8, 1233.
38 Gallardo-Donaire, J. and Martin, R. (2013). *J. Am. Chem. Soc.* 135: 9350–9353.
39 Ueda, S. and Nagasawa, H. (2008). *Angew. Chem. Int. Ed.* 47: 6411–6413.
40 Guin, S., Ghosh, T., Rout, S.K. et al. (2011). *Org. Lett.* 13: 5976–5979.
41 Chiba, S., Zhang, L., and Lee, J.-Y. (2010). *J. Am. Chem. Soc.* 132: 7266–7267.
42 Tnay, Y.L., Chen, C., Chua, Y.Y. et al. (2012). *Org. Lett.* 14: 3550–3553.
43 Li, M., Xie, Y., Ye, Y. et al. (2014). *Org. Lett.* 16: 6232–6235.
44 Wang, H., Xu, W., Wang, Z. et al. (2015). *J. Org. Chem.* 80: 2431–2435.
45 Cai, Z.-J., Wang, S.-Y., and Ji, S.-J. (2012). *Org. Lett.* 14: 6068–6071.
46 Wang, Y.-F., Chen, H., Zhu, X., and Chiba, S. (2012). *J. Am. Chem. Soc.* 134: 11980–11983.
47 Deb, M.L., Dey, S.S., Bento, I. et al. (2013). *Angew. Chem. Int. Ed.* 52: 9791–9795.
48 Modak, A., Dutta, U., Kancherla, R. et al. (2014). *Org. Lett.* 16: 2602–2605.
49 Zhang, G., Zhao, Y., and Ge, H. (2013). *Angew. Chem. Int. Ed.* 52: 2559–2563.
50 Wu, X., Wang, M., Zhang, G. et al. (2015). *Chem. Sci.* 6: 5882–5890.

12

Green Oxidation of Sulfide to Sulfoxide and Sulfone

Yiming Li and Xuefeng Jiang

School of Chemistry and Molecular Engineering, East China Normal University, Shanghai, China

Organic sulfoxides and sulfones are important compounds that widely exist in natural products [1, 2], pharmaceuticals [3, 4], flavors [5–7], and so on. Due to their importance in chemical and related studies, the syntheses of them have been widely investigated. Traditionally, oxidants with high oxidation states, such as $NaIO_4$, $K_2S_2O_8$, Oxone, and so on, or peroxy compounds, such as ^{m}CPBA, ^{t}BuOOH, were used to synthesize them from sulfides. Transformations with those reagents suffered from contamination and selectivity. As a greener and milder oxidant, hydrogen peroxide has attracted great interest as a powerful reagent for selective production of sulfoxides or sulfones, even on large scales [8]. Molecular oxygen is wildly regarded as a sustainable, cheap, mild, and stable oxidant, it has been utilized in industrial radical oxidation processes on scales of millions of tons [9, 10], such as in the oxidation of *p*-xylene to terephthalic acid, autoxidation of cumene en route to phenol and acetone in the Hock process, and radical-chain autoxidation of cyclohexane to a mixture of cyclohexanone and cyclohexanol. The oxidation of sulfide with oxygen has been studied over the past decades, typical reactions under mild conditions are summarized here. The related transformation with H_2O_2 and other oxidants, which has been outlined many times [111–113], is not covered.

12.1 Sulfoxide

12.1.1 Achiral Oxidization

The achiral organic sulfoxides can not only used as multiple functional compounds, but also as important intermediates in various bonds formation [11–14] and rearrangement reactions including the Pummerer [15–20] and Mislow–Evans rearrangements [21]. Moreover, sulfoxides can functionalize as ligands [22–24] for realizing a large set of important transformations. Beyond organic synthesis, sulfoxidation is also attractive from the viewpoint of industrial water treatment [114], warfare agent disposal [115], and such like. The significances mentioned here highlight the necessity for sulfonization and sulfidation, especially with green and mild procedures. As an oxygen source with the highest atom efficiency, compared to hydrogen peroxide [25–31], halogen

Green Oxidation in Organic Synthesis, First Edition. Edited by Ning Jiao and Shannon S. Stahl.

derivatives [32], and some novel oxygen-atom-transferred reagents [33], molecular oxygen is involved in many transition-metal-catalyzed, photocatalyzed, and enzyme-catalytic [34] transformations.

12.1.1.1 Transition-Metal-Catalyzed Processes

In 1995, Alper et al. reported selective oxidation of dialkyl or arylalkyl sulfides to corresponding sulfoxides with palladium dimer $[Pd(P^tBu_2H)(u\text{-}P^tBu_2)]_2$ **1**, in which benzyl, allyl, amide, nitro, carboxylic acid, and imine substituents were tolerated (Scheme 12.1) [35]. It is noteworthy that the transformation of (+)-biotin derivative **2** is stereospecific. The oxidation of dibenzothiophene **3** and diarylsulfides phenothiazines **4** were unsuccessful. The subsequent transformations with chiral ligands, such as (*R*)-(+)-2,2-bis(diphenylphosphino)-l,l-binaphthyl (BINAP), (2*S*,4*S*)-(-1-l-(*tert*-butoxycarbonyl)-4-(diphenylphosphino)-2-[(diphenylphosphino)methyllpyrrolidine (BPPM), and (−)-dimethyl 2,3-*O*-benzylidene-L-tartrate, did not induce enantioselectivity.

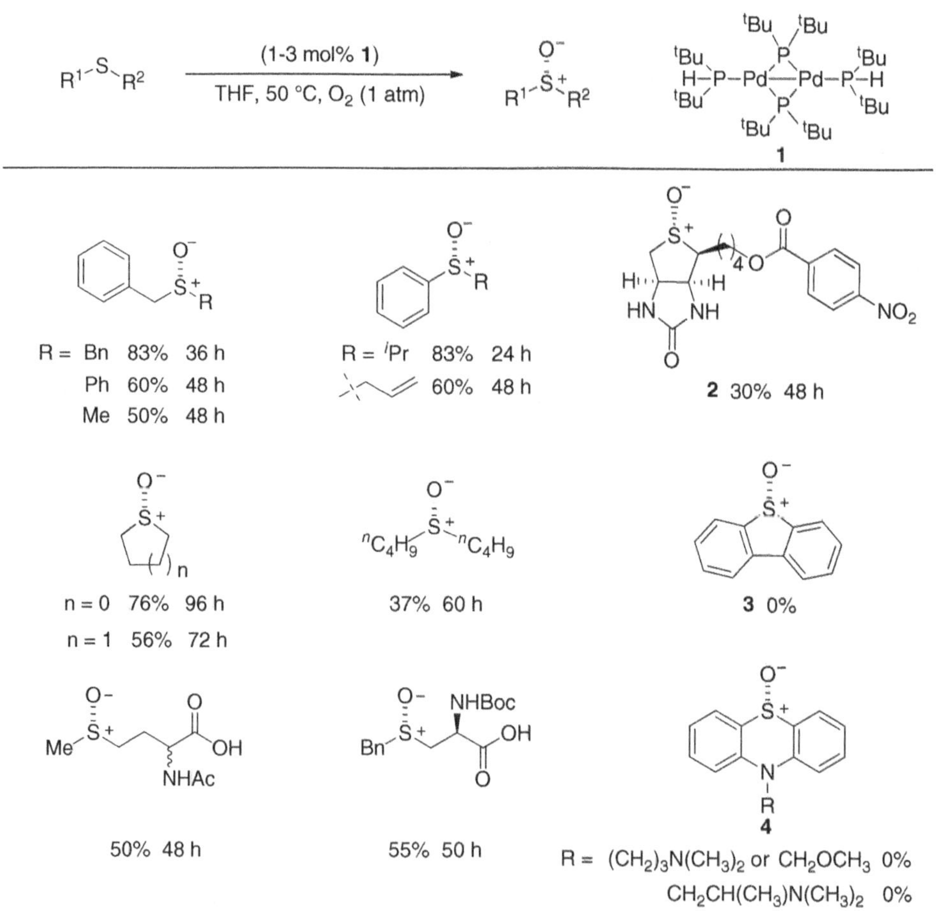

Scheme 12.1 Palladium-catalyzed sulfoxidation from sulfides.

In 2007, Ji et al. showed a highly efficient selective oxygenation method via bubbling oxygen to the system with ruthenium (III) *meso*-tetraphenylporphyrin chloride (Ru(TPP)Cl) (0.01 mmol%) as the catalyst and isobutyraldehyde as an oxygen acceptor (Scheme 12.2) [36]. Remarkably, the turnover number could reach up to 92 000 on a 20 mmol scale (Scheme 12.3).

In addition to noble-metal-catalyzed systems, the selective oxidation of sulfides with earth abundant metals have also been developed. In 2001, Martín et al. demonstrated a cocatalyst system with $Fe(NO_3)_3$ and $FeBr_3$, which achieved the oxidation of dialkyl and arylalkyl sulfide to sulfoxide at room temperature with air under normal pressure without sulfone being detected, and in which several functional groups could be tolerated, including Br, NO_2, CN, and CO_2H. (Scheme 12.4) [37]. Control experiments showed that sole $FeBr_3$ could not produce desired sulfoxide, while sole $Fe(NO_3)_3$ could help to convert 26% of thioanisole. The authors showed that in situ generated NO_2 or bromides/bromine couple controlled by Fe(III) might be the active oxidant species.

In 2012, He et al. found that the efficient oxygenation of sulfide with $Fe(acac)_2$ could be achieved at 100 °C under 2 MPa pressure (O_2), in which polyethylene glycol (PEG) worked as a green solvent and might play an important role in stabilizing the proposed

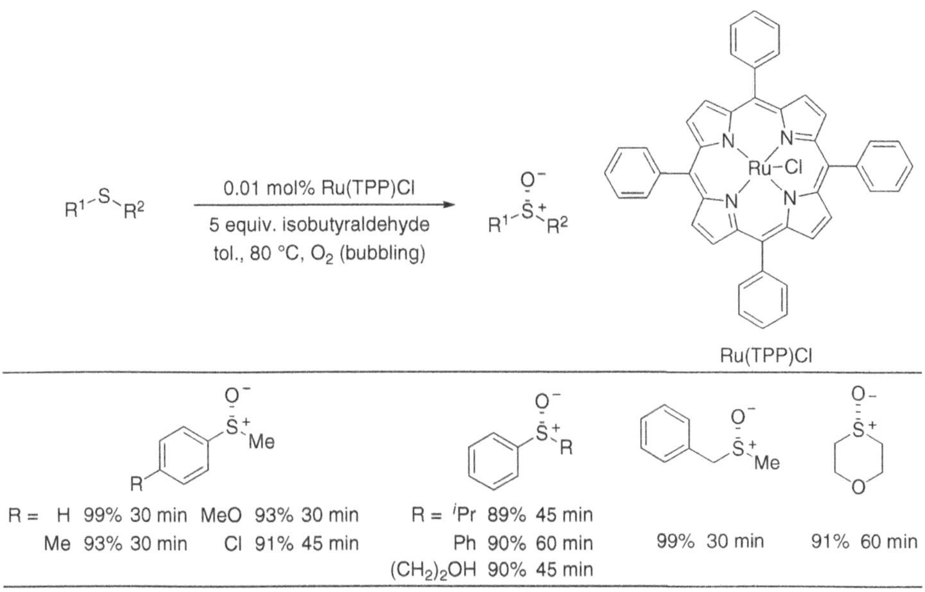

Scheme 12.2 Ru(TPP)Cl-catalyzed sulfoxidation from sulfides.

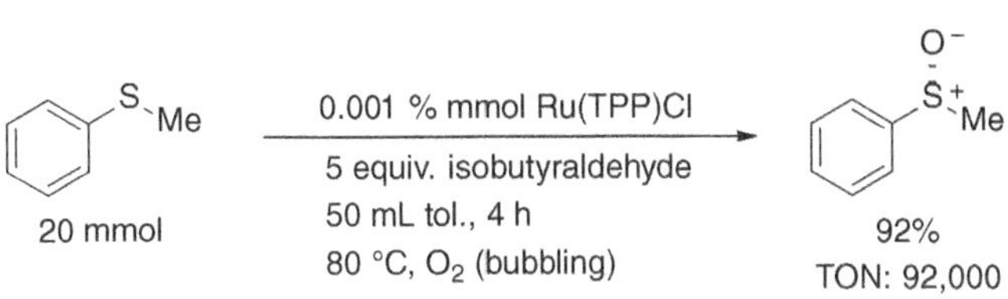

Scheme 12.3 Ru(TPP)Cl-catalyzed oxygenation of thioanisole on a 20 mmol scale.

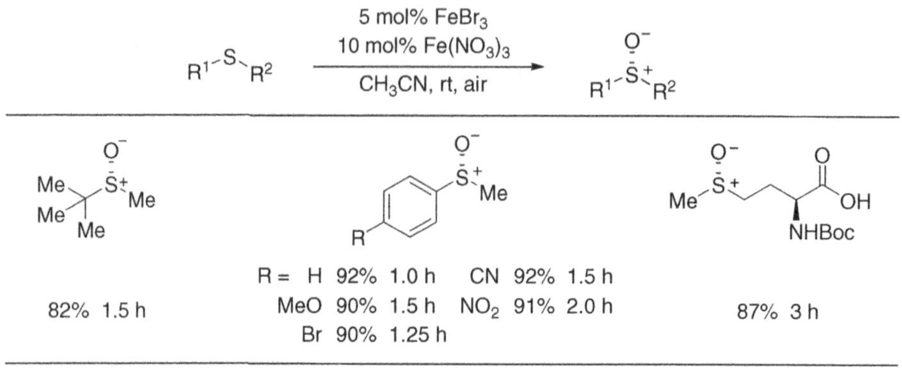

Scheme 12.4 Iron-catalyzed sulfoxidation from sulfide under air.

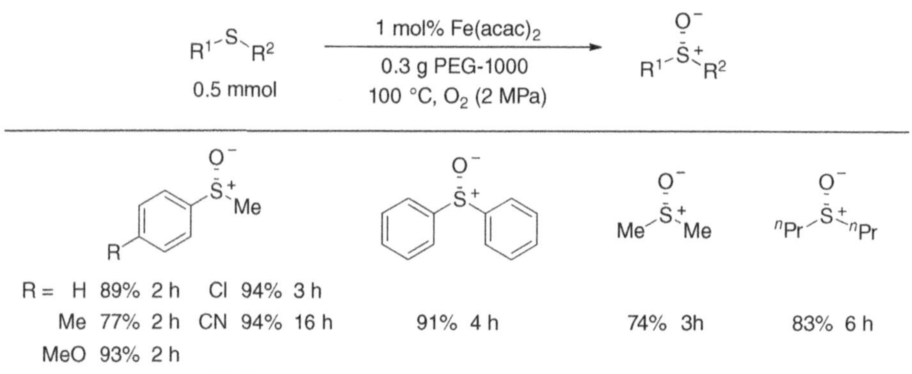

Scheme 12.5 $Fe(acac)_2$-catalyzed oxygenation of thioanisole with PEG-1000.

Fe(IV)-oxo species (Scheme 12.5) [38]. Control experiments exhibited that no product was obtained without catalyst and PEG or with only 1 atm of air, which indicated that catalyst, oxygen, and PEG are all prerequisites for performing the reaction smoothly. The substrates construction showed that diakyl, arylalkyl, and even diaryl sulfide could be efficiently transformed. Moreover, sulfones could also be produced with a longer reaction time. When 10 mol% of 2,2,6,6-tetramethylpiperidinooxy (TEMPO) was introduced into the standard system, the corresponding sulfoxidation process was totally inhibited. Combing with the references, the authors proposed a possible mechanism (Scheme 12.6): initially, Fe(III)-superoxo **5** was generated from Fe(II) and oxygen, which experienced dimerization and delivered Fe(III)–$(O)_4$–Fe(III) **6**. Then, the release of one molecule of oxygen gave birth to new iron complexes **7**, which yielded the Fe(IV)-oxo species **8** via O–O bond hemolysis. Finally, oxygen-atom transfer between **8** and sulfides produced sulfoxides and regenerated Fe(II). When $Fe(acac)_3$ was introduced into standard system instead of $Fe(acac)_2$, a similar result could be obtained, which indicated that Fe(IV)-oxo **8** could also be derived from Fe(III) via Fe(III)-O_2 **9**.

In 2015, He et al. developed an upgraded methodology with $Fe(NO_3)_3 \cdot 9H_2O$ as the catalyst and 2,2,2-trifluoroethanol (TFE) as the solvent. The relevant oxygenation of sulfide could be realized at 80 °C under 0.2 MPa pressure (Scheme 12.7) [39].

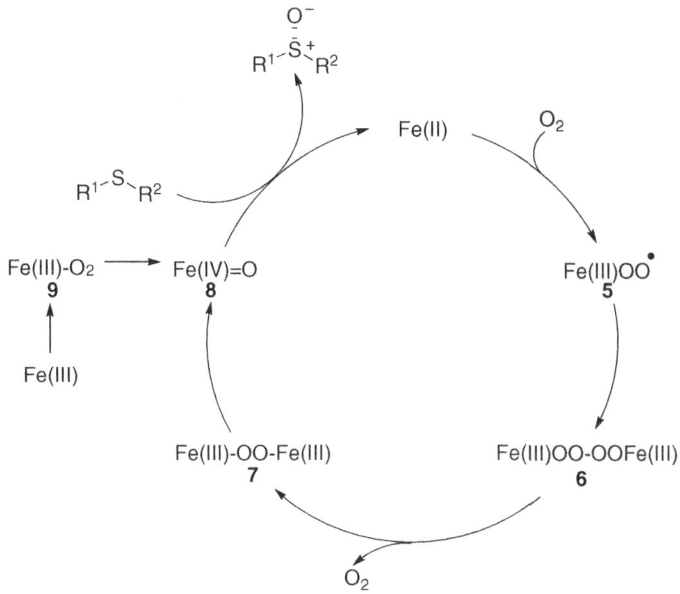

Scheme 12.6 Proposed mechanism for $Fe(acac)_2$-catalyzed oxygenation of thioanisole with PEG-1000.

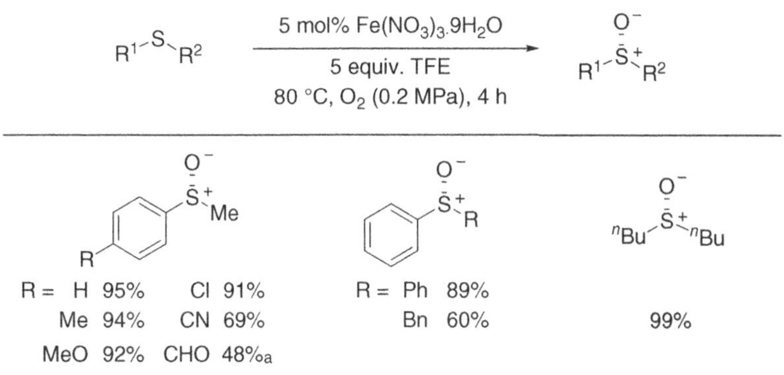

[a]from 4-$HOCH_2PhSMe$

Scheme 12.7 $Fe(NO_3)_3$-catalyzed oxygenation of sulfides.

The combination of $Fe(NO_3)_3 \cdot 9H_2O$ and TFE was proposed to generate trace amounts of HNO_3 analogs, which help to oxidize sulfides via intermediate **10**. The sulfoxide was produced after releasing hydrogen nitrite promoted by Fe(III). Then, the hydrogen nitrate (HNO_3) was regenerated through the oxidation of HNO_2 with the help of iron(III) and oxygen (Scheme 12.8).

In 2014, Hu et al. developed a sulfoxidation method with a $Mn(OAc)_2$/[C_{12}mim][NO_3] system, which could be recycled without loss of catalytic efficiency at least six times (Scheme 12.9) [40]. Mn(IV) oxide was proposed as a critical catalytic intermediate, which is generated from Mn(II) and transfer oxygen atom to sulfide.

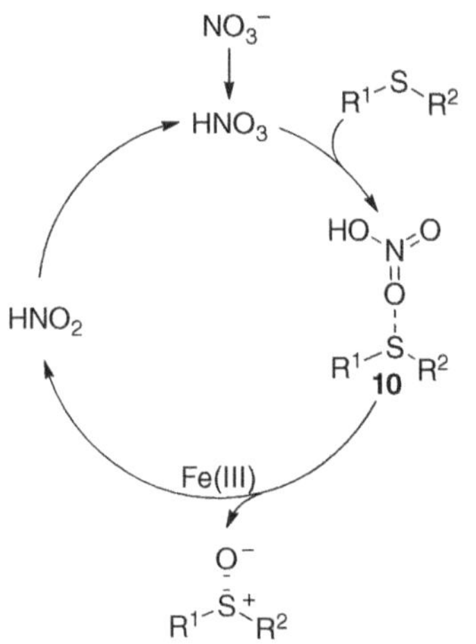

Scheme 12.8 Proposed mechanism for $Fe(NO_3)_3$-catalyzed oxygenation of sulfides.

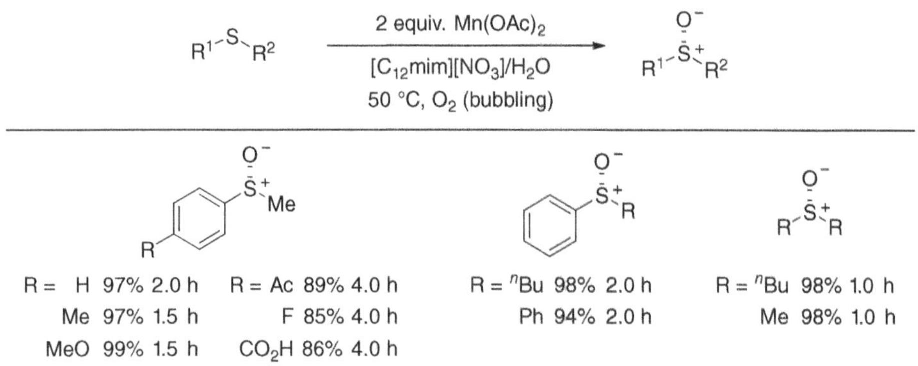

Scheme 12.9 $Mn(OAc)_2$/[C_{12}mim][NO_3]-catalyzed oxygenation of sulfides.

12.1.1.2 Photocatalyzed Processes

Aiming to decrease energy consumption in chemical transformations, visible-catalyzed systems were developed for realizing water splitting, CO_2 reduction, and multifarious organic transformations. Since 1962, the photooxidation of sulfides has been extensively studies with energy-inefficient light sources, such as the halogen and xenon lamps [41–48]. Recently, a low-wattage condensed fluorescent lamp (CFL), light-emitting diode (LED), and photoelectrosynthesis model cell [49] were utilized to realize the oxygenation of sulfide. Distinguished by active oxygen species, general photooxygenation modes of sulfide are listed as shown in Scheme 12.10 [42, 50, 116]. On the one hand, photosensitized energy transfer between a triplet oxygen $^3O_2$ and active photocatalyst (PC*) generated singlet oxygen $^1O_2$ (route a), which was then attacked by sulfide and released zwitterionic intermediate **11**. The final atom transfer between **11** and sulfide produced sulfoxide. On the other hand, sulfide could be oxidized to sulfide radical cation **12**, the following separate interaction with triplet oxygen (route b) or oxygen radical anion (route c) could deliver **13** and **11**. Further studies by Albini and co-workers [51, 52] showed that interaction of sulfide radical cation with superoxide resulted mainly in

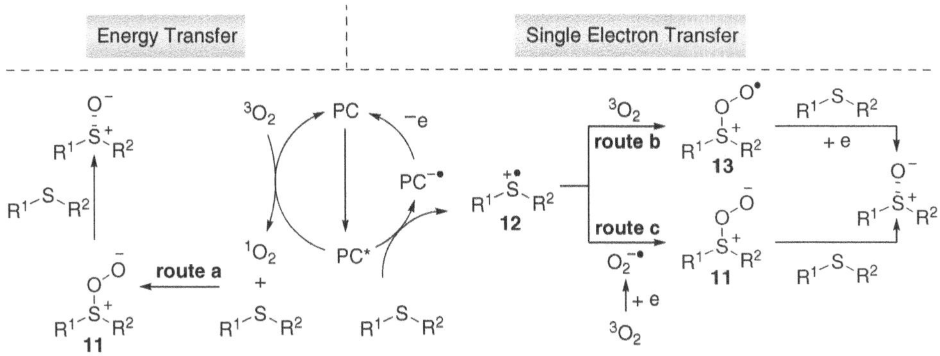

Scheme 12.10 General catalytic modes of photooxygenation of sulfides.

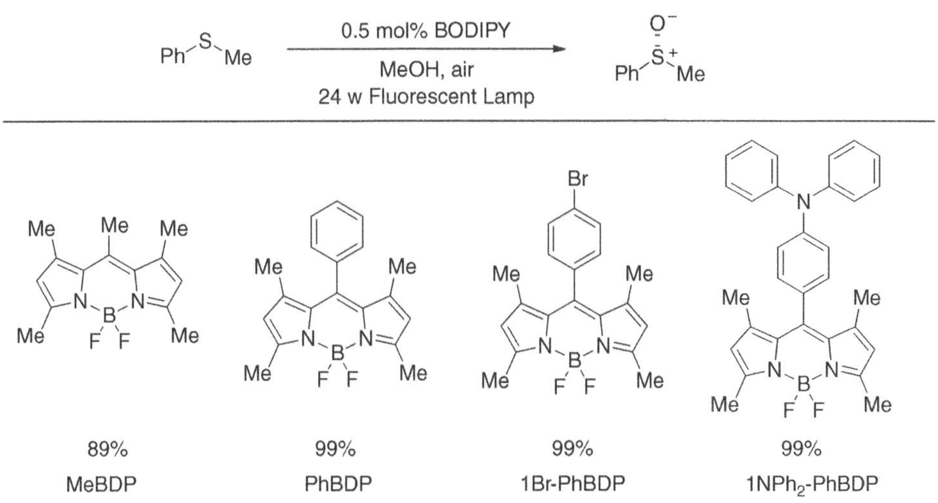

Scheme 12.11 BODIPY-catalyzed oxygenation of sulfides.

unproductive back electron transfer (BET) and electron-transferred sulfoxidation most likely proceeds via the reaction of sulfide radical cation with triplet oxygen.

In 2011, Xie et al. investigated the oxidation of thioanisole under visible light with 4,4-difluoro-4-bora-3a,4a-diaza-s-indacene (BODIPY) dyes that have excellent thermal and photochemical stability, high fluorescence quantum yield and good solubility. Four BODIPY dyes were prepared and delivered into methanol under air, which produced the desired sulfoxide in excellent yields. Singlet oxygen ($^1O_2$) was proposed as the active oxygen species (Scheme 12.11) [53].

In 2013, Xie et al. further exploited a batch of iodo-BODIPY and finally reduced the reaction time from 24 to 3 hours. The singlet-oxygen generation quantum yield studies of BODIPYs exhibited that HBDP, MeBDP, and PhBDP have negligible singlet-oxygen generation efficiencies. Dramatic increases of 0.81, 0.64, and 0.83 for HBDP-2I, MeBDP-2I, and PhBDP-2I, respectively, were observed when two iodine atoms were incorporated (Scheme 12.12) [54]. This was attributed to the distribution of the

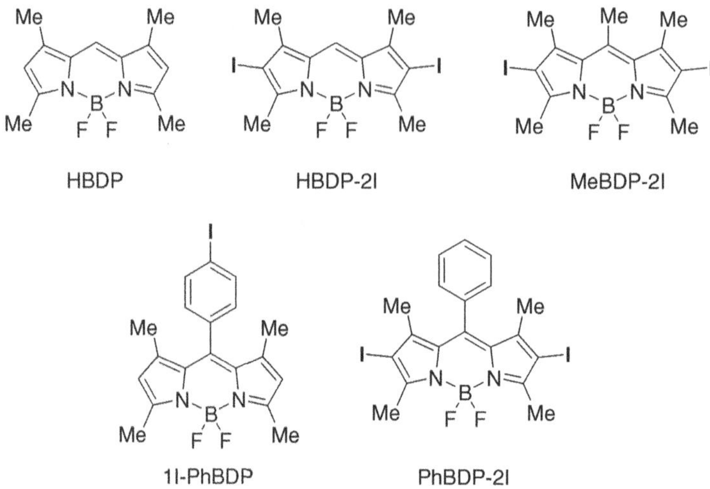

Scheme 12.12 Iodo-BODIPYs.

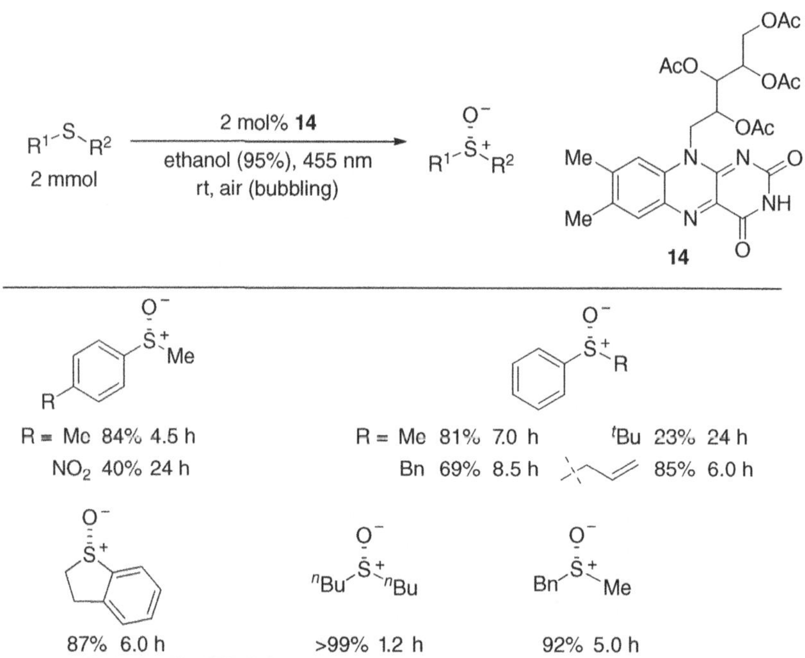

Scheme 12.13 Tetra-*O*-acetylriboflavin-catalyzed oxygenation of sulfides.

electron cloud caused by the presence of iodine atoms, which was calculated via theoretical studies on the frontier molecular orbitals based on the ground state structures.

In 2012, a photooxidation of dialkyl and arylalkyl sulfide to sulfoxide was completed by Cibulka et al. with vitamin B_2 derivative tetra-*O*-acetylriboflavin **14** in 95% ethanol (Scheme 12.13) [55]. It is worth mentioning that the quantum yield of this system is

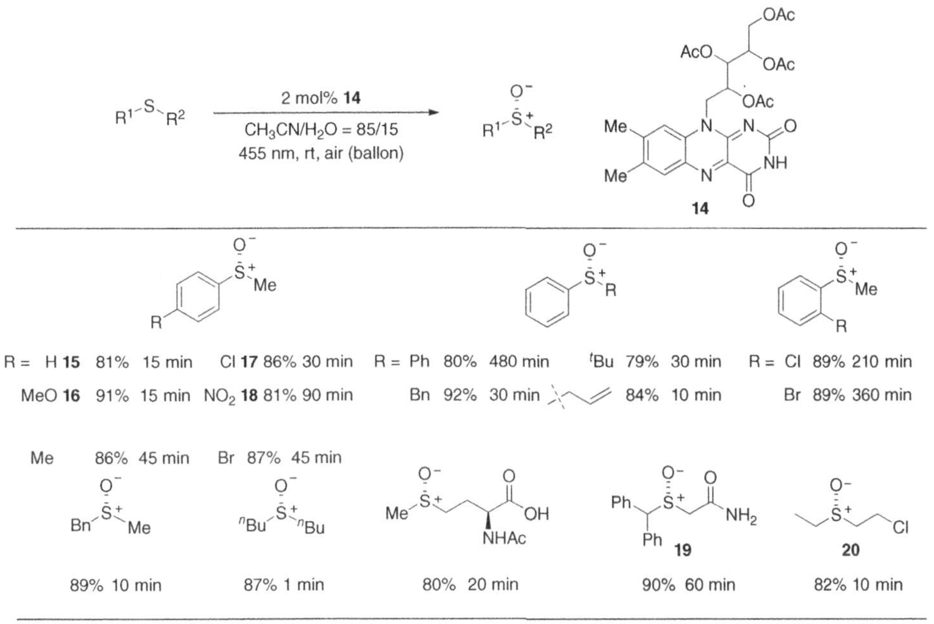

Scheme 12.14 Tetra-*O*-acetylriboflavin-catalyzed oxygenation of sulfides in CH_3CN/H_2O.

up to 0.7. Mechanistic studies revealed that thioanisol was transformed faster in deuterated methanol than in methanol (10 minutes, 90% vs. 25% conversion), which helped to show that singlet oxygen ($^1O_2$) might be the active species due to the longer existence of $^1O_2$ in deuterated solvent [56].

The same group reported an updated method that utilized acetonitrile/water (85/15) instead of aqueous ethanol (95%) as the solvent (Scheme 12.14) [57]. Comparatively, the related conversions needed shorted reaction times (1–480 minutes) and lower catalyst loading (0.1–2 mol%), which could be attributed to the high photocatalyst stability and extremely high quantum yields of **14** (up to 1.3, the theoretical highest value is 2). Beyond common substrates, psychostimulant Modafinil **19** with amide and active methylene could be efficiently produced. The detoxification of mustard gas derivatives **20** was accomplished in 10 minutes. Combining efficiency and mildness of this system with availability and biodegradability of catalyst, this procedure has potential applicability. To gain insight into the mechanism, effects of singlet-oxygen quencher $Co(acac)_3$ addition on the relative rate of photosulfoxidation of *p*-substituted thioanisols were investigated. The non-zero rate achieved with $Co(acac)_3$ showed that electron transfer pathways were also involved in the generation of **15**, **16**, and **17** with estimated relative contributions of 75, 34, and 28%, respectively. The almost zero limiting rate observed with **18** indicated predominating participation of the singlet-oxygen mechanism. Importantly, an analogous experiment on thioanisole oxidation in an ethanol-water (95 : 5) mixture confirmed only the singlet-oxygen mechanism occurs in this medium. These results indicated that both energy transfer induced singlet oxygen and electron transfer processes might participate in the corresponding translation.

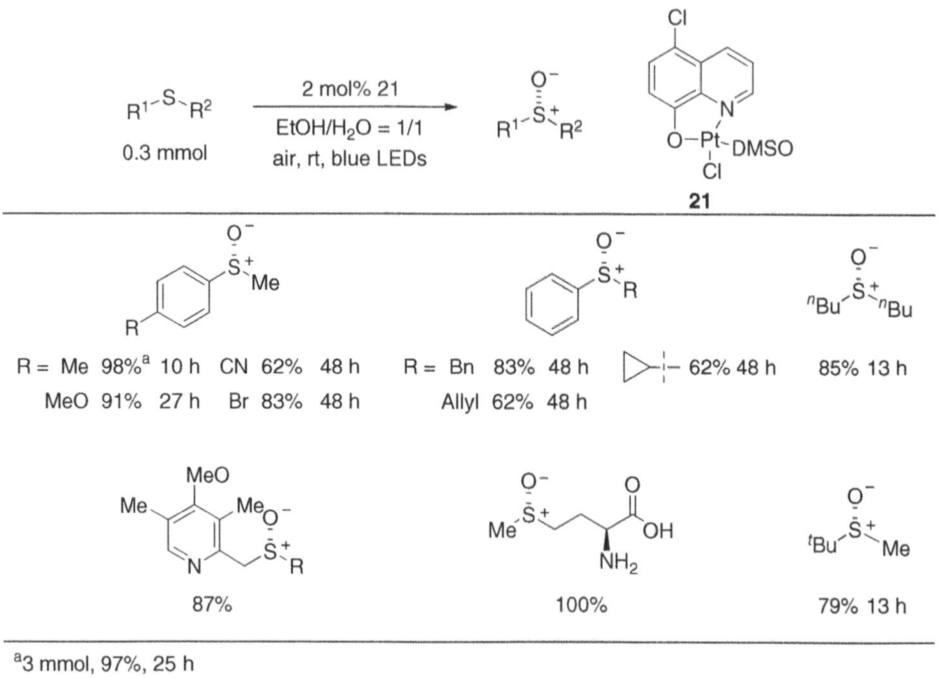

Scheme 12.15 Pt-complex-catalyzed sulfoxidation from sulfides.

In 2016, a variety of coordination complexes based on platinum(II) and hydroxyquinoline ligands were developed by Cabrera and Alemán et al., which were utilized for sulfoxidation from sulfide (Scheme 12.15) [58]. Multifarious functional groups are tolerated, such as nitrile, bromo, benzyl, allyl, pyridinyl, unprotected amine, and acid groups. In addition to batch processes, the flow experiments were conducted with much higher efficiency (Scheme 12.16). A mechanistic study with DABCO, a scavenger of singlet oxygen, did not induce obvious change, the same as the one with deuterated methanol. In contrast, the experiment with an oxygen radical anion quencher benzoquinone resulted in only 3% conversion within 2 hours. These experiments helped to prove the $O_2^{-\bullet}$ rather than $^1O_2$ was the key active oxygen species in this system.

In addition to homogeneous systems, the heterogeneous ones could also be used to realize the selective oxygenation of sulfide. In 2014, Zhang et al. utilized the high internal phase emulsion polymerization technique and micropore engineering to synthesize high surface area porous conjugated polymer B-(Boc-CB)$_2$-BT as heterogeneous photocatalysts (Scheme 12.17), which realized the selective oxygenation of sulfide under the irradiation of blue LEDs (Scheme 12.18) [59]. The insoluble polymer was immobilized in a glass column, irradiated with light, which could continuously catalyze the selective oxygenation within a circulatory flow system.

In 2014, Zhan et al. constructed an anionic indium porphyrin framework UNLPF-10 (UNLPF: University of Nebraska–Lincoln porous framework) composed of close-packed Williams β-tetrakaidecahedral cages using an octatopic ligand linked with 4-connected [In(COO)$_4$]-SBUs (secondary building units). Considering its rich photophysical and

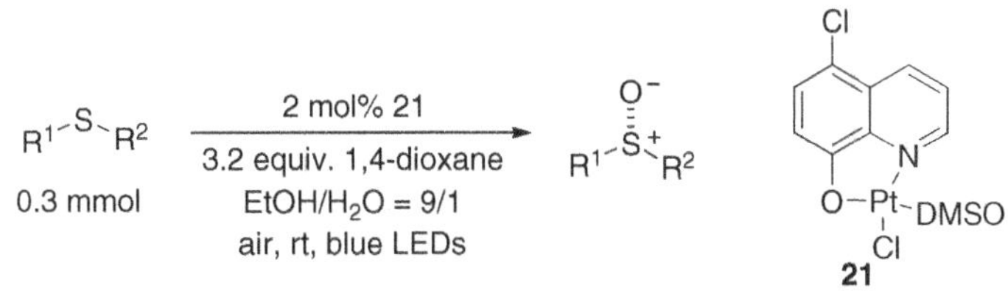

sulfoxide		solvent flow rate (mL/min)	O_2 flow rate (mL/min)	residence time (min)	conv. (%)
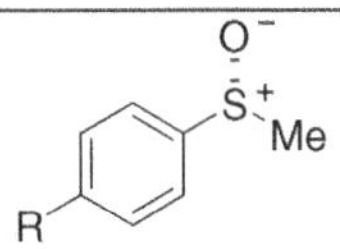	R = Me	0.04	0.33	11.0	100
	Br	0.01	0.04	82.0	50
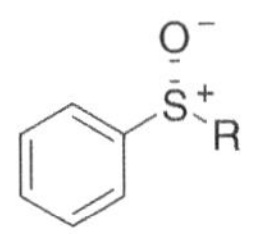	R = Bn	0.01	0.09	40.0	94
	Allyl	0.01	0.10	37.0	76
	Cyclopropyl	0.01	0.09	39.5	78
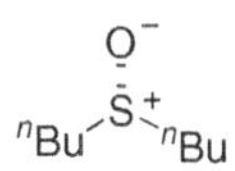		0.04	0.35	10.5	99
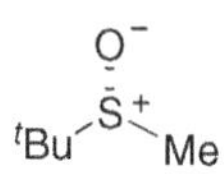		0.02	0.34	11.5	100
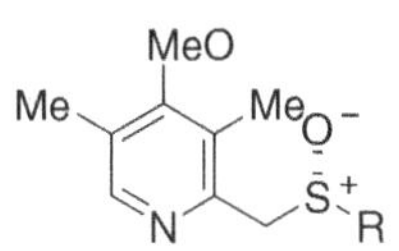		0.01	0.17	23.0	100

Scheme 12.16 Pt-complex-catalyzed sulfoxidation from sulfide using continuous flow processes.

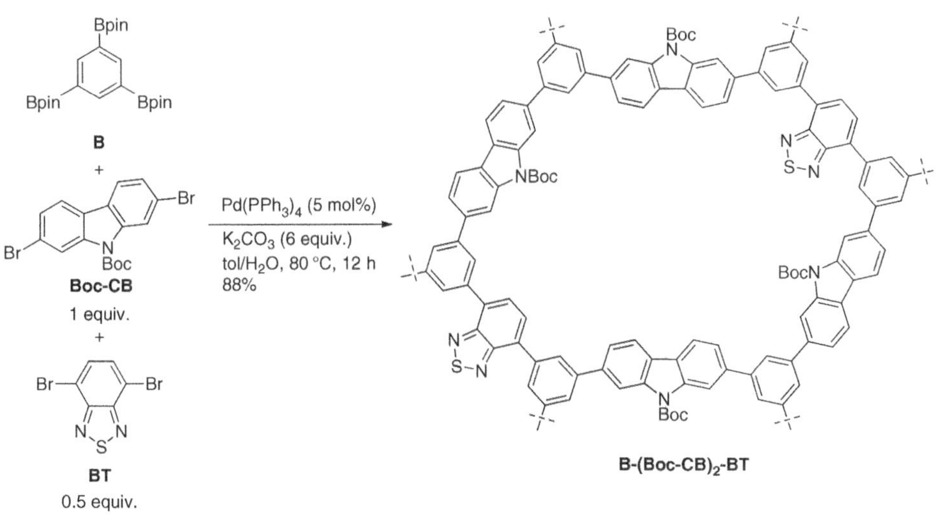

Scheme 12.17 Synthesis of high surface area porous conjugated polymers B-(Boc-CB)$_2$-BT.

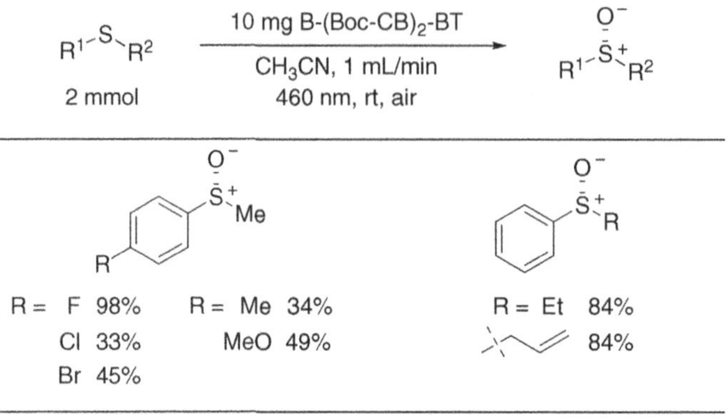

Scheme 12.18 B-(Boc-CB)$_2$-BT-catalyzed oxygenation of sulfides.

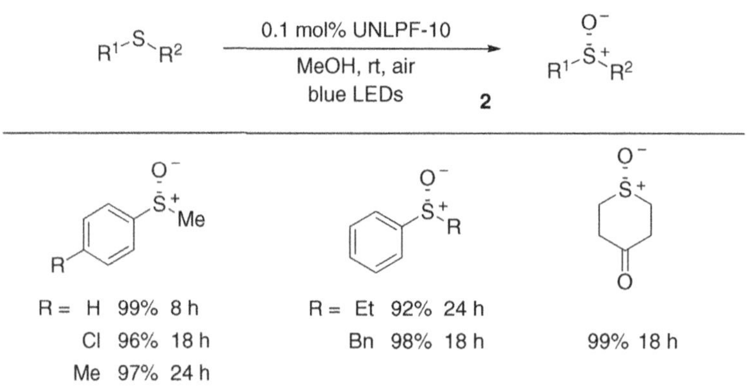

Scheme 12.19 UNLPF-catalyzed construction of β-keto sulfoxides.

photochemical properties, it was utilized to selectively oxidize some arylalkyl and dialkyl sulfides with air and blue LEDs (Scheme 12.19) [60].

In addition to the sole oxidation of sulfides, some cascade transformations have been achieved as well. In 2014, Yadav et al. described an efficient difunctionalization approach to β-keto sulfoxides with alkenes, thiophenols, and air catalyzed by the dye eosin Y [61]. The alkenes could be alkyl- and phenyl-substituted with various functional groups, such as bromo, nitro, trifluoromethyl, and so on, while the sulfur source is relatively limited (Scheme 12.20). Mechanistic studies showed that radical inhibitors TEMPO (2,2,6,6-tetramethyl-1-piperidinyloxy) or DPPH (2,2-diphenyl-1-picrylhydrazyl) could totally quench the reactions. In contrast, the singlet-oxygen quenchers, 2,3-dimethyl-2-butene or DABCO, did not have this ability. Moreover, the KI/starch indicator revealed the existence of H_2O_2 that might be generated from superoxide radical anion. Combining those experiments with the previous ones discussed, the authors proposed a possible mechanism (Scheme 12.21): the reductive quenching process between active eosin Y and thiophenol yield thiophenol radical cation **22**. The following deprotonation and addition to alkene produced a new radical **24**, and it then captured triplet oxygen and

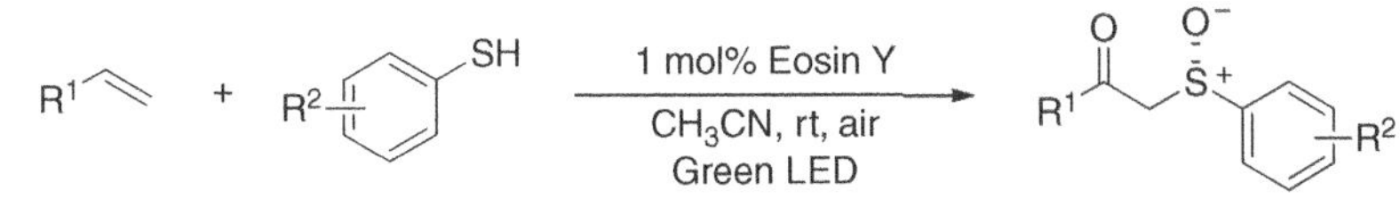

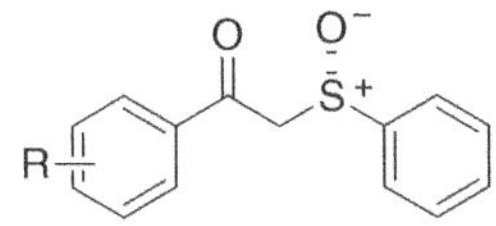

R = H	88%	2.0 h	2-CF_3	74%	3.5 h	4-MeO	86%	2.0 h
2-Me	89%	2.0 h	3-MeO	87%	3.0 h	4-Br	80%	3.0 h
2-MeO	90%	2.0 h	3-NO_2	74%	4.5 h	4-CF_3	70%	2.5 h
2-Br	80%	2.5 h						

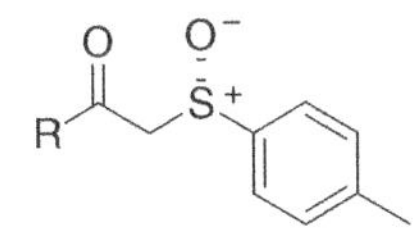

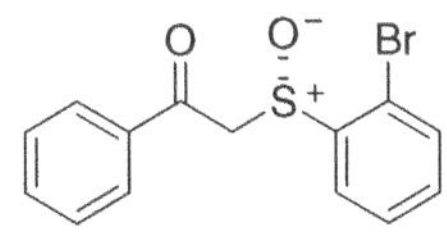

R = ^{i}Pr	69%	6.5 h	^{n}C$_8$H$_{17}$	60%	8.0 h	79%	2.5 h
^{n}C$_5$H$_{11}$	64%	6.0 h	Ph	80%	2.5 h		

Scheme 12.20 Eosin Y-catalyzed construction of β-keto sulfoxides.

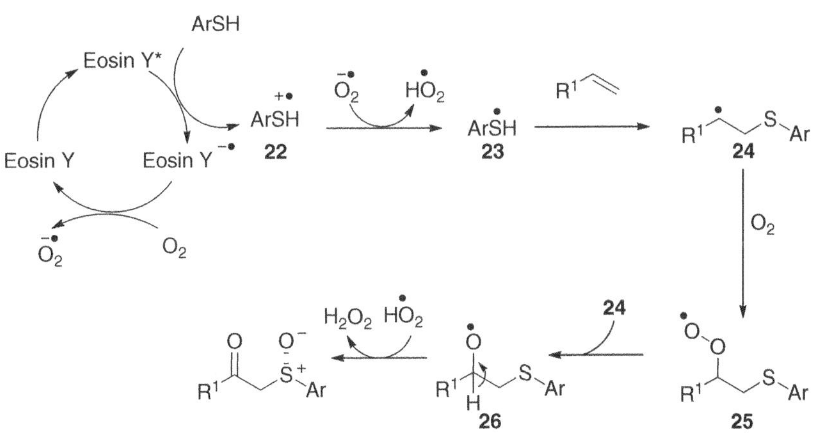

Scheme 12.21 Proposed mechanism for eosin Y-catalyzed construction of β-keto sulfoxides.

produced the peroxy radical **25**. The following oxygen-atom transfer to **24**, deprotonation, and oxygenation finally delivered β-keto sulfoxides.

Recently, a similar cascade thio-ene and sulfoxidation reaction catalyzed by Rose Bengal (RB) in EtOH/H_2O in air was developed by Wang et al. [62], which produced alkyl-substituted sulfoxides instead of β-keto sulfoxides (Scheme 12.22). Control experiments showed that sulfide might be an important intermediate, whereas disulfide is not (Scheme 12.23). Stern–Volmer experiments showed that the emission intensity of

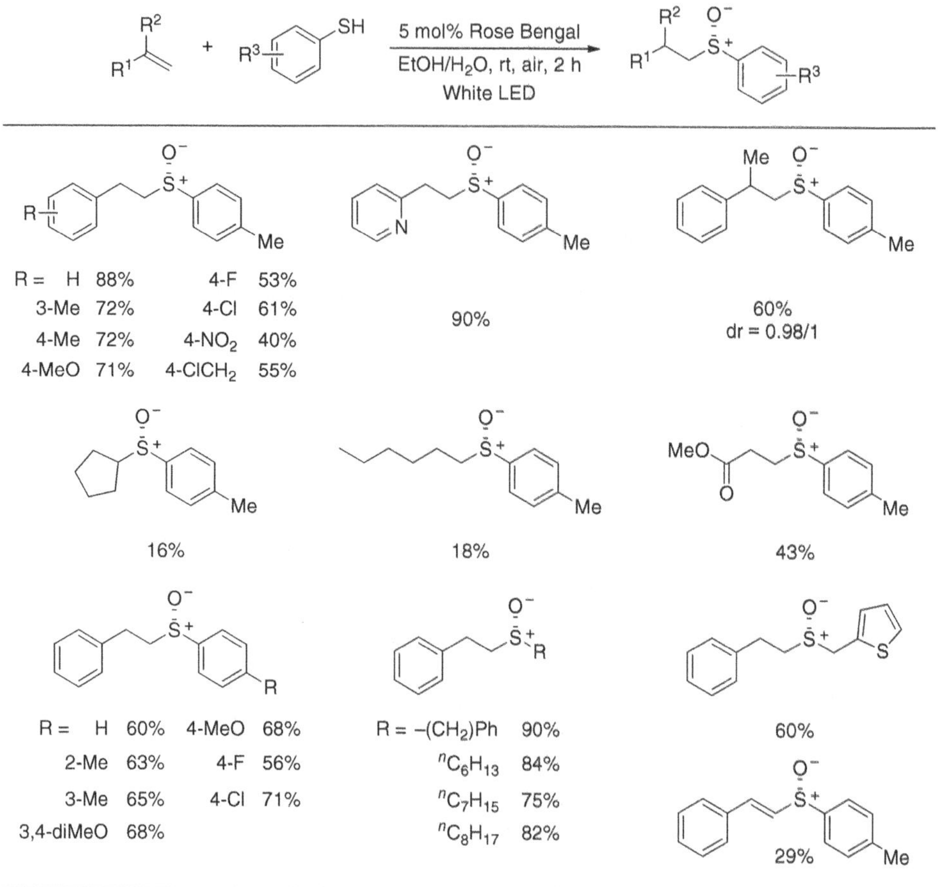

Scheme 12.22 Results for RB-catalyzed cascade thio-ene and sulfoxidation.

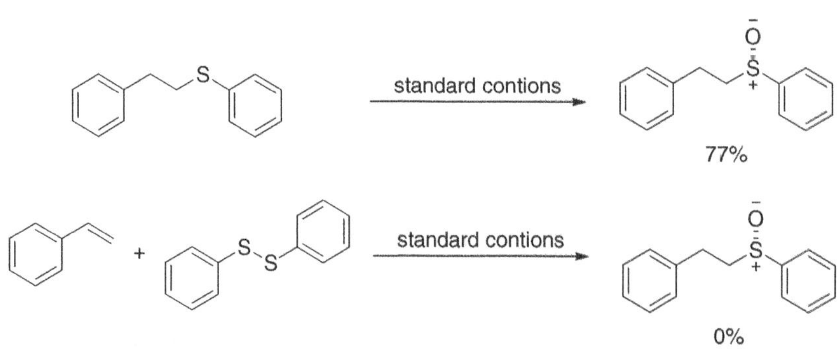

Scheme 12.23 Control experiments for RB-catalyzed cascade thio-ene and sulfoxidation.

excited RB was diminished with the addition of thiophenol. However, styrene did not have this ability. A possible mechanism was proposed based on these results as shown in Scheme 12.24 [62]: excited RB* was obtained from its ground state under visible light irradiation, which then experienced a single electron transfer with thiol and afforded

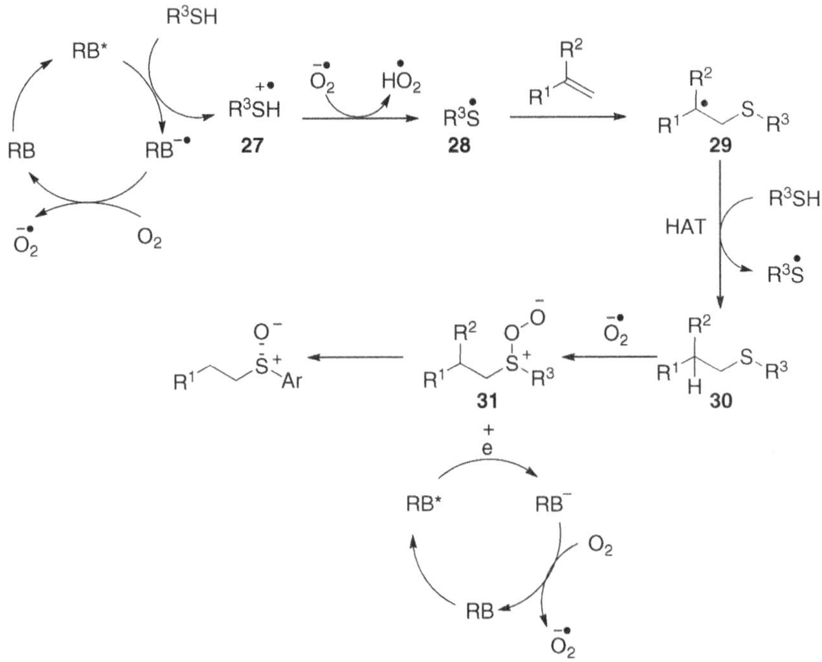

Scheme 12.24 Proposed mechanism for RB-catalyzed cascade thio-ene and sulfoxidation.

the radical cation and $RB^{-\bullet}$ radical anion. The interaction between $RB^{-\bullet}$ and triplet oxygen generated the ground state RB and $O2^{-\bullet}$. The following deprotonation of radical cation **27** by $O2^{-\bullet}$ released a thiyl radical **28**, which further added to alkene and produced the alkyl radical **29**. Subsequently, hydrogen atom transfer between thiol and alkyl radical **29** delivered sulfide intermediate **30**. Last, the photooxygenation of sulfide produced the targeted sulfoxide through intermediate **31**.

Beyond molecular oxygen, water could also be utilized as the oxygen source for sulfidation processes from sulfides. In 2015, a two chromophore-catalyst dyad containing a Ru-aqua complex **32** or its chloro **33** form were synthesized for oxidizing sulfide into sulfoxide with water as the oxygen source and $[Co(NH_3)_5Cl]Cl_2$ as a sacrificial electron acceptor (Scheme 12.25) [63]. Both catalysts showed significant activities with TONs of up to 1000 and high product selectivity (>99%). It is worth mentioning that nitro- and amine- groups could be tolerated in this system. Isotopic labeling experiments with H_2O^{18} demonstrated the oxygen-atom transfer from the water to the organic substrate, an active intermediate Ru(IV) = O was proposed as a key intermediate. On the basis of previous reports and the previous results, a proposed catalytic mechanism was presented (Scheme 12.26), The excited-state *Ru(II) phot **32** moiety can be quenched by the sacrificial electron acceptor Co(III) and generated Ru(III)phot Ru(II)cat-H_2O **33**, which provides a driving force for oxidizing the neighboring Ru(II)cat-H_2O moiety to generate **34** through subsequent proton-coupled electron transfer (PCET) processes. Then, another similar process delivered Ru(II)phot Ru(IV)cat = O **37**, the subsequent oxygen-atom transfer released sulfoxides. However, a mechanism involving successive two-electron two-proton oxidation processes cannot be entirely ruled out.

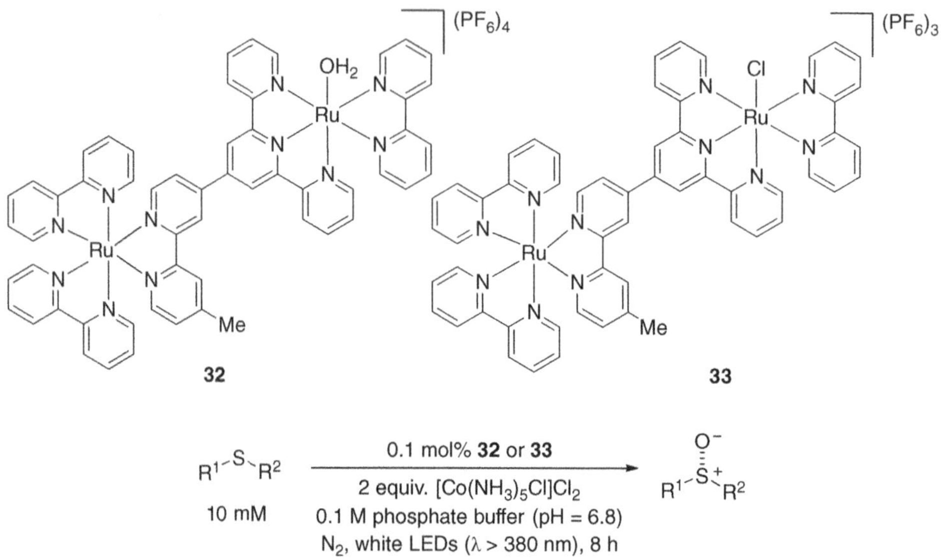

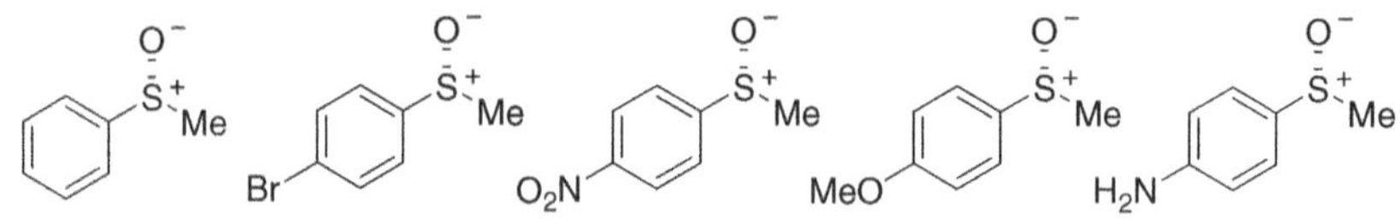

TON with **32**	532	237	167	645	512
TON with **33**	745	314	265	709	567
				1000[a]	

[a]20 mM Co(III) was added, illuminated for another 4 h.

Scheme 12.25 Ru-Ru dyad-catalyzed sulfoxidation from sulfides.

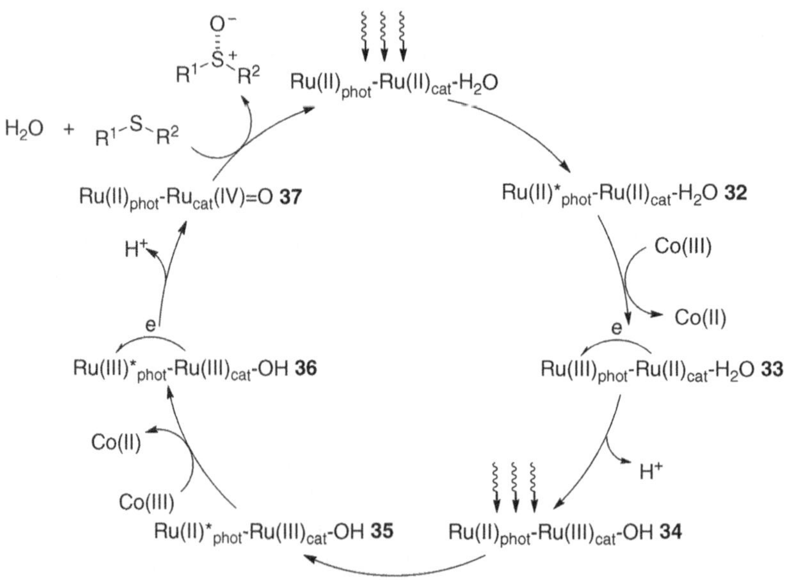

Scheme 12.26 Proposed mechanism for Ru-Ru dyad-catalyzed sulfoxidation from sulfides.

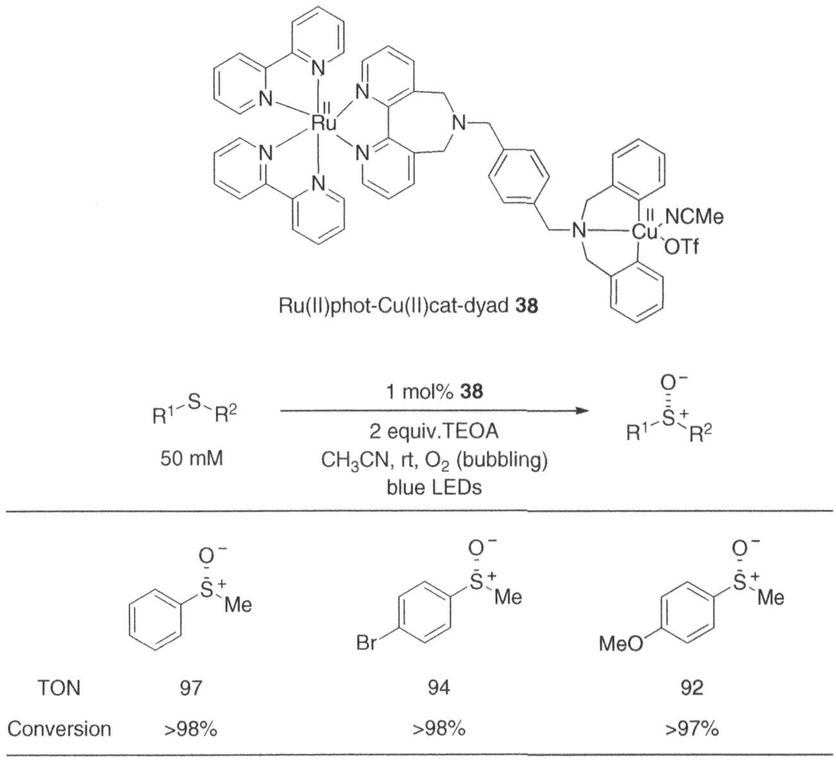

Scheme 12.27 Ru^{II}_{phot}-Cu^{II}_{cat}-dyad-catalyzed sulfoxidation from sulfides.

In 2015, Torelli and Hamelin et al. disclosed a $Ru(II)_{phot}$-$Cu(II)_{cat}$-dyad-catalyzed oxygenation of sulfide with O_2 under light irradiation. Control experiments demonstrated the importance of light, catalyst, electron donor (triethanolamine, TEOA), and the synergistic effect between Ru and Cu center in the dyad, which is more efficient compared to the biomolecular system in this study (Scheme 12.27) [64]. The EPR (electron paramagnetic resonance) experiments helped to confirmed the existence of Cu(I) species, besides, the involvement of copper oxygen-activated species was verified when 1 : 10 mixture of the reduced Cu(I) catalyst and 4-bromothioanisol produced sulfoxide in 70% yield with respect to Cu(I) catalyst in anhydrous acetonitrile in air. Besides, nanosecond time resolved absorption spectroscopy under catalytic conditions supported an oxidative quenching of the ^{3}MLCT state leading to a photoinduced electron transfer to the Cu(II) from the ruthenium photosensitizer in its excited state. Based on detailed mechanistic studies, a proposed mechanism was explained as shown in Scheme 12.28: photon absorption by $Ru(II)_{phot}$-$Cu(II)_{cat}$ yielded photoexcited $Ru(II)^*_{phot}$-$Cu(II)_{cat}$, which generated $Ru(III)_{phot}$-$Cu(I)_{cat}$ through intramolecular electron transfer. Subsequent reduction of Ru(III) to Ru(II) by TEOA and dioxygen activation generated $Ru(II)_{phot}$-$Cu(II)_{cat}$-O_2, this oxygen adduct finally oxidized sulfides to sulfoxides.

12.1.1.3 Others

In 2003, Murahashi et al. found that lumiflavin **39** can be utilized to catalyze the oxidation of sulfides and amines with molecular oxygen or even air in the presence of

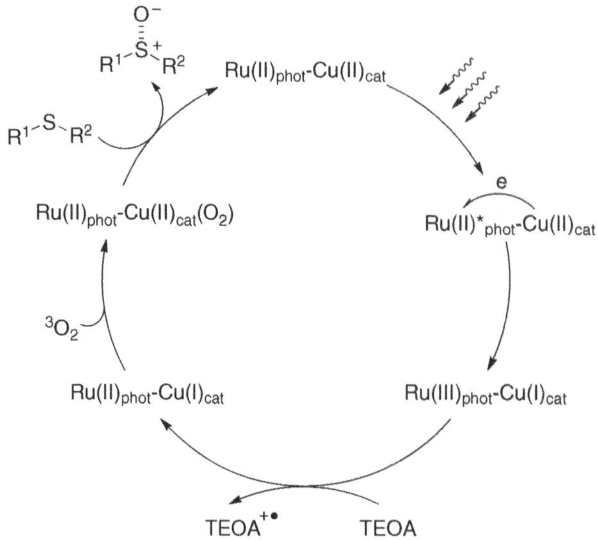

Scheme 12.28 Proposed mechanism for Ru^{II}_{phot}-Cu^{II}_{cat}-dyad-catalyzed sulfoxidation from sulfide.

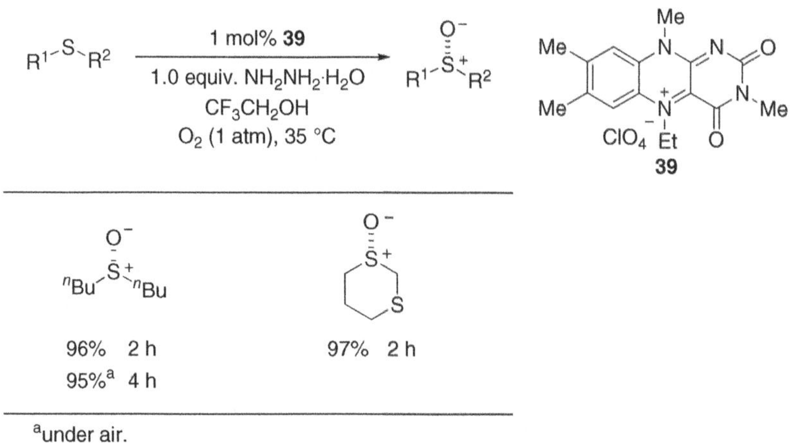

Scheme 12.29 Flavin-catalyzed aerobic oxidations.

hydrazine monohydrate in TFE, giving the corresponding oxides along with water and molecular nitrogen (Scheme 12.29) [65]. Recently, Imada et al. realized the oxygenation of thioanisol with newly designed peptide-containing N5-unmodified neutral flavins **40** under similar conditions (Scheme 12.30), where the possible mechanism is shown in Scheme 12.31 [66]. The catalytic cycle was initiated by reducing Fl-Pep-NH_2NH_2, which was generated from the combination of Fl-Pep and NH_2NH_2, with another molecule of NH_2NH_2 or NH=NH to afford FlH_2-Pep-NH_2NH_2 and diazene NH=NH or N_2. Then, molecular oxygen can be inserted into FlH_2-Pep-NH_2NH_2 to give FlOOH-Pep-NH_2NH_2 **41**, the **41** oxidized sulfide to sulfoxide and **41** is transformed to FlOH-Pep-NH_2NH_2 **42**, which finally undergoes dehydration to regenerate Fl-Pep-NH_2NH_2.

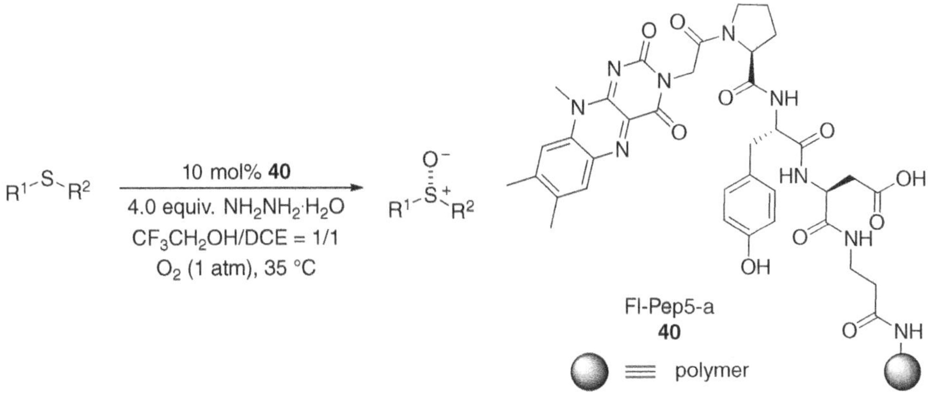

Scheme 12.30 Flavin-peptide-catalyzed aerobic oxidations.

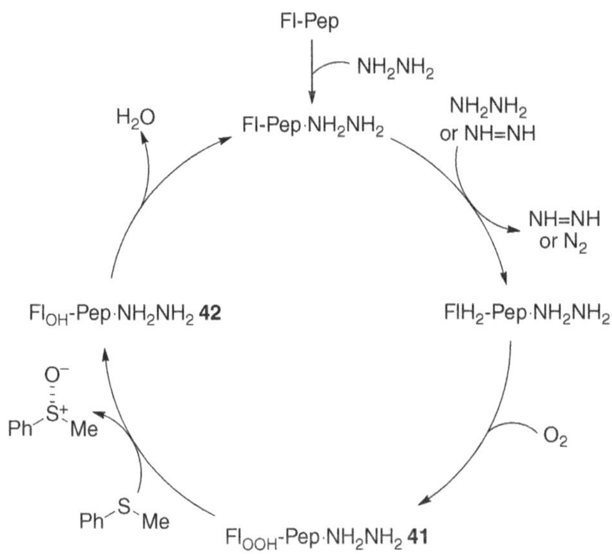

Scheme 12.31 Proposed mechanism for flavin-catalyzed aerobic oxidations.

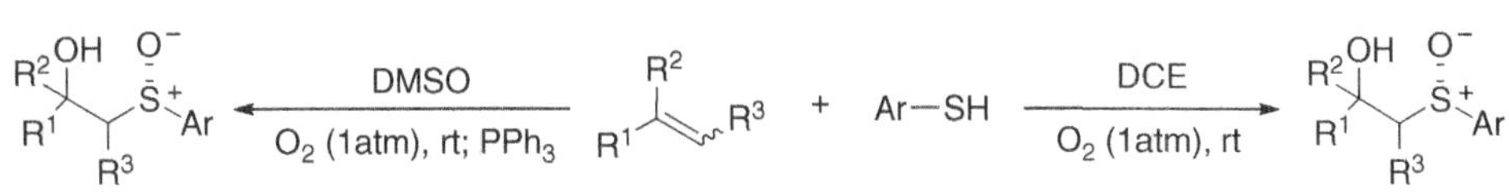

Scheme 12.32 Solvent-enabled tunable sulfoxidation and sulfidation.

In 2016, Lei et al. realized the selective synthesis of β-oxy sulfoxides and β-hydroxy sulfides from thiol, alkene, and oxygen via tuning solvents at room temperature without an external catalyst (Scheme 12.32) [67]. The library of β-oxy sulfoxides is shown in Scheme 12.33. The authors monitored the generating process of **43** with in situ IR, and no inductive period and obvious reaction intermediate were observed. However, the control experiments demonstrated that hydroperoxy sulfide **44** generated in DMSO

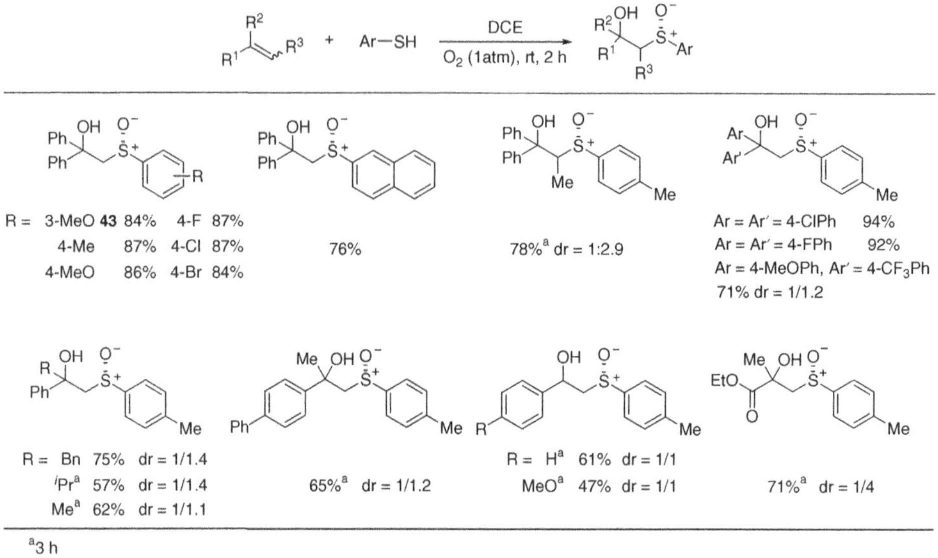

Scheme 12.33 The construction of β-oxy sulfoxides from alkenes, thiophenols, and oxygen.

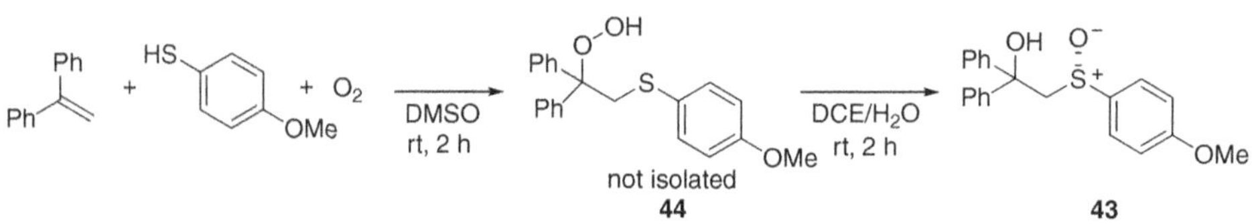

Scheme 12.34 Switchable synthesis of β-hydroxy sulfoxide.

could further transform to β-oxy sulfoxides **43** (Scheme 12.34). The in situ IR and NMR studies of generating process of β-hydroxy sulfide also showed the presence of hydroperoxy sulfide. First, a possible mechanism was proposed as in Scheme 12.35. Second, autoxidation of thiophenol generated a sulfur radical. Then, addition of sulfur radical to alkene afforded a carbon-centered radical **45**, which could experience a reaction with dioxygen and transformation into peroxyl radical **46**. Subsequently, an intermolecular hydrogen abstraction process between **46** and thiophenol delivered hydroperoxy sulfide **47**. Finally, **47** underwent a redox process with or without assistance of solvent-bonding, providing β-hydroxy sulfoxides and β-hydroxy sulfides, respectively.

In 2017, Chi et al. developed an one-pot cascade thio-ene/-yne reaction and sulfoxidation reaction with *N*-fluorobenzenesulfonimide (NFSI), which constructed alkyl-/alkenyl-substituted sulfoxide (Schemes 12.36 and 12.37) [68]. A control experiment with TEMPO totally suppressed the transformation, which indicated the radical property of this system. An ^{18}O isotope labeling experiment suggested the oxygen atom of sulfoxide came from the trace amount of water in the solvent. Reaction with 30 mol% of NFSI only produced sulfide, the generated sulfide **48** could be transformed to sulfoxide **49** under standard conditions (Scheme 12.38). These results showed that NFSI worked as both a radical initiator and oxidant. The authors presented a proposed mechanism as shown in Scheme 12.39: the single electron transfer between NFSI and thiophenol generated thiophenol radical cation, the following deprotonation delivered sulfur

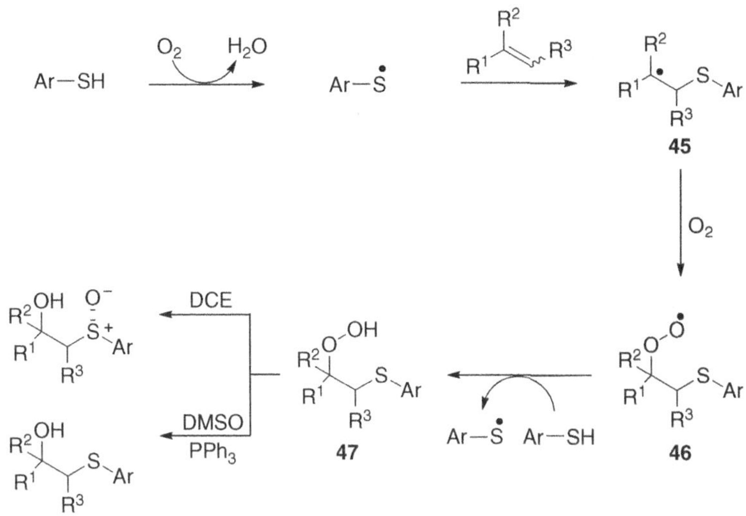

Scheme 12.35 Proposed mechanism for solvent-enabled tunable sulfoxidation and sulfidation.

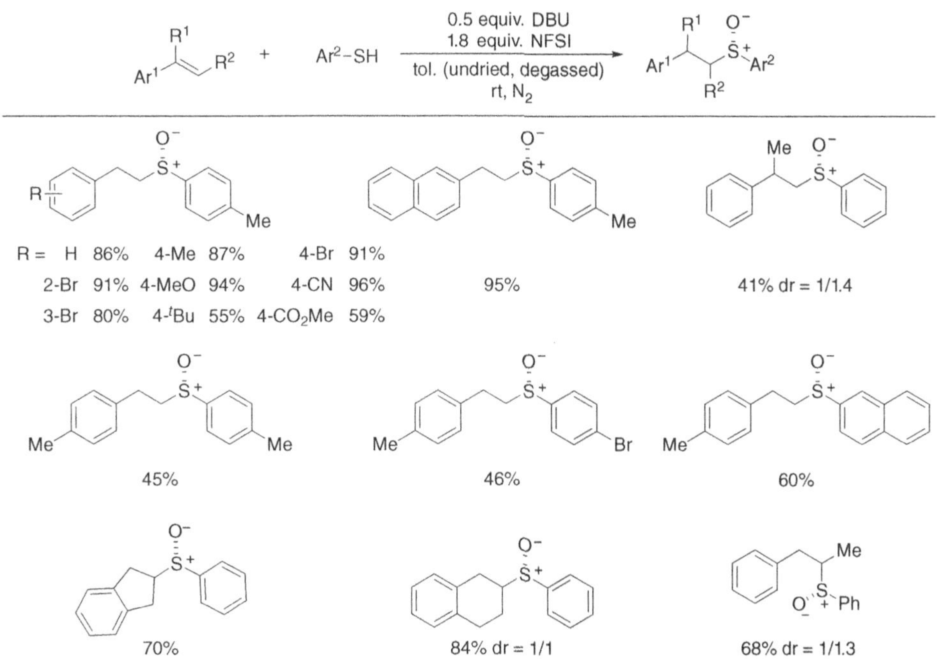

Scheme 12.36 NFSI-promoted cascade addition and oxygenation between thiophenol and alkenes.

radical. Then, the thio-ene process produced new carbon radical **50**, which experienced hydrogen atom transfer with another thiophenol and released sulfide **51**. Next, the possible nucleophilic addition of the sulfide to the electrophilic fluorine atom of NFSI afforded intermediate **52**. Substitution of the fluorine atom of **52** by water forms **53**, which was then transformed to sulfoxide through a proton transfer process.

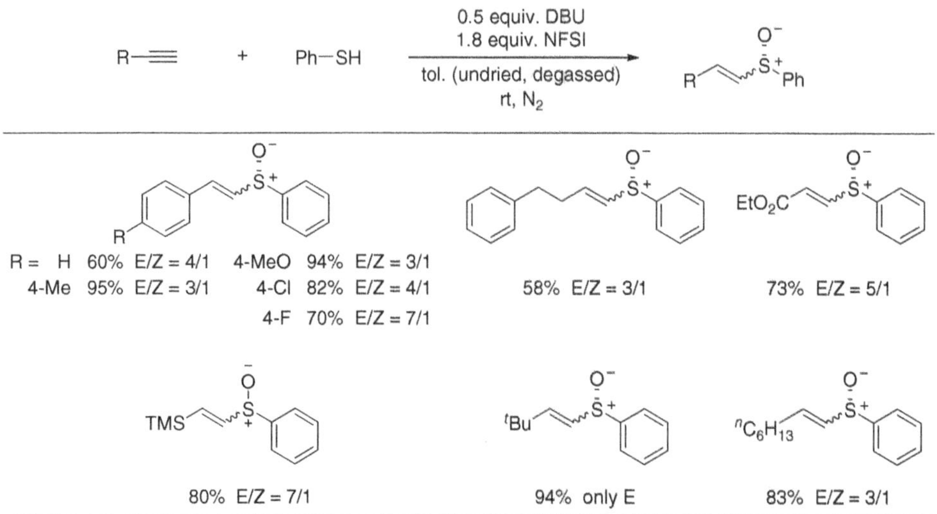

Scheme 12.37 NFSI-promoted cascade addition and oxygenation between thiophenols and alkynes.

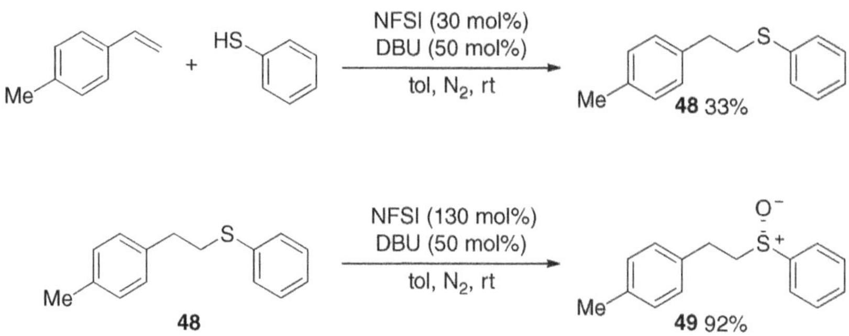

Scheme 12.38 Formation and oxidation of sulfides enabled by NFSI.

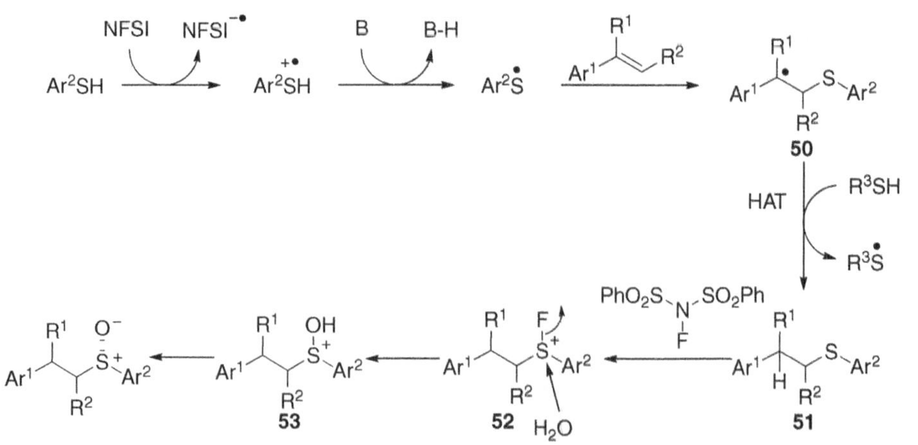

Scheme 12.39 Proposed mechanism for NFSI-promoted cascade addition and oxygenation between thiophenols and alkenes.

12.1.2 Chiral Oxidation

Chiral sulfoxides are important motifs of pharmaceuticals and ligands. In addition, extensive usage of asymmetrical organic sulfoxides as chiral auxiliaries and intermediates is developed. The demand for ligand developments and active compounds applications promoted the flourish of enantioselective sulfoxidation, especially in catalytic ways. Traditionally, H_2O_2 and other peroxides are utilized as important green oxidants under a variety of transition-metal-catalyzed [69–75] and biocatalyzed [76] conditions. Promising and underdeveloped asymmetrical oxygenation of sulfide with molecular oxygen is an alternative route, the related studies are summarized here.

In 1995, Mukaiyama et al. introduced an enantioselective oxygenation of sulfide with molecular oxygen catalyzed by chiral manganese (III)-salen complexes promoted by pivalaldehyde (Scheme 12.40) [77]. Moderate enantioselectivities were obtained, also, electron-rich and -poor groups at different positions of aryl sulfides were tolerated.

In 2010, Katsuki et al. utilized Ru(NO)-salen complexes to catalyze asymmetric oxygenation of sulfide in the presence of water under visible light irradiation (halogen lamp) at room temperature (Scheme 12.41) [78]. Chiral aryl alkyl sulfoxides and 2-substituted 1,3-disulfoxides were obtained with moderate to excellent yields and enantioselectivities, and dialkyl sulfide and stereohindered aryl alkyl organic sulfoxides were produced with relative low yields and moderate enantioselectivities. Remarkably, this aerobic oxygen-atom transfer reaction did not require a sacrificial reductant. Experiments in a $^{18}O_2$ atmosphere revealed that ^{18}O incorporation increased from 76 to 82% as the reaction proceeded. This result suggested that most of the incorporated oxygen atom originates from molecular oxygen but some originated from another oxygen source. An experiment with $H_2{}^{18}O$ under $^{16}O_2$ atmosphere showed that ^{18}O incorporation value of the sulfoxide product was 70% at 11% conversion, and it diminished to 43% at a 94% conversion. Experiments with $H_2{}^{18}O$ and $^{18}O_2$ indicated an aqua ligand coordinated with the ruthenium ion served as a proton transfer agent for the oxygen activation process (Scheme 12.42, step 1), and it was recycled and used as the proton

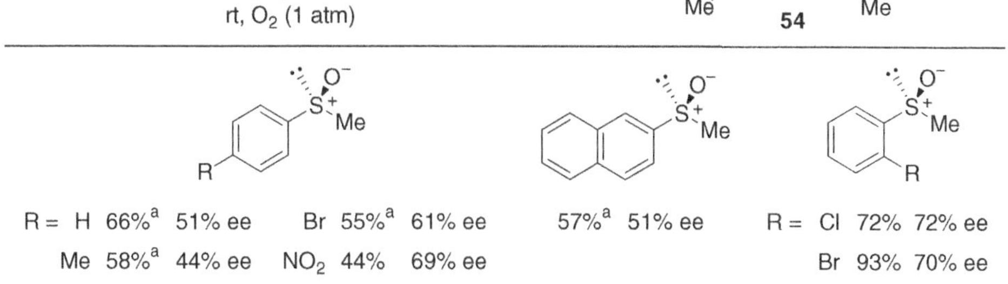

[a]Another 3 equiv. pivalaldehyde was added during reaction.

Scheme 12.40 Asymmetrical sulfoxidation of sulfide with O_2 catalyzed by Mn(III).

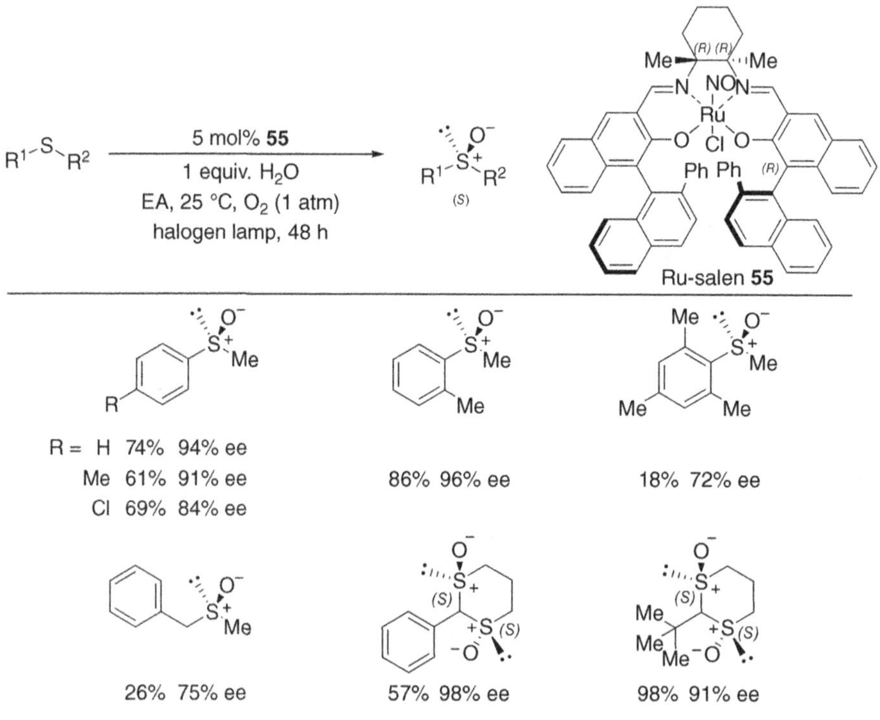

Scheme 12.41 Ru-salen-catalyzed asymmetrical sulfoxidation.

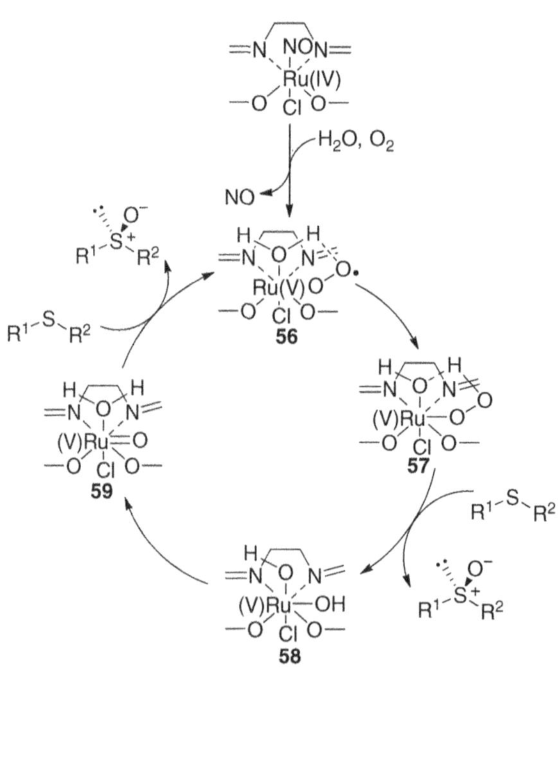

Scheme 12.42 Proposed mechanism of Ru-salen-catalyzed asymmetrical sulfoxidation.

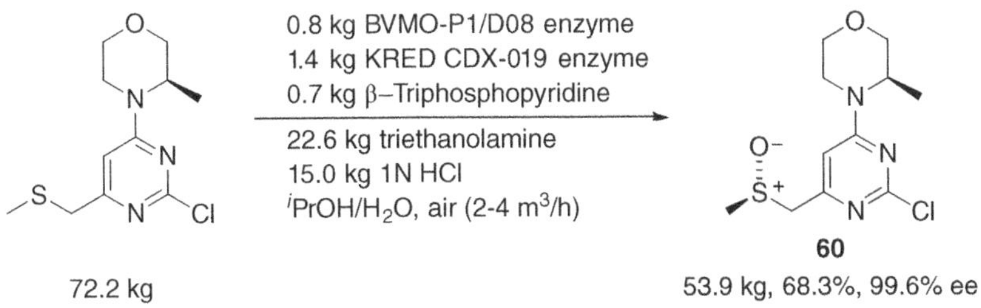

Scheme 12.43 Enzyme-catalyzed asymmetrical construction of **60** on a kilogram scale.

transfer mediator during the process. The hydroperoxide **57** transferred an oxygen to sulfide in asymmetrical way. The generated Ru-OH **58** experienced intermolecular hydrogen atom transfer and produced Ru=O **59**, which could further transfer an oxygen atom to sulfide and regenerated active catalyst **56**.

In 2017, scientists in AstraZeneca utilized a Baeyer–Villiger monooxygenase enzyme to synthesize an asymmetrical sulfoxide drug intermediate **60** on a kilogram scale, in which the efficient gas–liquid mass transfer of oxygen was crucial to obtaining a high yield. Compared to the original unselective m-CPBA oxidation and chromatography with an approximately 50% yield, this process has obvious advantages in cost, safety, and environmental factors (Scheme 12.43) [79].

12.2 Sulfone

Sulfones are an important class of organic molecules due to their prominent biological activities [80–82] and their wide chemical [83, 84] and physical applications [85–87]. They can be used as versatile building blocks in a variety of transformations, including classical reactions such as Julia olefination and various rearrangements [18, 19, 88], and multiple organocatalytic reactions [89–91]. What is more, sulfonizations are important processes, sometimes known as oxidative desulfuration [27, 92, 93]. In common, H_2O_2 was used to realize the sulfonization from sulfide in uncatalytic and catalytic ways [31, 94–104]. Reactions in this section focus on the ones with molecular oxygen as an oxidant.

In 1982, Iyanagi et al. reported a kinetic study on enzymatic *S*-oxygenation of different arylmethylsulfoxides to sulfones with purified cytochrome P450 as the catalyst, NADPH (nicotinamide adenine dinucleotide phosphate) as the electron sacrifice, and molecular oxygen as the oxidant, in which a linear correlation was observed between log(Vmax)'s (kinetics of oxidation of NADPH in the reconstituted system of purified cytochrome P-450) and the one-electron oxidation potentials of sulfoxides, suggesting the single electron transfer process occurred during the oxygenation of sulfoxides (Scheme 12.44) [105].

In 1996, Noible et al. reported a Co(acac)$_2$/Co(AAEMA)$_2$-catalyzed (AAEMA = deprotonated form of 2-(acetocetoxy)ethyl methacrylate) sulfonization of dimethylsulfide, dibutylsulfide, and diphenylsulfide under homogeneous and heterogeneous conditions using oxygen or air as the oxidant with ^{i}valcraldehyde as an additive (Scheme 12.45) [106]. The heterogeneous Co(AAEMA)$_2$ was prepared by reaction of Co(AAEMA)$_2$ with

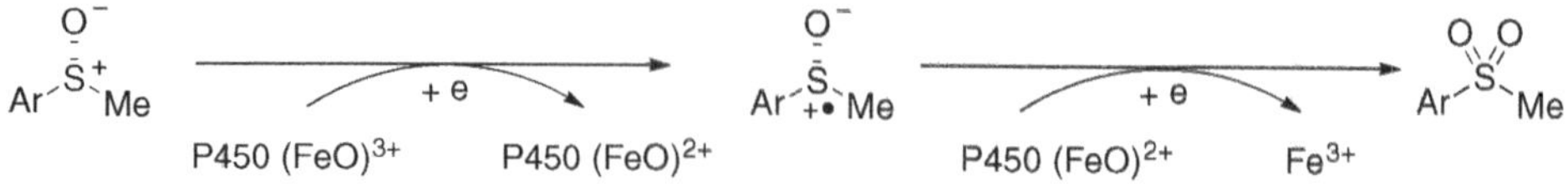

Scheme 12.44 P450-catalyzed oxygenation of sulfoxides.

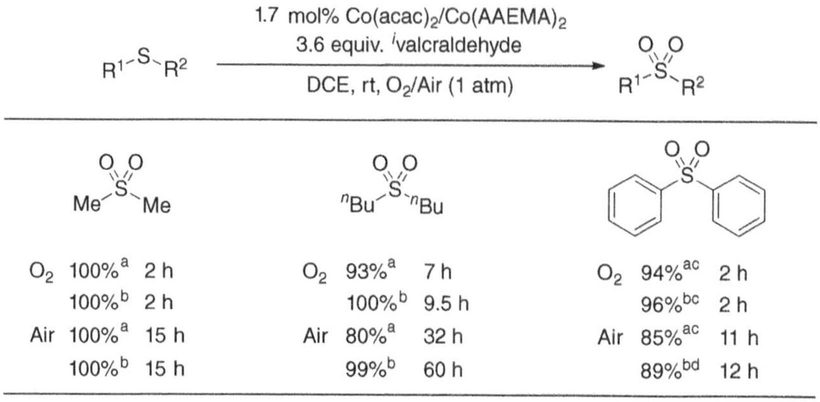

[a]$Co(acac)_2$ as catalyst, [b]$Co(AAEMA)_2$ as catalyst, [c] 5.4 equiv. *i*valcraldehyde, [d]7.2 equiv. *i*valcraldehyde

Scheme 12.45 Co(II)-catalyzed homogeneous asymmetrical sulfonization.

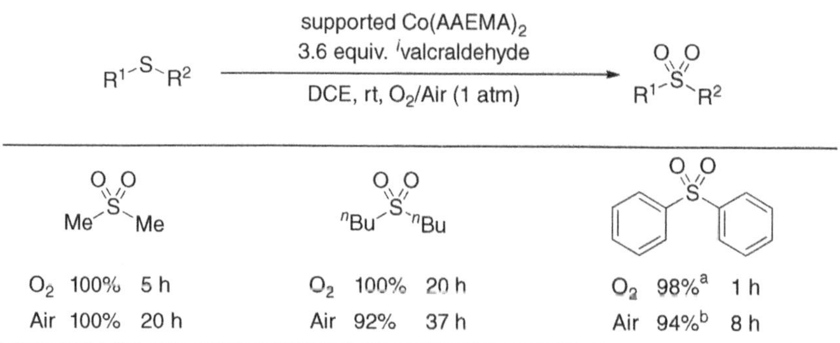

[a]5.4 equiv. *i*valcraldehyde, [b]7.2 equiv. *i*valcraldehyde

Scheme 12.46 Co(II)-catalyzed heterogeneous asymmetrical sulfonization.

N,N-dimethylacrylamide and *N,N'*-methylenebisacrylamide in *N,N*-dimethylformamide, its application in the construction of sulfones were efficient as well (Scheme 12.46). The supported complexes were recyclable with negligible loss of efficiency and did not suffer from metal leaching. In addition to sulfones, this system has also been utilized for the selective sulfoxidation, which is mainly controlled by reducing reaction times.

In 1996, Maikap et al. found that sulfide could easily translate to sulfone with catalytic amount of cobalt(II) salophen, molecular oxygen, and 2-methyl-1-propanol. Interestingly, control experiments showed that the reaction could also proceed efficiently without Co(II) (Scheme 12.47) [107]. The construction of compound library was conducted under metal

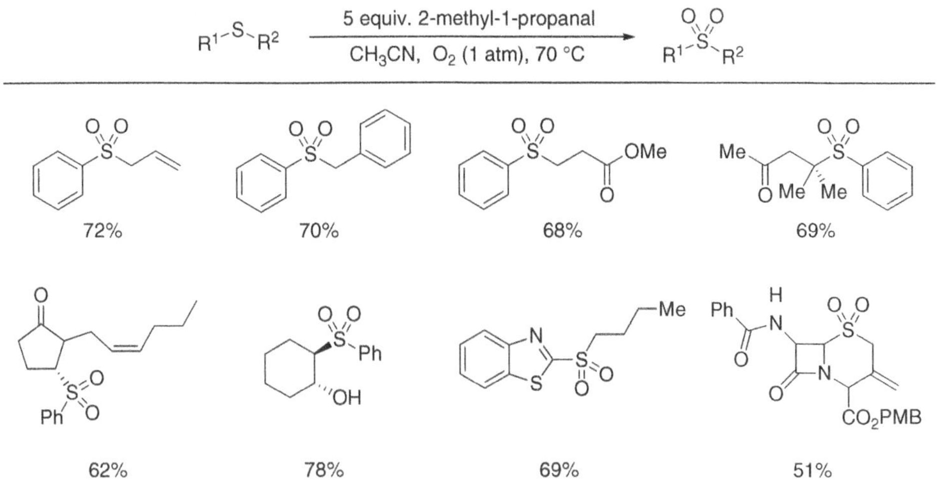

Scheme 12.47 Alcohol-promoted sulfonization.

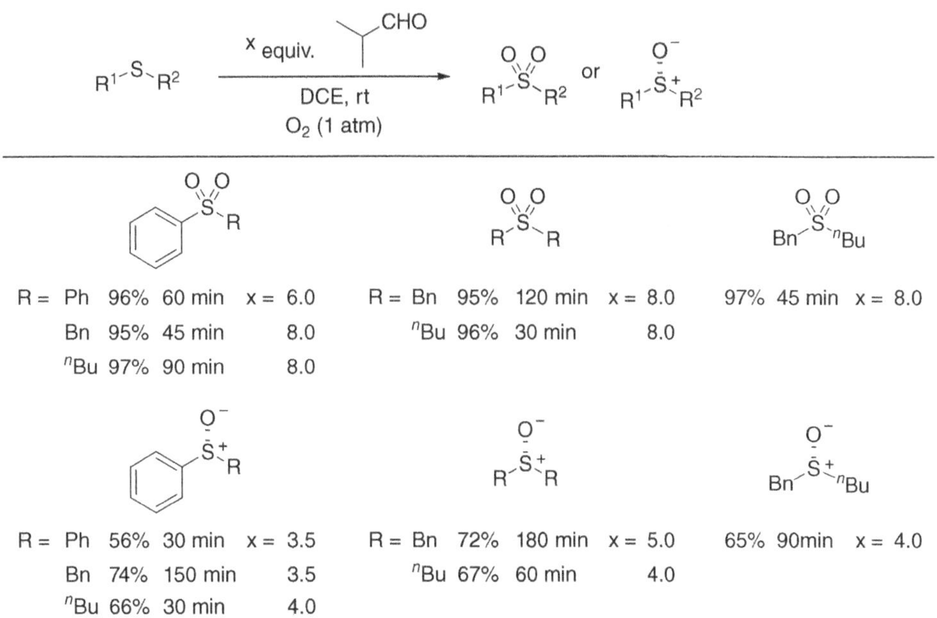

Scheme 12.48 Aldehyde-promoted sulfonization and sulfoxidation.

free conditions under normal pressure at 70 °C. Different sensitive groups, such as allyl, benzyl, ester, ketone, alkene, amide, and nitrogen-containing heterocycle, were tolerated.

In 1998, Joshi et al. demonstrate a simple process for the oxidation of sulfide to sulfone with molecular oxygen as the oxidant and isobutyraldehyde as the additive, in which typical diaryl, arylalkyl, and dialkyl sulfides could be transformed into sulfone within 2 hours with an almost quantitative yield (Scheme 12.48) [108]. It was thought that the reaction was initiated with the autoxidation of aldehyde and the generation of acyl radical. The following combination of molecular oxygen and acyl radical produced a peroxyacyl radical, which in turn abstracted a hydrogen atom from aldehyde forming peroxyacid and releasing another acyl radical in a chain propagation. The sulfone and sulfoxide were obtained via oxidation by peroxyacid.

In 2003, Choudary et al. developed an osmate exchanged layered double hydroxide (LDH–OsO_4) as a recyclable catalyst and achieved sulfonization with it in an oxygen atmosphere at 55 °C (Scheme 12.49) [109]. The catalyst could be reused four times without significant decrease in activity. Besides, this catalyst showed better activity than resin-OsO_4 and $K_2OsO_4 \cdot 2H_2O$ in the oxidation of thioanisole. Detailed studies were performed to understand the interaction mode between osmium center and sulfur-containing species, in which the X-ray photoelectron spectroscopy (XPS) of transient surface intermediate LDH-OsO_4 suggested the possible existence of Os-sulfone complex along with the unreacted LDH-OsO_4 in a +VI oxidation state. XPS studies of LDH–OsO_4^* did not show the existence of an Os—S bond, which combined with kinetic and thermogravimetric analysis/differential thermoanalysis/mass spectrometry (TGA–DTA–MS) studies indicated the concerted 3 + 1 cycloaddition through the delivery of two oxygen atoms at the same time. Continuous monitoring of reaction system showed that sulfoxide was not generated during the whole process. The concise proposed mechanism is shown in Scheme 12.50.

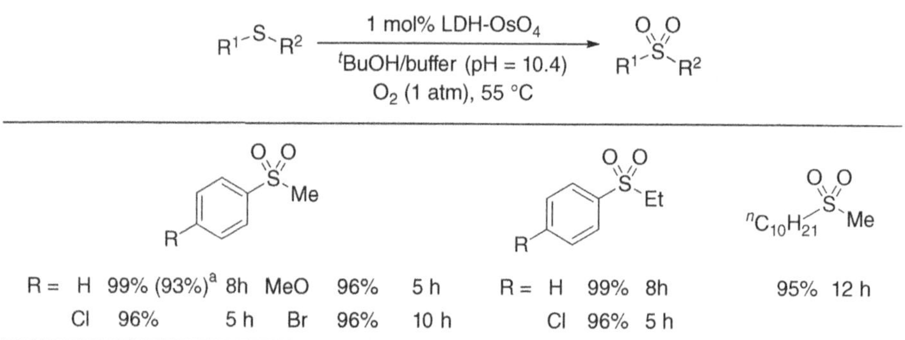

Scheme 12.49 LDH-OsO_4-catalyzed sulfonization.

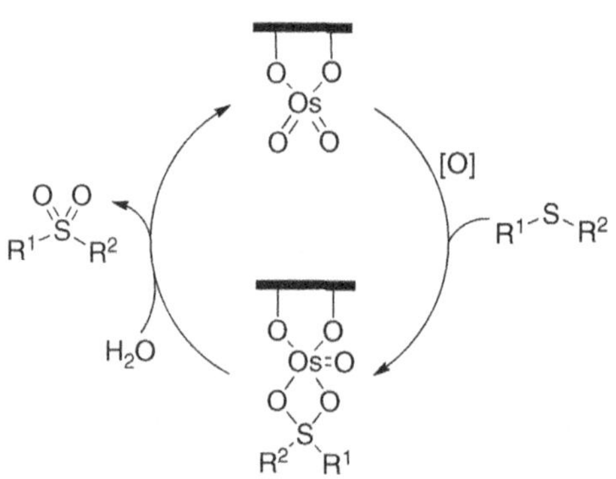

Scheme 12.50 Proposed mechanism of LDH-OsO_4-catalyzed sulfonization.

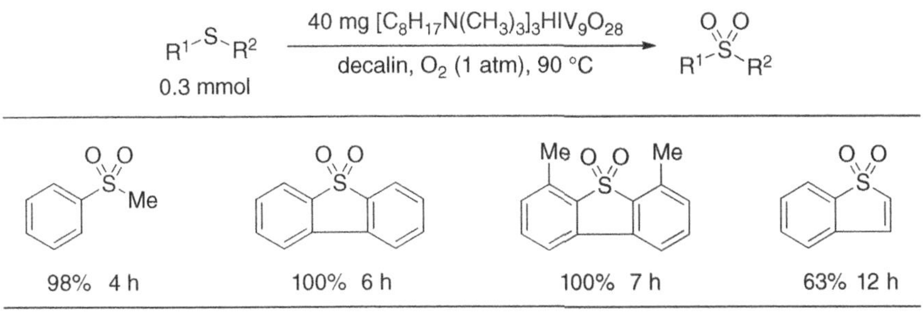

Scheme 12.51 Vanadoperiodate-catalyzed sulfonization and sulfoxidation.

In 2015, Li et al. have discovered a monoiodine(VII) derivative of the decavanadate $[C_8H_{17}N(CH_3)_3]_3HIV_9O_{28}$, which exhibited high efficiency on the oxidation of dibenzothiophene and analogs to sulfone with molecular oxygen at 90 °C (Scheme 12.51) [110]. The catalyst could be reused at least four times with conservation of activity. EPR studies with DMPO (5,5-dimethyl-1-pyrroline N-oxide) as a trap revealed the existence of oxygen radical anion when vanadoperiodate and molecular oxygen interacted at >60 °C. UV-Vis spectroscopy indicated the existence of vanadium superoxo bond. The following oxygen-atom transfer to sulfides released sulfones.

In conclusion, the selective oxygenation of organic sulfides with oxygen or air under mild conditions has been summarized. Previous studies mainly focused on the unsymmetrical oxygenation of dialkyl and arylalkyl sulfides; compatible sulfoxidation strategies with diarylsulfides still need to be developed. The asymmetrical sulfoxidation and sulfonization are still in their infancy. The TON and TOF of current strategies are mostly not yet high enough for industrial application. New catalytic methodologies with higher efficiency and novel technologies with higher-lever safety would encourage the development of this area and finally realize green industrial application.

References

1 Jacob, C. (2006). *Nat. Prod. Rep.* 23: 851–863.
2 Block, E. (1992). *Angew. Chem. Int. Ed.* 31: 1135–1178.
3 Ilardi, E.A., Vitaku, E., and Njardarson, J.T. (2014). *J. Med. Chem.* 57: 2832–2842.
4 Feng, M., Tang, B., Liang, S.H., and Jiang, X. (2016). *Curr. Top. Med. Chem.* 16: 1200–1216.
5 Cheng, H. (2010). *Crit. Rev. Food Sci.* 50: 938–950.
6 Khanum, F., Anilakumar, K.R., and Viswanathan, K.R. (2004). *Crit. Rev. Food Sci.* 44: 479–488.
7 Sun, B.G. (2007). *Sulfur-Containing Perfume Chemistry*. Beijing Science Press.
8 Caron, S., Dugger, R.W., Ruggeri, S.G. et al. (2006). *Chem. Rev.* 106: 2943–2989.
9 Hruszkewycz, D.P., Miles, K.C., Thiel, O.R., and Stahl, S.S. (2017). *Chem. Sci.* 8: 1282–1287.
10 Tomás, R.A.F., Bordado, J.C.M., and Gomes, J.F.P. (2013). *Chem. Rev.* 113: 7421–7469.
11 Pulis, A.P. and Procter, D.J. (2016). *Angew. Chem. Int. Ed.* 55: 9842–9860.
12 Jones-Mensah, E., Karki, M., and Magolan, J. (2016). *Synthesis* 48: 1421–1436.

13 Wu, X.F. and Natte, K. (2016). *Adv. Syn. Cata.* 358: 336–352.
14 Cremlyn, R.J. (1996). *An Introduction to Organosulfur Chemistry*. Chichester: Wiley.
15 Bur, S.K. and Padwa, A. (2004). *Chem. Rev.* 104: 2401–2432.
16 Feldman, K.S. (2006). *Tetrahedron* 62: 5003–5034.
17 Smith, L.H.S., Coote, S.C., Sneddon, H.F., and Procter, D.J. (2010). *Angew. Chem. Int. Ed.* 49: 5832–5844.
18 Schaumann, E. (2007). *Sulfur-Mediated Rearrangements I*. Berlin Heidelberg: Springer-Verlag.
19 Schaumann, E. (2007). *Sulfur-Mediated Rearrangements II*. Berlin Heidelberg: Springer-Verlag.
20 Yorimitsu, H. (2017). *Chem. Rec.* 17, 1156–1167.
21 Rojas, C.M. (2015). *Molecular Rearrangements in Organic Synthesis*. Wiley.
22 King, A.W., Wang, L., and Rack, J.J. (2015). *Acc. Chem. Res.* 48: 1115–1122.
23 Sipos, G., Drinkel, E.E., and Dorta, R. (2015). *Chem. Soc. Rev.* 44: 3834–3860.
24 Otocka, S., Kwiatkowska, M., Madalińska, L., and Kiełbasiński, P. (2017). *Chem. Rev.* 117: 4147–4181.
25 Kaczorowska, K., Kolarska, Z., Mitka, K., and Kowalski, P. (2005). *Tetrahedron* 61: 8315–8327.
26 Liu, K., Ou, H., Shi, X. et al. (2014). *Chin. J. Org. Chem.* 34: 681–692.
27 Dai, C., Zhang, J., Huang, C., and Lei, Z. (2017). *Chem. Rev.* 117: 6929–6983.
28 Panda, M.K., Shaikh, M.M., and Ghosh, P. (2010). *Dalton Tran.* 39: 2428–2440.
29 Šturala, J., Boháčová, S., Chudoba, J. et al. (2015). *J. Org. Chem.* 80: 2676–2699.
30 Meninno, S., Parrella, A., Brancatelli, G. et al. (2015). *Org. Lett.* 17: 5100–5103.
31 Bahrami, K. (2006). *Tetrahedron Lett.* 47: 2009–2012.
32 Kowalski, P., Mitka, K., Ossowska, K., and Kolarska, Z. (2005). *Tetrahedron* 61: 1933–1953.
33 Stingl, K.A. and Tsogoeva, S.B. (2010). *Tetrahedron Asymmetry* 21: 1055–1074.
34 Watanabe, Y., Iyanagi, T., and Oae, S. (1980). *Tetrahedron Lett.* 21: 3685–3688.
35 Aldea, R. and Alper, H. (1995). *J. Org. Chem.* 60: 8365–8366.
36 Zhou, X.-T., Ji, H.-B., Cheng, Z. et al. (2007). *Bio. Med. Chem. Lett.* 17: 4650–4653.
37 Martín, S.E. and Rossi, L.I. (2001). *Tetrahedron Lett.* 42: 7147–7151.
38 Li, B., Liu, A.-H., He, L.-N. et al. (2012). *Green Chem.* 14: 130–135.
39 Li, X.-D., Ma, R., and He, L.-N. (2015). *Chin. Chem. Lett.* 26: 539–542.
40 Hu, Y.-L., Fang, D., and Xing, R. (2014). *RSC Adv.* 4: 51140–51145.
41 Schenck, G.O. and Krauch, C.H. (1962). *Angew. Chem.* 74: 510.
42 Clennan, E.L. (2001). *Acc. Chem. Res.* 34: 875–884.
43 Zen, J.M., Liou, S.L., Kumar, A.S., and Hsia, M.S. (2003). *Angew. Chem. Int. Ed.* 42: 577–579.
44 Zhang, P., Wang, Y., Li, H., and Antonietti, M. (2012). *Green Chem.* 14: 1904–1908.
45 To, W.P., Liu, Y., Lau, T.C., and Che, C.M. (2013). *Chem. Eur. J.* 19: 5654–5664.
46 Lang, X., Hao, W., Leow, W.R. et al. (2015). *Chem. Sci.* 6: 5000–5005.
47 Lang, X., Zhao, J., and Chen, X. (2016). *Angew. Chem. Int. Ed.* 55: 4697–4700.
48 Suzuki, K., Jeong, J., Yamaguchi, K., and Mizuno, N. (2016). *New J. Chem.* 40: 1014–1021.
49 Farràs, P., Di Giovanni, C., Clifford, J.N. et al. (2015). *Coordin. Chem. Rev.* 304–305: 202–208.
50 Baciocchi, E., Chiappe, C., Del Giacco, T. et al. (2009). *Org. Lett.* 11: 1413–1416.
51 Bonesi, S.M., Manet, I., Freccero, M. et al. (2006). *Chem. Eur. J.* 12: 4844–4857.

52 Bonesi, S.M., Fagnoni, M., and Albini, A. (2008). *Eur. J. Org. Chem.* 2008: 2612–2620.
53 Li, W., Xie, Z., and Jing, X. (2011). *Catal. Commun.* 16: 94–97.
54 Li, W., Li, L., Xiao, H. et al. (2013). *RSC Adv.* 3: 13417–13421.
55 Dad'ová, J., Svobodová, E., Sikorski, M. et al. (2012). *ChemCatChem* 4: 620–623.
56 Ogilby, P.R. and Foote, C.S. (1983). *J.Am. Chem. Soc.* 105: 3423–3430.
57 Neveselý, T., Svobodová, E., Chudoba, J. et al. (2016). *Adv. Syn. Catal.* 358: 1654–1663.
58 Casado-Sanchez, A., Gomez-Ballesteros, R., Tato, F. et al. (2016). *Chem. Commun.* 52: 9137–9140.
59 Wang, Z.J., Ghasimi, S., Landfester, K., and Zhang, K.A.I. (2014). *Chem. Commun.* 50: 8177–8180.
60 Johnson, J.A., Zhang, X., Reeson, T.C. et al. (2014). *J. Am. Chem. Soc.* 136: 15881–15884.
61 Keshari, T., Yadav, V.K., Srivastava, V.P., and Yadav, L.D.S. (2014). *Green Chem.* 16: 3986–3992.
62 Cui, H., Wei, W., Yang, D. et al. (2017). *Green Chem.* 19: 3520–3524.
63 Li, T.-T., Li, F.-M., Zhao, W.-L. et al. (2015). *Inorg. Chem.* 54: 183–191.
64 Iali, W., Lanoe, P.H., Torelli, S. et al. (2015). *Angew. Chem. Int. Ed.* 54: 8415–8419.
65 Imada, Y., Iida, H., Ono, S., and Murahashi, S.-I. (2003). *J. Am. Chem. Soc.* 125: 2868–2869.
66 Arakawa, Y., Yamanomoto, K., Kita, H. et al. (2017). *Chem. Sci.* 8: 5468–5475.
67 Wang, H., Lu, Q., Qian, C. et al. (2015). *Angew. Chem. Int. Ed.* 55: 1094–1097.
68 Zhang, Y., Wong, Z.R., Wu, X. et al. (2017). *Chem. Commun.* 53: 184–187.
69 Fernández, I. and Khiar, N. (2003). *Chem. Rev.* 103: 3651–3706.
70 Burke, A.J. and Carreiro, E.P. (2013). Homogeneous catalytic applications. In: *Comprehensive Inorganic Chemistry II*, 2e, vol. 6 (ed. K. Poeppelmeier), 309–382. Amsterdam: Elsevier.
71 Carmen Carreno, M., Hernandez-Torres, G., Ribagorda, M., and Urbano, A. (2009). *Chem. Commun.* 6129–6144.
72 Konstantin, P.B. and Evgenii, P.T. (2012). *Curr. Org. Chem.* 16: 1215–1242.
73 Legros, J., Dehli, J.R., and Bolm, C. (2005). *Adv. Syn. Catal.* 347: 19–31.
74 Wojaczyńska, E. and Wojaczyński, J. (2010). *Chem. Rev.* 110: 4303–4356.
75 Bryliakov, K.P. (2017). *Chem. Rev.* 117: 11406–11459.
76 Grogan, G. (2015). 3.5.2 Oxidation at sulfur. In: *Biocatalysis in Organic Synthesis*, vol. 3 (ed. K. Faber, W.-D. Fessner and N. Turner), 286–311. Stuttgart, Germany: Thieme.
77 Imagawa, K., Nagata, T., Yamada, T., and Mukaiyama, T. (1995). *Chem. Lett.* 24: 335–336.
78 Tanaka, H., Nishikawa, H., Uchida, T., and Katsuki, T. (2010). *J. Am. Chem. Soc.* 132: 12034–12041.
79 Goundry, W.R.F., Adams, B., Benson, H. et al. (2017). *Org. Process Res. Dev.* 21: 107–113.
80 Meadows, D.C. and Gervay-Hague, J. (2006). *Med. Res. Rev.* 26: 793–814.
81 Butawan, M., Benjamin, L.R., and Bloomer, J.R. (2017). *Nutrients* 9.
82 Chen, X., Hussain, S., Parveen, S. et al. (2012). *Curr. Med. Chem.* 19: 3578–3604.
83 Tilstam, U. (2012). *Org. Process Res. Dev.* 16: 1273–1278.
84 Bäckvall, J.-E., Chinchilla, R., Nájera, C., and Yus, M. (1998). *Chem. Rev.* 98: 2291–2312.
85 Liu, Y.-L. (2012). *Polym. Chem.* 3: 1373–1383.
86 Dizman, C., Tasdelen, M.A., and Yagci, Y. (2013). *Polym. Int.* 62: 991–1007.
87 Kausar, A., Zulfiqar, S., and Sarwar, M.I. (2014). *Polym. Rev.* 54: 185–267.
88 Henderson, A.R.P., Kosowan, J.R., and Wood, T.E. (2017). *Can. J. Chem.* 95: 483–504.

89 Dubbaka, S.R. and Vogel, P. (2005). *Angew. Chem. Int. Ed.* 44: 7674–7684.

90 Nielsen, M., Jacobsen, C.B., Holub, N. et al. (2010). *Angew. Chem. Int. Ed.* 49: 2668–2679.

91 Alba, A.-N.R., Companyo, X., and Rios, R. (2010). *Chem. Soc. Rev.* 39: 2018–2033.

92 Kulkarni, P.S. and Afonso, C.A.M. (2010). *Green Chem.* 12: 1139–1149.

93 Chandra Srivastava, V. (2012). *RSC Adv.* 2: 759–783.

94 García-Gutiérrez, J.L., Fuentes, G.A., Hernández-Terán, M.E. et al. (2006). *Appli. Catal. A Gen.* 305: 15–20.

95 Al-Maksoud, W., Daniele, S., and Sorokin, A.B. (2008). *Green Chem.* 10: 447–451.

96 Karmee, S.K., Greiner, L., Kraynov, A. et al. (2010). *Chem. Commun.* 46: 6705–6707.

97 Zhou, X., Lv, S., Wang, H. et al. (2011). *Appli. Catal. A Gen.* 396: 101–106.

98 Tumula, V.R., Bondwal, S., Bisht, P. et al. (2012). *React. Kinet. Mech. Catal.* 107: 449–466.

99 Widger, L.R., Siegler, M.A., and Goldberg, D.P. (2013). *Polyhedron* 58: 179–189.

100 Hu, Y.-L., Liu, X.-B., and Fang, D. (2014). *Catal. Sci. Tech.* 4: 38–42.

101 Afrasiabi, R., Jalilian, F., Yadollahi, B., and Farsani, M.R. (2014). *Inorg. Chem. Commun.* 50: 113–116.

102 Ahammed, S., Kundu, D., Siddiqui, M.N., and Ranu, B.C. (2015). *Tetrahedron Lett.* 56: 335–337.

103 Bogolubsky, A.V., Moroz, Y.S., Mykhailiuk, P.K. et al. (2015). *ACS Comb. Sci.* 17: 348–354.

104 Gu, Q., Wen, G., Ding, Y. et al. (2017). *Green Chem.* 19: 1175–1181.

105 Watanabe, Y., Iyanagi, T., and Oae, S. (1982). *Tetrahedron Lett.* 23: 533–536.

106 Dell'Anna, M.M., Mastrorilli, P., and Nobile, C.F. (1996). *J. Mol. Catal. A: Chem.* 108: 57–62.

107 Khanna, V., Maikap, G.C., and Iqbal, J. (1996). *Tetrahedron Lett.* 37: 3367–3370.

108 Rao, T.V., Sain, B., Kumar, K. et al. (1998). *Synthetic Commun.* 28: 319–326.

109 Choudary, B.M., Reddy, C.R.V., Prakash, B.V. et al. (2003). *Chem. Commun.* 754–755.

110 Tang, N., Jiang, Z., and Li, C. (2015). *Green Chem.* 17: 817–820.

111 Baird, C.P. and Rayner, C.M. (1998). *J. Chem. Soc.*, Perkin Trans. 1: 1973.

112 Procter, D.J. (1999). *J. Chem. Soc.*, Perkin Trans. 1: 641–667.

113 Procter, D.J. (2001). *J. Chem. Soc.*, Perkin Trans. 1: 335–354.

114 Tobien, T., Cooper, W.J., Nickelsen, M.G., Pernas, E., O'Shea, K.E., and Asmus, K-D. (2000). *Environ. Sci. Technol.* 34: 1286–1291.

115 Yang, Y.C., Baker, J.A., and Ward, J.R. (1992). *Chem. Rev.* 92: 1729–1743.

116 Wang, Y., Li, Y., and Jiang, X. (2018). *Chem. Asian J.* 13: 2208–2242.

13

Oxidative Couplings with C—H Bonds Forming C—P and C—S Bonds

Tieqiao Chen[1] and Li-Biao Han[2]

[1] *College of Materials and Chemical Engineering, Hainan University, Haikou, China*
[2] *National Institute of Advanced Industrial Science and Technology (AIST), Tsukuba, Japan*

13.1 Introduction

Organophosphorus and organosulfur compounds are important chemicals in biology, medicinal chemistry, agro-chemistry, material chemistry, and organic synthesis [1]. However, synthesis of those compounds usually relies on the classical methods [2, 3]. For example, the nucleophilic substitutions of phosphorus-halogen compounds PX with RMgX or RLi, and the reactions of trialkyl phosphites with alkyl halides at high temperature (Michaelis–Arbuzov reaction), are generally used to prepare organophosphorus compounds [1, 2a]. The transition metal-catalyzed cross couplings of P(O)—H compounds with organohalides are also widely used for the synthesis of organophosphorus compounds [2c, d]. Cross coupling of organohalides with thiols are also generally used for the formation of C—S bonds [2e–g]. Simpler and cleaner methods for their preparation like recently developed C—P and C—S bonds formation catalyzed by metals are highly desirable [3].

$$\text{C–H} + \underset{\text{Z = P, S}}{\text{Z–H}} \xrightarrow[-2\,[\text{H}]]{} \text{C–Z} \qquad (13.1)$$

Oxidative coupling between C—H bonds and Z—H bonds is a straightforward and atom-economic method for constructing C—Z bonds [4]. This transformation avoids the need for prefunctionalization of the starting materials and is a reliable alternative to the classical coupling reactions with salt elimination. During the past several decades, this strategy has been widely employed in organic synthesis and great advances have been made. This chapter deals with the oxidative couplings of C—H compounds with thiols and P(O)—H compounds (Eq. 13.1). The oxidative couplings of C—H compounds with disulfides forming C—S bonds are also discussed.

Green Oxidation in Organic Synthesis, First Edition. Edited by Ning Jiao and Shannon S. Stahl.

13.2 Copper-Mediated Oxidative Coupling Forming C—P and C—S Bonds

13.2.1 Copper-Mediated Oxidative Coupling Forming C—P Bonds

In 2009, we reported a copper-catalyzed aerobic oxidative coupling of terminal alkynes with H-phosphonates, generating alkynylphosphonates (Scheme 13.1) [5]. In the presence of a catalytic amount of Cu/base, both aromatic and aliphatic terminal alkynes coupled readily with diisopropylphosphonate. The synthetic value was further demonstrated by the synthesis of phosphonates bearing a norethindrone and nucleotide fragment. However, secondary phosphine oxide like diphenylphosphine oxide did not work under similar conditions because of the severe oxidation of the diphenylphosphine oxide to diphenylphosphinic acid. The oxidative coupling of terminal alkynes with H-phosphonates could be conducted by using a copper catalyst immobilized on SiO_2-supported carbene with similar results [6]. This immobilized catalyst could be easily separated from products and reused. The cheap and stable $CuSO_4.5H_2O$ could also mediate the oxidative coupling efficiently [7]. Later, Zhao modified the copper catalytic system and achieved the cross oxidative coupling of diphenylphosphine oxide with phenylacetylene by adding diphenylphosphine oxide to the reaction mixture dropwise [8]. In 2016, Radivoy reported a base-free Cu_2O-catalyzed aerobic oxidative coupling between terminal alkynes and H-phosphonates [9].

The copper-catalyzed oxidative coupling was also applicable to sp^2C—P bond formation. In 2013, Yang discovered a copper-catalyzed phosphorylation of α-amino carbonyl compounds with diphenylphosphine oxide (Scheme 13.2) [10]. Thus, by using 10 mol% $Cu(OAc)_2.H_2O$ as a catalyst and 2 equiv. TBHP as an oxidant, the oxidative coupling proceeded smoothly in MeCN at 30 °C to produce the corresponding imidoylphosphine oxides in 20–91% yields.

Later, Yang disclosed a dehydrocoupling between indoles and P(O)—H compounds (Scheme 13.3) [11]. This transformation did not require additional oxidants with dihydrogen gas as the sole by-product. A variety of phosphorylated indoles were produced in good yields. P-chiral 3-phosphoindoles could also be prepared *stereo*-selectively.

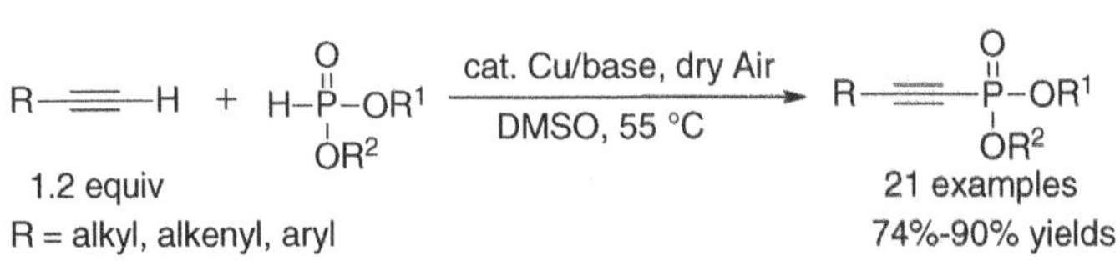

Scheme 13.1 Cu-catalyzed phosphorylation of terminal alkynes.

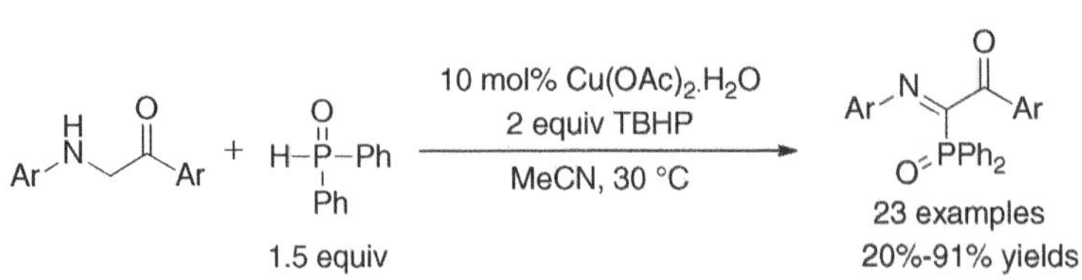

Scheme 13.2 Cu-catalyzed phosphorylation of α-amino carbonyl compounds.

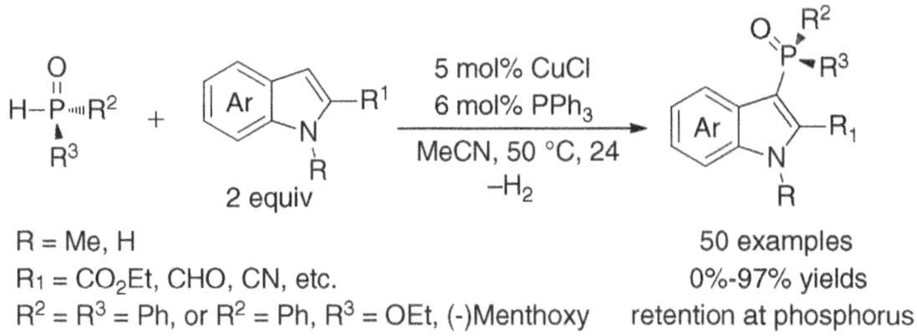

Scheme 13.3 Cu-catalyzed phosphorylation of indoles.

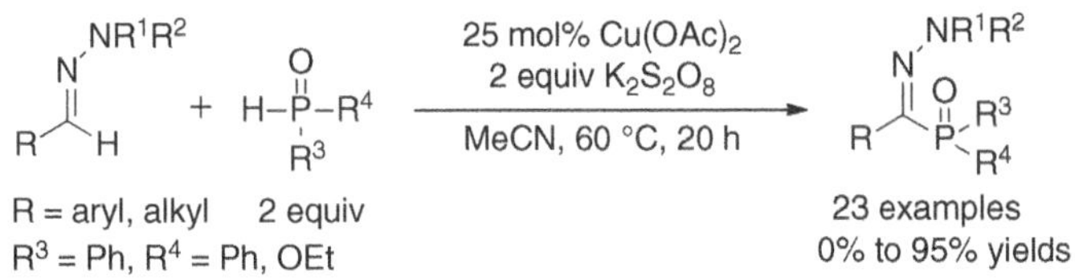

Scheme 13.4 Cu-catalyzed phosphorylation of *N,N*-dialkylhydrazones.

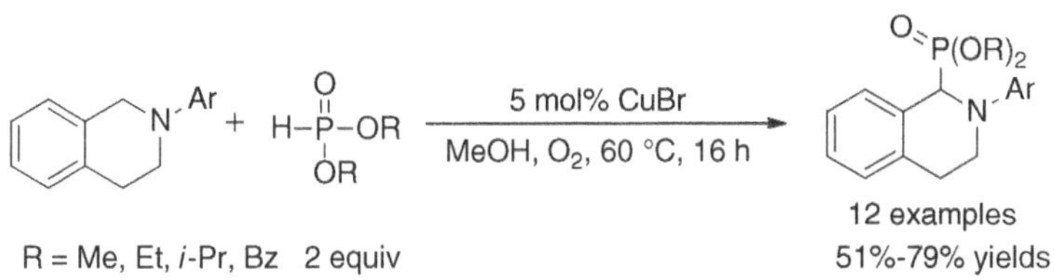

Scheme 13.5 Cu-catalyzed phosphorylation of *N*-aryl tetrahydroisoquinolines.

By using the strategy, a drug candidate IDX899 was synthesized under a clinical trial, displaying its synthetic value well.

By employing 2 equiv. $K_2S_2O_8$ as an oxidant, the copper-catalyzed phosphorylation of *N,N*-dialkylhydrazones with hydrogen phosphoryl compounds also took place in MeCN at 60 °C regioselectively (Scheme 13.4) [12]. This reaction could produce (*E*)-α-iminophosphine oxides in 0–95% yields. Functional groups like methoxy, F, Cl, Br, I, CF_3, CN, and even free hydroxyl groups were tolerable under the reaction conditions. However, although secondary phosphine oxides and H-phosphinates worked well, H-phosphonates was not applicable to this reaction.

Copper could also be employed as a catalyst for the construction of sp^3C—P bonds via oxidative coupling. In 2009, Li described an aerobic oxidative sp^3C—H/P(O) —H coupling through copper catalysis to produce α-aminophosphonates (Scheme 13.5) [13]. Mediated by 5 mol% CuBr, *N*-aryl tetrahydroisoquinolines were phosphorylated readily with H-phosphonates in MeOH at 60 °C under O_2 atmosphere, affording the expected coupling products in 51–79% yields. This reaction used dioxygen as an oxidant and provided a clean method for the synthesis of α-aminophosphonates. Later, Li modified the copper catalytic system and extended the substrate scope of C—H compounds to glycine derivatives in a CuI/TBHP oxidative system [14].

In 2015, Lei developed an efficient copper-catalyzed sp^3C—P bond-forming reaction based on radical/radical cross coupling strategy (Scheme 13.6) [15]. This reaction used CuCl as a catalyst and PCy_3 as a ligand. Various *β*-ketophosphonates including those bearing functional groups were produced by oxidative coupling of aryl ketone *O*-acetyloximes with secondary phosphine oxides, and the yields ranged from 47 to 86%. In addition to secondary phosphine oxides, H-phosphonates, and H-phosphinates were also applicable to this reaction with relatively low yields.

13.2.2 Copper-Mediated Oxidative Coupling Forming C—S Bonds

Oxidative coupling mediated by copper was applicable to C—S bonds formation. In 2006, Yu demonstrated the direct *ortho*-thiolation of 2-phenylpyridine with PhSH and dimethyl disulfides in the presence of 1 equiv. $Cu(OAc)_2$ to produce the coupling products in 40 and 51% yields, respectively (Scheme 13.7) [16]. Despite the low yields, this reaction represented a pioneering early work on the copper-catalyzed oxidative coupling forming C—S bonds.

Later, Fukuzawa disclosed a direct thiolation of benzoxazoles through copper catalysis (Scheme 13.8) [17]. Thus, catalyzed by 20 mol% CuI/2,2′-bipyridine, various benzoxazoles were coupled with diaryl disulfides under dioxygen atmosphere to produce the corresponding 2-benzoxazole thioethers. Electron-rich disulfides afforded the coupling products in good yields, whereas the electron-deficient ones hardly worked in the

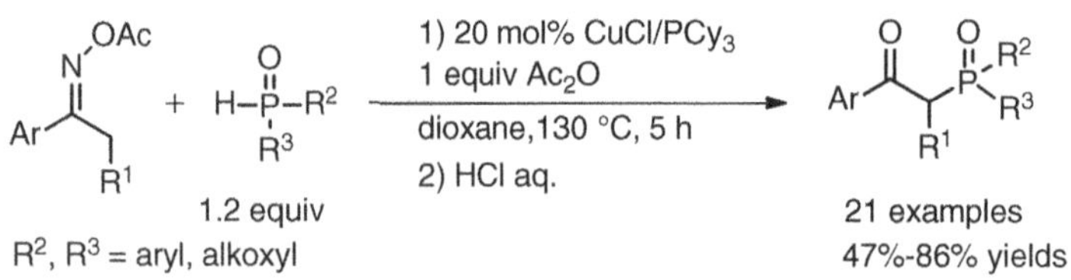

Scheme 13.6 Cu-catalyzed synthesis of *β*-ketophosphoryl compounds.

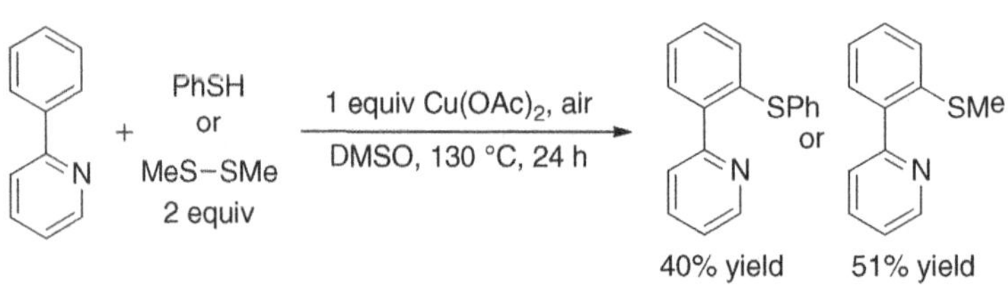

Scheme 13.7 Cu-catalyzed *ortho*-thiolation of 2-phenylpyridine.

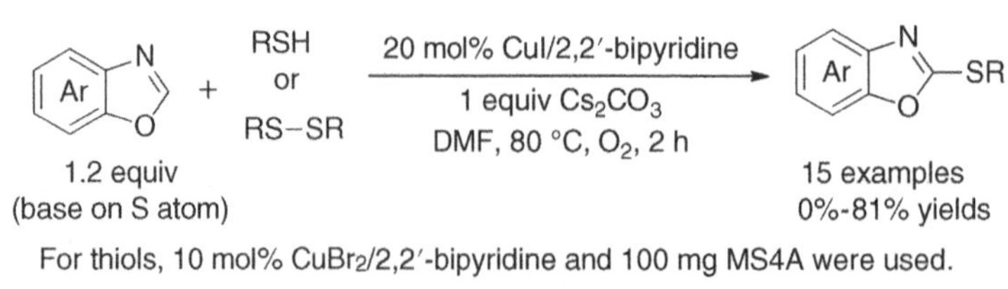

Scheme 13.8 Cu-catalyzed thiolation of benzoxazoles.

catalytic system. Under similar reaction conditions, the oxidative coupling of benzoxazoles with thiophenols also took place. In 2011, Liu disclosed a direct thiolation of thiazoles with thiols using stoichiometric CuI/2,2-bipyridyl under a dioxygen atmosphere [18]. Both aliphatic and electron-rich aromatic thiols were applicable to this reaction, affording the corresponding coupling products in good to high yields. Heterocyclic compounds, that is, oxazoles [19], imidazoles [19], oxadiazoles [19], pyrrole [20], thiazoles [21, 22], xanthines [23], and imidazo[1,2-a]pyridines [24], were also directly thiolated regioselectively with thiols or disulfides by employing a catalytic or stoichiometric amount of copper complexes under similar reaction conditions.

In 2010, Cheng reported a CuI-catalyzed thiolation of electron-rich benzenes with disulfides (Scheme 13.9) [25]. Aliphatic, aromatic, electron-rich, and electron-deficient disulfides all were applicable to this reaction, furnishing the corresponding sulfides in good to excellent yields. However, the substrate scope of C—H compounds seemed to be limited to tri- and di-methoxybenzenes. Later, Yu demonstrated a copper-catalyzed double thiolation of pentafluorobenzene with diaryl disulfides by C—H and C—F activation under a strong alkaline reaction condition [26]. This reaction most frequently gave a mixture of mono-arylthiolation (via C—H bond or C—F bond cleavage) and bisarylthiolation products.

Aryl trifluoromethyl sulfides commonly occur in pharmaceuticals and agrochemicals. The introduction of trifluoromethylthio group into functional molecules is important and attracts chemists' attention [27]. In 2012, Daugulis accomplished the direct trifluoromethylthiolation of carboxylic acid derivatives with bis(trifluoromethyl) disulfide by using auxiliary-assisted strategy through copper catalysis (Scheme 13.10) [28]. Thus, by using 8-aminoquinoline as an auxiliary group, a variety of carboxylic acid derivatives having both electron-withdrawing and electron-donating groups were thiolated under the optimal reaction conditions. Other aliphatic and aromatic disulfides were also applicable to this reaction giving similar results. By elevating the temperature to 130 °C, benzyl amine derivatives bearing a picolinic acid directing group also coupled with various disulfides in the presence of over-stoichiometric $Cu(OAc)_2$, yielding the corresponding γ-thiolated products in moderate yields. Compared with pyridine-directed

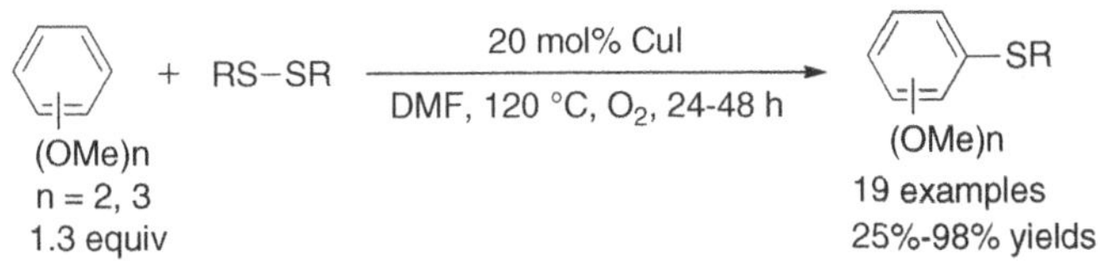

Scheme 13.9 Cu-catalyzed thiolation of benzenes.

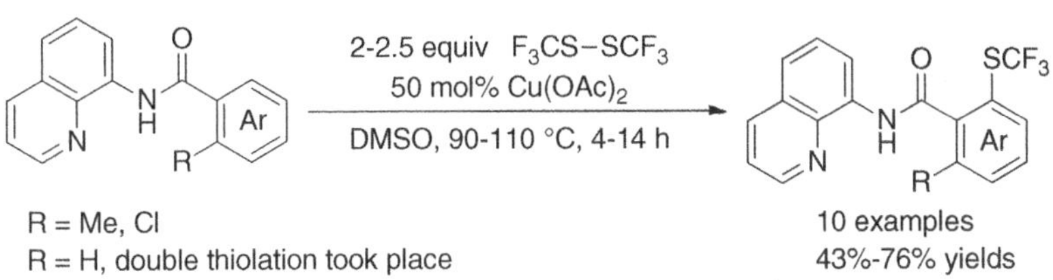

Scheme 13.10 Cu-mediated trifluoromethylthiolation of aryl amides.

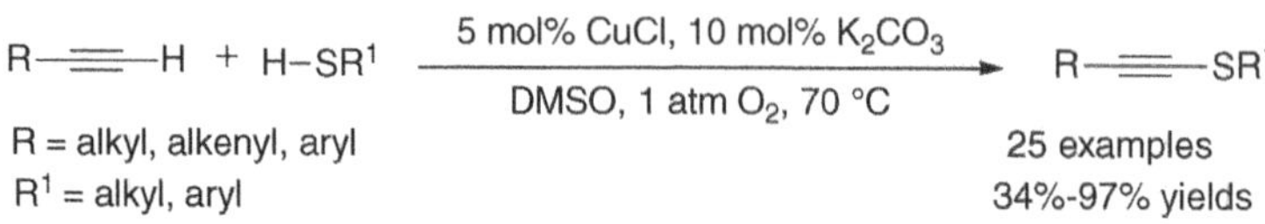

Scheme 13.11 Cu-catalyzed thiolation of terminal alkynes.

thiolation reported by Yu [16], the directing groups employed in this reaction are removable, facilitating functionalization of molecules much more. Despite over-stoichiometric disulfides being used, this reaction represented an important example in direct thiolation of C—H bonds. By using the 8-aminoquinoline-directed strategy, Liu later achieved the direct thiolation of carboxylic acid derivatives with thiols by addition of 2 equiv. Ag_2CO_3 as an oxidant via copper catalysis [29]. Very recently, a one-pot reaction of 8-aminoquinoline, benzoyl chlorides, and thiophenols forming sulfenylated benzamides was reported [30]. Interestingly, by using K_2CO_3 as a base, double thiolated products were afforded, whereas mono-thiolated products were preferred with $NaHCO_3$ under similar reaction conditions.

The oxidative coupling forming *sp*C—S bonds could also be mediated by copper (Scheme 13.11) [31]. The optimal reaction conditions very resembled that of copper-catalyzed aerobic oxidative phosphorylation of terminal alkynes [5]. Thus, catalyzed by 5 mol% CuCl and 10 mol% K_2CO_3, 4-methylphenylacetylene reacted readily with thiophenol under dioxygen atmosphere to produce the corresponding alkynyl sulfide in 97% yields. A wide arrange of alkynyl sulfides including those bearing functional groups could be generated in 34–97% yields under the reaction conditions. The synthetic value was further displayed by preparation of sulfur-rich π-conjugated polymers. This reaction could also be mediated by a copper catalyst immobilized on SiO_2-supported amine with similar results. This immobilized catalyst could be easily separated from the products and reused [32].

13.3 Silver-Mediated Oxidative Coupling Forming C—P and C—S Bonds

13.3.1 Silver-Mediated Oxidative Coupling Forming C—P Bonds

Silver complexes were also used to mediate the oxidative coupling forming C—P bonds. An early work was reported by Wan in 2012 (Scheme 13.12) [33]. Using 3 equiv. AgOAc, the oxidative coupling of indoles with dialkyl phosphonates proceeded readily in DCE at 90 °C, producing the bioactive 1*H*-indol-2-yl-phosphonates in 18–71% yields,

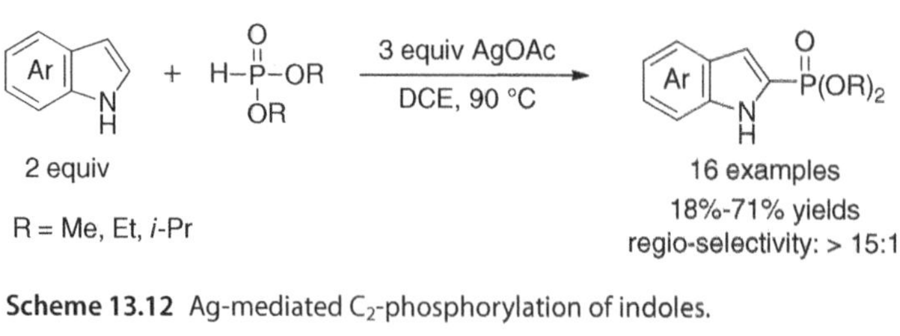

Scheme 13.12 Ag-mediated C_2-phosphorylation of indoles.

respectively. This reaction showed high regioselectivity, the C_2-phosphorylated products were preferred with over 94% selectivity. Subsequently, the phosphorylation of furans, thiophenes, thioazoles, pyrroles, and pyridines with dialkyl phosphonates was also accomplished with a catalytic amount of $AgNO_3$ and 4 equiv. $K_2S_2O_8$ as an oxidant [34]. This oxidative coupling took place at room temperature in a biphasic CH_2Cl_2/H_2O system generating the heteroarylphosphonates in moderate to good yields. Kim also investigated the oxidative coupling between dialkyl phosphonates and substituted pyrroles through silver catalysis [35]. Thiazolo [3,2-b]-1,2,4-triazoles could also be oxidatively phosphorylated with dialkyl phosphites under similar reaction conditions [36].

In 2015, an oxidative coupling of thiazoles with secondary phosphine oxides generating 2-diarylphosphoryl thiazoles was reported by Wen (Scheme 13.13) [37]. By heating a mixture of 3 equiv. diarylphosphine oxides, thiazoles and 1 equiv. $AgNO_3$ in MeCN at 90 °C, the expected phosphine oxides were produced in 19–92% yields. In 2016, Zou developed an efficient $Ag_2CO_3/Mg(NO_3)_2$ catalytic system for C—H phosphorylation of indoles with secondary phosphine oxides [38]. Interestingly, when C_2-position was occupied, C_3—H bonds were phosphorylated, whereas C_2-phosphorylated products could be produced if the C_3-position was unavailable.

In the same year, Qu demonstrated a silver-mediated oxidative coupling of coumarins with dialkyl phosphonates (Scheme 13.14) [39]. The reaction was catalyzed by 5 mol% $AgNO_3$ using 0.5 equiv. $Mg(NO_3)_2.6H_2O$ as an oxidant. Various coumarin-3-yl phosphonates were produced regioselectively, and the yields ranged from 0 to 77%. Under the reaction conditions, 2-quinolinones also served well as substrates and coupled readily with dialkyl phosphonates to give the corresponding C_3-phosphorylated products in moderate yields. By the aid of microwave irradiation, β-aryl-α,β-unsaturated carbonyl compounds were also successfully phosphorylated, selectively affording the α-C-phosphorylated *trans*-alkenes [40]. The oxidative phosphorylation of ketene dithioacetals with P(O)-H compounds was also reported [41]. Interestingly, with H-phosphonates, this transformation could take place with 3 equiv. $AgNO_3$ as the oxidant, whereas 3 equiv. $K_2S_2O_8$ was required with secondary phosphine oxides. Mediated by 5 mol% $AgNO_3$ and 40 mol% TEMPO, the simple styrenes could also be phosphorylated

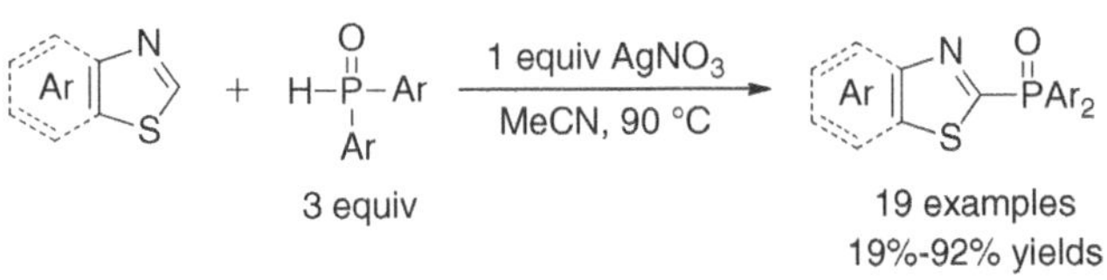

Scheme 13.13 Ag-mediated phosphorylation of thiazoles.

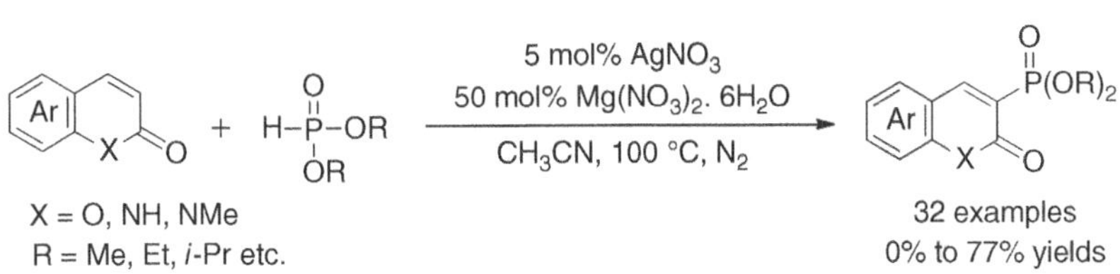

Scheme 13.14 Ag-mediated phosphorylation of coumarins and 2-quinolinones.

in toluene at 100 °C with 2 equiv. $K_2S_2O_8$ as the oxidant, furnishing the β-aryl-(*E*)-alkenylphosphoryl compounds in 51–81% yields [42].

In 2015, Ag_2O-mediated oxidative coupling forming sp^2C—P bonds was demonstrated with the use of 2 equiv. $K_2S_2O_8$ as an oxidant (Scheme 13.15) [43]. This reaction featured high regioselectivity with *para*-C-H being phosphorylated. Various formyl and acylphenylphosphonates were produced in good to high yields. Subsequently, Wu disclosed another elegant Ag-mediated regioselective phosphorylation of C—H bonds [44]. In the presence of 2.5 equiv. Ag_2CO_3, 2 equiv. $K_2S_2O_8$ and 2 equiv. NaOH, a variety of 8-aminoquinoline amides coupled readily with dialkyl phosphonates, regioselectively generating the C_5-phosphorylated products in good yields.

The oxidative phosphorylation of terminal alkynes could also be mediated by silver (Scheme 13.16) [45]. By heating a mixture of 2 equiv. Ag_2CO_3, 1.5 equiv. terminal alkynes and secondary phosphine oxides in DMSO at 120 °C for 15 hours, a various alkynylphosphine oxides including those bearing functional groups were produced in 40–71% yields. Dialkyl phosphonates did not work under the reaction conditions. Despite the limitation, this reaction is a good complement of the copper-catalyzed oxidative phosphorylation of terminal alkynes [5].

13.3.2 Silver-Mediated Oxidative Coupling Forming C—S Bonds

Compared with oxidative coupling forming C—P bonds, only limited examples of silver-mediated C—S bond-forming reactions were reported via oxidative coupling. In 2012, Gao disclosed a silver-catalyzed, copper-mediated thiolation between heteroarenes and thiols (Scheme 13.17) [46]. The oxidative coupling took place in DMF at 120 °C using

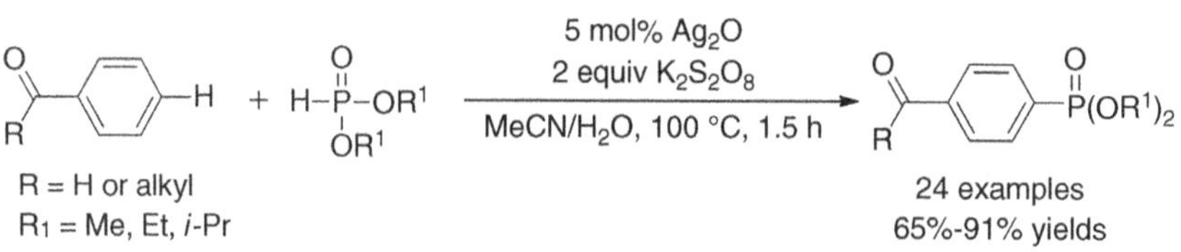

Scheme 13.15 Ag-mediated *para*-phosphorylation of bezenes.

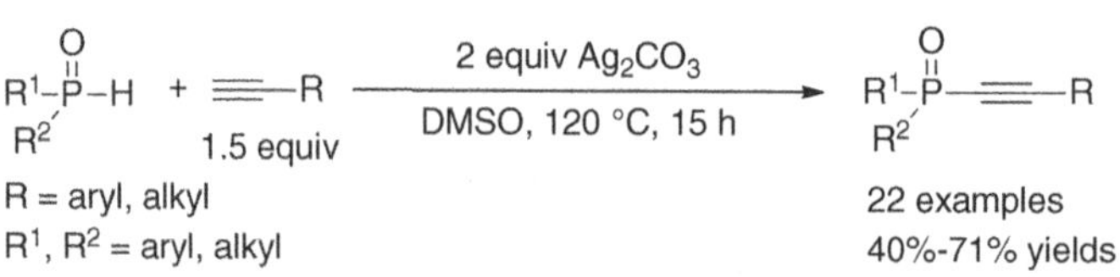

Scheme 13.16 Ag-mediated phosphorylation of terminal alkynes.

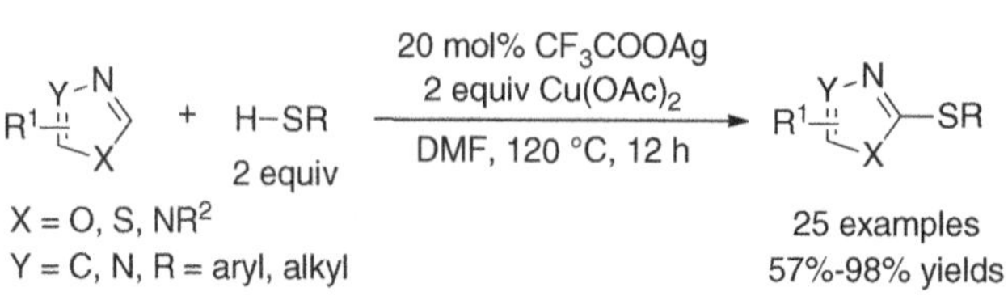

Scheme 13.17 Ag-mediated thiolation of heteroarenes.

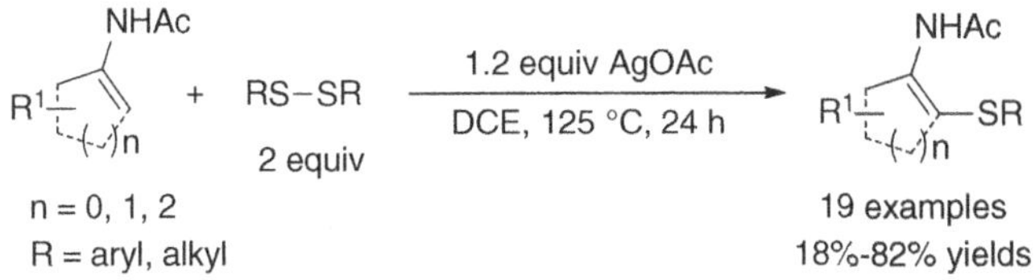

Scheme 13.18 Ag-mediated thiolation of enamides.

CF_3COOAg as a catalyst and 2 equiv. $Cu(OAc)_2$ as an oxidant, a wide arrange of heteroaryl sulfides could be produced in 57–98% yields. Interestingly, when the reaction was performed in DMSO, the solvent could participate in the reaction as the source of thiomethyl group and reacted with heteroarenes at an elevated temperature (140 °C), generating the heteroaryl methyl sulfides in good yields.

The selective C—H thiolation of enamides with disulfides could also be promoted by silver (Scheme 13.18) [47]. A variety of enamides could couple with disulfides mediated by 1.2 equiv. AgOAc in DCE at 125 °C to give the expected thiolated olefins in 18–82% yields. Aromatic disulfides showed high reactivity, whereas the reaction of aliphatic disulfides proceeded sluggishly under these reaction conditions. Employing AgOAc (0.2 equiv.)/dppp (0.24 equiv.) as the catalyst, $(NH_4)_2S_2O_8$ (3.0 equiv.) as the oxidant, and Bu_4NBF_4 (1.0 equiv.) as the additive, the thiolation of quinones with diaryl disulfides was also accomplished in DMSO at room temperature, affording the corresponding quinonyl aryl thioethers in 42–88% yields [48].

13.4 Palladium-Catalyzed Oxidative Coupling Forming C—P and C—S Bonds

13.4.1 Palladium-Catalyzed Oxidative Coupling Forming C—P Bonds

The first palladium-catalyzed dehydrocoupling reaction forming C—P bonds was reported in 2011 (Scheme 13.19) [49]. In the presence of 5 mol% $Pd(OAc)_2$, secondary phosphine oxides bearing a biphenyl group underwent intra-C—H/P(O)—H coupling in THF at 65 °C, producing the corresponding dibenzophosphole oxides in 61–95% yields. By using the strategy, a ladder-type dibenzophosphole oxide was also synthesized. The transformation did not require additional oxidant and thus avoids the oxidation of the starting P(O)—H compounds to phosphinic acids, providing a clean and facile protocol to synthesize dibenzophosphole oxides.

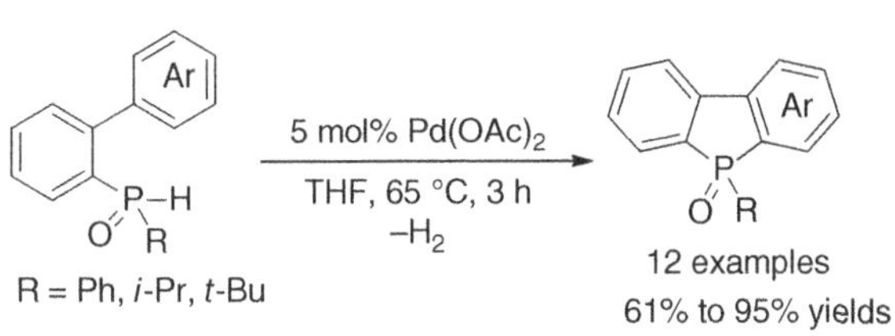

Scheme 13.19 Palladium-catalyzed intra-C—H/P(O)—H coupling.

Later, Li reported an efficient interoxidative coupling between dialkyl phosphonates and benzothiazoles (Scheme 13.20) [50]. This reaction took place in the presence of 3 equiv. $K_2S_2O_8$, 5 mol% $Pd(OAc)_2$ and 30 mol% proline, producing the C_2-phosphorylated products in good to moderate yields. By using bipyridyl as the ligand, benzoxazoles were also phosphonated. Under similar reaction conditions, Wu achieved the phosphorylation of coumarins with dialkyl phosphites [51]. This reaction featured high regioselectivity that only the C_3-phosphorylated products were produced. The phosphonation of imidazo[2,1-b]thiazoles was also accomplished in a similar catalytic system, regioselectively yielding the C_5-phosphonated products in moderate to good yields [52].

In 2013, Yu developed an elegant oxidative coupling of benzenes with P(O)-H compounds based on the heterocycle-directed *ortho*-palladation strategy (Scheme 13.21) [53]. Both secondary phosphine oxides and H-phosphonates were applicable to this reaction, coupling with *N*-heterocycles to produce the expected products in 0–80% yields. Meanwhile, Murakami also reported a similar reaction [54]. By using the group-directed strategy, azoarenes also were phosphorylated with H-phosphonates [55].

The *sp*C—P bonds could also be constructed via oxidative coupling. In 2015, we described an efficient method for the synthesis of alkynylphosphoryl compounds using $Pd(OAc)_2/AgBF_4$ via *sp*C—H/P(O)—H coupling (Scheme 13.22) [56]. This reaction overcame the substrate limitation of the copper-catalyzed aerobic oxidative phosphorylation of terminal alkynes [5], and all three kinds of hydrogen phosphoryl compound, that is, H-phosphonates, H-phosphinates, and secondary phosphine oxides, coupled

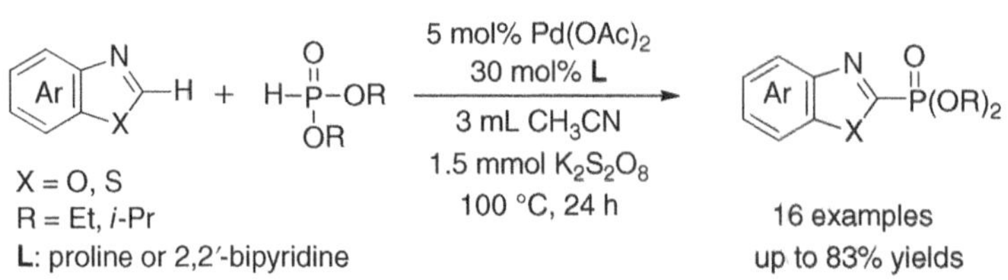

Scheme 13.20 Palladium-catalyzed phosphonation of benzothiazoles and benzoxazoles.

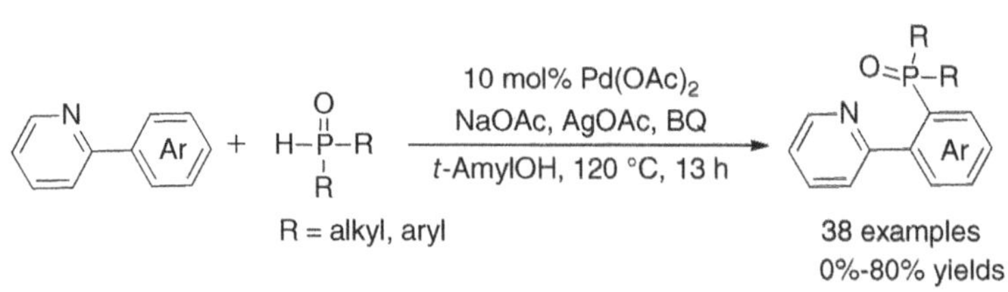

Scheme 13.21 Palladium-catalyzed oxidative coupling of benzenes with P(O)—H compounds.

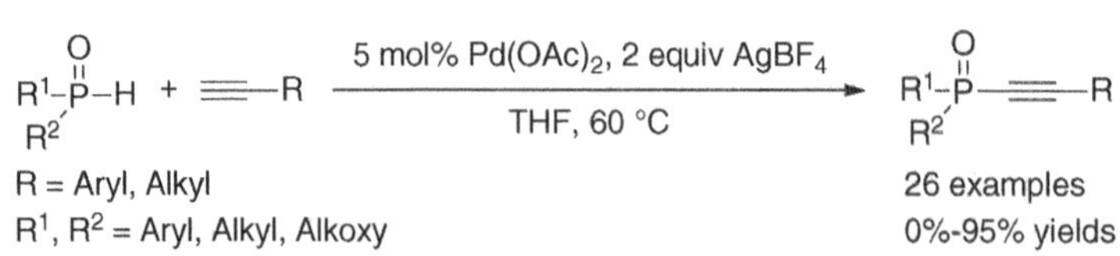

Scheme 13.22 Palladium-catalyzed phosphorylation of terminal alkynes.

readily with aromatic and aliphatic terminal alkynes. P-chiral alkynylphosphoryl compounds were also prepared *stereo*-specifically under the reaction conditions [57]. This transformation provided a general method for the synthesis of alkynylphosphoryl compounds. However, 2 equiv. expensive $AgBF_4$ was required in this reaction, limiting its application in organic synthesis. Very recently, we also developed a silver-free *sp*C—H/P(O)—H cross coupling [58]. Thus, in the presence of 2.5 mol% $Pd(OAc)_2/PPh_3$, various terminal alkynes reacted readily with H-phosphine oxides at 80 °C in the presence of acrylonitrile. This reaction could be performed on a gram scale and applied to the phosphorylation of bioactive ethisterone.

13.4.2 Palladium-Catalyzed Oxidative Coupling Forming C—S Bonds

The oxidative coupling forming C—S bonds can also be achieved by palladium catalysis. An early study of palladium-catalyzed oxidative coupling was reported by Doi in 2008 (Scheme 13.23) [59]. By using 10 mol% $PdCl_2$ or $PdCl_2$(COD) as a catalyst, a variety of thioenols could be converted into the corresponding benzo[*b*]thiophenes in 13–89% yields. The reaction provided an efficient method for the synthesis of multi-substituted benzo[*b*]thiophenes.

In the same year, this group reported a similar intracyclization of thiobenzanilides (Scheme 13.24) [60]. Thus, in the presence of 10 mol% $PdCl_2$(COD), 50 mol% CuI and 2 equiv. Bu_4NBr, various thiobenzanilides underwent intracyclization to produce the expected 2-substituted benzothiazoles in good to high yields. This reaction featured high functional group tolerance; methoxy, F, Cl, Br, I, ester, nitro, and nitrile all survived under the reaction conditions. Later, the reaction conditions were further optimized. Catalyzed by 10 mol% $PdCl_2$, the transformation could proceed smoothly in DMSO by addition of 50 mol% CsF under O_2 atmosphere [61]. By slightly tuning the reaction conditions, the reaction could also take place in water, providing a more practical protocol to benzothiazoles [62]. As a parallel work, Batey also described a similar aerobic oxidative intracyclization of *N*-arylthioureas through Pd/Mn co-catalysis in 2009 [63]. A variety of 2-aminobenzothiazoles were produced in good to moderate yields.

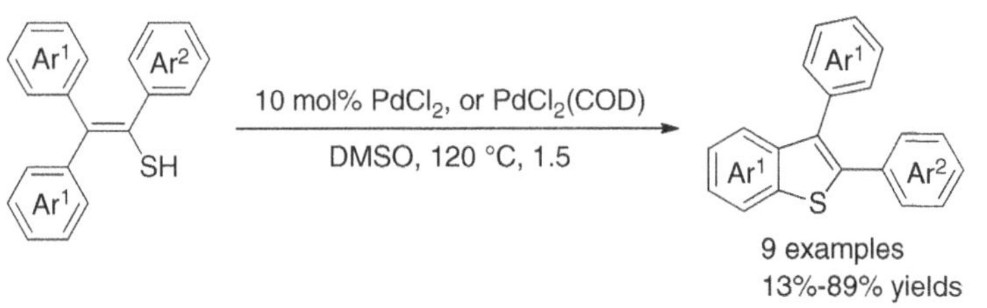

Scheme 13.23 Palladium-catalyzed cyclization of thioenols via intra-C—H/S—H coupling.

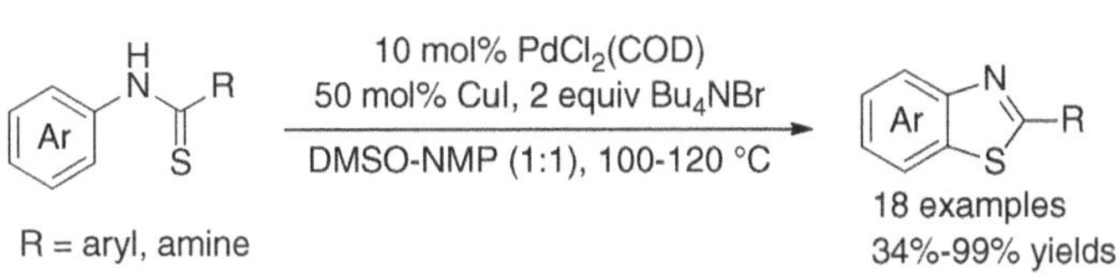

Scheme 13.24 Palladium-catalyzed cyclization of thiobenzanilides via intra-C—H/S—H coupling.

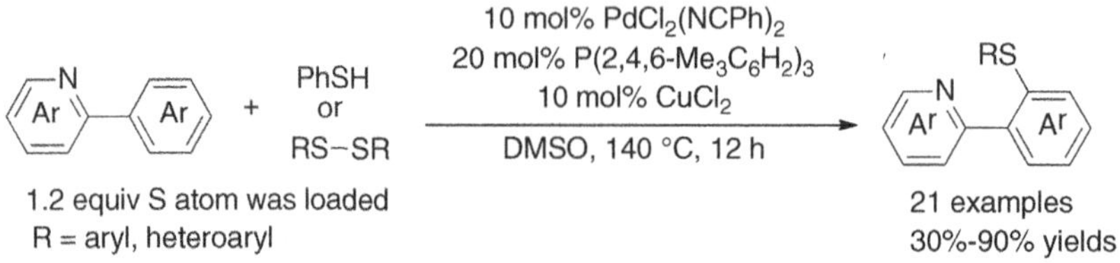

Scheme 13.25 Palladium-catalyzed direct thiolation of arenes.

In 2012, Liu employed this strategy for the synthesis of sugar-functionalized benzothiazoles from the corresponding glycosyl thiourea substrates [64].

In 2014, Nishihara communicated a direct thiolation of arenes with disulfides or thiols by employing the *N*-heterocycle-directed *ortho*-palladation strategy through Pd/Cu co-catalysis (Scheme 13.25) [65]. This reaction could produce the *ortho*-thiolated products including those bearing functional groups in 30–90% yields. Both sulfenyl moieties of the disulfide were incorporated into the thiolated products, and thus this transformation could complete with only 0.6 equiv. disulfides.

13.5 Other Metal-Mediated Oxidative Coupling Forming C—P and C—S Bonds

13.5.1 Oxidative Coupling Forming C—P Bonds

Mn complexes could mediate the oxidative coupling forming C—P bonds. The first example was reported in 2006 by Ishii [66]. Using 5 mol% $Mn(OAc)_2$ and 1 mol% $Co(OAc)_2$, the aerobic oxidative coupling of benzenes with diethyl phosphonates occurred in acetic acid at 45 °C under a mixed gas of O_2 (0.5 atm) and N_2 (0.5 atm) to produce the corresponding arylphosphonates in 59% yield (Scheme 13.26). When substituted benzenes were loaded, a mixture of regioisomers was usually given. Later, the phosphorylation of heterocyclic compounds, that is, thiazoles, pyrroles, and furans, with dialkyl phosphites was achieved in acetic acid using 3 equiv. $Mn(OAc)_3$ as an oxidant [67]. A wide arrange of heteroaryl phosphonates with functional groups like carbonyl, ether, alkoxy, ester, and formyl groups were selectively produced in high yields. Simple arenes could also be oxidatively phosphorylated with diethyl phosphonates under the reaction conditions [68]. Compared with the work reported by Ishii [66], this reaction showed higher regioselectivity. With electron-rich arenes, *ortho*-phosphorylated products were preferred, whereas *para*-phosphorylated ones were given if one

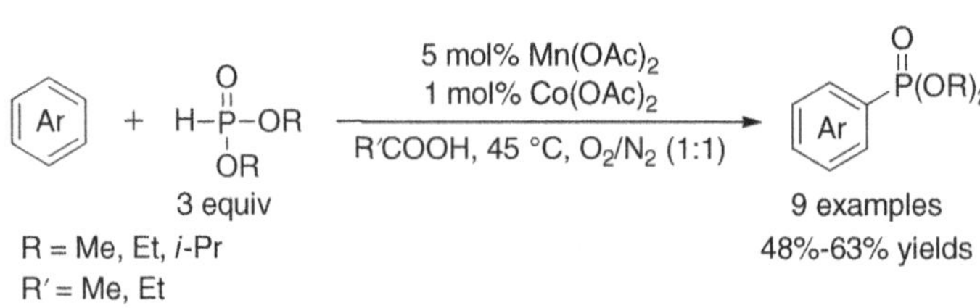

Scheme 13.26 Mn/Co-catalyzed phosphorylation of enamides.

ortho-position was occupied. With electron-deficient arenes bearing ester and nitrile groups, *para*-C—H bonds were usually phosphorylated. The phosphorylation of electron-rich arenes with diphenylphosphine oxide also took place under the reaction conditions. The valuable bioactive imidazo[1,2-a]pyridine-3-phosphonates [69], indole-2-phosphonates [69], and deazapurine-8-phosphonates [70] could also be regioselectively prepared in good yields via similar $Mn(OAc)_3$-mediated oxidative coupling.

The β-aryl-*α*,β-unsaturated compounds were also phosphorylated regioselectively at the α-position mediated by $Mn(OAc)_3$ (Scheme 13.27) [71]. Thus, in the presence of 3 equiv. $Mn(OAc)_3$, aryl alkenes bearing conjugated groups like ketone, ester, amide, and nitro reacted with dialkyl phosphonates in acetic acid to produce the corresponding (*E*)-alkenylphosphonates in moderate to good yields (37–87%). Phosphonates bearing functional groups such as coumarins [72], flavones [72], and uracil [73] were also synthesized under similar reaction conditions.

Interestingly, this Mn-mediated oxidative coupling is applicable to the phosphorylation of [60]fullerene [74, 75]. Thus, refluxing a mixture of 2 equiv. dialkyl phosphites, [60]fullerene, and 2 equiv. $Mn(OAc)_3$ in chlorobenzene for 50 minutes generated a *meso* and racemic isomeric mixture of phosphorylated singly-bonded fullerene dimers in 28–34% yields [74]. The hydrophosphorylated fullerenes ([60]fullerene/$Mn(OAc)_3$/dialkyl phosphites = 1 : 2 : 10) and acetoxyphosphorylated fullerenes ([60]fullerene/$Mn(OAc)_3$/dialkyl phosphites = 1 : 10 : 2) could also be produced selectively under similar reaction conditions by adjusting the ratio of the starting materials [75].

In 2014, Montchamp demonstrated a Mn-catalyzed oxidative coupling of arenes with P(O)—H compounds (Scheme 13.28) [76]. All the three kinds of hydrogen phosphoryl compounds, that is, H-phosphonates, H-phosphinates, and secondary phosphine oxides, were applicable to this reaction. Various aryl phosphoryl compounds including those bearing functional groups were produced in 37–94% yields. With a P-chiral P(O)—H compounds, the oxidative coupling took place stereospecifically with the configuration at phosphorus retained.

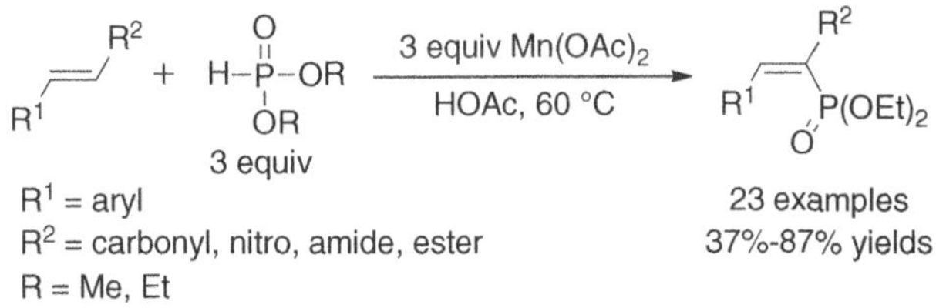

Scheme 13.27 Mn-mediated phosphorylation of alkenes.

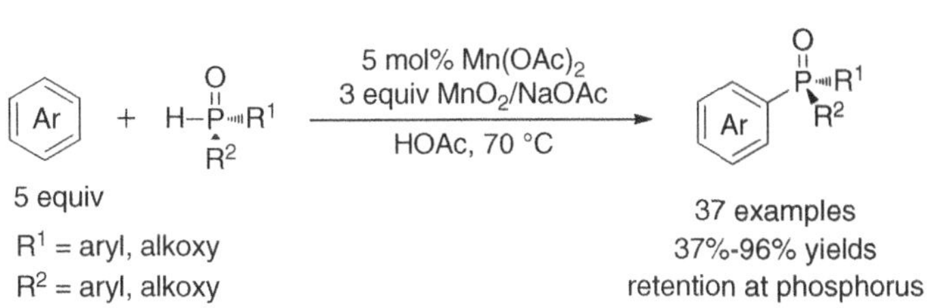

Scheme 13.28 Mn-mediated phosphorylation of alkenes.

The oxidative coupling forming C—P bonds could also be accomplished through iron catalysis. In 2009, Ofial disclosed an iron-catalyzed oxidative coupling of *N,N*-dimethylanilines with H-phosphonates forming sp^3C—P bonds (Scheme 13.29) [77]. The reaction was catalyzed by a catalytic amount of $FeCl_2$ using 2.5 equiv. *t*-BuOOH as an oxidant. Anilines with both electron-donating and electron-withdrawing groups at the benzene ring served as the right substrates to produce the corresponding *α*-aminophosphonates in 57–84% yields. Under similar reaction conditions, the double phosphorylated products could also be generated by increasing the amount of dialkyl phosphonates [78]. The oxidative phosphorylation of *N*-aryl tetrahydroisoquinolines with dialkyl phosphonates was also achieved through iron catalysis [79].

In addition to copper and iron, gold also served as a good catalyst for the aerobic oxidative coupling of *N*-aryl tetrahydroisoquinolines with P(O)—H compounds (Scheme 13.30) [80]. Catalyzed by 6 mol% gold catalyst, both H-phosphonates and secondary phosphine oxides coupled readily with various *N*-aryl tetrahydroisoquinolines at 60 °C in MeCN under air atmosphere to produce the *α*-aminophosphonates in good to high yields. By using 5 mol% MoO_3 as a catalyst, the oxidative phosphorylation of *N*-aryl tetrahydroisoquinolines with dialkyl phosphonates also took place in methanol at 60 °C under dioxygen atmosphere to give *α*-aminophosphonates in 86–95% yields, respectively [81]. Worth noting is that only 1.1 equiv. dialkyl phosphonates were used in the reaction.

The oxidative coupling of *N*-aryl tetrahydroisoquinolines with P(O)-H compounds forming sp^3C—P bonds was also realized by using visible light catalysts. In 2011, Rueping discovered a light-mediated oxidative coupling forming C—P bonds by using a photoredox ruthenium/bipyridyl or iridium/bipyridyl catalyst [82]. The transformation was carried out in a biphasic solution of toluene and water. A variety of α-aminophosphonates were produced in moderate to high yields through the oxidative coupling between *N*-aryltetrahydroisoquinolines and H-phosphonates (3 equiv.). Due to severe hydrolysis, dimethyl phosphonate was not applied to this reaction. This oxidative coupling also proceeded in MeCN with a palladium/porphyrin photocatalyst [83]. Under the reaction conditions, dimethyl phosphonate worked well.

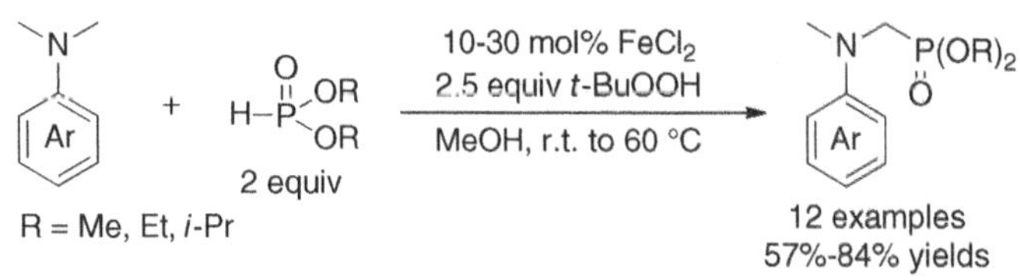

Scheme 13.29 Iron-catalyzed phosphorylation of *N,N*-dimethylanilines.

Scheme 13.30 Au-catalyzed phosphorylation of *N*-aryl tetrahydroisoquinolines.

Similar oxidative coupling also occurred in methanol mediated by an iridium photocatalyst immobilized on a polymer [84]. Under the reaction conditions, both secondary phosphine oxides and H-phosphonates coupled readily with *N*-aryltetrahydroisoquinolines. Eosin Y was also an efficient photocatalyst for the oxidative phosphorylation of *N*-aryltetrahydroisoquinolines with H-phosphonates [85].

13.5.2 Oxidative Coupling Forming C—S Bonds

Nickel catalysts were applicable to C—S bond-forming reactions via oxidative coupling. In 2015, Liu developed a nickel-catalyzed thiolation of benzamides based on heterocycle-directing strategy (Scheme 13.31) [86]. In the presence of 5 mol% $NiCl_2$, 10 mol% PhCOOH and 2 equiv. Ag_2CO_3, benzamides bearing both electron-donating and electron-withdrawing groups at the benzene ring readily coupled with diaryl disulfides in DCE at 140 °C to give the corresponding thiolated products in 0–95% yields. Functional groups such as methoxy, F, Cl, Br, and CF_3 groups were compatible. It seems that only diaryl disulfides were applicable to the reaction. Meanwhile, Shi reported a parallel work, achieving the reaction in DMSO at 140 °C by using 10 mol% $NiCl_2.6H_2O$, 20 mol% BIOL and 2 equiv. CF_3COOK [87]. It is worth noting that this reaction did not use an oxidant. The nickel-catalyzed oxidative coupling of 1,2,3-triazole *N*-oxides with thiols generating 4-thiolated 2-aryl-1,2,3-triazoles was also reported in the same year [88].

Iron complexes were also applied for oxidative coupling forming C—S bonds. In 2012, Lei reported an iron-catalyzed radical cyclization of thoibenzanilides (Scheme 13.32) [89]. A lot of 2-substituted benzothiazoles were synthesized in 0–97% yields by the intracyclization. Functional groups such as methoxy, Br, I, and nitro groups were tolerant well under the reaction conditions. This transformation could also be mediated by a photocatalytic system of 1 mol% $Ru(bpy)_3(PF_6)_2$ and 1 equiv. DBU (1,8-diazabicycloundec-7-ene) using molecule dioxygen as an oxidant [90].

In 2016, Glorius and co-workers used cobalt catalysis to achieve the 2-thiolation of indoles with thiols (Scheme 13.33) [91]. This reaction showed high regioselectivity, and

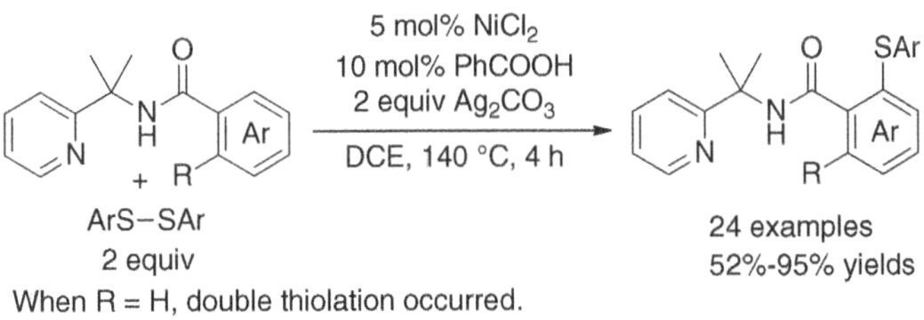

Scheme 13.31 Nickel-catalyzed C—H thiolation of benzamides.

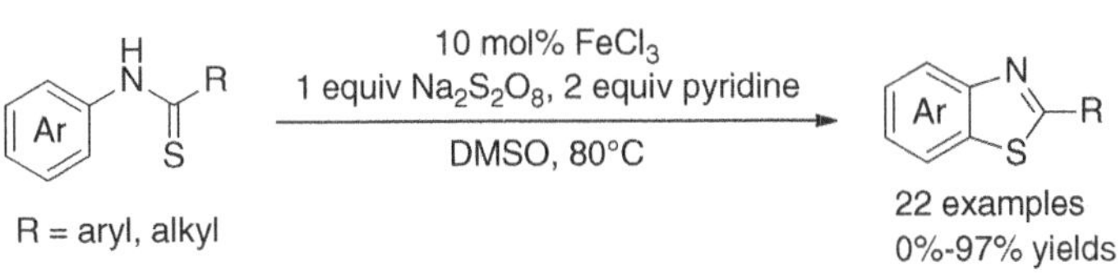

Scheme 13.32 Iron-catalyzed cyclization of thiobenzanilides.

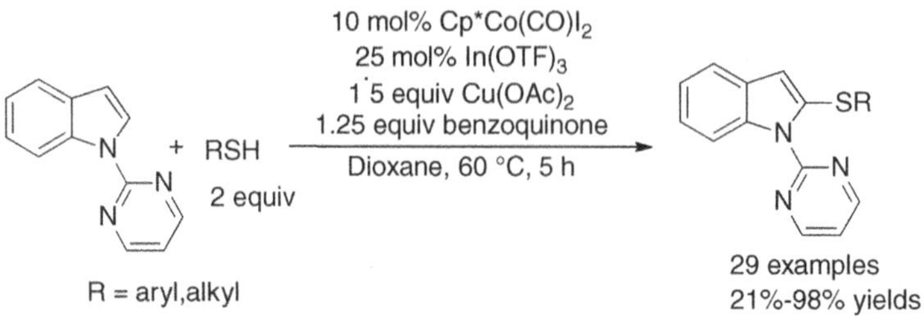

Scheme 13.33 Cobalt-catalyzed 2-thiolation of indoles.

only 2-thiolated products were produced under the reaction conditions. Aromatic thiols worked well, however, aliphatic thiols proceeded sluggishly. The oxidative coupling of the 3-thiolation of indoles with thiols was also achieved by Lei through electrocatalysis [92]. Various indoles coupled readily with thiophenols with dihydrogen evolution, producing the expected thiolated indoles in good to high yields.

13.6 Transition Metal-Free Oxidative Coupling Forming C—P and C—S Bonds

Despite transition metal catalysis has been extensively used in chemical bond-forming reactions, the necessary removal of the metals from the products in pharmaceuticals and material chemistry is costly. The transition metal-free reactions provided a solution. A few examples of oxidative couplings generating C—P and C—S bonds were reported in recent years, and we highlight them next.

In 1962, Jason found that naphthalene could couple with diethyl phosphonate to give a low yield of phosphorylated products at 130–150 °C with use of a stoichiometric amount of *t*-butyl peroxide [93]. The reaction represented the first example on metal-free oxidative coupling forming C—P bonds. In 2012, Wan disclosed the oxidation phosphorylation of *N*-aryl tetrahydroisoquinolines with dialkyl phosphonates under transition metal free conditions (Scheme 13.34) [94]. This transformation used 1.1 equiv. DDQ as the oxidant and took place in MeCN at room temperature to produce the corresponding coupling products in 51–99% yields. By elevating the temperature to 80 °C, the oxidative phosphorylation also proceeded in DCE under air atmosphere [95]. The sp^3C—P bond-forming reaction could also be accomplished with SO_2Cl_2 [96], AIBN [97], triarylaminium salts [98], and iodine [99].

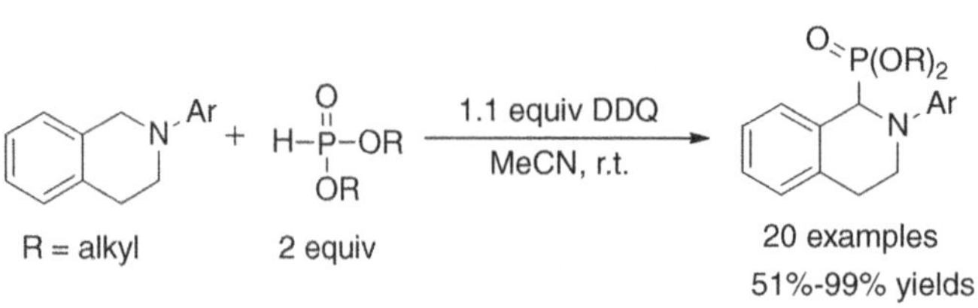

Scheme 13.34 Transition metal-free phosphorylation of *N*-aryl tetrahydroisoquinolines.

In 2012, Wei demonstrated an I_2/DMSO catalytic system for the 3-thiolation of indoles with disulfides (Scheme 13.35) [100]. In the presence of 5 mol% iodine and 3 equiv. DMSO, various indoles coupled with disulfides readily in DMC at 40 °C, selectively yielding the 3-thiolated indoles in 0–97% yields. Both sulfenyl groups in disulfides were incorporated into indoles. Aromatic disulfides worked well, whereas aliphatic ones hardly gave the expected products under the reaction conditions. By switching the solvent to DCE and elevating the temperature to 60 °C, the oxidative coupling of indoles with thiols also occurred with similar results in the I_2/DMSO catalytic system [101]. The 3-thiolation of indoles with thiols was also achieved by using I_2/O_2 in MeCN [102] or bovine serum albumin (BSA)–I_2/O_2 in water [103]. This transformation also took place by using NCS [104], NaOH/air [105, 106], and Cs_2CO_3/air [107].

The oxidative coupling between electron-rich arenes with thiols was also realized through iodine catalysis. In 2015, Parumala and Peddinti reported that by using 5 mol% I_2 and 3 equiv. DMSO, electron-rich chemicals like phenols, trimethoxy substituted benzenes and *N*,*N*-dimethylanilines reacted with thiophenols at 80 °C to give the *ortho*-thiolated products (if the *para*-position was occupied) in high yields [108]. Also applicable to this reaction were 4-hydroxycoumarins and ketene dithioacetals. Wang reported a similar reaction in I_2/DTBP catalytic system in the same year [109]. Under microwave irradiation, the *para*-thiolation of *N*,*N*-dimethylanilines with disulfides could complete at 110 °C in only 10 minutes [110]. The thiolated flavones [111], imidazo[1,2-a]pyridines [112, 113], pyrazoles [114], and pyrazolones [115] could also be prepared selectively through iodine catalysis. The bioactive benzo[*b*]thiophenes were also synthesized via iodine-catalyzed aerobic oxidative intracyclization [116].

In 2013, Deng disclosed an iodine catalytic system preparing *ortho*-arylsulfenylphenols by oxidative coupling of cyclohexanones with thiophenols (Scheme 13.36) [117]. The transformation used inexpensive iodine as a catalyst and dioxygen as an oxidant. Many functional groups like fluoro, chloro, bromo, hydroxyl, acetamido, and ester groups were well tolerated. This reaction represented a green method for synthesis of *ortho*-thiolated phenols starting from cheap cyclohexanones. It should be noted that by

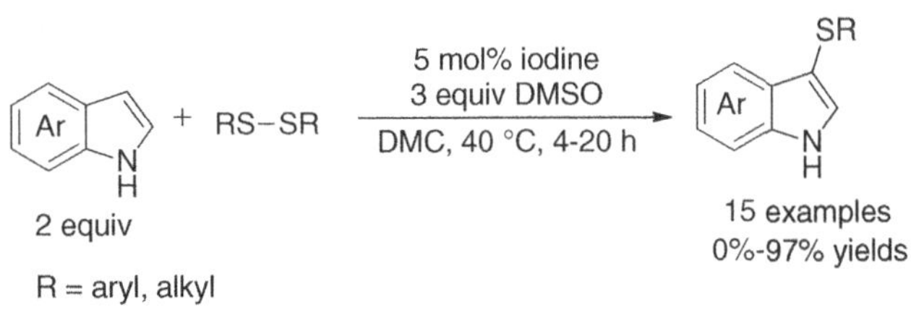

Scheme 13.35 Iodine-catalyzed phosphorylation of *N*-aryl tetrahydroisoquinolines.

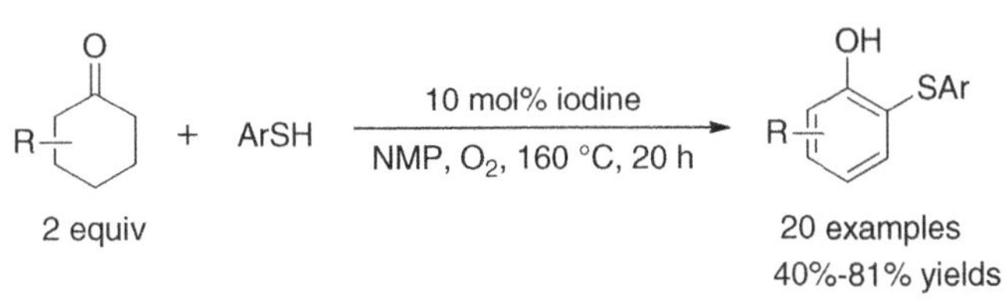

Scheme 13.36 Iodine-catalyzed oxidative coupling of cyclohexanone with thiophenols.

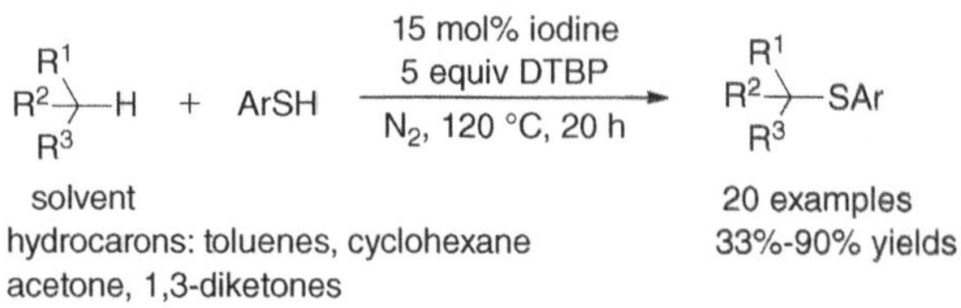

Scheme 13.37 Iodine-catalyzed oxidative coupling of hydrocarbons with thiophenols.

using NCS as an oxidant, the oxidative coupling of ketones with thiols proceeded in CH_2Cl_2 at room temperature to generate the β-thiolated ketones in good yields [118].

The oxidative coupling forming sp^3C—S bonds were also realized via iodine catalysis. In 2014, Lei developed a I_2/DTBP catalytic system for thiolation of hydrocarbons like toluenes, cyclohexane, acetone, and 1,3-diketones with thiophenols forming sp^3C—S bonds selectively (Scheme 13.37) [119]. This reaction took place at 120 °C with the starting hydrocarbons as the solvent. A various aryl alkyl sulfides were produced and the yields ranged from 33 to 90%. Later, Lei screened the conditions of oxidative coupling between 1,3-diketones with thiophenols, accomplishing the reaction in ethyl acetate with 1.5 equiv. 1,3-diketones under similar reaction conditions [120]. Notably, mediated by 1 equiv. Cs_2CO_3, the thiolation of 1,3-dicarbonyl compounds with thiophenols (2 equiv.) could take place in DMF at room temperature under air atmosphere to produce the corresponding α-sulfenylated carbonyl compounds in 50–95% yields [121].

In addition to iodine catalysis, there are other C—S bond-forming reactions via oxidative coupling [122–127]. For example, by using TBHP as an oxidant, the thiolation of hexane [122], acetone [123] and N-Me in amides [124] with disulfides constructing sp^3C—S bonds could take place under catalyst-free conditions. The electron-rich di or trimethoxybenzenes could couple with thiophenols to generate diaryl sulfides with use of NCS as an oxidant [125]. The α-thiolation of β-naphthols with thiophenols could be promoted by piperidine/DMSO under air atmosphere [126].

13.7 Conclusion

In summary, oxidative C—H/Z—H (Z = S, P) coupling forming C—Z bonds is a potentially powerful method for the synthesis of organosulfur and organophosphorus compounds, and significant advances have been made in recent years. Unfortunately, some severe drawbacks are also associated with these reactions; that is, narrow substrate scopes, over-stoichiometric oxidants, and overdose of C—H or Z—H compounds. These drawbacks need to be conquered in the future, in order to make the oxidative coupling reactions really useful.

References

1 (a) Corbridge, D.E.C. (2013). *Phosphorus: Chemistry, Biochemistry and Technology*, 6e. London: CRC Press. (b) Quin, L.D. (2000). *A Guide to Organophosphorus Chemistry*. New York: Wiley Interscience. (c) Kukhar, V.P. and Hudson, H.R. (eds.) (2000). *Aminophosphonic and Aminophosphinic Acids: Chemistry and Biological Activity*.

Chichester: Wiley. (d) Cremlyn, R.J. (1997). *An Introduction to Organosulfur Chemistry*. Mishawaka, USA: Wiley. (e) Mellah, M., Voituriez, A., and Schulz, E. (2007). *Chem. Rev.* 107: 5133–5209. (f) Murphy, A.R. and Frecher, J.M.J. (2007). *Chem. Rev.* 107: 1066–1096.
2 (a) Bhattacharya, A.K. and Thyagarajan, G. (1981). *Chem. Rev.* 81: 415–430. (b) der Jeught, S.V. and Stevens, C.V. (2009). *Chem. Rev.* 109: 2672–2702. (c) Demmer, C.S., Krogsgaard-Larsen, N., and Bunch, L. (2011). *Chem. Rev.* 111: 7981–8006. (d) Schwan, A.L. (2004). *Chem. Soc. Rev.* 33: 218–224. (e) Hartwig, J.F. (2008). *Acc. Chem. Res.* 41: 1534–1544. (f) Ley, S.V. and Thomas, A.W. (2003). *Angew. Chem. Int. Ed.* 42: 5400–5449. (g) Kondo, T. and Mitsudo, T. (2000). *Chem. Rev.* 100: 3205–3220.
3 (a) Xu, Q. and Han, L.-B. (2011). *J. Organomet. Chem.* 696: 130–140. (b) Tappe, F.M.J., Trepohl, V.T., and Oestreich, M. (2010). *Synthesis* 3037–3062. (c) Montchamp, J.-L. (2014). *Acc. Chem. Rev.* 47: 77–87. (d) Chen, T., Zhang, J.-S., and Han, L.-B. (2016). *Dalton Trans.* 45: 1483–1489. (e) Shen, C., Zhang, P., Sun, Q. et al. (2015). *Chem. Rev. Soc.* 44: 291–314.
4 (a) Li, C.-J. (2009). *Acc. Chem. Res.* 42: 335–344. (b) Girard, S.A., Knauber, T., and Li, C.-J. (2014). *Angew. Chem. Int. Ed.* 53: 74–100. (c) Yeung, C.S. and Dong, V.M. (2011). *Chem. Rev.* 111: 1215–1292. (d) Waterman, R. (2013). *Chem. Soc. Rev.* 42: 5629–5641.
5 Gao, Y., Wang, G., Chen, L. et al. (2009). *J. Am. Chem. Soc.* 131: 7956–7567.
6 Liu, P., Yang, J., Li, P., and Wang, L. (2011). *Appl. Organomet. Chem.* 25: 830–835.
7 Qu, Z., Chen, X., Yuan, J. et al. (2012). *Can. J. Chem.* 90: 747–752.
8 Liu, L., Wu, Y., Wang, Z. et al. (2014). *J. Org. Chem.* 79: 6816–6822.
9 Moglie, Y., Mascaró, E., Gutierrez, V. et al. (2016). *J. Org. Chem.* 81: 1813–1818.
10 Yang, B., Yang, T.-T., Li, X.-A. et al. (2013). *Org. Lett.* 15: 5024–5207.
11 Zhou, A.-X., Mao, L.-L., Wang, G.-W., and Yang, S.-D. (2014). *Chem. Commun.* 50: 8529–8532. The work was subsequently highlighted by ChemSusChem, see: He, K.-H., and Li, Y., *ChemSusChem* 2014, *7*, 2788–2790.
12 Xu, P., Wu, Z., Zhou, N., and Zhu, C. (2016). *Org. Lett.* 18: 1143–1145.
13 Baslé, O. and Li, C.-J. (2009). *Chem. Commun.* 4124–4126.
14 Zhi, H., Ung, S.P.-M., Liu, Y. et al. (2016). *Adv. Synth. Catal.* 358: 2553–2557.
15 Ke, J., Tang, Y., Yi, H. et al. (2015). *Angew. Chem. Int. Ed.* 54: 6604–6607.
16 Chen, X., Hao, X.-S., Goodhue, C.E., and Yu, J.-Q. (2006). *J. Am. Chem. Soc.* 128: 6790–6791.
17 Fukuzawa, S., Shimizu, E., Atsuumi, Y. et al. (2009). *Tetrahedron Lett.* 50: 2374–2376.
18 Ranjit, S., Lee, R., Heryadi, D. et al. (2011). *J. Org. Chem.* 76: 8999–9007.
19 Zhou, A.-X., Liu, X.-Y., Yang, K. et al. (2011). *Org. Biomol. Chem.* 9: 5456–5462.
20 Alves, D., Lara, R.G., Contreira, M.E. et al. (2012). *Tetrahedron Lett.* 53: 3364–3368.
21 Rosario, A.R., Casola, K.K., Oliveira, C.E.S., and Zeni, G. (2013). *Adv. Synth. Catal.* 355: 2960–2966.
22 Inomata, H., Toh, A., Mitsui, T., and Fukuzawa, S. (2013). *Tetrahedron Lett.* 54: 4729–4731.
23 He, Z., Luo, F., Li, Y., and Zhu, G. (2013). *Tetrahedron Lett.* 54: 5907–5910.
24 Cao, H., Chen, L., Liu, J. et al. (2015). *RSC Adv.* 5: 22356–22360.
25 Zhang, S., Qian, P., Zhang, M. et al. (2010). *J. Org. Chem.* 75: 6732–6735.
26 Yu, C., Zhang, C., and Shi, X. (2012). *Eur. J. Org. Chem.* 1953–1959.
27 Leroux, F., Jeschke, P., and Schlosser, M. (2005). *Chem. Rev.* 105: 827–856.
28 Tran, L.D., Popov, I., and Daugulis, O. (2012). *J. Am. Chem. Soc.* 134: 18237–18240.
29 Yan, X.-B., Gao, P., Yang, H.-B. et al. (2014). *Tetrahedron* 70: 8730–8736.
30 Liu, Y., Huang, M., and Wei, L. (2017). *Asian J. Org. Chem.* 6: 41–43.

31 Yang, Y., Dong, W., Guo, Y., and Rioux, R.M. (2013). *Green Chem.* 15: 3170–3175.
32 Fang, Z., He, W., Cai, M. et al. (2015). *Tetrahedron Lett.* 56: 6463–6467.
33 Wang, H., Li, X., Wu, F., and Wan, B. *Synthesis* 2012: 941–945.
34 Xiang, C.-B., Bian, Y.-J., Mao, X.-R., and Huang, Z.-Z. (2012). *J. Org. Chem.* 77: 7706–7710.
35 Kim, S.H., Kim, K.H., Lim, J.W., and Kim, J.N. (2014). *Tetrahedron Lett.* 55: 531–534.
36 Liu, W., Wang, S., Li, Z. et al. (2015). *Synlett* 26: 2561–2564.
37 Zhang, H.-J., Lin, W., Wu, Z. et al. (2015). *Chem. Commun.* 51: 3450–3453.
38 Sun, W.-B., Xue, J.-F., Zhang, G.-Y. et al. (2016). *Adv. Synth. Catal.* 358: 1753–1758.
39 Yuan, J.-W., Li, Y.-Z., Yang, L.-R. et al. (2015). *Tetrahedron* 71: 8178–8186.
40 Yuan, J.-W., Li, Y.-Z., Mai, W.-P. et al. (2016). *Tetrahedron* 72: 3084–3091.
41 Zhu, L., Yu, H., Guo, Q. et al. (2015). *Org. Lett.* 17: 1978–1981.
42 Gui, Q., Hu, L., Chen, X. et al. (2015). *Chem. Commun.* 51: 13922–13924.
43 Huang, X.-F., Wu, Q.-L., He, J.-S., and Huang, Z.-Z. (2015). *Org. Biomol. Chem.* 13: 4466–4472.
44 Sun, M., Sun, S., Qiao, H. et al. (2016). *Org. Chem. Front.* 3: 1646–1650.
45 Wang, T., Chen, S., Shao, A. et al. (2015). *Org. Lett.* 17: 118–121.
46 Dai, C., Xu, Z., Huang, F. et al. (2012). *J. Org. Chem.* 77: 4414–4419.
47 Yang, L., Wen, Q., Xiao, F., and Deng, G.-J. (2014). *Org. Biomol. Chem.* 12: 9519–9523.
48 Zhang, C., McClure, J., and Chou, C.J. (2015). *J. Org. Chem.* 80: 4919–4927.
49 Kuninobu, Y., Yoshida, T., and Takai, K. (2011). *J. Org. Chem.* 76: 7370–7376.
50 Hou, C., Ren, Y., Lang, R. et al. (2012). *Chem. Commun.* 48: 5181–5183.
51 Mi, X., Huang, M., Zhang, J. et al. (2013). *Org. Lett.* 15: 6266–6269.
52 Liu, W., Wang, S., Yao, H. et al. (2015). *Tetrahedron Lett.* 56: 6100–6103.
53 Feng, C.-G., Ye, M., Xiao, K.-J. et al. (2013). *J. Am. Chem. Soc.* 135: 9322–9325.
54 Li, C., Yano, T., Ishida, N., and Murakami, M. (2013). *Angew. Chem. Int. Ed.* 52: 9801–9804.
55 Hong, G., Mao, D., Wu, S., and Wang, L. (2014). *J. Org. Chem.* 79: 10629–10635.
56 Yang, J., Chen, T., Zhou, Y. et al. (2015). *Chem. Commun.* 51: 3549–3551.
57 Yang, J., Chen, T., Zhou, Y. et al. (2015). *Organometallics* 34: 5095–5098.
58 Zhang, J.-Q., Chen, T., Zhang, J.-S., and Han, L.-B. (2017). *Org. Lett.* https://doi.org/10.1021/acs.orglett.7b02389.
59 Inamoto, K., Arai, Y., Hiroya, K., and Doi, T. (2008). *Chem. Commun.* 5529–5531.
60 Inamoto, K., Hasegawa, C., Hiroya, K., and Doi, T. (2008). *Org. Lett.* 10: 5147–5150.
61 Inamoto, K., Hasegawa, C., Kawasaki, J. et al. (2010). *Adv. Synth. Catal.* 352: 2643–2655.
62 Inamoto, K., Nozawa, K., and Kondo, Y. (2012). *Synlett* 23: 1678–1682.
63 Joyce, L.L. and Batey, R.A. (2009). *Org. Lett.* 11: 2792–2795.
64 Shen, C., Xia, H., Yan, H. et al. (2012). *Chem. Sci.* 3: 2388–2393.
65 Iwasaki, M., Iyanaga, M., Tsuchiya, Y. et al. (2014). *Chem. Eur. J.* 20: 2459–2462.
66 Kagayama, T., Nakano, A., Sakaguchi, S., and Ishii, Y. (2006). *Org. Lett.* 8: 407–409.
67 Mu, X.-J., Zou, J.-P., Qian, Q.-F., and Zhang, W. (2006). *Org. Lett.* 8: 5291–5293.
68 Xu, W., Zou, J.-P., and Zhang, W. (2010). *Tetrahedron Lett.* 51: 2639–2643.
69 Yadav, M., Dara, S., Saikam, V. et al. (2015). *Eur. J. Org. Chem.* 2015: 6526–6533.
70 Sabat, N., Slavětínská, L.P., Klepetářová, B., and Hocek, M. (2016). *J. Org. Chem.* 81: 9507–9514.
71 Pan, X.-Q., Zou, J.-P., Zhang, G.-L., and Zhang, W. (2010). *Chem. Commun.* 46: 1721–1723.

72 Zhang, P., Jiang, Y.-J., Zou, J.-P., and Zhang, W. (2012). *Synthesis* 2012: 1043–1050.
73 Kim, S.H., Kim, S.H., Lim, C.H., and Kim, J.N. (2013). *Tetrahedron Lett.* 54: 1697–1699.
74 Wang, G.-W., Wang, C.-Z., Zhu, S.-E., and Murata, Y. (2011). *Chem. Commun.* 47: 6111–6113.
75 Wang, G.-W., Wang, C.-Z., and Zou, J.-P. (2011). *J. Org. Chem.* 76: 6088–6094.
76 Berger, O. and Montchamp, J.-L. (2014). *Chem. Eur. J.* 20: 12385–12388.
77 Han, W. and Ofial, A.R. (2009). *Chem. Commun.* 6023–6025.
78 Han, W., Mayer, P., and Ofial, A.R. (2010). *Adv. Synth. Catal.* 352: 1667–1676.
79 Cai, J., Liu, Y., Jiang, Y., and Yang, Y. (2017). *Phosphorus Sulfur Silicon* 192: 1068–1073.
80 Xie, J., Li, H., Xue, Q. et al. (2012). *Adv. Synth. Catal.* 354: 1646–1650.
81 Alagiri, K., Devadig, P., and Prabhu, K.R. (2012). *Tetrahedron Lett.* 53: 1456–1459.
82 Rueping, M., Zhu, S., and Koenigs, R.M. (2011). *Chem. Commun.* 47: 8679–8681.
83 To, W.-P., Liu, Y., Lau, T.-C., and Che, C.-M. (2013). *Chem. Eur. J.* 19: 5654–5664.
84 Yoo, W.-J. and Kobayashi, S. (2014). *Green Chem.* 16: 2438–2442.
85 Hari, D.P. and König, B. (2011). *Org. Lett.* 13: 3852–3855.
86 Yang, K., Wang, Y., Chen, X. et al. (2015). *Chem. Commun.* 51: 3582–3585.
87 Yan, S.-Y., Liu, Y.J., Liu, B. et al. (2015). *Chem. Commun.* 51: 4069–4072.
88 Zhu, J., Chen, Y., Lin, F. et al. (2015). *Org. Biomol. Chem.* 13: 3711–3720.
89 Wang, H., Wang, L., Shang, J. et al. (2012). *Chem. Commun.* 48: 76–78.
90 Cheng, Y., Yang, J., Qu, Y., and Li, P. (2012). *Org. Lett.* 14: 98–101.
91 Gensch, T., Klauck, F.J.R., and Glorius, F. (2016). *Angew. Chem. Int. Ed.* 55: 11287–11291.
92 Wang, P., Tang, S., Huang, P., and Lei, A. (2017). *Angew. Chem. Int. Ed.* 56: 3009–3013.
93 Jason, E.F. and Fields, E.K. (1962). *J. Org. Chem.* 27: 1402–1405.
94 Wang, H., Li, X., Wu, F., and Wan, B. (2012). *Tetrahedron Lett.* 53: 681–683.
95 Dhineshkumar, J., Samaddar, P., and Prabhu, K.R. (2017). *ACS Omega* 2: 4885–4893.
96 Tanoue, A., Yoo, W.-J., and Kobayashi, S. (2014). *Org. Lett.* 16: 2346–2349.
97 Alagiri, K., Devadig, P., and Prabhu, K.R. (2012). *Chem. Eur. J.* 18: 5160–5164.
98 Huo, C., Wang, C., Wu, M. et al. (2014). *Org. Biomol. Chem.* 12: 3123–3128.
99 Dhineshkumar, J., Lamani, M., Alagiri, K., and Prabhu, K.R. (2013). *Org. Lett.* 15: 1092–1095.
100 Ge, W. and Wei, Y. (2012). *Green Chem.* 14: 2066–2070.
101 Yi, S., Li, M., Mo, W. et al. (2016). *Tetrahedron Lett.* 57: 1912–1916.
102 Liu, X., Cui, H., Yang, D. et al. (2016). *Catal. Lett.* 146: 1743–1748.
103 Saima, Eqqubal, D., Lavekar, A.G., and Sinha, A.K. (2016). *Org. Biomol. Chem.* 14: 6111–6118.
104 Schlosser, K.M., Krasutsky, A.P., Hamilton, H.W. et al. (2004). *Org. Lett.* 6: 819–821.
105 Zhang, X., Zhou, X., Xiao, H., and Li, X. (2013). *RSC Adv.* 3: 22280–22284.
106 Liu, Y., Zhang, Y., Hu, C. et al. (2014). *RSC Adv.* 4: 35528–35530.
107 Song, S., Zhang, Y., Yeerlan, A. et al. (2017). *Angew. Chem. Int. Ed.* 56: 2487–2491.
108 Parumala, S.K.R. and Peddinti, R.K. (2015). *Green Chem.* 17: 4068–4072.
109 Yan, K., Yang, D., Sun, P. et al. (2015). *Tetrahedron Lett.* 56: 4792–4795.
110 Saba, S., Rafique, J., and Braga, A.L. (2016). *Catal. Sci. Technol.* 6: 3087–3098.
111 Zhao, W., Xie, P., Bian, Z. et al. (2015). *RSC Adv.* 5: 59861–59864.
112 Rafique, J., Saba, S., Rosário, A.R., and Braga, A.L. (2016). *Chem. Eur. J.* 22: 11854–11862.
113 Hiebel, M.-A. and Berteina-Raboin, S. (2015). *Green Chem.* 17: 937–944.
114 Yang, D., Sun, P., Wei, W. et al. (2016). *Org. Chem. Front.* 3: 1457–1461.

115 Siddaraju, Y. and Prabhu, K.R. (2017). *Org. Biomol. Chem.* 15: 5191–5196.
116 Fan, W., Chen, K.-Y., Chen, Q.-P. et al. (2017). *Org. Biomol. Chem.* 15: 6493–6499.
117 Liao, Y., Jiang, P., Chen, S. et al. (2013). *Green Chem.* 15: 3302–3306.
118 Yadav, J.S., Reddy, B.V.S., Jain, R., and Baishya, G. (2008). *Tetrahedron Lett.* 49: 3015–3018.
119 Yuan, J., Ma, X., yi, H. et al. (2014). *Chem. Commun.* 50: 14386–14389.
120 Cao, H., Yuan, J., Liu, C. et al. (2015). *RSC Adv.* 5: 41493–41496.
121 Chen, Q., Wang, X., Wen, C. et al. (2017). *RSC Adv.* 7: 39758–39761.
122 Zhao, J., Fang, H., Han, J. et al. (2014). *Adv. Synth. Catal.* 356: 2719–2724.
123 Yan, G., Borah, A.J., Wang, L. et al. (2015). *Tetrahedron Lett.* 56: 4305–4307.
124 Tang, R.-Y., Xie, Y.-X., Xie, Y.-L. et al. (2011). *Chem. Commun.* 47: 12867–12869.
125 Reddy, M.R., Kumar, G.S., and Meshram, H.M. (2016). *Tetrahedron Lett.* 57: 3622–3624.
126 Xiao, F., Tian, J., Xing, Q. et al. (2017). *ChemistrySelect* 2: 428–431.
127 Li, Y., Liu, D., Liu, C., and Lei, A. (2016). *Chem. Asian J.* 11: 2246–2249.

14

The Recent Developments of Photocatalytic Oxidation

Wei Ding, Liang-Qiu Lu, and Wen-Jing Xiao

College of Chemistry, Central China Normal University, Wuhan, China

14.1 Introduction

The development of green and sustainable methods for "effective" synthesis of fine chemicals is an important goal for synthetic organic chemists. A century ago, Giacomo Ciamician speculated that a clean, cost-effective, and environmentally benign photochemical transformation had the capacity to replace high-energy synthetic process [1]. However, the inability of most organic compounds to absorb visible light has limited the use of photochemical synthesis. Hence, utilizing visible-light-absorbing inorganic and organic compounds as photocatalysts (PCs) that sensitize organic molecules may allow scientists to overcome this barrier. The most commonly employed visible-light photocatalysts are the polypyridyl complexes of ruthenium and iridium, such as Ru(bpy: 2,2'-bipyridine), Ir(ppy: 2-phenylpyridine; dtbbpy: 4,4'-di-tert-butyl-2,2'-bipyridine), and organic dyes (i.e. Eosin Y and Rose Bengal) (Scheme 14.1a) [2]. Photocatalysts can absorb visible light to generate a photo-excited state (PC*), which usually has a long lifetime and remarkable redox properties. Over the past decade, we have witnessed a rapidly growing interest in the field of visible light photoredox catalysis. From the viewpoints of green and sustainable chemistry, visible-light-induced photocatalysis has been emerged as a powerful tool for promoting useful organic transformations, because of its low cost, easy availability, and environmental friendliness [3]. Three distinctive pathways, namely, reductive quenching, oxidative quenching, and energy transfer (ET), are frequently used to describe the photocatalytic mechanism (Scheme 14.1b).

Oxidative reactions are an important process in Nature and constitute one of the most fundamental chemical transformations in organic synthesis [4]. In this context, great efforts have been devoted to the search for new and more efficient oxidants for the oxidative transformations. Over the past decade, molecular oxygen has been widely utilized as an ideal terminal oxidant for oxidation reactions because of its natural, highly environmentally friendly, and inexpensive characteristics, and therefore offers attractive academic and industrial prospects [5]. Recently, visible light photocatalytic oxidation processes using oxygen as a terminal oxidant under mild reaction conditions have gained substantial and rapid development as a result of the inherent features of green chemistry and sustainability [6]. In these transformations, there are two kinds of reaction

Green Oxidation in Organic Synthesis, First Edition. Edited by Ning Jiao and Shannon S. Stahl.

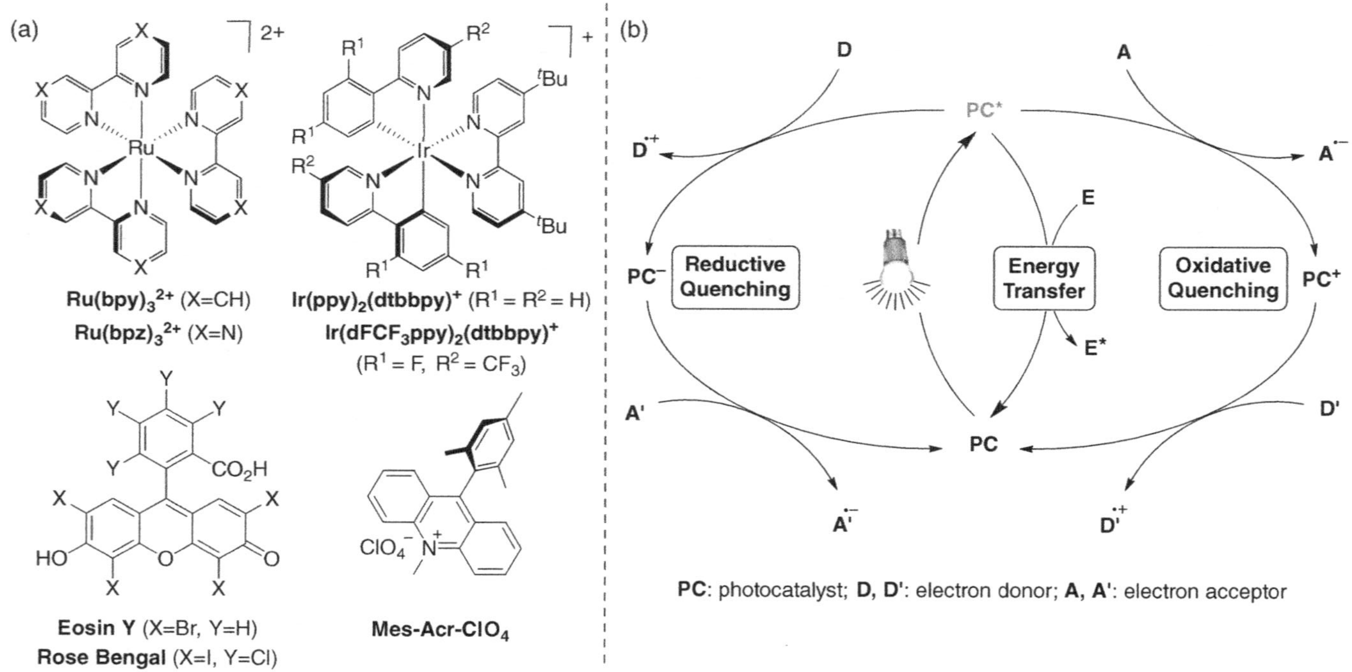

Scheme 14.1 Commonly used photocatalysts (PC) and mechanism profiles.

models: one uses only molecular oxygen as an oxidant to accomplish the photoredox catalytic cycle (*oxidation reaction*), another uses oxygen not only as an oxidizing agent but also as a terminal oxygen atom source incorporated into the target molecule to construct oxygen-containing compounds (*oxygenation reaction*). In this chapter, we will highlight the recent advances of visible-light-mediated aerobic oxidation and oxygenation reactions.

14.2 Visible-Light-Induced Photocatalytic Aerobic Oxidation

14.2.1 Photocatalytic Aerobic Oxidation of Amine Compounds

The amine motif is a kind of good electron donor due to the nitrogen atom possessing lone pair electron, and they are the most common reductive quenchers for the photoredox catalysis, particularly tertiary amines. Visible-light-induced photocatalytic oxidation reactions of amines using oxygen as an oxidant have attracted significant interest from organic chemists. In the photoredox catalytic cycle, amine compounds can be easily oxidized by the excited photocatalyst into an amino radical cation intermediate (Scheme 14.2) [7]. Subsequently, oxygen can oxidize the low-valence photocatalyst (PC^-) to complete the photoredox cycle with the formation of superoxide radical anion, which may then abstract a hydrogen atom from the amino radical cation to generate the key iminium ion intermediate. The iminium ion may be trapped by nucleophiles to realize the oxidative functionalization of amines. On the other hand, the amino radical cation intermediate can also undergo a β-scission via a homolytic cleavage of C—C bond as well as concomitant formation of both the iminium ion and α-amino radical.

The first elegant report of amine oxidative functionalization under visible-light-induced photocatalytic aerobic conditions appeared at the beginning of 2010, when Stephenson and co-workers achieved an aza-Henry reaction via the addition of nitromethane to a catalytically generated iminium ion (Eq. 14.1) [8]. This reaction proceeded in high chemical yields by using $Ir(ppy)_2(dtbbpy)PF_6$ (1 mol%) as the photocatalyst and air as the terminal oxidant. In 2005, Brasholz found that aza-Henry reaction products (1-[nitromethyl]-2-aryl-1,2,3,4-tetrahydroisoquinolines) could undergo the further photocatalytic oxidation by using 1-aminoanthraquinone as photocatalyst (Eq. 14.2) [9].

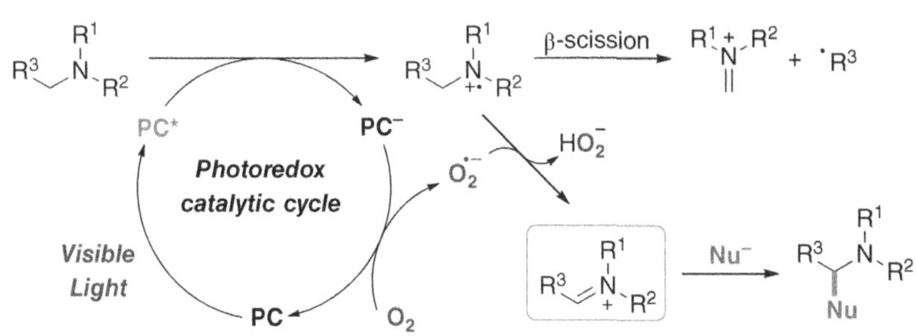

Scheme 14.2 Main pathways of photocatalytic amine oxidation to iminium ions.

Under visible light irradiation, this reaction affording valuable tetracyclic 12-nitro-indolo[2,1-a]isoquinolines in moderate to good yields via an photocatalytic aerobic dehydrogenation/6π-cyclization/oxidation cascade.

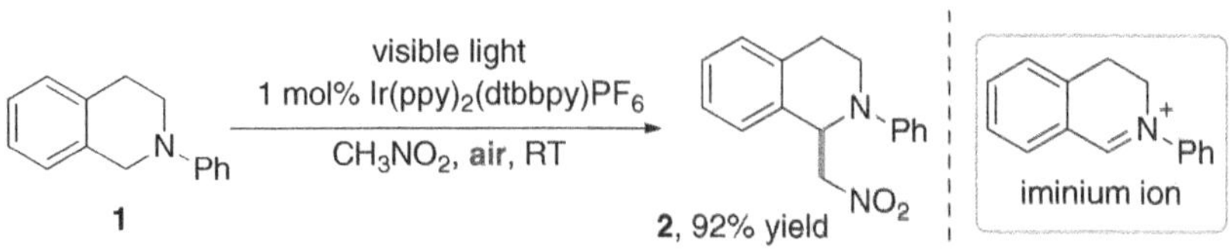

(14.1)

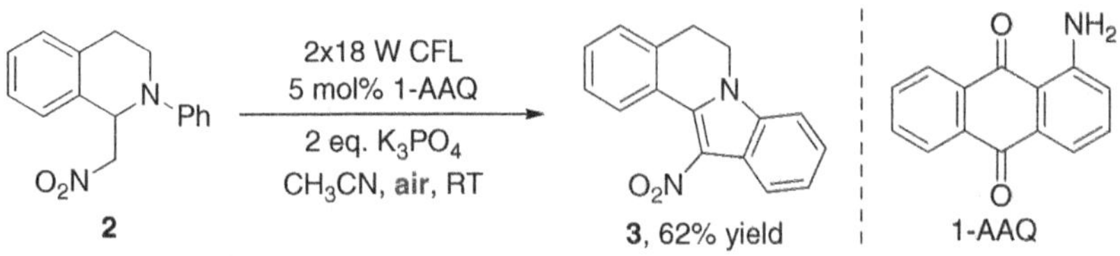

(14.2)

Subsequently, the tertiary amines aerobic oxidation functionalization under visible light irradiation were developed rapidly (Scheme 14.3). Rueping, König, Wu, Tan, and Zhou's groups have used the photocatlytic aerobic conditions to realize the addition reactions to the catalytically generated iminium ion by a broad range of appropriate nucleophiles, for instance, ketones [10d, e], potassium cyanide [10a, b], terminal alkynes [10f], the Ruppert–Prakash reagent ($TMSCF_3$) [10c], indoles [10g], and ethyl diazoacetate [10h]. In 2015, Feng and colleagues used the similar aerobic conditions to finish a visible-light-promoted three-component Ugi-type reaction of tertiary amines with isonitriles and carboxylic acids [11]. Alternatively, asymmetric aerobic oxidative functionalizations of tertiary amines were also investigated. Jiang's group reported an enantioselective aerobic oxidative olefination of *N*-aryl-tetrahydroisoquinolines with acrolein through a cooperative photoredox and asymmetric catalysis where *β*-ICD was used as a chiral Lewis base catalyst [12]. This reaction likely proceeds via the formation of a nucleophilic chiral enol intermediate that undergoes asymmetric addition to the iminium ion.

In 2012, Wang and co-workers developed an innovative strategy for the generation of an iminium ion intermediate through a visible-light-induced photocatalytic oxidative C—C bond cleavage pathway using oxygen as an oxidant. As for the general way of forming an iminium ion, when vicinal diamines were used as substrates, they could also be oxidized by an excited photocatalyst into the amine radical cation, in which a *β*-nitrogen atom is in close vicinity to a N-radical cation that would induce reorganization of electron density and lead to the cleavage of the C—C bond as well as concomitant formation of the iminium ion and *α*-amino radical. The iminium ion and *α*-amino radical intermediates have been employed in photoredox aza-Henry reactions and radical photopolymerizations (Scheme 14.4) [13].

Xiao and co-workers have reported an intramolecular variant of these iminium trapping reactions, in which an intramolecular oxidative cyclization reaction of chiral diamines was disclosed under visible-light-induced photocatalytic conditions using oxygen as an oxidant (Eq. 14.3) [14a]. Under irradiation with visible light, the diamines **15** are converted to iminium ions, which upon intramolecular trapping of the iminium fragment by a nitrogen anion yield the highly substituted tetrahydroimidazoles **16** in

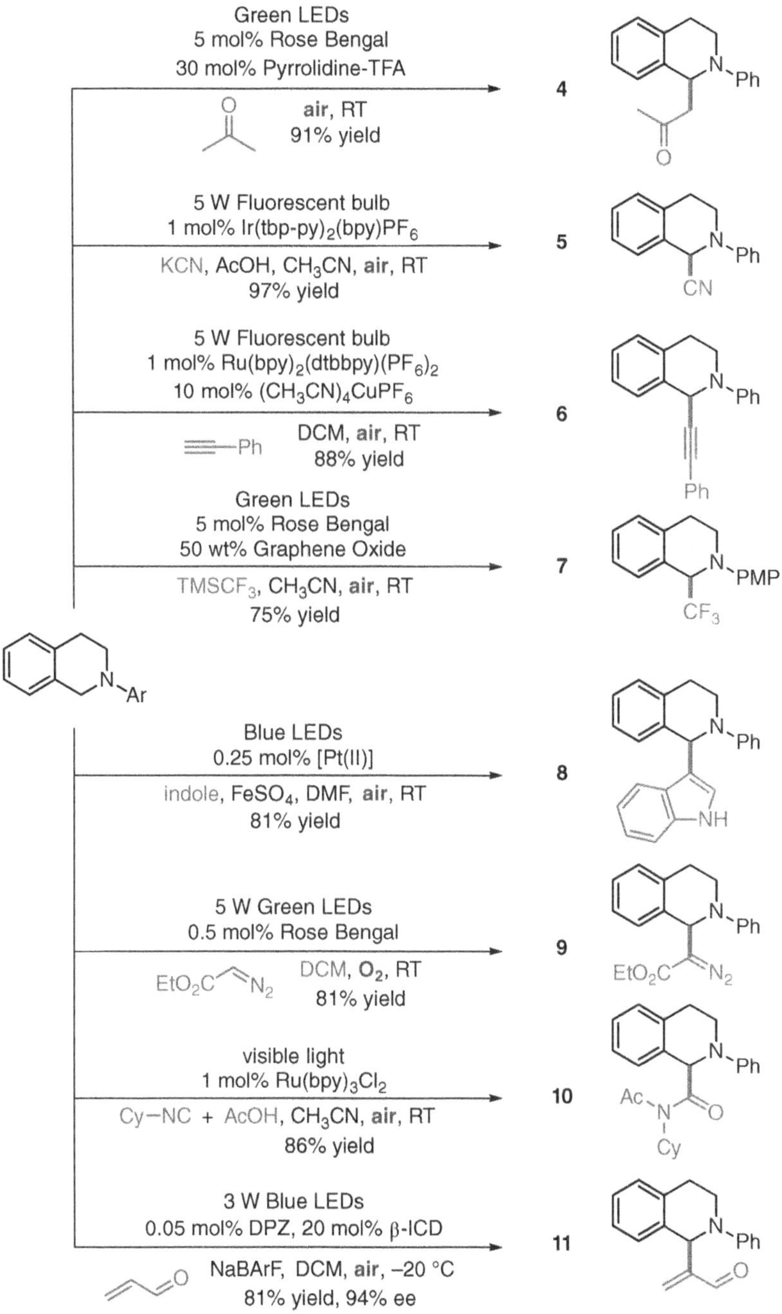

Scheme 14.3 Photocatalytic aerobic oxidative functionalization of *N*-phenyl tetrahydroisoquinoline.

good diastereoselectivity. Due to steric repulsion, intramolecular nucleophilic attack at the Re face was thought to be favorable in comparison to attack at the Si face. Subsequently, these authors have shown that the same strategy may be extended to the construction of isoquino[2,1-a][3,1]oxazine and isoquino[2,1-a]pyrimidine derivatives

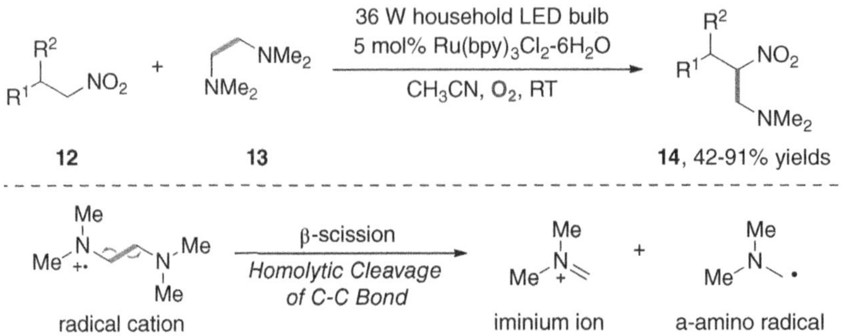

Scheme 14.4 Photocatalytic oxidative cleavage of the C—C bond.

via the attack of pendent oxygen or nitrogen nucleophiles on photoredoxgenerated iminium ions (Eq. 14.4) [14b].

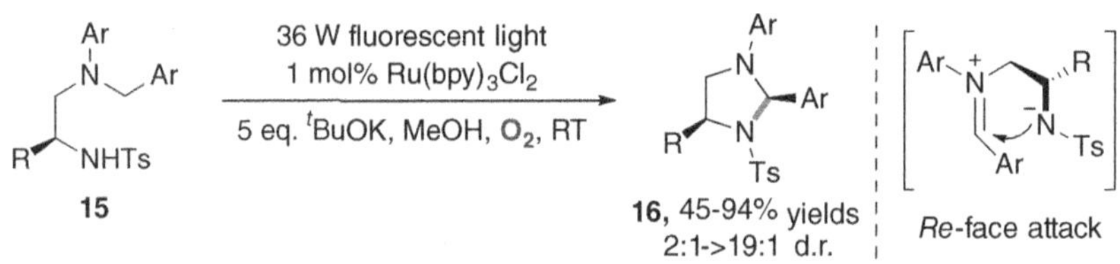

(14.3)

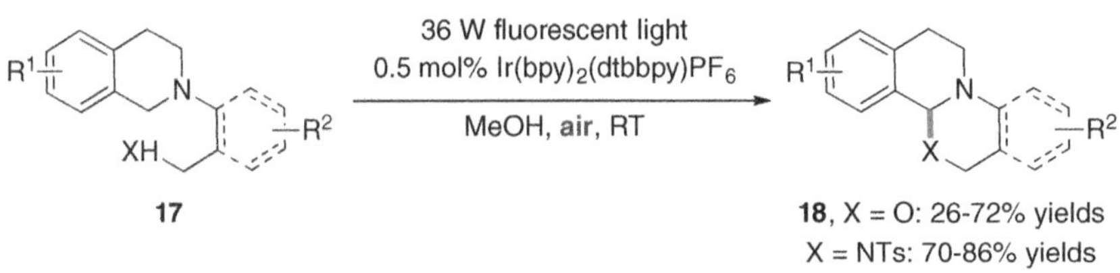

(14.4)

Iminium ions generated via aerobic photoredox catalysis have additionally been harnessed to perform cycloadditions. In 2011, Xiao and co-workers utilized a visible-light-mediated oxidation/[3 + 2] cycloaddition/aromatization sequence for the construction of pyrrolo[2,1-a]isoquinolines **21** (Scheme 14.5) [15a]. A potential mechanism featuring an azomethine ylide as the key intermediate was proposed by the authors. Initially, tetrahydroisoquinoline ester **19** is oxidized by the excited photocatalyst (**Ru[II]***) to give the amino radical cation **22**. Hydrogen atom abstraction, presumably by superoxide radical anion generated during catalyst turnover, then furnishes the iminium ion **23**. At this point, the generated HO_2^- may deprotonate the α-position of the ester group to afford the azomethine ylide **24**, a 1,3-dipole. Cycloaddition with a variety of dipolarophiles, followed by aromatization, delivers the desired product **21**.

Not long thereafter, Rueping's group independently reported a similar [3 + 2] cycloaddition reaction (Eq. 14.5) [15b]. And Maurya have expanded this strategy to cycloaddition reaction of tetrahydro-*β*-carbolines (Eq. 14.6) [15c]. In addition, Itoh and co-workers used methylene blue to replace the ruthenium complex as photocatalyst to realize the same transformation [15d].

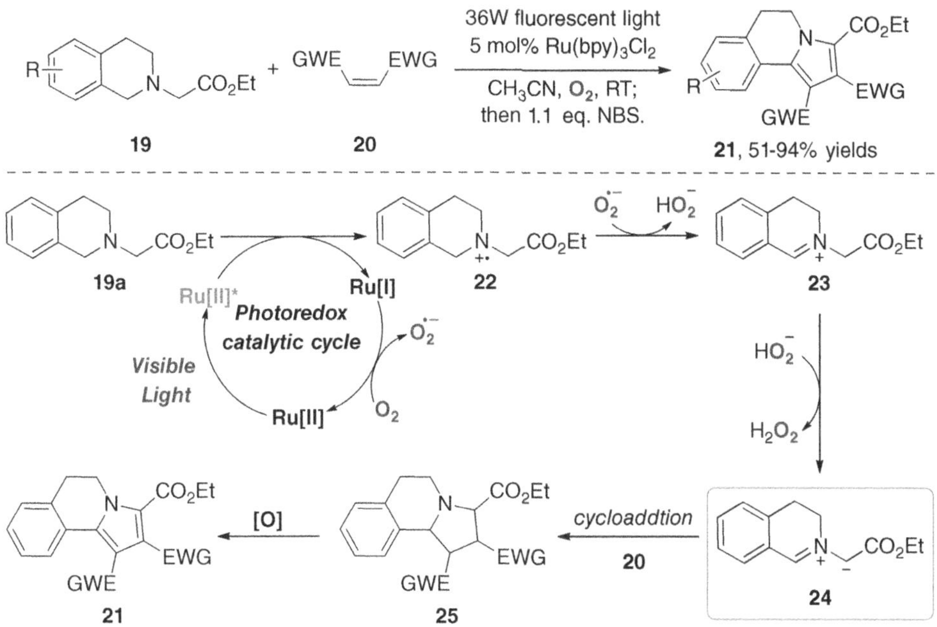

Scheme 14.5 Photocatalytic oxidative [3 + 2] cycloaddition reaction.

Rueping's work

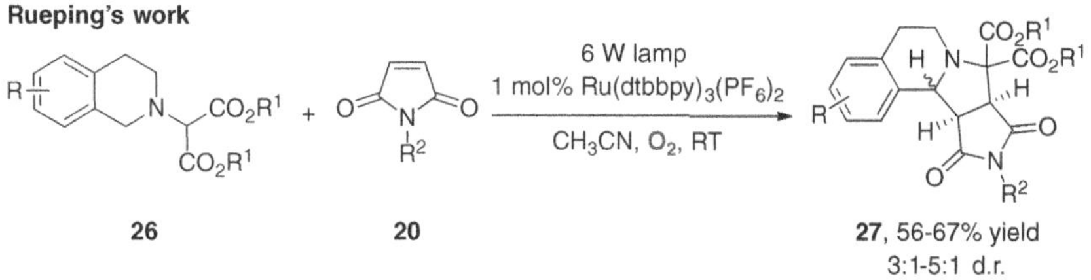

(14.5)

Maurya's work

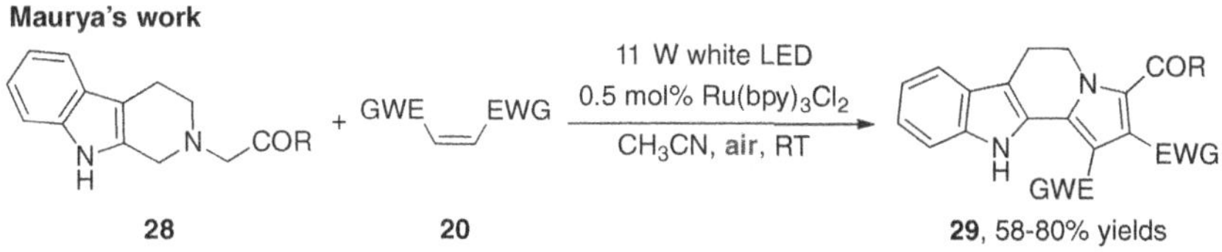

(14.6)

As the photocatalytic oxidation reactions of tertiary amines (see Scheme 14.2), secondary amines could also be oxidized to generate iminium ion intermediate under visible-light-induced photocatalytic aerobic conditions. However, the iminium ion intermediate generated from secondary amines is not as stable as that from tertiary amines, which can undergo deprotonation to generate imine (Scheme 14.6). The photocatalytically generated imine intermediates have been applied into a series of nucleophilic addition and cycloaddition reactions.

In 2012, Li and co-workers developed a new method toward the functionalization of secondary amines with indoles through visible-light-induced aerobic oxidation (Scheme 14.7) [16a]. In this work, they established a general methodology for the

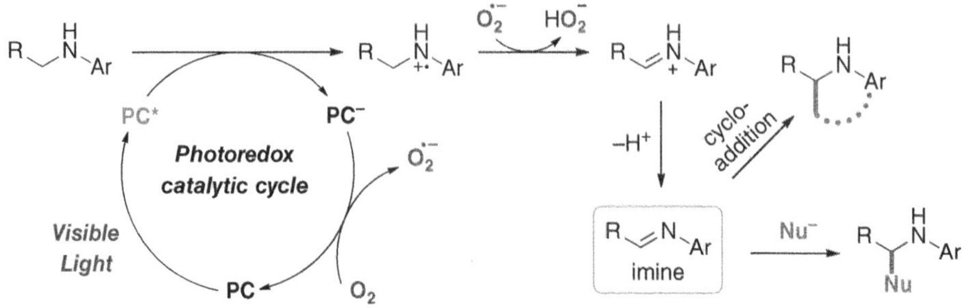

Scheme 14.6 Main pathways of photocatalytic secondary amine oxidation to imine.

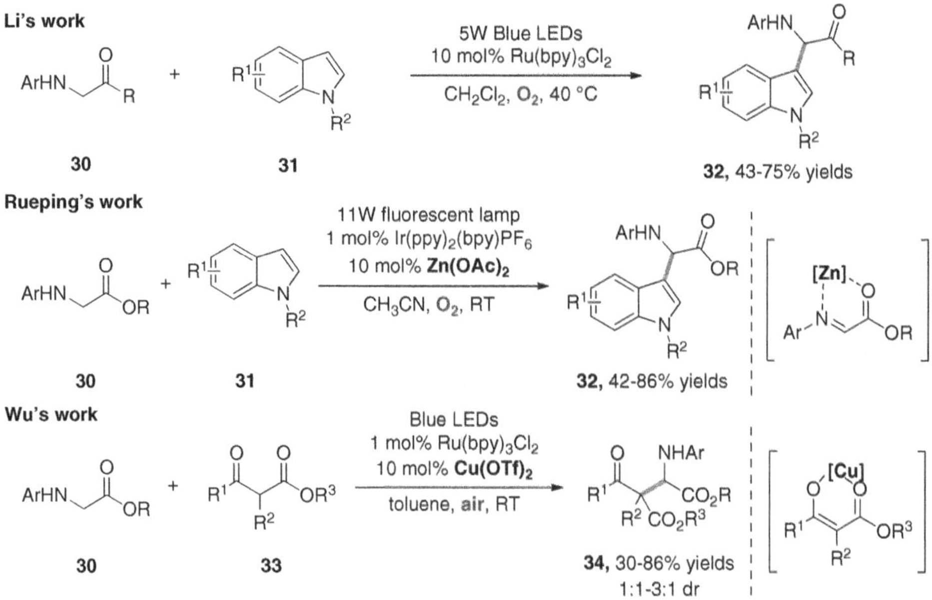

Scheme 14.7 Photocatalytic oxidative functionalization of secondary amines.

construction of a number of functionalized 2-(1H-indol-3-yl)-2-aminocarbonyl compounds from α-amino carbonyls and indoles using O_2 as the oxidant under visible-light irradiation. At the same time, Rueping's group reported a similar transformation of peptides and glycine esters through a combination of visible-light photoredox catalysis and Lewis acid catalysis (Scheme 14.7) [16b]. The addition of the Lewis acid $Zn(OAc)_2$ was essential for C—H arylation due to its capacity to activate the electrophilic imine intermediate. Soon after, Wu changed the nucleophiles from indoles to *β*-keto esters to achieve a aerobic oxidative alkylation of secondary amines via dual photoredox and copper catalysis (Scheme 14.7) [16c].

Sulfonylmethyl isocyanide is a useful synthon as a 1,3-dipole. In early 2014, Xiao's group disclosed a visible-light-induced photocatalytic oxidative [3 + 2] cycloaddition reaction between secondary amines and isocyanides (Scheme 14.8) [17a]. In this photocatalytic cascade, readily accessible glycine derivatives **30** were oxidized to imines

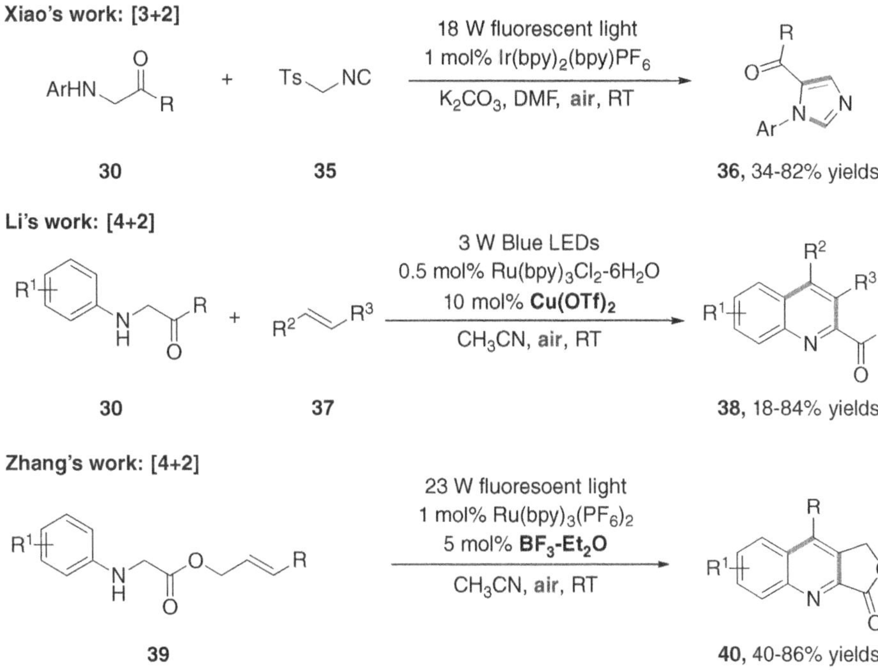

Scheme 14.8 Photocatalytic oxidative cycloaddition of secondary amines.

and then underwent a formal [3 + 2] cycloaddition/proton transfer/aromatization process to yield various biologically significant 1,5-disubstituted imidazoles. Recently, Li and Zhang reported a formal oxidative [4 + 2] cycloaddition reaction of secondary amines and unactivated alkenes through a combination of aerobic photoredox catalysis and Lewis acid catalysis, affording a range of substituted quinolines with moderate to good yields (Scheme 14.8) [17b]. At the same time, the similar intramolecular [4 + 2] cycloaddition reaction of secondary amines **39**, bearing alkene fragments, to deliver quinoline-fused lactones, was developed by Zhang's group (Scheme 14.8) [17c].

In 2012, Zheng and co-workers reported a new photocatalytic oxidative transformation of secondary amines based on the reactivity of nitrogen-centered radical cations [18]. In this process, an amino radical cation undergoes electrophilic addition to a tethered alkene instead of transformation into an iminium ion (Scheme 14.9) [18a]. It is important that the addition of silica gel significantly accelerates the process, perhaps through the absorption of O_2, facilitating single-electron oxidation, and acting as a source of protons. It is worth mentioning that the styryl anilines **43**, tethering an oxygen or nitrogen nucleophile, were transformed into fused indolines via interception of the photogenerated benzylic cation by the tethered nucleophile [18b].

Generally, primary amines are much less reactive for photocatalytic oxidation than tertiary amines and secondary amines, probably due to the relative higher oxidation potential. In 2015, Itoh and co-workers developed a photooxidative reaction of primary benzylamine using molecular oxygen as an oxidant under visible-light irradiation to form highly desirable imines (Eq. 14.7, left) [19a]. Soon after, Tao's group used dual

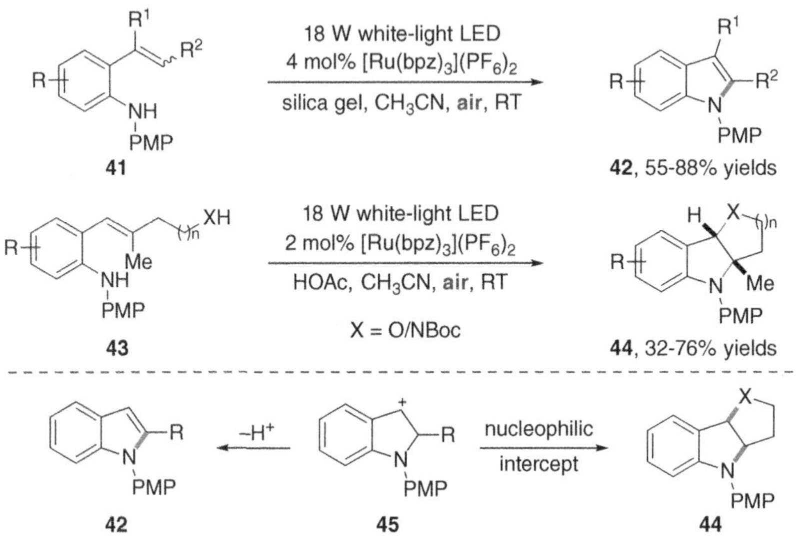

Scheme 14.9 Photocatalytic oxidative cyclization of secondary amines.

catalysis combining visible-light photocatalysis and copper catalysis to accomplish efficient aerobic oxidation of benzylamine to nitriles (Eq. 14.7, right) [19b].

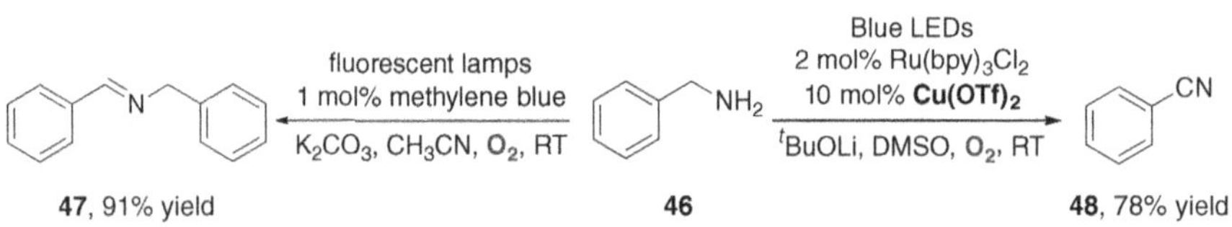

(14.7)

Based on these reports on the aerobic photocatalytic oxidation of amines, Rueping and co-workers utilized *N*-substituted hydroxylamine **49** as the starting materials for the generation of the nitrone intermediate under visible-light-induced aerobic oxidation conditions (Eq. 14.8) [20a]. Following intermolecular [3 + 2] cycloaddition with alkenes, the nitrone intermediate can efficiently be transformed into valuable isoxazolidine. The similar oxidation of hydrazine has been developed by Zhu's group to generate reactive diazene intermediate, which could attack Michael acceptor **53** and cyclize to a polysubstituted pyrazole in excellent yield (Eq. 14.9) [20b].

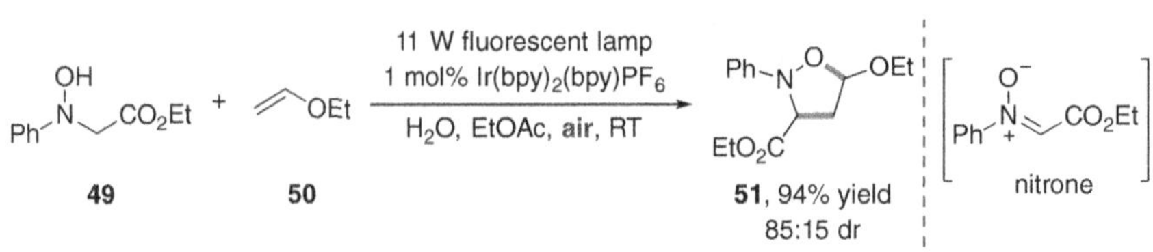

(14.8)

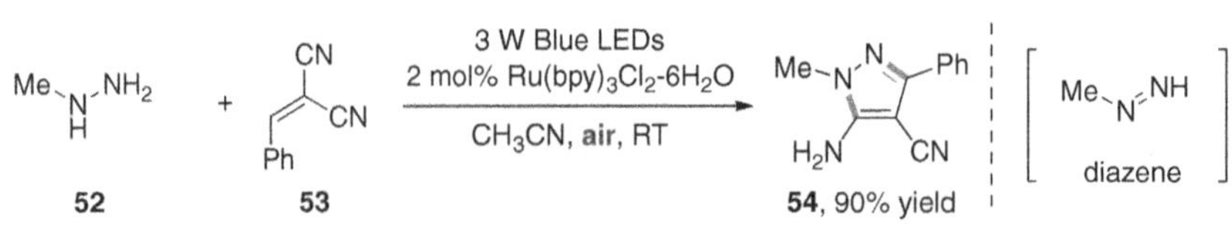

(14.9)

14.2.2 Merging Photocatalytic Aerobic Oxidation with Metal-Catalyzed C—H Activation

Transition metal-catalyzed C—H oxidative functionalization has proven to be a powerful tool for the construction of complex organic molecules. However, in order to complete the full catalytic cycle, stoichiometric amounts of Cu(II) or Ag(I) salts have often been applied to reoxidize and regenerate the active catalyst. Recently, the development of dual catalysis concept, merging visible light photoredox catalysis and transition metal catalysis, may enable the use of an oxygen molecule as a terminal oxidant to reoxidize a metal catalyst [21]. The following general mechanism can be proposed (Scheme 14.10): after C—H activation by the catalyst $M^{(n+2)}X_2$ (**A**), a metal–aryl complex **B** is formed. Insertion of the olefinic compound, which yields the metal complex **C**, and *β*-hydride elimination give a metal hydride **D**. Then, the in situ generated metal hydride or low-valent metal complexes can be reoxidized by either the photoredox catalyst or the in situ formed superoxide anion yielding the catalytically active metal species.

Rueping has utilized the dual catalysis strategy, using air as a terminal oxidant, in oxidative C—H olefinations of aryl amides [22a] and *o*-(2-pyridyl) phenols [22b] catalyzed by rhodium and ruthenium, respectively (Scheme 14.11).

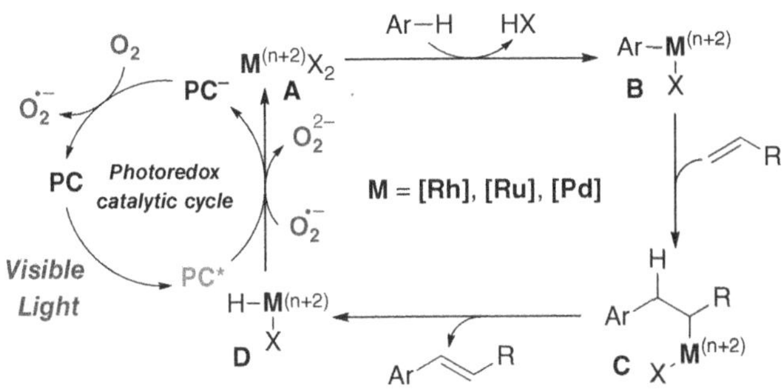

Scheme 14.10 General mechanism for metal-catalyzed C—H functionalization using combined photoredox catalysis.

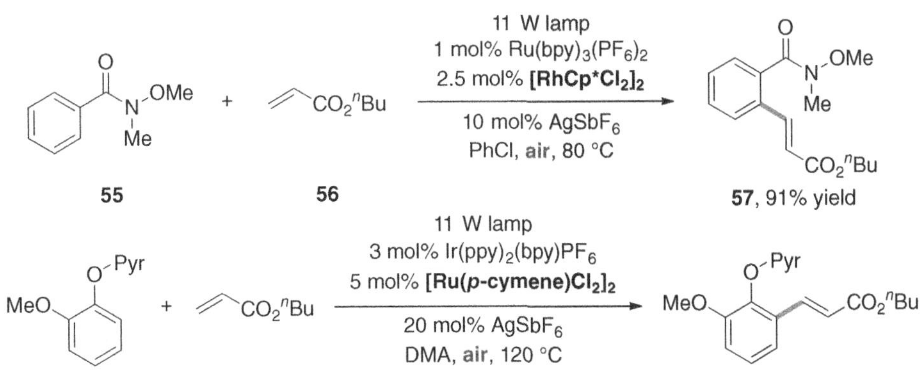

55 56 57, 91% yield

58 56 59, 75% yield

Scheme 14.11 Rh/Ru-catalyzed olefination in the presence of photoredox catalysis.

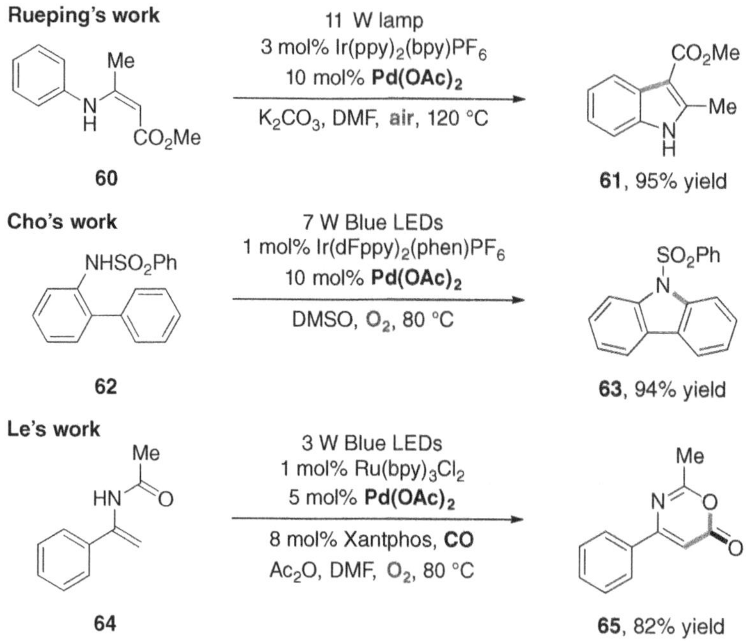

Scheme 14.12 Pd-catalyzed cyclization in the presence of photoredox catalysis.

Subsequently, Rueping showed that the same strategy could be extended to the palladium catalyzed intramolecular oxidative C—H arylation of enamides enabling the synthesis of indoles (Scheme 14.12) [23a]. Wu's group has developed a new photocatalytic oxidative radical cyclization strategy to achieve the same transformation for the synthesis of indoles without adding palladium catalyst [23d]. Then, Cho's group has successfully utilized the merged visible light photoredox and palladium catalysis in the intramolecular C—H bond oxidative amination of N-substituted 2-amidobiaryls to synthesize carbazoles [23b]. Recently, an aerobic intramolecular oxidative carbonylation of enamides by combining palladium and photoredox catalysis has been reported by Lei and co-workers [23c]. In this reaction, the authors directly used oxygen as an oxidant to provide a mild and environmentally friendly method for the synthesis of 1,3-oxazin-6-ones.

14.2.3 Photocatalytic Aerobic Oxidative Construction of C–Heteroatom Bond

The regiospecific formation of the C–heteroatom bond is an important transformation in organic synthesis for the construction of complex organic compounds. Visible-light-induced photocatalytic aerobic reactions have proven to be extremely powerful and versatile synthetic tools for this purpose. In 2014, Lei and co-workers developed the first visible-light-induced decarboxylative amidation of α-keto acids, providing a novel and efficient route to various amides in good yields. One of the key features of this work was the use of O_2 as a green terminal oxidant (Scheme 14.13) [24a]. Soon after, the photocatalytic oxidative amidation of aromatic aldehydes was reported by Leow's group, which provides an economical, green, and mild alternative for the formation of C—N bond by using phenazine ethosulfate as a metal-free visible-light photocatalyst and air as the sole

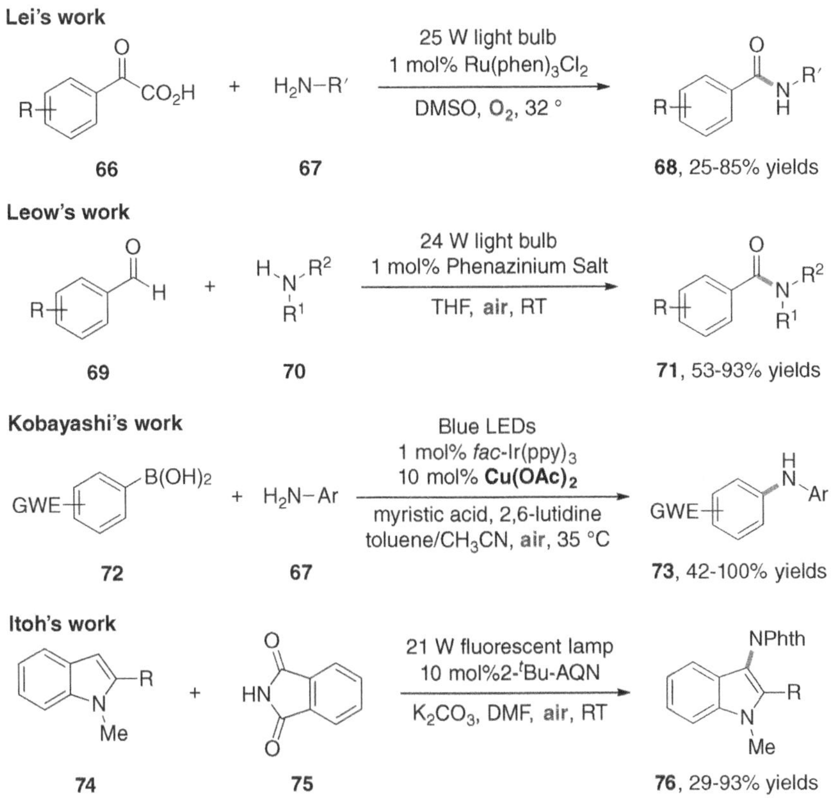

Scheme 14.13 Photocatalytic oxidative construction of the C—N bond.

oxidant [24b]. In 2015, Kobayashi utilized the merged visible light photoredox and copper catalysis to achieve an aerobic oxidative Chan–Lam coupling reaction of anilines and electron deficient aryl boronic acids, which were found to be poor substrates in traditional Chan–Lam coupling reactions [24c]. Very recently, Itoh and co-workers used phthalimide as a nitrogen-centered radical source for the direct C—H oxidative amination of indoles through a visible-light-mediated photocatalytic aerobic oxidation method [24d].

Direct C—H amination of arenes provides straightforward and streamlined access to aromatic amine moieties, which are widely found in pharmaceuticals, agrochemicals, and organic materials. In 2015, Nicewicz and co-workers developed an elegant TEMPO and acridinium photooxidant-based dual catalytic system for site-selective aryl C—H oxidative amination reactions by using oxygen as an oxidant (Eq. 14.10) [25a]. Latterly, the authors changed amine substrates to trimethylsilyl cyanide to realize the similar aryl C—H oxidative cyanation reactions, providing a novel and efficient route to various aromatic nitriles (Eq. 14.11) [25b].

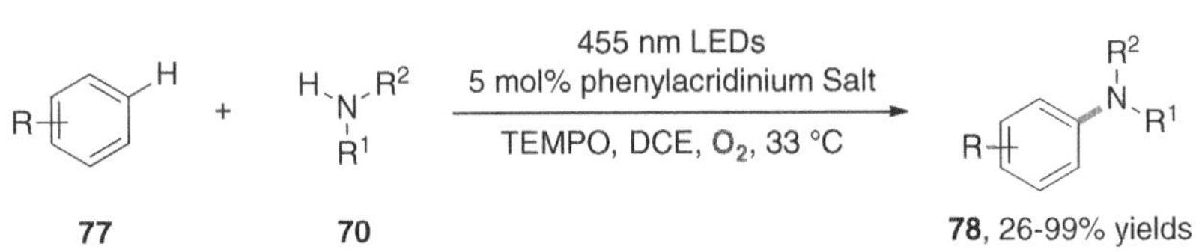

(14.10)

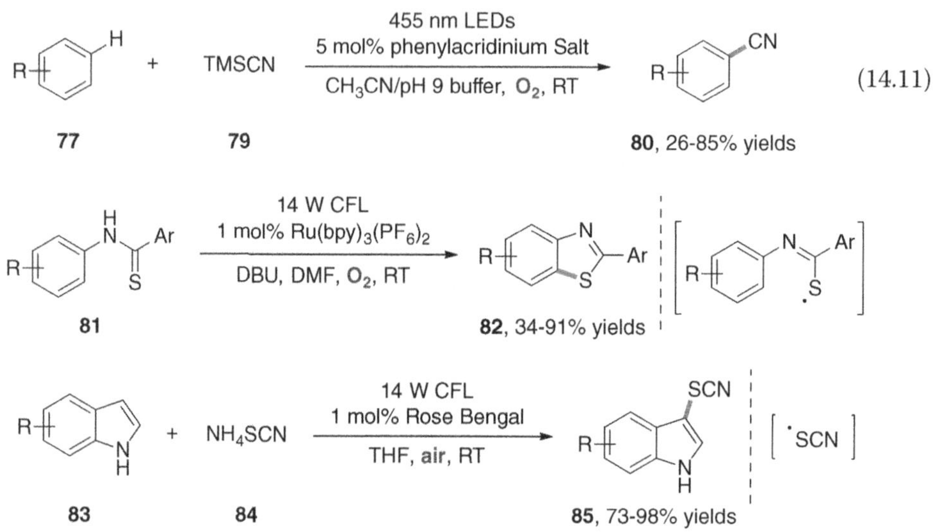

Scheme 14.14 Photocatalytic oxidative construction of C—S bond.

In 2012, Li's group developed a mild and convenient approach for the aerobic oxidative construction of C—S bond via visible-light photocatalysis (Scheme 14.14) [26a]. This method allows for the synthesis of a range of 2-substituted benzothiazoles through an intramolecular radical cyclization in good yields using oxygen as the terminal oxidant. Two years later, the authors used visible-light-promoted aerobic oxidation to achieve a direct C3 thiocyanation of indoles with ammonium thiocyanate via a photocatalytic radical addition [26b].

Visible-light photoredox catalysis has emerged as a powerful platform for the generation of various S-radical species and methodology development toward the synthesis of diverse S-containing compounds. In 2015, Noël and co-workers reported a visible-light-induced photocatalytic aerobic oxidation of thiols to disulfides by using Eosin Y as a photocatalyst, and a continuous flow procedure was also developed to accelerate this oxidative process, enabling the preparation of products with high purity in a timeframe of minutes (Eq. 14.12) [27a]. Zhang and Li utilized the similar photocatalytic aerobic oxidation to achieve the first direct S—P coupling reaction between thiols and diarylphosphine oxide, affording a range of S—P(O) coupling products in moderate to excellent yields (Eq. 14.13) [27b].

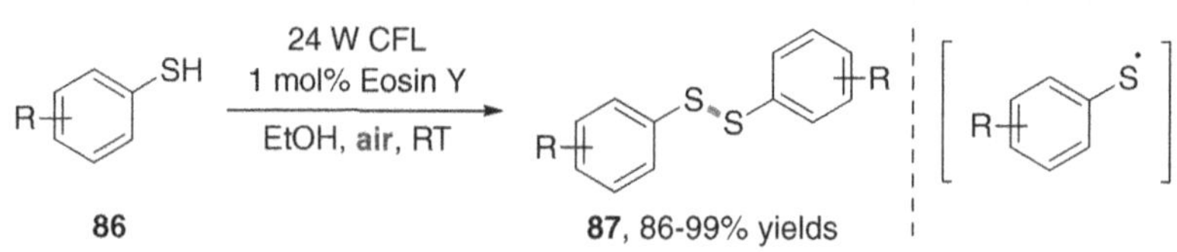

(14.12)

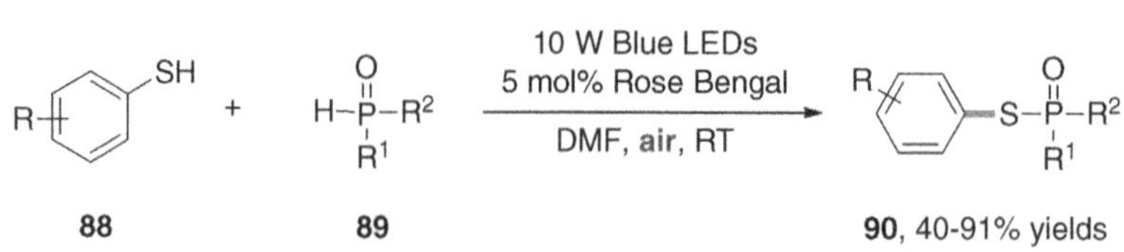

(14.13)

14.3 Visible-Light-Induced Photocatalytic Aerobic Oxygenation

Oxygen-containing groups, such as alcohol, carbonyl, and carboxyl, are attractive synthetic targets due to their wide presence in a great number of natural products and biologically active molecules, and their role as useful synthons in organic synthesis. Thus, the preparation of O-containing compounds has always been an important topic. The incorporation of oxygen atoms from molecular oxygen into substrates via a visible-light-induced photocatalysis is one of the most attractive approaches. For visible light photocatalytic aerobic oxygenation reactions, there are mainly three pathways to incorporate molecular oxygen into the product (Scheme 14.15): (i) radical addition of molecular oxygen ($^3O_2$) to alkyl radicals, which was generated by photoredox catalysis; (ii) generation of superoxide radical anion through a single-electron transfer (SET) process; and (iii) generation of reactive singlet oxygen ($^1O_2$) through ET. Several comprehensive reviews of this field have been published thus far [6]. Consequently, in this part, we will focus mainly on highlighting the new advances in this field by reaction type.

14.3.1 Photocatalytic Aerobic Oxygenation Via Radical Addition

Visible-light photoredox catalysis has become one of the most efficient and sustainable methods to generate radical intermediates from a great number of readily available starting materials. This emerging strategy brings a powerful platform to construct C—C and C-heteroatom bonds. Some photogenerated alkyl radical intermediates are easily trapped by ground state oxygen to form the hydroperoxide compounds, which can then translate to carbonyls or alcohols followed by the elimination of water or reduction (Scheme 14.16).

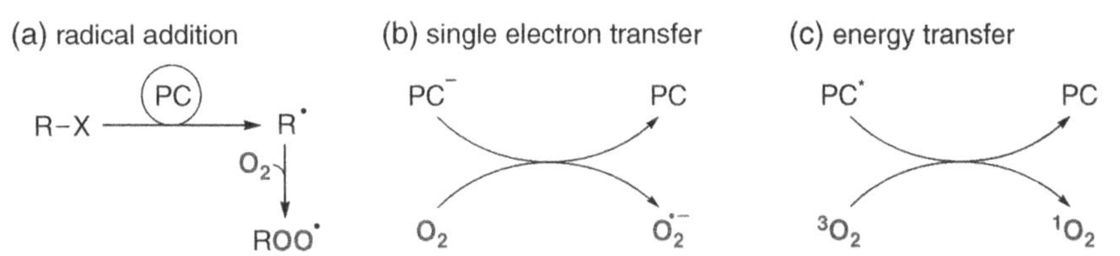

Scheme 14.15 General models for photocatalytic oxygenation.

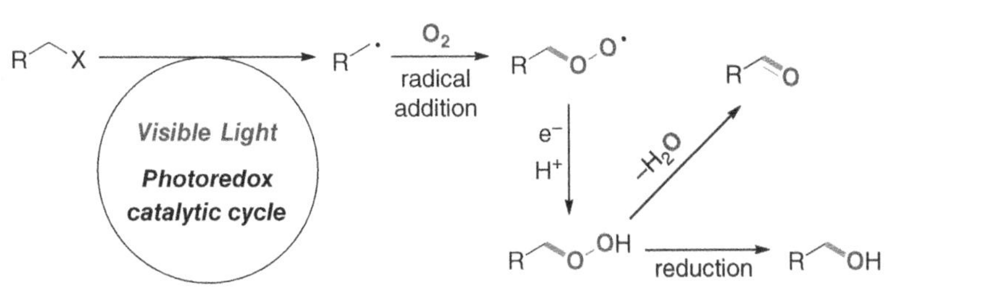

Scheme 14.16 General mechanism for photocatalytic aerobic oxygenation via radical addition.

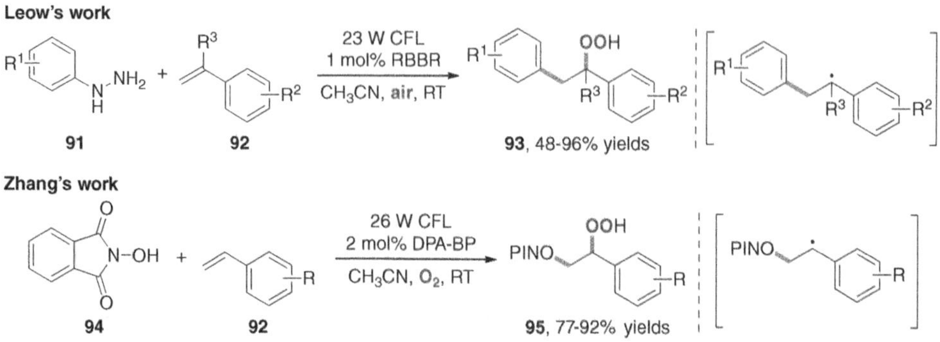

Scheme 14.17 Photocatalytic synthesis of hydroperoxides.

In 2015, Leow and co-workers reported a visible-light-induced photocatalytic hydroperoxyarylation reaction of styrenes with aryl hydrazines and air, affording a variety of hydroperoxides in moderate to good yields (Scheme 14.17) [28a]. In this reaction, the aryl radicals can be generated from the photocatalytic aerobic oxidation of aryl hydrazines, and then these aryl radicals can undergo the hydroperoxyarylation reaction. Soon after, Zhang's group utilized photoinduced proton-coupled electron transfer process to generate a phthalimide-*N*-oxyl (PINO) radical from *N*-hydroxyphthalimide (NHPI), realizing the ***dioxygenation*** of styrenes [28b]. Recently, Adimurthy found that this dioxygenation reaction could also occur under photocatalyst-free condition [28c].

Subsequently, in 2016, Liang and co-workers developed a mild approach for the construction of 4-aryl tetealones from styrene derivatives via visible light photoredox catalytic oxidative cyclodimerization–oxygenation (Eq. 14.14) [29]. And when 2-phenylmalononitrile (PMN) was used as a co-catalyst, the conversion efficiency improved remarkably.

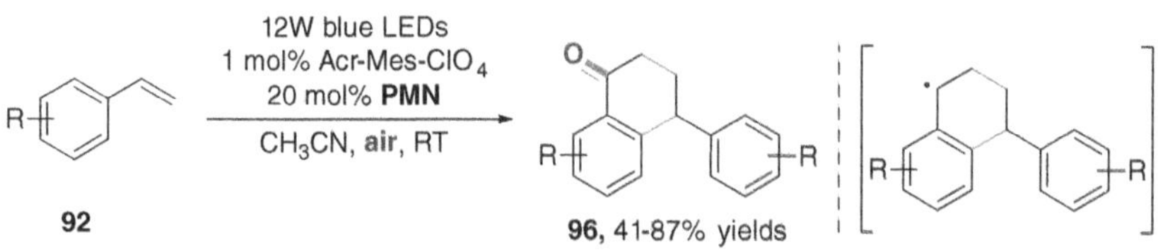

(14.14)

The photocatalytic aerobic oxygenation of allenes was reported by Kumaraswamy's group, which provided a visible-light-promoted phenylseleno-oxygenation reaction of activated allenes with diphenyl diselenide and air under photocatalyst-free condition (Eq. 14.15) [30a]. Mechanistic studies indicated the SET achieved between diphenyl diselenide and allenes by irradiation of visible light. In 2017, Wang utilized electron-rich aromatic disulfides as photocatalysts to achieve a visible-light-mediated aerobic oxidative cleavage of C=C bonds, affording a mild method for the conversion of aromatic olefins into ketones and aldehydes [30b].

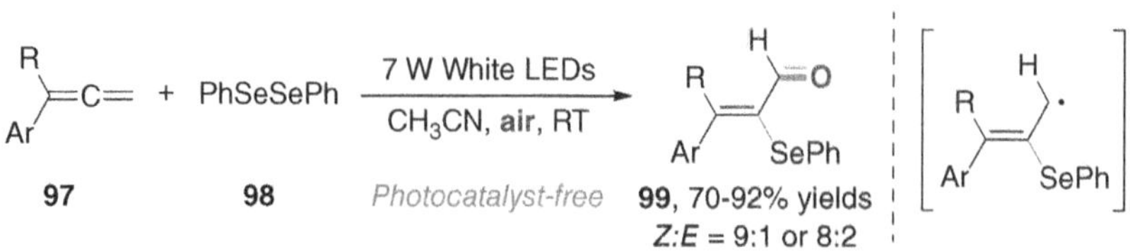

(14.15)

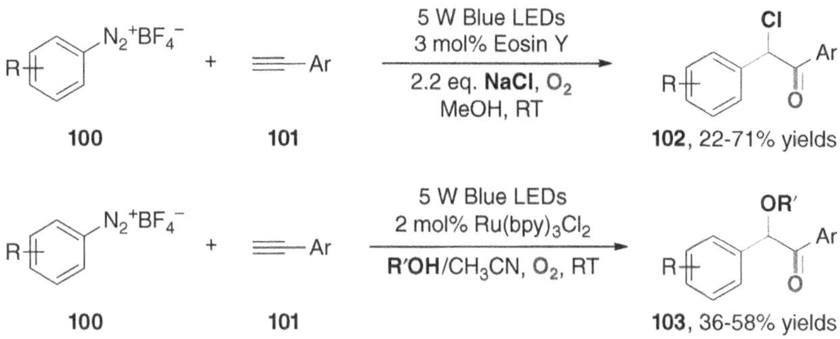

Scheme 14.18 Photocatalytic aerobic functionalization of alkynes.

Aryl diazonium salts are an important source of aryl radicals in photoredox catalysis. In 2016, Ni and Wang applied photocatalytic generated aryl radical to trigger a visible-light-induced four-component domino reaction of alkynes, Aryl diazonium salts, oxygen, and a nucleophile (NaCl or ROH), providing a direct approach to synthesize α-chloro or α-alkoxy aryl ketones (Scheme 14.18) [31].

Subsequently, Wang and co-workers developed an efficient methodology to build oxazole skeleton via a three-component cyclization reaction of 2*H*-azirines, alkynyl bromides, and molecular oxygen under visible-light photoredox catalysis (Eq. 14.16) [32].

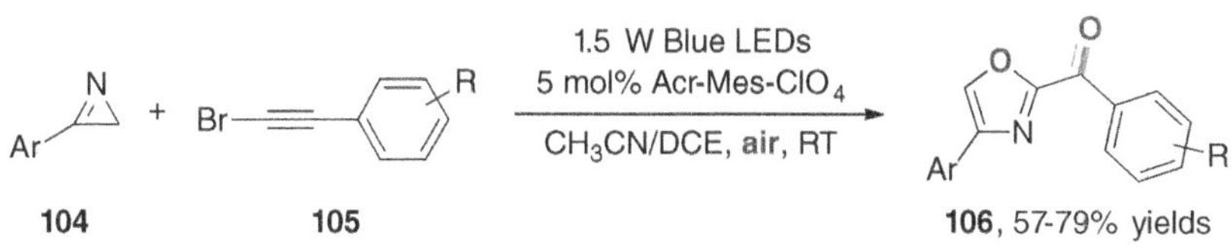

(14.16)

In 2014, Zhou and co-workers reported a visible-light-induced intramolecular oxidative cyclization of *ortho*-alkynylated *N,N*-dialkylamines using air as the oxidant, affording various 2-aryl-3-acylindoles in good yields (Scheme 14.19) [33a]. The plausible mechanism of this reaction is involving the addition of *α*-amino alkyl radicals to alkynes, followed by C—O bond formation. Using *ortho*-alkynylated carboxylic acids **109** as substrates, Xia utilized a photoredox-assisted decarboxylative generated alkyl radical to achieve a similar oxidative cyclization reaction, which enables the formation of a variety of 3-acylindoles, 3-acylbenzofurans, and 3-acylbenzothiophenes under mild conditions [33b]. Very recently, Xia's group demonstrated a dual copper and visible light photoredox catalyzed oxidative cyclization of aromatic enamines with alkynes for the synthesis of multisubstituted quinolone derivatives [33c].

Recently, Wang and co-workers developed a photoinduced oxidative formylation of *N,N*-dimethylanilines and *N,N*-di(prop-2-yn-1-yl)anilines with molecular oxygen under photocatalyst-free condition [34]. Mechanistic studies indicated that both the starting materials and products act as photocatalysts to trigger this transformation.

Natarajan's group described a one-pot method for the synthesis of electronically diverse aryl glyoxals from acetoarylones via bromination and visible-light-induced photocatalytic oxidative carbonylation sequence by using air as an oxidant (Eq. 14.17) [35].

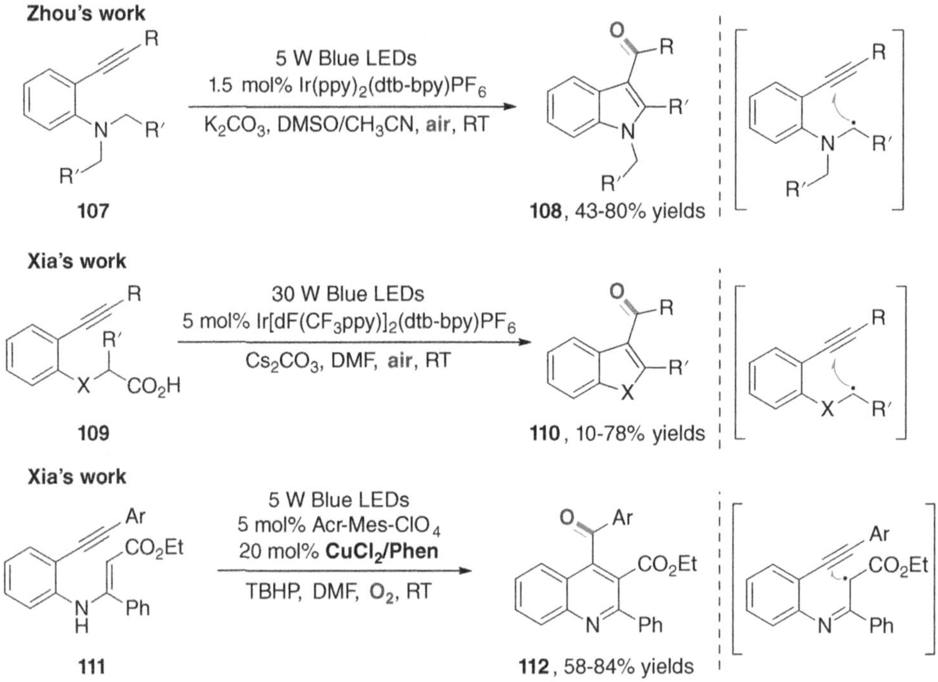

Scheme 14.19 Photocatalytic aerobic oxidative cyclization of alkynes.

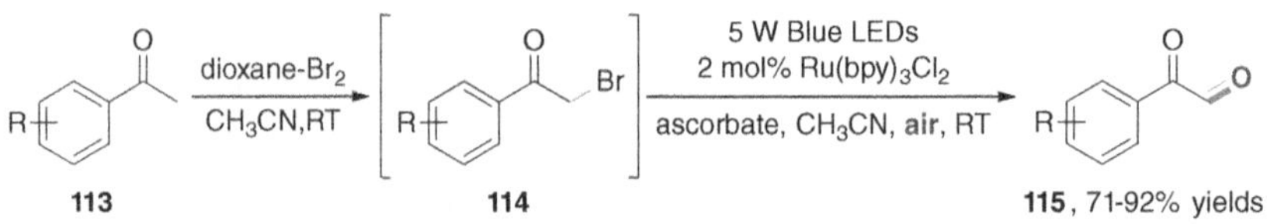

(14.17)

As part of their ongoing studies on the photocatalytic *N*-centered radical reaction [36a], Xiao and Chen developed a cooperative TEMPO/photocatalytic system for direct conversion of N—H bonds into N-centered radicals, enabling a general and selective oxidative radical oxyamination of β,γ-unsaturated hydrazones (Scheme 14.20) [36b]. In this protocol, PPh_3 was employed as a reductant to reduce the hydroperoxide intermediates into the hydroxylation products. At the same time, Xiao and Lu applied the capture of photocatalytic decarboxylative generated alkyl radical by molecular oxygen and a subsequent reduction by sodium borohydride to achieve the conversion of a variety of readily available carboxylic acids to alcohols in moderate to high yields [36c].

14.3.2 Photocatalytic Aerobic Oxygenation via Superoxide Radical Anion

As described in Section 14.2, oxygen acts as an oxidation to complete the photoredox cycle with the formation of a superoxide radical anion, and this kind of highly active species might have Lewis basicity. Therefore, the superoxide radical anion could react with the appropriate acidic components, such as arylboronic acids, to realize photocatalytic aerobic oxygenation reactions (Scheme 14.21).

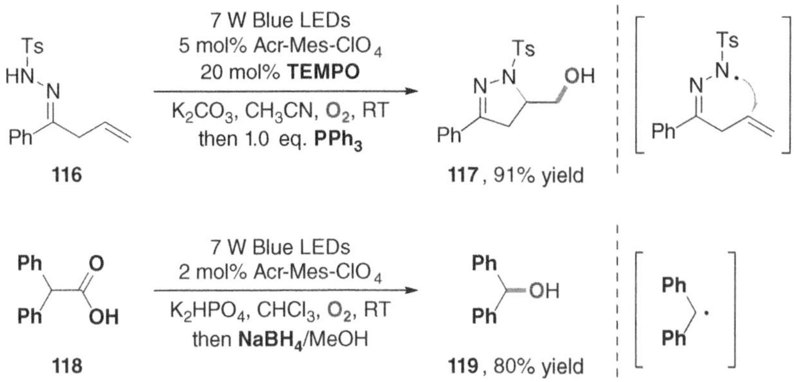

Scheme 14.20 Photocatalytic aerobic oxidative hydroxylation.

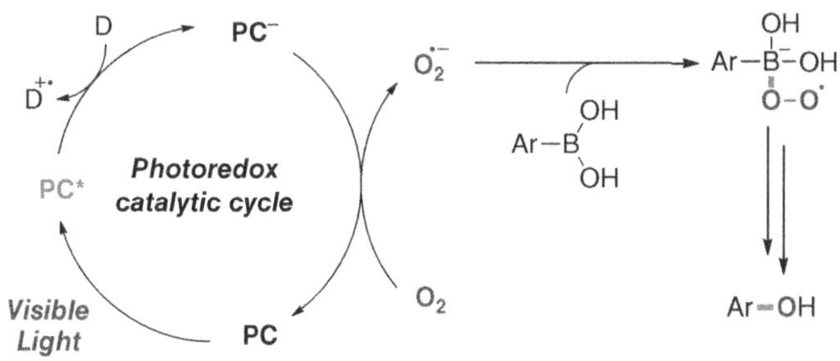

Scheme 14.21 Photocatalytic aerobic oxidative hydroxylation of arylboronic acids.

Xiao and co-workers have made significant progress in the visible-light-mediated aerobic oxidative hydroxylation of arylboronic acids [37a]. After Xiao's report, Xu et al. used water-soluble *N*-methyl-3(10*H*)-acridones as photocatalyst to realize the aerobic oxidative hydroxylation of arylboronic acids in water, giving an eco-friendly access to phenols (Eq. 14.18) [37b]. In 2016, Fu's group developed an efficient visible-light photoredox borylation of aryl halides with bis-(pinacolato)diboron. Then, the authors applied one-pot, two-step, including borylation of aryl halides and oxidative hydroxylation of arylboronic esters to achieve the visible-light photoredox conversion of aryl halides into phenols (Eq. 14.19) [37c].

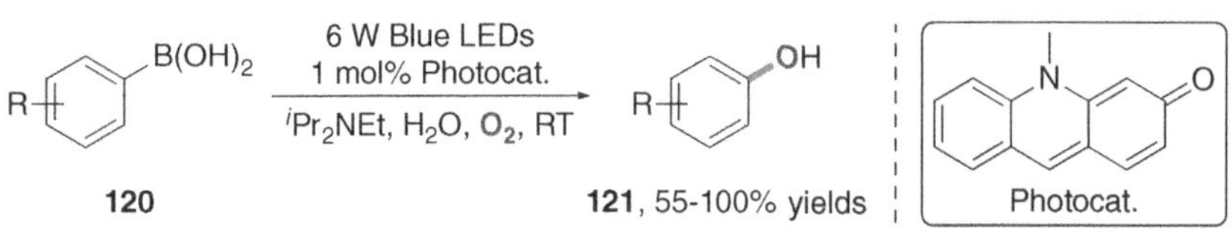

(14.18)

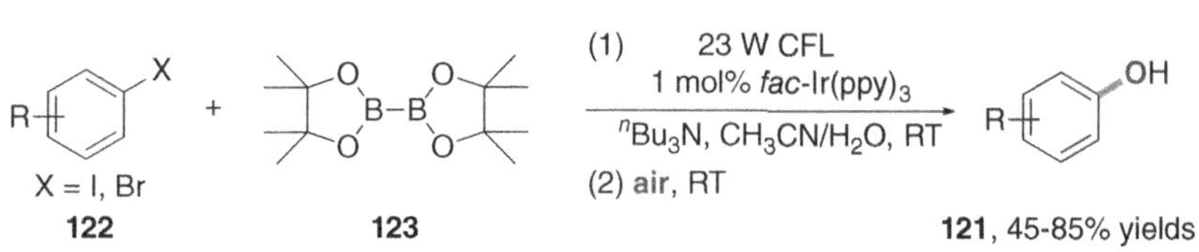

(14.19)

In 2015, Xiao's group developed a visible-light-induced phototandem catalysis strategy based on a single iridium photocatalyst, which consists of two mechanistically distinct catalytic cycles, namely photocatalytic ET and SET. This protocol allows a rapid and efficient construction of 3-ester-3-hydroxy-2-oxindole derivatives from a wide range of diazoamides through a cyclization/aerobic oxidation sequence (Eq. 14.20) [38].

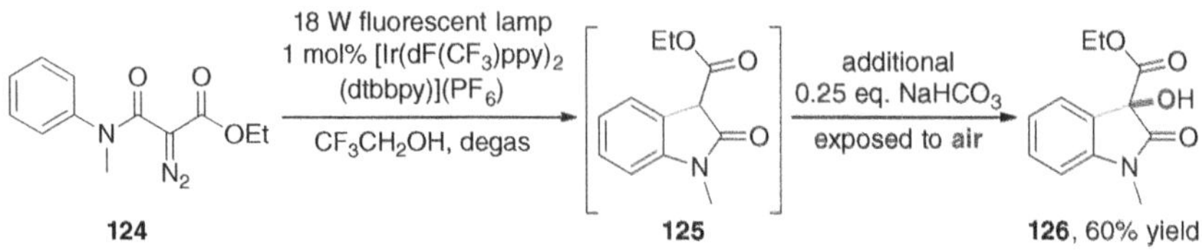

(14.20)

14.3.3 Photocatalytic Aerobic Oxygenation via Singlet Oxygen

Although molecular oxygen is considered an ideal green oxidant, the reactivity of ground state oxygen, (also known as triplet state, $^3O_2$), is often low. However, this reactivity can be increased dramatically by converting triplet oxygen ($^3O_2$) to singlet oxygen ($^1O_2$). Singlet molecular oxygen plays a significant role in several biochemical processes. Furthermore, chemists have utilized photogenerated molecular $^1O_2$ as an oxygen source for several synthetic transformations. Singlet oxygen can be generated by triplet-triplet ET from an excited photosensitizer (Scheme 14.22). The highly reactive singlet oxygen has been applied to visible light photocatalytic cycloadditions with alkenes and asymmetric α-hydroxylations of carbonyls.

In 2014, Wang's group reported a visible light photocatalytic C—C bond cleavage of enamines to produce amide products through a singlet oxygen ($^1O_2$)-initiated [2 + 2] cycloaddition and ring fragmentation (Scheme 14.23) [39a]. In addition, asymmetric [2 + 4] cycloaddition of 3-substituted 2-pyridones and singlet oxygen was carried out by Bach and co-workers with combination of visible light photocatalysis and organocatalysis [39b]. This reaction could offer various chiral 3-hydroxypyridine-2,6-diones via a template-mediated type II photooxygenation and an acid-catalyzed rearrangement, by using tetraphenylporphyrin (TPP) as photocatalyst and chiral amide **132** as the organocatalyst. The asymmetric induction model assumes the binding of the 3-substituted 2-pyridones to the catalyst **132** through a double hydrogen bond.

In 2004, Córdova developed an asymmetric oxidative α-hydroxylation of aldehydes with the combination of photocatalysis and enamine catalysis (Scheme 14.24) [40a]. While the TPP used as a photosensitizer under the irradiation of a 250-W high-pressure

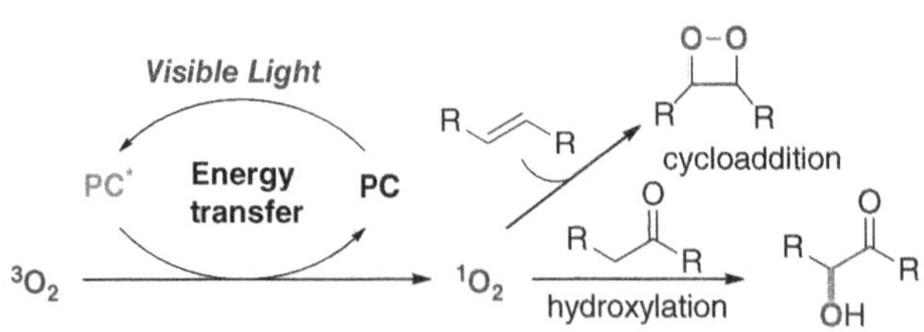

Scheme 14.22 General mechanism for photocatalytic aerobic oxygenation via $^1O_2$.

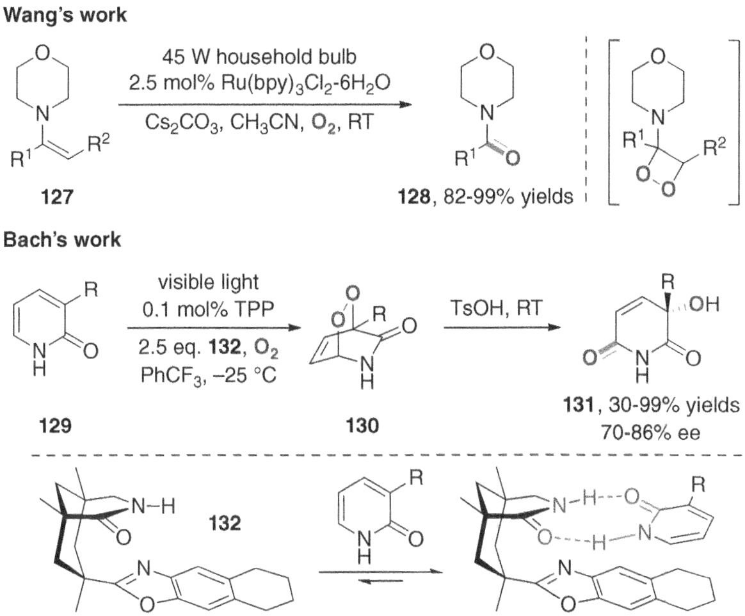

Scheme 14.23 Photocatalytic cycloaddition of singlet oxygen.

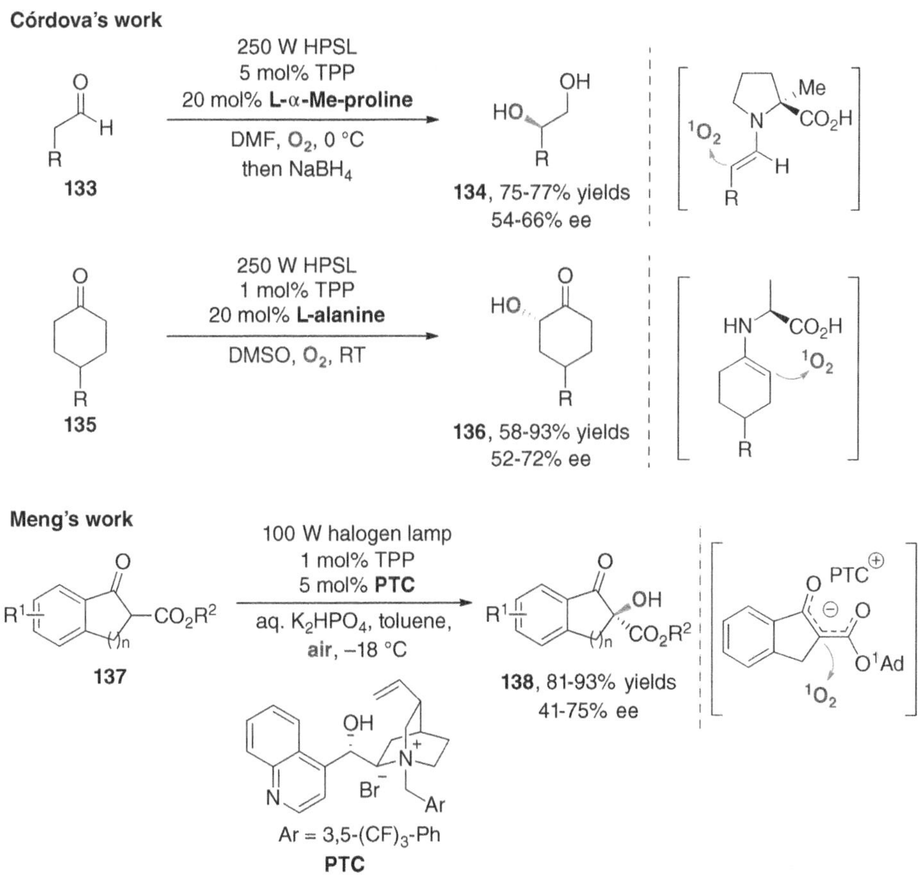

Scheme 14.24 Photocatalytic asymmetric α-hydroxylation of carbonyls via $^1O_2$.

sodium lamp, the aldehydes could be oxidized by molecular oxygen in the presence of the chiral proline derivative to give 1,2-diols after in situ reduction with $NaBH_4$. Soon after, these authors demonstrated a similar protocol to achieve asymmetric α-photooxygenation of cyclic ketones, yielding α-hydroxylated ketones [40b]. The key step of this reaction is the enantioselective reaction between the nucleophilic chiral enamine intermediate and singlet oxygen. Chiral phase-transfer catalysts have been received much attention in asymmetric organic transformations. The combination of chiral phase-transfer catalysis with photocatalysis has been reported by Meng's group, which can promote the oxidation reaction of the compounds with oxygen to afford the hydroxylation products in moderate enantioselectivity [40b]. The authors proposed that the reaction proceeded through a contact ion pair out of the quaternary ammonium ion of the chiral phase-transfer catalyst and the *β*-ketoester, which then reacts with singlet oxygen in a stereocontrolled fashion to form the final hydroxylated product.

Recently, Xiao and co-workers have successfully developed a novel family of visible-light-responsive chiral ligands by grafting a photosensitizer to chiral bisoxazoline ligands [41]. As shown in Scheme 14.25, the design chiral ligands **141** can be synthesized from readily available reagents **139** and **140** through an esterification operation. And UV/vis spectra reveal that these chiral ligands have a maximum absorption of visible light at roughly 400 nm wavelengths. Complexation of this ligand with $Ni(acac)_2$ results in a new kind of chiral bifunctional photocatalysts, which was powerful for the asymmetric oxidation *α*-hydroxylation of *β*-ketoesters by using oxygen as the green oxidant and visible light as the ideal driving force. In this bifunctional photocatalyst, the thioxanthone component could function as an efficient energy-transfer photosensitizer,

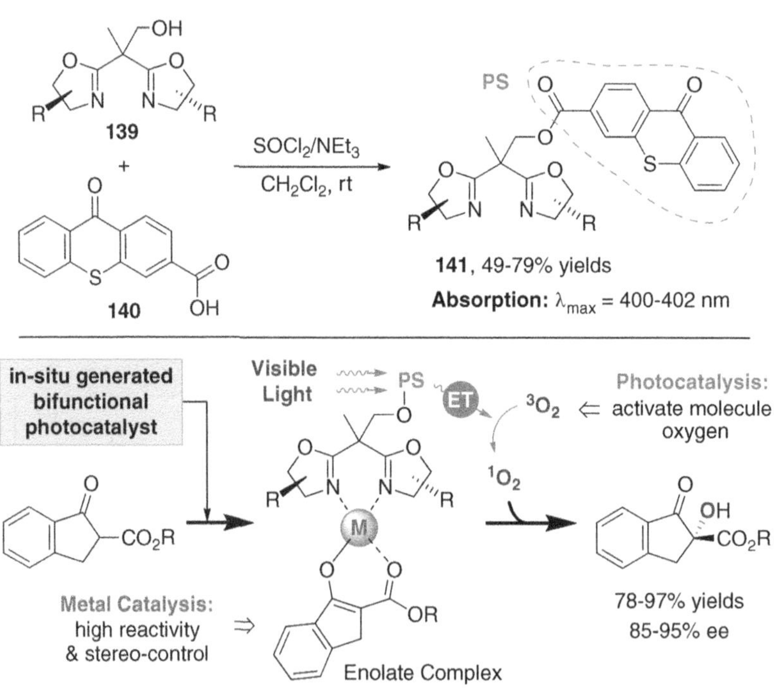

Scheme 14.25 Synthesis of new chiral ligands and asymmetric photooxygenation of a *β*-keto ester.

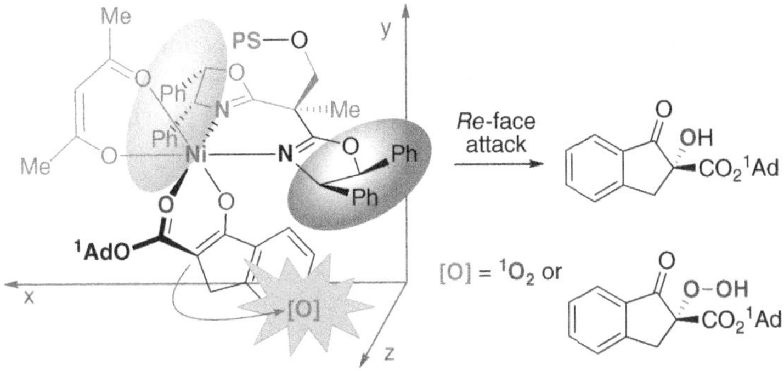

Scheme 14.26 Proposed asymmetric induction model.

with its ground state absorbing visible light and its triplet state then sensitizing molecule oxygen to generate $^1O_2$. The chiral metal complex derived from the BOX component can serve as an efficient catalyst to activate β-keto carbonyls and then control the enantioselectivity of the C—O bond-formation step. Using this protocol, products containing the α-hydroxy-β-dicarbonyl motif are produced in high yields and with excellent enantiopurities.

Based on the experiment results and previous studies, the authors proposed a possible stereoinduction model for this reaction (Scheme 14.26). Because the ^{1}Ad group is located away from the chiral ligand to avoid possible steric constraints, the Si face of enol-formed β-keto ester is blocked by the back phenyl groups. Then, the attack of oxidants from the Re face of enol seems relatively favorable, resulting in the hydroxylation product with the R-configuration.

14.4 Summary and Outlook

Since 2008, visible-light-induced photocatalysis has proved to be a powerful tool for the formation of new chemical bonds in synthetic chemistry that has attracted research interest from all around the world. Through this strategy, aerobic oxidative reactions under mild reaction conditions have gained substantial and rapid development and application. In these transformations, there are two kinds of reaction models: one is an oxidation reaction, the other is an oxygenation reaction. In this chapter, we have briefly introduced the recent advances of visible-light-mediated aerobic oxidation and oxygenation reactions. Spectroscopic measurements and electrochemistry techniques have been used widely and have played an increasingly important role in mechanistic studies. As noted, almost all the described reactions take place at room temperature under irradiation from low-energy light sources. The development of such approaches will undoubtedly make tremendous contributions to both researches in synthetic methodology and in industrial applications.

Despite these tremendous advances, several challenges in this field still remain to be solved. *First*, the general substrates employed are mainly aryl compounds or structures that contain heteroatoms. The oxidation of inherent C—H bonds in simple alkanes has rarely been studied. *Second*, the control of enantioselectivity remains a major problem.

Third, more spectroscopic measurements and electrochemistry techniques need to be incorporated into this research. Such work will provide insights into reaction mechanisms and allow for the rational design of new reactions. *Last, but not least,* further applications of this reaction methodology to the syntheses of natural products and pharmaceuticals are required to show their full potential.

References

1 Ciamician, G. (1912). *Science* 36: 385–394.
2 (a) Tucker, J.W. and Stephenson, C.R. (2012). *J. Org. Chem.* 77: 1617–1622. (b) Ravelli, D. and Fagnoni, M. (2012). *ChemCatChem* 4: 169–171. (c) Hari, D.P. and König, B. (2014). *Chem. Commun.* 50: 6688–6699. (d) Shaw, M.H., Twilton, J., and MacMillan, D.W. (2016). *J. Org. Chem.* 81: 6898–6926. (e) Romero, N.A. and Nicewicz, D.A. (2016). *Chem. Rev.* 116: 10075–10166. (f) Day, J.I., Teegardin, K., Weaver, J., and Chan, J. (2016). *Org. Process Res. Dev.* 20: 1156–1163.
3 (a) Zeitler, K. (2009). *Angew. Chem. Int. Ed.* 48: 9785–9789. (b) Narayanam, J.M. and Stephenson, C.R.J. (2011). *Chem. Soc. Rev.* 40: 102–113. (c) Xuan, J. and Xiao, W.-J. (2012). *Angew. Chem. Int. Ed.* 51: 6828–6838. (d) Prier, C.K., Rankic, D.A., and MacMillan, D.W. (2013). *Chem. Rev.* 113: 5322–5363. (e) Schultz, D.M. and Yoon, T.P. (2014). *Science* 343: 985–993. (f) Karkas, M.D., Porco, J.A. Jr., and Stephenson, C.R. (2016). *Chem. Rev.* 116: 9683–9747.
4 Bäckvall, J.E. (2004). *Modern Oxidation Methods.* Weinheim, Germany: Wiley-VCH.
5 (a) Stahl, S.S. (2004). *Angew. Chem. Int. Ed.* 43: 3400–3420. (b) Mallat, T. and Baiker, A. (2004). *Chem. Rev.* 104: 3037–3058. (c) Punniyamurthy, T., Velusamy, S., and Iqbal, J. (2005). *Chem. Rev.* 105: 2329–2363. (d) Piera, J. and Backvall, J.E. (2008). *Angew. Chem. Int. Ed.* 47: 3506–3523. (e) Shi, Z., Zhang, C., Tang, C., and Jiao, N. (2012). *Chem. Soc. Rev.* 41: 3381–3430.
6 (a) Cheng, X., Hu, X., and Lu, Z. (2017). *Chin. J. Org. Chem.* 31: 251–266. (b) Bian, C., Singh, A.K., Niu, L. et al. (2017). *Asian J. Org. Chem.* 6: 386–396.
7 (a) Shi, L. and Xia, W. (2012). *Chem. Soc. Rev.* 41: 7687–7697. (b) Xie, J., Jin, H., Xu, P., and Zhu, C. (2014). *Tetrahedron Lett.* 55: 36–48. (c) Beatty, J.W. and Stephenson, C.R. (2015). *Acc. Chem. Res.* 48: 1474–1484. (d) Liu, W., Zheng, X., Zeng, J., and Cheng, P. (2017). *Chin. J. Org. Chem.* 37: 1–19.
8 Condie, A.G., Gonzalez-Gomez, J.C., and Stephenson, C.R. (2010). *J. Am. Chem. Soc.* 132: 1464–1465.
9 Rusch, F., Unkel, L.-N., Alpers, D. et al. (2015). *Chem. Eur. J.* 21: 8336–8340.
10 (a) Rueping, M., Zhu, S., and Koenigs, R.M. (2011). *Chem. Commun.* 47: 12709–12711. (b) Hari, D.P. and König, B. (2011). *Org. Lett.* 13: 3852–3855. (c) Pan, Y., Wang, S., Kee, C.W. et al. (2011). *Green Chem.* 13: 3341–3344. (d) Pan, Y., Kee, C.W., Chen, L., and Tan, C.-H. (2011). *Green Chem.* 13: 2682–2685. (e) Liu, Q., Li, Y.-N., Zhang, H.-H. et al. (2012). *Chem. Eur. J.* 18: 620–627. (f) Rueping, M., Koenigs, R.M., Poscharny, K. et al. (2012). *Chem. Eur. J.* 18: 5170–5174. (g) Zhong, J.-J., Meng, Q.-Y., Wang, G.-X. et al. (2013). *Chem. Eur. J.* 19: 6443–6450. (h) Xiao, T., Li, L., Lin, G. et al. (2014). *Org. Lett.* 16: 4232–4235.
11 (a) Rueping, M. and Vila, C. (2013). *Org. Lett.* 15: 2092–2095. (b) Chen, Y. and Feng, G. (2015). *Org. Biomol. Chem.* 13: 4260–4265.

12 Wei, G., Zhang, C., Bureš, F. et al. (2016). *ACS Catal.* 6: 3708–3712.

13 Cai, S., Zhao, X., Wang, X. et al. (2012). *Angew. Chem. Int. Ed.* 51: 8050–8053.

14 (a) Xuan, J., Cheng, Y., An, J. et al. (2011). *Chem. Commun.* 47: 8337–8339. (b) Xuan, J., Feng, Z.-J., Duan, S.-W., and Xiao, W.-J. (2012). *RSC Adv.* 2: 4065–4068.

15 (a) Zou, Y.-Q., Lu, L.-Q., Fu, L. et al. (2011). *Angew. Chem. Int. Ed.* 50: 7171–7175. (b) Rueping, M., Leonori, D., and Poisson, T. (2011). *Chem. Commun.* 47: 9615–9617. (c) Chandrasekhar, D., Borra, S., Kapure, J.S. et al. (2015). *Org. Chem. Front.* 2: 1308–1312. (d) Fujiya, A., Tanaka, M., Yamaguchi, E. et al. (2016). *J. Org. Chem.* 81: 7262–7270.

16 (a) Wang, Z.-Q., Hu, M., Huang, X.-C. et al. (2012). *J. Org. Chem.* 77: 8705–8711. (b) Zhu, S. and Rueping, M. (2012). *Chem. Commun.* 48: 11960–11962. (c) Gao, X.-W., Meng, Q.-Y., Xiang, M. et al. (2013). *Adv. Synth. Catal.* 355: 2158–2164.

17 (a) Deng, Q.-H., Zou, Y.-Q., Lu, L.-Q. et al. (2014). *Chem.-Asian J.* 9: 2432–2435. (b) Yang, X., Li, L., Li, Y., and Zhang, Y. (2016). *J. Org. Chem.* 81: 12433–12442. (c) Dong, W., Hu, B., Gao, X. et al. (2016). *J. Org. Chem.* 81: 8770–8776.

18 (a) Maity, S. and Zheng, N. (2012). *Angew. Chem. Int. Ed.* 51: 9562–9566. (b) Morris, S.A., Nguyen, T.H., and Zheng, N. (2015). *Adv. Synth. Catal.* 357: 2311–2316. (c) Morris, S.A., Wang, J., and Zheng, N. (2016). *Acc. Chem. Res.* 49: 1957–1968.

19 (a) Itoh, A., Okada, A., Yuasa, H. et al. (2015). *Synlett* 26: 1705–1709. (b) Tao, C., Wang, B., Sun, L. et al. (2017). *Org. Biomol. Chem.* 15: 328–332.

20 (a) Hou, H., Zhu, S., Pan, F., and Rueping, M. (2014). *Org. Lett.* 16: 2872–2875. (b) Ding, Y., Zhang, T., Chen, Q.-Y., and Zhu, C. (2016). *Org. Lett.* 18: 4206–4209.

21 Fabry, D.C. and Rueping, M. (2016). *Acc. Chem. Res.* 49: 1969–1979.

22 (a) Fabry, D.C., Zoller, J., Raja, S., and Rueping, M. (2014). *Angew. Chem. Int. Ed.* 53: 10228–10231. (b) Fabry, D.C., Ronge, M.A., Zoller, J., and Rueping, M. (2015). *Angew. Chem. Int. Ed.* 54: 2801–2805.

23 (a) Zoller, J., Fabry, D.C., Ronge, M.A., and Rueping, M. (2014). *Angew. Chem. Int. Ed.* 53: 13264–13268. (b) Choi, S., Chatterjee, T., Choi, W.J. et al. (2015). *ACS Catal.* 5: 4796–4802. (c) Liu, K., Zou, M., and Lei, A. (2016). *J. Org. Chem.* 81: 7088–7092. (d) Liu, W.-Q., Lei, T., Song, Z.-Q. et al. (2017). *Org. Lett.* 19: 3251–3254.

24 (a) Liu, J., Liu, Q., Yi, H. et al. (2014). *Angew. Chem. Int. Ed.* 53: 502–506. (b) Leow, D. (2014). *Org. Lett.* 16: 5812–5815. (c) Yoo, W.J., Tsukamoto, T., and Kobayashi, S. (2015). *Angew. Chem. Int. Ed.* 54: 6587–6590. (d) Yamaguchi, T., Yamaguchi, E., and Itoh, A. (2017). *Org. Lett.* 19: 1282–1285.

25 (a) Romero, N.A., Margrey, K.A., Tay, N.E., and Nicewicz, D.A. (2015). *Science* 349: 1326–1330. (b) McManus, J.B. and Nicewicz, D.A. (2017). *J. Am. Chem. Soc.* 139: 2880–2883.

26 (a) Cheng, Y., Yang, J., Qu, Y., and Li, P. (2012). *Org. Lett.* 14: 98–101. (b) Fan, W., Yang, Q., Xu, F., and Li, P. (2014). *J. Org. Chem.* 79: 10588–10592.

27 (a) Talla, A., Driessen, B., Straathof, N.J.W. et al. (2015). *Adv. Synth. Catal.* 357: 2180–2186. (b) Sun, J.G., Yang, H., Li, P., and Zhang, B. (2016). *Org. Lett.* 18: 5114–5117.

28 (a) Chen, Y.-H., Lee, M., Lin, Y.-Z., and Leow, D. (2015). *Chem.-Asian J.* 10: 1618–1621. (b) Luo, J. and Zhang, J. (2016). *J. Org. Chem.* 81: 9131–9137. (c) Samanta, S., Ravi, C., Joshi, A. et al. (2017). *Tetrahedron Lett.* 58: 721–725.

29 Wei, D., Li, Y., and Liang, F. (2016). *Adv. Synth. Catal.* 358: 3887–3896.

30 (a) Kumaraswamy, G., Vijaykumar, S., Ankamma, K., and Narayanarao, V. (2016). *Org. Biomol. Chem.* 14: 11415–11425. (b) Deng, Y., Wei, X.J., Wang, H. et al. (2017). *Angew. Chem. Int. Ed.* 56: 832–836.

31 Niu, T.-F., Jiang, D.-Y., Li, S.-Y. et al. (2016). *Chem. Commun.* 52: 13105–13108.

32 Chen, L., Li, H., Li, P., and Wang, L. (2016). *Org. Lett.* 18: 3646–3649.

33 (a) Zhang, P., Xiao, T., Xiong, S. et al. (2014). *Org. Lett.* 16: 3264–3267. (b) Gao, F., Wang, J.-T., Liu, L.-L. et al. (2017). *Chem. Commun.* 53: 8533–8536. (c) Xia, X.-F., Zhang, G.W., Wang, D., and Zhu, S.-L. (2017). *J. Org. Chem.* 82: 8455–8463.

34 (a) Yang, S., Li, P., Wang, Z., and Wang, L. (2017). *Org. Lett.* 19: 3386–3389. (b) Ji, W., Li, P., Yang, S., and Wang, L. (2017). *Chem. Commun.* 53: 8482–8485.

35 Natarajan, P., Manjeet, N., Kumar, S., and Devi, K.M. (2017). *Tetrahedron Lett.* 58: 658–662.

36 (a) Chen, J.-R., Hu, X.-Q., Lu, L.-Q., and Xiao, W.-J. (2016). *Chem. Soc. Rev.* 45: 2044–2056. (b) Hu, X.-Q., Chen, J., Chen, J.-R. et al. (2016). *Chem. Eur. J.* 22: 14141–14146. (c) Song, H.-T., Ding, W., Zhou, Q.-Q. et al. (2016). *J. Org. Chem.* 81: 7250–7255.

37 (a) Zou, Y.-Q., Chen, J.-R., Liu, X.-P. et al. (2012). *Angew. Chem. Int. Ed.* 51: 784–788. (b) Xie, H.-Y., Han, L.-S., Huang, S. et al. (2017). *J. Org. Chem.* 82: 5236–5241. (c) Jiang, M., Yang, H., and Fu, H. (2016). *Org. Lett.* 18: 5248–5251.

38 Xia, X.-D., Ren, Y.-L., Chen, J.-R. et al. (2015). *Chem.-Asian J.* 10: 124–128.

39 (a) Wang, D., Li, J., Cai, S. et al. (2014). *Synlett* 25: 1626–1628. (b) Wiegand, C., Herdtweck, E., and Bach, T. (2012). *Chem. Commun.* 48: 10195–10197.

40 (a) Córdova, A., Sundén, H., Engqvist, M. et al. (2004). *J. Am. Chem. Soc.* 126: 8914–8915. (b) Sundén, H., Engqvist, M., Casas, J. et al. (2004). *Angew. Chem. Int. Ed.* 43: 6532–6535. (c) Lian, M., Li, Z., Cai, Y. et al. (2012). *Chem.-Asian J.* 7: 2019–2023.

41 Ding, W., Lu, L.-Q., Zhou, Q.-Q. et al. (2017). *J. Am. Chem. Soc.* 139: 63–66.

15

The Recent Developments of Electrochemical Oxidation

Jun-ichi Yoshida, Ryutaro Hayashi, and Akihiro Shimizu

Department of Synthetic Chemistry and Biological Chemistry, Graduate School of Engineering, Kyoto University, Kyoto, Japan

15.1 Introduction

Electrochemical oxidation serves as a straightforward and powerful method for oxidizing organic molecules. In principle, the electrochemical processes proceed by direct manipulation of electrons on the surface of the electrode, and consequently there is no need for using stoichiometric amounts of chemical oxidants, which often causes a problem of hazardous wastes. Therefore, electrochemical oxidation serves as an environmentally benign method for oxidation. Notably, it also serves as a tunable method for oxidizing organic compounds to generate a wide range of reactive intermediates that are difficult to generate by conventional chemical methods.

The electrochemical oxidation of organic compounds [1] has a long history and various applications to the synthesis of organic compounds [2, 3] have been developed both in academia and industry [4]. For example, the Kolbe coupling reaction [5], which was reported in 1848, serves as a powerful method for generating alkyl radicals that effectively undergo coupling reactions to give dimers. This reaction has been used for many years and still serves as a useful tool for carbon–carbon bond formation via free radical intermediates. Electrochemical oxidation is also effective for generation of carbocations that react with nucleophiles to achieve functionalization and carbon–carbon bond formation.

However, the power of the electrochemical method has still been underestimated by synthetic organic chemists presumably because there are so many variables accompanying the method, such as electrode materials, supporting electrolytes, cell design, cell voltage, current density, and so on. For chemists who are not familiar with the electrochemical method, the lack of knowledge and experiences on such variables seems to be the obstacle to use the method in laboratory and industrial synthesis. Recently, advantages of the electrochemical methods have been, however, reevaluated particularly from a view point of generation of reactive species that are difficult to achieve by chemical methods under mild conditions. This chapter aims to illustrate the utility of the electrooxidative method in organic synthesis, but due to the limitation of space, we mainly focus on the development of new techniques of the electrochemical method.

Green Oxidation in Organic Synthesis, First Edition. Edited by Ning Jiao and Shannon S. Stahl.

Particularly, we focus on the electrochemical C—H functionalization reactions including the formation of carbon-heteroatom bonds and carbon–carbon bonds because the field of oxidative C—H functionalization has recently witnessed remarkable progress and has received significant research interest from a wide range of synthetic organic chemists. Again, due to the limitation of space we highlighted selected examples that have been reported recently to give an indication of the potential of this exciting ongoing field.

15.2 Modern Techniques of Electrochemical Oxidation

There are many factors governing the processes that occur on and near the surface of electrodes, and extensive studies have been made to provide the information for understanding, designing, and developing electrochemical redox reactions. In this section, let us briefly touch on some modern tactics and strategies of controlling electrochemical redox reactions [6], although various conventional methods to control the electrochemical processes are also useful.

15.2.1 Electroauxiliary

Activating groups and directing groups have often been used for controlling the reactivity of substrate molecules and reaction pathways in organic synthesis [7]. A similar method has also been developed in organic electrochemistry. A functional group that activates substrate molecules toward the electron transfer and controls the reaction pathway is called an electroauxiliary. Electroauxiliaries enable highly selective electron transfer driven reactions that are difficult to achieve or impossible by conventional electrolysis of naked substrates. For example, triorganosilyl groups serve as electroaxuiliaries for the oxidation of heteroatom compounds. The electrochemical oxidation of organic compounds usually proceeds by the initial electron transfer from the HOMO (highest occupied molecular orbital) of a substrate molecule to the anode. The major part of the HOMO of heteroatom compounds, such as amines and their derivatives, is usually the nonbonding *p* orbital of the heteroatom (lone pair) and oxidative electron transfer. The introduction of a silyl group on the carbon adjacent to a heteroatom decreases the HOMO level by virtue of orbital interaction between C—Si σ orbital and the nonbonding *p* orbital of the heteroatom [8, 9]. Therefore, the silyl group significantly decreases the oxidation potential and makes the oxidation much easier. Also, the C—Si bond in the resulting radical cation is selectively cleaved without affecting C—H bonds making the subsequent process highly selective.

An example shown in Scheme 15.1 demonstrates the power of a silyl group as an electroauxiliary in the anodic oxidation of carbamates [10]. The oxidation potential of the silyl-substituted carbamate (b) is much less positive than that without the silyl-substituent (a). Another point is regioselectivity. Although two regioisomers are produced by anodic oxidation of a substrate without a silyl group (a), the anodic oxidation of a silyl-substituted substrate gives rises to regioselective introduction of a methoxy group onto the carbon to which the silyl group has been attached (b). Arylthio groups also serve as electroauxiliaries for oxidation of heteroatom compounds, although the principle is different (c) [11, 12]. The initial electron transfer occurs mainly from the *p* orbital of the sulfur. In the resulting radical cation, the C—S bond is cleaved selectively.

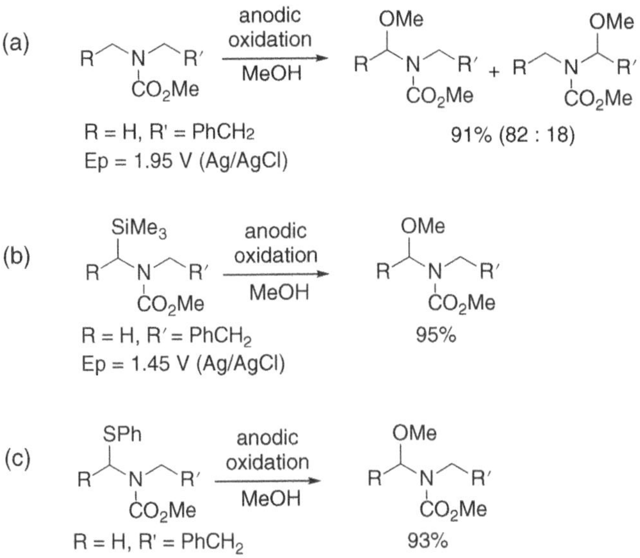

Scheme 15.1 Using silyl and thio groups as an electroauxiliary in the anodic oxidation.

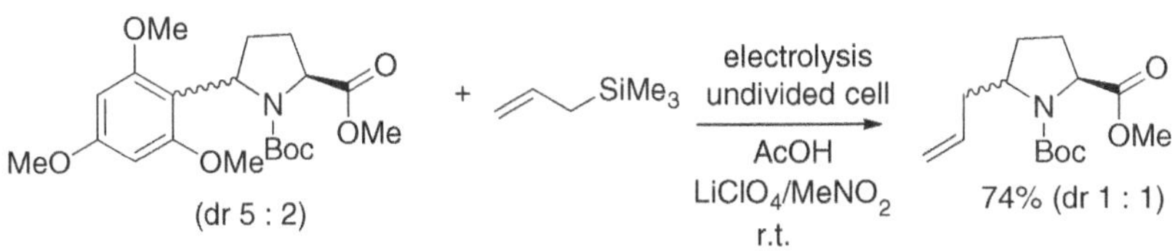

Scheme 15.2 Functionalization of proline derivatives.

The 2,4,6-trimethoxyphenyl group can also be used as an electroauxiliary as reported by Chiba and co-workers. This electroauxiliary has been effectively used for functionalization of proline derivatives as shown in Scheme 15.2 [13].

Silyl groups are also effective as electroauxiliaries for the oxidation of π-electron systems such as aromatic compounds and olefins. For example, the introduction of a silyl at the benzylic position decreases the oxidation potential of aromatic compounds and a nucleophile is introduced at the benzylic position via selective cleavage of the C—Si bond.

15.2.2 Low Temperature Electrolysis

Lowering reaction temperature is one of the simplest methods to increase the selectivity [14]. Generally, the rate of the reaction decreases and the lifetime of reactive species increases with a decrease in the reaction temperature. However, until a few decades ago, it was believed to be difficult to carry out electrochemical reactions at low temperatures such as −78 °C, although such low temperature reactions were popular in the field of organometallic chemistry. This is presumably because the conductivity of the solution decreases with a decrease in the temperature. However, the recent technical developments enable the electrolysis at very low temperatures, such as −78 °C. The choice of supporting electrolyte and solvent is crucial to performing low temperature electrolysis.

It was also believed that organic electrolysis cannot be carried out without moisture because organic solvents suitable for electrolysis are usually hygroscopic. However, modern techniques of organic chemistry including vacuum line techniques and syringe techniques enable the electrochemical reactions without the influence of moisture. An electrochemical cell is dried under vacuum by using a vacuum line and after cooling a substrate solution is introduced to an electrochemical cell by syringe techniques. The electrolysis is carried out under an atmosphere of dried inert gas. Thus, we can carry out electrochemical reactions at low temperatures without moisture. This technique is particularly beneficial in the case of generating and accumulating highly reactive intermediates, such as organic cations that are thermally unstable and sensitive to moisture [15]. A picture of typical divided cell that enables electrolysis at low temperatures under inert atmosphere is shown in Figure 15.1.

Based on the low temperature electrolysis technique, Yoshida et al. developed the "cation pool" method [16] and the indirect "cation pool" method [17, 18]. For example, the anodic oxidation of *N*-(methoxycarbonyl)pyrrolidine in Bu_4NClO_4/CH_2Cl_2 at −72 °C gives a solution of the corresponding *N*-acyliminium ion, which can be characterized by NMR and IR. A nucleophile such as allyltrimethylsilane is then added to obtain a desired product.

They also developed the indirect "cation pool" method. In this method an active reagent is electrochemically generated and accumulated in solution. In the next step, the reagent is reacted with a precursor of an organic cation. In the third step, a solution of the resulting organic cation is reacted with a nucleophile to obtain the desired product.

Chiba et al. developed a different method for generating and accumulating reactive organic cations in solution by electrolysis using the $LiClO_4/CH_3NO_2$ or $LiClO_4/C_2H_5NO_2$ system at higher temperatures such as 0 °C [19]. Recent successful applications to the synthesis of azanucloeoside derivatives are noteworthy [20, 21].

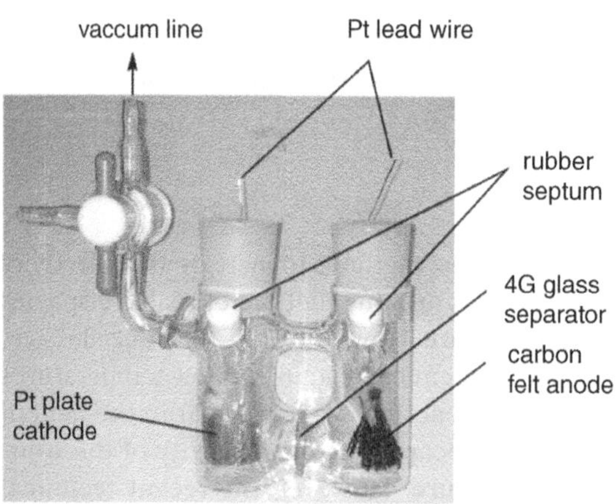

Figure 15.1 A typical divided cell for low temperature electrolysis without moisture.

15.2.3 Electrochemical Flow Microreactors

Flow microreactors have emerged as an enabling technology for organic synthesis [22–24]. The characteristic features of flow microreactors include the precise reaction time control based on flow as well as extremely fast mixing and heat transfer based on micro space. Electrochemical flow microreactors solve inherent problems in conducting organic electrochemical reactions on a preparative scale [25–28]. For example, high surface-to-volume ratio of micro space is advantageous for increasing the efficiency of the processes that occur on the surface of the electrodes. A short distance between the electrodes in a micro space can solve the problem of low conductivity of organic solvents. Also, the flow nature is quite effective for controlling extremely fast reactions of highly reactive species. Such species can be generated and be transferred immediately to the next reactor where the next reagent or catalyst is added to produce desired products before the reactive species decompose (Figure 15.2).

15.2.4 Reaction Media

Sometimes the choice of a suitable reaction media is the key to the success of desired reactions. This is also true for the electrochemical method, because solvation and behavior of ions often play a major role in electron transfer processes in solution. For example, Waldvogel and co-workers reported that a mixture of $(CF_3)_2CHOH$ and CH_3OH is quite effective for cross-coupling reactions of aniline derivatives [29]. Presumably, the hydrogen bond network is responsible for high selectivity. Also, ionic liquids [30, 31] that have been used as new reaction media in a variety of organic reactions can be used as solvents for electrolysis as well. Ionic liquids often have wider potential windows than common organic solvents.

Chiba et al. developed a unique [2 + 2] cycloaddition reactions [32] and Diels–Alder reactions [33] mediated by radical cations using a $LiClO_4/CH_3NO_2$ system. The entropic effect in such media is essential for achieving such reactions [34].

15.2.5 Supporting Electrolytes

The choice of supporting electrolyte is also often crucial. Electrochemical oxidation of organic compounds usually proceeds by the initial formation of radical cations.

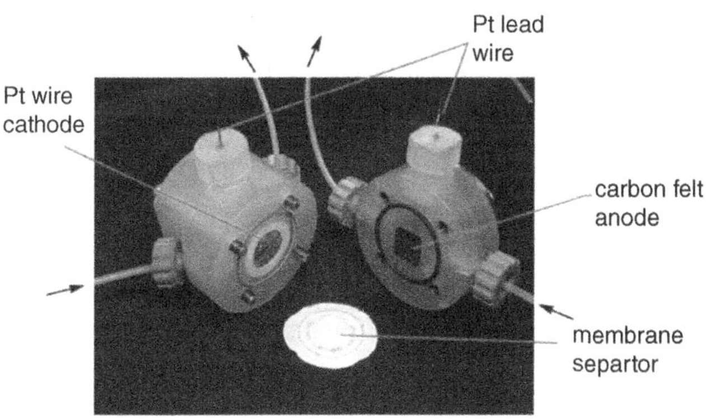

Figure 15.2 An electrochemical flow microreactor.

The counter anion should be the anionic part of the supporting electrolyte, and therefore, the nature of the counter anion should affect the reactivity of the radical cation. Similarly, the counter anions affect the stability and reactivity of cationic species generated in the anodic oxidation. Recently, solid-supported electrolytes [35–37] have been developed for the easy separation from products after electrolysis. This approach is beneficial from a view point of green chemistry because solid-supported electrolytes can be, in principle, recovered and reused. Moreover, the use of flow micro electrochemical cells enables electrolysis without intentionally added electrolytes [38]. In this case charged species generated by the electrolysis such as H^+ serve as charge carriers.

15.2.6 Electrode Materials

The choice of appropriate electrode materials is often crucial for the success of desired electrochemical reactions. Sometimes, it is not too much to say that the surface of the electrode is a catalyst for the electrochemical reactions. Recently, relatively new electrode materials such as reticulated vitreous carbon (RVC) [39] and boron doped diamond (BDD) [40] have often been used for organic electrosynthesis. Although RVC has been used largely in small scale electroanalytical studies, chemists have realized its advantageous properties for electrosynthesis such as high surface-to-volume ratio and minimal reactivity over a wide range of reaction conditions. Consequently, RVC has been used for electrosynthesis both in research laboratories and in industry. BBD has the widest solvent window among the common electrode materials. From the view point of organic chemistry, BBD has a special reactivity toward organic compounds. In many cases oxy radicals such as hydroxyl radicals generated by water oxidation play a role as mediator in the oxidation of organic compounds [41].

15.2.7 Mediators

Indirect electrochemical reactions using mediators play an important role in electrosynthesis [42]. In contrast to direct electrolysis where the electron transfer occurs directly between the electrode surface and substrate molecules, in indirect electrochemical oxidation an oxidizing agent (mediator) is produced by the oxidation of its precursor on the electrode surface and the oxidized form of the mediator reacts with substrate molecules in a bulk solution. In the in-cell method the oxidized form of the mediator is continuously regenerated by the electrolysis, whereas in the ex-cell method the electrochemical generation and the chemical reaction with the oxidized form of the mediator are carried out separately. Recently, a new type of redox mediator based on the iodine(I)/iodine(III) redox couple has been developed (Scheme 15.3) [43]. Notably, by tethering the redox-active iodophenyl moiety to an alkylammonium group, the supporting electrolyte is

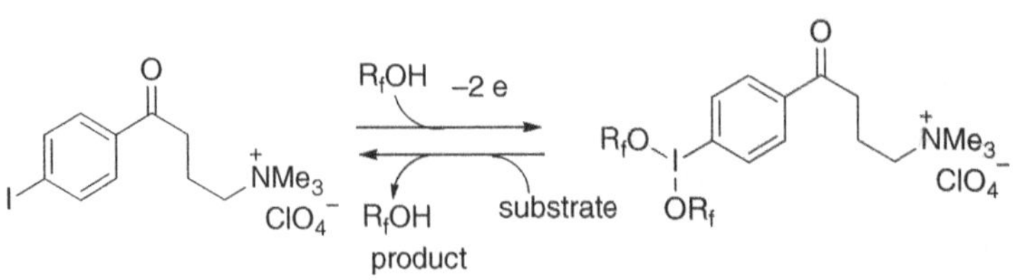

Scheme 15.3 Redox mediator based on the iodine(I)/iodine(III) redox.

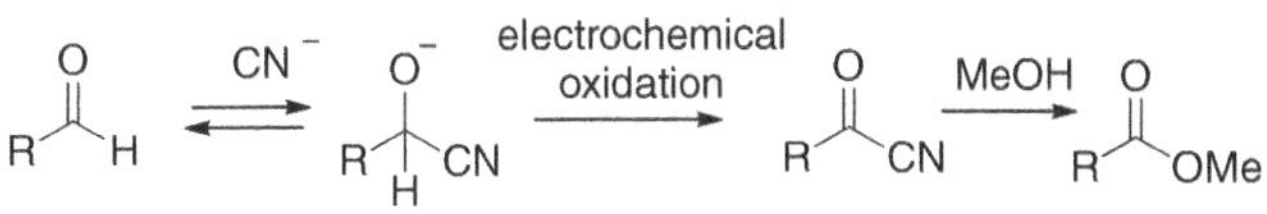

Scheme 15.4 Electrochemical oxidation of aldehydes.

merged with the mediator. This new approach enables easy recovery and reuse of both mediator and supporting electrolyte components in electrosynthesis.

15.2.8 Organocatalysis

Recently, the use of organocatalysts in electrochemical reactions have received significant research interest [44]. Usually, organocatalysts are not directly participate in the electrochemical reactions, but an intermediate generated by the reaction of a substrate and an organocatalyst participates in the electrochemical reaction. A very early example of an organocatalyzed electrochemical oxidation is the oxidation of aldehydes to esters mediated by cyanide developed by Chiba and co-workers (Scheme 15.4) [45].

15.3 Oxidative Functionalization of C—H Bonds

15.3.1 Functionalization of Aromatic C—H Bonds

Electrochemical C—H functionalization of aromatic compounds has been known about for many years. Usually electronegative elements or electron-deficient functional groups are introduced to aromatic rings.

In contrast, the introduction of electron-rich functional groups is difficult because of the following two major reasons. One is that sources of such functional groups are usually more easily oxidized than aromatic substrates. For example, the use NH_3 as a nucleophile to synthesize aniline derivatives usually leads to oxidation of NH_3. The other one is overoxidation. Aromatic compounds bearing electron-rich functional group is more easily oxidized than parent aromatic substrates. For example, aniline derivatives are more easily oxidized than their parent aromatic compounds. Therefore, direct conversion of benzene to aniline is a kind of a "dream reaction." Various methods have been developed for amination of C—H bonds of aromatic compounds [46]. Although the photoredox method has made significant contribution to this fields recently [47], the contribution of the electrochemical method is also significant.

Morofuji, Shimizu, and Yoshida developed electrochemical C—H amination using pyridine as a nitrogen source (Scheme 15.5) [48]. The anodic oxidation of electron-rich aromatic compounds in the presence of pyridine gives *N*-arylpyridinium ions (Zincke salts).

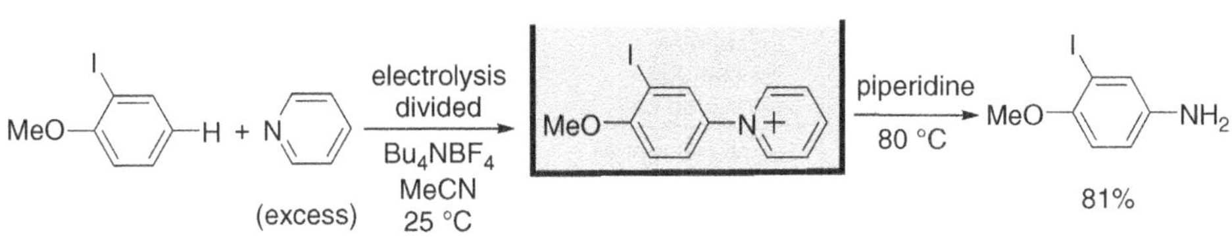

Scheme 15.5 Electrochemical C—H amination using pyridine as a nitrogen source.

A strong electron-withdrawing effect caused by a positive charge avoids overoxidation. In the next step, the treatment of *N*-arylpyridinium ions with piperidine gives desired aromatic compounds having an NH_2 group. This method was successfully applied to synthesis of 1,4-benzoxazin-3-ones [49].

An intramolecular version of the electrochemical C—H amination leads to the effective ortho-functionalization [50]. For example, the reaction of 2-pyrimidyloxybenzene gives 2-aminobenzoxazoles.

Based on a similar concept, imidazole moiety can be introduced on aromatic rings (Scheme 15.6) [51]. *N*-Mesylimidazole serves as the best nitrogen nucleophile, and the mesyl protecting group on the electrochemically generated *N*-arylimidazolium ions is removed after electrolysis to give desired *N*-arylimidazoles.

The introduction of functional amino groups such as 2- and 3-amino- and 2- and 3-oxy-alkyl groups can be introduced based on the heterocyclization approach prior to the electrolysis (Scheme 15.7) [52]. The heterocycles formed by heterocyclization are used as nitrogen nucleophiles in the electrochemical C—H amination. After the electrolysis the heterocycles are opened to give functionalized alkylamino-substituted aromatic compounds.

Zeng and Little developed the electrochemical oxidative amination of benzoxazoles using Bu_4NI as a redox catalyst (Scheme 15.8) [53]. They suggested that the electrogenerated I^+ plays a key role.

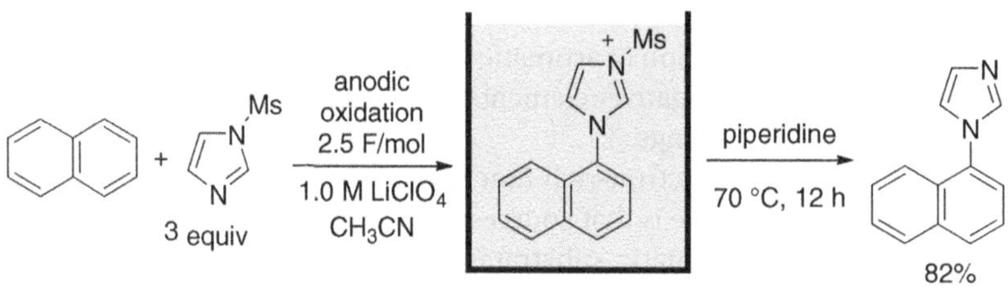

Scheme 15.6 Electrochemical C—H amination using imidazole.

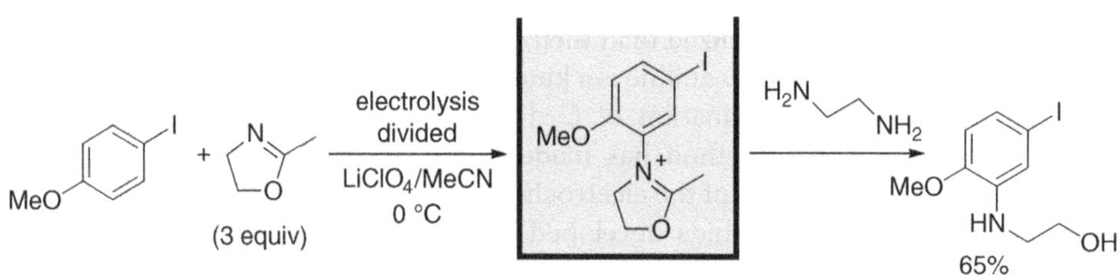

Scheme 15.7 Electrochemical C—H amination using 4,5-dihydrooxazole.

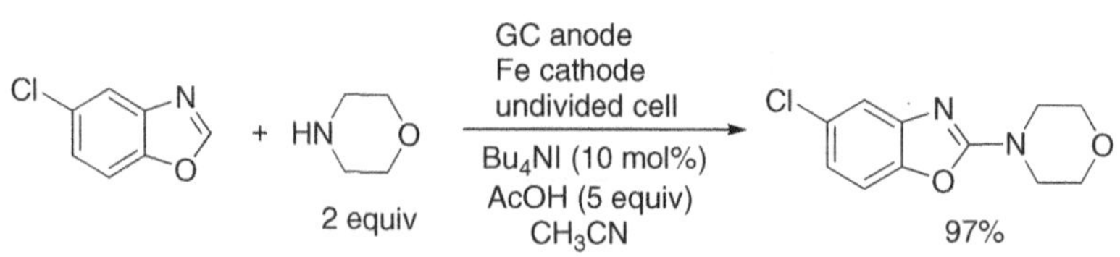

Scheme 15.8 Electrochemical oxidative amination of benzoxazoles.

Xu et al. reported the electrochemical intramolecular C—H amination (Scheme 15.9) [54]. A mechanism involving the formation of the amidinyl radical followed by radical cyclization has been suggested.

Oxygen functionalities can be introduced to aromatic rings by using the electrochemical method. For example, Mei, Zhang, and their co-workers reported Pd-catalyzed electrochemical acetoxylation of aromatic C—H bonds using acetate anion as a nucleophile (Scheme 15.10) [55]. A mechanism involving C—H bond insertion of Pd coordinated by the nitrogen atom followed by electrochemical oxidation of Pd(II) to Pd(IV) and the subsequent reductive elimination was suggested.

An intramolecular version of the electrochemical introduction of an oxygen functionality to aromatic C—H bonds was also achieved by Waldvogel et al. Benzoxazoles can be synthesized from readily available anilides (Scheme 15.11) [56]. An amidyl radical intermediate that resonates to aromatic carbon radical is postulated.

Aromatic thiols can also be used as nucleophiles for electrooxidative functionalization of aromatic C—H bonds. Lei and co-workers reported the electrochemical coupling of electron-rich aromatic and heteroaromatic compounds such as *N*-methylindole with thiophenols (Scheme 15.12) [57]. They postulated a mechanism involving coupling of the radical cation of an aromatic compound and an arylthio radical followed by proton elimination.

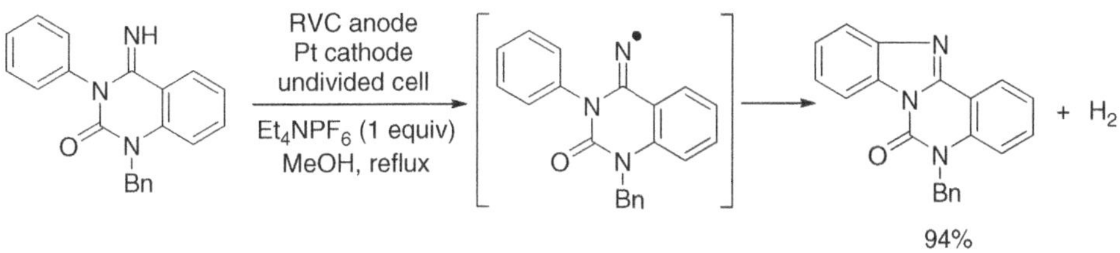

Scheme 15.9 Electrochemical intramolecular C—H amination.

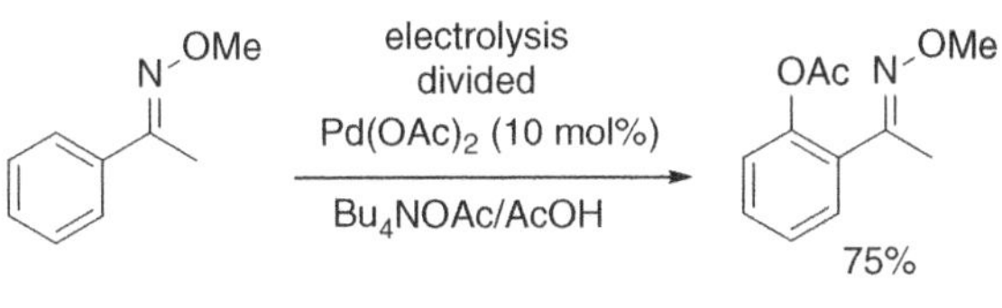

Scheme 15.10 Pd-catalyzed electrochemical acetoxylation of aromatic C—H bonds.

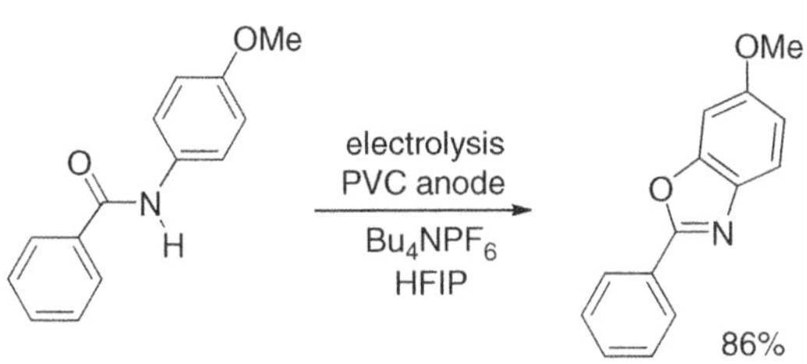

Scheme 15.11 Electrochemical C—O cyclization.

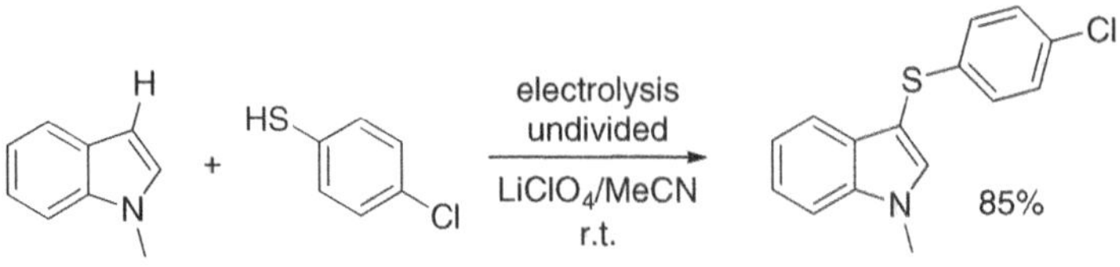

Scheme 15.12 Electrochemical C—S bond formation.

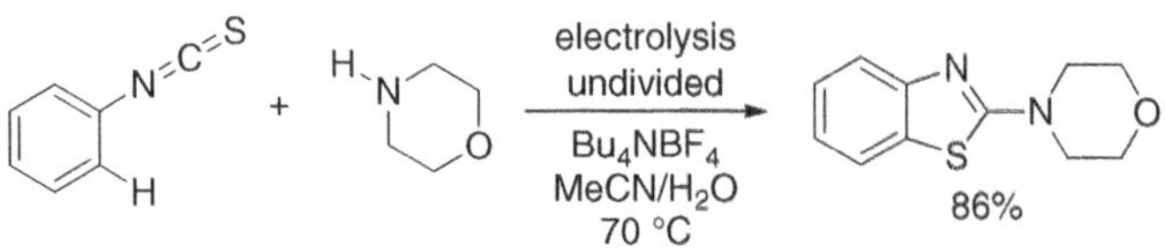

Scheme 15.13 Electrochemical reactions of phenyl isothiocyanates and amines.

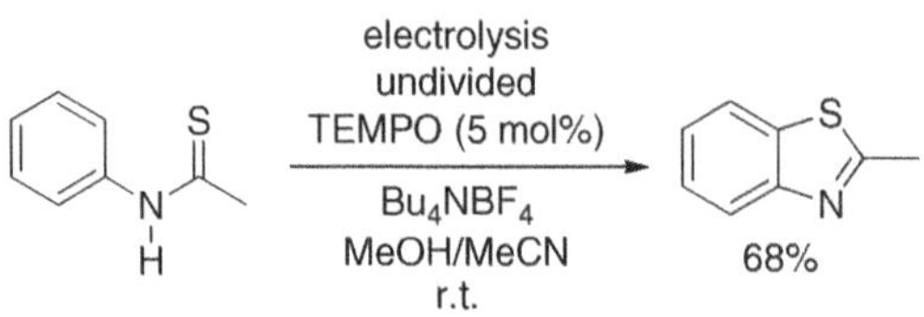

Scheme 15.14 Electrochemical C—S cyclization.

Lei et al. also reported the electrochemical reactions of phenyl isothiocyanates and amines such as morpholine to give 2-aminobenzothiazoles (Scheme 15.13) [58]. The reaction seems to be initiated by nucleophilic attack of an amine to the carbon of the isothicyanate moiety. The resulting S anion is oxidized on the surface of the electrode to give the S radical, which intramolecularly attacks the aromatic ring. Subsequent oxidation and proton elimination afford 2-aminobenzothiazoles.

Xu and co-workers developed TEMPO-catalyzed intramolecular thiolation of aromatic C—H bonds (Scheme 15.14) [59]. The electrooxidatively generated TEMPO cation attacks the S and the resulting S—O bond is homolytically cleaved to generate the S radical that reacts with the aromatic ring. This reaction serves as a useful method for synthesizing thiazoles from amino-substituted aromatic compounds.

Oxidative C—H/C—H cross-coupling serves as the most atom economical way of connecting two aromatic rings. Various methods using chemical oxidants [60], transition metal catalysts [61], photoredox systems [62], and mechanochemistry [63] have been developed, and electrochemical oxidation also serves as a powerful method. Although the early work based on direct electrochemical oxidation of a mixture of two different aromatic compounds suffers from low yields of the cross-coupling products presumably because of unselective oxidation of aromatics and overoxidation [64], the radical cation pool method enables the selective cross-coupling (Scheme 15.15) [65]. In this method, an aromatic substrate is anodically oxidized in a divided cell at low temperatures such as −78 °C to accumulate the radical cation. In the case of naphthalene, its radical cation can be accumulated in solution as a dimer. In the next step, the reaction partner is added after DME (dimethoxyethane) was added to solubilize the radical cation.

Because the coupled radical cation should be further oxidized by another radical cation, two equivalents of the radical cation is needed to obtain the desired cross-coupling product in high yields. In the following case, the cross-coupled product was obtained in a 91% yield based on the amount of pentamethylbenzene by using 4.4 equivalent of naphthalene.

Efficient oxidative coupling of arenes via electrochemical regeneration of 2,3-dichloro-5,6-dicyano-1,4-benzoquinone (DDQ) under mild reaction conditions has been developed by Hilt and co-workers (Scheme 15.16) [66]. *Ortho*-terphenylenes can be effectively coupled as shown in the following example. The application of this reaction to hexaphenylbenzene gave a coronene product in a 95% yield.

Waldvogel et al. developed an efficient method for cross-coupling of phenols and aromatic compounds using a BDD electrode (Scheme 15.17) [67]. A mechanism involving abstraction of phenolic hydrogen by an electrogenerated alkoxy radical has been proposed. The resulting phenoxy radical couples with an aromatic coupling partner. The coupled radical undergoes hydrogen abstraction reaction with an alkoxy radical

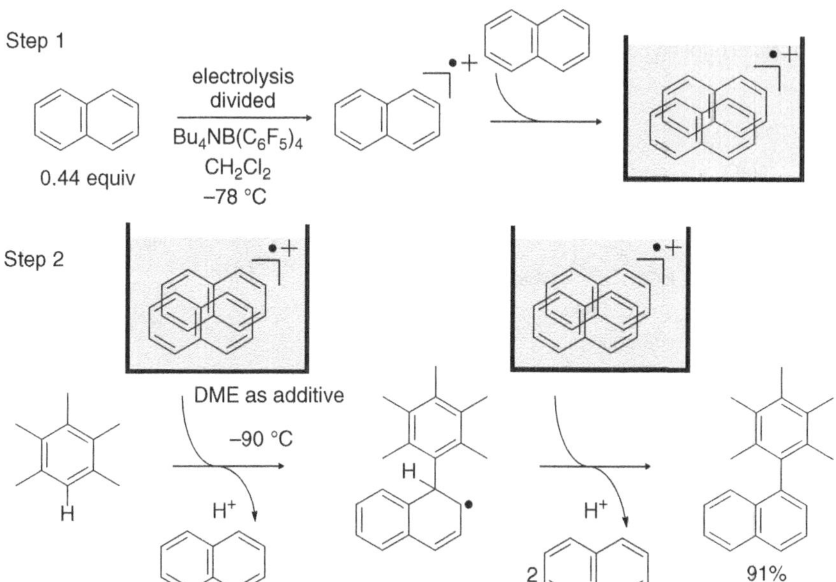

Scheme 15.15 Electrochemical oxidative C—H/C—H cross couplings.

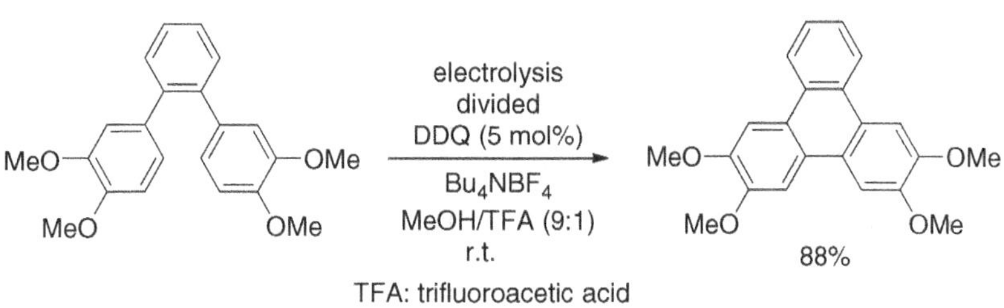

Scheme 15.16 Electrochemical triphenylene core construction.

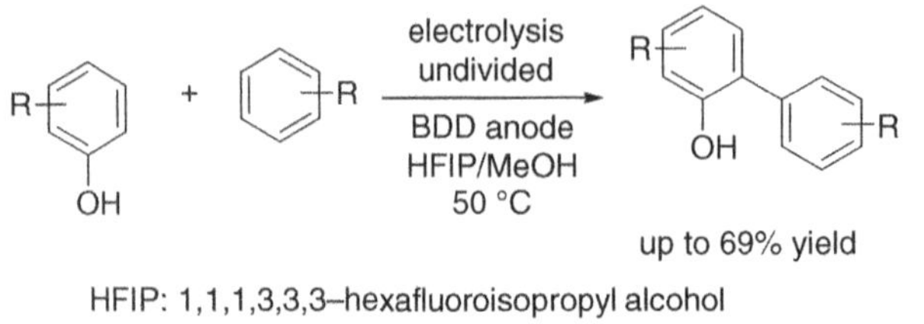

Scheme 15.17 Cross-coupling of phenols and aromatic compounds.

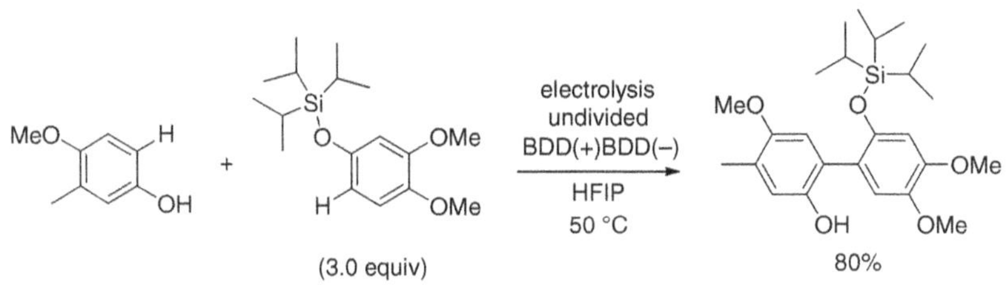

Scheme 15.18 Cross-coupling between protected phenols and non-protected phenols.

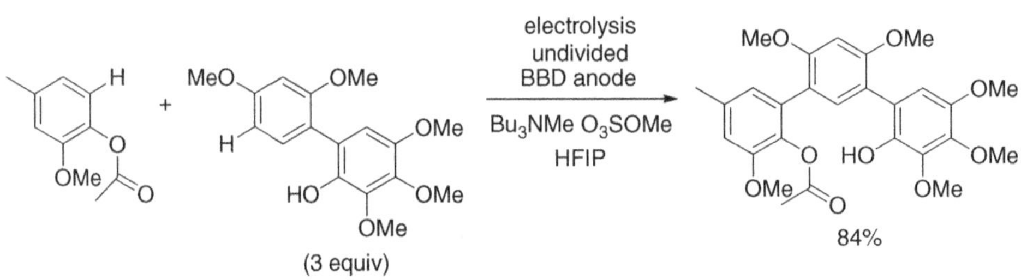

Scheme 15.19 Synthesis of non-symmetric *meta*-terphenyl-2,2″-diols.

to give the final product. Various biphenols [68] and terphenyl-2,2″-diols were successfully synthesized.

Selective coupling of protected phenols and non-protected phenols was achieved by Waldvogel et al. (Scheme 15.18) [69]. The reaction serves as a useful method for synthesizing partially protected non-symmetric 2,2′-biphenols. The use of a bulky silyl group on one of the hydroxyl moiety seems to retard the oxidation of the product.

Synthesis of non-symmetric *meta*-terphenyl-2,2″-diols was achieved by anodic C—C cross-coupling reactions of phenol derivative and phenol-arene derivatives, which were prepared by anodic cross-coupling (Scheme 15.19) [70].

The anodic cross-coupling of phenols has been successfully extended to effective cross-coupling of two aniline derivatives (Scheme 15.20) [29].

Kakiuchi et al. developed Pd-catalyzed cross-coupling reactions of two arenes using anodic oxidation in the presence of I_2 [71]. The bond formation takes place at the less-sterically hindered *ortho* position as shown in Scheme 15.21.

Anodic Pd/benzoquinone-catalyzed aromatic C—H/vinylic C—H cross-coupling was developed by Amatore and Jutand (Scheme 15.22) [72]. Benzoquinone was effectively recycled by the electrochemical process. A mechanism involving C—H insertion of

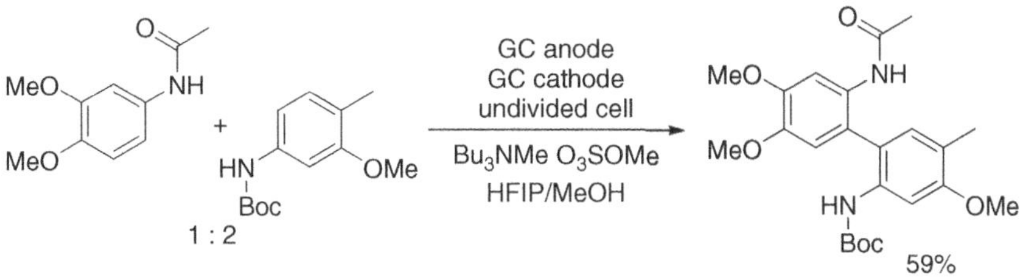

Scheme 15.20 Cross-coupling of two aniline derivatives.

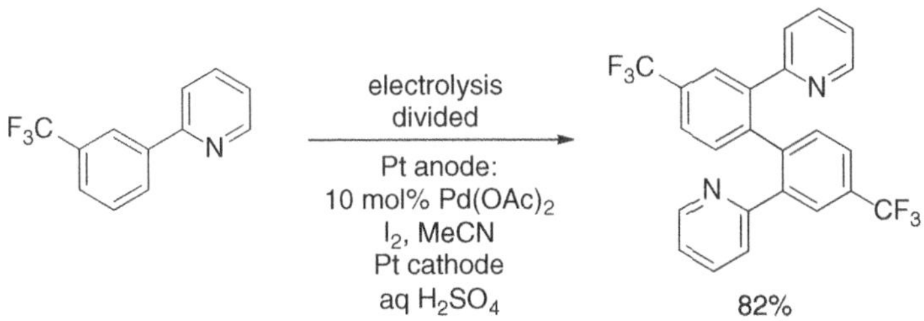

Scheme 15.21 Cross-coupling of two arenes.

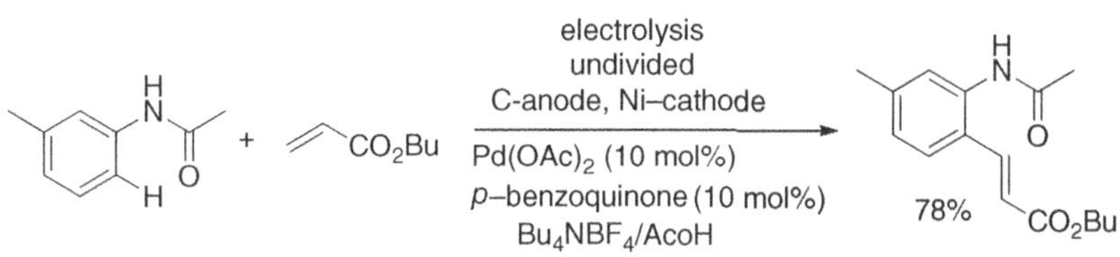

Scheme 15.22 Electrochemical Heck reactions.

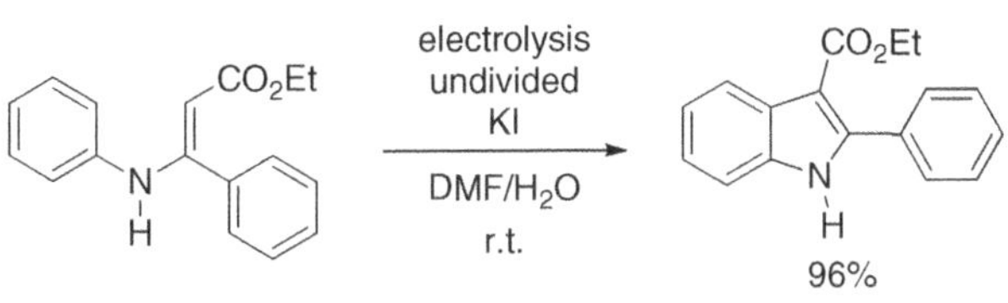

Scheme 15.23 Electrochemical intramolecular cyclization of enamines to indoles.

Pd(II)$(OAc)_2$ followed by carbopalladation of an alkene was proposed. Heck-type elimination gives the desired cross-coupled products and HPd(II)OAc. Reductive elimination of HOAc gives Pd(0), which is oxidized to regenerate Pd(II).

Lei and co-workers developed intramolecular aromatic C—H/vinylic C—H coupling mediated by KI using electrochemistry (Scheme 15.23) [73]. A mechanism involving electrogenerated I^+ that reacts with the enamine moiety of the substrate was suggested. The resulting N—I bond collapses to give an N radical. The carbon radical, which is the resonance form the N radical, reacts with the phenyl group on the nitrogen atom to give the five-membered ring. This method offers a simple and efficient method for the synthesis of indoles.

Jorgensen and co-workers reported organocatalyzed anodic coupling of *N*-tosyl-4-hydroxylaniline derivatives and aliphatic aldehydes (Scheme 15.24) [74]. A mechanism involving the formation of enamine derivatives derived from aldehydes and an organocatalyst, which nucleophilically attacks the electrooxidatively generated *N*-tosyl-4-iminocyclohexa-2,5-dien-1-ones followed by hydrolysis and cyclization to form hemiacetals, was proposed. The products were obtained in high enantioselectivities.

15.3.2 Functionalization of Alkenyl C—H Bonds

Electrochemical oxidation of chalcone oximes to give 3,5-disubstituted isoxazoles was developed by Zeng and co-workers (Scheme 15.25) [75]. A mechanism involving the electrooxidatively generated radical of the oxime was proposed.

15.3.3 Functionalization of Allylic and Benzylic C—H Bonds

Electrochemical allylic oxidation was developed by Shono and co-workers in the 1970s [76]. Recently, Baran and co-workers developed a new method of electrochemical allylic C—H oxidation using tetrachloro-*N*-hydroxyphthalimide (Cl_4NHPI) as a mediator (Scheme 15.26) [77]. Because *t*-BuOOH was used as a nucleophile, α,β-unsaturated ketones were obtained as final products. The method could be applied to regio- and

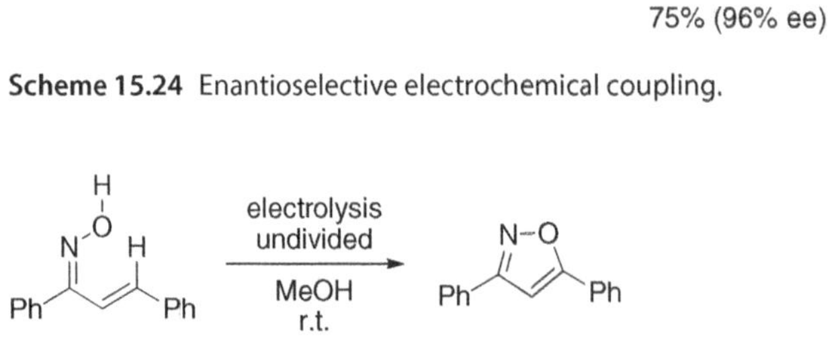

Scheme 15.24 Enantioselective electrochemical coupling.

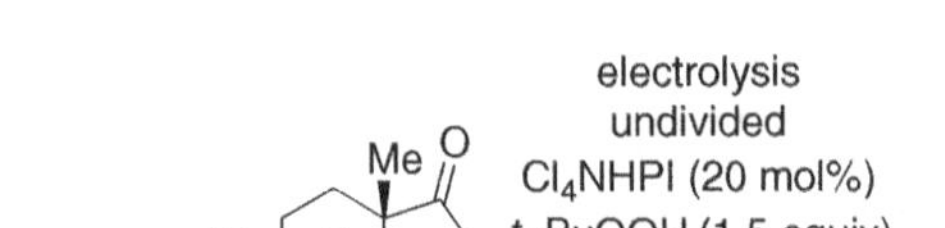

Scheme 15.25 Electrochemical oxidation of chalcone oximes.

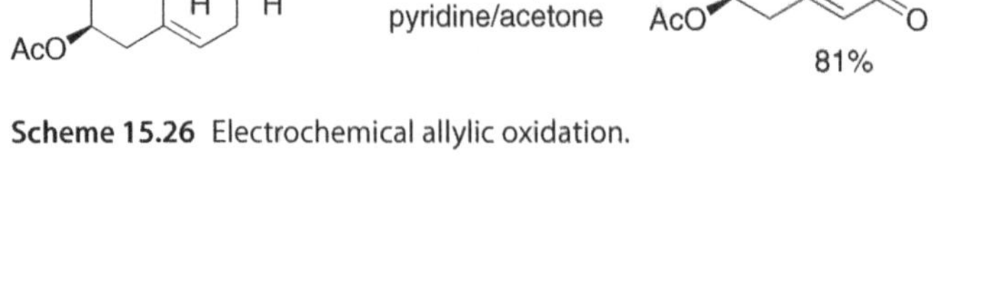

Scheme 15.26 Electrochemical allylic oxidation.

chemoselective allylic C—H oxidation of various complex organic molecules including terpenoids.

Anodic oxidation of benzylic C—H has been known for many years as anodic oxidation of side-chains of aromatic compounds [78]. Recently, Yoshida and co-workers reported anodic benzylic oxidation using dimethylsulfoxide (DMSO) as a nucleophile (Scheme 15.27) [79]. For example, anodic oxidation of diarylmethanes at −78 °C gives diarylcarbenium ion pools, which can be characterized by NMR. The addition of DMSO to the pool gives a pool of diarylmethoxysulfonium ions, which can be characterized by NMR. Treatment of triethylamine gives the corresponding diaryl ketones via Swern–Moffatt type oxidation.

Simple benzylic cations cannot be accumulated as cation pools because of their instability. However, the electrolysis in the presence of DMSO directly gives the benzyloxysulfonium ions, which can be converted to benzaldehyde by the treatment with triethylamine. Treatment of electrogenerated benzyloxysulfonium ions with aq. NaOH or MeOH/H_2O at 0 °C gives the corresponding benzyl alcohols [80].

Amination of benzylic C—H bonds can also be accomplished electrochemically (Scheme 15.28) [81]. Anodic oxidation of toluene derivatives in the presence of *N*-tosyldiphenylsulfilimine gives the corresponding benzylaminosulfonium ions. The treatment with tetrabutylammonium iodide under non-electrolytic conditions gives the corresponding *N*-tosylbenzylamines.

Imidazole moiety can be introduced by the anodic benzylic C—H amination [51]. The anodic oxidation of benzylic substrates in the presence of *N*-mesylimidazole followed

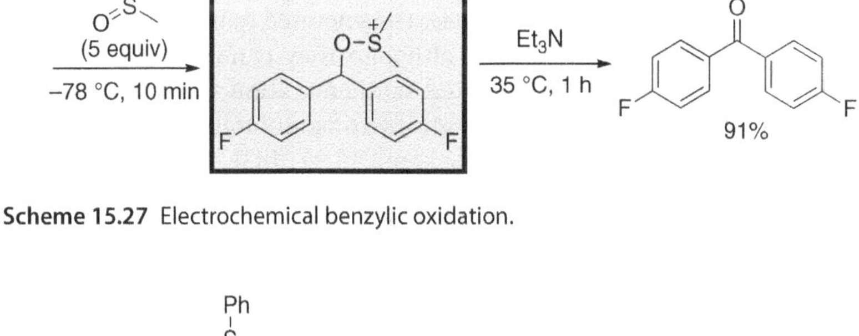

Scheme 15.27 Electrochemical benzylic oxidation.

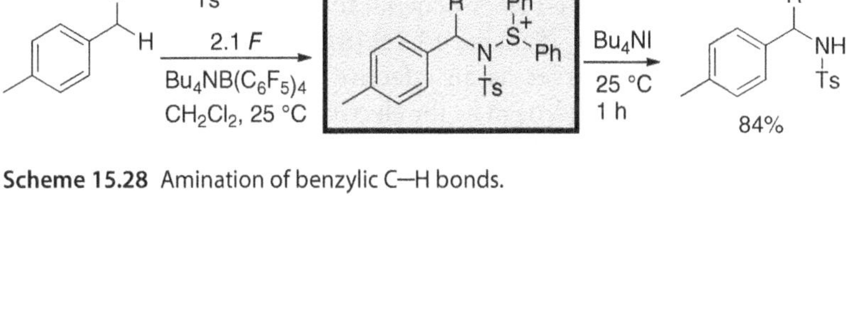

Scheme 15.28 Amination of benzylic C—H bonds.

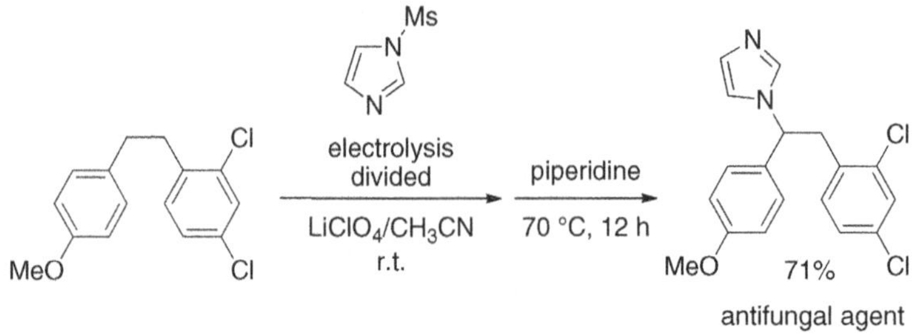

Scheme 15.29 Anodic benzylic C—H amination.

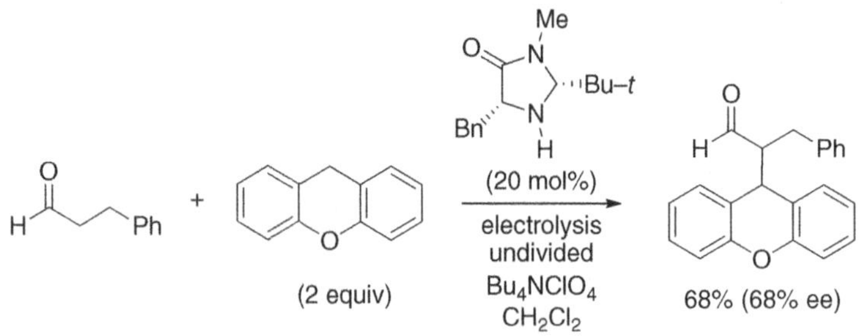

Scheme 15.30 Organocatalyzed anodic benzylic C—H amination.

by treatment with piperidine gives the corresponding *N*-benzylimidazoles. Primary, secondary, and tertiary benzylic C—H bonds can be functionalized efficiently. The following synthesis of an antifungal agent demonstrates the power of the present method, because the benzylic position adjacent to the more electron-rich aromatic ring was selectively oxidized without affecting the other benzylic position as shown in Scheme 15.29.

Jang and co-workers reported an interesting anodic coupling of xanthene and aliphatic aldehydes in the presence of an organocatalyst (Scheme 15.30) [82]. They suggested a mechanism involving coupling of electrogenerated xanthene radical and the electrogenerated enamine radical cations, although they cannot be ruled out a mechanism involving coupling of electrogenerated xanthene cation and enamines.

Luo and co-workers reported coupling of tetrahydroisoquinolines and aliphatic ketones in the presence of an organocatalyst (Scheme 15.31) [83]. They postulated a mechanism involving the coupling of the electrogenerated iminium cations and the enamine derived from the ketones.

15.3.4 Functionalization of Aliphatic C—H Bonds

C—H bonds α to a carbonyl group of ketones can be functionalized by the electrochemical method. Wang and co-workers reported electrochemical oxidation of acetophenones mediated by KI in the presence of O_2 (Scheme 15.32) [84]. Aryl α-ketoesters are the final products and *p*-nitrophenol serves as an effective catalyst. They proposed a mechanism involving abstraction of hydrogen by the electrogenerated I radical followed by the reaction of α-keto radical with O_2. The use of NH_4OAc as a nucleophile and *t*-$BuNH_2$ as a base gives α-ketoamides [85].

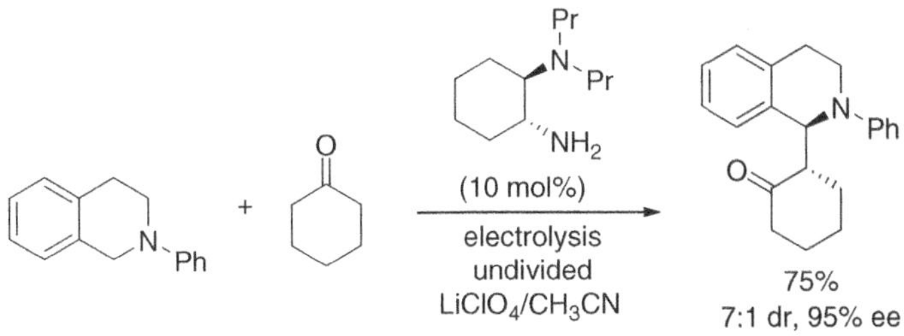

Scheme 15.31 Enantioselective coupling of tetrahydroisoquinolines and aliphatic ketones.

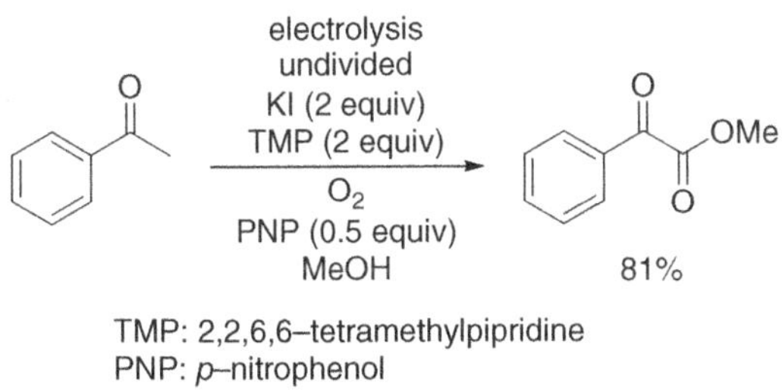

Scheme 15.32 Electrochemical oxidation of acetophenones.

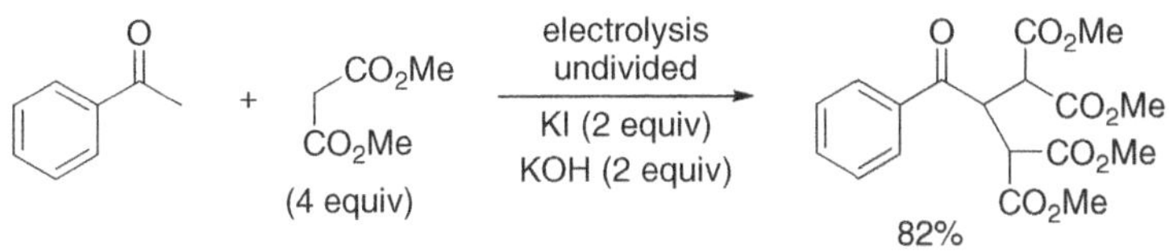

Scheme 15.33 Electrochemical difunctionalization of aryl ketones with malonate esters.

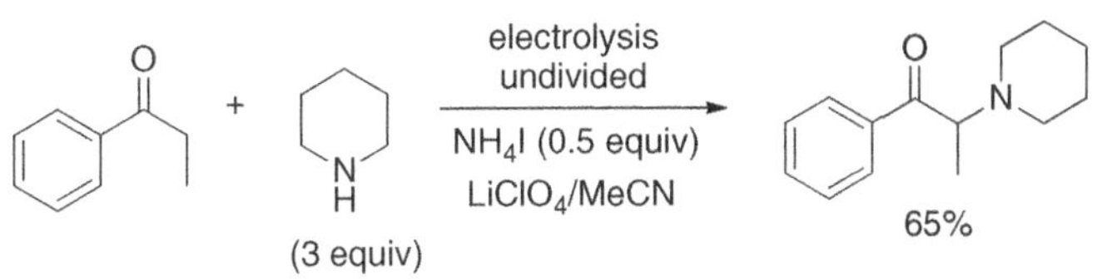

Scheme 15.34 Electrochemical α-amination of ketones.

The use of primary or secondary amines in EtOH leads to the formation of the corresponding α-ketoamides. In this case Bu_4NI is used. The intramolecular version was also accomplished [86].

Zha, Wang, and co-workers also reported difunctionalization of aryl ketones with malonate esters mediated by KI (Scheme 15.33) [87]. In this case the reactions are carried out without oxygen. A mechanism involving the formation of α,α-diiodoketones by the electrogenerated I radical followed by double alkylation with malonate esters was proposed.

The use of aryl alkyl ketones having longer chain and *sec*-amines as nucleophiles leads to the formation of alkyl α-aminoalkyl ketones as reported by Zeng et al. (Scheme 15.34) [88]. In this case the use of NH_4I is effective.

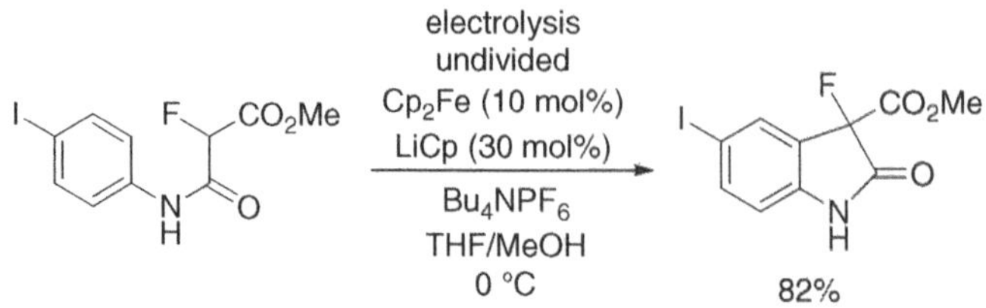

Scheme 15.35 Electrochemical synthesis of C3-fluorinated oxindoles.

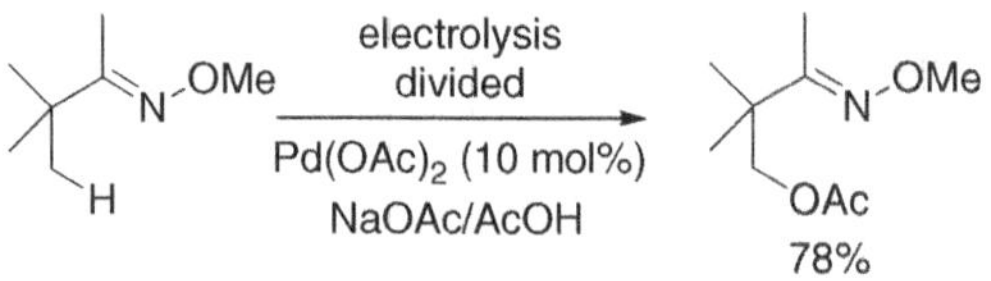

Scheme 15.36 Electrochemical Pd-catalyzed C—H bond oxidation.

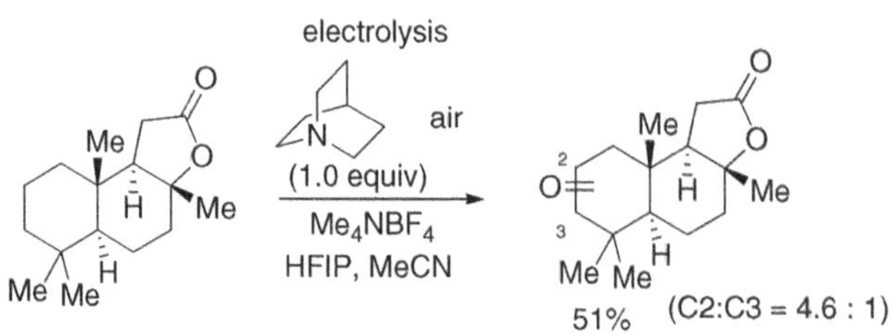

Scheme 15.37 Electrochemical C—H bond oxidation.

Electrooxidative functionalization of C—H bonds α to carbonyl groups can be used for cyclization. Xu and co-workers reported an interesting method for the synthesis of C3-fluorinated oxindoles (Scheme 15.35) [89]. The electrogenerated carbon radical α to the carbonyl group adds to the aromatic ring and further oxidation gives the final product. Ferrocene serves as an effective mediator.

15.3.5 Functionalization of Unactivated C—H Bonds

Mei and co-workers reported that C—H bonds β to C=N of oximes can be functionalized by the anodic oxidation (Scheme 15.36) [90]. The reaction is mediated by a catalytic amount of Pd(II), and oxygen functionalities such as OAc, OTs, and OMe can be introduced. Pd(II) insertion to the C—H bond is facilitated by the coordination of N. The resulting Pd(II) species is anodically oxidized to Pd(IV), which undergoes reductive elimination to give the product.

Baran and co-workers developed an effective method for conversion of an unactivated CH_2 group to a carbonyl group (Scheme 15.37) [91]. A mechanism of this quinuclidine mediated electrochemical C—H oxidation involves abstraction of hydrogen by electrooxidatively generated quinuclidine radical cation to generate a carbon radical that reacts with O_2 to afford the corresponding ketone.

15.4 Oxidative Functionalization of C—C Multiple Bonds

Electrochemical oxidative functionalization of carbon–carbon double bonds can be categorized into two types. The first one is initiated by electrochemical oxidation of carbon-carbon double bonds, which is followed by the attack of nucleophiles to give products. This type has been known about for many years, although various kinds of products are obtained depending on substrate structures and reaction conditions [92]. Sometimes allylic oxidation is the major reaction pathway (vide supra). The second one involves the reaction of electrooxidatively generated reactive species with carbon–carbon double bond.

The anodic oxidation of aryl-substituted olefins in the presence of DMSO developed by Yoshida and co-workers serves as an example of the first type (Scheme 15.38) [79]. For example, the electrochemical oxidation of 4,4′-dichloro-*trans*-stilbene in the presence of DMSO at 0 °C gave dialkoxysulfonium ion, which was characterized by NMR. After the electrolysis, treatment with triethylamine gave the corresponding 1,2-diketone in 83% yield via the Swern–Moffatt type oxidation process. The electrochemical oxidation of alkenes bearing a nucleophilic group such as tosylamino group in an appropriate position leads to the formation of cyclized product (Scheme 15.39). The treatment of dialkoxysulfonium ions with aq. NaOH leads to the formation of *vic* diols with high diastereoselectivity [80].

Electrochemical oxidation of electron-rich alkenes such as enol ethers accompanied by intramolecular attack of nucleophiles to give cyclized products has been extensively studied by Moeller and co-workers (Scheme 15.40) [93]. The tosylamino and hydroxyl groups serve as nucleophiles for the attack to the radical cation of alkenes.

The carbon–carbon double bond of indoles can be electrooxidatively difunctionalized. Harran and co-workers developed an interesting macrocyclization based on [3 + 2] annulation between phenol and indole shown in Scheme 15.41 [94]. The oxygen atom

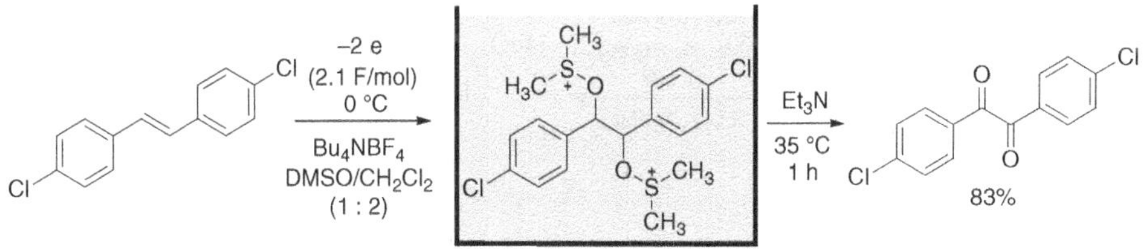

Scheme 15.38 Anodic oxidation of aryl-substituted olefins.

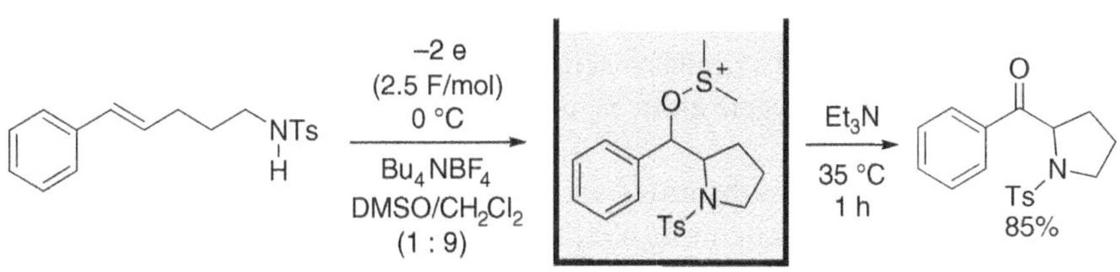

Scheme 15.39 Electrochemical oxidation of alkenes.

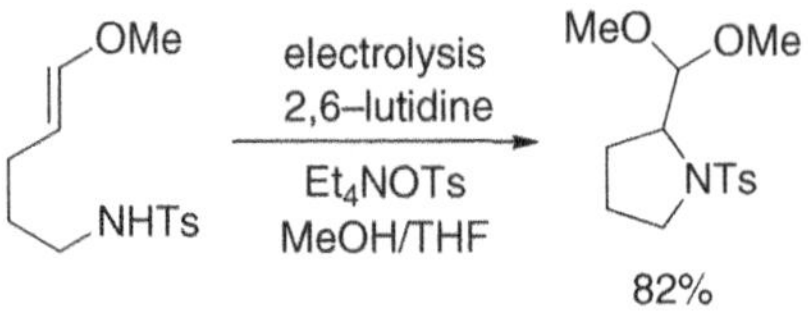

Scheme 15.40 Electrochemical oxidation of electron-rich alkenes.

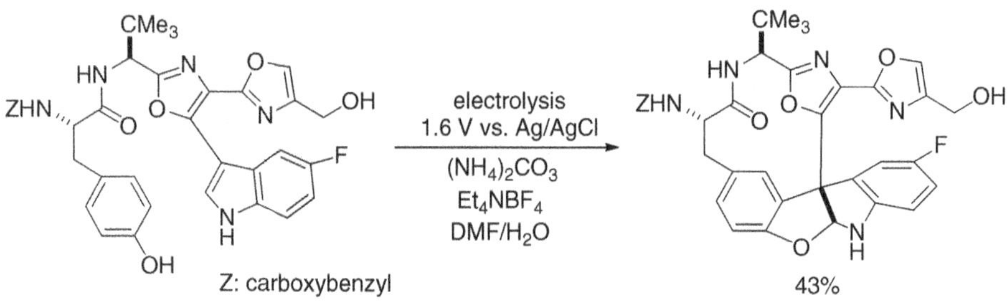

Scheme 15.41 Electrochemical intramolecular [3 + 2] annulation.

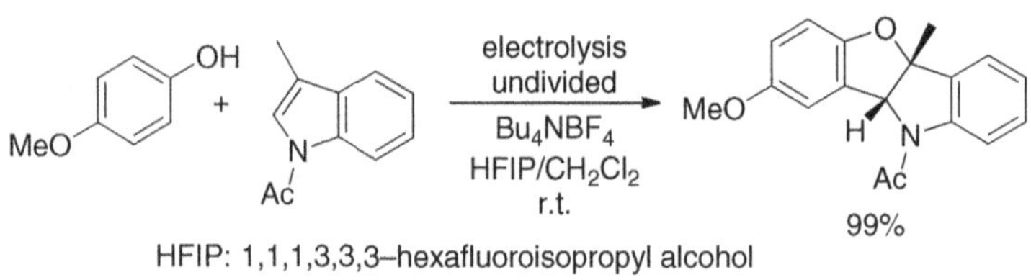

Scheme 15.42 Electrochemical intermolecular [3 + 2] annulation.

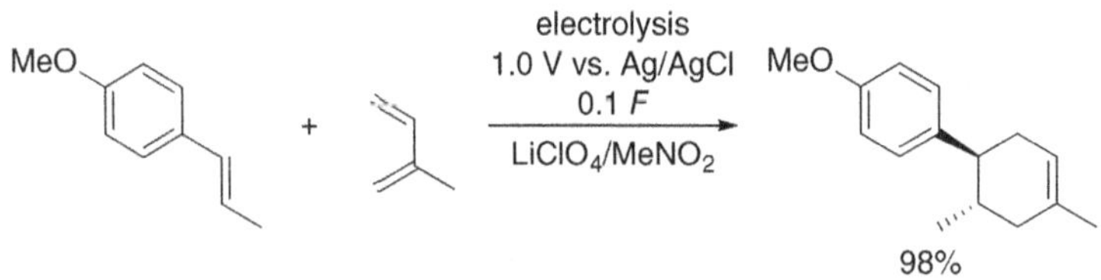

Scheme 15.43 Electrochemical Diels–Alder reactions.

attacks the carbon next to the nitrogen of the indole structure. Scalable synthesis of a diazoamide-based drug candidate was achieved using this method.

Lei and co-workers developed a similar reaction using 3-methyl-*N*-acetylindoles [95]. In this case, however, the oxygen atom is introduced to 3-position of the indole (Scheme 15.42).

Diels–Alder reactions of aryl-substituted alkenes with 1,3-dienes reported by Chiba and co-workers is interesting (Scheme 15.43) [33]. The electrolysis is carried out under the controlled potential conditions using a catalytic amount of electricity. An EC-backward-E mechanism was proposed for this electrocatalytic reactions.

Alkene difunctionalization with halogen and chalcogen cations stabilized by DMSO developed by Yoshida and co-workers serves as an example of the second type (Scheme 15.44). For example, the electrochemical oxidation of Bu_4NI (or I_2) in DMSO/CH_2Cl_2 (1:9 v/v) using Bu_4NBF_4 as a supporting electrolyte at −78 °C gives a solution of I^+ stabilized by DMSO. The reaction with *cis*-5-decene at 0 °C gives *β*-iodoalkoxysulfonium ion that can be characterized by NMR. The treatment with triethylamine affords the corresponding *β*-iodo ketone in 85% yield [96]. The treatment with aq. NaOH gives the corresponding iodohydrin, and the use of NaOMe as a base leads to the formation of the epoxide [97].

Zeng and Little reported aziridination of alkenes with *N*-aminophthimide mediated by Bu_4NI using an undivided cell (Scheme 15.45) [98]. They proposed a radical mechanism.

Yudin et al. also reported aziridination of olefins with *N*-aminophthalimide. In this case the anodic oxidation should be carried out under the controlled potential conditions (Scheme 15.46) [99].

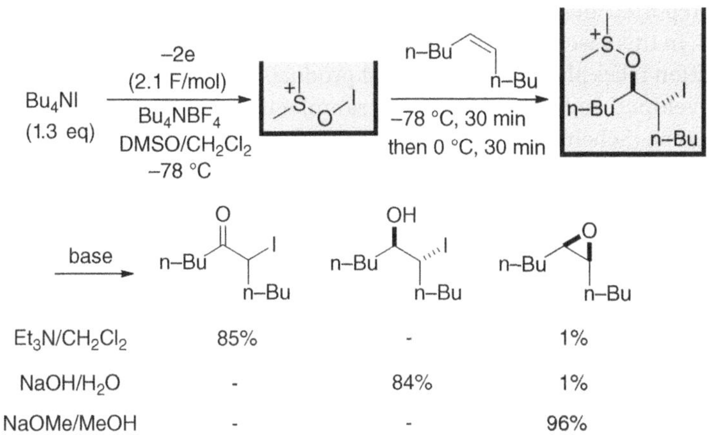

Scheme 15.44 Electrochemical alkene difunctionalization with halogen.

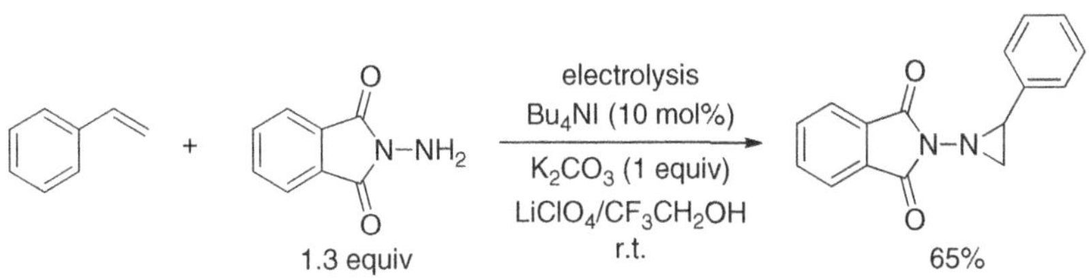

Scheme 15.45 Zeng and Little's electrochemical aziridination.

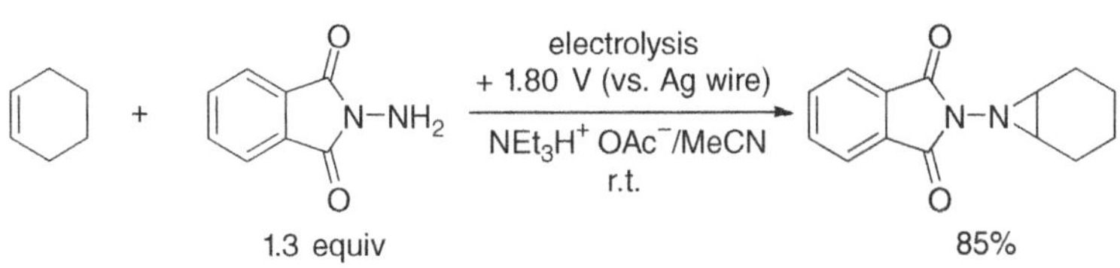

Scheme 15.46 Yudin's electrochemical aziridination.

Lin and co-workers developed Mn-catalyzed electrochemical dichlorination of olefins using $MgCl_2$ as the chlorine source (Scheme 15.47) [100]. They proposed a mechanism involving the anodic oxidation of Mn(II)Cl to Mn(III)Cl, which transfers Cl radical to the olefin. The use of indene as a substrate gives the *trans*-dichlorinated compound in high diastereoselectivity.

Lin and co-workers also reported electrochemical diazidation of olefins catalyzed by $MnBr_2$ (Scheme 15.48) [101]. Because the products can be reduced to vicinal diamines, this reaction serves as a powerful method for making vicinal diamines. A mechanism involving transfer of N_3 radical from electrogenerated Mn(II)-N_3 to olefin is proposed.

The method of the second type can also be used for cyclization. For example, Moeller and co-workers reported the intramolecular anodic coupling of amides and alkenes (Scheme 15.49) [102]. A mechanism involving the attack of the electrooxidatively generated amidyl radicals onto an appropriately positioned carbon–carbon double bond was suggested. Electrochemical oxidation of the cyclized radical to the cation followed by trapping with methanol gives the final product.

Xu and co-workers reported intramolecular anodic coupling of carbamates and alkenes (Scheme 15.50) [103]. In this case, the cyclized radicals are further oxidized and subsequent proton elimination takes place to give the final products.

Zeng and Sun developed iodine mediated intramolecular anodic coupling of sulfonamides and styrene (Scheme 15.51) [104]. The reaction seems to proceed by the attack of anodically generated I_2 or I^+ on the carbon–carbon double bond followed by intramolecular nucleophilic attack of the sulfonamide moiety. The nucleophilic

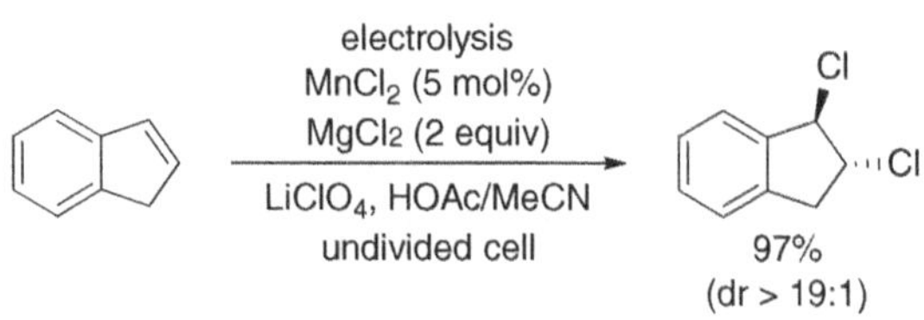

Scheme 15.47 Mn-catalyzed electrochemical dichlorination of olefins.

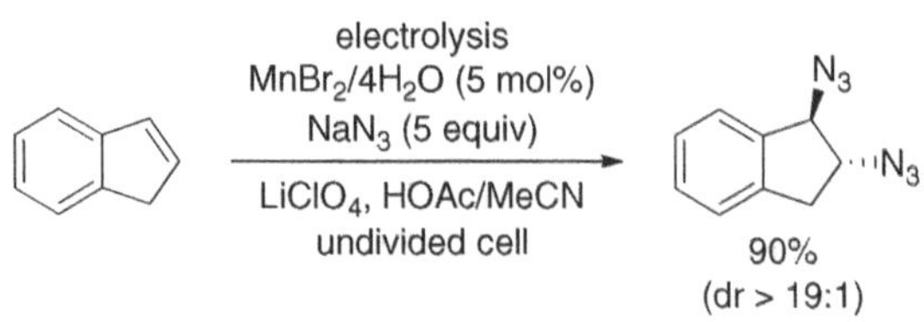

Scheme 15.48 Mn-catalyzed electrochemical diazidation of olefins.

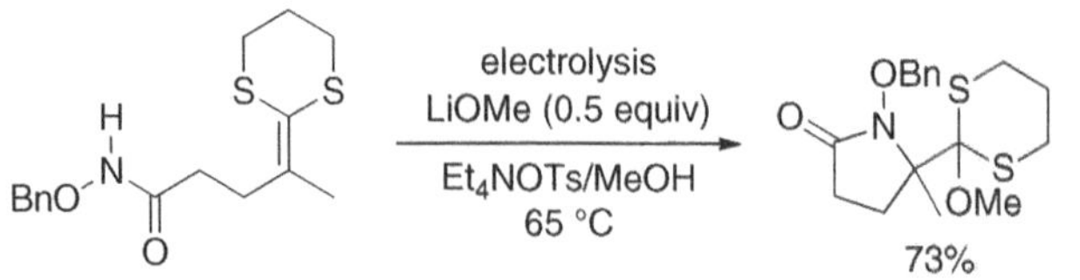

Scheme 15.49 Intramolecular anodic coupling of amides and alkenes.

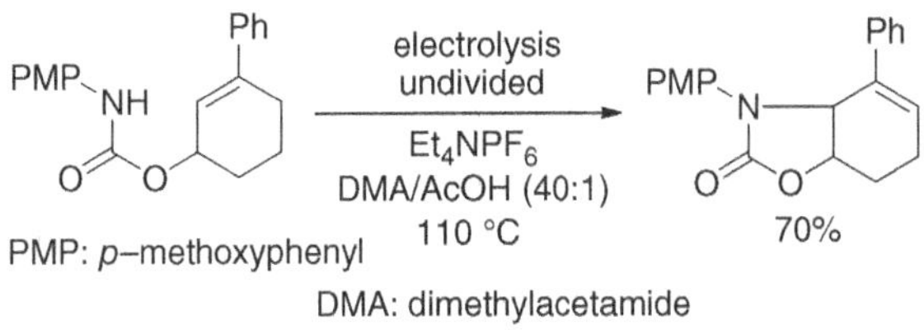

Scheme 15.50 Anodic coupling of carbamates and alkenes.

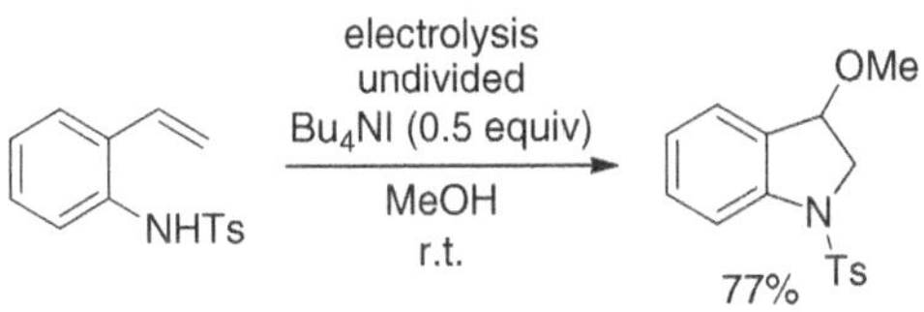

Scheme 15.51 Intramolecular anodic coupling of sulfonamides and styrene.

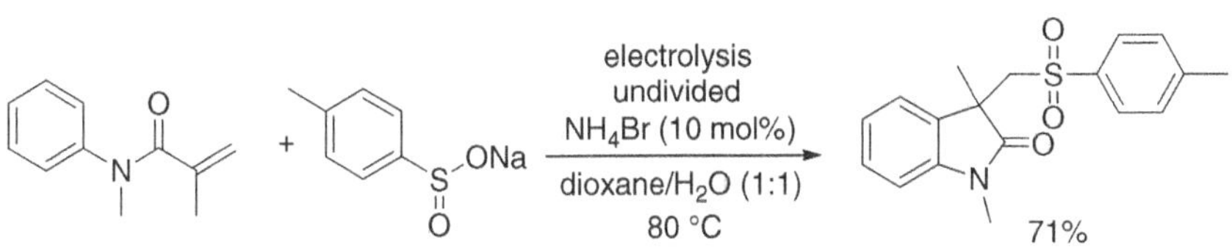

Scheme 15.52 Electrochemical oxidation of *N*-arylacrylamides and sodium *p*-methylbenzenesulfinate.

substitution by the cathodically generated methoxide releases the I^- that serves as a precursor of I_2 or I^+. Therefore, a catalytic amount of I^- is sufficient for the reaction. Notably, the reaction proceeds smoothly without using supporting electrolyte such as $LiClO_4$, although the cell voltage is a little bit higher.

The oxindole structure can be synthesized by Br-mediated electrochemical oxidation of a mixture of *N*-arylacrylamides and sodium *p*-methylbenzenesulfinate as reported by Zeng and Sun (Scheme 15.52) [105]. The reaction of sulfinate and electrogenerated Br_2 gives sulfonyl bromide. The O—Br bond is cleaved homolytically, and the resulting oxygen-centered radical resonates to the more stable sulfonyl radical, which adds to the carbon–carbon double bond. Thus-generated carbon radical adds to the phenyl ring on the nitrogen atom to make the five-membered ring.

Carbon–carbon triple bonds can also be used for intramolecular anodic coupling. Xu and Lu reported the intramolecular anodic coupling of a urea and a carbon–carbon triple bond followed by the intramolecular reaction with an aromatic ring (Scheme 15.53) [106]. This transformation catalyzed by Cp2Fe serves as a useful method for constructing indoles.

The intramolecular anodic coupling involving carbon–carbon triple bonds was successfully applied to construction of polycyclic aromatic compounds as reported by Xu and co-workers (Scheme 15.54) [107]. A mechanism involving domino radical cyclization was suggested.

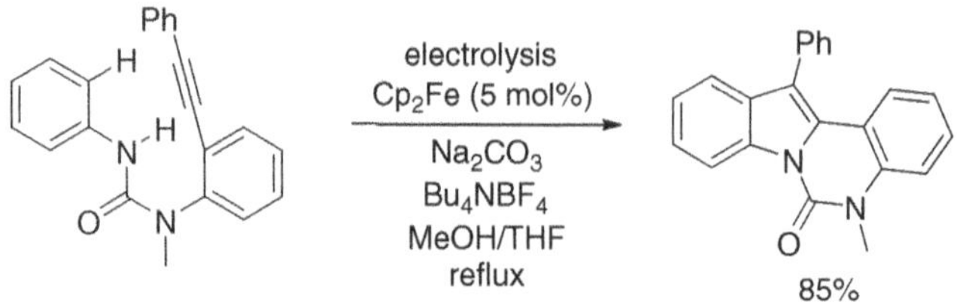

Scheme 15.53 Anodic construction of indoles.

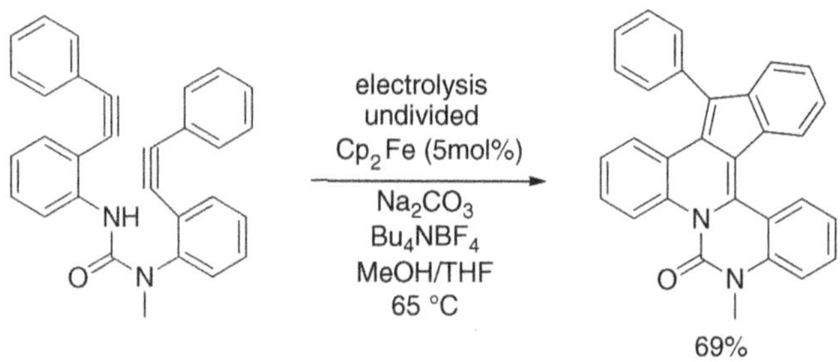

Scheme 15.54 Anodic construction of polycyclic aromatic compounds.

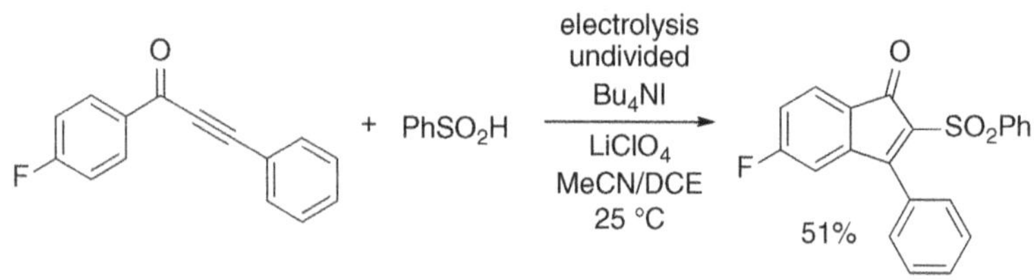

Scheme 15.55 Electrooxidative arylsulfonylation of aromatic ynones with sulfinic acids.

Sulfonyl radical is also effective for functionalization of carbon–carbon triple bonds. Lei and co-workers reported electrooxidative arylsulfonylation of aromatic ynones with sulfinic acids (Scheme 15.55) [108]. The reaction serves as an efficient method for synthesis of indenone derivatives.

15.5 Oxidation of Functional Groups

The direct electrochemical oxidation of alcohols to ketones and aldehydes is rather difficult, although such processes are very important from a viewpoint of chemical synthesis. It is well-known that organic nitroxyls including TEMPO (2,2,6,6-tetramethyl-1-piperidine-*N*-oxy) effectively mediate the process. However, their use in energy conversion has been rather limited because of high cell voltages. Stahl and his coworker reported that the use of (2,2′-bipyridine)Cu complex as a cocatalyst decreases the cell voltage dramatically [109]. They proposed a mechanism involving the anodic oxidation of Cu(I) to Cu(II), which reacts with an alcohol to give an alkoxyCu(II) (Scheme 15.56). The subsequent reaction with TEMPO gives an aldehyde regenerating Cu(I). Thus, Cu(II) acts as a one-electron oxidant, while TEMPO acts as an electron-proton acceptor.

Scheme 15.56 Proposed mechanism.

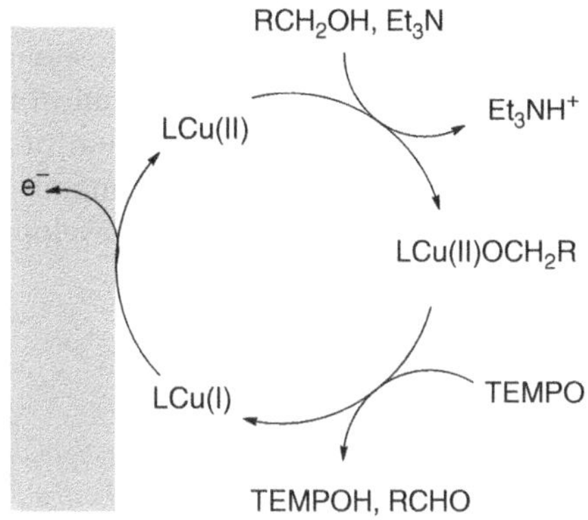

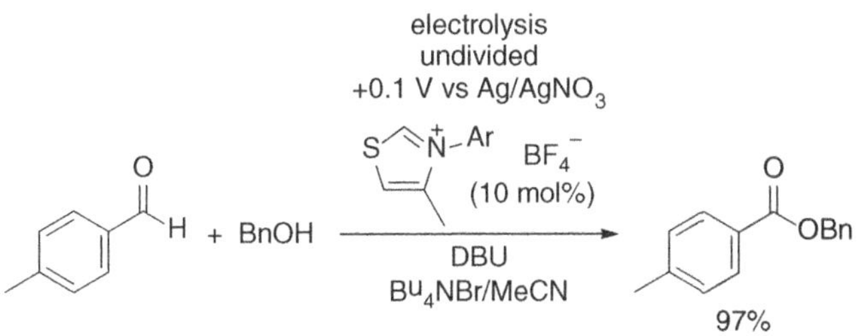

Scheme 15.57 Anodic oxidation of aldehydes to esters.

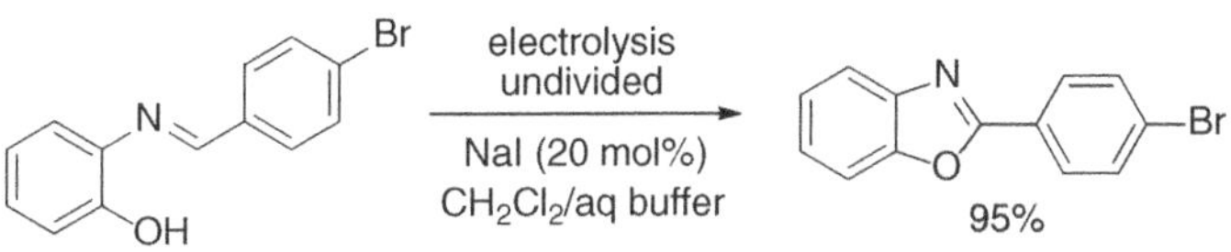

Scheme 15.58 Anodic oxidation of aldimines.

Boydston and co-workers developed a method for anodic oxidation of aldehyde to esters mediated by organocatalysts (Scheme 15.57) [110]. They postulated a mechanism involving anodic oxidation of the Breslow intermediate generated from an aldehyde and the catalyst followed by nucleophilic attack of an alcohol. The method is applicable to non-aromatic aldehyde as well. The use of diorgano disulfides (RSSR) instead of alcohols leads to the formation of thioesters [111]. In this case RS^- generated by the cathodic reduction of RSSR serve as nucleophiles.

Zeng and co-workers reported the anodic oxidation of aldimines mediated by NaI. In this case the intramolecular attack of OH gives oxazole rings (Scheme 15.58). Because the starting materials can be easily prepared from 2-aminophenols and aldehydes, the present reaction provides a useful method for synthesizing benzoxazoles.

15.6 Outlook

The examples discussed here are representative of recently developed electrooxidative transformations and they demonstrate, together with many other examples, the power of the electrochemical method in modern organic synthesis. Because of the limitation of space, this chapter mainly focused on C—H functionalizations, which have been extensively studied recently. We should keep in mind, however, that there are many other aspects of the electrochemical method that serve as effective environmentally benign methods for the oxidation of organic compounds. They will hopefully work together to meet the great demands of laboratory organic synthesis and chemical production in industry for the development of a sustainable society in the near future.

References

1 Saveant, J.-M. (2016). *ChemElectroChem* 3: 1967–1977.
2 (a) Moeller, K.D. (2000). *Tetrahedron* 56: 9527–9554. (b) Yoshida, J., Kataoka, K., Horcajada, R., and Nagaki, A. (2008). *Chem. Rev.* 108: 2265–2299. (c) Frontana-Uribe, B.A., Little, R.D., Ibanez, J.G. et al. (2010). *Green Chem.* 12: 2099–2119. (d) Schäfer, H.J. (2011). *C. R. Chim.* 14: 745–765.
3 Recent reviews: (a) Horn, E.J., Rosen, B.R., and Baran, P.S. (2016). *ACS Cent. Sci.* 2: 302–308. (b) Yan, M., Kawamata, Y., and Baran, P.S. (2017). *Angew. Chem. Int. Ed.* 56: 2–9. (c) Yan, M., Kawamata, Y., and Baran, P.S. (2017). *Chem. Rev.* 117: 13230–13319. (d) Chiba, K. and Okada, Y. (2017). *Curr. Opin. Electrochem.* 2: 53–59.
4 Sequeira, C.A.C. and Santos, D.M.F. (2009). *J. Braz. Chem. Soc.* 20: 387–406.
5 (a) Kolbe, H. (1848). *Ann. Chem. Pharm.* 68: 339–341. (b) Kolbe, H. (1849). *Ann. Chem. Pharm.* 69: 257–294.(c) Vijh, A.K. and Conway, B.E. (1967). *Chem. Rev.* 67: 623–664.
6 Yoshida, J., Kataoka, K., Horcajada, R., and Nagaki, A. (2008). *Chem. Rev.* 108: 2265–2299.
7 Yoshida, J. (2014). Electroauxiliary. In: *Encyclopedia of Applied Electrochemistry* (ed. G. Kreysa, K. Ota and R.F. Savinell), 386–392. Heidelberg: Springer.
8 Yoshida, J., Maekawa, T., Murata, T. et al. (1990). *J. Am. Chem. Soc.* 112: 1962–1970.
9 Yoshida, J. and Nishiwaki, K. (1998). *J. Chem. Soc. Dalton Trans.* 2589–2596.
10 Yoshida, J. and Isoe, S. (1987). *Tetrahedron Lett.* 28: 6621–6624.
11 Yoshida, J., Sugawara, M., and Kise, N. (1996). *Tetrahedron Lett.* 37: 3157–3160.
12 Sugawara, M., Mori, K., and Yoshida, J. (1997). *Electrochem. Acta* 42: 1995–2003.
13 Shoji, T., Kim, S., Yamamoto, K. et al. (2014). *Org. Lett.* 16: 6404–6407.
14 Yoshida, J. and Suga, S. (2002). *Chem. Eur. J.* 8: 2651–2658.
15 Yoshida, J., Ashikari, Y., Matsumoto, K., and Nokami, T. (2013). *J. Synth. Org. Chem. Jpn.* 71: 1136–1144.
16 Yoshida, J., Suga, S., Suzuki, S. et al. (1999). *J. Am. Chem. Soc.* 121: 9546–9549.
17 Suga, S., Matsumoto, K., Ueoka, K., and Yoshida, J. (2006). *J. Am. Chem. Soc.* 128: 7710–7711.
18 Matsumoto, K., Ueoka, K., Suzuki, S. et al. (2009). *Tetrahedron* 65: 10901–10907.
19 Kim, S., Hayashi, K., Kitano, Y. et al. (2002). *Org. Lett.* 4: 3735–3737.
20 Kim, S., Shoji, T., Kitano, Y., and Chiba, K. (2013). *Chem. Commun.* 49: 6525–6527.
21 Shoji, T., Kim, S., and Chiba, K. (2017). *Angew. Chem., Int. Ed.* 56: 4011–4014.
22 (a) Yoshida, J. (2005). *Chem. Commun.* 4509–4516. (b) Yoshida, J. (2009). *Interface* Summer: 40–45.
23 Watts, K., Gattrell, W., and Wirth, T. (2011). *Beilstein J. Org. Chem.* 7: 1108–1114.
24 Atobe, M., Tateno, H., and Matsumura, Y. (2017). *Chem. Rev.* https://doi.org/10.1021/acs.chemrev.7b00353.

25 Löwe, H. and Ehrfeld, W. (1999). *Electrochim. Acta* 44: 3679.
26 Paddon, C.A., Pritchard, G.J., Thiemann, T., and Marken, F. (2002). *Electrochem. Commun.* 825–831.
27 Suga, S., Okajima, M., Fujiwara, K., and Yoshida, J. (2001). *J. Am. Chem. Soc.* 123: 7941–7942.
28 Horii, D., Fuchigami, T., and Atobe, M. (2007). *J. Am. Chem. Soc.* 129: 11692–11693.
29 Schulz, L., Enders, M., Elsler, B. et al. (2017). *Angew. Chem. Int. Ed.* 56: 4877–4881.
30 Kathiresan, M. and Velayutham, D. (2015). *Chem. Commun.* 51: 17499–17516.
31 Fuchigami, T. and Inagi, S. (2011). *Chem. Commun.* 47: 10211–10223.
32 Chiba, K., Miura, T., Kim, S. et al. (2001). *J. Am. Chem. Soc.* 123: 11314–11315.
33 Okada, Y., Yamaguchi, Y., Ozaki, A., and Chiba, K. (2016). *Chem. Sci.* 7: 6387–6393.
34 Imada, Y., Yamaguchi, Y., Shida, N. et al. (2017). *Chem. Commun.* 53: 3960–3963.
35 Tajima, T. and Fuchigami, T. (2005). *J. Am. Chem. Soc.* 127: 2848–2849.
36 Ogumi, Z., Nishio, K., and Yoshizawa, Z. (1981). *Electrochim. Acta* 26: 1779–1782.
37 Jorissen, J. (1996). *Electrochim. Acta* 41: 553–562.
38 Horcajada, R., Okajima, M., Suga, S., and Yoshida, J. (2005). *Chem. Commun.* 1303–1305.
39 Friedrich, J.M., Ponce-de-León, C., Reade, G.W., and Walsh, F.C. (2004). *J. Electroanal. Chem.* 561: 203–217.
40 Macpherson, J.V. (2015). *Phys. Chem. Chem. Phys.* 17: 2935–2949.
41 Enache, T.A., Chiorcea-Paquim, A.-M., Fatibello-Filho, O., and Oliveira-Brett, A.M. (2009). *Electrochem. Commun.* 11: 1342–1345.
42 Francke, R. and Little, R.D. (2014). *Chem. Soc. Rev.* 43: 2492–2521.
43 Broese, T. and Francke, R. (2016). *Org. Lett.* 18: 5896–5899.
44 Ogawa, K.A. and Boydston, A. (2015). *J. Chem. Lett.* 44: 10–16.
45 Chiba, T., Okimoto, M., Nagai, H., and Takata, Y. (1982). *Bull. Chem. Soc. Jpn.* 55: 335–336.
46 Louillat, M.-L. and Patureau, F.W. (2014). *Chem. Soc. Rev.* 43: 901–910.
47 (a) Romero, N.A., Margrey, K.A., Tay, N.E., and Nicewicz, D.A. (2015). *Science* 349: 1326–1330. (b) Zheng, Y.-W., Chen, B., Ye, P. et al. (2016). *J. Am. Chem. Soc.* 138: 10080–10083.
48 Morofuji, T., Shimizu, A., and Yoshida, J. (2013). *J. Am. Chem. Soc.* 135: 5000–5003.
49 Wesenberg, L.J., Herold, S., Shimizu, A. et al. (2017). *Chem. Eur. J.* 23: 12096–12099.
50 Morofuji, T., Shimizu, A., and Yoshida, J. (2015). *Chem. Eur. J.* 21: 3211–3214.
51 Morofuji, T., Shimizu, A., and Yoshida, J. (2014). *J. Am. Chem. Soc.* 136: 4496–4499.
52 Morofuji, T., Shimizu, A., and Yoshida, J. (2015). *J. Am. Chem. Soc.* 137: 9816–9819.
53 Gao, W.-J., Li, W.-C., Zeng, C.-C. et al. (2014). *J. Org. Chem.* 79: 9613–9618.
54 Zhao, H.-B., Hou, Z.-W., Liu, Z.-J. et al. (2017). *Angew. Chem. Int. Ed.* 56: 587–590.
55 Li, Y.-Q., Yang, Q.-L., Fang, P. et al. (2017). *Org. Lett.* 19: 2905–2908.
56 Gieshoff, T., Kehl, A., Schollmeyer, D. et al. (2017). *Chem. Commun.* 53: 2974–2977.
57 Wang, P., Tang, S., Huang, P., and Lei, A. (2017). *Angew. Chem. Int. Ed.* 56: 3009–3013.
58 Wang, P., Tang, S., and Lei, A. (2017). *Green Chem.* 19: 2092–2095.
59 Qian, X.-Y., Li, S.-Q., Song, J., and Xu, H.-C. (2017). *ACS Catal.* 7: 2730–2734.
60 Dohi, T. and Kita, Y. (2008). *Angew. Chem. Int. Ed.* 47: 1301–1304.
61 (a) Stuart, D.R. and Fagnou, K. (2007). *Science* 316: 1172–1175. (b) Yang, Y., Lan, J., and You, J. (2017). *Chem. Rev.* 117: 8787–8863.
62 Hernandez-Perez, A.C. and Collins, S.K. (2016). *Acc. Chem. Res.* 49: 1557–1565.

63 Lou, S.-J., Mao, Y.-J., Xu, D.-Q. et al. (2016). *ACS Catal.* 6: 3890–3894.
64 Nyberg, K. (1971). *Acta Chem. Scand.* 25: 3770–3776.
65 Morofuji, T., Shimizu, A., and Yoshida, J. (2012). *Angew. Chem. Int. Ed.* 51: 7259–7262.
66 Röse, P., Emge, S., König, C.A., and Hilt, G. (2017). *Adv. Synth. Catal.* 359: 1359–1372.
67 (a) Kirste, A., Nieger, M., Malkowsky, I.M. et al. (2009). *Angew. Chem. Int. Ed.* 15: 2273–2277. (b) Kirste, A., Schnakenburg, G., Stecker, F. et al. (2010). *Angew. Chem. Int. Ed.* 49: 971–975. (c) Kirste, A. and Waldvogel, S.R. (2012). *J. Am. Chem. Soc.* 134: 3571–3576.
68 Elsler, B., Schollmeyer, D., Dyballa, K.M. et al. (2014). *Angew. Chem. Int. Ed.* 53: 5210–5213.
69 Wiebe, A., Schollmeyer, D., Dyballa, K.M. et al. (2016). *Angew. Chem. Int. Ed.* 55: 11801–11805.
70 Lips, S., Wiebe, A., Elsler, B. et al. (2016). *Angew. Chem. Int. Ed.* 55: 10872–10876.
71 Saito, F., Aiso, H., Kochi, T., and Kakiuchi, F. (2014). *Organometallics* 33: 6704–6707.
72 Amatore, C., Cammoun, C., and Jutand, A. (2007). *Adv. Synth. Catal.* 349: 292–296.
73 Tang, S., Gao, X., and Lei, A. (2017). *Chem. Commun.* 53: 3354–3356.
74 Jensen, K.L., Franke, P.T., Nielsen, L.T. et al. (2010). *Angew. Chem. Int. Ed.* 49: 129–133.
75 Xiao, H.-L., Zeng, C.-C., Tian, H.-Y. et al. (2014). *J. Electroanal. Chem.* 727: 120–124.
76 Shono, T. and Ikeda, A. (1972). *J. Am. Chem. Soc.* 94: 7892–7898.
77 Horn, E.J., Rosen, B.R., Chen, Y. et al. (2016). *Nature* 533: 77–81.
78 Examples of electrochemical benzylic C–H transformation: (a) Hoormann, D., Kubon, C., Jörissen, J. et al. (2001). *J. Electroanal. Chem.* 507: 215–225. (b) Tajima, T. and Fuchigami, T. (2005). *Angew. Chem. Int. Ed.* 44: 4760–4763. (c) Tajima, T. and Fuchigami, T. (eds.) (2005). *Angew. Chem.* 117: 4838–4841. (d) Okajima, M., Soga, K., Nokami, T. et al. (2006). *Org. Lett.* 8: 5005–5007.
79 Ashikari, Y., Nokami, T., and Yoshida, J. (2011). *J. Am. Chem. Soc.* 133: 11840–11843.
80 Ashikari, Y., Nokami, T., and Yoshida, J. (2012). *Org. Lett.* 14: 938–941.
81 Hayashi, R., Shimizu, A., Song, Y. et al. (2017). *Chem. Eur. J.* 23: 61–64.
82 Ho, X.-H., Mho, S.-i., Kang, H., and Jang, H.-Y. (2010). *Eur. J. Org. Chem.* 4436–4441.
83 Fu, N., Li, L., Yang, Q., and Luo, S. (2017). *Org. Lett.* 19: 2122–2125.
84 Zhang, Z., Su, J., Zha, Z., and Wang, Z. (2013). *Chem. Eur. J.* 19: 17711–17714.
85 Zhang, Z., Su, J., Zha, Z., and Wang, Z. (2013). *Chem. Commun.* 49: 8982–8984.
86 Qian, P., Su, J.-H., Wang, Y. et al. (2017). *J. Org. Chem.* 82: 6434–6440.
87 Gao, H., Zha, Z., Zang, Z. et al. (2014). *Chem. Commun.* 50: 5034–5036.
88 Liang, S., Zeng, C.-C., Tian, H.-Y. et al. (2016). *J. Org. Chem.* 81: 11565–11573.
89 Wu, Z.-J. and Xu, H.-C. (2017). *Angew. Chem. Int. Ed.* 56: 4734–4738.
90 Yang, Q.-L., Li, Y.-Q., Ma, C. et al. (2017). *J. Am. Chem. Soc.* 139: 3293–3298.
91 Kawamata, Y., Yan, M., Liu, Z. et al. (2017). *J. Am. Chem. Soc.* 139: 7448–7451.
92 Möller, K.-C. and Schäfer, H.J. (1997). *Electrochim. Acta* 42: 1971–1978.
93 (a) Sutterer, A. and Moeller, K.D. (2000). *J. Am. Chem. Soc.* 122: 5636–5637. (b) Xu, H.-C. and Moeller, K.D. (2008). *J. Am. Chem. Soc.* 130: 13542–13543. (c) Xu, H.-C. and Moeller, K.D. (2010). *J. Am. Chem. Soc.* 132: 2839–2844.(d) Xu, C.-H. and Moeller, K.D. (2010). *Angew. Chem. Int. Ed.* 49: 8004–8007.
94 Ding, H., DeRoy, P.L., Perreault, C. et al. (2015). *Angew. Chem. Int. Ed.* 54: 481–4822.
95 Liu, K., Tang, S., Huang, P., and Lei, A. (2017). *Nat. Commun.* 8: 775.
96 Ashikari, Y., Shimizu, A., Nokami, T., and Yoshida, J. (2013). *J. Am. Chem. Soc.* 135: 16070–16073.

97 Shimizu, A., Hayashi, R., Ashikari, Y. et al. (2015). *Beilstein J. Org. Chem.* 11: 242–248.
98 Chen, J., Yan, W.-Q., Lam, C.M. et al. (2015). *Org. Lett.* 17: 986–989.
99 Siu, T. and Yudin, A.K. (2002). *J. Am. Chem. Soc.* 124: 530–531.
100 Fu, N., Sauer, G.S., and Lin, S. (2017). *J. Am. Chem. Soc.* 139: 15548–15553.
101 Fu, N., Sauer, G.S., Saha, A. et al. (2017). *Science* 357: 575–579.
102 Xu, H.-C., Campbell, J.M., and Moeller, K.D. (2014). *J. Org. Chem.* 79: 379–391.
103 Xiong, P., Xu, H.-H., and Xu, H.-C. (2017). *J. Am. Chem. Soc.* 139: 2956–2959.
104 Liang, S., Zeng, C.-C., Luo, X.-G. et al. (2016). *Green Chem.* 18: 2222–2230.
105 Jiang, Y.-Y., Liang, S., Zeng, C.-C. et al. (2016). *Green Chem.* 18: 6311–6319.
106 Hou, Z.-W., Mao, Z.-Y., Zhao, H.-B. et al. (2016). *Angew. Chem. Int. Ed.* 55: 9168–9172.
107 Hou, Z.-W., Mao, Z.-Y., Song, J., and Xu, H.-C. (2017). *ACS Catal.* 7: 5810–5813.
108 Wen, J., Shi, W., Zhang, F. et al. (2017). *Org. Lett.* 19: 3131–3134.
109 Badalyan, A. and Stahl, S.S. (2016). *Nature* 535: 406–410.
110 Rinney, E.E., Ogawa, K.A., and Boydston, A.J. (2012). *J. Am. Chem. Soc.* 134: 12374–12377.
111 Ogawa, K.A. and Boydston, A.J. (2014). *Org. Lett.* 16: 1928–1931.

16

The Recent Developments of Enzymatic Oxidation

Severin Wedde, Marc Biermann, Ji Eun Choi, Keiko Oike, Nadine Zumbrägel, and Harald Gröger

Chair of Organic Chemistry I, Faculty of Chemistry, Bielefeld University, Bielefeld, Germany

16.1 Introduction and Overview of Enzymatic Oxidations

Oxidation is one of the most important transformations in organic chemistry [1–3]. Besides "classic chemical" non-catalytic and chemocatalytic methodologies, biocatalysis also turned out to provide attractive solutions for challenging oxidation reactions [4]. Among the representatives of enzyme class I (oxidoreductases) various types of enzymes have been applied in organic-synthetic biotransformations enabling efficient organic syntheses in many cases. Thus, numerous oxidoreductases are now used in organic synthesis. These types of applied biocatalytic reactions include, for example, dehydrogenation, hydroxylation, epoxidation, Baeyer–Villiger oxidation, and sulfoxidation reactions [4]. A major advantage of many of these biocatalysts consists of the use of molecular oxygen as an economically attractive and environmentally friendly, non-toxic oxidation agent. In addition, the tremendous progress in protein engineering techniques led to the design of tailor-made mutants of oxidoreductases with improved catalytic properties [5, 6]. The availability of efficient oxidation biocatalysts has allowed the development of a range of efficient transformations. Furthermore, industrial applications have been realized [7]. Representative examples, which are landmarks in the field of industrial biocatalytic oxidation, are the transformation of cephalosporin C into 7-ACA by means of a D-amino acid oxidase as well as hydroxylation of steroids with monooxygenases.

In the following, an overview of selected application areas of oxidoreductases in the field of organic synthesis, clustered according to the applied enzyme classes, will be given.

16.2 Oxidation with Dehydrogenases

16.2.1 Introduction and Overview

Dehydrogenases are enzymes that reversibly catalyze redox reactions by utilizing hydride acceptors/donors, such as nicotinamide cofactors (Scheme 16.1). However, processes for the oxidation of alcohols or amines are rare. On the one hand, in the case

Green Oxidation in Organic Synthesis, First Edition. Edited by Ning Jiao and Shannon S. Stahl.

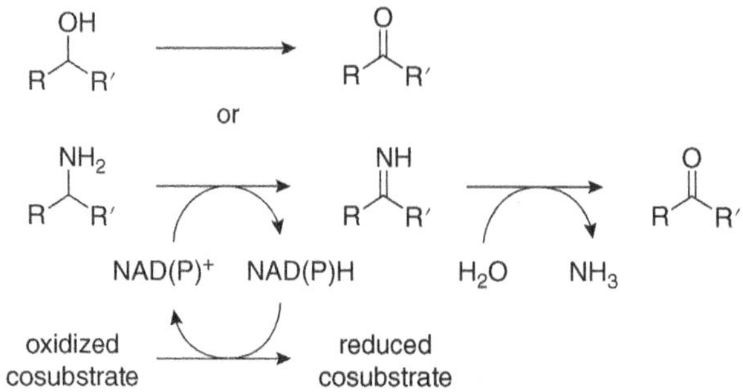

Scheme 16.1 General transformation catalyzed by dehydrogenases with cofactor regeneration.

of secondary alcohols and primary amino groups bound to a secondary carbon center, a stereogenic center is destructed, converting a chiral sp^3 into a prochiral sp^2 center, which leads to the loss of valuable stereogenic information. On the other hand, despite the name, the preferred direction of most of these enzymes is the reduction that is caused by the thermodynamically unfavorable break up of aromaticity in the cofactor. In addition, product inhibition is often observed due to the more lipophilic product, which is more tightly bound to the active center than the hydrophilic substrate [8].

Cofactor regeneration systems allow the equilibrium to be shifted toward the oxidative direction. They also make it possible to use catalytic amounts, since cofactors contribute a large extent to the cost of the process [9]. Many different systems have been developed to efficiently regenerate the oxidized cofactor. Generally, such systems can be substrate and enzyme coupled [10]. The substrate-coupled approach, which can be considered a "biocatalytic equivalent" to the Oppenauer oxidation, uses the same enzyme that is responsible for the oxidation also for reduction of a cosubstrate, thus simultaneously reoxidizing the cofactor. The enzyme-coupled system utilizes a second enzyme to perform the re-oxidation of the reduced form of the cofactor, NAD(P)H. For this purpose, a second alcohol dehydrogenase [11–13], a lactate dehydrogenase (LDH) [14], a glutamate dehydrogenase (GluDH) [15–17], an NAD(P)H oxidase [18–25], or a laccase [26–30] have been used. Recently, porphyrin-based complexes [31, 32] and electrochemical systems [33–35] have also been applied.

16.2.2 Dehydrogenation of Alcohols

16.2.2.1 Primary Alcohols

The oxidation of primary alcohols has been used for a long time in the form of acetic acid fermentation from ethanol, which is a two-step process. An alcohol dehydrogenase oxidizes ethanol to formaldehyde, which is subsequently oxidized to acetic acid by an aldehyde dehydrogenase [36]. In fact, most microorganisms possess aldehyde dehydrogenases (ALDH) that make whole-cell catalytic selective oxidations to aldehydes challenging. There are, however, exceptions like *Janibacter terrae*, which was successfully applied to the oxidation of aliphatic and especially allylic primary alcohols with negligible over-oxidation (Scheme 16.2) [12, 37].

While isolated enzymes can be used to avoid over-oxidation, Molinari et al. were also able to prevent over-oxidation by utilizing a two-phase system consisting of water

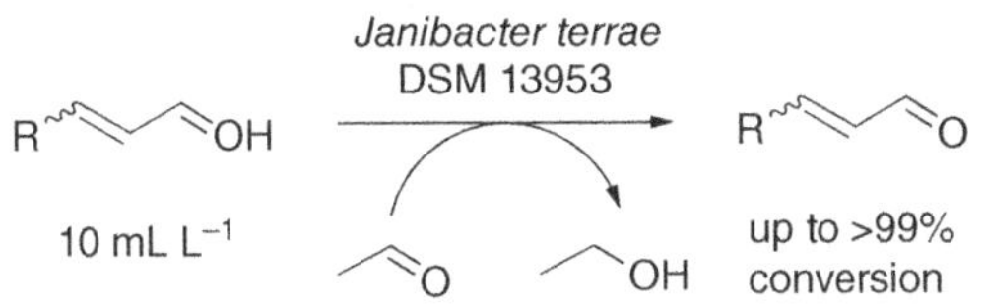

Scheme 16.2 Oxidation of allylic alcohols with *Janibacter terrae* [12, 37].

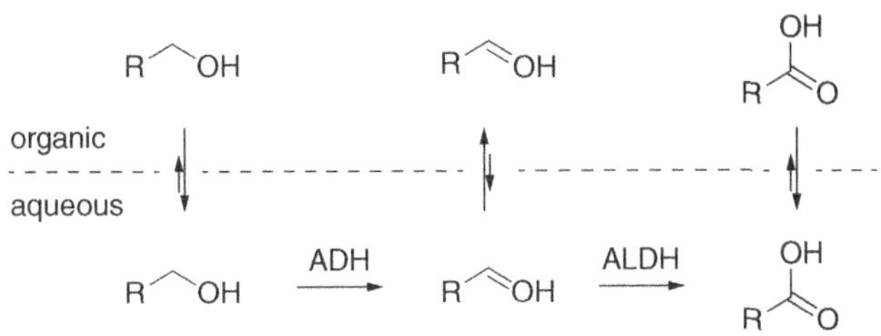

Scheme 16.3 Concept of preventing over-oxidation in the presence of aldehyde dehydrogenases (ALDH) by using a two-phase solvent system.

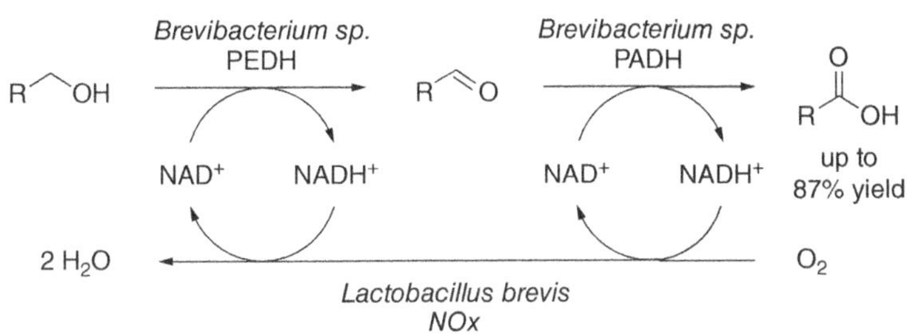

Scheme 16.4 Oxidation of primary alcohols to carboxylic acids with molecular oxygen [40–42].

and isooctane. Upon oxidation of the hydrophilic alcohol, the more lipophilic aldehyde was extracted into the organic phase, preventing oxidation by ALDH in the aqueous phase (Scheme 16.3) [38, 39].

Although, in many cases, over-oxidation is not desired, Ohta et al. were interested in combining alcohol and ALDH from *Brevibacterium* sp. with an NADH oxidase from *Lactobacillus brevis* for environmentally benign oxidation using molecular oxygen as the oxidant (Scheme 16.4). A range of aliphatic (e.g. 1-hexanol) and aromatic primary alcohols (e.g. 2-phenylethanol) were successfully converted into their respective carboxylic acids with an up to 87% yield [40–42].

Extending on this concept, Kara et al. achieved the complete oxidation of methanol to carbon dioxide by means of an alcohol dehydrogenase (ADH), a formaldehyde dismutase (FDM), and a formate dehydrogenase (FDH) (Scheme 16.5). The system could be used for cofactor regeneration by utilizing all three from methanol available reducing equivalents [43].

A major focus in biocatalysis in the recent past was based on the synthesis of lactones since they provide easy access to polyesters or polyamides. One possible way is to first selectively oxidize one of the hydroxyl groups of α,ω-diols to the corresponding aldehyde.

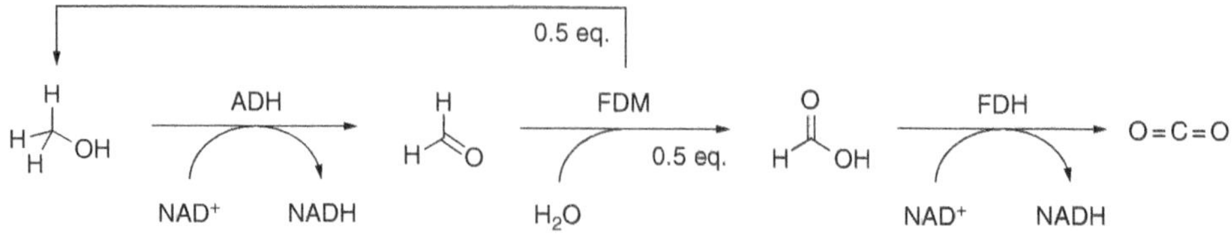

Scheme 16.5 Complete oxidation of methanol to carbon dioxide by a three-enzyme cascade consisting of an alcohol dehydrogenase (ADH), a formaldehyde dismutase (FDM) and a formate dehydrogenase (FDH) [43].

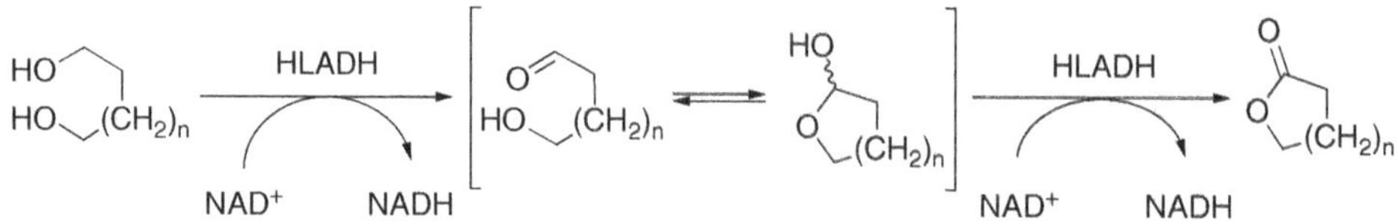

Scheme 16.6 Oxidative lactonization of α,ω-diols via intermediate hemi-acetal formation [28, 30, 44–46].

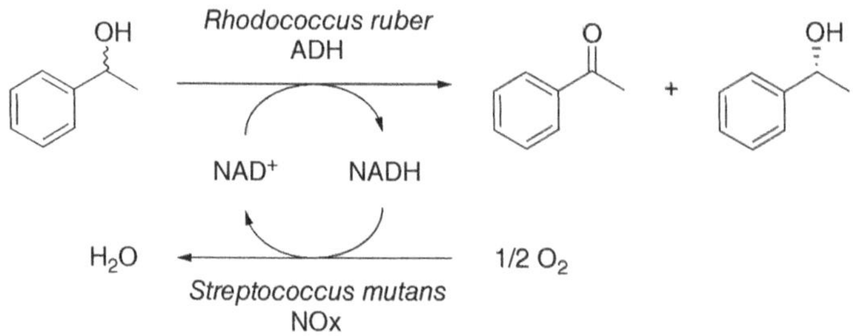

Scheme 16.7 Kinetic resolution of *rac*-2-phenylethanol [54].

After a spontaneous cyclization, the resulting intramolecular hemi-acetal can be oxidized to the lactone. Both oxidations are catalyzed by *horse liver* alcohol dehydrogenase (HLADH) in combination with different cofactor regeneration systems (Scheme 16.6) [28, 30, 44–46].

16.2.2.2 Secondary Alcohols

In the past, there were mainly two areas of interest for the dehydrogenase-catalyzed oxidation of secondary alcohols. First, they permit the regioselective oxidation of polyols (e.g. sugars) without the need for protecting groups [47–50]. Secondly, even though a stereogenic center is destructed, the enzyme can still act stereoselectively on its substrate, enabling a kinetic resolution of chiral secondary alcohols, in which one of the enantiomers is selectively oxidized while the second enantiomer remains unchanged. There are several examples for the oxidative kinetic resolution of secondary alcohols, which limits the yield to 50% [51–54]. DSM reported the kinetic resolution of 1-phenylethanol using an ADH from *Rhodococcus ruber* and an NAD(P)H oxidase from *Streptococcus mutans* and also discussed the possibilities to scale up the reaction to 1 l considering oxygen intake for NOx-based cofactor recycling (Scheme 16.7) [54].

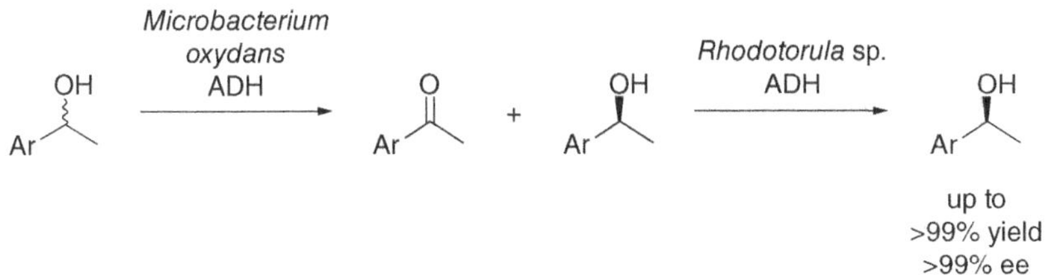

Scheme 16.8 Dynamic kinetic resolution of *rac*-2phenylethanol [58].

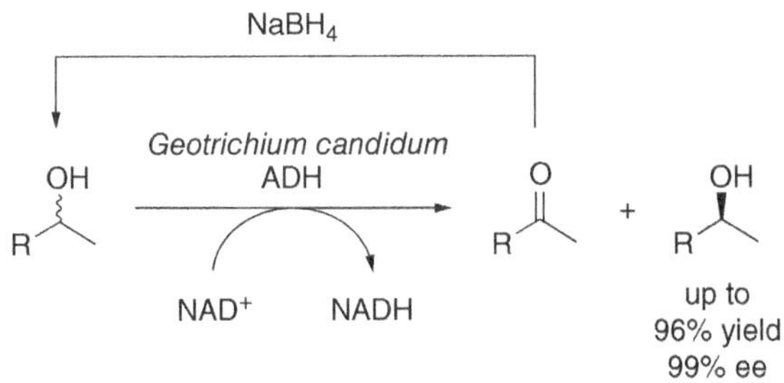

Scheme 16.9 Deracemization of secondary alcohols be means of a selective oxidation in combination with an unselective reduction [59].

By using *meso*-substrates, the limited conversion can be circumvented, increasing the theoretical yield to 100% [55]. Several possibilities for the dynamic kinetic resolution of chiral secondary alcohols have been examined in the past [56]. Beside systems that utilize unselective chemical oxidation (e.g. TEMPO oxidation) in combination with stereoselective enzymatic reduction [29, 57], such a dynamic kinetic resolution can also be set up by using an oxidative ADH. The most straight forward way is the combination of a stereoselective ADH for the oxidation and a stereo-complementary ADH for the reduction as used by Li et al. for the deracemization of 1-phenylethanol (Scheme 16.8). They obtained up to >99% *ee* and >99% yield for (*S*)-1-phenylethanol using (*R*)-selective *Microbacterium oxydans* for the oxidation and (*S*)-selective *Rhodotorula* sp. for the reduction [58].

In fact, Tanaka et al. showed that similar to the aforementioned combination of unselective oxidation and selective reduction, only one of the steps need to be selective. In their study, *Geotrichum candidum* and the unselective chemical reducing agent sodium borohydride were combined (Scheme 16.9) [59].

Alternatively, Voss et al. found *Rhodococcus* sp. to unselectively oxidize the alcohol while reducing the ketone selectively [60]. Musa and Karume et al. developed mutants of *Thermoanaerobacter ethanolicus* secondary alcohol dehydrogenase (TeSADH), which can act as racemization catalysts for a range of chiral secondary alcohols. By combining it with a stereoselective acylation, catalyzed by *Candida antarctica* lipase B (CALB), they could achieve a dynamic kinetic resolution toward acylated (*R*)-1-phenylethanol derivatives (Scheme 16.10) [61–63].

ADHs for alcohol oxidation have recently found more use in the context of enzymatic cascades. Usually, cascades have the advantage that a subsequent transformation of the ketone or consumption of NAD(P)H drives the redox-equilibrium of the ADH toward

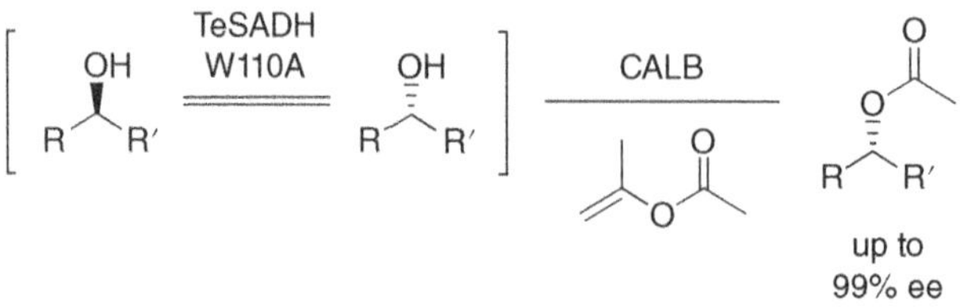

Scheme 16.10 Deracemization of secondary alcohols using a TeSADH mutant as racemization catalyst [61–63].

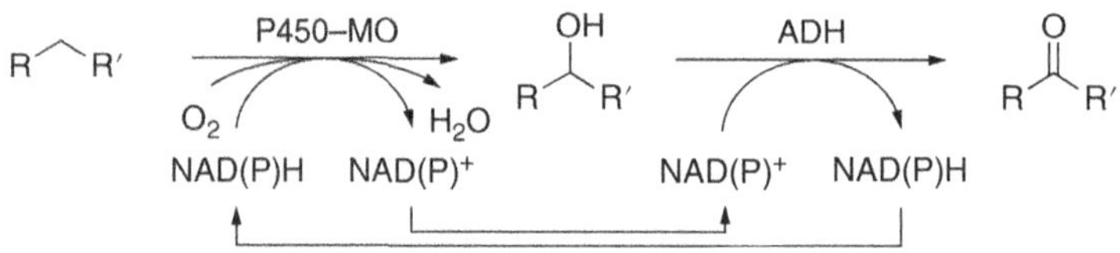

Scheme 16.11 General concept of alkane oxidation using a combination of P450 monooxygenase (P450-MO) and alcohol dehydrogenase (ADH).

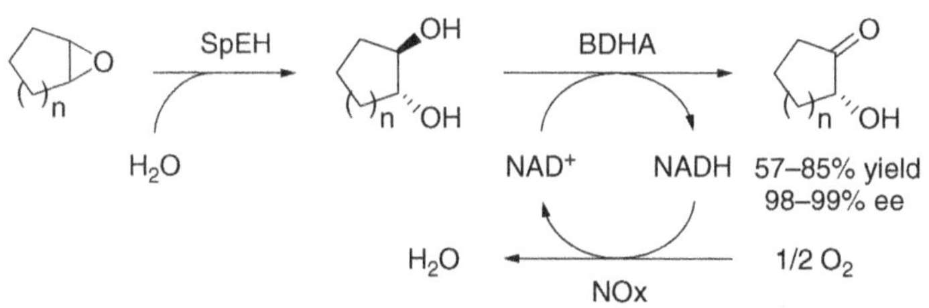

Scheme 16.12 Hydrolysis of epoxides by epoxide hydrolase (SpEH) and kinetic resolution of the resulting *gem*-diols via selective oxidation to α-hydroxy ketones by butanediol dehydrogenase (BDHA) [68].

the less favorable oxidative direction. Through P450 monooxygenases (MOs), which functionalize alkanes by hydroxylation, alcohols have become more readily available, whereas in the past most alcohols where made from the corresponding carbonyl compounds. Since P450-MOs reduce oxygen to water by oxidizing NAD(P)H to NAD(P)$^+$, they can be used in conjunction with an ADH catalyzed alcohol oxidation to oxidize alkanes to ketones without the need for external cofactor regeneration (Scheme 16.11). Examples include the oxidation of heptane to 2-heptanone, as well as cyclohexane and cyclooctane to cyclohexanone and cyclooctanone, respectively [64–67].

The combination of an epoxide hydrolase with an ADH as shown by Zhang et al. leads to chiral α-hydroxy ketones via geminal diols. *Meso*-epoxides, such as cyclohexene oxide, were converted into their respective (*R*)-α-hydroxy ketone with 85% conversion and 99% *ee* (Scheme 16.12). Furthermore, (*R*)-α-hydroxytetralone was obtained with 49% conversion and 99% *ee*. Despite the α-keto correlation, they did not observe racemization of the products [68].

In another interesting cascade, an ADH catalyzed oxidation of secondary alcohols was combined with either reductive amination by means of an amino acid dehydrogenase (AADH) [69, 70] or a transamination by means of a transaminase (TA) [71]. The latter is elegant in that it additionally includes an AADH (alanine dehydrogenase, ALD) for regeneration of cofactors and cosubstrates (Scheme 16.13). Since the ADH-AADH cascade is limited to amino acids, the ADH-TA-ALD cascade also extends the substrate scope to aliphatic alcohols.

Combining this concept with the aforementioned system of P450-MO and ADH, Both et al. extended the cascade by synthesizing the alcohol from ethylbenzene by using a monooxygenase. As an initial study, they obtained 26% overall conversion to (*R*)-phenylethylamine at 97.5% *ee* using whole-cell catalysts expressing all necessary enzymes (Scheme 16.14) [72].

16.2.3 Dehydrogenation of α-Amino Acids

Just like alcohol dehydrogenases, AADHs have been predominantly used in reductive, rather than oxidative direction; that is, the stereoselective synthesis of amino acids from keto acids. Therefore, only a few of examples for the reverse reaction are available. One application is the synthesis of rare D-amino acids by using native L-amino acid dehydrogenases (L-AADH) as shown by Hummel et al. for the kinetic resolution of DL-*tert*-leucine using an L-leucine dehydrogenase (L-LeuDH). In this process D-*tert*-Leucine was obtained at >99% *ee* (Scheme 16.15) [73].

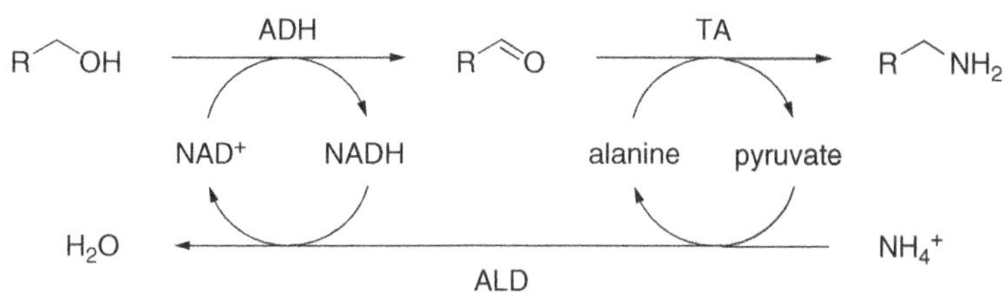

Scheme 16.13 Redox-neutral transformation of an alcohol to an amine by means of an alcohol dehydrogenase (ADH), a transaminase (TA) and an alanine dehydrogenase (ALD) [71].

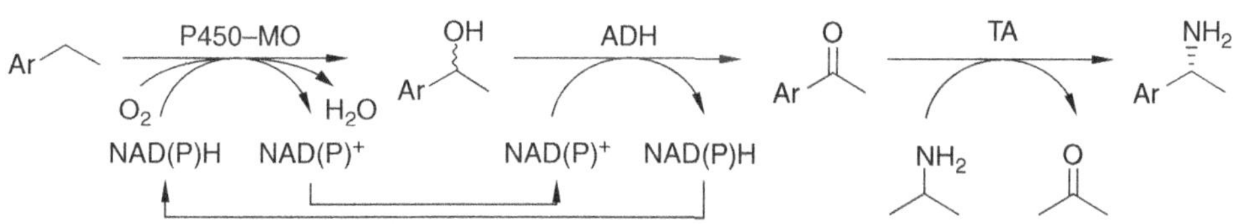

Scheme 16.14 Stereoselective amination of ethylbenzene derivatives via a combination of P450 monooxygenase (P450-MO), alcohol dehydrogenase (ADH) and transaminase (TA) [72].

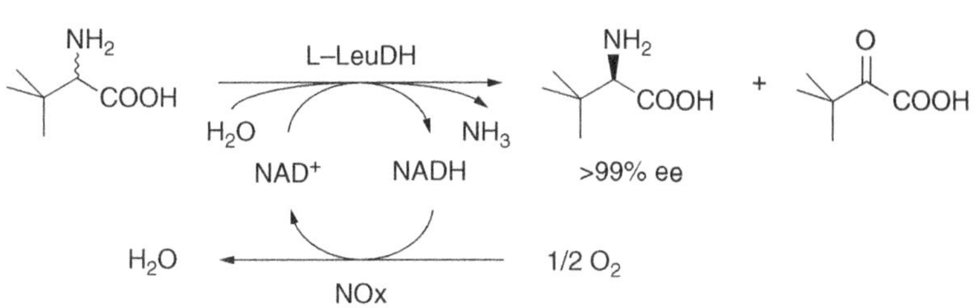

Scheme 16.15 Kinetic resolution of DL-*tert*-leucine via L-leucine dehydrogenase (LeuDH) and cofactor regeneration based on NADH oxidase (NOx) [73].

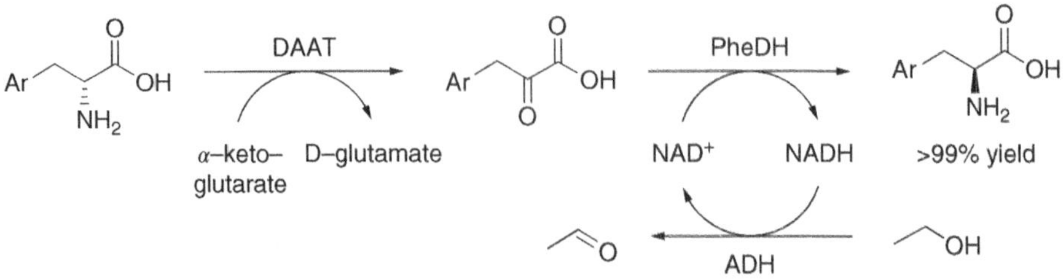

Scheme 16.16 Stereoinversion of D-*p*-bromophenylalanine to the L-enantiomer via *para*-bromophenylpyruvate using a combination of D-amino acid transaminase (DAAT), phenylalanine dehydrogenase (PheDH) and alcohol dehydrogenase [74].

As a model system for the artificial cofactor regeneration using iron–porphyrin complexes and molecular oxygen, Greschner et al. used the oxidative deamination of L-glutamate to α-ketoglutarate observing up to >99% conversion [32]. Khorsand et al. demonstrated the stereoinversion of D-*p*-bromophenylalanine combining a D-transaminase (D-TA) and L-AADH via *p*-bromophenylpyruvate. Additionally, an ADH was used for cofactor regeneration of the L-AADH oxidizing ethanol to acetaldehyde (Scheme 16.16) [74].

16.3 Oxidation with Monooxygenases

16.3.1 Introduction and Overview

Monooxygenases (EC 1.13.x.x and EC 1.14.x.x) activate molecular oxygen (O_2) via formation of an active peroxy-intermediate for catalysis of the subsequent regio- and stereoselective insertion of a single oxygen atom into an organic substrate (Scheme 16.17) [75–78]. Activation of the oxygen is achieved by the transfer of electrons from a donor, which can be the substrate itself (internal monooxygenases) or an external electron donor like the coenzyme NAD(P)H (external monooxygenases) [76, 79, 80]. The electron transfer from the donor to the oxygen can be mediated by cofactor(s) bound to the enzyme like heme, flavin, pterin, or metal ions (cofactor-dependent monooxygenases) or the substrate itself (cofactor-independent monooxygenases) [76, 77].

$$\text{substrate} + \text{donor–H} + O_2 + H^+ \xrightarrow[\text{}]{\text{monooxygenase (cofactor)}} \text{substrate–O} + \text{donor} + H_2O$$

donor–H = substrate: internal monooxygenase

donor–H = coenzyme: external monooxygenase

Scheme 16.17 General reaction equation for the oxidation of a substrate by an internal or external monooxygenase respectively.

Monooxygenases are widely present in nature and can be found in eukaryotes (mammals, insects, plants, fungi, protist), prokaryotes (bacteria, archaea), and even encoded in viruses (mimivirus) [81, 82]. The monooxygenases are classified into the following families based on the specific cofactor they use for catalysis (Figure 16.1):

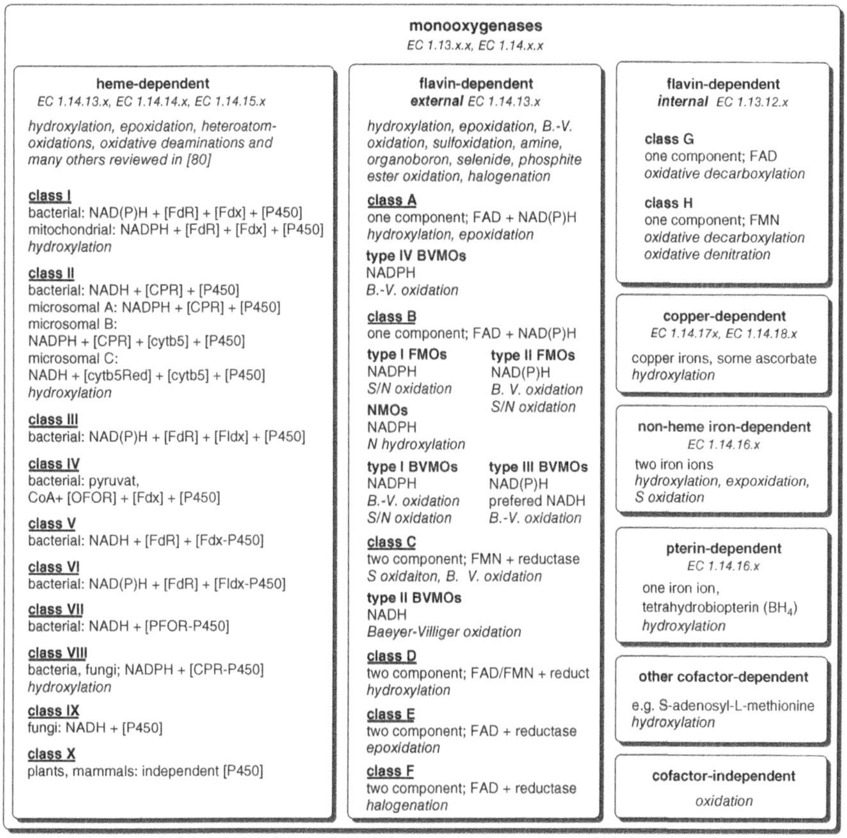

Fdx: iron–sulfur-cluster; FdR: ferredoxin reductase (FAD); CPR: cytochrome P450 reductase (FAD, FMN); Fldx: flavodoxin (FMN); OFOR: 2-oxoacid:ferredoxin oxidoreductase (thiamine pyrophosphate, [4Fe–4S] cluster); PFOR: phthalate-family oxygenase reductase (FMN, [2Fe–2S] cluster); FMO: flavin containing monooxygenase; NMO: *N*-hydroxylating monooxygenase; BVMO: Baeyer-Villiger monooxygenase

Figure 16.1 Overview of the different families and subfamilies of monooxygenases and the diverse oxidation reactions they catalyze. Categorization into families is based on the cofactor-dependency, whereas further subdivision is based on structural differences and/or the type of reaction [76–78,101–106].

heme-, flavin-, copper-, non-heme iron-, pterin-, as well as other cofactor-dependent and cofactor-independent monooxygenases [76, 77]. As a result of the functional interaction of the cofactor and the enzyme, reactions proceed according to different reaction mechanisms, which have been reviewed elsewhere [77].

In accordance to Scheme 16.17 external monooxygenases theoretically need equimolar amounts of cofactor. However, NAD(P)H can be consumed by heme-dependent monooxygenases via different shunt pathways under formation of toxic hydrogen peroxide or water without oxygenation of the substrate [84, 85]. This results in a reduced coupling efficiency, which is defined as the percentage of electrons from NAD(P)H used for product formation [86, 87]. Uncoupling is observed for non-native substrates and depends on the fitting of the substrate to the active site [88], which can be improved by protein engineering [89]. This undesired side-reaction has a tremendous impact on the economic efficiency and feasibility as it leads to the loss of reducing equivalents, increased oxygen demand and lowered specific activity per product [88, 90]. In contrast, in flavin-dependent monooxygenases side reactions are suppressed [91]. The application of external monooxygenases suffers from the consumption of cofactors and the need for cofactor regeneration. Thus, monooxygenases are frequently used in reaction cascades with other enzymes. Different strategies for cofactor regeneration have been already developed [92–95]. The application of whole cells is preferable to (semi)purified enzymes as the cellular cofactor pool usually obviates the need for external cofactor addition. Nevertheless, sometimes host cell proteins cause the formation of undesired side products [96, 97]. From an industrial point of view, internal monooxygenases are of special interest as they are independent of any cofactors. However, they are rare and show a very limited substrate scope [77].

The most prominent and best characterized family of monooxygenases with more than 11 000 gene products are the heme-dependent monooxygenases [75, 81, 82, 98]. Also known as cytochrome P450 monooxygenases or CYPs, they were historically named after their distinct absorption band with bound carbon monoxide at 450 nm – at the time, their function was unknown [99, 100]. The P450s are subdivided into 10 classes according the structure of their electron transport system (multicomponent, single chain, membrane-bound, soluble) (Figure 16.1) [101]. Mammalian P450s are involved in drug and xenobiotic metabolism, steroid hormone synthesis, fat-soluble vitamin metabolism, and conversion of polyunsaturated fatty acids to biologically active molecules. They also play a major role in insects and plants in the synthesis of secondary metabolites [81]. Last, in microorganisms they are involved in the degradation of fatty acids and alkanes as well as the synthesis of special fatty acid-based molecules such as secondary metabolites or signal molecules [78, 107]. Thus, it is hardly surprising that P450 monooxygenases catalyze diverse reactions on a broad range of substrates, like hydrocarbon and aromatic hydroxylation, *N*-hydroxylation, alkene and arene epoxidation, alkyne oxidation, *N*-oxidation, *S*-oxidation, aldehyde, and alcohol oxidation, oxidative deamination and dehalogenation as well as many other reactions, reviewed in Ref. [80]. Nevertheless, recombinant overexpression of P450s in soluble manner is not trivial, as many of these monooxygenases are multicomponent, membrane-bound enzymes [90, 108] – strategies for recombinant expression are described in Refs [109, 110].

With several hundred described and characterized enzymes the flavin-dependent monooxygenases represent the second largest family of monooxygenases, which are known to cover a wide range of different oxidation reactions in a highly regio- and/or enatioselective manner, like hydroxylation and Baeyer–Villiger monooxygenation. They characteristically employ a purely organic cofactor for oxygenation, either flavin mononucleotide (FMN) or flavin adenine dinucleotide (FAD). The flavin can function as a prosthetic group, tightly bound to the enzyme or as a coenzyme [76, 102]. Based on their diverse structure and function, flavin-dependent monooxygenases are divided into eight subfamilies (group A–H) [102], which themselves are partly further subdivided [76, 102, 103] (Figure 16.1). A detailed overview of the different types of monooxygenases and the oxidation reactions they catalyze is shown in Figure 16.1, which can be used as a tool for the selection of an enzyme (group) for a desired reaction.

Among the diverse repertoire of oxidations catalyzed by monooxygenases from a synthetic point of view the most frequently applied oxidative reactions are (enantioselective) hydroxylations, epoxidations, and Baeyer–Villiger oxidations (Scheme 16.18) [112]. Thus, the focus will be on these oxidations.

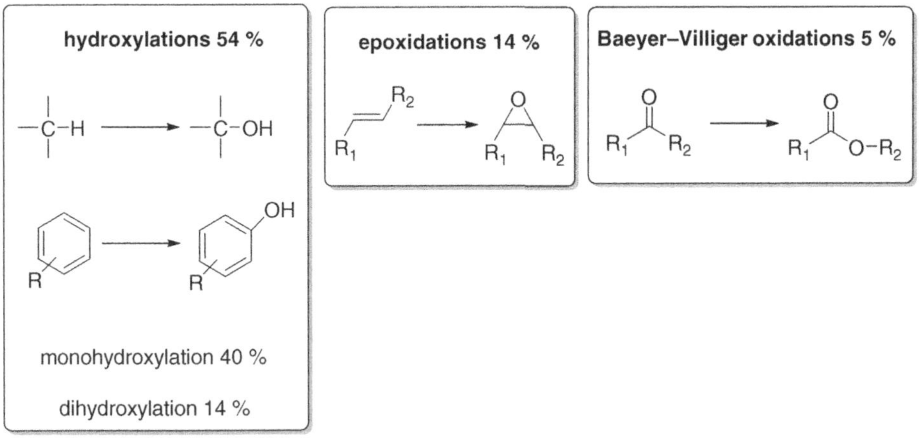

Scheme 16.18 Synthetically important enzymatic reactions with monooxygenases [101].

16.3.2 Hydroxylation

The stereoselective oxyfunctionalization of hydrocarbons like alkanes, arenes, and at benzylic position by means of chemocatalysis is still a challenging research area [113], thus making enzymatic approaches an attractive alternative [114]. One of the first observations of an enzymatic hydroxylation was the conversion of *n*-octane to 1-octanol by a non-heme di-iron monooxygenase (AlkBGT) from the soil bacterium *Pseudomonas oleovorans* GPo1 (later renamed *Pseudomonas putida* GPo1) in 1963 (Scheme 16.19, upper part) [117]. In contrast to the application of crude extracts, growing *P. oleovorans* showed no accumulation of 1-octanol [118], which was found due to utilization of *n*-octane as a sole energy and carbon source: 1-octanol is further metabolized by AlkJ to octanal, followed by AlkH mediated conversion to octanoic acid, to produce acyl-CoA (AlkK) for final degradation in the β-oxidation cycle [119].

Recombinant expression of an alkene hydroxylation system gene cluster in a *P. putida* strain enabled the whole-cell catalytic production of 1-octanol [118], whose production was estimated in 1998 to cost $6.00 to 10.00 per kg in a batch or a fed-batch process [120]. Alternatively, merely the alkane monooxygenase AlkBGT was recombinantly expressed in *Escherichia coli* W3110 for the hydroxylation of *n*-octane (Scheme 16.19, middle) [116]. As substrate mass transfer into cells was found to be limiting, alkL also from *Pseudomonas putida* GPo1, which putatively forms a hydrophobic pore in the outer membrane, was coexpressed. AlkL successfully facilitated the substrate transfer over the membrane and thereby increased the productivity of the whole-cell catalyst several fold, especially for longer chain alkanes (C_9–C_{16}) (Scheme 16.19, lower) [116, 121], but also for limonene [122] and dodecanoic acid [123]. Due to its importance, the alkL gene is patented by Evonik for a number of biocatalytic and fermentative processes [124, 125]. In the late 1970s the yeast *Candida guillermondii* (*Meyerozyma guilliermondii*) was found to hydroxylate long-chain alkanes (preferably C_{16}–C_{18}) to the corresponding primary alcohols. Further investigations revealed that a two component, NADPH-dependent cytochrome P450 monooxygenase was responsible for the hydroxylation [115, 126], whose mechanism also has been described [80, 127]. Meanwhile a broad

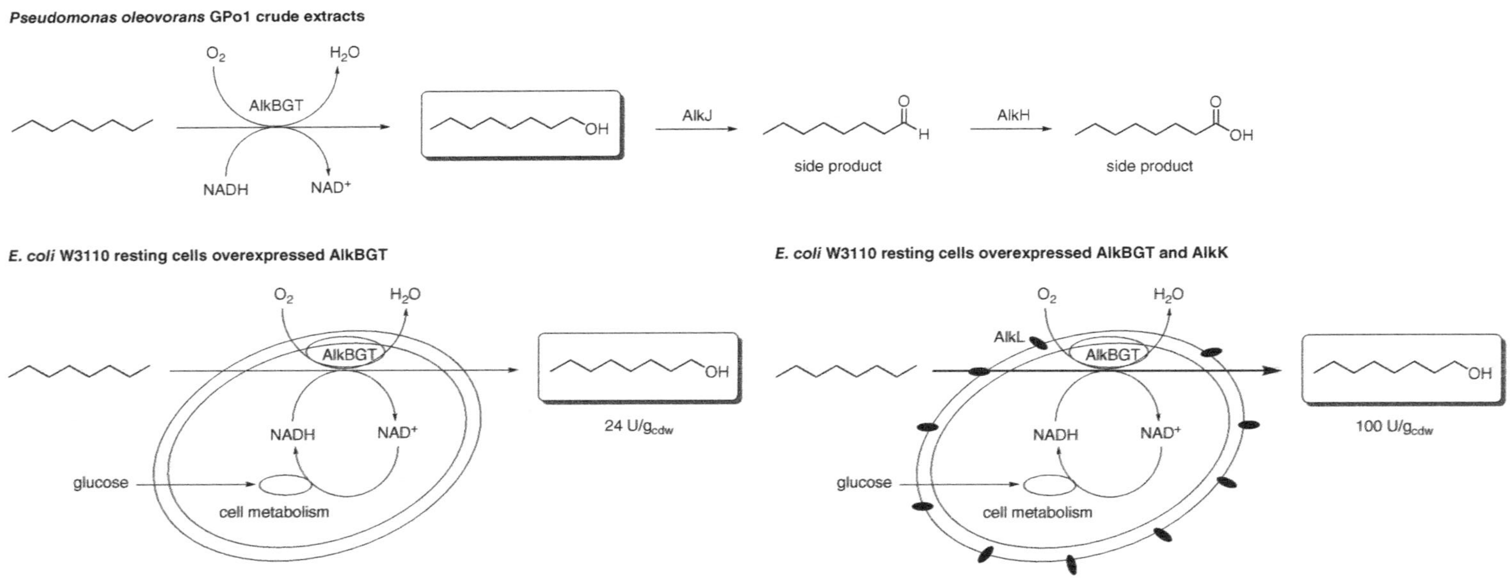

Scheme 16.19 *Upper part*: Biocatalytic conversion of octane to octanol by crude extract from *Pseudomonas oleovorans* GPo1 show side product formation due to further oxidation of 1-octanol by AlkJ toward octanal and by AlkH toward octanoic acid [115]. *Middle part*: Recombinant expression of the monooxygenase AlkBGT from *P. oleovorans* in *Escherichia coli* W3110 enables hydroxylation of octane to 1-octanol without side product formation. Nevertheless, this approach was limited by the substrate transfer over the outer membrane. *Lower*: By simultaneous overexpression of AlkBGT and AlkL, a putative membrane pore from *P. oleovorans*, in *E. coli*, catalytic activity could be increased fourfold [116].

range of natural and engineered monooxygenases are known, which altogether form a powerful toolbox for the hydroxylation of C_1–C_{22} *n*-alkanes to the corresponding primary and/or secondary alcohol(s) (Scheme 16.20) [66, 132–134].

Among the P450s the NADPH-dependent P450$_{BM3}$ from *Bacillus megaterium* (CYP102A1; class VIII) is one of the most popular, as it is soluble, self-sufficient (consists of a reductase domain and a P450 domain on the same peptide chain) [135–137] and thus can be successfully expressed recombinant, has a high catalytic turnover rate [138] and a high coupling efficiency of NADPH for product formation [139] – for a review, see Ref. [85]. P450$_{BM3}$ hydroxylates medium-chain (C_{12}–C_{20}) fatty acids ω-*n* ($n = 1$–3) as native substrates [85], but hydroxylates C_6–C_{10} alkanes merely with low activity and does not convert short-chain alkanes (<C_6) [140, 141]. Thus different P450$_{BM3}$ mutants were generated by directed evolution (random mutagenesis, recombination, and screening), which show increased activity toward alkanes (C_3–C_8) [134, 142], even with different enantioselectivity for octane [130, 131, 141] or high coupling efficiencies like 95% for propane [140]. Alternatively, the wild-type P450$_{BM3}$ was enabled to convert propane and butane to the corresponding 2-alkanols by the addition of perfluoro carboxylic acids (PFs) as inert dummy substrates (decoy molecules) [139, 143]. Nevertheless, for propane the coupling efficiency of 18% is very low [139]. Although coupling efficiency could be increased to 48% by high concentration of propane (0.5 MPa) in an autoclave (Scheme 16.21, left) [139], it is significantly lower than for the mutant. Interestingly, the pressure has an influence on the regioselectivity [139]. Working under high pressure is not desirable with respect to safety issues, thus an alternative approach is to utilize *n*-butane in liquid form without pressure at decreased temperatures. It was found, that P450$_{BM3}$(19A12) retains sufficiently active even at temperatures of 0 °C, at which the highest product concentration of 0.16 g l^{-1} of 2-butanol was obtained (Scheme 16.21, right) [144].

The selective conversion of the smallest and most inert alkane to the liquid fuel methanol is still challenging [147, 148] and can be performed by two different types of methane monooxygenases at ambient temperatures: the particulate methane monooxygenase (pMMO), a copper-dependent membrane protein, and the soluble methane monooxygenase (sMMO), an iron-dependent cytosolic protein [149]. Even after decades of research, biocatalytic application of these enzymes is restricted by a number of drawbacks, especially by the hurdle of functional heterologous expression of these complex multicomponent enzymes [143, 149, 150]. Thus, different P450s were used as a starting point for directed evolution and rational design [140, 151–153]. Despite these efforts, ethane is the smallest accepted alkane. However, perfluoro carboxylic acids as chemically inert guests enabled P450-mediated methane hydroxylation [143], as well as the use of iodosylbenzene as the oxidant, but with significantly lower turnover numbers (TONs) [150].

Besides primary alcohols, the production of chiral secondary alcohols as building blocks for chemical and pharmaceutical industry from cheaply available alkanes via monooxygenases offers an alternative synthetic route to ADH-based biocatalysis [145]. Enantioselectivity can be improved by directed evolution [131] or by oxidation of the unwanted enantiomer by an ADH to the corresponding ketone. This strategy was successfully applied for the production of *(S)*-2-octanol by a multicomponent P450 monooxygenase (CYP154A8, FdR/YkuN, or PdR/Pdx, each from different species)

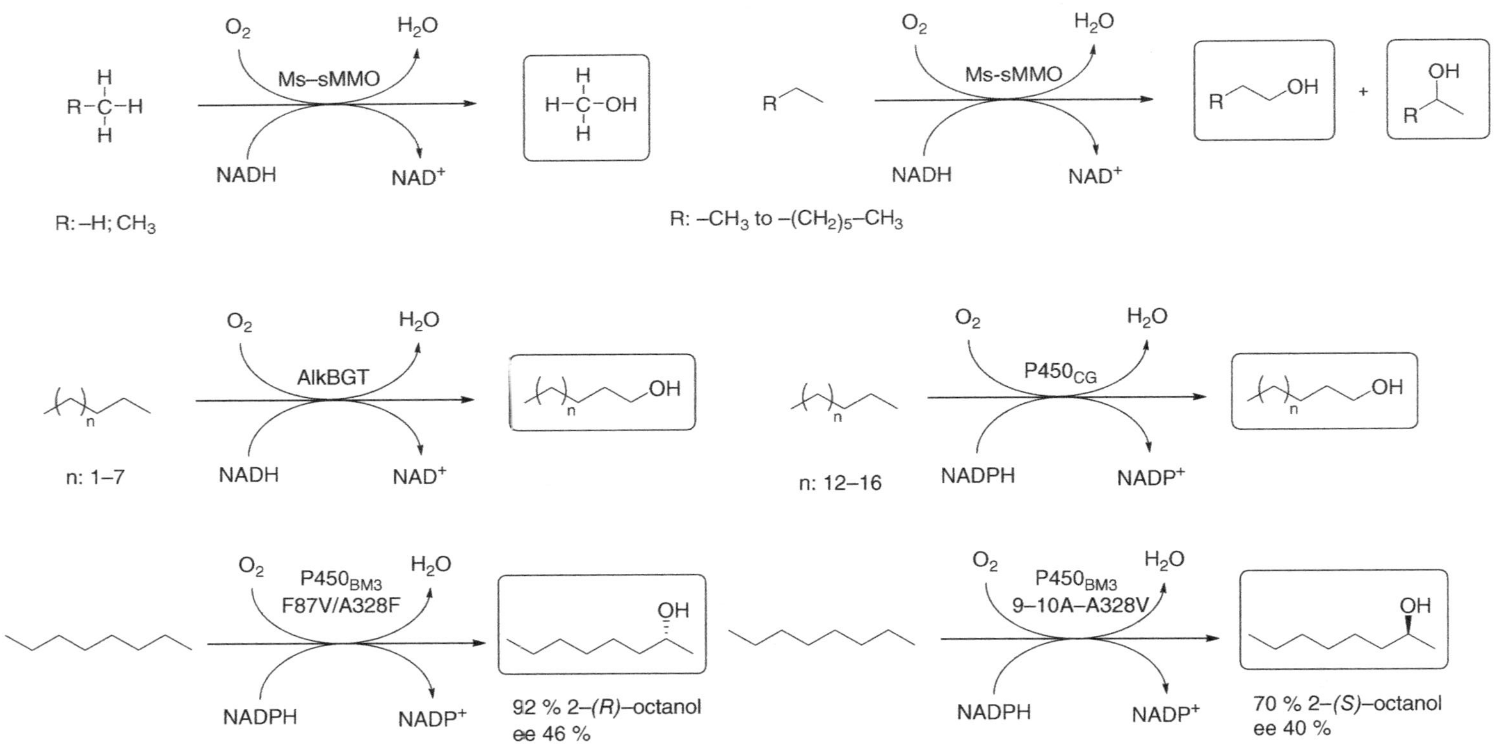

Scheme 16.20 Selected examples of different monooxygenases, which are able to hydroxylate *n*-alkanes of different side-chains to the primary or/and secondary alcohols: soluble methane monooxygenase from *Methylococcus capsulatus* hydroxylates gaseous methane and ethane to the corresponding alcohols, as well as C_3–C_8 alkanes to a mixture of 1-alkanols and 2-alkanols [128]; AlkBGT from *Pseudomona oleovorans* hydroxylates C_5–C_{12} alkanes exclusively to the coressponding 1-alkanols [129] as well as P450 from *Candida guillermondii* ($P450_{CG}$), which prefers C_{16}–C_{18} alkanes [115, 126]; In literature, different mutants of *Bacillus megaterium* P450s are described showing different substrate spectra and enantioselectivities (the mutant F87V/A328F produces preferably 2-(*R*)-octanol [130], whereas 9-10A-A328V predominantly 2-(*S*)-octanol [131]).

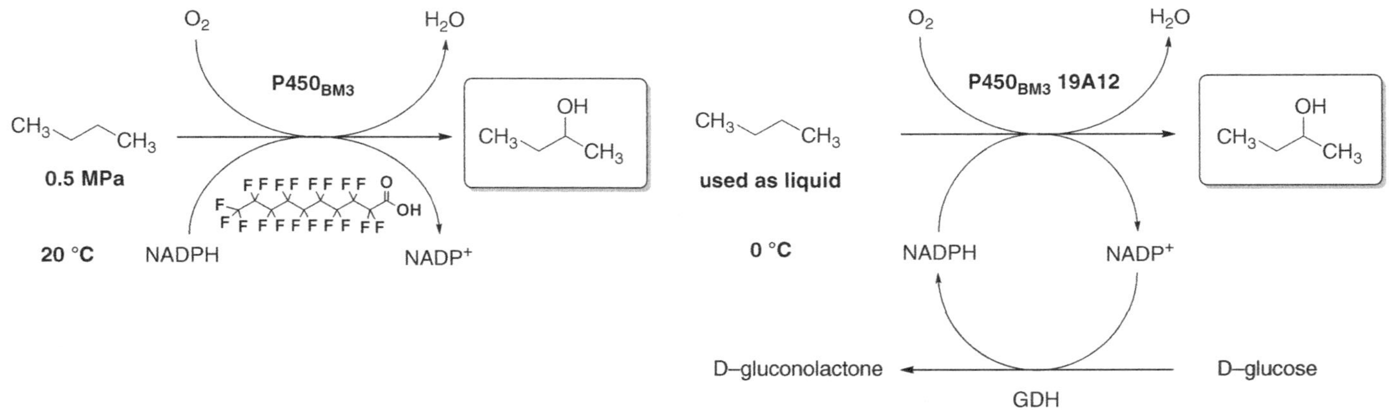

Scheme 16.21 *Left:* Synthesis of 2-butanol by $P450_{BM3}$ wild-type under high pressure (0.5 MPa at 20 °C) with the addition of the decoy molecule PFC10 [139]. *Right*: Application of the engineered mutant $P450_{BM3}$ 19A12 enables production of 2-bunatol at normal pressure at 0 °C, at which temperature *n*-butane is a liquid, with cofactor regeneration via glucose dehydrogenase (GDH) [144].

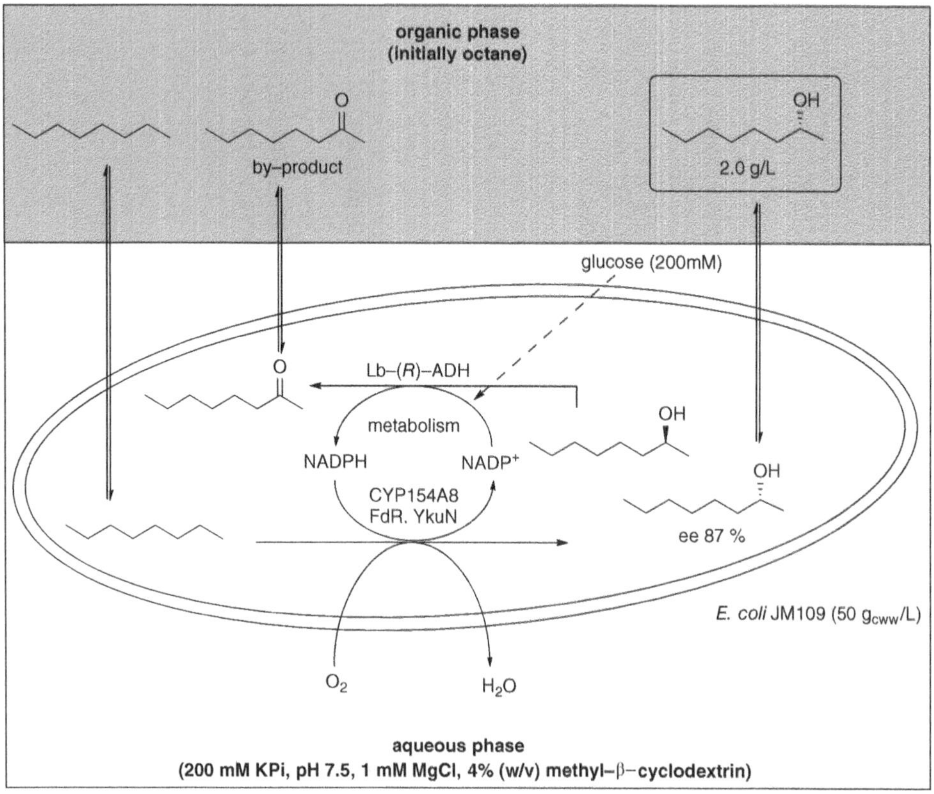

Scheme 16.22 Production of enantiomerically pure (*S*)-2-octanol by hydroxylation of octane via CYP154A8 and the redox partners FdR and YkuN with an *ee* of 87% using resting *Escherichia coli* whole-cell catalysts in a two-phase system. In the two-phase system octane acts as a substrate reservoir and simultaneously as an organic solvent for extraction of the products. The slightly formed (*R*)-2-octanol is converted to octanone via an *R*-selective ADH from *Lactobacillus kefir*. The cofactor is recycled by metabolizing glucose and also to some extent by the ADH-reaction. The addition of methyl-β-cyclodextrin stabilizes the monoxygenase [145, 146].

combined with *R*-selective Lb-ADH, which simultaneously facilitates cofactor regeneration, in a buffer-octane two-phase system (Scheme 16.22) [145, 146].

In 1965, the hydroxylation of cycloalkanes to their corresponding cycloalkanols (C_3–C_8) by a bacterium isolated from soil was observed [154] and further examples followed, like the alkane hydroxylase system from *Gordonia* sp. TF6 containing a non-heme monooxygenase AlkB [130, 133, 155]. In contrast to the flavin-dependent monooxygenases, which accept only aromatic substrates with an activating hydroxyl or amino group, P450s can hydroxylate also non-activated aliphatic or aromatic compounds [76]. Thus, isolated $P450_{BM3}$ mutants were utilized in cyclohexane-buffer biphasic system for the hydroxylation of cyclohexane. Cofactor regeneration was realized via a formate dehydrogenase (FDH) (Scheme 16.23, left) [155]. Alternatively, cofactor regeneration can be realized without an additional cosubstrate in a reaction integrated manner, enabling the production of the industrially relevant building blocks cyclohexanone, cyclooctanone, and cyclododecanone from the corresponding cycloalkanes, by combining a $P450_{BM3}$-mutant for hydroxylation and an alcohol dehydrogenase (ADH) for the subsequent formation of the ketones (Scheme 16.23, right) [64, 156].

Selective hydroxylation of aromatic compounds, which are used as precursors for pharmaceuticals, can be obtained by non-heme monooxygenases (di-iron, pterin, and flavin dependent), tyrosinase and P450s, reviewed in [157]. Due to its high activity, $P450_{BM3}$ was applied for hydroxylation of different benzenes [158–160] and was optimized by protein engineering ($P450_{BM3}$ R47S/Y51W/I401M) for the *o*-hydroxylation of different monosubstituted benzenes especially halogenated benzenes, whose corresponding phenols are important synthons [161].

Polyphenol oxidases (PPO; also called tyrosinases), which are copper containing monooxygenases, catalyze the *ortho*-hydroxylation of monophenols to the corresponding catechols, which are used as raw material for the synthesis of pesticides, drugs, and perfumes, which are hardly achievable via chemical synthesis [15, 162, 163]. From an industrial point of view, PPOs are highly interesting, as they do not require cofactors and can be applied in organic media [75] (for the catalytic mechanism, see [164]). Nevertheless, processes are limited by the subsequent oxidation of the products by the same enzyme to *ortho*-quinones (Scheme 16.24, upper) [145, 165, 68] and product inhibition [163]. Quinone formation can be prevented by applying a flavin containing NADH-dependent phenol monooxygenase (HbpA) from *Pseudomonas azelaica* HBP 1 in a recombinant *E. coli* strain (Scheme 16.24, lower) [166]. Apart from substituted catechols, which are as precursors in the manufacture of synthetic flavors, antioxidants, antihypertensive drugs, inks, plastics, and polymers [153], halocatechols are of special interest as they are used as intermediates for cross-coupling and halogen-metal exchange reactions in the synthesis of functionalized catechol building blocks [19]. However, biocatalysis struggles with the cell toxicity of phenols and catechols, thus, the substrate needs to be fed and the product must be removed *in situ* [166, 167]. Using these process strategies different 3-alkyl, 3-aryl, and 3-halocatechols could be produced on a gram scale with space–time yields up to $0.45\,g\,l^{-1}\,h^{-1}$ [166]. This approach is commercialized in a similar way by Fluka (now Sigma-Aldrich) on a 300 L reactor scale. *In situ* product removal is realized by continuous adsorption on the solid resin Amberlite™ XAD-4 and polymerization of dissolved 3-substituted catechol is prevented by a residence time of 10 hours [112]. The 4-halocatechols can be obtained via a 4-hydroxyphenylacetate 3-hydroxylase [169].

Catechols can be produced alternative from toluene via dihydroxylation by toluene *o*-monooxygenase (TOM) of *Burkholderia* (*Pseudomonas*) cepacia G4 [170] from benzene and toluene via successive hydroxylation by phenol hydroxylase from *Arthrobacter* sp. W1 (Scheme 16.25) [171] and combination of toluene-3-monooxygenase and phenol hydroxylase in *Burkholderia* (*Pseudomonas*) *picketti* PKO1 [172] as well as from *o*-xylene via toluene/o-xylene monooxygenase (ToMO) of *Pseudomonas stutzeri* OX1 [173].

With a world market of around 250 tons per annum, the catecholamine L-DOPA (3,4-dihydroxy-L-phenylalanine), which is used in the treatment of Parkinson's disease, is a very important representative [174, 175]. L-DOPA is frequently produced from L-tyrosine via different types of monooxygenases: biotransformations applying tyrosinases struggle with quinone formation (discussed in detail in [176]) and tyrosine hydroxylase from *Asperigillus terreus* [177] with oxidation of L-DOPA to L-2,3,4-trihydroxyphenylalanine (Scheme 16.26, upper) [178].

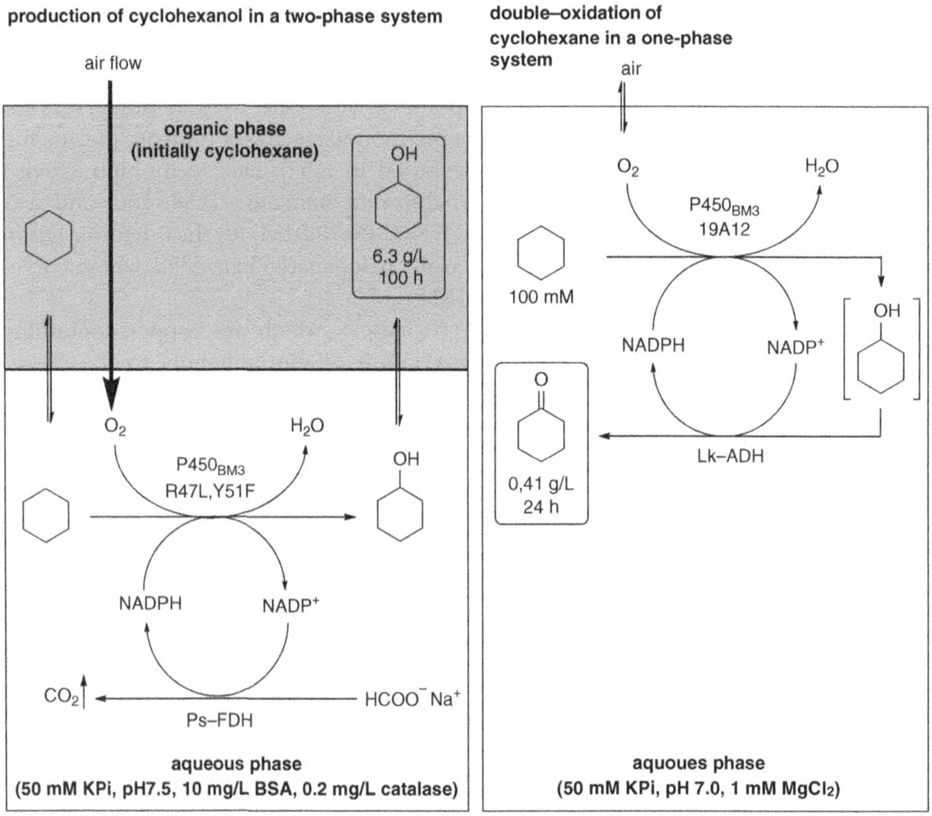

Scheme 16.23 *Left*: Oxidation of cyclohexane toward cyclohexanol in a biphasic system, consisting of a cyclohexane phase, which function as substrate pool and simultaneously as an organic solvent for *in situ* extraction of the product, and an aqueous phase, in which the enzymatic biotransformation takes place. A sufficient supply of the monooxygenase reaction with oxygen is realized by air flushing. The cofactor is regenerated via an FDH under consumption of formate. Addition of BSA and catalase stabilized the P450 [155]. *Right*: Combination of monooxygenase catalyzed hydroxylation of cyclohexane with subsequent alcohol dehydrogenase-catalyzed oxidation of the intermediate cyclohexanol to cyclohexanone in a pure aqueous system. As the cofactor is recycled within the reaction cascade, there is no need for an additional substrate [156].

In contrast, *p*-hydroxyphenylacetat 3-hydroxylase (PHAH) from *E. coli* W shows no side reactions and yielded 48 mM (9,4 g l^{-1}) L-DOPA, when applied as resting cells fed with L-tyrosine [178]. Alternatively, L-tyrosine can also be produced in a previous step from L-phenylalanine by a phenylalanine hydroxylase, a pteridine-dependent monooxygenase [179] or from glucose by a metabolically engineered *E. coli* strain (320 mg l^{-1}; 51.7 mg L-DOPA/g glucose) [180]. Potentially, the reaction cascade might be further extended by a DOPA decarboxylase to yield dopamine. Dopamine can be further oxidized to noradrenaline, a drug used, for example, for the treatment of anaphylaxis, by dopamine β-hydroxylase (Scheme 16.26, lower), a copper-dependent monooxygenase using ascorbate as an electron donator [181].

In addition, monooxygenases are of great significance in the synthesis of steroids – terpenoid lipids consisting of a carbon skeleton with four basal rings (A, B, C, and D)

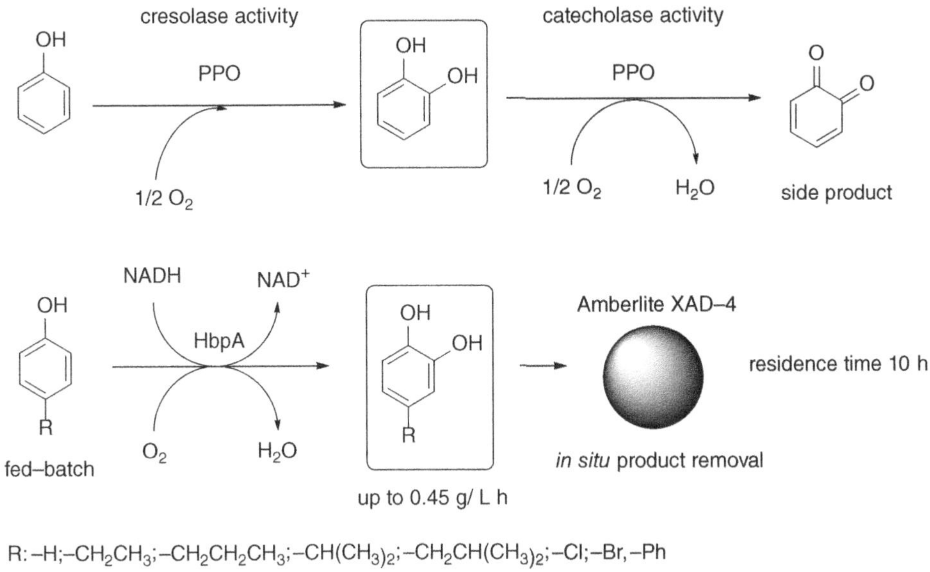

Scheme 16.24 *Above*: Polyphenol oxidase (PPO) from the mushroom *Agaricus bisporus* hydroxylates phenol via the cresolase activity to catechol. Due to catecholase activity the catechol is also partially further oxidized to quinone [165]. *Below*: Application of the NADH-dependent phenol monooxygenase from *Pseudomonas azelaica* HBP 1 (HbpA) enables the exclusive production of the catechol as well as 3-alky-, 3-aryl-, and 3-halocatechols, from the corresponding monophenols. Due to toxicity of the phenols and catechols, the substrate needs to be fed and the products removed by adsorption on a resin [112, 166, 167].

generally arranged in a 6-6-6-5 fashion. They represent the second largest category of the most commercialized medical products with more than 300 approved steroid drugs, encompassing a worldwide market of US$ 10 billion and more than one million tons per annum [182–184]. As efficient regio- and stereoselective functionalization of steroids is hardly achievable via chemical synthesis, without monooxygenase reactions many potent steroidal medications in the field of endocrinology, oncology, rheumatology, and gynecology would be hardly available on production scale at an economic price [183, 185, 186]. Furthermore, upon functionalization of steroids, hydroxylation is of special significance as it often improves biological activity, like in the case of 7α-hydroxy derivatives of dehydroepiandrosterone (DHEA), which exhibit a several fold higher immunoprotective and immunoregulatory properties compared to non-hydroxylated DHEA [183, 187]. Steroid biocatalysis suffers from the fact that most of the steroid hydroxylases are multicomponent membrane-bound P450 monooxygenases, such as the mammal P450s, and the repertoire of soluble P450s, like CYP106A2 from *B. megaterium*, is very limited [108, 182, 184, 188, 189]. Thus, historically wild-type microorganisms were used for steroid hydroxylation, in which actinobacteria are the most prominent and efficient [185]. Nevertheless, these organisms show limited substrate uptake, product degradation, product toxicity or significant by-product formation [190] and low expression levels. Thus, a lot of effort has been made in the recombinant expression of membrane-bound P450s [109, 110] – without [191] and with coexpression [192] of the natural redox partners in an appropriate ratio [193, 194]. These approaches unlock the

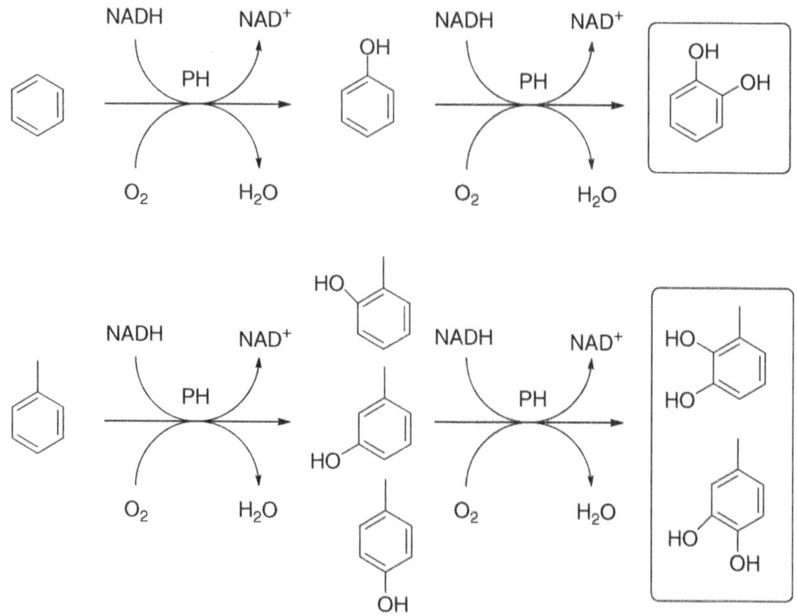

Scheme 16.25 Dihydroxylation of benzene and toluene to the corresponding catechols by phenol hydroxylase from *Arthrobacter* sp. W1 [171].

potential of mammalian P450s for biocatalytic application, as the use of mammalian cells for biocatalytic processes is not an option. For nearly all carbon atoms of the steroid skeleton, except for positions C4, C10, and C13, different microorganisms and enzymes are known to hydroxylate diverse substrates with different efficiency (Scheme 16.27) [182, 183, 185].

Among these reactions, the 7α-, 9α-, 11α-, 11β-, 15 α-, 16α-, and 17α-hydroxylations are of particular industrial importance [183, 184, 196]. In microbial steroid C11-hydroxylations it is observed that 11β-hydroxylation reactions generally proceed at lower yields and with more side reactions than 11α-hydroxylations, which might be due to a higher steric hindrance of the axial compared to the equatorial position [184]. As mentioned above, CYP106A2 from *B. megaterium* is the most prominent soluble steroid hydroxylase, showing a productivity of $5.5\,g\,l^{-1}\,day^{-1}$ for 15β-OH testosterone and 15β-OH progesterone [197]. Thus, it was subject to protein engineering for shifting the regioselectivity from C15 to C11 to increase the 11α-OH progesterone production compared with the formation of 15β-OH progesterone [184]. CYP102A1 from *B. megaterium* is not a steroid hydroxylase, but very attractive as it is self-sufficient and thus, was transformed by enlargement of the active site to a steroid hydroxylase enabling the hydroxylation of testosterone [198], progesterone [199], and norethisterone [200].

The first industrial application of microbial steroid hydroxylation is the conversion of progesterone to 11α-hydroxyprogesterone by *Rhizopus* sp., established in the 1950s by Pharmacia and Upjohn (now Pfizer) for the production of cortisone and derivatives (Scheme 16.28). [201–203]. Due to the microbial 11-hydroxylation of progesterone the production process of cortisone was dramatically simplified from 37 synthetic steps to 11 and thereby decreased the manufacturing costs from \$200 to \$6 per gram [204].

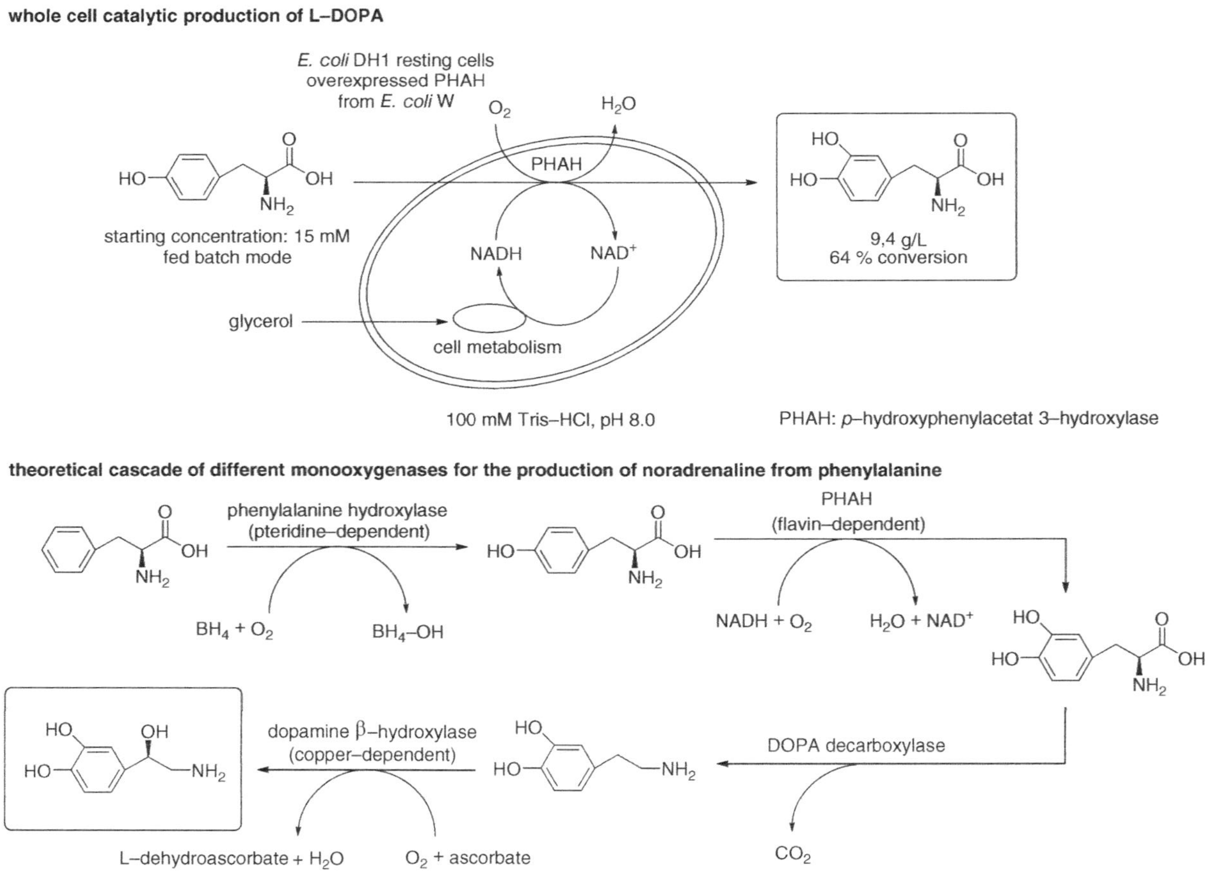

Scheme 16.26 *Above*: Production of L-DOPA by hydroxylation of L-tyrosine by resting *Escherichia coli* whole-cell catalyst containing overexpressed PHAH [178]. *Below*: Theoretical cascade for the production of noradrenaline from L-phenylalanine by combination of three different monooxygenases and a DOPA carboxylase. Recycling of the ascorbate can be realized via a NADPH- or NADH-dependent dehydroascorbate reductase. Regeneration of NAD(P)H can be obtained, for example, by an ADH transforming isopropanol to acetone.

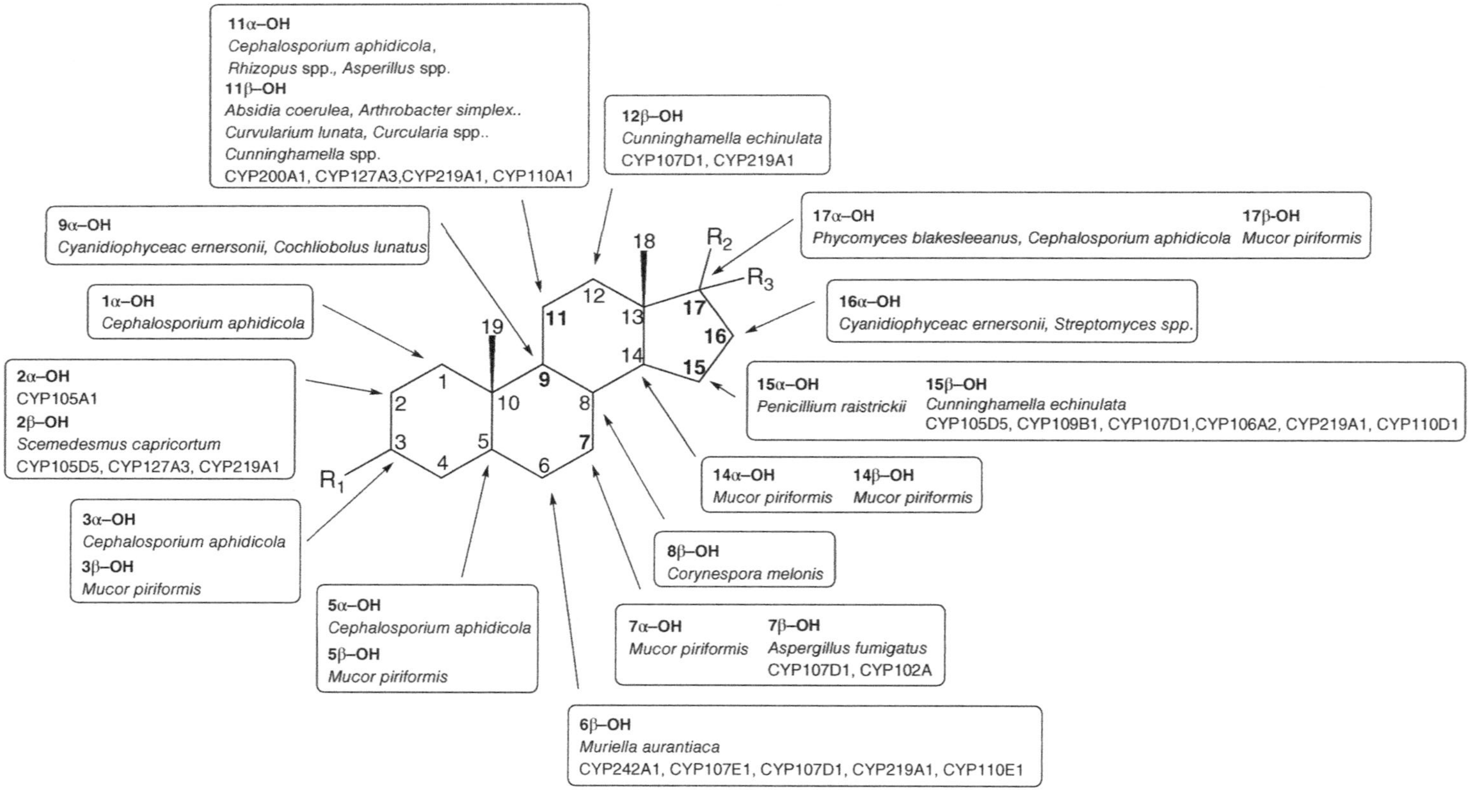

R_1: =O, –OH R_2: =O, –OH, $-CH_2OCH_3$ R_3: –OH, –H

Scheme 16.27 Known microorganisms and P450s for hydroxylation of nearly all carbon atoms of the steroid nucleus, except for position C4, C10, C13. The 7α-, 9α-, 11α-, 11β-, 15 α-, 16α-, and 17α-hydroxylation are of special industrial interest. Source: Information taken from [182, 183, 185, 195].

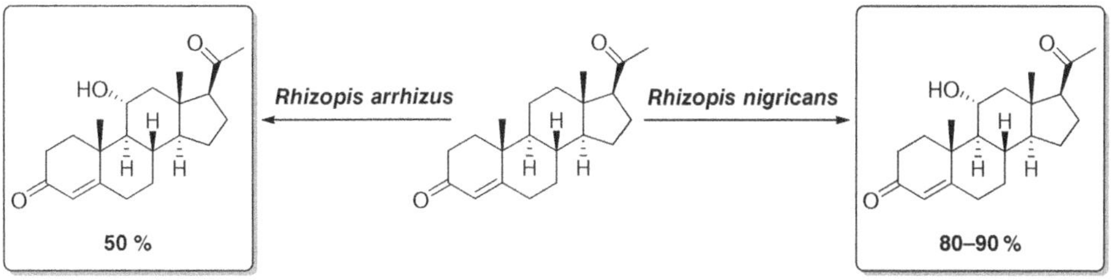

Scheme 16.28 Biotransformation of progesterone to 11α-hydroxyprogesterone developed in the 1950s by Pharmacia and Upjohn. Efficiency of the hydroxylation depends on the used whole-cell catalyst [201–203].

In addition, the 11β-hydroxylation of Reichstein's compound S by *Streptomyces fradiae* and *Cunninghamella blakesleeana* to hydrocortisone, the major adrenal glucocorticoid of mammals and an important intermediate of steroidal drug synthesis [200], was observed at Upjohn in the 1950s [205, 206]. At the same time at Squibb (today, Bristol-Myers-Squibb) the 11α-hydroxylation of Reichstein's compound S by *Aspergillus niger* was identified [207, 208]. Bayer Schering Pharma AG (now Bayer Pharma AG) utilizes *Curvularia* sp. at a scale of ~100 tons per year for the 11β-hydroxylation of Reichstein's compound S, a process that has no chemical equivalent [90, 209, 210]. The advances in recombinant expression of membrane-bound P450s enabled the preparative scale production of 6 β-hydroxytestosterone from testosterone with recombinantly expressed human CYP3A4 and NADPH-P450 reductase in *E. coli* [193]. Even a recombinant enzyme cascade (13 genes) was assembled in *Saccharomyces cerevisiae* encompassing also three steroid hydroxalases (bovine: CYP17A1; human: CYP11B1, CYP21A1) for the production of hydrocortisone from a simple carbon source [211]. This approach was considered for commercialization by Sanofi-Aventis [78].

Further industrial applications are the 6α-hydroxylation of compactin (ML-236B) to form the blockbuster cholesterol-lowering drug pravastatin by *Streptomyces carbophilus* containing CYP105A3 [212, 213] and by *Mucor hiemalis* industrially applied by Daiichi Sankyo and Bristol-Myers-Squibb [90].

16.3.3 Epoxidation of Non-activated Alkenes

One of the first observed biological epoxidations with industrial potential was the conversion of 1-octene to 1,2-epoxyoctane by *Pseudomonas putidas* (ATCC 29347), formerly *P. oleovorans*, in 1963 by Shell (Scheme 16.29, upper) [218]. Investigations revealed that a non-heme, three component, iron-dependent, soluble monooxygenase (alkB, alkG, alkT) is responsible for the epoxidation. [119, 219–222] As the same enzyme also catalyzes the terminal hydroxylation of alkanes [222, 223], besides 1,2-epoxyoctane also 7-octen-1-ol and 2-oxiranpentanol were obtained [214]. However, in the absence of any steric constraints, olefin epoxidation is kinetically more favorable than hydroxylation [220, 224]. The industrial relevance of biocatalytic epoxidation is elucidated by the intense efforts made by Exxon and Esso in the 1970: isolation of strains with improved epoxidation rate [225, 226], investigation of the substrate scope (C_6–C_{12} 1-alkenes and 1,7-octadiene), the optimal pH, temperature, substrate concentration and biomass loading for the application of resting cells [227]

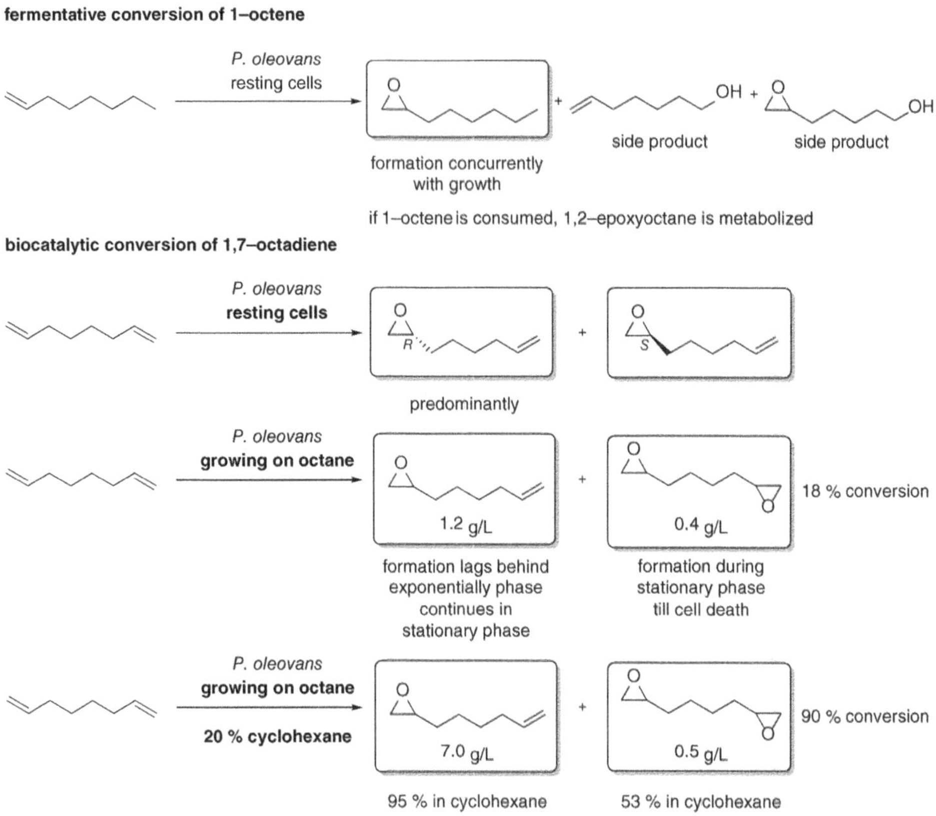

Scheme 16.29 *Above*: fermentative conversion of 1-octene to 1,2-epoxyoctane via *P.oleovorans* growing on 1-octene. As alkBGT also terminally hydroxylate alkanes 7-octen-1-ol and 2-oxiranpentanol are formed as side products. If 1-octene as the carbon source is consumed, 1,2-epoxyoctane is metabolized. *Below*: biocatalytic conversion of 1,7-octadiene with resting cells results stereoselectively in the formation of preferentially (*R*)-(+)-7,8-epoxy-1-octene. In a fermentative process *P. oleovans* produces 7,8-epoxy-1-octene, and 1,2,7,8-diepoxyoctane. Productivity can be improved by the addition of 20% cyclohexane as second phase. *Source:* Information taken from [214–217].

and the finding that cell viability is necessary for substrate transport and epoxidation [215]. The epoxidation of 1,7-octadiene by resting cells give preferentially (*R*)-(+)-7,8-epoxy-1-octene [216], whereas the application of growing cells also led to the formation of 1,2,7,8-diepoxyoctane in a growth phase-dependent manner (Scheme 16.29, lower) [215]. As cell growth is inhibited by 7,8-epoxy-1-octene in concentrations above $0.8\,g\,l^{-1}$ [215] conversion in the pure aqueous system is limited to 18%, but could be increased to 90% by the addition of an organic phase of 20% cyclohexane for *in situ* product removal [217].

Alkane-growing microorganisms, like *P. oleovorans*, are often able to epoxidize alkenes due to the activity of P450s or non-heme iron-dependent monooxygenases. As these enzymes generate a highly reactive iron-oxygen intermediate, which can attack not only carbon–carbon double bonds leading to epoxidation (mechanism described in [86]) but also unactivated C—H bonds, unintended hydroxylations are obtained [228, 229]. In contrast, monooxygenases from alkene utilizers generally do not hydroxylate alkanes and show higher enantioselectivity, like the alkene monooxygenase (Xamo) from *Xanthobacter* strain Py2 [230, 231]. Some alkene-utilizing bacteria are also reported to successfully expoxidate the gaseous and volatile ethene, propene, 1-butene, *cis*- and *trans*-2-butene [228]. Although alkene utilizers show a higher robustness against the cell toxic epoxyalkanes than alkane utilizers, the application of such alkene utilizers is problematic, as the epoxides are frequently intermediates of a decomposition pathway [230, 232]. Thus, recombinant expression is recommended. Examples are the recombinant expression of the *xylMA* genes from the TOL plasmid of *P. putida* mt-2 [233] and the styrene monooxygenase StyAB from styrene-degrading *Pseudomonas sp.* VLB120 for the selective oxidation of styrene to the corresponding *(S)*-styrene oxide [234], an important building block in the synthesis of optically active compounds like pharmaceuticals [235]. In a fermentative fed-batch process using a two-phase system consisting of a defined growth medium and bis(2-ethylhexyl)-phthalate as carrier solvent $11\,g^{-1}l$ of enantiopure *(S)*-styrene oxide was obtained after 10 hours [233]. Production of *(S)*-styrene oxide was also up-scaled to a 30 l fed-batch process with a product yield of 388 g [236]. Furthermore, asymmetric epoxidation of different styrene derivatives is described, showing that the conversion rates depend on the position and nature of the styrene substituent [237–239].

On an industrial scale *Nocordia coralline* B 276 is applied by Nippon Mining for the stereoselective epoxidation of different terminal and sub-terminal alkenes (short-chain gaseous C_3–C5, short-chain liquid C_6–C_{12} and long-chain $C_{>12}$ olefins) [112]. In contrast to StyAB, the monooxygenase from *Nocordia coralline* B 276 produces predominantly the *(R)*-epoxides. Due to the high toxicity of the produced epoxides for C_6–C_{12} substrates resting cells are used, whereas for the long-chain substrates growing cells are used. In both cases a two-phase system with alkanes as cosolvent is used for *in situ* product removal [112].

16.3.4 Baeyer–Villiger Oxidation

The cleavage of C—C bonds with the concomitant insertion of an oxygen atom by peroxy acids was firstly reported by von Baeyer and Villiger in 1899 [240]. In contrast to the toxic and hazardous peroxy acids, whose application is challenging in terms of safety issues especially on large-scale, Baeyer–Villiger monooxygenases (BVMOs) are able to catalyze the oxidation of ketones to the corresponding esters or lactones under mild conditions in an environmentally friendly way [241, 242].

The first indication of an enzymatic Baeyer–Villiger oxidation was observed in 1948 in a microbial degradation pathway of the steroid A-ring [243, 244], whereas the first BVMO – the 2,5-diketocamphane monooxygenase (2,5-DKCMO) from *Pseudomonas* sp., which catalyzes in the degradation pathway of camphor the lactonization of 2,5-diketocamphane – was identified and characterized in 1964 [245, 246]. In the last decade, the number of known BVMOs rapidly increased, although not all could be successfully recombinantly expressed (no type II BVMOs [104, 105, 242]). The substrate range of BVMOs comprises cyclic ketones (e.g. cyclohexanone monooxygenase, cyclopentanone monooxygenase), linear ketones (e.g. acetone monooxygenase, methyl ketone monooxygenase), aromatic ketones (e.g. phenylacetone monooxygenase), and aryl-aliphatic ketones (e.g. 4-hydroxyacetophenone monooxygenase) or large ring cyclic ketones (e.g. cyclododecanone monooxygenase and cyclopentadecanone monooxygenase) [247].

One of the best characterized BVMOs is the flavin- and NADPH-dependent cyclohexanone monooxygenase (CHMO) from *Acinetobacter* sp. NCIMB 9871 [248], which is esteemed for its ability to oxidize a wide range of aliphatic and small aromatic substrates (over 100 non-native substrates [249, 250]) in a regioselective manner and in many cases with high enantioselectivity [242, 249, 251, 252]. Originally, it was found to be responsible for the conversion of cyclohexanone to ε-caprolactone in the metabolism of *Acinetobacter* sp. NCIMB 9871 [253, 254]. The product ε-caprolactone belongs to the most important industrial bulk chemicals, which is produced on a multi-10 000 tons scale utilizing stoichiometric amounts of peracetic acid as oxidizing agent, having only a modest selectivity (85–90%) [3, 255].

The first milestone for the biocatalytic production, which requires only molecular oxygen as a reagent, was the successful cloning and recombinant expression of the CHMO from *Acinetobacter* sp. NCIMB 9871 in *E. coli* [256], since the wild-type organisms metabolize ε-caprolactone [253, 254]. Since that time, a lot of efforts have been made to improve the productivity. In a fermentative biotransformation with feeding of glucose as an energy source, recombinant *Corynebacterium glutamicum* produced $16.0\,g\,l^{-1}$ ε-caprolactone with volumetric productivity of $2.3\,g\,l^{-1}\,h^{-1}$ [257]. In this case, cofactor regeneration is realized by the cellular carbon metabolism. As cofactor recycling is a critical factor regarding economic feasibility, another approach is to combine the CHMO with an ADH for a cascade integrated cofactor recycling, simultaneously enabling the use of cyclohexanol as a readily available starting material [258–260].

However, the productivity of the process suffers from the low stability of the CHMO, which is inactivated by high concentrations of the substrate and the product. Thus, different mutants [252, 260–262] were engineered, for instance by introduction of stabilizing disulfide bonds [252, 260, 262], and different sources were screened for more stable BVMOs [263]. In addition, recombinant *E. coli* whole-cell catalysts can be encapsulated in alginate for biotransformation in a SpinChem-reactor facilitating consecutive batch reactions with reduced loss of activity [264].

Nevertheless, the CHMO also show a substrate and product inhibition, which were addressed both by a fed-batch approach [259, 264, 265] as well as different *in situ* product removal strategies [260, 265–267]. In a batch process $11.2\,g\,l^{-1}$ ε-caprolactone were obtained by extraction over a semipermeable PDMS-membrane by methylcyclohexane (99% conversion) [267]. The oligomers were obtained by polymerization via lipase A in the aqueous reaction medium [260], whereas the polymer polycaprolactone (PCL) – a final product made from its monomer – can be directly obtained by polymerization via lipase B in the methylcyclohexane compartment with a polymer fraction having a degree of polymerization of commercial PCL [267]. The reaction cascade was also extended by a P450 monooxygenase to obtain cyclohexanol by hydroxylation from cyclohexane [268]. Alternatively, ε-caprolactone can be hydrolyzed in the pure aqueous medium by lipase B under basic conditions to 6-hydroxyhexanoate, which after acidification and extraction can be polymerized enzymatically [265, 266]. CHMO and ADH were also combined by a two-substrate approach (cyclohexanone and 1,6-hexandiol) with opposite cofactor demand, leading in both parallel reactions to the formation of ε-caprolactone (Scheme 16.30) [269].

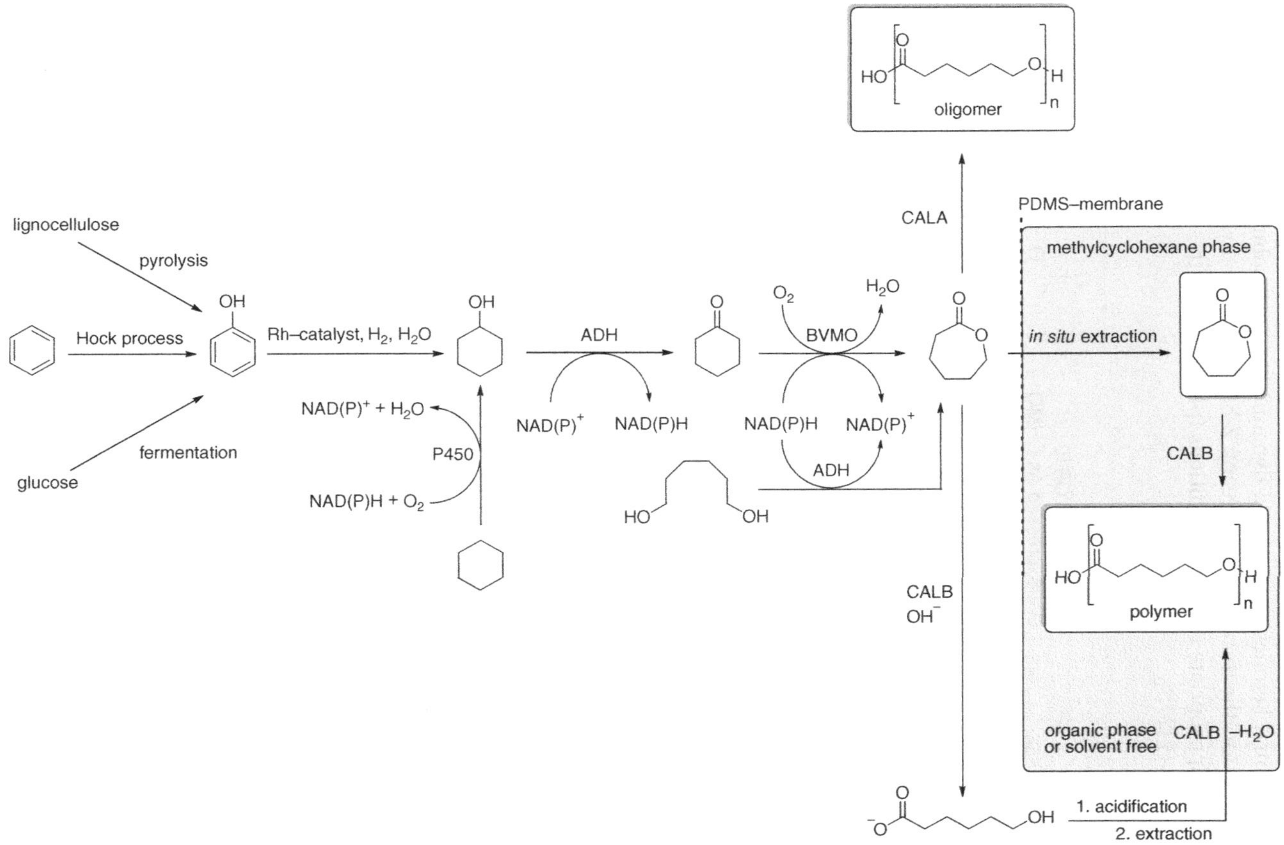

Scheme 16.30 Overview of the different (chemo)enzymatic cascades to couple the BVMO catalyzed oxidation of cylohexanone toward ε-caprolactone: (a) Beginning with P450 catalyzed hydroxylation of cyclohexane to obtain cyclohexanol, which is further oxidized to cyclohexanone *in situ* and ε-caprolactone. Suitable combinations of enzymes enable reaction internal cofactor regeneration [268]; (b) Starting from cyclohexanone (2 equivalents) and 1,6-hexandiol (1 equiv.) to yield ε-caprolactone (3 equiv.) [269]; (c) Phenol as a starting point with the prospective of coming from a biorenewable source is hydrogenated by Rh-catalyst. The cyclohexanol-crude product can be directly applied in the enzymatic double oxidation [267]. Product inhibition can be addressed by (a) oligomerization via lipase A from *Candida antarctica* in an aqueous system [260], (b) hydrolysis of ε-caprolactone via lipase B from *Candida antarctica* in an basic aqueous system yielding 6-hydroxyhexanoate [265, 266] or (c) by *in situ* extraction over a semipermeable PDMS-membrane [267] and (d) by combining *in situ* extraction with polymerization in the methylcyclohexane phase [267].

Although the conversion of cyclohexanone to ε-caprolactone was the first described CHMO oxidation, to date profitability of its biocatalytic production is still behind chemical processes, whereas the production of chiral lactones by BVMOs in the field of fine chemicals has a lower entrance barrier due to advantageous regio- and stereoselectivity of these enzymes [270–272]. Thus, BVMOs have been applied for desymmetrization of prochiral ketones as well as in dynamic kinetic resolutions, which was successfully shown for the racemic mixture of 2-benzyloxy-methylcyclopentanone yielding, under the appropriate reaction conditions (pH 9), 85% of the *(R)*-lactone with 96% *ee* (Scheme 16.31) [273].

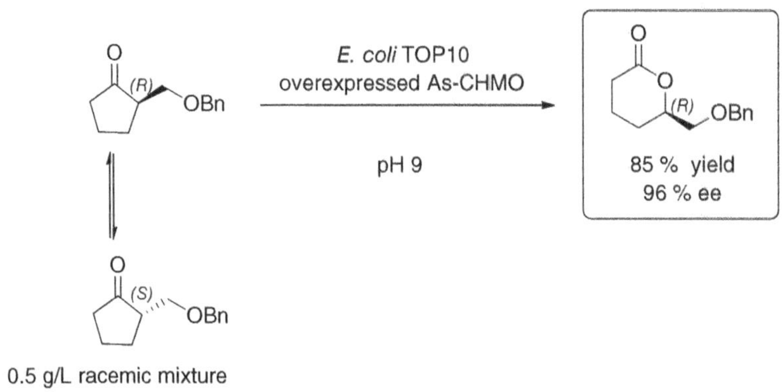

Scheme 16.31 Dynamic kinetic resolution of racemic 2-benzyloxy-methylcyclopentanone by CHMO toward (*R*)-6-benzyloxymethyl-tetrahydropyran-2-one [273].

The high synthetic potential of BVMOs is also exemplified by the oxidation of bicyclo[3.2.0]hept-2-en-6-one by *Acinetobacter* TD63 whole cells, a hydrolase deficient mutant of *Acinetobacter sp.* [274] to its corresponding regioisomeric lactones (−)-(1*S*,5*R*)-2-oxabicyclo[3.3.0]oct-6-en-3-one – which is a precursor of Corey lactone, one of the most important key intermediates in pharmaceutically relevant prostaglandin synthesis [275] – and (−)-(1*R*,5*S*)-3-oxabicyclo[3.3.0]oct-6-en-2-one (Scheme 16.32, upper) [277]. The expected "normal" lactone is formed by migration of the more substituted carbon atom, whereas the "non-typical" lactone is generated due to the migration of the less substituted one [278].

As *Acinetobacter calcoaceticus* belongs to class 2 organisms [279], application of this natural strain in industrial processes is not desirable. Thus, recombinant whole-cell catalysts, which also enable the overexpression of the CHMO, are the best choice. Investigations of different process parameters revealed a substrate and product inhibition, as well as aeration as limiting key parameters [272, 276]. These limitations were addressed by the combination of an increased biomass loading, improved aeration by a designed vessel and addition of an adsorbent resin Optipore L-493, enabling the increase of the substrate concentration from 1 to 20 g l^{-1} [276]. As the substrate also adsorbed on the resin, it functions as a substrate reservoir and simultaneously as an *in situ* product removal strategy (Scheme 16.32, lower) [276]. This process strategy could be successfully up-scaled to a 50 L bioreactor [272]. In contrast to many other reactions a biphasic approach showed only limited success [276]. Further investigations indicated that the cell membrane, which prevents cofactor leaching, is responsible for mass transfer

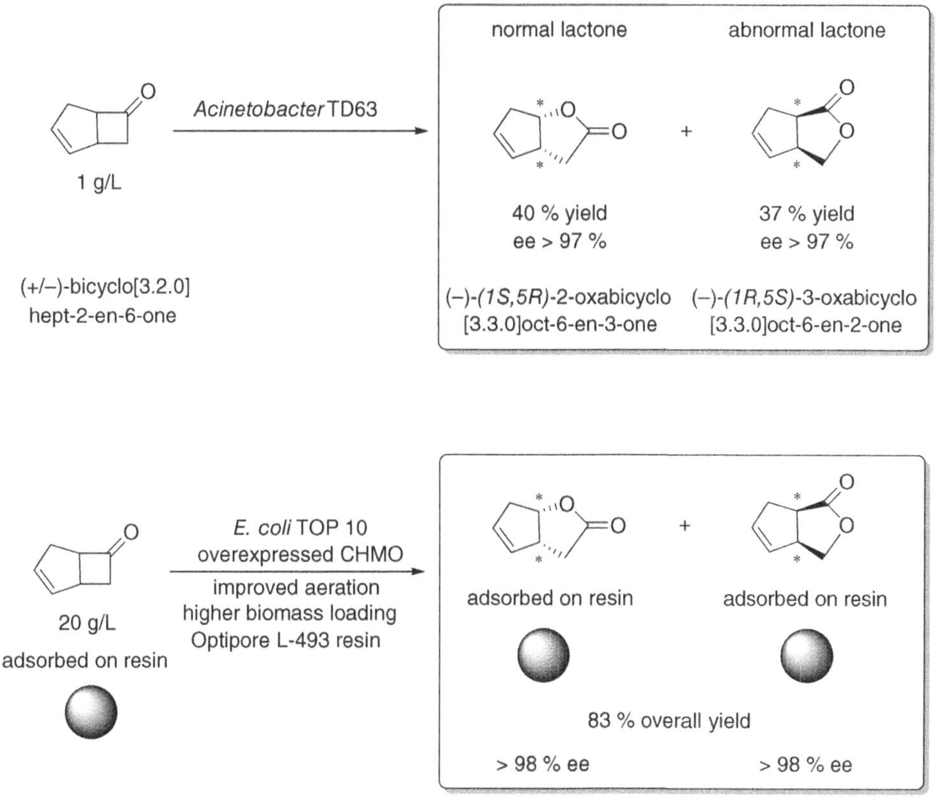

Scheme 16.32 *Upper*: Oxidation of bicyclo[3.2.0]hept-2-en-6-one to its corresponding regioisomeric lactones by *Acinetobacter* TD63 grown on cyclohexanediol to (–)-*(1S,5R)*-2 oxabicyclo[3.3.0]oct- 6-en-3-one and (–)-*(1R,5S)*-3-oxabicyclo[3.3.0]oct-6-en-2-one [272]. *Lower*: Productivity could be increased by the application of *Escherichia coli* whole-cell catalyst, higher biomass loading, improved aeration technique and *in situ* substrate feeding and product removal via an adsorbent resin [276].

limitations of substrate and products [280]. Furthermore, an optimal substrate concentration between 0.2–0.4 $g\,l^{-1}$ was detected, which could be maintained in a fed-batch process on a 200 l pilot-plant scale [280, 281].

16.4 Oxidation with Dioxygenases

16.4.1 Peroxide Formation

Reactive and unstable peroxo-species are intermediates in dioxygenase reactions in which both atoms of oxygen are transferred into the product. These hydro- or endoperoxides can be transformed afterward in enzymatic or non-enzymatic reductions or rearrangements [282].

Since dioxygen is chemically unreactive in the absence of a suitable catalyst, it needs to be activated. Transition metal ions that contain unpaired electrons, like metal cofactors in dioxygenases, can activate dioxygen in three different ways: Orbital overlap with a metal ion, single electron transfer and reaction with a substrate radical [282–292].

Orbital overlap with a metal ion: The unpaired electrons in the π^* orbitals of the dioxygen are able to overlap with those of the metal ion, if there is a complexation of the dioxygen to the transition metal. The orbital overlap of dioxygen and the metal ion then allows the reaction with a singlet organic reagent provided that the total number of unpaired electrons is constant [282, 293].

Single electron transfer: The metal ion in dioxygenases is able to transfer a single electron to bound oxygen, as the transition metals have two sequential available oxidation states (Fe(II)/Fe (III)). When $^3O_2$ oxygen accepts a single electron superoxide is formed, another possible oxygen activation [283]. Superoxide can afterwards undergo a variety of one- or two-electron chemical reactions [283, 284]. According to the redox potential this route is energetically unfavorable. Indeed, redox enzymes are able to influence the redox potential of bound cofactors [285]. This route of activation has, for example, been found in hemoglobin [286] and could also be used by metalloenzymes. An example is the indoleamine 2,3-dioxygenase, which catalyzes the oxidative cleavage of C=C bond in indolamine derivatives like tryptamine and serotonine. This dioxygenase can either be activated by binding of superoxide to the ferric enzyme or by binding of dioxygen to the ferrous enzyme [287].

Moreover, a single electron transfer can also be catalyzed by organic reagents that are able to build a stable radical intermediate, like flavin cofactor. By a single electron transfer, the flavin cofactor activates dioxygen in order to form a stable flavin semiquinone and superoxide that is recombined to a hydroperoxide intermediate (Scheme 16.33) [288].

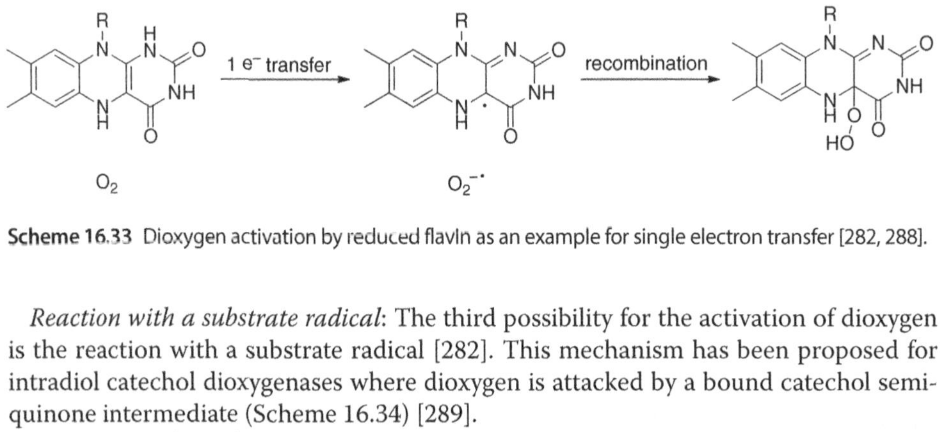

Scheme 16.33 Dioxygen activation by reduced flavin as an example for single electron transfer [282, 288].

Reaction with a substrate radical: The third possibility for the activation of dioxygen is the reaction with a substrate radical [282]. This mechanism has been proposed for intradiol catechol dioxygenases where dioxygen is attacked by a bound catechol semiquinone intermediate (Scheme 16.34) [289].

The proposed mechanism for lipoxygenases [290, 291] and cyclo oxygenases [292] also proceeds via formation of a substrate radical before reaction with dioxygen.

Moreover, a small number of dioxygenases are known to need neither a metal ion nor an organic cofactor to activate dioxygen. In these cases, a substrate carbanion can directly react with dioxygen accessing a stable substrate radical (Scheme 16.35) [282].

After activation of dioxygen and the formation of hydro- or endoperoxide intermediates, these peroxides can react in different ways, for example as nucleophiles or electrophiles. Moreover, cleavage of the peroxide bond can proceed via migration of the C—C bond or via a homolytic cleavage [282].

Hydroperoxides are excellent nucleophiles. An adjacent carbonyl group can, for example, undergo a nucleophilic attack by a hydroperoxide under formation of a

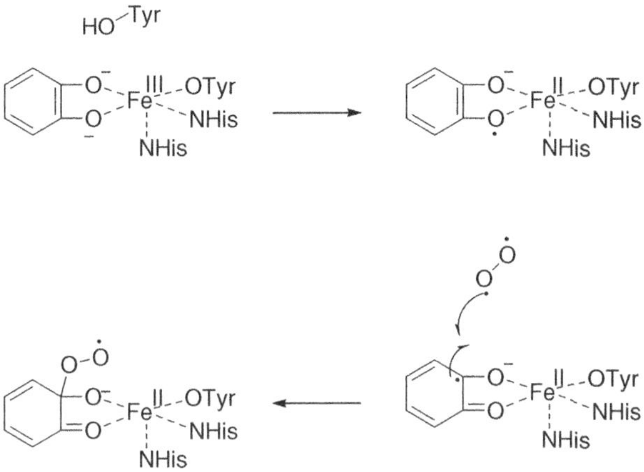

Scheme 16.34 Proposed mechanism for the activation of dioxygen in intradiol catechol dioxygenases [282, 289].

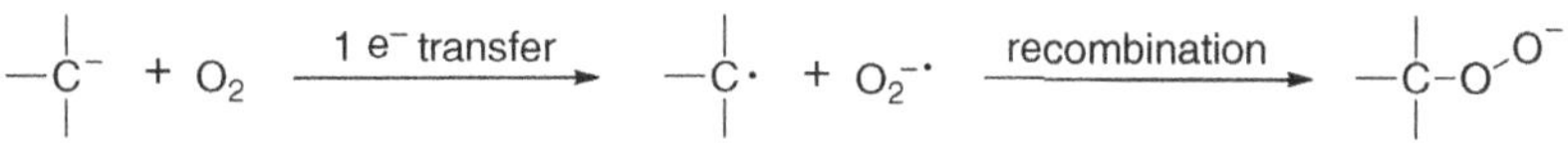

Scheme 16.35 Reaction of carbanion with dioxygen [282].

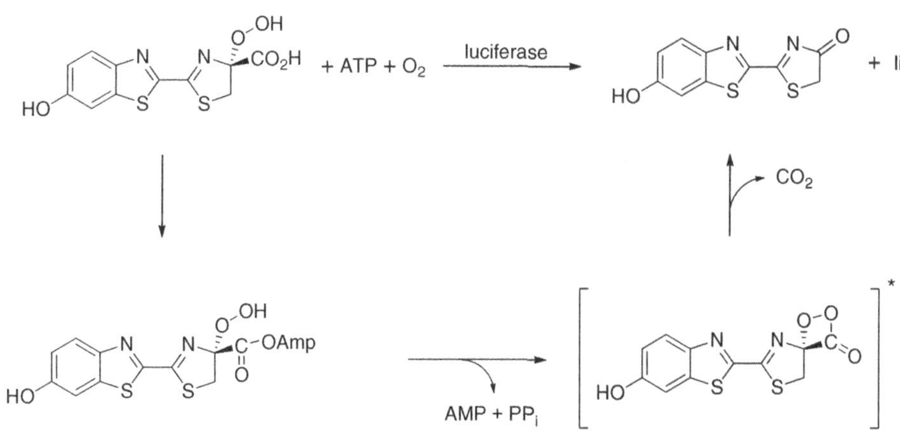

Scheme 16.36 Firefly luciferase catalyzed reaction with dioxetane intermediate [282, 294].

dioxetane intermediate [282]. The mechanism of the firefly luciferase enzyme, for example, proceeds via the formation of a dioxetane intermediate (Scheme 16.36) [294].

On the other hand, hydroperoxide can also act as oxidizing agent, when being attacked by a nucleophile [282]. An example is the flavin hydroperoxide intermediate in *p*-hydroxybenzoate hydroxylase that reacts with a phenol nucleophile (Scheme 16.37) [288].

Under acid catalysis, 1,2-rearrangement in which a C—C bond migrates onto the electron deficient oxygen of a peroxide can occur. This reaction is known as Criegee rearrangement (Scheme 16.38) [282, 295]. A Criegee rearrangement is, for example, proposed for proximal hydroperoxides in extradiol catechol dioxygenases [289].

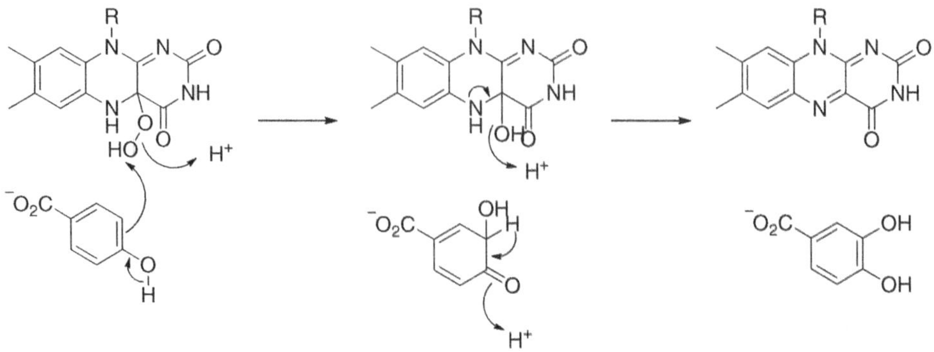

Scheme 16.37 Reaction of flavin hydroperoxide (electrophile) with phenol nucleophile [282, 288].

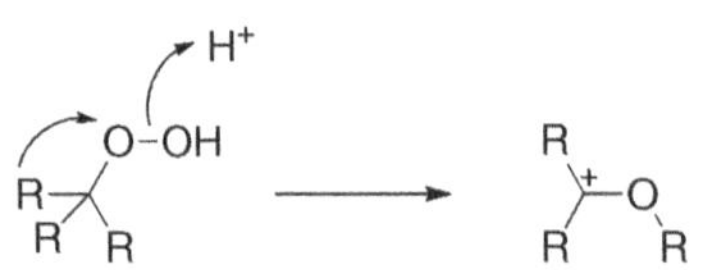

Scheme 16.38 1,2-rearrangement as an example of O-O-cleavage [282, 295].

Another possibility is the homolytic cleavage of the peroxide bond in the presence of transition metal catalysts; for example, iron salts. The emerging alkoxy radical can afterwards fragment via homolytic cleavage of an adjacent C—C bond. An iron(III) oxo complex would be the corresponding second product; for example, if the cleavage of the peroxide bond is initiated by complexation to iron(II). These iron(III) oxo complexes are proposed intermediates in non-heme iron-dependent metalloenzyme-catalyzed reactions among others [296]. An example for the homolytic cleavage of the peroxide bond is given in Scheme 16.39 [297].

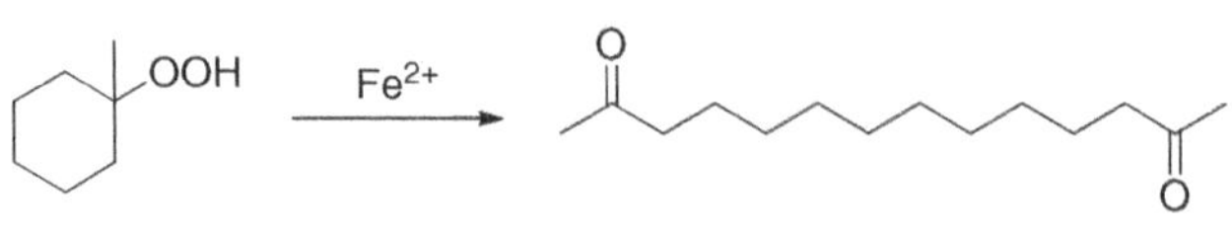

Scheme 16.39 Example of homolytic cleavage of the peroxide bond of organic hydroperoxides [282, 297].

16.4.2 Dihydroxylation of Aromatic Compounds

Most of aromatic compounds are monocyclic aromatic hydrocarbons including benzene, toluene, ethylbenzene and xylenes (BTEX), and polycyclic aromatic hydrocarbons (PAHs) containing benzene rings and two or three cyclic rings such as naphthalene, phenanthrene, anthracene, pyrene, and chrysenes [298]. Aromatic compounds are found in petroleum, which is important energy resource and raw material for industrial purpose, but also considered as pollutants. Therefore, bioremediation or biodegradation of petroleum pollutants has been focusing on improving biodegradation rate and reducing toxicity or petroleum in environment. To develop an effective biodegradation method that is low cost, safe, and does not need secondary waste treatment, screening of allochthonous microorganisms has been introduced [299, 300]. Two genes, ahdA1b-1 encoding polycyclic aromatic hydrocarbon ring-hydroxylating dioxygenase (PAH-RHDα) and xyLE encoding catechol-2,3-dioxygenase (C23O), was expressed from a copper-tolerant phenanthrene (PHE)-degrading bacterium, strain *Sphingobium* sp. PHE-1. These enzymes degrade phenanthrene up to $600\,mg\,l^{-1}$ and both related genes were induced in phenanthrene containing medium [299].

Apart from *in situ* bioremediation methods, biocatalysts are currently studied for degrading aromatic compounds. Dioxygenases are oxygenases that can introduce both oxygen atoms of molecular oxygen into a substrate [301]. Dioxygenases use four electrons or two electrons from a substrate and two from an external donor. Rieske-type iron-dependent dioxygenases are mononuclear non-heme iron-dependent dioxygenases, which catalyze *cis*-dihydroxylation reaction using NADH and two oxygen atoms as cosubstrates to produce *cis*-dihydrodiol products [302]. The previous study using benzoate 1,2-dioxygenase showed that it utilized benzoate, 4-fluorobenzoate (4-FB), 3,5-difluorobenzoate (3,5-FB), 3,4,5-trifluorobenzoate (3,4,5-FB) as substrates and showed production yield to 47 ± 3, 62 ± 5, 63 ± 7, $51 \pm 8\%$, respectively [302]. In the first step, substrate and O_2 are bound, forming a reactive species in the next step, and finally an electron transfer from the Rieske cluster proceeds. The resulting Fe(III)-peroxo-aryl radical species then readily accepts an electron from the Rieske cluster to complete the cis-dihydroxylation reaction (Scheme 16.40) [302].

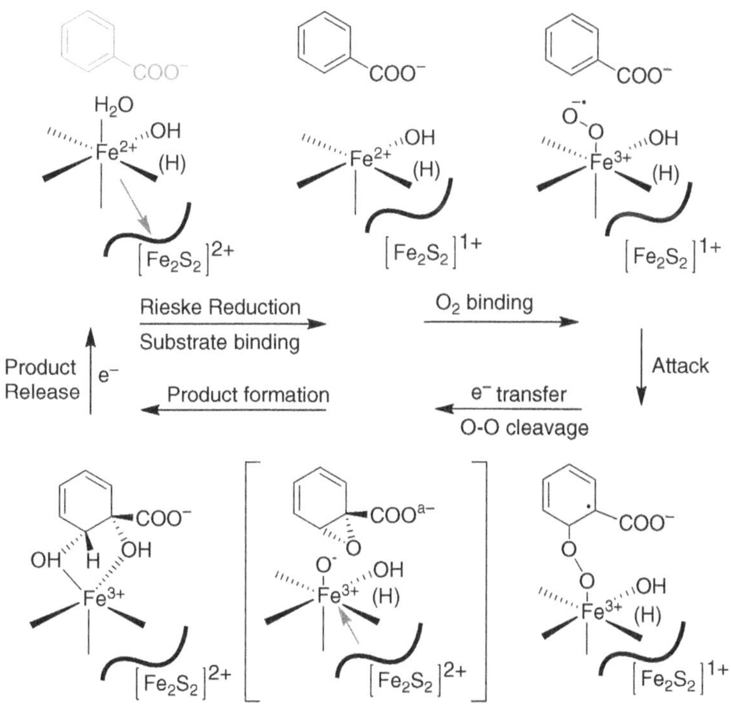

Scheme 16.40 Reaction cycle for benzoate 1,2-dioxygenase [302].

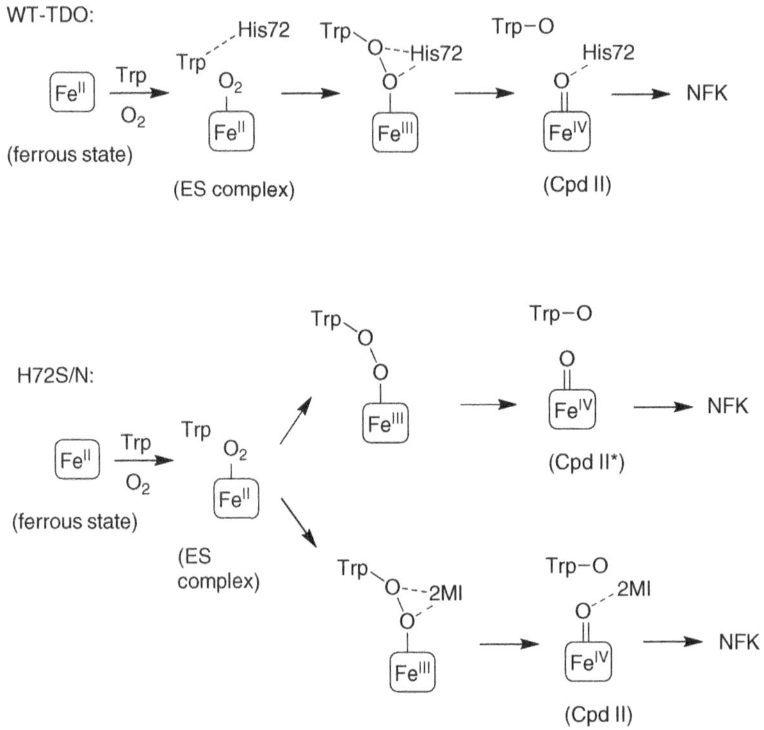

Scheme 16.41 Proposed role of His72 in catalysis during the dioxygenase reaction. WT-TDO: wild-type tryptophan 2,3-dioxygenase [303]. Trp; tryptophan, NFK; N-formylkynurenine, Cdp II; Fe(IV)O intermediate, 2MI; methylimidazole.

For a deeper understanding of the oxidation mechanism of Rieske-type iron-dependent dioxygenases, site directed mutagenesis was performed [303]. Mutant proteins H72S, H72N, and Q73F were generated in *Cupriavidus metallidurans* tryptophan 2,3-dioxygenase (cmTDO) and the function of distal His72 was studied by measuring spectroscopic characterizations, enzymatic kinetic analysis, and chemical rescue assays with L-tryptophan as a substrate. Exchanging His72 to Ser/Asn decreased the K_{cat} value to approximately 5% and 0.5% of that of the wild-type enzyme, respectively. This study revealed that His72 is not only essential for substrate binding but is assumed to stabilize the oxygen-bound heme intermediates (Scheme 16.41).

Another type of Rieske-type iron-dependent dioxygenase is the naphthalene dioxygenase, which consists of an NAD(P)H-oxidoreductase, a ferredoxin and an oxygenase. Recently, a study of the degradation pathway of high molecular weight-PAHs by biocatalyst has been published [111]. Naphthalene dioxygenase from *Sphingomonas* CHY-1 is currently reported to have the broadest specificity of all dioxygenases for high molecular weight (HMW) PAHs. This dioxygenase catalyzed the oxidation of fluoranthene and chrysene to *cis*-7,8-dihydroxy-7,8-dihydrofluoranthene and *cis,cis*-3,4,9,10-tetrahydroxy-3,4,9,10-tetrahydrochrysene, respectively (Figure 16.2). PAH substrates such as benz[a]anthracene, benzo[a]pyrene, and pyrene converted to *cis*-1,2-dihydrodiol, *cis*-4,5-dihydrodiol, and *cis*-9,10-dihydrodiol as main products (Figure 16.2). The substrate preference of the naphthalene dioxygenase follows the order from fluoranthene, benz[a]anthracene, chrysene, benzo[a]pyrene, and pyrene.

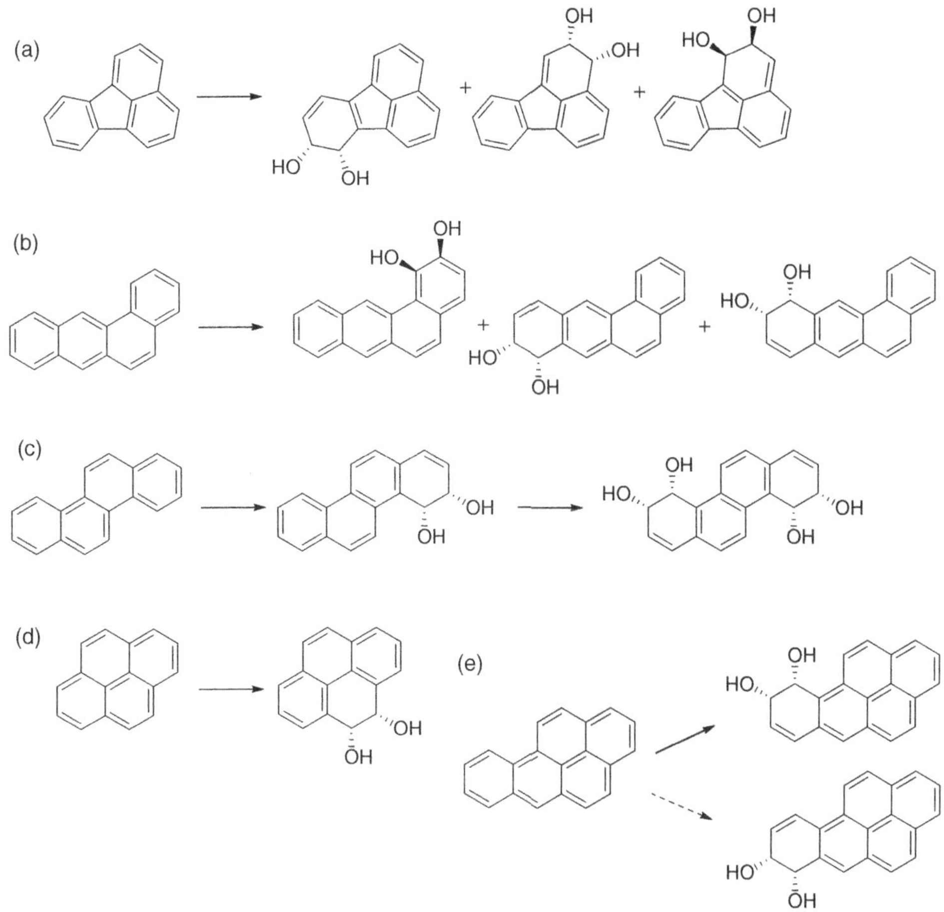

Figure 16.2 Oxidation products generated by *Sphingomonas* CHY-1 RHD from four- and firve-ring PAHs. The dihydrodiols identified as dioxygenation products of fluoroanthene (a), benz[a]anthracene (b), chrysene (c), pyrene (d), and benzo[a]pyrene (e) are presented [111]. Reproduced with permission of John Wiley & Sons.

Another type of non-heme iron enzymes containing the 2-his-1-carboxylate facial triad (two histidine residues and one aspartate or glutamate residue arranged at the vertices of one face of the iron-coordinating octahedron) can catalyze the oxidative ring cleavage of aromatic compounds. The reaction mechanism of oxidative cleavage of 2-aminophenol using 2-aminophenol 1,6-dioxygenase (APD) was studied by quantum and molecular mechanics analysis [304]. Accordingly, in the initial step Fe—O_2 adducts (Fe(III) – superoxo or substrate radical–Fe(II) – superoxo) are formed, in which a proton transfer from the substrate to His195 coupled an electron transfer from the substrate to Fe is involved. Subsequently, a proton transfer from His195 to the proximal oxygen of the dioxygen ligand coupled with the transfer of an electron from the substrate to the dioxygen takes place. Some aromatic compounds that have cross-linked aromatic compounds such as lignin are broken down into individual aromatic molecules by biocatalysts [301]. Oxygenases were utilized to couple ring cleavage with oxygen insertion and ring-opening of aromatic compounds. Metapyrocatechase was isolated from *Pseudomonas* species and was tested toward catalyzing the oxygenation of catechol to

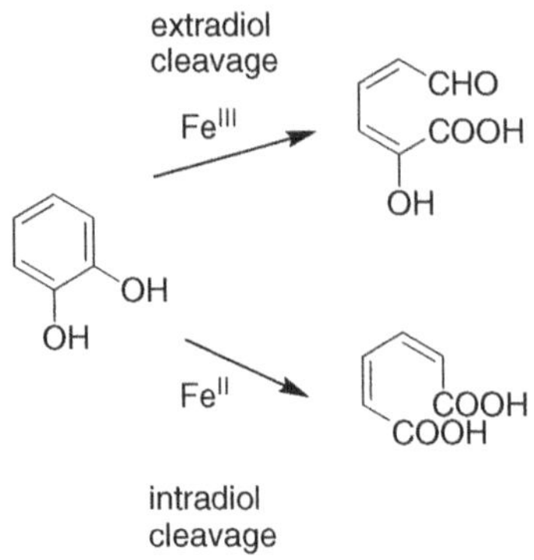

Figure 16.3 Distinct natural catechol dioxygenase activities [301]. Reproduced with permission of John Wiley & Sons.

a-hydroxymucomic semialdehyde [111]. By means of this enzyme, oxygen could be inserted to catecholic substrates at positions near the two phenolic hydroxyl groups. A similar study was performed with a 3,4-dihydroxyphenylacetate-2,3-dioxygenase (homo-proteocatechuate 2,3-dioxygenase), which catalyze oxidation of 3,4-dihydroxyphenylacetate to 2-hydroxy-5-carboxymethylmuconate semialdehyde [305]. Another catechol aromatic ring-degrading enzyme, protocatechuate-3,4-dioxygenase (3,4-PCD), is also known for converting 3,4-dihydroxybenzoate to 3-carboxy-*cis,cis*-muconate in a reaction similar to that of the catechol 1,2-dioxygenase [306]. Intradiol and extradiol dioxygenases utilize a non-heme Fe center that can use triplet dioxygen to react with substrates and have a third iron ligand site near the substrate ligand sites to form the oxygen intermediate. The reaction with catechol dioxygenase is shown in Figure 16.3.

16.5 Oxidation with Oxidases

16.5.1 Resolution of Amines via Amine Oxidases

Amine oxidases are enzymes catalyzing the oxidative deamination of amino moieties. They can be found in bacteria, fungi, and mammals including humans.

Monoamine oxidases catalyze the conversion of amines into the corresponding imines. Simple amines such as butylamine, amylamine, and benzylamine are oxidized to imines using wild-type monoamine oxidases with molecular oxygen as an oxidant (Scheme 16.42) [307]. In 1995, the flavin-dependent monoamine oxidase N (MAO-N, EC 1.4.3.4) was isolated from *Aspergillus niger* [308] and later overexpressed in *E. coli* [309].

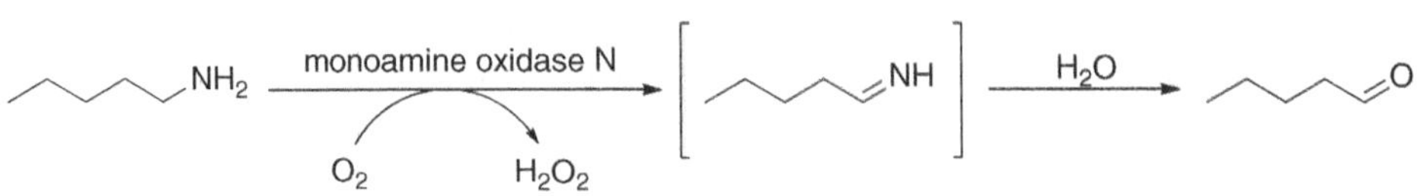

Scheme 16.42 The conversion of an amine (e.g. hexylamine) into the corresponding aldehyde using monoamine oxidase N.

Tetrahydro-β-carbolines (THBCs) are important bioactive alkaloids in nature. Examples include harmicine, reserpine, ajmalicine, and yohimbine (Scheme 16.43)

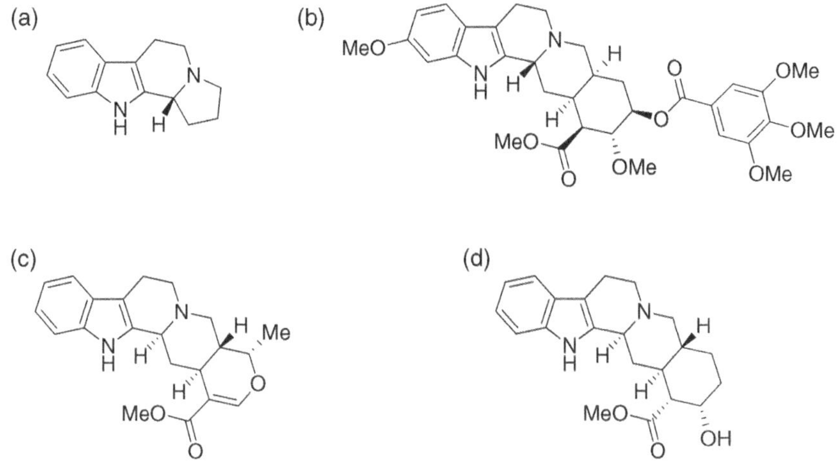

Scheme 16.43 The examples of alkaloids with the β-carboline backbone. (a) Harmicine, (b) reserpine, (c) ajmalicine, and (d) stimulant yohimbine.

[310]. A chemoenzymatic deracemization of various racemic alkaloids with the β-carboline backbone was investigated using the combination of MAO-N and the non-selective reductant NH_3BH_3, leading to 99% *ee* and up to 93% isolated yield (Figure 16.4) [310]. Interestingly, the MAO-N variants exhibited different enantioselectivities with various racemic alkaloids containing the β-carboline backbone. Moreover, docking simulations using MAO-N showed (*R*)-selectivity for various racemic alkaloids with the β-carboline backbone.

1,2,3,4-Tetrahydroquinoline (THQ) is an optically active substance of interest in the field of pharmaceuticals and agrochemicals [311, 312]. Li et al. made various mutants of amine oxidases using 11 amino acids placed within a 5 Å distance from the cyclohexanone substrate and screened them with various amines as substrates [313]. A cyclohexylamine oxidase (CHAO) mutant (T198FL199SM226F) from *Brevibacterium oxidans* IH-35A expressed in *E. coli* turned out to be the preferred enzyme for the conversion of racemic 2-methyl-THQ into (*R*)-2-methyl-THQ showing 76% yield and 98% *ee* (Scheme 16.44) [313].

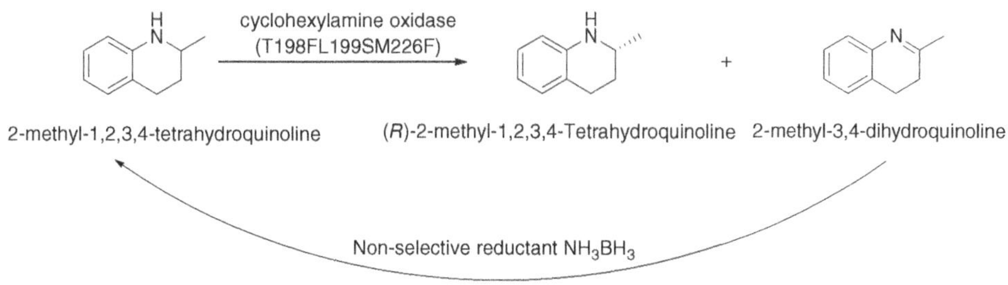

Scheme 16.44 The conversion of 2-methyl-1,2,3,4-tetrahydroquinoline into (*R*)-2-methyl-1,2,3,4-tetrahydroquinoline using cyclohexylamine oxidase mutant (T198FL199SM226F) from *B. oxidans* IH-35A expressed in *Escherichia coli*.

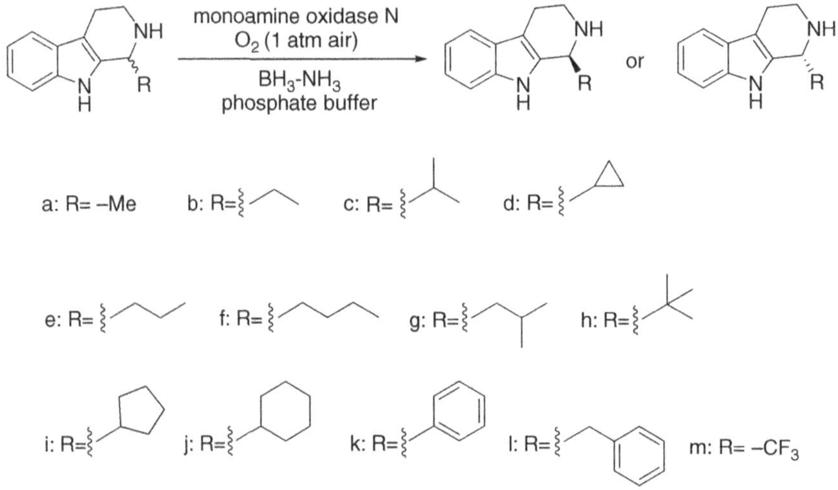

Entry	Time (h)	*ee* (%)	
		MAO-N D9	MAO-N D11
a	24	>99 (*R*)	99 (*R*)
b	84	36 (*R*)	25 (*R*)
c	72	40 (*S*)	33 (*S*)
d	48	86 (n.d.)	35 (n.d.)
e	48	99 (*S*)	>99 (*S*)
f	48	96 (*S*)	99 (*S*)
g	48	58 (*S*)	>99 (*S*)
h	48	62 (*S*)	96 (*S*)
i	48	85 (*S*)	96 (*S*)
j	48	80 (*S*)	97 (*S*)
k	72	30 (*S*)	92 (*S*)
l	48	-	-
m	48	-	-

Figure 16.4 Enantioselectivity of monoamine oxidase (MAO-N) N D9 and D11 with various racemic alkaloids with the β-carboline backbone [310]. Reproduced with permission of the American Chemical Society.

16.5.2 Keto Acid Synthesis Using L-Amino Acid Oxidases

Amino acid oxidases (AAOs) are enzymes that oxidize amino acids to keto acids, releasing ammonium and hydrogen peroxide. There are two main categories of AAOs depending on the chirality of the amino acid as substrate, namely L-amino acid oxidases (EC 1.4.3.2, L-AAOs or LAOs) and D-amino acid oxidases (EC 1.4.3.3, D-AAOs or DAOs). L-AAOs are flavoenzymes containing non-covalently bound FAD, which catalyze the stereospecific oxidative deamination of L-amino acids to the corresponding imino acids as intermediate to α-keto acids and produce ammonia and hydrogen peroxide. L-AAOs are widely distributed in nature. The most studied L-AAOs are from snake venoms [314–316], but also those of fungal [317, 318], bacterial [319–321], and algal origin [322] have been reported. L-Lysine oxidase (LO, EC 1.4.3.14) catalyzes the oxidation of the α,ω-amino group of L-lysine [307]. LO from *Trichoderma viride* and *N*-methyl-L-amino acid dehydrogenase (NMAADH)

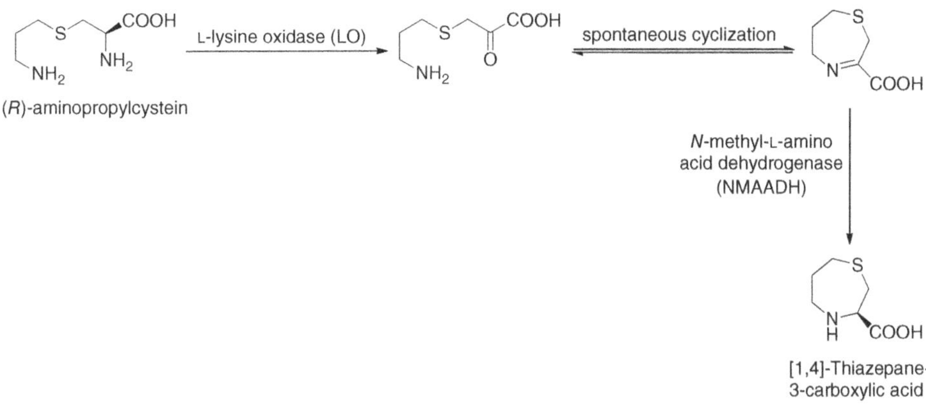

Scheme 16.45 The reaction from aminopropylcystein to [1,4]-thiazepane-3-carboxylic acid by the combination of L-Lysine oxidase and *N*-methyl-L-amino acid dehydrogenase.

from *Pseudomonas putida* ATCC12633 expressed in *E. coli* convert's aminopropylcystein into [1,4]-thiazepane-3-carboxylic acid in a one-pot system (Scheme 16.45) [323]. First, the α,ω-diamino acid (aminopropylcystein) is oxidized to the α-keto acid by LO from *T. viride* (Scheme 16.45). Next, the α-keto acid spontaneously converts into the imino acid by cyclization (Scheme 16.45). Finally, the imine is reduced to L-cyclic amino acid (CAA) by NMAADH from *P. putida* ATCC12633 expressed in *E. coli* using NADPH as a cofactor (Scheme 16.45). D-AAO and NMAADH can also obtain L-CAA from racemic CAA.

16.5.3 Keto Acid Synthesis Using D-AAO

D-AAOs are flavoproteins and catalyze the stereospecific oxidative deamination of D-amino acid to the corresponding α-keto acids, ammonium, and peroxide. The enzymes widely exist in various organisms such as bacteria [324], fungi [325], plants [326], and humans [327]. D-AAOs have been intensively studied because these enzymes have been considered to be model flavo-oxidase catalysts in basic research. In addition, the enzymes are already used for industrial and biotechnological processes [328], especially for the production of 7-aminocephalosporanic acid (7-ACA).

7-ACA is an important intermediate in many industrial processes to produce semi-synthetic cephalosporin antibiotics. Besides the chemical synthesis of 7-ACA, enzymatic approaches have been studied [329, 330]. Lopez-Gallego et al. have developed a one-pot system for the generation of 7-ACA from cephalosporin C (CPC) with D-AAO, catalase and glutaryl acylase (GAG) as immobilized enzymes (Scheme 16.46) [331]. By D-AAO coimmobilized with catalase, this one-pot system was carried out without hydrogen peroxide. After amino acid oxidation, the intermediate α-ketoadipyl-7-aminocephalosporanic acid (α-ketoadipyl-7-ACA) was hydrolyzed in the second reaction forming glutaric acid as a side-product (Scheme 16.47). Furthermore, Lopez.-Gallego et al. reported an improved hydrolysis with mutated glutaryl acylase using a mutant of glutaryl acylase in the absence of hydrogen peroxide for hydrolysis of α-ketoadipyl-7-ACA (Scheme 16.47) [332]. The double mutant Y178F + F375H of glutaryl acylase from *Pseudomonas* SY-77 was identified as the preferred mutant. Whereas the

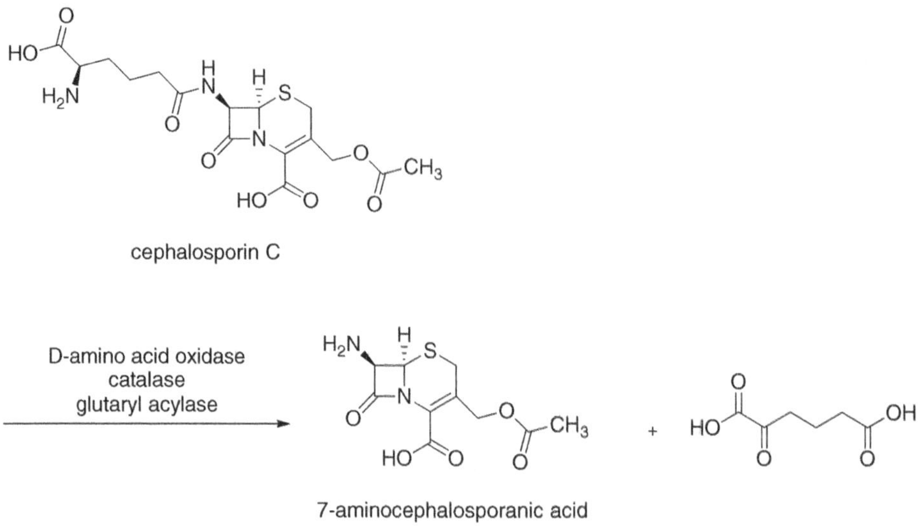

Scheme 16.46 The conversion of cephalosporin C into 7-aminocephalosporanic acid using D-AAO, catalase, and glutaryl acylase.

wild-type enzyme showed 40% yield of 7-ACA, this mutant led to 80% yield of 7-ACA [332]. However, an alternative enzymatic one-step process for the conversion of CPC into 7-ACA using a cephalosporin C acylase was introduced later [333].

16.5.4 Deracemization of Racemic Amino Acids Using AAOs

Deracemization is the conversion of a racemic mixture of a compound into an enantiomerically enriched form, in which one of its enantiomers can be obtained with a theoretical yield of 100%.

(*S*)-2-Amino-3-(6-o-tolylpyridin-3-yl)propanoic acid is an important intermediate in the production of an antidiabetic drug candidate. A formal deracemization of *rac*-2-amino-3-(6-o-tolylpyridin-3-yl)propanoic acid yielding the (*S*)-2-amino-3-(6-o-tolylpyridin-3-yl) propanoic acid has been realized by three different (chemo-)enzymatic cascade reactions (Scheme 16.48) [334]. (*R*)-2-amino-3-(6-o-tolylpyridin-3-yl)propanoic acid was selectively oxidized to the corresponding keto acid by (*R*)-AAO from *Trigonopsis variabilis* expressed in *E. coli* and then the keto acid was transferred to the (*S*)-amino acid by (*S*)-aminotransferase from *Burkholderia* sp. expressed in *E. coli* (Scheme 16.48a).

By means of this method, the (*S*)-amino acid was produced in 73% yield and with 99.9% *ee*. Regarding the second method for producing the (*S*)-amino acid, the (*R*)-amino acid was also oxidized to the keto acid by (*R*)-AAO from *T. variabilis* expressed in *E. coli* and then the keto acid was converted into the (*S*)-amino acid by (*S*)-amino acid dehydrogenase from *Sporosarcina ureae* using NADH as a cofactor (Scheme 16.48b). In this case, the (*S*)-amino acid was produced in 54% yield and with >99% *ee*. In addition to these two enzymatic cascades, a chemoenzymatic dynamic kinetic resolution of the racemic amino acid has also been reported (Scheme 16.48c). The (*S*)-amino acid was obtained using Celite-immobilized (*R*)-AAO and chemical imine reduction with borane. This process yielded the (*S*)-amino acid with an isolated yield of 68% at >99% *ee*.

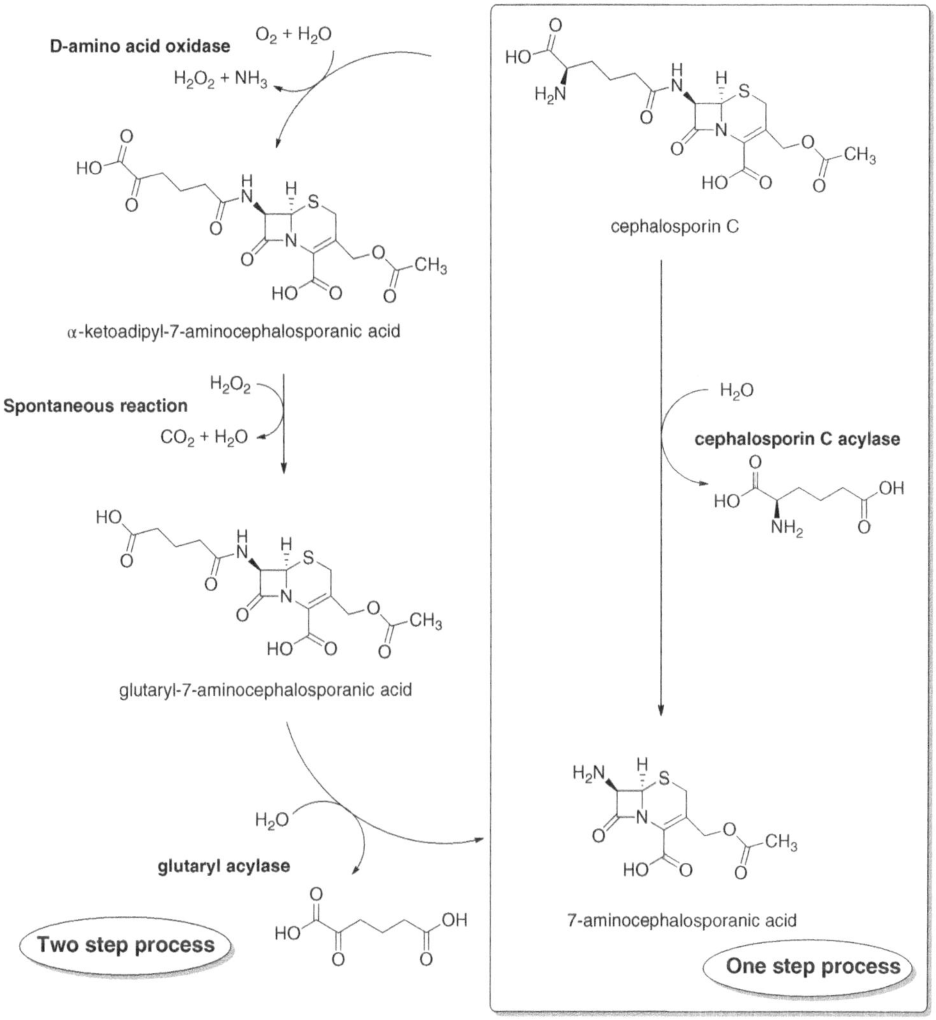

Scheme 16.47 The two-step (left) or one step (right) conversion of cephalosporin C into 7-aminocephalosporanic acid.

16.6 Oxidation with Artificial Oxidizing Enzymes

16.6.1 Oxidation with Phytase-Vanadate Complexes

Vanadium, a heavy metal with rather low toxicity, is abundantly available in nature and thus present in various metalloenzymes [335, 336]. In the field of organic synthesis, the number of chiral vanadium catalysts is steadily increasing [337–339].

Biocatalysts containing vanadium have already been applied in organic synthesis. One concept is to use vanadate-containing metalloenzymes and apply them for the catalysis of reactions that are not done physiologically. For example, vanadate-containing haloperoxidases were found to catalyze asymmetric sulfoxidations. A disadvantage is that only small substrates can enter the active site of

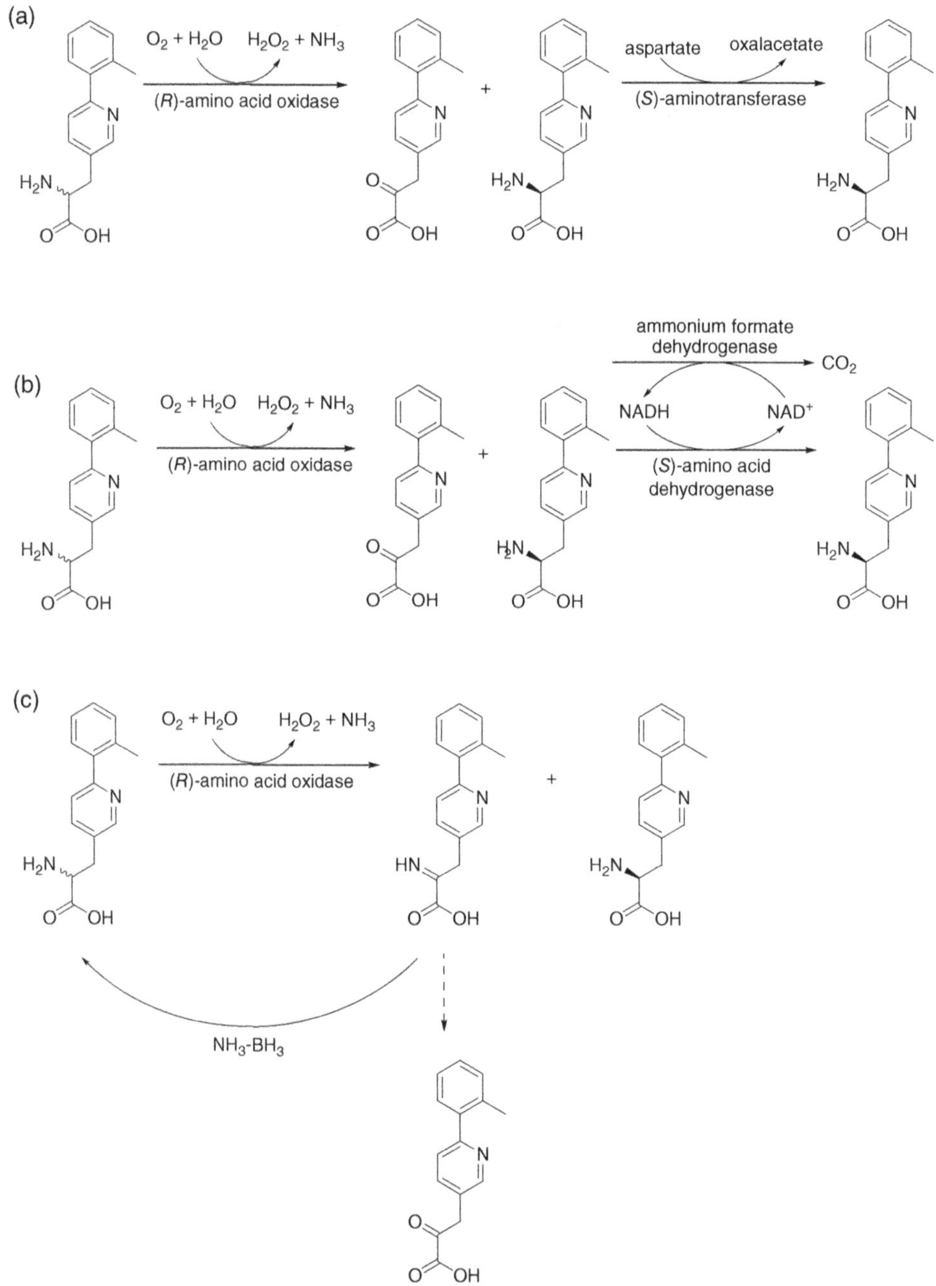

Scheme 16.48 The conversion of racemic 2-amino-3-(6-o-tolylpyridin-3-yl)propanoic acid into (*S*)-2-amino-3-(6-o-tolylpyridin-3-yl)propanoic acid (a) using (*R*)-AAO and (*S*)-aminotransferase, (b) using (*R*)-AAO and (*S*)-amino acid dehydrogenase, and (c) by dynamic resolution using (*R*)-AAO and chemical reduction of imine.

vanadium-dependent haloperoxidases leading to low conversion rates and only moderate enantioselectivity [340, 341].

A promising novel approach is the development of an artificial vanadium biocatalyst, which was done by Sheldon's group [342–345]. It was shown that vanadium chloroperoxidases and (membrane-bound) acid phosphatases are structurally closely related to each other and that the apoenzyme of vanadium choloroperoxidase shows phosphatase-like activity [346, 347]. Because vanadate is also known to be a potent inhibitor of acid phosphatases [348, 349], as well as related phytases [350, 351], Sheldon's group concluded that the incorporation of a vanadate ion into the active site of these enzymes should give artificial peroxidases. Phytase is a commercially available, heat-stable and inexpensive enzyme, since it is used as feed additive [352–354]. Moreover, the active site of a phytase can accept larger substrates compared to haloperoxidases. The incorporation of vanadate into phytase generated a semisynthetic peroxidase that is catalyzing the sulfoxidation of thioanisole (Scheme 16.49) [342–345].

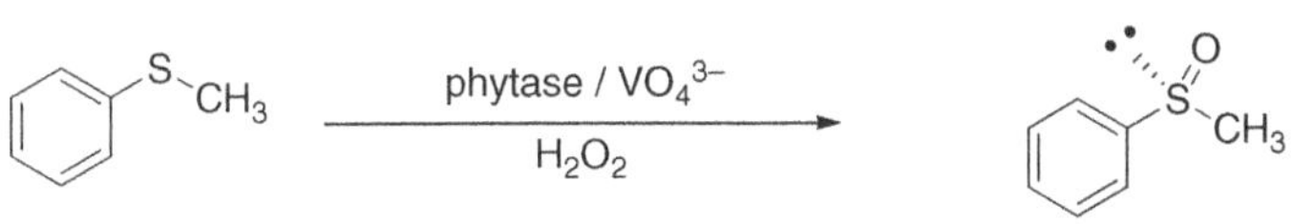

1

(*S*)-**2**

> 99% conversion
> 99% selectivity (sulfoxide)
66% *ee*

Scheme 16.49 Asymmetric sulfoxidation by phytase-vanadate complex established by the Sheldon's group [342–345].

At a vanadate concentration of 15 μM, thioanisole (**1**) could successfully be converted to the corresponding sulfoxide (*S*)-**2** with >99% conversion, >99% selectivity for the sulfoxide and an enantiomeric excess of 66%. Due to an increase in vanadate concentration, slow oxidation to the corresponding sulfone was observed. In a fed-batch operation with vanadate-phytase complex over 3 days, thioanisole could also be successfully converted to the corresponding sulfoxide with 50% enantiomeric excess [342–345].

Moreover, only the incorporation of a vanadate ion into the active site of phytase generated a semisynthetic peroxidase with a 13-fold increase in enzyme activity. The pH optimum of the phytase-vanadate complex is 5.0, which is consistent with the pH optimum for the physiological phytase reaction (hydrolysis of phytic acid) [342–345].

In the proposed mechanism (Scheme 16.50) for the sulfoxidation catalyzed by vanadium-incorporated phytase, the peroxide is bound alongside to the vanadium in the vanadium-phytase complex. In the next step, the peroxovanadium(V) species is nucleophilically attacked by the sulfur atom of the sulfide facilitating an asymmetric sulfoxidation under regeneration of the vanadate-phytase complex [345].

To get an insight into the scope of sulfoxidations catalyzed by vanadate-incorporated phytase, a series of *meta*- and *para*-substituted thioanisole derivatives as well as ethyl phenyl sulfide were examined by Sheldon's group. Electron-donating substituents showed an increase in the sulfoxidation rate, whereas electron-withdrawing substituents decreased the sulfoxidation rate. This is consistent with the rate-limiting nucleophilic attack of the sulfur atom on the electrophilic peroxo-species [342–345].

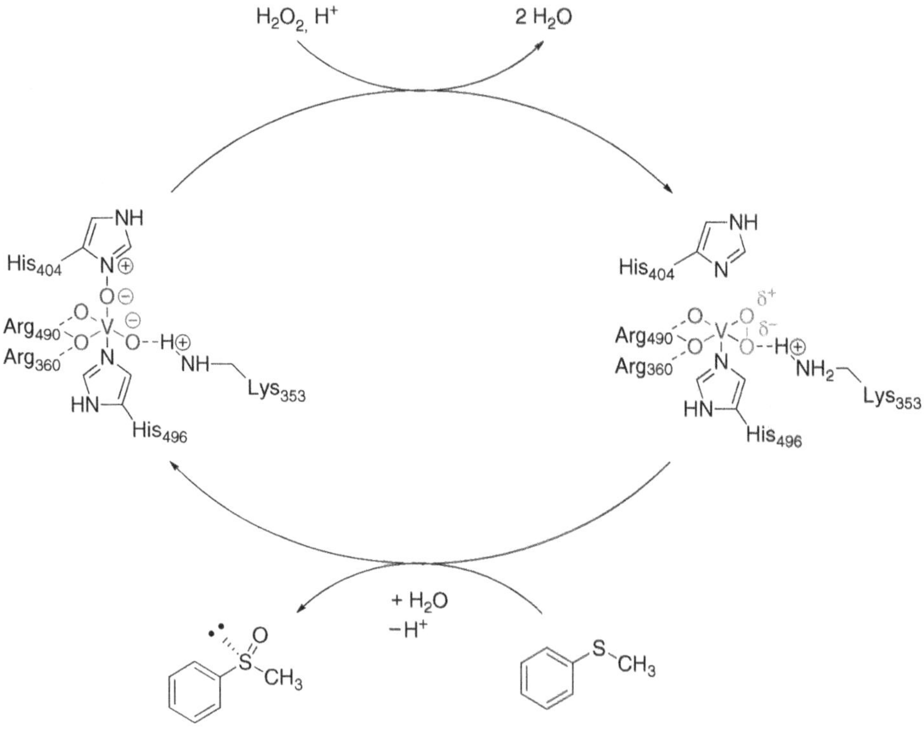

Scheme 16.50 Proposed mechanism for the sulfoxidation by vanadium-incorporated phytase. Numbering of amino acid residues according to vanadium chloroperoxidase from *Curvularia inequalis* [345].

Even though the enantioselectivities that have been obtained in sulfoxidations catalyzed by the phytase-vanadate complex are still in a moderate range, the concept of designing artificial metalloenzymes by incorporation of a vanadate ion into an inexpensive enzyme like phytase is promising. To increase the enantioselectivities as well as the substrate scope protein engineering could be applied in future work.

16.6.2 Oxidation with Iron-Porphyrin Complexes

Cytochrome P450 monooxygenases are oxygen-activating enzymes that have an active site consisting of a heme group with iron embedded in the center linked to the protein via a sulfur bridge to a cystein residue [355]. Over 450 different P450 enzymes were found in mammalian tissue and organ as a well as in plants, bacteria, yeast, and insects [296]. Dioxygen in the catalytic cycle of cytochrome P450 is activated in the presence of two protons and two electrons, which are provided by NADPH. The reactive iron(V) oxospecies in cytochrome P450 is responsible for hydroxylation of substrates (Scheme 16.51). Iron(III)- or manganese(III) porphyrin in cytochrome P450 are known to be able to activate dioxygen with oxygen donating agents. A prosthetic group, iron(III) protoporphyrin IX, is located at the active site and is bound to the apoprotein through cysteinate axial ligand of iron.

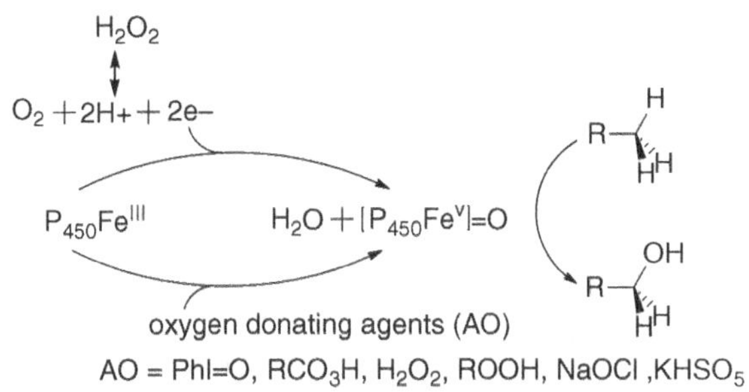

Scheme 16.51 Oxidation of substrates catalyzed by cytochrome P450.

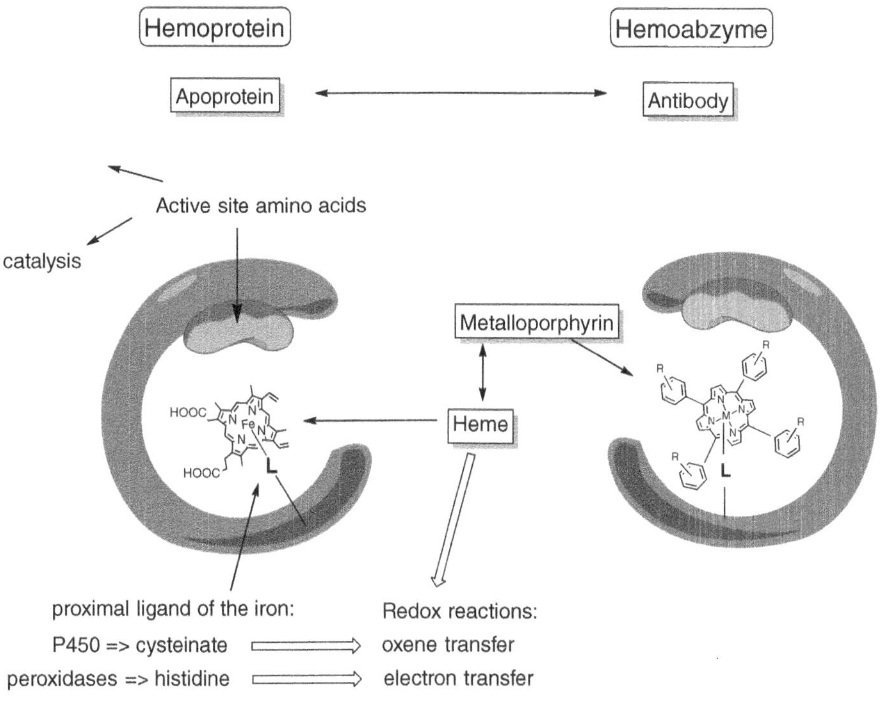

Figure 16.5 Schematic view of the structure of hemoproteins and their anti-metalloporphyrin antibodies mimics: hemoabzymes [356]. Reproduced with permission of the American Chemical Society.

The first generation of artificial hemoprotein containing an iron-porphyrin complex was referred to as hemoabzymes, which was created by the non-covalent association of a synthetic iron-porphyrins with monoclonal antibodies against these cofactors. The scheme of production of hemoabzymes was well explained in a study that obtained a fourth set of monoclonal anti-porphyrin antibodies, using as a hapten iron(III)-α,α,α,β-meso-tetrakis(ortho-carboxyphenyl) porphyrin, α,α,α,β-Fe(ToCPP) (Figure 16.5) [83, 356]. The second generation are the so-called hemozymes, which were constructed by non-covalent association of a non-relevant protein with a tetrarylporphyrin. To produce new artificial hemoproteins, it was necessary to design ideal porphyrin hapten, which is the binding site for the metalloporphyrin and hydrophobic pocket, and axial ligand or the iron [357].

Monoclonal antibodies, which are specific for a presumed transition-state analog for porphyrin metalation, N-methylmesoporphyrin IX, were prepared in 1990 [358, 359] The iron mesoporphyrin IX-antibody complex showed activity for the oxidation of pyrogallol, hydroquinone, *o*-dianisidine, and 2,2′-azino-bis-(3-ethylbenzothiazolined-sulfonic acid) (ABTS) by H_2O_2 [359]. The catalytic efficiency of the antibody complex was four times higher than the one of the free ion(III) mesoporphyrin. In detail, the corresponding value for the antibody complex is $274\,M^{-1}S^{-1}$ compared to $64\,M^{-1}S^{-1}$ for the free mesoporphyrin.

Another artificial porphyrin-antibody complex 3A3-MP8 was prepared by microperoxidase 8 (MP8), which is a heme octapetide having a proximal ligand at the iron atom consisting of histidine 18 and its corresponding antibody MP8 [83]. Porphyrin-antibody complex 3A3-MP8 was able to catalyze the nitration of phenol by NO_2 in the presence of H_2O_2 [360]. The total amount of nitrophenols was 14.5 ± 0.7 mM in the presence of 3A3-MP8 as a catalyst and 6.8 ± 0.5 mM in the presence of MP8 as a catalyst, which suggested that the heme of MP8 enhances the activity. This result could be explained by a protection of the heme of MP8 by antibody 3A3 against oxidative degradations. In addition, 3A3-MP8 complex induced a regioselectivity of the reaction toward 2-nitrophenol instead of 4-nitrophenol (Scheme 16.52). The 3A3-MP8 complex was found to catalyze the oxidation of thioanisole by H_2O_2 in a stereoselective fashion with an enantiomeric excess of 45% *ee* in favor of the *R*-enantiomer (Scheme 16.52) [361]. Furthermore, the oxidation of *N*-monosubstituted hydroxylamines in the presence of 3A3-MP8 characterized as antibody-porphyrin-iron(II)-metabolite complex, 3A3-MP8-FeII-RNO. The formation of such complexes was more difficult with the bulky and hydrophobic *N*-(1-*p*-chlorophenyl-propyl)-hydroxylamine than with small and hydrophilic hydroxylamines, such as isopropylhydroxylamine, because the antibody on the distal face of MP8 caused a partial steric hindrance (Scheme 16.52) [361].

A new artificial hemoprotein, Fe[T*p*CPP]-Xln10A, consisting of *β*-1,4-endoxylanase or xylanase A (Xln10A) containing Fe(III)-meso-tetra-*p*-carboxyphenylporphyrin (Fe(T*p*CPP))1-Fe or *meso*-tetra(4-sulfonatophenyl)porphyrinatoiron(III) (FeTpSPP)-Xln10A were used for oxidation of thioanisole by H_2O_2 [362, 363]. Both Fe(T*p*CPP)- and (FeT*p*SPP)-Xln10A showed lower yields and TONs compared to free metalporphyrins (Table 16.1). However, with (FeT*p*SPP)-Xln10A and imidazole an increased enantiomeric excess of 40% *ee* in favor of the *S*-sulfoxide was reached.

Another example of an hemozyme is myoglobin showing peroxidase activity, which was developed by the insertion of modified hemes into mutated sperm whale myoglobin [364]. In this study, new myoglobin was produced by mutation of an amino acid at the heme distal side and hydrophobic substrate binding site near the heme pocket to give peroxidase activity. An artificially created moiety that was linked at each terminal of heme-propionate side chain in protohemin IX contains one benzene ring and four carboxylate groups. Mutant rMb(H64D) showed dramatic degradation rate for bisphenol A, which was 35-fold higher than that of the wild-type. Also, the K_{cat}/K_m value of the 2-methoxyphenol oxidation was remarkably increased with usage of hybridization of a mutant and a modified heme. The highly oxidized artificial heme is supposed to enhance the binding affinity of substrates within an artificially created substrate binding domain.

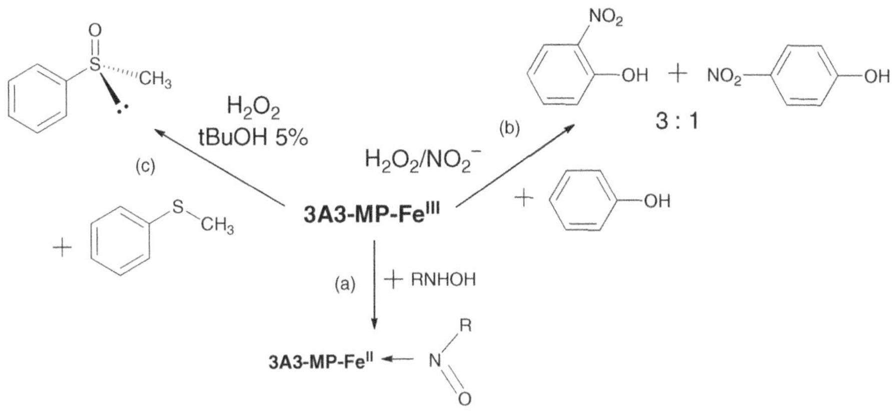

Scheme 16.52 (a) Formation of iron(II)–nitrosoalkane complexes by oxidation of N-substituted hydroxylamines by the 3A3-MP8 complex, (b) nitration of phenol by H_2O_2/NO_2^- catalyzed by the artificial metalloenzyme 3A3-MP8 and (c) enantioselective sulfoxidation of thioanisole by 3A3-MP8 [357].

Table 16.1 Yield, turnover number and enantiomeric excesses obtained for the oxidation of thioanisole into sulfoxide by H_2O_2 catalyzed by (FeTpSPP) and complexed with Xln10A in 50 mM phosphate buffer pH 7.4 [363].

Buffer	Catalyst	Yield (%)	Turnover number	*ee* (%)
50 mM phosphate pH 7.4	(FeTpSPP)	33 ± 2	57 ± 2	–
	(FeTpSPP) + ImH	93 ± 3	162 ± 3	–
	(FeTpSPP)-Xln10A	24 ± 3	42 ± 3	36 ± 3
	(FeTpSPP)-Xln10A + ImH	85 ± 4	145 ± 3	40 ± 3

16.7 Summary and Outlook

In summary, a broad range of enzyme classes have already been demonstrated to be of high synthetic relevance for oxidation reactions in organic chemistry. Such biocatalysts, for example dehydrogenases, monoxygenases, dioxygenases, and oxidases, efficiently catalyze a broad set of synthetically useful chemical reactions, for example transformation of alcohols into their carbonyl counterparts, hydroxylation of hydrocarbon moieties in both, functionalized and unfunctionalized molecules (such as steroids or alkanes), epoxidation, Bayer–Villiger reactions, or peroxide formations. What makes such biocatalytic oxidations particularly interesting as a solution for organic synthesis is the use of molecular oxygen as a readily available, non-toxic and environmentally friendly oxidation agent, which can be utilized in many cases. At the same time, safety issues have been considered accordingly even when utilizing an atmospheric oxygen level. Further advantages of biocatalytic oxidations are the excellent selectivities that can be often achieved (underlined by, e.g. the impressive regio- and diastereoselectivities when hydroxylating steroid molecules). In spite of many successful synthetic examples and, in part, industrial applications of such types of biocatalytic oxidation reactions, a current challenge often still consists of achieving a high process efficiency. Thus, the development of such processes represents a task for the future, and combination of protein engineering and process engineering strategies can be seen as a promising strategy to overcome hurdles that still exist today in the field of enzymatic oxidations leading to novel and further improved oxidation processes by means of enzyme catalysis.

References

1 Bäckvall, J.-E. (ed.) (2010). *Modern Oxidation Methods*, 2e. Weinheim: Wiley-VCH.
2 Mizuno, N. (ed.) (2009). *Modern Heterogeneous Oxidation Catalysis - Design, Reactions and Characterization*. Weinheim: Wiley-VCH.
3 For industrial examples, see: Weissermel, K. and Arpe, H.-J. (2003). *Industrial Organic Chemistry*, 4e. Weinheim: Wiley-VCH.
4 Drauz, K., Gröger, H., and May, O. (eds.) (2012). *Enzyme Catalysis in Organic Synthesis*, 3e, vol. 1-3. Weinheim: Wiley-VCH.
5 Reetz, M.T. (2012). *Enzyme Catalysis in Organic Synthesis*, 3e, vol. 3 (ed. K. Drauz, H. Gröger and O. May), 119–190. Weinheim: Wiley-VCH.
6 Romero, P.A. and Arnold, F.H. (2009). *Nat. Rev. Mol. Cell Biol.* 10: 866–876.
7 Liese, A., Seelbach, K., and Wandrey, C. (2006). *Industrial Biotransformations*. Weinheim: Wiley-VCH.
8 Lemière, G.L., Lepoivre, J.A., and Alderweireldt, F.C. (1985). *Tetrahedron Lett.* 26: 4527–4528.
9 Tufvesson, P., Lima-Ramos, J., Nordblad, M., and Woodley, J.M. (2011). *Org. Process Res. Dev.* 15: 266–274.
10 Weckbecker, A., Gröger, H., and Hummel, W. (2010). *Adv. Biochem. Eng. Biotechnol.* 120: 195–242.
11 Lavandera, I., Kern, A., Resch, V. et al. (2008). *Org. Lett.* 10: 2155–2158.
12 Orbegozo, T., Lavandera, I., Fabian, W.M.F. et al. (2009). *Tetrahedron* 65: 6805–6809.
13 Rodriguez, C., Lavandera, I., and Gotor, V. (2012). *Curr. Org. Chem.* 16: 2525–2541.
14 Garvie, E.I. (1980). *Microbiol. Rev.* 44: 106–139.
15 Veronese, F.M., Degani, Y., Nyc, J.F. et al. (1974). *J. Biol. Chem.* 249: 7922–7941.
16 Veronese, F.M., Boccu, E., and Conventi, L. (1975). *Biochim. Biophys. Acta* 377: 217–228.
17 Peterson, P.E., Pierce, J., and Smith, T.J. (1997). *J. Struct. Biol.* 120: 73–77.
18 Geueke, B., Riebel, B., and Hummel, W. (2003). *Enzym. Microb. Technol.* 32: 205–211.
19 Riebel, B.R., Gibbs, P.R., Wellborn, W.B., and Bommarius, A.S. (2003). *Adv. Synth. Catal.* 345: 707–712.
20 Hummel, W. and Riebel, B. (2003). *Biotechnol. Lett.* 25: 51–54.
21 Riebel, B.R., Gibbs, P.R., Wellborn, W.B., and Bommarius, A.S. (2002). *Adv. Synth. Catal.* 344: 1156–1168.
22 Rocha-Martín, J., Vega, D., Bolivar, J.M. et al. (2011). *BMC Biotechnol.* 11: 101.
23 Lountos, G.T., Riebel, B.R., Wellborn, W.B. et al. (2004). *Acta Crystallogr.* Sect. D Biol. Crystallogr. 60: 2044–2047.
24 Findrik, Z., Šimunović, I., and Vasić-Rački, D. (2008). *Biochem. Eng. J.* 39: 319–327.
25 Wu, X., Kobori, H., Orita, I. et al. (2012). *Biotechnol. Bioeng.* 109: 53–62.

26 Aksu, S., Arends, I.W.C.E., and Hollmann, F. (2009). *Adv. Synth. Catal.* 351: 1211–1216.
27 Ferrandi, E.E., Monti, D., Patel, I. et al. (2012). *Adv. Synth. Catal.* 354: 2821–2828.
28 Könst, P., Kara, S., Kochius, S. et al. (2013). *ChemCatChem* 5: 3027–3032.
29 Díaz-Rodríguez, A., Ríos-Lombardía, N., Sattler, J.H. et al. (2015). *Catal. Sci. Technol.* 5: 1443–1446.
30 Kara, S., Spickermann, D., Schrittwieser, J.H. et al. (2013). *ACS Catal.* 3: 2436–2439.
31 Maid, H., Böhm, P., Huber, S.M. et al. (2011). *Angew. Chem. Int. Ed.* 50: 2397–2400.
32 Greschner, W., Lanzerath, C., Reß, T. et al. (2014). *J. Mol. Catal. B Enzym.* 103: 10–15.
33 Hollmann, F. and Schmid, A. (2004). *Biocatal. Biotransform.* 22: 63–88.
34 Kochius, S., Magnusson, A.O., Hollmann, F. et al. (2012). *Appl. Microbiol. Biotechnol.* 93: 2251–2264.
35 Kochius, S., Park, J.B., Ley, C. et al. (2013). *J. Mol. Catal. B Enzym.* 103: 94–99.
36 Raspor, P., Goranovic, D., and Goranovič, D. (2008). *Crit. Rev. Biotechnol.* 28: 101–124.
37 Orbegozo, T., de Vries, J.G., and Kroutil, W. (2010). *Eur. J. Org. Chem.* 3445–3448.
38 Molinari, F., Aragozzini, F., Cabral, J.M.S., and Prazeres, D.M.F. (1997). *Enzym. Microb. Technol.* 20: 604–611.
39 Gandolfi, R., Ferrara, N., and Molinari, F. (2001). *Tetrahedron Lett.* 42: 513–514.
40 Miyamoto, K., Hirano, J.I., and Ohta, H. (2004). *Biotechnol. Lett.* 26: 1385–1388.
41 Hirano, J.I., Miyamoto, K., and Ohta, H. (2007). *Appl. Microbiol. Biotechnol.* 76: 357–363.
42 Hirano, J.I., Miyamoto, K., and Ohta, H. (2008). *Tetrahedron Lett.* 49: 1217–1219.
43 Kara, S., Schrittwieser, J.H., Gargiulo, S. et al. (2015). *Adv. Synth. Catal.* 357: 1687–1691.
44 Boratyński, F., Kiełbowicz, G., and Wawrzeńczyk, C. (2010). *J. Mol. Catal. B Enzym.* 65: 30–36.
45 Gargiulo, S., Arends, I.W.C.E., and Hollmann, F. (2011). *ChemCatChem* 3: 338–342.
46 Rauch, M., Schmidt, S., Arends, I.W.C.E. et al. (2017). *Green Chem.* 13: 226–265.
47 Landis, B.H., McLaughlin, J.K., Heeren, R. et al. (2002). *Org. Process Res. Dev.* 6: 547–552.
48 De Muynck, C., Pereira, C., Soetaert, W., and Vandamme, E. (2006). *J. Biotechnol.* 125: 408–415.
49 Habe, H., Shimada, Y., Yakushi, T. et al. (2009). *Appl. Environ. Microbiol.* 75: 7760–7766.
50 Yuan, J., Wu, M., Lin, J., and Yang, L. (2016). *J. Biosci. Bioeng.* 122: 10–16.
51 Fogagnolo, M., Giovannini, P.P., Guerrini, A. et al. (1998). *Tetrahedron* 9: 2317–2327.
52 Edegger, K., Mang, H., Faber, K. et al. (2006). *J. Mol. Catal. A Chem.* 251: 66–70.
53 Kagohara, E., Pellizari, V.H., Comasseto, J.V. et al. (2008). *Food Technol. Biotechnol.* 46: 381–387.
54 Brummund, J., Sonke, T., and Müller, M. (2015). *Org. Process Res. Dev.* 19: 1590–1595.
55 Kochius, S., Paetzold, M., Scholz, A. et al. (2014). *J. Mol. Catal. B Enzym.* 103: 61–66.
56 Gruber, C.C., Lavandera, I., Faber, K., and Kroutil, W. (2006). *Adv. Synth. Catal.* 348: 1789–1805.
57 Méndez-Sánchez, D., Mangas-Sánchez, J., Lavandera, I. et al. (2015). *ChemCatChem* 7: 4016–4020.
58 Li, Y.L., Xu, J.H., and Xu, Y. (2010). *J. Mol. Catal. B Enzym.* 64: 48–52.
59 Tanaka, T., Iwai, N., Matsuda, T., and Kitazume, T. (2009). *J. Mol. Catal. B Enzym.* 57: 317–320.
60 Voss, C.V., Gruber, C.C., and Kroutil, W. (2007). *Tetrahedron: Asymmetry* 18: 276–281.
61 Musa, M.M., Patel, J.M., Nealon, C.M. et al. (2015). *J. Mol. Catal. B Enzym.* 115: 155–159.
62 Karume, I., Musa, M.M., Bsharat, O. et al. (2016). *RSC Adv.* 6: 96616–96622.

63 Karume, I., Takahashi, M., Hamdan, S.M., and Musa, M.M. (2016). *ChemCatChem* 8: 1459–1463.
64 Müller, C.A., Weingartner, A.M., Dennig, A. et al. (2016). *J. Ind. Microbiol. Biotechnol.* 43: 1641–1646.
65 Müller, C.A., Akkapurathu, B., Winkler, T. et al. (2013). *Adv. Synth. Catal.* 355: 1787–1798.
66 Müller, C.A., Dennig, A., Welters, T. et al. (2014). *J. Biotechnol.* 191: 196–204.
67 Staudt, S., Burda, E., Giese, C. et al. (2013). *Angew. Chem. Int. Ed.* 52: 2359–2363.
68 Zhang, J., Wu, S., Wu, J., and Li, Z. (2015). *ACS Catal.* 5: 51–58.
69 Fan, C., Xu, G.-C., Ma, B. et al. (2015). *J. Biotechnol.* 195: 67–71.
70 Resch, V., Fabian, W.M.F., and Kroutil, W. (2010). *Adv. Synth. Catal.* 352: 993–997.
71 Klatte, S. and Wendisch, V.F. (2014). *Bioorg. Med. Chem.* 22: 5578–5585.
72 Both, P., Busch, H., Kelly, P.P. et al. (2016). *Angew. Chem. Int. Ed.* 55: 1511–1513.
73 Hummel, W., Kuzu, M., and Geueke, B. (2003). *Org. Lett.* 5: 3649–3650.
74 Khorsand, F., Murphy, C.D., Whitehead, A.J., and Engel, P.C. (2017). *Green Chem.* 19: 503–510.
75 Burton, S.G. (2003). *Trends Biotechnol.* 21: 543–549.
76 van Berkel, W.J., Kamerbeek, N.M., and Fraaije, M.W. (2006). *J. Biotechnol.* 124: 670–689.
77 Torres Pazmiño, D.E., Winkler, M., Glieder, A., and Fraaije, M.W. (2010). *J. Biotechnol.* 146: 9–24.
78 Bernhardt, R. (2006). *J. Biotechnol.* 124: 128–145.
79 Ruckpaul, K. and Rein, H. (eds.) (1989). *Basis and Mechanism of Regulation of Cytochrome P450*, 1–55. Berlin: Akademie-Verlag.
80 Sono, M., Roach, M.P., Coulter, E.D., and Dawson, J.H. (1996). *Chem. Rev.* 96: 2841–2887.
81 Estabrook, R.W. (2003). *Drug Metab. Dispos.* 31: 1461–1473.
82 Nelson, D.R. (2009). *Hum. Genomics* 4: 59–65.
83 Ricoux, R., Sauriat-Dorizon, H., Girgenti, E. et al. (2002). *J. Immunol. Methods* 269: 39–57.
84 Denisov, I.G., Makris, T.M., Sligar, S.G., and Schlichting, I. (2005). *Chem. Rev.* 105: 2253–2277.
85 Whitehouse, C.J.C., Bell, S.G., and Wong, L.-L. (2012). *Chem. Soc. Rev.* 41: 1218–1260.
86 Paulsen, M.D. and Ornstein, R.L. (1992). *J. Comput. Aided Mol. Des.* 6: 449–460.
87 Hamdane, D., Zhang, H., and Hollenberg, P. (2008). *Photosynth. Res.* 98: 657–666.
88 Loida, P.J. and Sligar, S.G. (1993). *Biochemistry* 32: 11530–11538.
89 Fasan, R., Chen, M.M., Crook, N.C., and Arnold, F.H. (2007). *Angew. Chem. Int. Ed.* 46: 8414–8418; (2007). Angew. Chem. 119: 8566–8570.
90 van Beilen, J.B., Duetz, W.A., Schmid, A., and Witholt, B. (2003). *Trends Biotechnol.* 21: 170–177.
91 Palfey, B.A. and McDonald, C.A. (2010). *Arch. Biochem. Biophys.* 493: 26–36.
92 Zhao, H. and van der Donk, W.A. (2003). *Curr. Opin. Biotechnol.* 14: 583–589.
93 Hollmann, F., Hofstetter, K., and Schmid, A. (2006). *Trends Biotechnol.* 24: 163–171.
94 Wu, H., Tian, C., Song, X. et al. (2013). *Green Chem.* 15: 1773–1789.
95 Hummel, W. and Gröger, H. (2014). *J. Biotechnol.* 191: 22–31.
96 Holland, H.L. and Weber, H.K. (2000). *Curr. Opin. Biotechnol.* 11: 547–553.
97 Leresche, J.E. and Meyer, H.-P. (2006). *Org. Process Res. Dev.* 10: 572–580.
98 Danielson, P.B. (2002). *Curr. Drug Metab.* 3: 561–591.
99 Omura, T. and Sato, R. (1962). *J. Biol. Chem.* 237: 1375–1376.

100 Klingenberg, M. (1958). *Arch. Biochem. Biophys.* 75: 376–386.
101 Hannemann, F., Bichet, A., Ewen, K.M., and Bernhardt, R. (2007). *Biochim. Biophys. Acta* 1770: 330–344.
102 Huijbers, M.M.E., Montersino, S., Westphal, A.H. et al. (2014). *Arch. Biochem. Biophys.* 544: 2–17.
103 Riebel, A., Fink, M.J., Mihovilovic, M.D., and Fraaije, M.W. (2014). *ChemCatChem* 6: 1112–1117.
104 Beam, M.P., Bossermann, M.A., Noinja, N. et al. (2009). *Biochemistry* 48: 4476–4487.
105 Willetts, A., Joint, I., Gilbert, J.A. et al. (2012). *Microb. Biotechnol.* 5: 549–559.
106 Willets, A. (1997). *Trends Biotechnol.* 15: 55–62.
107 van Bogaert, I.N., Groeneboer, S., Saerens, K., and Soetaert, W. (2011). *FEBS J.* 278: 206–221.
108 Eiben, S., Kaysser, L., Maurer, S. et al. (2006). *J. Biotechnol.* 124: 662–669.
109 Julsing, M.K., Cornelissen, S., Bühler, B., and Schmid, A. (2008). *Curr. Opin. Chem. Biol.* 12: 177–186.
110 Guengerich, F.P., Parikh, A., Johnson, E.F. et al. (1997). *Drug Metab. Dispos.* 25: 1234–1241.
111 Jouanneau, Y., Meyer, C., and Duraffourg, N. (2016). *Appl. Microbiol. Biotechnol.* 100: 1253–1263.
112 Liese, A., Seelbach, K., and Wandrey, C. (2000). *Industrial Biotransformations.* Weinheim: Wiley-VCH.
113 Roduner, E., Kaim, W., Sarkar, B. et al. (2013). *ChemCatChem* 5: 82–112.
114 Fujii, T., Narikawa, T., Kato, J., and Otake, H. (2005). Protein related to hydroxylation of alkane and gene encoding the same. JP2005058053, filed 13 September 2005 and issued 29 March 2007.
115 Schunck, W.H., Riege, P., Blasif, R. et al. (1978). *Acta. Biol. Med. Ger.* 37: 3–7.
116 Julsing, M.K., Schrewe, M., Cornelissen, S. et al. (2012). *Appl. Environ. Microbiol.* 78: 5724–5733.
117 Baptist, J.N., Gholson, R.K., and Coon, M.J. (1963). *Biochim. Biophys. Acta* 69: 40–47.
118 Bosetti, A., van Beilen, J.B., Preusting, H. et al. (1992). *Enzym. Microb. Technol.* 14: 702–708.
119 van Beilen, J.B., Wubbolts, M.G., and Witholt, B. (1994). *Biodegradation* 5: 161–174.
120 Mathys, R.G., Schmid, A., and Witholt, B. (1999). *Biotechnol. Bioeng.* 64: 459–477.
121 Grant, C., Deszcz, D., Wei, Y.-C. et al. (2014). *Sci. Rep.* 5844: 1–9.
122 Cornelissen, S., Julsing, M.K., Volmer, J. et al. (2013). *Biotechnol. Bioeng.* 110: 1282–1192.
123 Scheps, D., Malca, S.H., Richter, S.M. et al. (2013). *Microb. Biotechnol.* 6: 694–707.
124 Poetter, M., Schmid, A., Andreas. et al. (2010). Biocatalytic oxidation process with alkl gene product. WO2011131420, filed 20 April 2010 and issued 27 October 2011.
125 Schafferm, S., Gielen, J., Decker, N. et al. (2011). Biotechnological synthesis process of organic compounds with the aid of an alkl gene product. WO2013024111, filed 15 August 2011 and issued 21 February 2013.
126 Müller, H.G., Schunck, W.H., Riege, P., and Honeck, H. (1979). *Acta. Biol. Med. Ger.* 38: 345–349.
127 Ortiz de Montellano, P.R. (2010). *Chem. Rev.* 110: 932–948.
128 Colby, J., Stirling, D.I., and Dalton, H. (1977). *Biochem. J.* 165: 395–402.
129 Grant, C., Woodley, J.M., and Baganz, F. (2011). *Enzym. Microb. Technol.* 48: 480–486.

130 Weber, E., Seifert, A., Antonovici, M. et al. (2011). *Chem. Commun.* 47: 944–946.
131 Peters, M.W., Meinhold, P., Glieder, A., and Arnold, F.H. (2003). *J. Am. Chem. Soc.* 125: 13442–13450.
132 van Beilen, J.B. and Funhoff, E.G. (2005). *Curr. Opin. Biotechnol.* 16: 308–314.
133 Fujii, T., Narikawa, T., Takeda, K., and Kato, J. (2004). *Biosci. Biotechnol. Biochem.* 68: 2171–2177.
134 Glieder, A., Farinas, E.T., and Arnold, F.H. (2002). *Nat. Biotechnol.* 20: 1135–1139.
135 Miura, Y. and Fulco, A.J. (1974). *J. Biol. Chem.* 249: 1880–1888.
136 Narhi, L.O. and Fulco, A.J. (1986). *J. Biol. Chem.* 26: 7160–7169.
137 Narhi, L.O. and Fulco, A.J. (1987). *J. Biol. Chem.* 262: 6683–6690.
138 Ost, T.W.B., Miles, C.S., Murdoch, J. et al. (2000). *FEBS Lett.* 486: 173–177.
139 Kawakami, N., Shoji, O., and Watanabe, Y. (2013). *Chem. Sci.* 4: 2344–2348.
140 Fasan, R., Meharenna, Y.T., Snow, C.D. et al. (2008). *J. Mol. Biol.* 383: 1069–1080.
141 Farinas, E.T., Schwaneberg, U., Glieder, A., and Arnold, F.H. (2001). *Adv. Synth. Catal.* 343: 601–606.
142 Fasan, R., Chen, M.M., Crook, N.C., and Arnold, F.H. (2007). *Angew. Chem.* 119: 8566–8570; (2007). Angew. Chem. Int. Ed. 46: 8414–8418.
143 Zilly, F.E., Acevedo, J.P., Augustyniak, W. et al. (2011). *Angew. Chem. Int. Ed.* 50: 2720–2724.
144 Staudt, S., Müller, C.A., Marienhagen, J. et al. (2012). *Beilsstein J. Org. Chem.* 8: 186–191.
145 Tieves, F., Erenburg, I.N., Mahmoud, O., and Urlacher, V.B. (2016). *Biotechnol. Bioeng.* 113: 1845–1852.
146 von Bühler, C.J. and Urlacher, V.B. (2014). *Chem. Commun.* 50: 4089–4091.
147 Labinger, J.A. and Bercaw, J.E. (2002). *Nature* 417: 507–514.
148 Zakaria, Z. and Kamarudin, S.K. (2016). *Renew. Sust. Energ. Rev.* 65: 250–261.
149 Sazinsky, M.H. and Lippard, S.J. (2015). *Sustaining Life on Planet Earth: Metalloenzymes Mastering Dioxygen and Other Chewy Gases*, vol. 15 (ed. M.H. Kroneck and M.E.S. Torres), 205–256. Springer International Publishing.
150 Chen, M.M., Coelho, P.S., and Arnold, F.H. (2012). *Adv. Synth. Catal.* 354: 964–968.
151 Xu, F., Bell, S.G., Lednik, J. et al. (2005). *Angew. Chem. Int. Ed.* 44: 4029–4032.
152 Meinhold, P., Peters, M.W., Chen, M.M.Y. et al. (2005). *ChemBioChem* 6: 1765–1768.
153 Chen, M.M., Snow, C.D., Vizcarra, C.L. et al. (2012). *Protein Eng. Des. Sel.* 25: 171–178.
154 Ooyama, J. and Foster, J.W. (1965). *Antonie Leeuwenhoek* 31: 45–65.
155 Maurer, S.C., Kühnel, K., Kaysser, L.A. et al. (2005). *Adv. Synth. Catal.* 347: 1000–1098.
156 Staudt, S., Burda, E., Giese, C. et al. (2013). *Angew. Chem. Int. Ed.* 52: 2359–2363; (2013). Angew. Chem. 125: 2415–2419.
157 Ullrich, R. and Hofrichter, M. (2007). *Cell. Mol. Life Sci.* 64: 271–293.
158 Dennig, A., Marienhagen, J., Ruff, A.J. et al. (2012). *ChemCatChem* 4: 771–773.
159 Whitehouse, C.J., Yang, W., Yorke, J.A. et al. (2010). *ChemBioChem* 11: 2549–2556.
160 Whitehouse, C.J., Rees, N.H., Bell, S.G., and Wong, L.L. (2011). *Chem. Eur. J.* 17: 6862–6868.
161 Dennig, A., Lülsdprf, N., Liu, H., and Schwaneberg, U. (2013). *Angew. Chem. Int. Ed.* 52: 8459–8462.
162 Burton, S.G. (2003). *Curr. Org. Chem.* 7: 1317–1331.
163 Rescigno, A., Sollai, F., Pisu, B. et al. (2002). *J. Enzym. Inhib. Med. Chem.* 17: 207–218.
164 Guzik, U., Hupert-Kocurek, K., and Wojcieszyńska, D. (2014). *Molecules* 19: 8995–9018.

165 Kazandjian, R.Z. and Klibanov, A.M. (1985). *J. Am. Chem. Soc.* 107: 5448–5450.
166 Held, M., Suske, W., Schmid, A. et al. (1998). *J. Mol. Catal. B Enzym.* 5: 87–93.
167 Held, M., Panke, S., Kohler, H.-P. et al. (1999). *Bioworld* 5: 3–6.
168 Boshoff, A., Burton, M.H., and Burton, S.G. (2003). *Biotechnol. Bioeng.* 83: 1–7.
169 Coulombel, L., Nolan, L.C., Nikodinovic, J. et al. (2011). *Appl. Microbiol. Biotechnol.* 89: 1867–1875.
170 Newman, L.M. and Wackett, L.P. (1995). *Biochemistry* 34: 14066–14076.
171 Ma, F., Shi, S.-N., Sun, T.-H. et al. (2013). *Appl. Microbiol. Biotechnol.* 97: 5097–5103.
172 Bertoni, G., Bolognese, F., Galli, E., and Barbieri, P. (1996). *Appl. Environ. Microbiol.* 62: 3704–3711.
173 Olsen, R.H., Kukor, J.J., and Kaphammer, B. (1994). *J. Bacteriol.* 176: 3749–3756.
174 Patil, S.A., Apine, O.A., Surwase, S.N., and Jadhav, J.P. (2013). *APD* 2: 81–87.
175 Ghisalba, O., Meyer, H.-P., and Wohlgemuth, R. (2010). *Encyclopedia of Industrial Biotechnology: Bioprocess, Bioseparation, and Cell Technology* (ed. M.C. Flickinger), 1–18. Wiley.
176 Min, K., Park, K., Park, D.-H., and Yoo, Y.J. (2015). *Appl. Microbiol. Biotechnol.* 99: 575–584.
177 Chattopadhyay, S. and Das, A. (1990). *FEMS Microbiol. Lett.* 72: 195–200.
178 Lee, J.-Y. and Xun, L. (1998). *Biotechnol. Lett.* 20: 479–482.
179 Gunsalus, I.C., Pederson, T.C., and Sligar, S.G. (1975). *Annu. Rev. Biochem.* 44: 377–407.
180 Muñoz, A.J., Hernández-Chávez, G., de Anda, R. et al. (2011). *J. Ind. Microbiol. Biotechnol.* 38: 1845–1852.
181 Kuhar, M.J., Couceyro, P.R., and Lambert, P.D. (1999). *Basic Neurochemistry Molecular, Cellular and Medical Aspects*, 6e (ed. G.J. Siegel, B.W. Agranoff, R.W. Albers, et al.), 243–262. Philadelphia: Lippincott Williams & Wilkins.
182 Tong, W.-Y. and Dong, X. (2009). *Recent Pat. Biotechnol.* 3: 141–153.
183 Donova, M.V. and Egorova, O.V. (2012). *Appl. Microbiol. Biotechnol.* 94: 1423–1447.
184 Bureik, M. and Bernhardt, R. (2007). *Modern Biooxidation: Enzymes, Reactions and Applications* (ed. R.D. Schmid and V.B. Urlacher), 155–176. Wiley-VCH.
185 Donova, M.V. (2007). *Appl. Biochem. Microbiol.* 43: 1–14.
186 Constable, D.J.C., Dunn, P.J., Hayler, J.D. et al. (2007). *Green Chem.* 9: 411–420.
187 Janeczkoa, T., Dmochowska-Gładysza, J., Kostrzewa-Susłowa, E. et al. (2009). *Steroids* 74: 657–661.
188 Berg, A., Gustaffson, J.-A., and Ingelmann-Sundberg, M. (1976). *J. Biol. Chem.* 251: 2831–2838.
189 Kumar, S. (2010). *Expert Opin. Drug Metab. Toxicol.* 6: 115–131.
190 Weis, R., Winkler, M., Schittmayer, M. et al. (2009). *Adv. Synth. Catal.* 351: 2140–2146.
191 Zehentgruber, D., Dragan, C.-A., Bureik, M., and Lütz, S. (2010). *J. Biotechol.* 146: 179–185.
192 Petric, S., Hakki, T., Bernhardt, R. et al. (2010). *J. Biotechnol.* 150: 428–437.
193 Vail, R.B., Homann, M.J., Hanna, I., and Zaks, A. (2005). *J. Ind. Microbiol. Biotechnol.* 32: 67–74.
194 Parikh, A., Gillam, E.M.J., and Guengerich, F.P. (1997). *Nat. Biotechnol.* 15: 784–788.
195 Carballeira, J.D., Quezada, M.A., Hoyos, P. et al. (2009). *Biotechnol. Adv.* 27: 686–714.
196 Fernandes, P., Cruz, A., Angelova, B. et al. (2003). *Enzym. Microb. Technol.* 32: 688–705.

197 Zehentgruber, D., Hannemann, F., Bleif, S. et al. (2010). *ChemBioChem* 11: 713–721.

198 van Vugt-, B.M.A., Lussenburg, M.C., Damsten, D.M. et al. (2006). *Biochem. Biophys. Res. Commun.* 346: 810–818.

199 Lewis, J.C., Mantovani, S.M., Fu, Y. et al. (2010). *ChemBioChem* 11: 2502–2505.

200 Reinen, J., Vredenburg, G., Klaering, K. et al. (2015). *J. Mol. Catal. B Enzym.* 121: 64–74.

201 Peterson, D.H. and Murray, H.C. (1952). *J. Am. Chem. Soc.* 74: 1871–1872.

202 Murray, H.C. and Peterson, D.H. (1952). Oxygenation of steroids by mucorales fungi. US Patent 2, 602,769, filed 23 February 1952 and issued 8 July 1952.

203 Peterson, D.H., Murray, H.C., Eppstein, S.H. et al. (1952). *J. Am. Chem. Soc.* 75: 5933–5936.

204 May, S.W. and Padgette, S.R. (1983). *Nat. Biotechnol.* 1: 677–686.

205 Collingsworth, D.R., Brunner, M.P., and Haines, W.J. (1952). *J. Am. Cem. Soc.* 74: 2381–2382.

206 Hanson, F.R., Mann, K.M., Nielson, E.D. et al. (1953). *J. Am. Chem. Soc.* 75: 5369–5370.

207 Fried, J., Thoma, R.W., Gerke, J.R. et al. (1952). *J. Am. Chem. Soc.* 74: 3962–3963.

208 Hogg, J.A. (1992). *Steroids* 57: 593–616.

209 Petzoldt, K., Annen, K., Laurent, H. et al. (1982). Process for the preparation of 11- beta-hydroxy steroids. Schering Aktiengesellschaft, US Patent No. US 4353985.

210 Urlacher, V.B. and Eiben, S. (2006). *Trends Biotechnol.* 24: 324–330.

211 Szczebara, F.M., Chandelier, C., Villeret, C. et al. (2003). *Nat. Biotechnol.* 21: 143–149.

212 Park, J.W., Lee, J.K., Kwon, T.J. et al. (2003). *Biotechnol. Lett.* 25: 1827–1831.

213 Serizawa, N. and Matsuoka, T. (1991). *Biochim. Biophys. Acta* 1084: 35–40.

214 May, S.W. and Abbott, B.J. (1973). *J. Biol. Chem.* 248: 1725–1730.

215 Schwartz, R.D. and McCoy, C.J. (1976). *Appl. Environ. Microbiol.* 31: 78–82.

216 May, S.W. and Schwartz, R.D. (1974). *J. Am. Chem. Soc.* 96: 4031–4032.

217 Schwartz, R.D. and McCoy, C.J. (1977). *Appl. Environ. Microbiol.* 34: 47–49.

218 van der Linden, A.C. (1963). *Biochim. Biophys. Acta* 77: 157–159.

219 Peterson, J.A., Basu, D., and Coon, M.J. (1996). *J. Biol. Chem.* 241: 5162–5163.

220 May, S.W. and Abbott, B.J. (1972). *Biochem. Biophys. Res. Commun.* 48: 1230–1234.

221 Ruettinger, R.T., Griffith, G.R., and Coon, M.J. (1977). *Arch. Biochem. Biophys.* 183: 528–537.

222 Kok, M., Oldenhuis, R., van der Linden, M.P. et al. (1989). *J. Biol. Chem.* 264: 5435–5441.

223 van Beilen, J.B., Kingma, J., and Witholt, B. (1994). *Enzym. Microb. Technol.* 16: 904–911.

224 Stirling, D.I. and Dalton, H. (1979). *FEMS Microbial. Lett.* 5: 315–318.

225 Schwartz, R.D. (1973). *Appl. Microbiol.* 25: 574–577.

226 Schwartz, R.D. and McCoy, C.J. (1973). *Appl. Microbiol.* 26: 217–218.

227 Abbott, B. and Hou, C.T. (1973). *Appl. Microbiol.* 26: 86–91.

228 Zhou, N.-Y., Jenkins, A., Chan Kwo Chion, C.K.N.C., and Leak, D.J. (1999). *Appl. Environ. Microbiol.* 65: 1589–1595.

229 Farinas, E.T., Alcalde, M., and Arnold, F.H. (2004). *Tetrahedron* 60: 525–528.

230 van Ginkel, C.G., Welten, H.G.J., and de Bont, J.A.M. (1987). *Appl. Environ. Microbiol.* 53: 2903–2907.

231 van Ginkel, C.G. and de Bont, J.A.M. (1986). *Arch. Microbiol.* 145: 403–407.

232 Breuer, M., Ditrich, K., Habicher, T. et al. (2004). *Angew. Chem. Int. Ed.* 43: 788–824.

233 Wubbolts, M.G., Favre-Bulle, O., and Witholt, B. (1996). *Biotechnol. Biotechnol. Bioeng.* 52: 301–308.

234 Panke, S., Wubbolts, M., Schmid, A., and Witholt, B. (2000). *Biotechnol. Bioeng.* 69: 91–100.
235 Furuhashi, K. (1995). *Chirality in Industry* (ed. A.N. Collins, G.N. Sheldrake and J. Crosby), 167–186. Chichester: Wiley.
236 Panke, S., Held, M., Wubbolts, M.G. et al. (2002). *Biotechnol. Bioeng.* 80: 33–41.
237 Lin, H., Liu, Y., and Wu, Z.-L. (2011). *Tetrahedron Asymmetry* 22: 134–137.
238 Bernasconi, S., Orsini, F., Sello, G. et al. (2000). *Terahedron Lett.* 41: 9157.
239 Wubbolts, M.G., Hoven, J., Meigert, B., and Witholt, B. (1994). *Enzym. Microb. Technol.* 16: 887–894.
240 Baeyer, A. and Villiger, V. (1899). *Ber. Dtsch. Chem. Ges.* 32: 3625–3633.
241 Brink, G.T., Arends, I., and Sheldon, R. (2004). *Chem. Rev.* 104: 4105–4124.
242 Riebel, A., Dudek, H.M., de Gonzalo, G. et al. (2012). *Appl. Microbiol. Bitotechnol.* 95: 1479–1489.
243 Turfitt, G.E. (1948). *Biochem. J.* 42: 376–383.
244 Laskin, A.I., Grabowich, P., De Lisle Meyers, C., and Fried, J. (1964). *J. Med. Chem.* 7: 406–409.
245 Conrad, H.E., DuBus, R., Namtvedt, M.J., and Gunsalus, I.C. (1965). *J. Biol. Chem.* 240: 495–503.
246 Taylor, D.G. and Trudgill, P.W. (1986). *J. Bacteriol.* 165: 489–497.
247 Leisch, H., Morley, K., and Lau, P.C.K. (2011). *Chem. Rev.* 111: 4165–4222.
248 Donoghue, N.A., Norris, D.R., and Trudgill, P.W. (1976). *Eur. J. Biochem.* 63: 175–192.
249 Stewart, J.D. (1998). *Curr. Org. Chem.* 2: 195–216.
250 Ottolina, G., Carrea, G., Colonno, S., and Rückemann, A. (1996). *Tetrahedron Asymmetry* 7: 1123–1136.
251 Fraaije, M.W., Kamerbeek, N.M., van Berkel, W.J.H., and Janssen, D.B. (2002). *FEBS Lett.* 518: 43–47.
252 van Beek, H.L., Wijma, H.J., Fromont, L. et al. (2014). *FEBS Open Bio.* 4: 168–174.
253 Norris, D.B. and Trudgill, P.W. (1971). *Biochem. J.* 121: 363–370.
254 Donoghue, N.A. and Trudgill, P.W. (1975). *Eur. J. Biochem.* 60: 1–7.
255 Labet, M. and Thielemans, W. (2009). *Chem. Soc. Rev.* 38: 3484.
256 Chen, Y.C., People, O.P., and Walsh, C.T. (1988). *J. Bacteriol.* 170: 781–789.
257 Doo, E.-H., -Lee, W.-H., Seo, H.-S. et al. (2009). *J. Biotechnol.* 142: 164–169.
258 Mallin, H., Wulf, H., and Bornscheuer, U.T. (2013). *Enzym. Microb. Technol.* 53: 283–287.
259 Staudt, S., Bornscheuer, U.T., Menyes, U. et al. (2013). *Enzym. Microb. Technol.* 53: 288–292.
260 Schmidt, S., Scherkus, C., Manschiol, J. et al. (2015). *Angew. Chem. Int. Ed.* 54: 2784–2787.
261 Opperman, D.J. and Reetz, M.T. (2010). *ChemBioChem* 11: 2589–2596.
262 Schmidt, S., Genz, M., Balke, K., and Bornscheuer, U.T. (2015). *J. Biotechnol.* 214: 199–211.
263 Romero, E., Castellanos, J.R., Mattevi, A., and Fraaije, M.W. (2016). *Angew. Chem. Int. Ed.* 55: 15852–15855.
264 Mallin, H., Muschiol, J., Byström, E., and Bornscheuer, U.T. (2013). *ChemCatChem* 5: 3529–3532.
265 Scherkus, C., Schmidt, S., Bornscheuer, U.T. et al. (2016). *ChemCatChem* 8: 3446–3452.
266 Scherkus, C., Schmidt, S., Bornscheuer, U.T. et al. (2017). *Biotechnol. Bioeng.* 114: 1215–1221.

267 Wedde, S., Rommelmann, P., Scherkus, C. et al. (2017). *Green Chem.* 19: 1286–1290.

268 Pennec, A., Hollmann, F., Smit, M.S., and Opperman, D.J. (2015). *ChemCatChem* 7: 236–239.

269 Bornadel, A., Hatti-Kaul, R., Hollmann, F., and Kara, S. (2015). *ChemCatChem* 7: 2442–2445.

270 Bes, M.T., Villa, R., Roberts, S.M. et al. (1996). *J. Mol. Catal. B Enzym.* 1: 127–134.

271 Mihovilovic, M.D., Müller, B., and Stanetty, P. (2002). *Eur. J. Org. Chem.* 22: 3711–3730.

272 Hilker, I., Wohlgemuth, R., Alphand, V., and Furstoss, R. (2005). *Biotechnol. Bioeng.* 92: 702–710.

273 Berezina, N., Alphand, V., and Furstoss, R. (2002). *Tetrahydron: Asymmetry* 13: 1953–1955.

274 Davey, J.F. and Trudgill, P.W. (1977). *Eur. J. Biochem.* 74: 115–127.

275 Miyaji, K., Ohara, Y., Takahashi, Y. et al. (1991). *Tetrahedron Lett.* 32: 4557–4560.

276 Simpson, H.D., Alphand, V., and Furstoss, R. (2001). *J. Mol. Catal. B Enzym.* 16: 101–108.

277 Alphand, V., Archelas, A., and Furstoss, R. (1989). *Tetrahedron Lett.* 30: 3663–3664.

278 Mihovilovic, M.D. and Kapitán, P. (2004). *Tetrahedron Lett.* 45: 2751–2754.

279 Doig, S.D., O'Sullivan, L.M., Patel, S. et al. (2001). *Enzyme Microb. Techol.* 28: 265–274.

280 Doig, S.D., Simpson, H., Alphand, V. et al. (2003). *Enzym. Microb. Technol.* 32: 347–355.

281 Baldwin, C.V.F., Wohlgemuth, R., and Woodley, J.M. (2008). *Org. Process Res. Dev.* 12: 660–665.

282 Bugg, T.D. (2003). *Tetrahedron* 59: 7075–7101.

283 Sawyer, D.T. and Valentine, J.S. (1981). *Acc. Chem. Res.* 14: 393–400.

284 Lee-Ruff, E. (1977). *Chem. Soc. Rev.* 6: 195.

285 Stephens, P.J., Jollie, D.R., and Warshel, A. (1996). *Chem. Rev.* 96: 2491–2514.

286 Winterhalter, K.H. (1976). *Chimia* 30: 9.

287 Taniguchi, T., Sono, M., Hirata, F. et al. (1979). *J. Biol. Chem.* 254: 3288–3294.

288 Massey, V. (1994). *J. Biol. Chem.* 269: 22459–22462.

289 Que, L. and Ho, R.Y.N. (1996). *Chem. Rev.* 96: 2607–2624.

290 Nelson, M.J. and Cowling, R.A. (1990). *J. Am. Chem. Soc.* 112: 2820–2821.

291 de Groot, J., Garssen, G.J., Vliegenthart, J., and Boldingh, J. (1973). *Biochim. Biophys. Acta, Lipids Lipid Metab* 326: 279–284.

292 Hamberg, M. and Samuelsson, B. (1967). *J. Biol. Chem.* 242: 5336–5343.

293 Jones, R.D., Summerville, D.A., and Basolo, F. (1979). *Chem. Rev.* 79: 139–179.

294 McCapra, F. (1976). *Acc. Chem. Res.* 9: 201–208.

295 Lowry, T.H. and Richardson, K.S. (1976). *Mechanism and Theory in Organic Chemistry*. New York: Harper & Row.

296 Sono, M., Roach, M.P., Coulter, E.D., and Dawson, J.H. (1996). *Chem. Rev.* 96: 2841–2888.

297 Hawkins, E.G.E. (1950). *Nature* 166: 69.

298 Varjani, S.J. (2017). *Bioresour. Technol.* 223: 277–286.

299 Song, M., Yang, Y., Jiang, L. et al. (2017). *Environ. Pollut.* 220: 1059–1067.

300 Zhang, G.-Y., Ling, J.-Y., Sun, H.-B. et al. (2009). *J. Hazard. Mater.* 172: 580–586.

301 Wang, Y., Li, J., and Liu, A. (2017). *J. Biol. Inorg. Chem.* https://doi.org/10.1007/s00775-017-1436-5.

302 Rivard, B.S., Rogers, M.S., Marell, D.J. et al. (2015). *Biochemistry* 54: 4652–4664.

303 Geng, J., Dornevil, K., and Liu, A. (2012). *J. Am. Chem. Soc.* 134: 12209–12218.
304 Dong, G., Lu, J., and Lai, W. (2016). *ACS Catal.* 6: 3796–3803.
305 Adachi, K., Takeda, Y., Senoh, S., and Kita, H. (1964). *Biochim. Biophys. Acta, Gen. Subj.* 93: 483–493.
306 Stanier, R.Y. and Ingraham, J.L. (1954). *J. Biol. Chem.* 236: 2223–2228.
307 Turner, N.J. (2011). *Chem. Rev.* 111: 4073–4087.
308 Schilling, B. and Lerch, K. (1995). *Biochim. Biophys. Acta* 1243: 529–537.
309 Sablin, S.O., Yankovskaya, V., Bernard, S. et al. (1998). *Eur. J. Biochem.* 253: 270–279.
310 Ghislieri, D., Houghton, D., Green, A.P. et al. (2013). *ACS Catal.* 3: 2869–2872.
311 Ghislieri, D., Green, A.P., Pontini, M. et al. (2013). *J. Am. Chem. Soc.* 135: 10863–10869.
312 Wang, T., Zhuo, L.G., Li, Z. et al. (2011). *J. Am. Chem. Soc.* 133: 9878–9891.
313 Li, G., Ren, J., Yao, P. et al. (2014). *Am. Chem. Soc.* 4: 902–908.
314 Ahn, M.Y., Lee, B.M., and Kim, Y.S. (1997). *Int. J. Biochem. Cell Biol.* 29: 911–919.
315 Souza, D.H.F., Eugenio, L.M., Fletcher, J.E. et al. (1999). *Arch. Biochem. Biophys.* 368: 285–290.
316 Du, X.-Y. and Clemetson, K.J. (2002). *Toxicon* 40: 659–665.
317 Kusakabe, H., Kodama, K., Kuninaka, A., and Yoshino, H. (1980). *J. Biol. Chem.* 255: 976–981.
318 Hahn, K., Neumeister, K., Mix, A. et al. (2016). *Appl. Microbiol. Biotechnol.* 101: 2853–2864.
319 Geueke, B. and Hummel, W. (2002). *Enzyme Microb. Technol.* 31: 77–87.
320 Geueke, B. and Hummel, W. (2003). *Protein Expr. Purif.* 28: 303–309.
321 Faust, A., Niefind, K., Hummel, W., and Schomburg, D. (2007). *J. Mol. Biol.* 367: 234–248.
322 Piedras, P., Pineda, M., Muñoz, J., and Cardenas, J. (1992). *Planta* 188: 13–18.
323 Yasuda, M., Ueda, M., Muramatsu, H. et al. (2006). *Tetrahedron: Asymmetry* 17: 1775–1779.
324 Takahashi, S., Furukawara, M., Omae, K. et al. (2014). *Appl. Environ. Microbiol.* 80: 7219–7229.
325 Yano, S., Ozaki, H., Matsuo, S. et al. (2012). *Adv. Biosci. Biotechnol.* 3: 7–13.
326 Gisby, M.F., Mudd, E.A., and Day, A. (2012). *Plant Physiol.* 160: 2219–2226.
327 Molla, G., Sacchi, S., Bernasconi, M. et al. (2006). *FEBS Lett.* 580: 2358–2364.
328 Pilone, M.S. and Pollegioni, L. (2002). *Biocatal. Biotransform.* 20: 145–159.
329 Matsuda, A., Matsuyama, K., Yamamoto, K. et al. (1987). *J. Bacteriol.* 169: 5815–5820.
330 Nikolov, A. and Danielsson, B. (1994). *Enzyme Microb. Technol.* 16: 1037–1041.
331 Lopez-Galleo, F., Batencor, L., Hidalgo, A. et al. (2005). *Adv. Synth. Catal.* 347: 1804–1810.
332 Lopez-Galleo, F., Batencor, L., Sio, C.F. et al. (2008). *Adv. Synth. Catal.* 350: 343–348.
333 Pollegioni, L., Rosini, E., and Molla, G. (2013). *Appl. Microbiol. Biotechnol.* 97: 2341–2355.
334 Chen, Y., Goldberg, S.L., Hanson, R.L. et al. (2011). *Org. Process Res. Dev.* 15: 241–248.
335 Rehder, D. (2010). *Chem. Unserer Zeit* 44: 322–331.
336 Bauer, G., Güther, V., Hess, H. et al. (eds.) (2000). *Vanadium and Vanadium Compounds*. Weinheim, Germany: Wiley-VCH Verlag GmbH & Co. KGaA.
337 Canali, L. and Sherrington, D.C. (1999). *Chem. Soc. Rev.* 28: 85–93.
338 Bolm, C. (2003). *Coord. Chem. Rev.* 237: 245–256.
339 Takizawa, S., Gröger, H., and Sasai, H. (2015). *Chem. Eur. J.* 21: 8992–8997.
340 ten Brink, H.B., Tuynman, A., Dekker, H.L. et al. (1998). *Inorg. Chem.* 37: 6780–6784.

341 Andersson, M., Willetts, A., and Allenmark, S. (1997). *J. Org. Chem.* 62: 8455–8458.
342 van de Velde, F., Arends, I.W., and Sheldon, R.A. (2000). *Top. Catal.* 13: 259–265.
343 van de Velde, F. and Könemann, L. (1998). *Chem. Commun.* 1891–1892.
344 Correia, I., Aksu, S., Adão, P. et al. (2008). *J. Inorg. Biochem.* 102: 318–329.
345 van de Velde, F., Könemann, L., van Rantwijk, F., and Sheldon, R.A. (2000). *Biotechnol. Bioeng.* 67: 87–96.
346 Bhattacharyya, S. and Tracey, A.S. (2001). *J. Inorg. Biochem.* 85: 9–13.
347 Neuwald, A.F. (1997). *Protein Sci.* 6: 1764–1767.
348 Lindquist, Y., Schneider, G., and Vihko, P. (1994). *Eur. J. Biochem.* 221: 139–142.
349 Hemrika, W., Renirie, R., Dekker, H.L. et al. (1997). *Proc. Natl. Acad. Sci. U. S. A.* 94: 2145–2149.
350 Vescina, C.M., Salice, V.C., Cortizo, A.M., and Etcheverry, S.B. (1996). *Biol. Trace Elem. Res.* 53: 185–191.
351 Mahajan, A. and Dua, S. (1997). *J. Agric. Food Chem.* 45: 2504–2508.
352 Ullah, A.H. and Mullaney, E.J. (1996). *Biochem. Biophys. Res. Commun.* 227: 311–317.
353 Nelson, T.S., Shieh, T.R., Wodzinski, R.J., and Ware, J.H. (1971). *J. Nutr.* 101: 1289–1293.
354 Pasamontes, L., Haiker, M., Wyss, M. et al. (1997). *Appl. Environ. Microbiol.* 63: 1696–1700.
355 Kumar, D., Tahsini, L., Visser, d. et al. (2009). *J. Phys. Chem. A* 113: 11713–11722.
356 de Lauzon, S., Mansuy, D., and Mahy, J.-P. (2002). *Eur. J. Biochem.* 269: 470–480.
357 Mahy, J.-P., Maréchal, J.-D., and Ricoux, R. (2015). *Chem. Commun.* 51: 2476–2494.
358 Cochran, A. and Schultz, P. (1990). *Science* 249: 781–783.
359 Cochran, A.G. and Schultz, P.G. (1990). *J. Am. Chem. Soc.* 112: 9414–9415.
360 Ricoux, R., Girgenti, E., Sauriat-Dorizon, H. et al. (2002). *J. Protein Chem.* 21: 473–477.
361 Ricoux, R., Lukowska, E., Pezzotti, F., and Mahy, J.-P. (2004). *Eur. J. Biochem.* 271: 1277–1283.
362 Ricoux, R., Dubuc, R., Dupont, C. et al. (2008). *Bioconjug. Chem.* 19: 899–910.
363 Ricoux, R., Allard, M., Dubuc, R. et al. (2009). *Org. Biomol. Chem.* 7: 3208–3211.
364 Sato, H., Hayashi, T., Ando, T. et al. (2004). *J. Am. Chem. Soc.* 126: 436–437.

Index

Green Oxidation in Organic Synthesis, First Edition. Edited by Ning Jiao and Shannon S. Stahl.

b

n

s

t

x

y

z

Wiley The manufacturer's authorized representative according to the EU General Product Safety Regulation is Wiley-VCH GmbH, Boschstr. 12, 69469 Weinheim, Germany, e-mail: Product_Safety@wiley.com.

Printed and bound by CPI Group (UK) Ltd, Croydon, CR0 4YY
15/09/2026
02217418-0001